AF710961

Advances in Understanding Landscape Influences on Freshwater Habitats and Biological Assemblages

Publication of this book was made possible with financial support from

Advances in Understanding Landscape Influences on Freshwater Habitats and Biological Assemblages

Edited by

Robert M. Hughes
Amnis Opes Institute, and Department of Fisheries and Wildlife, Oregon State University
2895 SE Glenn, Corvallis, Oregon 97331, USA

Dana M. Infante
Department of Fisheries and Wildlife, Michigan State University
1405 South Harrison Road, Suite 318, East Lansing, Michigan 48824, USA

Lizhu Wang
Great Lakes Regional Office, International Joint Commission
Post Office Box 32869, Detroit, Michigan, USA

Kai Chen
Department of Entomology, Nanjing Agricultural University
Weigang #1, Xuanwu District, Nanjing, Jiangsu, 210095, China

Bianca de Freitas Terra
Center for Agrarian and Biological Sciences, State University of Vale do Acaraú
Avenida da Universidade 850, Sobral, Ceará, Brazil

American Fisheries Society Symposium 90

Based on the Proceedings of the Symposia

"Advances in Understanding Landscape Influences on
Freshwater Habitats and Biological Assemblages"
American Fisheries Society Annual Meeting
Held in Atlantic City, New Jersey, USA
August 22–23, 2018

"Landscape Influences on Freshwater Habitats and Biological Assemblages"
Society for Freshwater Science Annual Meeting
Held in Detroit, Michigan, USA
May 23–24, 2018

American Fisheries Society
Bethesda, Maryland
2019

Suggested citation formats follow.

Entire book

Hughes, R. M., D. M. Infante, L. Wang, K. Chen, and B. F. Terra, editors. 2019. Advances in understanding landscape influences on freshwater habitats and biological assemblages. American Fisheries Society, Symposium 90, Bethesda, Maryland.

Chapter in book

Falke, J. A., B. M. Huntsman, and E. R. Schoen. 2019. Climatic variation drives growth potential of juvenile Chinook Salmon along a subarctic boreal riverscape. Pages 57–82 *in* R. M. Hughes, D. M. Infante, L. Wang, K. Chen, and B. F. Terra, editors. Advances in understanding landscape influences on freshwater habitats and biological assemblages. American Fisheries Society, Symposium 90, Bethesda, Maryland.

Cover photo by Steve Cline (U.S. Environmental Protection Agency, Corvallis, Oregon); over Bower's Rock State Park, Willamette River, Oregon.

© Copyright 2019 by the American Fisheries Society

All rights reserved. Photocopying for internal or personal use, or for the internal or personal use of specific clients, is permitted by the American Fisheries Society (AFS) provided that the appropriate fee is paid directly to Copyright Clearance Center (CCC), 222 Rosewood Drive, Danvers, Massachusetts 01923, USA; phone: 978-750-8400. Request authorization to make multiple copies for classroom use from CCC. These permissions do not extend to electronic distribution or long-term storage of multiple articles or to copying for resale, promotion, advertising, general distribution, or creation of new collective works. For such uses, permission or license must be obtained from AFS.

Printed in the United States of America on acid-free paper.

Library of Congress Control Number 2019944894

ISBN 978-1-934874-56-1
ISSN 0892-2284

American Fisheries Society Web site: www.fisheries.org

American Fisheries Society
425 Barlow Place, Suite 110
Bethesda, Maryland 20814
USA

This book is dedicated to Jim Omernik (U.S. Environmental Protection Agency [USEPA] and U.S. Geological Survey, retired), who provided landscape perspectives on interpreting spatial patterns in the National Eutrophication Survey data (particularly stream nutrient–land use relationships) and the National Acid Precipitation Assessment Program (i.e., national alkalinity patterns of streams and lakes, which were used to focus sampling in the most acid-sensitive regions). He then developed hierarchical ecoregion maps for the United States and North America based on natural and anthropogenic factors, as well as lake phosphorus maps for the north-central and northeastern United States. Those spatial frameworks led to the development of regionalized biological and water quality criteria by many state water quality agencies, and the ecoregions currently are key geographic strata for data interpretation in the USEPA's National Aquatic Resource Surveys and for other environmental research.

Contents

Preface

Over a 2017 lunch in East Lansing, Michigan, we (Robert Hughes, Dana Infante, and Lizhu Wang) discussed the idea of a book to follow up on *Landscape Influences on Stream Habitats and Biological Assemblages* (2006). To broaden the content and perspectives to other continents, we invited Kai Chen (China) and Bianca de Freitas Terra (Brazil) to serve as co-editors. To further expand the material, we organized symposia for the 2018 annual meetings of the Society for Freshwater Sciences (chaired by Dana Infante and Lizhu Wang) and the American Fisheries Society (AFS; chaired by Jared Ross). The title of the new book, *Advances in Understanding Landscape Influences on Freshwater Habitats and Biological Assemblages*, indicated the goals of including lentic as well as lotic content and updating the earlier volume. We all felt that an updated book was timely because of the many conceptual, database, and analytical advances in our field over the past decade, the expansion of landscape-aquatic science globally, and the critical need for improved landscape-aquatic management in the Anthropocene. To ensure publication prior to the joint AFS–The Wildlife Society meeting in September 2019, the editors followed a tight schedule of manuscript submission, peer review, and author revision, resulting in rejection of three manuscripts and withdrawals of several others. Book publication or symposium travel were supported by sponsorships from the AFS Fish Habitat, Fish Management, International Fisheries, and Water Quality sections. We are also extremely grateful to Debby Lehman for copy editing and 57 diligent manuscript reviewers for improving the clarity and quality of the 21 book chapters, which incorporate landscape–freshwater science advances from four continents.

Reviewers

Theodore Angradi
U.S. Environmental Protection Agency, Duluth, Minnesota, USA

Lee Baumgartner
Charles Sturt University, Albury-Wodonga, New South Wales, Australia

Sandra Bryce
InfoGraphics, Corvallis, Oregon, USA

Lilian Casatti
São Paulo State University, São Jose do Rio Preto, São Paulo, Brazil

Arthur Cooper
Michigan State University, East Lansing, Michigan, USA

Christopher Edge
University of Toronto, Toronto, Ontario, Canada

Michael Eggleton
University of Arkansas, Pine Bluff, Arkansas, USA

Peter Esselman
U.S. Geological Survey, Ann Arbor, Michigan, USA

Matthew Falcy
Oregon Department of Fish and Wildlife, Corvallis, Oregon, USA

Michael Gangloff
Appalachian State University, Boone, North Carolina, USA

Pedro Giovâni da Silva
Federal University of Minas Gerais, Belo Horizonte, Minas Gerais, Brazil

Robert Gresswell
U.S. Geological Survey, Bozeman, Montana, USA

Glenn Griffith
U.S. Geological Survey, Corvallis, Oregon, USA

Gretchen Hansen
University of Minnesota, Saint Paul, Minnesota, USA

Alan Herlihy
Oregon State University, Corvallis, Oregon, USA

Nathaniel Hitt
U.S. Geological Survey, Kearneysville, West Virginia, USA

Lucinda Johnson
University of Minnesota, Duluth, Minnesota, USA

Yoichiro Kanno
Colorado State University, Fort Collins, Colorado, USA

Jason Leach
Natural Resources Canada, Sault Ste. Marie, Ontario, Canada

Xueqin Liu
Institute of Hydrobiology, Chinese Academy of Sciences, Wuhan, Hubei, China

Diego Macedo
Federal University of Minas Gerais, Belo Horizonte, Minas Gerais, Brazil

Erika Martin
Emporia State University, Emporia, Kansas, USA

Dale McCullough
Columbia River Intertribal Fish Commission, Portland, Oregon, USA

Michael Meador
U.S. Geological Survey, Reston, Virginia, USA

Norman Mercado
Autonomous University of Morelos, Cuernavaca, Morelos, Mexico

Steve Miranda
U.S. Geological Survey, Mississippi State, Mississippi, USA

Matthew Mitro
Wisconsin Department of Natural Resources, Madison, Wisconsin, USA

Daisuke Nakano
Central Research Institute of the Electric Power Industry, Chiba, Japan

Junjiro Negishi
Hokkaido University, Hokkaido, Japan

Joseph Nohner
Michigan Department of Natural Resources, Lansing, Michigan, USA

Thierry Oberdorff
Research Institute for Development, Toulouse, France

Julian Olden
University of Washington, Seattle, Washington, USA

Anthony Olsen
U.S. Environmental Protection Agency, Corvallis, Oregon, USA

Allison Pease
Texas Tech University, Lubbock, Texas, USA

Brenda Pracheil
Oak Ridge National Laboratory, Oak Ridge, Tennessee, USA

Xiaodong Qu
China Institute of Water Resources and Hydropower Research, Beijing, China

Gordon Reeves
U. S. Forest Service, Corvallis, Oregon, USA

Wayne Robinson
Charles Sturt University, Thurgoona, New South Wales, Australia

Jared Ross
Michigan State University, East Lansing, Michigan, USA

Nancy Roth
Tetra Tech, Owings Mills, Maryland, USA

Victor Satoru Saito
Federal University of São Carlos, São Carlos, São Paulo, Brazil

Daniel Schindler
University of Washington, Seattle, Washington, USA

Stefan Schmutz
University of Natural Resources and Life Sciences, Vienna, Austria

Patricia Soranno
Michigan State University, East Lansing, Michigan, USA

Jason Thiem
Department of Primary Industries, Narrandera, New South Wales, Australia

Darren Thornbrugh
U.S. Forest Service, Milwaukee, Wisconsin, USA

Ralph Tingley
University of Missouri, Columbia, Missouri, USA

Ryan Utz
Chatham University, Pittsburgh, Pennsylvania, USA

Ian Waite
U.S. Geological Survey, Portland, Oregon, USA

Kyle Wellband
University of Laval, Quebec City, Quebec, Canada

Seth Wenger
University of Georgia, Athens, Georgia, USA

Christopher Wheeler
Tennessee Tech University, Cookeville, Tennessee, USA

Jodi Whittier
University of Missouri, Columbia, Missouri, USA

Jack Williams
Trout Unlimited, Medford, Oregon, USA

Taylor Woods
University of Tennessee, Knoxville, Tennessee, USA

Adam Yates
Western University and Canadian Rivers Institute, Waterloo, Ontario, Canada

Yixin Zhang
Xi'an Jiaotong-Liverpool University, Suzhou, Jiangsu, China

Symbols and Abbreviations

The following symbols and abbreviations may be found in this book without definition. Also undefined are standard mathematical and statistical symbols given in most dictionaries.

A	ampere
AC	alternating current
Bq	becquerel
C	coulomb
°C	degrees Celsius
cal	calorie
cd	candela
cm	centimeter
Co.	Company
Corp.	Corporation
cov	covariance
DC	direct current; District of Columbia
D	dextro (as a prefix)
d	day
d	dextrorotatory
df	degrees of freedom
dL	deciliter
E	east
E	expected value
e	base of natural logarithm (2.71828…)
e.g.	(exempli gratia) for example
eq	equivalent
et al.	(et alii) and others
etc.	et cetera
eV	electron volt
F	filial generation; Farad
°F	degrees Fahrenheit
fc	footcandle (0.0929 lx)
ft	foot (30.5 cm)
ft^3/s	cubic feet per second (0.0283 m^3/s)
g	gram
G	giga (10^9, as a prefix)
gal	gallon (3.79 L)
Gy	gray
h	hour
ha	hectare (2.47 acres)
hp	horsepower (746 W)
Hz	hertz
in	inch (2.54 cm)
Inc.	Incorporated
i.e.	(id est) that is
IU	international unit
J	joule
K	Kelvin (degrees above absolute zero)
k	kilo (10^3, as a prefix)
kg	kilogram
km	kilometer
l	levorotatory
L	levo (as a prefix)
L	liter (0.264 gal, 1.06 qt)
lb	pound (0.454 kg, 454g)
lm	lumen
log	logarithm
Ltd.	Limited
M	mega (10^6, as a prefix); molar (as a suffix or by itself)
m	meter (as a suffix or by itself); milli (10^{-3}, as a prefix)
mi	mile (1.61 km)
min	minute
mol	mole
N	normal (for chemistry); north (for geography); newton
N	sample size
NS	not significant
n	ploidy; nanno (10^{-9}, as a prefix)
o	ortho (as a chemical prefix)
oz	ounce (28.4 g)
P	probability
p	para (as a chemical prefix)
p	pico (10^{-12}, as a prefix)
Pa	pascal
pH	negative log of hydrogen ion activity

ppm	parts per million
qt	quart (0.946 L)
R	multiple correlation or regression coefficient
r	simple correlation or regression coefficient
rad	radian
S	siemens (for electrical conductance); south (for geography)
SD	standard deviation
SE	standard error
s	second
T	tesla
tris	tris(hydroxymethyl)-aminomethane (a buffer)
UK	United Kingdom
U.S.	United States (adjective)
USA	United States of America (noun)
V	volt
V, Var	variance (population)
var	variance (sample)
W	watt (for power); west (for geography)
Wb	weber
yd	yard (0.914 m, 91.4 cm)
α	probability of type I error (false rejection of null hypothesis)
β	probability of type II error (false acceptance of null hypothesis)
Ω	ohm
μ	micro (10^{-6}, as a prefix)
′	minute (angular)
″	second (angular)
°	degree (temperature as a prefix, angular as a suffix)
%	per cent (per hundred)
‰	per mille (per thousand)

American Fisheries Society Symposium 90:1–9, 2019
© 2019 by the American Fisheries Society

Advances in Understanding Landscape Influences on Freshwater Habitats and Biological Assemblages: Introduction

Robert M. Hughes*
Amnis Opes Institute, and Department of Fisheries and Wildlife, Oregon State University 2895 SE Glenn, Corvallis, Oregon 97331, USA

Dana M. Infante
Department of Fisheries and Wildlife, Michigan State University 1405 South Harrison Road, Suite 318, East Lansing, Michigan 48824, USA

Lizhu Wang
Great Lakes Regional Office, International Joint Commission Post Office Box 32869, Detroit, Michigan 48232, USA

Kai Chen
Department of Entomology, Nanjing Agricultural University Weigang #1, Xuanwu District, Nanjing, Jiangsu, 210095, China

Bianca de Freitas Terra
Center for Agrarian and Biological Sciences, State University of Vale do Acaraú Avenida da Universidade 850, Sobral, Ceará, Brazil

Previously, Hughes et al. (2006) synthesized the state of knowledge related to influences of landscapes on river ecosystems. That volume, containing contributions from 77 authors, highlighted challenges encountered when measuring and linking landscapes with river habitats, described approaches for resolving those challenges, and identified major knowledge gaps to direct future research. In addition, Johnson and Host (2010) provided a timeline from the 1970s to 2008 to generalize the significant contributions of the landscape approach to aquatic studies, summarized landscape approaches used in studying freshwater ecosystems, and synthesized the literature regarding studies of freshwater ecosystems conducted at or above the catchment extent. Those reviews have been invaluable in describing the usefulness of the landscape approach, including improving our understanding of how landscape factors influence freshwater habitats and biological assemblages, supporting science-based landscape-aquatic management, and informing policy to conserve freshwater ecosystems. Because of these benefits, the landscape approach has continued to be tested and applied in multiple contexts. As a result, our knowledge in understanding, measuring, and predicting how characteristics of freshwater habitats and biological assemblages are influenced by landscapes has advanced rapidly. Hence, there is a need to synthesize current understanding of landscape influences on freshwaters and identify new challenges and knowledge gaps to direct future research

* Corresponding author: hughes.bob@amnisopes.com

and its application in freshwater resource management and conservation.

Evolving Understanding of Landscape Influences on Freshwater Habitats and Biological Assemblages

Since the early recognition that rivers and streams are influenced by the landscapes through which they flow (Hynes 1975), landscape-aquatic system studies have emphasized habitat heterogeneity, connectivity, and scale (both extent and resolution or grain) in flowing waters (Allan 1995; Ward et al. 2002; Allan 2004). Before the 2000s, most early work investigating these relationships had been conducted on small- to medium-sized streams and their catchments. Other studies emphasized the hierarchical influences of large-extent drivers that shape waters at intermediate spatial extents and thereby create habitats suitable for biota at smaller extents (Frissell et al. 1986; Tonn 1990; Snelder and Biggs 2002; Gámez-Virués et al. 2015). Studies have applied that understanding to evaluate landscape influences on freshwater systems at regional, national, and continental extents for lakes (e.g., Soranno et al. 2009), streams (e.g., Esselman et al. 2003; Hill et al. 2017), and riverscapes (Fausch et al. 2002), and from a historical perspective (Harding et al. 1998; Brown et al. 2009).

Major Factors Contributing to Our Evolved Understanding

Several factors have contributed much to our evolving understanding of landscape influences on freshwater ecosystems and biological assemblages. These include advances in spatial frameworks, databases, analytical approaches, computing power, and stressor sources, along with the usefulness of the landscape approach for management (Infante et al. 2019, this volume). The development of spatial frameworks to account for hierarchical influences of landscape features on water bodies are applicable to both lotic and lentic systems (Soranno et al. 2009; Wang et al. 2011). Such frameworks identify fundamental spatial units for study (e.g., water body, buffer area, and catchment). Development of spatial frameworks has occurred together with development of large, computer-accessible regional, national, and multinational databases describing characteristics of freshwater systems and landscapes. Data are then attributed to units, providing an organized means for analyzing landscape-aquatic linkages at multiple spatial extents. These large databases have substantially advanced our capability to study landscape-aquatic system interactions across large areas (e.g., Hill et al. 2017, 2018).

We have also experienced advances in analytical understanding along with greater accessibility of novel statistical techniques. For example, Allan (2004) described the difficulty in accounting for covariation in natural and anthropogenic landscape features as a primary challenge related to quantifying landscape influences on river ecosystems. Today, we see greater application of different strategies that aid in teasing apart natural and anthropogenic influences, as described in this book and elsewhere (Infante et al. 2019).

Improvements in geographic information system technology and computing power have enabled a new era of landscape-aquatic studies. This capability makes it feasible to study landscape-aquatic interactions for entire nations or large river basins to develop new aquatic landscape concepts and to improve science-based management and decision making.

Ecologists and environmental scientists have increasingly recognized the importance of stressor linkages beyond a water body's catchment. These include the aquatic effects of climate change (Winfield et al. 2016; Isaac et al. 2018), groundwater connections (Falke et al. 2011; Warner et al. 2012; Perkin et al. 2017), and atmospheric dispersal of contaminants and nutrients (Baker et al. 1991; Peterson et al. 2007; Landers et al. 2008; Meire et al. 2012).

A final important factor contributing to our advance in understanding landscape influences on freshwaters has been the value of landscape information to support conservation and management of freshwater ecosystems across large spatial extents. Because information on characteristics and condition of freshwater sites are often not available consistently over large regions, the landscape approach is increasingly used to characterize features or assess conditions (Kuehne et al. 2017). The impetus for these regional-scale efforts may result from multiple issues that need to be addressed, such as protection of freshwater ecosystems and land use planning (Steel et al. 2010), which can be best integrated using a landscape approach.

These advancements are further described in detail throughout this book with the aims of (1) reporting on major advancements in the knowledge of the influences of landscape on freshwater ecosystems since Hughes et al. (2006) and (2) identifying major challenges and knowledge gaps to direct future research and improve freshwater resource management (Infante et al. 2019).

Structure of the Book and Key Findings

The next 19 chapters in this book are ordered by seven different groups with some degree of overlap, followed by a final synthesizing chapter. Yates et al. (2019, this volume) calls for spending additional effort at finding and interpreting census data, remote sensing images and visual surveys, in addition to widely available geographic information system data, for better evaluating the impacts of anthropogenic pressures on aquatic ecosystems.

The next three chapters employ markedly different riverscape approaches for assessing natural and anthropogenic limits on fish species. Davis et al. (2019, this volume) used Chinook Salmon *Oncorhynchus tshawytscha* riverscape genetics to determine that gene flow was limited more by migration-channel gradients than site-specific factors in an Oregon river basin. Falke et al. (2019, this volume) conducted spatially explicit riverscape modeling and bioenergetics to determine that warmer years increased juvenile Chinook Salmon growth potential in an Alaskan river basin. Examining riverscapes on the five largest Hawaiian islands, Martin et al. (2019, this volume) concluded that waterfalls excluded nonnative aquatic vertebrate and macroinvertebrate species except where there were high levels of anthropogenic disturbance.

Three chapters then focus on landscape–lake interactions but with differing objectives. Treating lakes like islands, Alofs (2019, this volume) used boosted regression trees to establish that nonnative Rainbow Smelt *Osmerus mordax* were most likely to invade as a function of site-specific chemical and physical habitat factors plus the transport and use of live bait by anglers. Allerton et al. (2019, this volume) showed that the relative locations of the number and area of connected lakes upgradient of a river site differ among river basins, likely altering the connectivity, stability, and resilience of river networks and the lakes themselves. Jacobson et al. (2019, this volume) applied general additive modeling to catchment land use and site and remotely sensed lake data to estimate historic and current lake coldwater habitat. They concluded that there have been substantial reductions in lake coldwater habitat as a function of climate change and eutrophication from agriculture.

The following two chapters are directed towards improved landscape-aquatic survey designs and sampling methodologies. Bruckerhoff and Gido (2019, this volume) used available data and randomization tests to evaluate the effects of uniform, random, and skewed site distributions on landscape-fish assemblage responses. They concluded that uniform site distributions and large numbers of sites (>50) were best for teasing out the subtle effects of landscape gradients and that uniformity could best be achieved through use of equal-random-stratified designs. In a multiyear study in Australia's Murray–Darling basin, Robinson et al. (2019, this volume) used such a survey design weighted by drainage area to ensure sufficient numbers of sites for smaller hydrologic units. They also evaluated alternative fish sampling methods and demonstrated that a standardized electrofishing protocol was preferable for assemblage-level assessments of sites and hydrologic units. Based on generalized additive modeling, they found that small-bodied species were collected more often in dry periods, whereas large-bodied species were less often collected.

The subsequent three chapters related landscape conditions to stream habitats. Ross et al. (2019, this volume) reported that natural landscape characteristics, and especially timber harvest practices, influenced multiple instream wood variables in Alaskan streams. Also studying Alaskan streams, Cooper et al. (2019, this volume) developed a cumulative landscape pressure index and determined that streams overall had very low threat levels, although those for anadromous fish were nearly twice as high, especially when predicting future threats from climate change and mining. Negishi (2019, this volume) studied multiple sites along a single river segment in Japan and found that nitrate, chlorophyll, and the abundance of larval benthic and hyporheic and adult benthic insects increased with percent cropland. However, hyporheic adults were maximized at intermediate levels of cropland, indicating the importance of assessing the vertical dimension of streambeds.

Five chapters used hydrologic units or regions to partition observed patterns in biotic responses to landscape conditions. Two studies from the same four hydrologic units in the Cerrado biome of Brazil indicated differing geologies, climate, and land use (Callisto et al. 2019; Pompeu et al. 2019; both this volume). Higher stream-site macroinvertebrate taxa richness and multimetric index scores were associated with lower proportions of catchment agriculture. However, fish species richness and diversity were highest in one of the hydrological units with a high amount of agriculture, and land use explained little of the variability in fish assemblage structure. Deweber et al. (2019, this volume) used land use and physiography to classify hydrologic units through use of multivariate regression tree analysis. Then, they tested the classification with available fish data. They found that the relationships between land use and various fish assemblage metrics differed among land use/physiography classes in the upper Mississippi River basin. Thornbrugh and Infante (2019) used existing hydrologic regions to assess the effects of land use and channel fragmentation on stream fish assemblages. Through use of cascade multivariate regression trees, they found differences in dominant pressures (e.g., fragmentation versus land use) among and within regions. Herlihy et al. (2019, this volume) took a markedly different approach for analyzing fish assemblage differences. Rather than classify landscapes, they clustered available fish data and determined greater group mean classification strength than that obtained from existing map classifications of ecoregions or hydrologic units (which were weaker than geopolitical regions).

The penultimate chapters employed landscape variables for predictive conservation modeling. Perkin et al. (2019, this volume) employed gradient forest analysis to assess the relative importance of catchment and riverscape pressures on fish distributions at network- and local-catchment extents. Key predictors differed by landscape extent but explained comparable levels of variability. They then used these models to predict threats to unsampled local- and network-catchments. Mostafavi et al. (2019, this volume) used nine different modeling approaches to predict known and potential distributions of nine sensitive fish species in Iran from site and climatic variables. Key predictor variables differed among the species. Although random forest and generalized boosting method had the highest performance and surface range envelopes the lowest, the authors recommended using ensemble modeling.

Based on the preceding chapters as well as selected literature, Infante et al. (2019) summarizes what the editors deem as the major advancements, challenges, and gaps in understanding as regards landscape influences on freshwater ecosystems. Finally, because of considerable confusion and misuse of terms, we add a glossary that defines and explains the choice of terms that we used in this book.

References

Allan, J. D. 1995. Stream ecology—structure and function of running waters. Chapman and Hall, London.

Allan, J. D. 2004. Landscapes and riverscapes: the influence of land use on stream ecosystems. Annual Review of Ecology, Evolution, and Systematics 35:257–284.

Allerton, M. L., N. E. Jones, S. J. Melles, and B. J. Schmidt. 2019. Ecologically relevant structural metrics for complex stream–lake networks. Pages 125–139 *in* R. M. Hughes, D. M. Infante, L. Wang, K. Chen, and B. F. Terra, editors. Advances in understanding landscape influences on freshwater habitats and biological assemblages. American Fisheries Society, Symposium 90, Bethesda, Maryland.

Alofs, K. M. 2019. Landscape approaches to understanding invasions in inland lakes: Rainbow Smelt as a case study. Pages 105–123 *in* R. M. Hughes, D. M. Infante, L. Wang, K. Chen, and B. F. Terra, editors. Advances in understanding landscape influences on freshwater habitats and biological assemblages. American Fisheries Society, Symposium 90, Bethesda, Maryland.

Baker, L. A., A. T. Herlihy, P.R. Kaufmann, and J. M. Eilers. 1991. Acidic lakes and streams in the United States: the role of acidic deposition. Science 252:1151–1154.

Brown, L. R., T. F. Cuffney, J. F. Coles, F. Fitzpatrick, G. McMahon, J. Steuer, A. H. Bell, and J. T. May. 2009. Urban streams across the USA: lessons learned from studies in 9 metropolitan areas. Journal of the North American Benthological Society 28:1051–1069.

Bruckerhoff, L. A., and K. B. Gido. 2019. Assessing site-selection strategies for modeling the influence of landscape factors on stream fish assemblages. Pages 159–177 *in* R. M. Hughes, D. M. Infante, L. Wang, K. Chen, and B. F. Terra, editors. Advances in understanding landscape influences on freshwater habitats and biological assemblages. American Fisheries Society, Symposium 90, Bethesda, Maryland.

Callisto, M., D. R. Macedo, M. S. Linares, and R. M. Hughes. 2019. Multistatus and multispatial scale assessment of landscape effects on benthic macroinvertebrates in the Neotropical savanna. Pages 275–302 *in* R. M. Hughes, D. M. Infante, L. Wang, K. Chen, and B. F. Terra, editors. Advances in understanding landscape influences on freshwater habitats and biological assemblages. American Fisheries Society, Symposium 90, Bethesda, Maryland.

Cooper, A. R., D. M. Infante, J. A. Ross, W. M. Daniel, K. Herreman, C. C. Estes, and G. E. Whelan. 2019. Contemporary and future landscape-based anthropogenic disturbances to Alaskan streams: implications for anadromous fish habitat conservation. Pages 229–253 *in* R. M. Hughes, D. M. Infante, L. Wang, K. Chen, and B. F. Terra, editors. Advances in understanding landscape influences on freshwater habitats and biological assemblages. American Fisheries Society, Symposium 90, Bethesda, Maryland.

Davis, C. D., C. W. Epps, R. L. Flitcroft, and M. A. Banks. 2019. Beyond isolation by distance: riverscape effects on genetic structure of fall-run Chinook Salmon. Pages 35–56 *in* R. M. Hughes, D. M. Infante, L. Wang, K. Chen, and B. F. Terra, editors. Advances in understanding landscape influences on freshwater habitats and biological assemblages. American Fisheries Society, Symposium 90, Bethesda, Maryland.

Deweber, J. T., L. Sleezer, and E. A. Frimpong. 2019. A new regionalization framework to quantify how physiography mediates the effect of land use on stream fishes. Pages 321–350 *in* R. M. Hughes, D. M. Infante, L. Wang, K. Chen, and B. F. Terra, editors. Advances in understanding landscape influences on freshwater habitats and biological assemblages. American Fisheries Society, Symposium 90, Bethesda, Maryland.

Esselman, P. C., D. M. Infante, L. Wang, A. Cooper, D. Wieferich, Y. P. Tsang, D. Thornbrugh, and W. W. Taylor. 2013. Regional fish community indicators of landscape disturbance to catchments of the conterminous United States. Ecological Indicators 26:163–173.

Falke, J. A., K. D. Fausch, R. Magelky, A. Aldred, D. S. Durnford, L. K. Riley, and R. Oad. 2011. The role of groundwater pumping and drought in shaping ecological futures for stream fishes in a dryland river basin of the western Great Plains, USA. Ecohydrology 4:682–697.

Falke, J. A., B. M. Huntsman, and E. R. Schoen. 2019. Climatic variation drives growth potential of juvenile Chinook Salmon along a subarctic boreal riverscape. Pages 57–82 *in* R. M. Hughes, D. M. Infante, L. Wang, K. Chen, and B. F. Terra, editors. Advances in understanding landscape influences on freshwater habitats and biological assemblages. American Fisheries Society, Symposium 90, Bethesda, Maryland.

Fausch, K. D., C. E. Torgersen, C.V. Baxter, and H. W. Li. 2002. Landscapes to riverscapes: bridging the gap between research and conservation of stream fishes. BioScience 52:483–498.

Frissell, C. A., W. J. Liss, C. E. Warren, and M. D. Hurley. 1986. A hierarchical framework for stream habitat classification: viewing streams in a watershed context. Environmental Management 10:199–214.

Gámez-Virués, S., D. J. Perović, M. M. Gossner, C. Börschig, N. Blüthgen, H. De Jong, N. K. Simons, A. M. Klein, J. Krauss, G. Maier, and C. Scherber. 2015. Landscape simplification filters species traits and drives biotic homogenization. Nature Communications 6:8568.

Harding, J. S., E. F. Benfield, P. V. Bolstad, G. S. Helfman, and E. B. D. Jones, III. 1998. Stream biodiversity: the ghost of land use past. Proceedings of the National Academy of Sciences of the United States of America 95:14843–14847.

Herlihy, A. T., J. C. Sifneos, R. M. Hughes, D. V. Peck, and R. M. Mitchell. 2019. Lotic fish assemblage clusters across the conterminous United States and their associations with environmental variables. Pages 385–408 *in* R. M. Hughes, D. M. Infante, L. Wang, K. Chen, and B. F. Terra, editors. Advances in understanding landscape influences on freshwater habitats and biological assemblages. American Fisheries Society, Symposium 90, Bethesda, Maryland.

Hill, R. A., E. W. Fox, S. G. Leibowitz, A. R. Olsen, D. J. Thornbrugh, and M. H. Weber. 2017. Predictive mapping of the biotic condition of conterminous U.S. rivers and streams. Ecological Applications 27:2397–2415.

Hill, R. A., M. H. Weber, R. M. Debbout, S. G. Leibowitz, and A. R. Olsen. 2018. The lake-catchment (LakeCat) dataset: characterizing landscape features for lake basins within the conterminous USA. Freshwater Science 37:208–221.

Hughes, R. M., L. Wang, and P. W. Seelbach, editors. 2006. Landscape influences on stream habitats and biological assemblages. American Fisheries Society, Symposium 48, Bethesda, Maryland.

Hynes, H. B. N. 1975. The stream and its valley. International Vereinigung für Limnologie Theoretische and Angewendte Verhandlungen 19:1–5.

Infante, D. M., L. Wang, R. M. Hughes, K. Chen, and B. F. Terra. 2019. Advancements, challenges, and gaps in understanding landscape influences on freshwater ecosystems. Pages 463–495 *in* R. M. Hughes, D. M. Infante, L. Wang, K. Chen, and B. F. Terra, editors. Advances in understanding landscape influences on freshwater habitats and biological assemblages. American Fisheries Society, Symposium 90, Bethesda, Maryland.

Isaac, D. J., C. H. Luce, D. L. Horan, G. L., Chandler, S. P. Wollrab, and D. E. Nagel. 2018. Global warming of salmon and trout rivers in the northwestern U.S.: road to ruin or path through purgatory? Transactions of the American Fisheries Society 147:566–585.

Jacobson, P. C., G. J. A. Hansen, L. G. Olmanson, K. E. Wehrly, C. L. Hein, and L. B. Johnson. 2019. Loss of coldwater fish habitat in glaciated lakes of the Midwestern United States after a century of land use and climate change. Pages 141–157 *in* R. M. Hughes, D. M. Infante, L. Wang, K. Chen, and B. F. Terra, editors. Advances in understanding landscape influences on freshwater habitats and biological assemblages. American Fisheries Society, Symposium 90, Bethesda, Maryland.

Johnson, L. B., and G. E. Host. 2010. Recent developments in landscape approaches for the study of aquatic ecosystems. Journal of the North American Benthological Society 29:41–66.

Kuehne, L. M., J. D. Olden, A. L. Strecker, J. J. Lawler, and D. M. Theobald. 2017. Past, present, and future of ecological integrity assessment for fresh waters. Frontiers in Ecology and the Environment 15:197–205.

Landers, D. H., S. L. Simonich, D. A. Jaffe, L. H. Geiser, D. H. Campbell, A. R. Schwindt, A. R., C. B. Schreck, M. L. Kent, W. D. Hafner, H. E. Taylor, K. J. Hageman, S. Usenko, L. K. Ackerman, J. E. Schrlau, N. L. Rose, T. F. Blett, and M. M. Erway. 2008. The fate, transport, and ecological impacts of airborne contaminants in western national parks (USA). U.S. Environmental Protection Agency, EPA/600/R-07/138, Corvallis, Oregon.

Martin, B., Y.-P. Tsang, R. W. Tingley, III, H. Cliverd, and D. M. Infante. 2019. Characterizing natural barriers to nonnative stream fauna in Hawai'i. Pages 83–104 *in* R. M. Hughes, D. M. Infante, L. Wang, K. Chen, and B. F. Terra, editors. Advances in understanding landscape influences on freshwater habitats and biological assemblages. American Fisheries Society, Symposium 90, Bethesda, Maryland.

Meire, R. O., S. E. Lee, Y. Yao, A.C. Targino, J.P.M. Torres, and T. Harner. 2012. Seasonal and altitudinal variations of legacy and current-use pesticides in the Brazilian tropical and subtropical mountains. Atmospheric Environment 59:108–116.

Mostafavi, H., Z. Kordjazi, R. Valavi, H. Shafizadeh-Moghadam, J. Kambouzia, and D. M. Infante. 2019. Ensemble modelling of sensitive stream fish species distributions in Iran: expanding knowledge to aid species conservation. Pages 441–462 *in* R. M. Hughes, D. M. Infante, L. Wang, K. Chen, and B. F. Terra, editors. Advances in understanding landscape influences on freshwater habitats and biological assemblages. American Fisheries Society, Symposium 90, Bethesda, Maryland.

Negishi, J. N. 2019. Does the proportion of crop fields in a small catchment predict coupled benthic-hyporheic invertebrate responses? Pages 255–274 *in* R. M. Hughes, D. M. Infante, L. Wang, K. Chen, and B. F. Terra, editors. Advances in understanding landscape influences on freshwater habitats and biological assemblages. American Fisheries Society, Symposium 90, Bethesda, Maryland.

Perkin, J. S., K. B. Gido, J. A. Falke, K. D. Fausch, H. Crockett, E. R. Johnson, and J. Sanderson. 2017. Groundwater declines are linked to changes in Great Plains stream fish assemblages. Proceedings of the National Academy of Sciences of the United States of America 114:7373–7378.

Perkin, J. S., J. C. Wellemeyer, and J. D. Fore. 2019. Multiscale fish assemblage distribution models to guide riverscape conservation planning. Pages 409–440 *in* R. M. Hughes, D. M. Infante, L. Wang, K. Chen, and B. F. Terra, editors. Advances in understanding landscape influences on freshwater habitats and biological assemblages. American Fisheries Society, Symposium 90, Bethesda, Maryland.

Peterson, S. A., J. Van Sickle, A.T. Herlihy, and R. M. Hughes. 2007. Mercury concentration in fish from streams and rivers throughout the western United States. Environmental Science and Technology 41:58–65.

Pompeu, P. S., C. G. Leal, D. R. Carvalho, N. T. Junqueira, M. A. Castro, and R. M. Hughes. 2019. Effects of catchment land use on stream fish assemblages in the Brazilian savanna. Pages 303–320 *in* R. M. Hughes, D. M. Infante, L. Wang, K. Chen, and B. F. Terra, editors. Advances in understanding landscape influences on freshwater habitats and biological assemblages. American Fisheries Society, Symposium 90, Bethesda, Maryland.

Robinson, W. A., M. Lintermans, J. H. Harris, and F. Guarino. 2019. A landscape-scale electrofishing monitoring program can evaluate fish responses to climatic conditions in the Murray–Darling River system, Australia. Pages 179–201 *in* R. M. Hughes, D. M. Infante, L. Wang, K. Chen, and B. F. Terra, editors. Advances in understanding landscape influences on freshwater habitats and biological assemblages. American Fisheries Society, Symposium 90, Bethesda, Maryland.

Ross, J. A., D. M. Infante, D. J. Martin, and K. Herreman. 2019. Influences of local and landscape characteristics, including past timber harvest practices, on large wood in streams draining young-growth Southeast Alaska catchments. Pages 203–227 *in* R. M. Hughes, D. M. Infante, L. Wang, K. Chen, and B. F. Terra, editors. Advances in understanding landscape influences on freshwater habitats and biological assemblages. American Fisheries Society, Symposium 90, Bethesda, Maryland.

Snelder, T. H., and B. J. F. Biggs. 2002. Multiscale river environmental classification for water resources management. Journal of the American Water Resources Association 38:1225–1239.

Soranno, P. A., K. E. Webster, K. S. Cheruvelil, and M. T. Bremigan. 2009. The lake landscape-context framework: linking aquatic connections, terrestrial features and human effects at multiple spatial scales. Verhandlungen der Internationalen Vereinigung für theoretische und angewandte Limnologie 30:695–700.

Steel, A., R. M. Hughes, S. Schmutz, S. Muhar, M. Poppe, C. Trautwein, M. Fukushima, H. Shimazaki, J. Young, B. Feist, A. Fullerton, and B. Sanderson. 2010. Are we meeting the challenges of landscape scale riverine research? A review. Living Reviews in Landscape Research [online serial] 4:1.

Thornbrugh, D. J., and D. M. Infante. 2019. Landscape effects on stream fishes: broadscale responses to anthropogenic land use across temperate mesic regions of the United States. Pages 351–383 *in* R. M. Hughes, D. M. Infante, L. Wang, K. Chen, and B. F. Terra, editors. Advances in understanding landscape influences on freshwater habitats and biological assemblages. American Fisheries Society, Symposium 90, Bethesda, Maryland.

Tonn, W. M. 1990. Climate change and fish communities: a conceptual framework.

Transactions of the American Fisheries Society 119:337–352.

Wang, L., D. M. Infante, P. Esselman, A. Cooper, D. Wu, W. Taylor, D. Beard, G. Whelan, and A. Ostroff. 2011. A hierarchical spatial framework and database for the national river fish habitat condition assessment. Fisheries 36:436–449.

Ward, J.V., F. Malard, and K. Tockner. 2002. Landscape ecology: a framework for integrating pattern and process in river corridors. Landscape Ecology 17:35–45.

Warner, N. R., R. B. Jackson, T.H. Darrah, S.G. Osborn, A. Down, K. Zhao, A. White, and A. Vengosh. 2012. Geochemical evidence for possible natural migration of Marcellus Formation brine to shallow aquifers in Pennsylvania. Proceedings of the National Academy of Sciences of the United States of America 109:11961–11966.

Winfield, I. J., C. Baigun, P. A. Balykin, B. Becker, Y. Chen, A. F. Filipe, Y. V. Gerasimov, A. L. Godinho, R. M. Hughes, J. D. Koehn, D. N. Kutsyn, V. Mendoza-Portillo, T. Oberdorff, A. M. Orlov, A. P. Pedchenko, F. Pletterbauer, I. G. Prado, R. Rösch, and S. J. Vatland. 2016. International perspectives on the effects of climate change on inland fisheries. Fisheries 41:399–405.

Yates, A. G., J. M. Culp, R. C. Bailey, and P. A. Chambers. 2019. Strengths and weaknesses of data sources for describing exposure of aquatic ecosystems to human activity. Pages 11–33 *in* R. M. Hughes, D. M. Infante, L. Wang, K. Chen, and B. F. Terra, editors. Advances in understanding landscape influences on freshwater habitats and biological assemblages. American Fisheries Society, Symposium 90, Bethesda, Maryland.

American Fisheries Society Symposium 90:11–33, 2019
© 2019 by the American Fisheries Society

Strengths and Weaknesses of Data Sources for Describing Exposure of Aquatic Ecosystems to Human Activity

Adam G. Yates*
Department of Geography, Western University and Canadian Rivers Institute
1151 Richmond Street, London, Ontario N6A 5C2, Canada

Joseph M. Culp
Environment and Climate Change Canada
Department of Geography and Environmental Studies, Wilfrid Laurier University
75 University Avenue West, Waterloo, Ontario N2L 3C5, Canada

Robert C. Bailey
University of Ontario Institute of Technology
2000 Simcoe Street North, Oshawa, Ontario L1G 0C5, Canada

Patricia A. Chambers
Environment and Climate Change Canada, Canada Centre for Inland Waters
867 Lakeshore Road, Post Office Box 5050, Burlington, Ontario L7R 4A6, Canada

Abstract.—Over the past decade, numerous studies have identified correlative relationships between aquatic biota and human activities at landscape scales. In addition to demonstrating the pervasive effects of these activities on aquatic biota, these findings have encouraged researchers to suggest that predictive relationships between human activities and aquatic biota could be used to enhance diagnostic power of biological assessments, predict future changes in species distributions, and inform land-use planning. However, to achieve these important goals, descriptions of human activities will need to become more detailed than the simple land use/land cover classifications frequently used. Our purpose is to highlight four sources of human activity data (existing geographic information system layers, census data, remotely sensed images, and visual landscape surveys) that can be used to increase the level of detail with which the human environment is described. Strengths and weaknesses of each data source are discussed and methods for adapting those data to aquatic studies are described by drawing on experiences from studies in the agricultural landscapes of southern Manitoba and southwestern Ontario, Canada. Based on the observations and lessons learned from our previous experiences, we make recommendations for how researchers can identify and apply the data sources that best meet their needs. We also discuss challenges and possible solutions for applying the described data sources as well as for improving data availability in the future. Moreover, we encourage aquatic researchers to allot more time to detailed description of human

* Corresponding author: adam.yates@uwo.ca

activities because we believe this to be an effective approach to improving our ability to predict the effects of human activity and thus better assist decision makers in protecting aquatic ecosystems.

Introduction

Initiated by Hynes' (1975) eloquent description of how "the valley rules the stream" and fueled by rapid advancements in geographic information systems (GIS), the number of aquatic ecology studies including a landscape component has substantially increased over the past few decades (Hughes et al. 2006; Johnson and Host 2010). Now, more than 40 years following the publication of "The Stream and Its Valley," the importance of catchment character has been unequivocally demonstrated (Allan 2004). Important drivers of the structure and function of aquatic biota include many natural characteristics of the catchment, such as geology, climate, elevation, and latitude (e.g., Richards et al. 1996; Wang et al. 2003; Chaves et al. 2005; Gerth et al. 2013). However, it may be the many studies demonstrating pervasive effects of human activities on aquatic ecosystems that have most changed the way aquatic ecosystems are perceived and managed (e.g., Wang et al. 2001; Morse et al. 2003; Bruns 2005; Nislow and Lowe 2006; Von Schiller et al. 2008; Vandewalle et al. 2010; Maloney and Weller 2011; Krynak and Yates 2018).

The knowledge that human activities occurring at the catchment scale can strongly influence the structure and function of aquatic ecosystems has led to the initiation of numerous aquatic monitoring programs in many jurisdictions—state/provincial (Ohio EPA 1987; Jones et al. 2007), national (Reynoldson et al. 1999; Paulsen et al. 2008), and international (European Water Framework Directive 2000)—in nearly every continent of the world (e.g., Reynoldson et al. 1999; Davies 2000; Ollis et al. 2006; Poquet et al. 2009; Jun et al. 2012). In addition to providing better understanding of the ecological condition of aquatic ecosystems worldwide, the large ecological data sets generated by these monitoring programs have prompted researchers to go beyond simple assessment of aquatic ecosystems towards predictive analyses aimed at guiding ecosystem management and land-use planning (Bailey et al. 2007). To generate these analyses, land use/land cover (LULC) layers depicting the nature and extent of human activity have been used. However, LULC descriptions homogenize what is often a diverse landscape into a small number of generalized categories (e.g., agriculture). As a result, a large proportion of variation in human activities known to affect aquatic biota (e.g., fertilizer application [Winter et al. 2002], housing density [Cookson and Schorr 2009], housing proximity [Macedo et al. 2014], type and age of agriculture or deforestation [Leal et al. 2016; Leitão et al. 2018], and road-stream crossings [Leal et al. 2016; Leitão et al. 2018]) may remain undocumented.

Using more comprehensive descriptors of human activity could improve predictive models in three ways. First, studies have demonstrated that including more detailed information can significantly improve the strength of human activity–biota relationships (Yates and Bailey 2010; Yates et al. 2014), suggesting that the predictive power and accuracy of models would be increased by using more such descriptions. Second, relationships between LULC and ecological condition may change over time as the nature of the activities (e.g., crop type, housing density) associated with a land-use type vary (Allan 2004). Models based on more thorough descriptions of human activities, especially those that include measures of management practices, are more likely to capture such changes. Third, and perhaps most important, including more detailed descriptors of human activity also aligns predictive models more closely with the scale at which human activities are managed. Planners and managers generally do not make decisions based on land cover, but rather on the extent and intensity of specific activities (e.g., housing and livestock densities, forest harvest intensity, water withdrawal rate, and fertilizer application rate). Inclusion of more in-depth descriptors of LULC will more readily facilitate translation of research findings into improved management strategies and policy.

The popularity of LULC data stems from its wide availability and the ease with which it can be applied. There are, however, a variety of sources that provide more comprehensive human activity data that can and should be more widely applied to develop more detailed GIS data sets that can be used in predictive models for stream assessment, land-use planning, and species-distributions processes. Therefore, the goal of this chapter is to describe four sources of human activity data (i.e., existing GIS layers, census data, remotely sensed images, and visual landscape surveys) available to aquatic scientists to increase the level of detail with which the human environment is described (Table 1). Strengths and weaknesses of each data source are discussed, and methods for adapting those data to aquatic studies are described by drawing on experiences from two studies in the agricultural landscapes of southwestern Ontario and southern Manitoba, Canada.

Sources of Human Activity Data

Existing Geographic Information System Layers

Geographic information systems have become a common component of aquatic studies in part because GIS software has become more user-friendly and increases in computing power have allowed researchers to rapidly analyze data sets covering very large extents (Johnson and Gage 1997). Use of GIS as an information archiving and management tool has also become standard practice in many public and private sectors, and as a result, there has been an enormous increase in the number and scope of GIS layers available, many of which can be of use to aquatic ecologists looking to describe human activities. Indeed, the ultimate end of data collected from the four sources described in this chapter will likely be incorporation into a GIS through processing the data into a GIS layer.

Creators and publishers of GIS layers include agencies at all levels of government (i.e., federal, state/provincial, and municipal), and indeed, these are the major sources of GIS data that most aquatic ecologist are familiar with and use in their research. Although the types of activities described by these agencies vary, GIS layers commonly generated by federal and state sources tend to be of greater geographic coverage (e.g., national or state/provincial) and coarser detail (e.g., transportation, land use). However, more detailed layers of government-regulated activities such as mining activities, wastewater treatment facilities, and forestry are often maintained by specific departments. Government data layers are usually publicly available and can often be downloaded directly through Web-based spatial data clearinghouses (e.g., Manitoba Land initiative [https://mli2.gov.mb.ca] and California Geoportal [http://http://portal.gis.ca.gov/geoportal]).

Less well known is the breadth of GIS layers maintained by nongovernment organizations. We have found that marketing boards, crop insurance agencies, agricultural federations, and other such organizations within the agricultural sector maintain highly detailed and current GIS layers describing a wide variety of activities (e.g., livestock farming) and management practices (e.g., fertilizer application rates). Corporations and federations in other sectors (e.g., forestry, oil and gas), as

Table 1. Characteristics and examples of GIS layers, census data, remote sensing images, and visual landscape surveys that are useful for increasing the detail with which human activities are described.

Data source	Accessibility	Resolution	Coverage	Example sources
Existing GIS layers	Government layers —Web access, no to minimal restrictions	Government layers —variable	Government layers —municipal to national	Government layers 1. Geogratis[a] 2. Ontario Land Information[b]
	Nongovernment layers —access through organization, highly restricted	Nongovernmental layers —fine	Nongovernmental layers —local to regional	Nongovernmental layers 1. Manitoba Agricultural Services Corporation[c]
Census data	Web access, data summarized to census-defined boundaries, some data may be suppressed	Coarse	National	1. American Census[d] 2. Canadian Census of Agriculture[e] 3. Portuguese Census of Agriculture[f]
Remotely sensed images	Web access, no restrictions, may be a purchasing fee, often available through government servers	Coarse to fine	Global	1. Global Earth[g] 2. Quickbird[h] 3. SPOT[i]
Visual landscape surveys	Not applicable	Activities are finely resolved	Determined by scope of survey	

[a] geogratis.cgdi.gc.ca
[b] https://www.ontario.ca/page/land-information-ontario
[c] www.masc.mb.ca/masc.nsf/mmpp_index.html
[d] http://factfinder.census.gov
[e] www.statcan.gc.ca/eng/ca2016
[f] www.ine.pt/xportal/xmain?xpid=INE&xpgid=ine_main
[g] www.google.com/earth
[h] www.digitalglobe.com/resources/satellite-information
[i] www.spot.com/

well as special interest groups, are also potential sources of GIS layers (e.g., forest management, protected lands). Yet despite existence of these data, their identification and use are hampered by the fact that these layers are private property and as such are not widely publicized, easily accessed, and consistent in quality.

Once the existence of such GIS layers is identified, there remain two major constraints to employing the data in research. First, and most important, are issues related to individual privacy. Many of these layers incorporate personal or corporate information gathered with specific permissions regarding the use and dissemination of that information. Lengthy negotiation processes are sometimes necessary to develop a data-sharing agreement that meets research requirements while maintaining confidentiality. Recoding data to eliminate personal identifiers and/or scaling the data to a larger unit (e.g., catchment rather than property boundaries) are effective ways of meeting an organization's privacy requirements while still maintaining the integrity of the data set for research purposes. Second, is the issue of spatial extent. Unlike government layers, which tend to cover large geographic jurisdictions, private layers tend to incorporate highly specific data that describe small and sometimes discontinuous areas. As a result, it may be necessary to gather layers from multiple organizations to generate a layer with coverage that matches the desired study area and descriptors that integrate with study objectives. When using existing GIS layers, any available metadata regarding the origins and specifics of the layer must be carefully examined. Doing so will increase the probability that the layer is portraying the desired information at the scale and level of accuracy required for the project. Geographic information system layers without metadata should be avoided unless the quality of the data can be confirmed.

Census Data

Census data are underused by aquatic ecology researchers (but see Wang et al. 2008; Diebel and Vander Zanden 2009; Falcone et al. 2009; Bartsch et al. 2015) despite being widely used in other research fields, including agricultural (Xiong et al. 2008), medical (Booth et al. 2017), and soil sciences (Goidts et al. 2009). Census data are a consistent and widely available source of human activity data that is reliably collected in many countries, usually at a frequency of 5–10 years. In addition, through the encouragement and support of the World Programme for the Census of Agriculture (FAO 2005), many nations also conduct censuses of agriculture (e.g., Canadian Census of Agriculture, www.statcan.gc.ca/eng/ca2016; U.S. Census of Agriculture, www.nass.usda.gov/AgCensus/). Although the descriptors collected vary and rarely include actual input values (e.g., kilograms of fertilizer applied per unit area), agricultural censuses do collect a wealth of data on numerous agricultural activities and management practices, including livestock numbers, crop types, fertilizer application, and tillage practices.

The strengths of census data as a source of human activity data stem largely from the fact that their collection and maintenance is legislated in most nations, leading to consistent and substantial expenditure of resources on the collection, processing, and publication of the data. Completion of surveys is usually mandatory and generates comprehensive databases at a national scale. Furthermore, the data are made widely available through easily used online interfaces (e.g., U.S. Census Bureau's American Factfinder, http://factfinder.census.gov), making it a straightforward process for researchers to access data for their study area.

Using census data for research of aquatic ecosystems is, however, constrained by confidentiality procedures aimed at protecting the privacy of contributing citizens. These procedures lead to the data being summarized to institutionally based boundaries (e.g., counties). Furthermore, in agricultural censuses, data from areas with small numbers of farmers can be suppressed. Although a few census programs provide services that can summarize data to meet specific project objectives, these services are often expensive and may still be subject to data suppression if activity levels are low (e.g., less than three farms reporting for the Canadian Census of Agriculture). Classification or area-weighted interpolations are necessary to adapt the census data to the catchment boundaries usually used for aquatic studies.

In our experience, the selection of a method for interpolating census data to study unit boundaries (e.g., catchment) is best determined based on the size of the study units and regional extent of the study. When study units are much larger than census divisions, such as large river basins, data from multiple divisions are aggregated through interpolation, and thus, the resulting analyses are susceptible to the modified areal unit problem (Openshaw 1984; Maas-Hebner et al. 2015). In particular, aggregation of census data invokes the scale problem component of the modified areal unit problem, which is associated with the variation in results that may be obtained when areal data are combined into increasingly larger area units for analysis. Although having no effect on the regional mean, regional variation can be reduced as divisions are successively aggregated, potentially reducing a researcher's ability to detect relationships between activity descriptions and ecological conditions (Jelinski and Wu 1996). The modified areal unit problem can be minimized through the use of area-based interpolation, particularly when interpolation uses dasymetric mapping techniques.

Area-based interpolation (Lam 1983) is commonly used to translate census data from census boundaries to hydrologic units or other study units of interest (e.g., ecoregions, catchments [Agriculture and Agri-food Canada 2008]). This method uses geographic overlays of the source (i.e., census boundaries) and target (e.g., catchment boundaries) layers to determine the proportion of the area in each study unit that is comprised of a particular census division (Figure 1). Values for human activities described by the census are then weighted based on the calculated proportions. For activities that are LULC-specific (e.g., area of agricultural land to which commercial fertilizers are applied), interpolated values may be improved substantially through dasymetric mapping techniques (Holt et al. 2004). Dasymetric mapping more accurately depicts patterns of geographic data by dividing the original source zones of a data set (e.g., census boundaries) into smaller spatial units using boundaries, usually LULC types, that better reflect the spatial distribution of the variable of interest. Using a LULC layer as a mask to dasymetrically model census data thus ensures that area-weighted proportions are calculated based on the area of a relevant land-cover type present in the study unit as opposed to the total study-unit area that overlaps the census division.

The accuracy of area-based interpolation can decline as the study-unit size decreases (Zoraghein et al. 2016). Declines in accuracy will be particularly rapid once the study-unit size becomes small enough that multiple study units are completely contained within a single census division and important variation in human activity among study units can no longer be well described (Figure 2). The decline in accuracy is usually greater for descriptors that are more haphazardly distributed across the landscape, such as number of livestock and number of dwellings. In such cases where study units are small, classification is likely to be the better interpolation method, especially if the study region is large

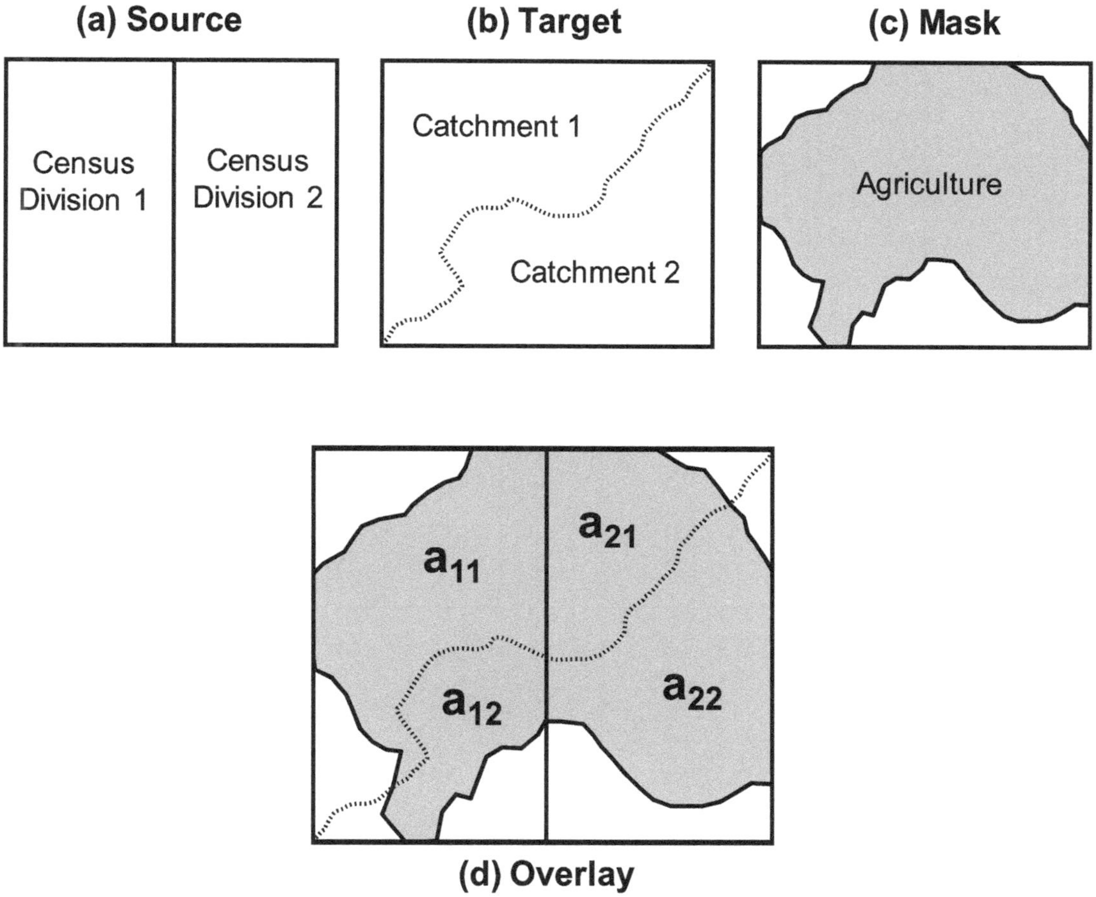

Figure 1. Example of area-based interpolation of census data to catchment boundaries. The **(a)** geographic boundaries of the source census data are compared to the **(b)** target catchment, as well as the **(c)** agriculture land cover mask. The **(d)** three layers are then overlain to determine area–weight proportion of the census data for each catchment relevant to land cover. (Adapted from Agriculture and Agri-food Canada 2008)

enough to incorporate geographically related gradients in human activity.

Classification can be achieved either through qualitative (e.g., expert judgment) or quantitative (e.g., cluster analysis) methods. The goal is to assign a categorical description (e.g., low, medium, or high) of the magnitude or extent of an activity in the census district within which the study unit is located. Using classification does, however, limit the use of the data to describing coarse differences in the extent of human activities. Moreover, classification increases the risk of committing the fallacy of division where in the case of census data, it may be incorrectly assumed that the properties of study units within a census division are accurately depicted by the entire census division. Indeed, we recommend that in cases where study catchments are substantively smaller than the census division, researchers are likely best served to use remotely sensed images or landscape surveys to collect human-activity information.

There are also automated methods of fusing census data to LULC classifications using regression-based statistics. These have been successfully used to combine crop land from LULC layers with reported crop census data

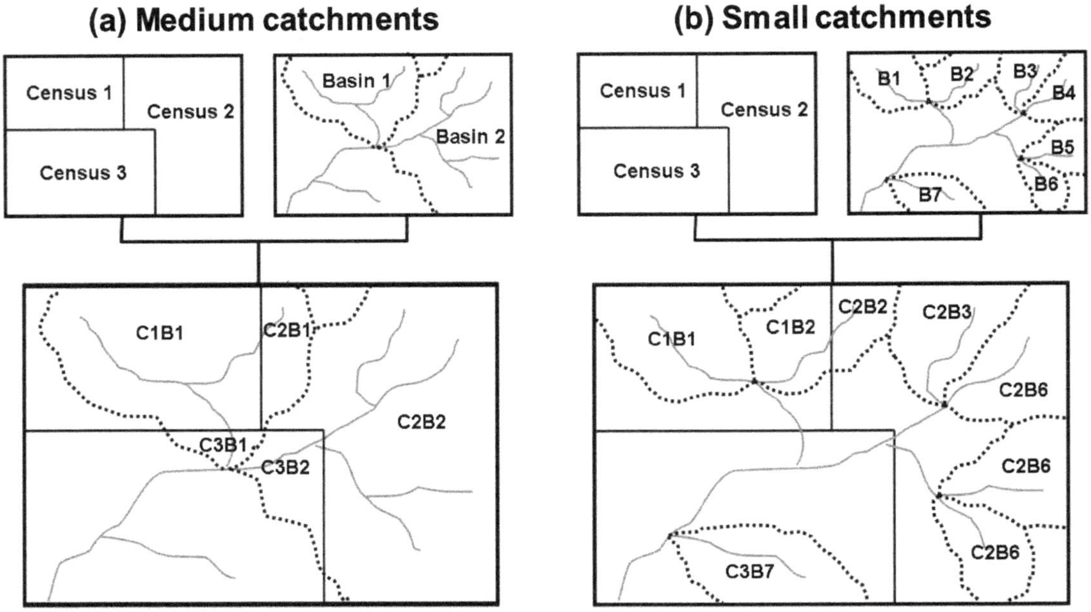

Figure 2. Comparison of **(a)** medium-sized and **(b)** small-sized catchments to demonstrate how interpolation of census data to catchment boundaries results in reduced variability among study catchments that are small enough for multiple units to be contained in the same census division.

(Ramankutty and Foley 1998; Cardille and Clayton 2007). However, whether these interpolation methods can be used for census data not directly linked to the area of cropland (e.g., livestock density) remains to be seen. Furthermore, to our knowledge, these methods have not yet been applied to the smaller geographical study units that aquatic ecologists typically study, and as such, the applicability of these methods may be limited at this point.

Remotely Sensed Images

The value of having a bird's-eye view of the landscape has long been recognized by aquatic ecologists. As such, remotely sensed images have been, and continue to be, regularly used in aquatic ecology studies, including some studies in which these images have augmented LULC to increase the detail of human-activity descriptions (e.g., Bryce et al. 1999; Davies et al. 2010; Yates and Bailey 2010). Indeed, because of the long history of aerial photography, aquatic ecologists have also effectively used air photos to understand legacy effects of past amounts and types of human activity (Harding et al. 1998).

A remote-sensing image is defined as an image produced by a recording device that is not in physical contact with the object under study. Remotely sensed images can include a variety of image types with many applications to aquatic sciences (Johnson and Gage 1997; Johnson and Host 2010). Particularly important to the scope of this chapter is the increasing use of satellite imagery to generate detailed LULC layers at high spatial and temporal resolutions, such as the annual crop inventory produced by the Agriculture and Agri-Food Canada (https://open.canada.ca/data/en/dataset/ba2645 d5-4458-414 d-b196-6303ac06c1c9). Commenced in 2009, this inventory uses images from several satellites (e.g., Landsat-5, Radarsat-2) to generate a nationwide inventory of crops at the field scale. For the purposes of this chapter, however, we are most interested in those images that are traditionally referred to as air photos, as it is these images that can be most easily used by aquatic ecologists to describe human activities not depicted by pre-existing LULC layers.

Air photos are one of the oldest methods of remotely sensing the character of the earth's surface and are still widely used today. Traditionally taken from airplanes, these images are now also captured by satellites (e.g., SPOT, Quickbird, and Worldview I and II) and can generate global coverage at very high resolutions. Image availability is, however, somewhat variable both in terms of coverage, resolution, and cost. Images taken from satellites can be expensive to obtain if the region of interest is large, but they are usually available for large geographic areas at a variety of resolutions (e.g., 2.5 to 20 m for SPOT Images) and in panchromatic (black and white) or color. Images taken from aircraft cover smaller acquisition areas than satellite images, but recent technology has resulted in high-resolution images (e.g., 30 cm). The advent of virtual globes (e.g., Google Earth, NASA World Wind, and Skyline Globe) has greatly increased the accessibility, as well as spatial and temporal coverage, of high-quality imagery, including street-level imagery, available to researchers (Yu and Gong 2012). Virtual globes also provide platforms that are increasingly GIS-integrated with which researchers can generate and share descriptions of human activities.

A wide variety of human activity information can be extracted from remotely sensed images, depending on the image being used. However, what activities can be seen is dependent upon the quality of the image and the time of year the image is taken (Figure 3). With images of coarser resolution, identification of large-scale human activities may be all that is achievable (e.g., mines, farmsteads, and golf courses). Finer resolution data have fewer limitations and can enable identification of smaller-scaled human activities (e.g., housing density, manure storage structures) and even management practices (e.g., riparian fencing, erosion control structures). Moreover, integrating fine-resolution images with multispectral images can enable extraction of additional activity descriptors. For example, high-resolution, multispectral images are frequently used to generate layers describing the degree of landscape perviousness (e.g., Lu and Weng 2009; Hu and Weng 2011; Chen et al. 2012).

The season in which an image was taken can also affect what information can be extracted. Many human activities, especially those related to management practices, are only visible at particular times of the year. Tillage practices and crop types are examples of two management decisions related to the same human activity that cannot be viewed during the same season. In this and other similar cases, it would be necessary to use two sets of images to generate a complete description of the activity.

The increasing availability, affordability, and capability of drones, or unmanned aerial systems (UAS), has added an additional tool for generating remotely sensed human-activity data (Tang and Shao 2015) that is increasingly being applied to aquatic-ecosystem management (Shahbazi et al. 2014). Indeed, UAS equipped with fine-resolution sensors can capture a scale of observations between visual landscape surveys and remote sensing images obtained via satellite and manned aircraft (Tang and Shao 2015). The use of UAS is spatially restricted compared to aerial photography and satellite imagery, with applications likely limited to the stream corridor and small headwater catchment scales, or to targeted, activity-specific assessments. However, UAS have the possibility of generating data at user-defined time intervals. Furthermore, as

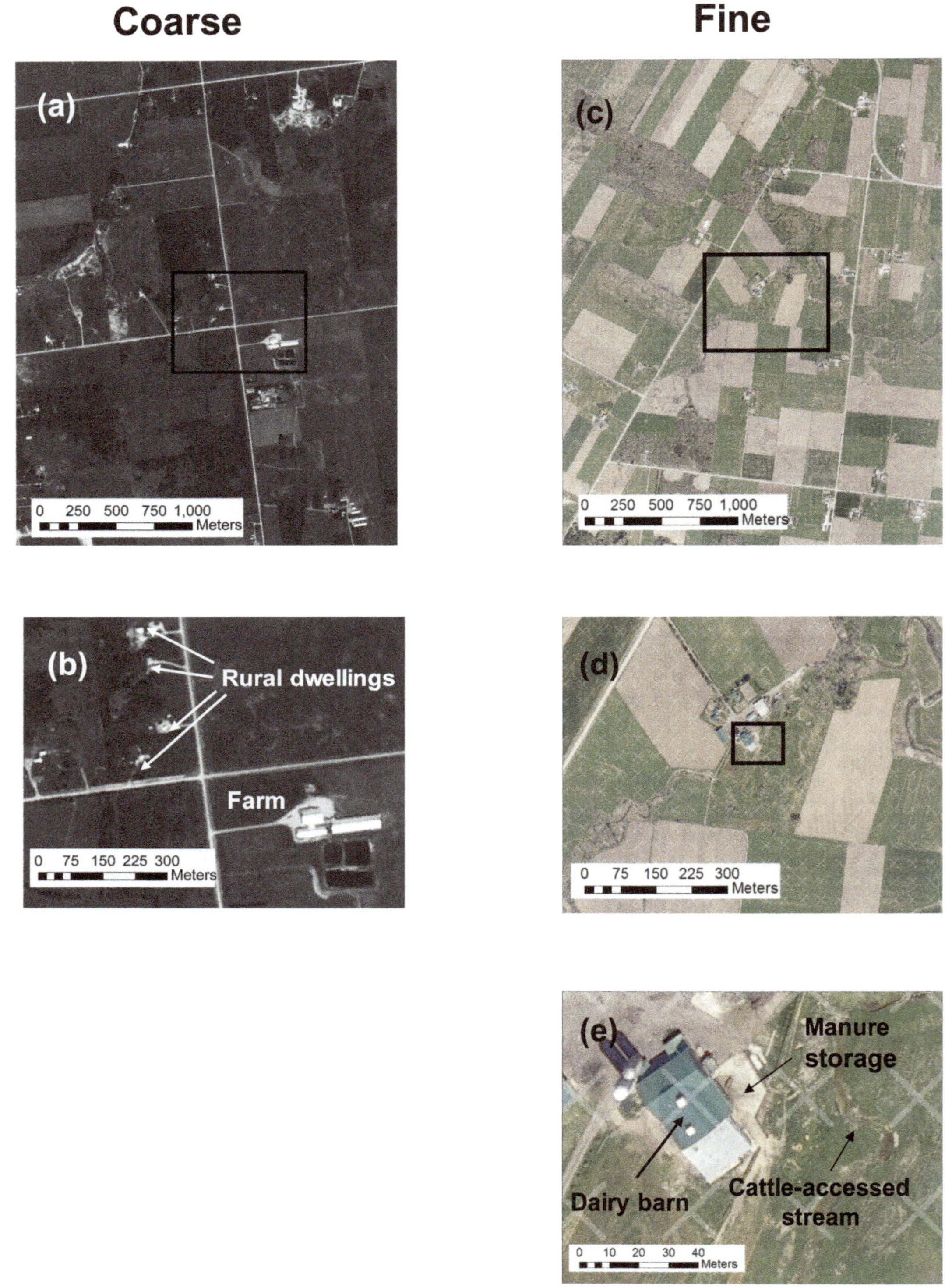

Figure 3. Remotely sensed images showing differences in human activity that can be resolved between coarse and fine images. **(a)** Coarse resolution enables **(b)** broader scale human activities to be described whereas **(c)** fine-resolution images allow delineation of **(d)** broad-scale and **(e)** fine-scale activities, as well as (e) management practices.

technological advancements allow the development of additional sensors (e.g., Zhang et al. 2016) to be deployed on UAS, the potential for capturing additional aspects of human activity beyond spatial extent will increase. There are, however, ongoing issues that limit applications of UAS for landscape description, including image-processing issues, and require additional research before UAS reach their full potential (Shahbazi et al. 2014).

Despite recent advancements in image accessibility and quality, as well as freely available viewing and GIS-integrated software, generating human-activity descriptions from remotely sensed images still requires substantial processing time as individual GIS layers must be digitized (either manually or through creation of an algorithm) from the images before they can be summarized. The time and resources required to effectively use remotely sensed images make early identification of human-activity data needs essential to ensure that data can be collected and processed within a research project's time frame. Moreover, early recognition of possible cost constraints associated with image acquisition and processing (e.g., technician salaries) is important to ensure that project budgets are sufficient.

Visual Landscape Surveys

Regardless of image quality, a great deal of human activity simply cannot be easily described from remotely sensed images. In these cases, ground-based survey techniques may be the only alternative given the limitations of using census data and/or lack of existing GIS layers. Most survey techniques (e.g., mail surveys, interviews) are too costly, time-consuming, and unpredictable with regards to response rate to employ for aquatic-ecology studies. Visual landscape surveys can, however, be an effective method of gathering human-activity data.

Most aquatic ecologists already conduct some level of visual landscape survey by recording descriptions of human activity, such as riparian land use, as part of ecological or habitat sampling. Indeed, several agencies have formalized such descriptions as part of habitat assessments (e.g., Barbour et al. 1999; Kaufmann et al. 1999; Parsons et al. 2004; Hughes and Peck 2008). Such surveys are almost exclusively conducted at the stream site scale. However, effective description of human activities can also be achieved for the catchment scale.

Visual landscape surveys at catchment or landscape scales are sometimes dubbed "windshield surveys" because the process of surveying human activities is often done by surveying the landscape from roadsides. Surveys can be conducted for relatively large areas (e.g., headwater to mid-order catchments) in reasonable amounts of time, although data collection for larger scale (e.g., drainage basin) projects is likely to be impractical under most circumstances. This technique is especially useful for assessing activities associated with high road densities, such as urban and agricultural activities, where viewing of human activities is relatively unobstructed. A drawback of using windshield surveys is the difficulty in generating complete coverages in hilly or wooded landscapes. Incorporating UAS into such surveys could enhance usability in these landscape settings. A second drawback of this method is that collected data require substantial processing to generate usable data. However, with the recent development of commercially available software, such as ArcPad (ESRI, Redlands California), spatially referenced survey data are increasingly efficient to record in the field and greatly reduces data processing time.

Industry-led mapping, such as Google Street View (https://mapstreetview.com/), is also making use of windshield survey data more cost-effective, particularly for study of urban areas where imagery is updated more frequently. However, even the less frequently collected imagery from rural areas may provide a low-cost method for aquatic scientists to conduct armchair windshield surveys. The potential for collection of human activity data from tools such as Google Street View is sure to expand in the future.

Logistical constraints associated with conducting windshield surveys will, in most cases, limit the application of this technique to studies with smaller geographic scopes (e.g., headwater catchments). Studies focusing on a specific human activity or requiring spatial positioning of activities (as opposed to regional summaries) are therefore most likely to use this source. Furthermore, windshield surveys are also an excellent way to groundtruth human activity estimates from census data or remotely sensed images as surveys can be conducted on a subset of the study units to establish error rates.

Case Studies

We use two ongoing studies in agricultural regions of Canada (Figure 4) to illustrate when and how the four different data sources can be used and to share our experiences in using these data. The first study was undertaken in southwestern Ontario to develop predictive relationships between ecological condition of streams and rural human activity for use in land-use planning models (Yates and Bailey 2006; Yates and Bailey 2010). The goal of the second study was to relate estimates of nutrients released from human activities to

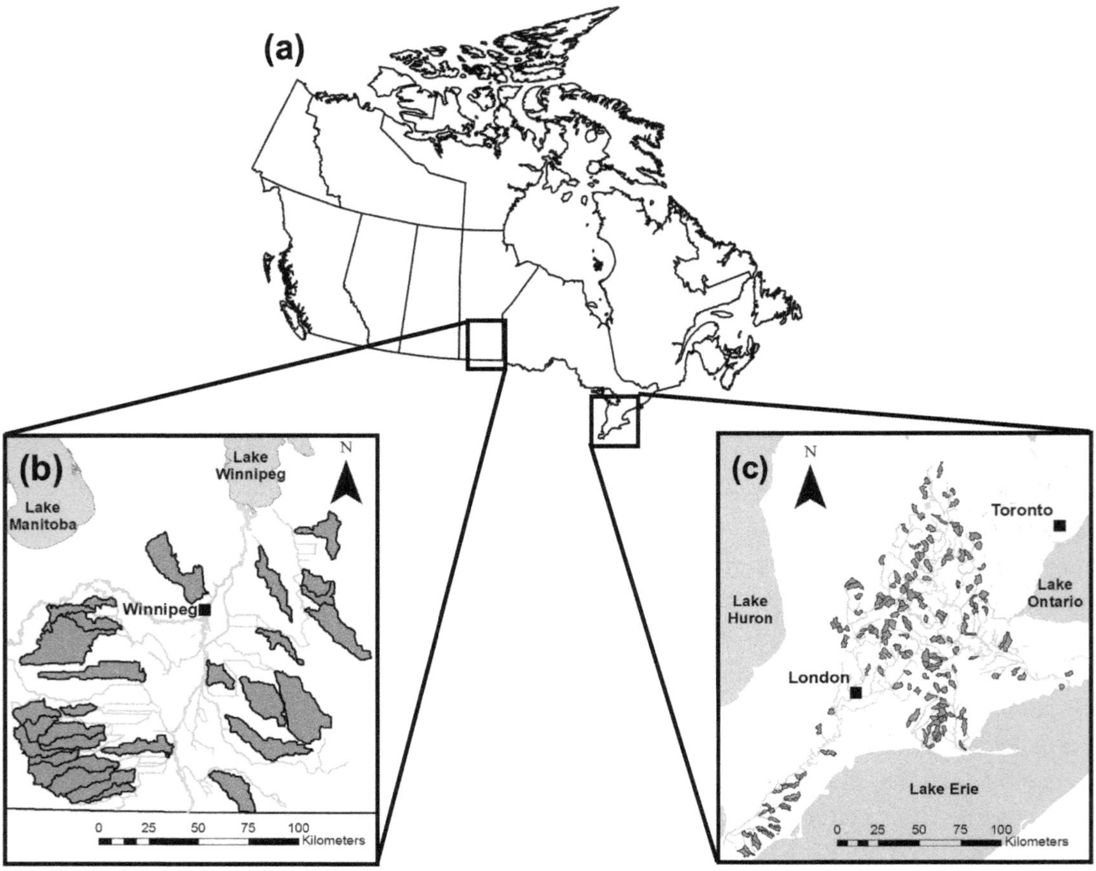

Figure 4. Locations of **(a)** case study areas in Canada and the associated study unit locations in **(b)** southern Manitoba and **(c)** southwestern Ontario.

the ecological condition of receiving streams as a function of the presence of agricultural beneficial management practices (BMPs). This study is ongoing in southern Manitoba as part of an initiative to understand and manage nonpoint source contributions of nutrients to tributary streams in the Red River valley (Yates et al. 2012).

Both studies required detailed descriptions of human activity, though for different purposes. In the southwestern Ontario study, detailed descriptions of the types and extents of each activity were needed to determine how each activity varied and covaried in relation to ecological condition. In contrast, the southern Manitoba study required human activity information detailed enough to generate estimates of the amount of anthropogenic phosphorus (P) and nitrogen (N) being released to each catchment. Although these projects are of similar geographic extents, the sizes of the study catchments were very different. The southwestern Ontario project focused on headwater streams with small catchments (6–30 km^2), whereas the southern Manitoba study units were medium-sized, permanent streams with catchment areas varying from 64 to 602 km^2.

Southwestern Ontario

Farm-level information is very difficult to obtain for the agricultural industry because of the large number of producers and minimal reporting of farm-management practices in most jurisdictions. As such, a major challenge for the southwestern Ontario project was adequate description of the types and extent of livestock production within each catchment. However, the production of several livestock types in Ontario is governed by marketing boards, which set production quotas. As a result, livestock marketing boards were identified as a potential source of human-activity information leading to a series of meetings to develop partnerships and negotiate data sharing. From initiation of contact to delivery of data, this process covered in some cases almost a year, underscoring the need for early identification and contact with potential data sources when private information is being requested.

Although we were able to obtain GIS layers that identified the location of major livestock producers, privacy restrictions prevented us from obtaining associated production levels (e.g., livestock numbers). Given that livestock farms in the region were known to vary widely in stock number, a measure of the livestock numbers within each study catchment was needed to capture among-catchment differences in livestock production. It was decided that livestock density would be estimated as the product of the area of livestock housing facilities on a farm and recommended stocking densities for livestock (see Yates and Bailey 2010 for full details). The area of livestock housing facilities on each farm in the study area was determined by measuring the area of all livestock housing facilities on each farm using high-resolution (30 cm) air photos and a GIS. This process was resource-intensive, requiring several months of GIS analyst time to measure the buildings of the more than 4,000 livestock producers in the study catchments. Furthermore, this analysis was only possible because of the high-quality air photo images available for the region that allowed an accurate assessment of farm buildings.

Livestock data derived as part of this study were used along with other descriptions of human activities in an analysis to generate a detailed human activity gradient using principal component analysis (Yates and Bailey 2010). The detailed activity gradient was found to be more strongly associated with multivariate descriptions of fish and benthic invertebrate assemblages than was an activity gradient generated from a traditional, coarse LULC data set (Yates and Bailey 2010). Indeed, the canonical R^2 value increased by

33% and 20% for fish and benthic invertebrates, respectively, resulting in the conclusion that the additional resources required to obtain a more detail description of human activity were merited.

The southwestern Ontario study also had the goal of establishing relationships between measures of ecological condition and tillage practices. However, study-unit size was considered too small and the application of the practice too related to farmer preferences (as opposed to geographic characteristics) to allow the use of census data. Furthermore, available remotely sensed images were taken in summer when tillage practices were indistinguishable and no existing GIS layers could be found for the study region. Given that the study region had a very high road density and was relatively flat with few wooded areas, it was determined that a windshield survey would be an effective method for collecting the tillage information. This dictated that the survey be undertaken several weeks after harvest for a subset of the headwater catchments (see Yates et al. 2006 for full details). This survey was conducted over a 2-week period using air photos to provide spatial reference for each cultivated field. Tillage practices for each field were recorded on the air photos. Following the survey, field boundaries were digitized in a GIS using a geo-rectified image of the air photos and attributed based on the surveyed tillage information.

In this case, Yates et al. (2006) found that tillage practices were associated with the condition of stream ecosystems because both nutrient concentrations and benthic macroinvertebrate assemblage composition were correlated with catchment-scale proportions of tillage types. Moreover, the use of management-level information regarding agricultural activity provided insight into catchment-scale controls of benthic macroinvertebrate diversity and taxa richness that were not detected using LULC data in a companion study of the same 33 streams (Rios and Bailey 2006). The contrast between the results of the two studies illustrates the importance of obtaining more detailed descriptions of human activity as a means of achieving greater predictive power for land-use planning models.

Southern Manitoba

One of the largest tasks faced in the southern Manitoba study was generating consistent, comparable estimates of the amounts of anthropogenic N and P being released in each of the study catchments. An important source of nutrients in southern Manitoba is the application of synthetic fertilizers to cultivated land (Bourne et al. 2002). However, generating accurate estimates of fertilizer application rates was difficult because of the large number of farmers and considerable regional variation in application rates among crop and soil types (Glozier et al. 2006). Fortunately, comprehensive records of fertilizer application rates are collected at the quarter section (one-fourth of a square mile or 2.59 km^2) level by the province's crop insurance agency (Manitoba Agricultural Services Corporation, www.masc.mb.ca/masc.nsf/index.html). Negotiations to access this data and to generate a data-sharing agreement took approximately 2 months. The final agreement resulted in complete sharing of the past 10 years of data with all released information being summarized to the catchment scale to ensure confidentiality. This agreement was unexpectedly terminated a year later when the corporation leadership changed, highlighting the difficulty associated with obtaining layers from private or nongovernment organizations.

The other substantial challenge in the southern Manitoba study was quantifying the frequency that BMPs were adopted in each area. Generating an estimate of these practices is difficult for large areas because of a lack of implementation records. One source of BMP implementation data in Canada is the

Canadian Census of Agriculture, which collects information regarding the use of several practices including zero tillage, riparian buffer zones, and crop rotation (www.statcan.gc.ca/eng/ca2016). Applying those data, however, required interpolation from census division boundaries to the boundaries of the study catchments. Because the study catchments were of substantial size, and often overlapped large portions of at least one census division, it was decided that area weighting was an appropriate method for estimating BMP implementation within the catchments. Area weighting to each study catchment was achieved by first converting the census descriptors into area-neutral units (i.e., proportion for area variables [e.g., area of crop land under zero tillage] or density for count variables [e.g., number of farms employing crop rotation]) using the areas of appropriate land cover types for each census division. The proportion of a catchment comprised by a particular land-cover type for each overlying census district was then determined. The resulting proportions were multiplied by the associated area-neutral BMP descriptors and summed for each catchment to provide a total for each BMP.

Unlike southwestern Ontario, where color images of high resolution were available, only coarser-resolution (2.5 m) panchromatic SPOT images were available for southern Manitoba. The quality of these images limited the activities that could be described, particularly the ability to identify building types. Groups of buildings and property lines could be discerned and were used to identify rural dwellings and farmsteads with the purpose of generating an estimate of the number of private septic systems in the study catchments. These estimates were then multiplied by the average number of household members for the region and nutrient production coefficients to generate estimates of N and P released from private septic systems.

Applying more detailed descriptions of human activities in rural southern Manitoba enabled a more complete prediction of the nutrient footprint of catchments across the Red River valley. These predictions were used to make decisions regarding site selection for studies aimed at identifying drivers of ecological conditions (Yates et al. 2013, 2014). These studies identified that regional variation in the presence and size of municipal wastewater treatment lagoons were the most important driver of ecosystem metabolism and that nutrients from these activities were entering aquatic food webs (Yates et al. 2013, 2014). Moreover, similar to the southern Ontario study, we found that some measures of aquatic ecological condition, particularly benthic macroinvertebrates, were not associated with general LULC descriptions but could be predicted by more specific descriptions of agricultural best management practices derived from census data (Yates et al. 2014). The combined results of the southern Manitoba studies further demonstrate the importance of having more detailed human activity descriptions to explain ecological variability in landscapes that appear to be highly homogeneous at coarse scales.

Challenges, Recommendations, and Management Implications

The availability of high-quality, GIS-ready data sources is an invaluable resource to aquatic scientists. Therefore, it is important that, as a group, aquatic scientists continue to support and promote development of Web-based spatial data clearinghouses and contribute layers that they have developed when it does not conflict with confidentiality agreements. Moreover, aquatic scientists need to advocate for policies that mandate and fund collection of more detailed human activity data and to ensure that data are collected and processed into a product that can be readily applied to the geographical scales at which aquatic scientists are working. An example of this would be to encourage policymakers associated with census agencies to enable summarization of raw data with user-defined boundaries. A Web-based tool, especially one linked to a virtual globe platform, would increase applicability of the data to aquatic research while still maintaining a high degree of confidentiality. Such efforts would go a long way to ensuring that policies and practices are put in place that mandate that current sources of information are maintained and updated as well as continuing the trend of increasing data availability experienced over recent decades.

A continuing challenge associated with applying detailed human activity information is that these more detailed sources are frequently more variable in time than are coarser LULC descriptions. For example, crop types and associated tillage and fertilizer practices vary at the farm scale from year to year according to crop rotations, commodity prices, and interannual variability in weather. Likewise, livestock densities can change rapidly, as illustrated by a census division in southern Manitoba that experienced an increase in the number of pigs from approximately 1.2 million to 1.8 million between the censuses of 2011 and 2016 (Statistics Canada 2016). As a result, spatial data describing human activities can quickly become out of date. Advances in computing power and remote sensing are alleviating this problem to some extent by providing more frequent images of landscapes that can be rapidly processed. For example, Agriculture and Agri-Food Canada's annual crop inventory now provides an annual, field-scale update of all crop types planted across Canada (www.agr.gc.ca/atlas/aci/). Similar GIS layers are being generated and maintained at yearly or monthly time scales for activities such as oil and gas development (e.g., Alberta Energy Regulator Land Development Package: www1.aer.ca/ProductCatalogue/217.html). Continued advocacy for the creation and maintenance of layers with high resolution is thus needed, as is partnership with industries that frequently generate and maintain layers updated at shorter time scales. Indeed, governments should be encouraged to work with industry, scientists, and other producers of high-resolution spatial data to generate mechanisms for the sharing of these data.

The four data sources described in this paper vary extensively in terms of their accessibility, coverage, and grain size (Table 1). As described previously, these characteristics are major determinants of the suitability of a data source for a particular study. In general, existing GIS layers are suitable for all studies, provided that the coverage of the layer is large enough or layers are available for adjoining jurisdictions. However, where studies cross large-scale political boundaries (e.g., international or state/provincial borders), researchers may find inconsistencies between layer properties such as resolution, time frame, and classification assignment and naming conventions for specific human activities. For example, provincial/state level LULC layers often differ between adjoining jurisdictions in terms of the number and naming convention of agricultural classification. In such cases, researchers may be required to adopt less regionally specific national layers or conduct processing routines to further generalize one or more of the state-level layers to the lowest common convention (e.g., combine pasture and cropland classes to generate a common agriculture class).

Coverage and consistency in that coverage are not usually an issue for census data, which are usually undertaken at the national scale, although international consistencies may arise. However, grain size helps determine the pertinence of the data for the study and the most useful method for interpolating the data to the study units. Grain size is also the constraining characteristic of remotely sensed images as it determines which human activities can be described. Visual landscape surveys are not constrained in the same ways, but coverage can be affected by road density and the ability to clearly view the landscape, as well as by the time and resources required to conduct the surveys for even relatively small areas. Researchers must consider these inherent characteristics when deciding which sources might be applicable to their study.

The data sources described in this chapter also comprise a gradient with regards to the level of effort that must be expended to ultimately integrate the data into a GIS. Existing GIS layers require relatively little processing time compared to census data, remotely sensed images, and windshield surveys that require progressively more effort. Based on these resource requirements, as well as the constraints of availability, grain size, and coverage discussed above, we propose that researchers adopt a decision tree for the process of collecting human-activity data (Figure 5). Existing GIS layers are the preferred source of data because of their ease of use and high level of detail. However, because existing GIS layers are also the most difficult to locate and obtain, it will often be necessary to use one or more of the remaining three sources. In these circumstances, we recommend that the suitability of census data be evaluated first, followed by remotely sensed images and windshield surveys. Although presented as a decision tree, it is important to remember that the characteristics of the human activities to be described should be the driving factor in the types of sources used. Moreover, for many research projects, it is likely that the use of multiple data sources and data sources of different types will be required to adequately describe all activities of interest. In these cases, additional time will need to be allotted to generate harmonized GIS databases that enable spatial analyses synthesizing the collected human activity descriptions. Furthermore, as was demonstrated by the case studies, we recommend that data sources be used in a complementary fashion either to provide further information about an activity or to ground-truth estimates. Remotely sensed images and windshield surveys are particularly well suited to these tasks.

Conclusion

With the exception of widely available, relatively coarse-grained, and quality-assured GIS layers, the assembly of georeferenced human-activity data requires substantial effort on the part of the researcher. Aquatic ecologists should not, however, shy away from using these sources simply because of the resource and time commitments required to extract critical human activity data. As aquatic scientists, we often devote months and substantial resources to the collection of ecological data, but only hours in the collection of human activity data. Instead, collection of human activity data needs to be viewed in a similar light to collection of ecological data and attributed similar amounts of resources. High-quality human-activity data are no less important to modeling efforts than are ecological data, and the information may be the source of the most easily obtained improvement in the predictive capacity of future models.

Acknowledgments

We thank E. Krynak for comments on an earlier version of this chapter as well as assistance with formatting and literature review. R. M. Hughes and three anonymous reviewers provided helpful comments that were incorporated into the final version of this chapter. Funding for this work was provided through Natural Science and Engineering Research Council of Canada Discovery grants to A. G. Yates and J. M. Culp, as well as funds from

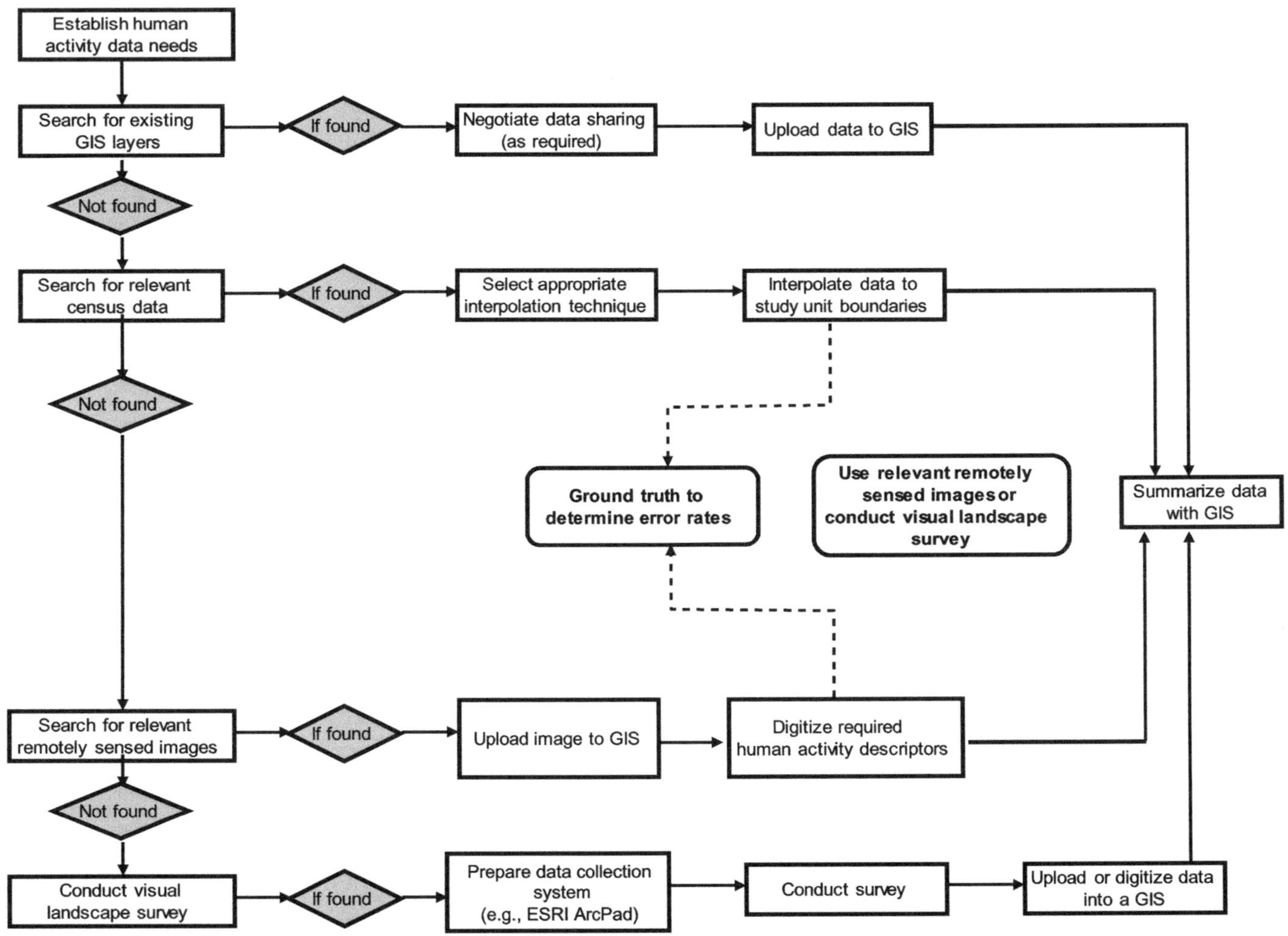

Figure 5. Framework for the identification, collection and processing of detailed human activity data from four data sources (i.e., existing GIS layers, census data, remotely sensed images, and visual landscape surveys) based on accessibility, grain size, coverage, and processing requirements.

Environment and Climate Change Canada's Lake Winnipeg Basin Initiative.

References

Agriculture and Agri-food Canada. 2008. Interpolated census of agriculture to soil landscapes, ecological frameworks and drainage areas of Canada: documentation and user guide. Agriculture and Agri-food Canada, AAFCAAC#1116052, Ottawa.

Allan, J. D. 2004. Landscapes and riverscapes: the influence of land use on stream ecosystems. Annual Review of Ecology Evolution and Systematics 35:257–284.

Bailey, R. C., T. B. Reynoldson, A. G. Yates, J. Bailey, and S. Linke. 2007. Integrating stream bioassessment and landscape ecology as a tool for landuse planning. Freshwater Biology 52:908–917.

Barbour, M.T., J. Gerritsen, B. D. Snyder, and J. B. Stribling. 1999. Rapid bioassessment protocols for use in streams and wadeable rivers: periphyton, benthic macroinvertebrates and fish. U.S. Environmental Protection Agency, EPA 841-B-99–002, Washington, D.C.

Bartsch, W. M., R. P. Axler, and G. E. Host. 2015. Evaluating a Great Lakes scale landscape stressor index to assess water quality in the St. Louis River Area of Concern. Journal of Great Lakes Research 41:99–110.

Booth, B. J., R. R. Jones, M. E. Turyk, S. Freels, D. M. Patel, L. T. Stayner, and M. H. Ward. 2017. Livestock and poultry density and childhood cancer incidence in nine states in the USA. Environmental Research 159:444–451.

Bourne, A., A. Armstrong, and G. Jones. 2002. A preliminary estimate of total nitrogen and total phosphorus loading to streams in Manitoba, Canada. Manitoba Water Quality Management Section, Manitoba Conservation Report No. 2002-04, Winnipeg.

Bruns, D. A. 2005. Macroinvertebrate response to land cover, habitat, and water chemistry in a mining-impacted river ecosystem: a GIS watershed analysis. Aquatic Sciences 67:403–423.

Bryce, S. A., D. P. Larsen, R. M. Hughes, and P. R. Kaufmann. 1999. Assessing relative risks to aquatic ecosystems: a mid-Appalachian case study. Journal of the American Water Resources Association 35:23–36.

Cardille, J. A., and M. K. Clayton. 2007. A regression tree-based method for integrating land-cover and land-use data collected at multiple scales. Environmental and Ecological Statistics 14:161–179.

Chaves, M. L., P. M. Chainho, and J. L. Costa. 2005. Regional and local environmental factors structuring undisturbed benthic macroinvertebrate communities in the Mondego River basin, Portugal. Archiv Für Hydrobiologie 163:497–523.

Chen, R., F. Fan, and Y. Deng. 2012. Extraction and analysis of impervious surfaces based on a spectral un-mixing method using Pearl River delta of China Landsat TM/ETM+ imagery from 1998 to 2008. Sensors 12:1846–1862.

Cookson, N., and M. S. Schorr. 2009. Correlations of watershed housing density with environmental conditions and fish assemblages in a Tennessee Ridge and Valley stream. Journal of Freshwater Ecology 24:553–561.

Davies, P. E. 2000. Development of a national river bioassessment system (AUSRIVAS) in Australia. Pages 113–124 *in* J. F. Wright, D. W. Sutcliffe, and M. T. Furse, editors. Assessing the biological quality of fresh waters: RIVPACS and other techniques. Freshwater Biological Association, Ambleside, UK.

Davies, P. J., I. A. Wright, S. J. Findlay, O. J. Jonasson, and S. Burgin. 2010. Impact of urban development on aquatic macroinvertebrates in south eastern Australia: degradation of in-stream habitats and comparison with non-urban streams. Aquatic Ecology 44:685–700.

Diebel, M. W., and M. J. Vander Zanden. 2009. Nitrogen stable isotopes in streams: effects of agricultural sources and transformations. Ecological Applications 19:1127–1134.

European Water Framework Directive 2000. Directive 2000/60/EC of the European Parliament and of the council of 23 October 2000 establishing a framework for community action in the field of water policy. Official Journal of the European Communities L 327: 1–72.

Falcone, J. M., D. M. Carlisle, and L. C. Weber. 2009. Quantifying human disturbance in watersheds: variable selection and performance of a GIS-based disturbance index for predicting the biological condition of perennial streams. Ecological Indicators 10:264–273.

FAO (Food and Agriculture Organization of the United Nations). 2005. A system of integrated agricultural censuses and surveys, volume 1: World Programme for the Census of Agriculture 2010. FAO, Rome. Available: www.fao.org/3/a-a0135e.pdf. (February 2019).

Gerth, W. J., A. T. Herlihy, and J. C. Sifneos. 2013. Large-scale macroinvertebrate assemblage patterns from least-disturbed wadeable stream sites across the 48 contiguous US states. Knowledge and Management of Aquatic Ecosystems [online serial] 408:02.

Glozier, N. E., J. A. Elliott, B. Holliday, J. Yarotski, and B. Harker. 2006. Water quality characteristics and trends in a small agricultural watershed: south Tobacco Creek, Manitoba, 1992–2001. Environment Canada, Ecological Sciences Division, Prairie and Northern Region, Saskatoon, Saskatchewan.

Goidts, E., B. van Weasmael, and K. Van Oost. 2009. Driving forces of soil organic carbon evolution at the landscape and regional scale using data from a stratified soil monitoring. Global Change Biology 15:2981–3000.

Harding, J. S., E. F. Benfield, P. V. Bolstad, G. S. Helfman, and E. B. D. Jones, III. 1998. Stream biodiversity: the ghost of land use past. Proceedings of the National Academy of Sciences of the United States of America 8:14843–14847.

Holt, J. B., C. P. Lo, and T. W. Hodler. 2004. Dasymetric estimation of population density and areal interpolation of census data. Cartography and Geographic Information Science 31:103–121.

Hu, X., and Q. Weng. 2011. Impervious surface area extraction from IKONOS imagery using an object-based fuzzy method. Geocarto International 26:3–20.

Hughes, R. M., and D. V. Peck. 2008. Acquiring data for large aquatic resource surveys: the art of compromise among science, logistics, and reality. Journal of the North American Benthological Society 27:837–859.

Hughes, R. M., L. Wang, and P. W. Seelbach, editors. 2006. Landscape influences on stream habitat and biological assemblages. American Fisheries Society, Symposium 48, Bethesda, Maryland.

Hynes, H. B. N. 1975. The stream and its valley. Verhandlungen der Internationale Vereinigung für Theoretische und Angewandte Limnologie 19:1–15.

Jelinski, D. E., and J. Wu. 1996. The modifiable areal unit problem and implications for landscape ecology. Landscape Ecology 11:129–140.

Johnson, L. B., and S. H. Gage. 1997. Landscape approaches to the analysis of aquatic ecosystems. Freshwater Biology 37:113–132.

Johnson, L. B., and G. E. Host. 2010. Recent developments in landscape approaches for the study of aquatic ecosystems. Journal of the North American Benthological Society 29:41–66.

Jones, F. C., K. M. Somers, B. Craig, and T. B. Reynoldson. 2007. Ontario benthos biomonitoring network: protocol manual. Ontario Ministry of the Environment, Toronto. Available: https://desc.ca/sites/default/files/OBBN2007finalapril18c.pdf. (February 2019).

Jun, Y., D. Won, S. Lee, D. Kong, and S. Hwang. 2012. A multimetric benthic macroinvertebrate index for the assessment of stream biotic integrity in Korea. International Journal of Environment, Resources and Public Health 9:3599–3628.

Kaufmann, P. R., P. Levine, E. G. Robison, C. Seeliger, and D. V. Peck. 1999. Quantifying physical habitat in wadeable streams. U.S. Environmental Protection Agency, EPA/620/R-99/003, Corvallis, Oregon

Krynak, E. M., and A. G. Yates. 2018. Benthic invertebrate taxonomic and trait associations with land use in an intensively managed watershed: implications for indicator identification. Ecological Indicators 93:1050–1059.

Lam, N. S. 1983. Spatial interpolation methods: a review. The American Cartographer 10:129–50.

Leal, C. G., P. S. Pompeu, T. A. Gardner, R. P. Leitão, R. M. Hughes, P. R. Kaufmann, J. Zuanon, F. R. de Paula, S. F. B. Ferraz, J. R. Thomson, R. MacNally, J. Ferreira, and J. Barlow. 2016. Multi-scale assessment of human-induced changes on Amazonian instream habitat. Landscape Ecology 31:1725–1745.

Leitão, R. P., J. Zuanon, D. Mouillot, C. G. Leal, R. M. Hughes, P. R. Kaufmann, S. Villéger, P. S. Pompeu, D. Kasper, F. R. de Paula, S. F. B. Ferraz, and T. Gardner. 2018. Disentangling the pathways of land use impacts on the functional structure of fish assemblages in Amazon streams. Ecography 41:219–232.

Lu, D., and Q. Weng. 2009. Extraction of urban impervious surfaces from an IKONOS image. International Journal of Remote Sensing 30:1297–1311.

Maas-Hebner, K. G., M. J. Harte, N. Molina, R. M. Hughes, C. B. Schreck, and J. A. Yeakley. 2015. Combining and aggregating environmental data for status and trends assessments: challenges and approaches. Environmental Monitoring and Assessment 187:278–295.

Macedo, D. R., R. M. Hughes, R. Ligeiro, W. R. Ferreira, M. Castro, N. T. Junqueira, D. R. O. Silva, K. R. Firmiano, P. R. Kaufmann, P. S. Pompeu, and M. Callisto. 2014. The relative influence of multiple spatial scale environmental predictors on fish and macroinvertebrate assemblage richness in cerrado ecoregion streams, Brazil, Landscape Ecology 29:1001–1016.

Maloney, K. O., and D. E. Weller. 2011. Anthropogenic disturbance and streams: land use and land-use change affect stream ecosystems via multiple pathways. Freshwater Biology 56:611–626.

Morse, C. C., A. D. Huryn, and C. Cronan. 2003. Impervious surface area as a predictor of the effects of urbanization on stream insect communities in Maine, U.S.A. Environmental Monitoring and Assessment 89:95–127.

Nislow, K. H., and W. H. Lowe. 2006. Influences of logging history and riparian forest characteristics on macroinvertebrates and Brook Trout (*Salvelinus fontinalis*) in headwater streams (New Hampshire, USA). Freshwater Biology 51:388–397.

Ohio EPA (Ohio Environmental Protection Agency). 1987. Biological criteria for the protection of aquatic life: volumes I–III. Ohio EPA, Columbus.

Ollis, D. J., H. F. Dallas, K. J. Esler, and C. Boucher. 2006. Bioassessment of the ecological integrity of river ecosystems using aquatic macroinvertebrates: an overview with a focus on South Africa. African Journal of Aquatic Science 31:205–227.

Openshaw, S. 1984. The modifiable areal unit problem. GeoBooks, Norwich, England.

Parsons, M., M. C. Thoms, and R. H. Norris. 2004. Development of a standardised approach to river habitat assessment in Australia. Environmental Monitoring and Assessment 98:109–130.

Paulsen, S. G., A. Mayio, D. V. Peck, J. L. Stoddard, E. Tarquinio, S. M. Holdsworth, J. Van Sickle, L. L. Yuan, C. P. Hawkins, A. T. Herlihy, P. R. Kaufmann, M. T. Barbour, D. P. Larsen, and A. R. Olsen. 2008. Condition of stream ecosystems in the US: an overview of the first national assessment. Freshwater Science 27:812–821.

Poquet, J. M., J. Alba-Tercedor, T. Punti, M. D. Sanchez-Montoya, S. Robles, M. Alvarez, C.

Zamora-Munoz, C. E. Sainz-Cantero, M. R. Vidal-Abarca, M. L. Suarez, M. Toro, A. M. Pujante, M. Rieradevall, and N. Prat. 2009. The MEDiterranean prediction and prediction and classification system (MEDPACS): an implementation of the RIVPACS/AUSRIVAS predictive approach for assessing Mediterranean aquatic macroinvertebrate communities. Hydrobiologia 623:153–171.

Ramankutty, N., and J. A. Foley. 1998. Characterizing patterns of global land use: an analysis of global croplands data. Global Biogeochemical Cycles 12:667–685.

Reynoldson, T. B., M. Bombardier, D. B. Donald, H O'Neill, D. M. Rosenberg, H. Shear, T. M. Tuominen, and H. H. Vaughan. 1999. Strategy for a Canadian aquatic biomonitoring network. National Water Research Institute, Contribution Number 99-248, Burlington, Ontario.

Richards, C., L. B. Johnson, and G. E. Host. 1996. Landscape-scale influence on stream habitats and biota. Canadian Journal of Fisheries and Aquatic Sciences 53(Supplement):295–311.

Rios, S. L., and R. C. Bailey. 2006. Relationship between riparian vegetation and stream benthic communities at three spatial scales. Hydrobiologia 553:153–160.

Shahbazi, M., J. Théau, and P. Ménard. 2014. Recent applications of unmanned aerial imagery in natural resource management. GIScience and Remote Sensing 51:339–365.

Statistics Canada. 2016. Table: 32-10-0426-01: Pigs on census day. Available: https://www150.statcan.gc.ca/t1/tbl1/en/tv.action?pid=3210042601&pickMembers%5B0%5D=1.1497. (February 2018).

Tang, L., and G. Shao. 2015. Drone remote sensing for forestry research and practices. Journal of Forestry Research 26:791–797.

Vandewalle, M., F. de Bello, M. P. Berg, T. Bolger, S. Dolédec, F. Dubs, C. K. Feld, R. Harrington, P. A. Harrison, S. Lavorel, P. M. da Silva, M. Moretti, J. Niemelä, P. Santos, T. Sattler, J. P. Sousa, M. T. Sykes, A. J. Vanbergen, and B. A. Woodcock. 2010. Functional traits as indicators of biodiversity response to land use changes across ecosystems and organisms. Biodiversity and Conservation 19:2921–2947.

Von Schiller, D., E. Marti, J. L. Riera, M. Ribot, J. C. Marks, and F. Sabater. 2008. Influence of land use on stream ecosystem function in a Mediterranean basin. Freshwater Biology 53:2600–2612.

Wang, L., T. Brenden, P. Seelbach, A. Cooper, D. Allan, R. Clark, Jr., and M. Wiley. 2008. Landscape based identification of human disturbance gradients and reference conditions for Michigan streams. Environmental Monitoring and Assessment 141:1–17.

Wang, L., J. Lyons, and P. Kanehl. 2001. Impacts of urbanization on stream habitat and fish across multiple spatial scales. Environmental Management 28:255–266.

Wang, L., J. Lyons, P. Rasmussen, P. Seelbach, T. Simon, M. Wiley, P. Kanehl, E. Baker, S. Niemela, and P. M. Stewart. 2003. Watershed, reach, and riparian influences on stream fish assemblages in the Northern Lakes and Forest Ecoregion, U.S.A. Canadian Journal of Fisheries and Aquatic Sciences 60:491–505.

Winter, J. G., K. M. Somers, P.J. Dillion, C. Paterson, and R. A. Reid. 2002. Impacts of golf courses on macroinvertebrate community structure in Precambrian Shield streams. Journal of Environmental Quality 31:2015–2025.

Xiong, W., I. Holman, D. Conway, E. Lin, and Y. Li. 2008. A crop model cross calibration for use in regional climate impacts studies. Ecological Modelling 213:365–380.

Yates, A. G., and R. C. Bailey. 2006. The stream and its altered valley: integrating landscape ecology into environmental assessments of agro-ecosystems. Environmental Monitoring and Assessment 114:257–271.

Yates, A. G., and R. C. Bailey. 2010. Improving the description of human activities potentially affecting rural stream ecosystems. Landscape Ecology 25:371–382.

Yates, A. G., R. C. Bailey, and J. A. Schwindt. 2006. No-till cultivation improves stream ecosystem quality. Journal of Soil and Water Conservation 61:14–19.

Yates, A. G., R. B. Brua, J. M. Culp, and P. A. Chambers. 2013. Multi-scaled drivers of rural prairie stream metabolism along human activity gradients. Freshwater Biology 58:675–689.

Yates, A. G., R. B. Brua, J. M. Culp, P. A. Chambers, and L. I. Wassenaar. 2014. Sensitivity of structural and functional indicators depends on type and resolution of anthropogenic activities. Ecological Indicators 45:274–284.

Yates, A. G., J. M. Culp, and P. A. Chambers. 2012. Estimating nutrient production from human activities in subcatchments of the Red River, Manitoba. Journal of Great Lakes Research 38(Supplement 3):106–114.

Yu, L., and P. Gong. 2012. Google Earth as a virtual globe tool for earth science applications at the global scale: progress and perspectives. International Journal of Remote Sensing 33:3966–3986.

Zhang, J., C. Yang, H. Song, W. C. Hoffmann, D. Zhang, and G. Zhang. 2016. Evaluation of an airborne remote sensing platform consisting of two consumer-grade cameras for crop identification. Remote Sensing 8:257.

Zoraghein, H., S. Leyk, M. Ruther, and B. P. Buttenfield. 2016. Exploiting temporal information in parcel data to refine small area population estimates. Computers, Environment and Urban Systems 58:19–28.

American Fisheries Society Symposium 90:35–56, 2019
© 2019 by the American Fisheries Society

Beyond Isolation by Distance: Riverscape Effects on Genetic Structure of Fall-Run Chinook Salmon

Chanté D. Davis*
Coastal Oregon Marine Experiment Station, Hatfield Marine Science Center
Department of Fisheries and Wildlife, Oregon State University
2030 SE Marine Science Drive, Newport, Oregon 97365, USA

Clinton W. Epps
Department of Fisheries and Wildlife, Oregon State University
Nash Hall Room 104, Corvallis, Oregon 97331, USA

Rebecca L. Flitcroft
U.S. Forest Service, Pacific Northwest Research Station
3200 SW Jefferson Way, Corvallis, Oregon 97331, USA

Michael A. Banks
Coastal Oregon Marine Experiment Station, Hatfield Marine Science Center
Department of Fisheries and Wildlife, Oregon State University
2030 SE Marine Science Drive, Newport, Oregon 97365, USA

Abstract.—Habitat fragmentation, land use practices, and flow impediments modify the natural course of rivers, disrupting connectivity and subsequently affecting dispersal and gene flow in aquatic organisms. Many of the relationships between the physical river network and the genetic structure of populations are not well understood. Riverscape genetics is a developing field that uses population genetic metrics to assess genetic structure within the context of the environmental variables that drive functional connectivity in a river network. Here, we applied an effective distance network approach to characterize the effects of hydrology in shaping neutral genetic population structure of fall-run Chinook Salmon *Oncorhynchus tshawytscha* within a small, coastal Oregon catchment. We evaluated whether gene flow was limited by (1) site-specific features occurring within spawning habitat, using a dissimilarity matrix, and (2) the cumulative effect of the environment accrued while traveling en route between reaches. We found that Chinook Salmon that spawned at higher elevations (site specific effects) after traversing steeper gradients (en-route effects) were more genetically distinct from individuals that traversed gradual gradients and spawned at lower elevations. This effect (isolation by resistance) was distinguishable from isolation by distance, which was not detected among spawning groups. Our study enhanced interpretation of habitat heterogeneity in constraining gene flow and spatial genetic structure among reaches within a small, coastal catchment. Given that smaller catchments may hold life history

* Corresponding author: davis.chanted@gmail.com

variation that is important to long-term population persistence, there is need to understand these relationships that maintain genetic diversity.

Introduction

Expanding quantitative genetic research in river systems by explicitly considering habitat heterogeneity and the physical configuration of river networks could clarify the evolutionary consequences of environmental change and habitat fragmentation in rivers on population persistence. Habitat fragmentation, land use practices, and barriers modify the natural courses of rivers disrupting functional connectivity within the riverscape (Ward and Stanford 1983). These alterations have subsequent effects on species dispersal ability and gene flow (Nehlsen et al. 1991; Yoshiyama et al. 1998; Brenkman et al. 2012). The genetic response resulting from the inability of individuals to disperse may include reduced genetic and life history variation (Yoshiyama et al. 1998). The ability to observe and assess genetic changes to life history traits is necessary to understand how natural and anthropogenic changes to functional connectivity may affect fish assemblages now and in the future.

The distribution of riverine habitat is driven by hydrologic, stochastic (e.g., fires and floods), and geomorphic (e.g., erosion and deposition) processes that create and reconfigure stream habitats over time (Reeves et al. 1995). Physical habitat diversity within riverscapes is characterized by differences in bed type, channel morphology, biological communities, and physical configuration, processes that are driven by natural and anthropogenic disturbance (Frissell et al. 1986; Montgomery 1999; Fausch et al. 2002; Borrett et al. 2014). River research that considers the spatial structure of the physical network in conjunction with disturbance and geomorphic processes may use network theory and graph theoretic approaches in analyses (Ganio et al. 2005; Wipfli et al. 2007). River network structure complicates traditional spatial statistics because of the directional covariation inherent in a river system that is formed and maintained by unidirectional water, sediment, and organic material flows (Wipfli et al. 2007; Wohl et al. 2019). However, advances in geospatial tools and availability of geospatial data now facilitate evaluation of hydrologic, geomorphic, and ecological changes throughout entire river networks (Benda et al. 2007; Peterson and Ver Hoef 2014; Roux et al. 2015). The spatial structure of rivers influences functional connectivity, but many of the relationships between the physical network and genetic structure are not well understood (Frissell et al. 1986; Benda et al. 2004; Peterson et al. 2013).

Riverscape genetics (RG), a counterpart to land- and seascape genetics, is a developing field with methodology that combines spatial statistics, population genetics, and landscape ecology to understand the role of habitat heterogeneity in shaping gene flow (Kanno et al. 2011; Selkoe et al. 2016). The term "riverscape" in RG is analogous to the use of "landscape" in landscape genetics (LG) and "seascape" in seascape genetics (SG). Therefore, it is defined here as "a hierarchically structured mosaic of differentially distributed habitat within freshwater environments" (Davis et al. 2018). Collectively, LG and SG strive to develop methodologies that move beyond explaining genetic structure as a function of geographic distance (isolation by distance [IBD]) by describing how specific habitat variables such as elevation, tree canopy coverage, or oceanic currents inhibit or facilitate dispersal (McRae 2006; Selkoe et al. 2010; Liggins et al. 2013). The methods developed in LG quantify the contribution of habitat variation along the path traveled en route (e.g., effective distance, path analysis; Figure 1) on spatial genetic variation among individuals and groups (Shirk et al. 2010; Zeller et al. 2012; Hall and Beissinger 2014). Similarly, seascape genetics research has adapted graph-theoretic models (Galindo et al. 2006; Riginos and Liggins 2013) to quantify relationships between hydrogeologic features and genetic distance or differences between populations (Banks et al. 2007; Selkoe et al. 2010; Johansson et al. 2015).

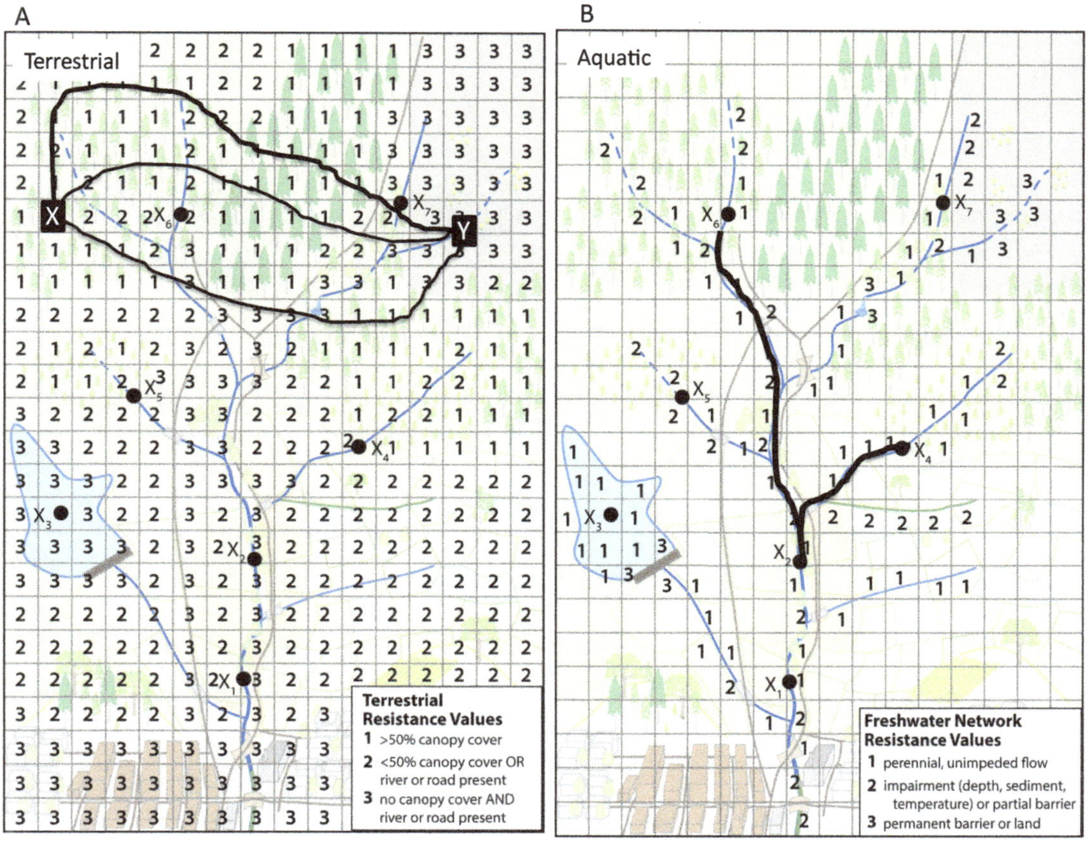

Figure 1. A hypothetical catchment has been transformed into a resistance surface that represents travel by a terrestrial organism across areas with different levels of **(A)** canopy cover and **(B)** travel by an aquatic organism through areas with varying water flow. Within each image, the origin and destination for potential dispersing aquatic and terrestrial organisms are labeled: seven reaches (black dots X_1–X_7) are labeled within the river and two sample locations (black squares X, Y) are on land. The resistance surface transforms an image into a grid of cells (raster image) that is used to test hypotheses about the influence of environmental variables on movement or gene flow. The values within each cell depict the cost of traveling caused by a covariate of interest (e.g., elevation, canopy cover, and water flow). Routes of travel between two points may be modeled (solid black lines) using least-cost path analysis, circuit theory, or other models, resulting in a calculation of effective distance. Effective distance value(s) are then used to understand how habitat complexity affects movement. However, the lack of alternative pathways within dendritic river networks poses a difficulty in the application of this landscape genetic method. Background images are modified from Davis et al. (2018).

Studies of RG have identified effects of isolation by resistance to understand the effects of climate variation and habitat heterogeneity (e.g., temperature, confluence number, discharge, and maximum stream gradient) within rivers on genetic variation of freshwater species (i.e., Olsen et al. 2010a, 2010b; Kanno et al. 2011; Galbraith et al. 2015; Pilger et al. 2015; Hand et al. 2016). The body of research in freshwater systems emphasizes the usefulness of RG approaches to explain genetic variation beyond geographic distance metrics (e.g., waterway distance). Unlike research in terrestrial and marine systems, RG has yet to incorporate the en-route effects of the heterogeneous habitat on dispersal beyond physical barriers like dams.

Linking genetic variation to riverscapes requires consideration of the functional connectivity within the network (Cote et al. 2009; Auerbach and Poff 2011; Flitcroft et al. 2014), conditions that alter functional connectivity, and the spatial configuration of the network (Ganio et al. 2005; Thorp et al. 2006; Blanchet et al. 2011; Peterson et al. 2013). Given the differences between terrestrial, marine, and freshwater habitats, Davis et al. (2018) reviewed the applicability of using LG (Figure 1) and SG (Figure 2) methods to explore the en-route effects of traversing rivers on genetic variation. The authors suggested an integrated network model where edges of the network are weighted using effective distance and nodes represent sampling areas (Figure 3). Both genetic diversity (e.g., allelic richness, heterozygosity) and genetic distance (e.g., Jost's *D*, fixation index) could be modeled as a function of the network, either as metrics of the sample site (e.g., genetic diversity as a function of reach characteristics) or difference-based metrics (genetic difference between pairs of reaches as a function of edge characteristics that link them). This approach would enable broad-scale (e.g., catchment, evolutionarily significant units [ESUs]) and fine-scale (e.g., within catchments, within reach) analysis (Figure 3).

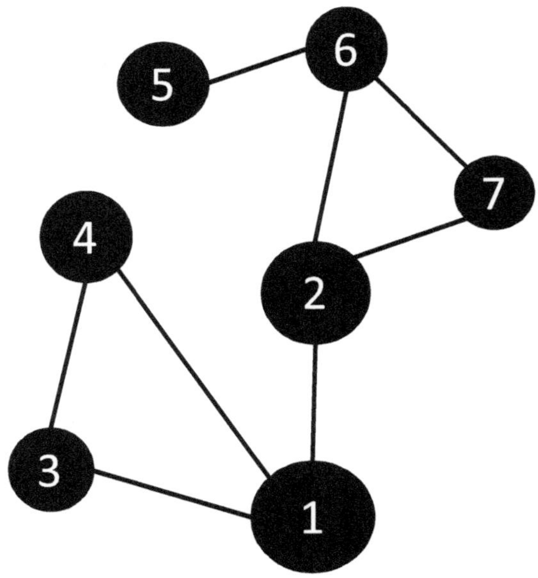

Figure 2. A hypothetical island system is modeled as a graph-theoretic network representing movement by marine organisms among islands based on the direction of ocean currents. Nodes (filled circles) represent individual islands, and differences in node size reflects genetic distance within each island population (larger nodes reflect greater genetic distance). Edges (black lines) connecting nodes reflect gene flow between island populations.

Here, we apply this concept of riverscape genetics to characterize the effects of riverscape variability in shaping the genetic structure of fall-run Chinook Salmon *Oncorhynchus tshawytscha* within a small coastal catchment. Chinook Salmon express diverse life histories that are characterized by the season of the adult return migration to spawning habitat (called a run). Life history diversity of Chinook Salmon in the Pacific Northwest is characterized by differences in return timing (O'Malley and Banks 2008; O'Malley et al. 2013), homing fidelity to natal spawning rivers (Waples 2001), and freshwater rearing strategies (Groot and Margolis 1991). Addition-

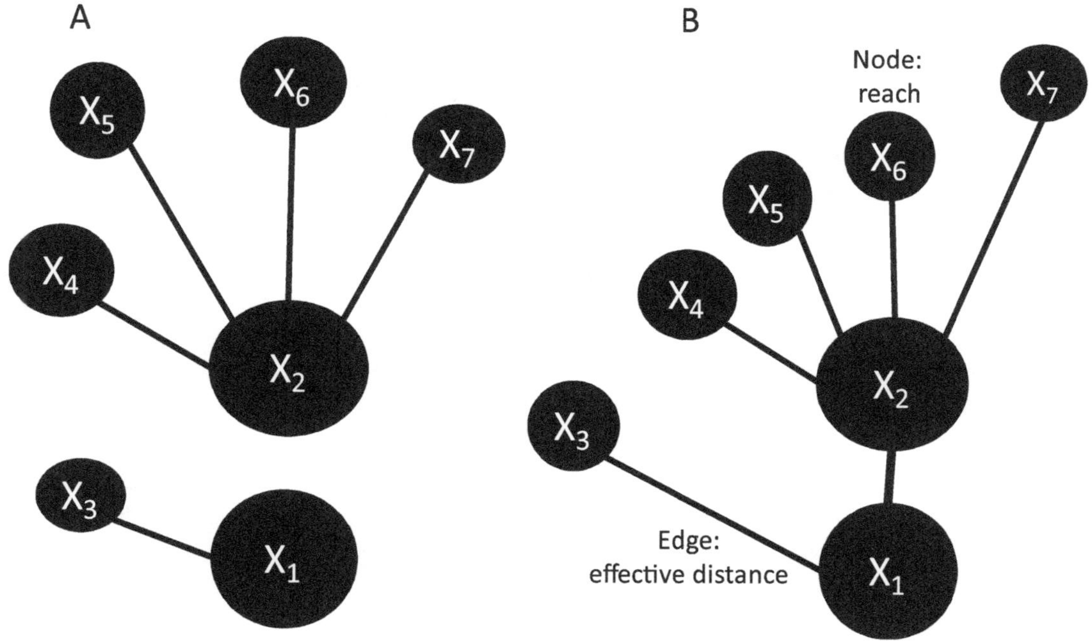

Figure 3. A riverscape genetics approach to evaluate the effects of environmental resistance on gene flow based on the hypothetical river system in Figure 1B. The number within each node corresponds to its location in Figure 1B, site X_3 is upstream of a permanent barrier (e.g., dam) and site X_7 is upstream of a semipermanent barrier (e.g., waterfall). Each node (circle) represents genetic distance within each reach (e.g., D_{est}, F_{ST}). Edges (black lines) reflect **(A)** geographic distance and **(B)** effective distance measured as the cumulative effect of a covariate of interest (e.g., gradient) on dispersal. Evaluating genetic distance between pairs of reaches as a function of (A) distance may result in no relationship. However, by evaluating genetic variation or genetic distance as a function of (B) riverscape characteristics that link them, the effects of the riverscape on genetic diversity at sites X_3 and X_7 become clearer.

ally, changes in genetic structure that reflect restricted gene flow resulting from decreased functional connectivity among essential habitat have been well documented among the Salmonidae (Dionne et al. 2008; Fraser et al. 2011; Kanno et al. 2011; Bradbury et al. 2014; Benavente et al. 2015). Identifying features of the riverscape that facilitate or impede dispersal would provide an ecological context for understanding how Chinook Salmon populations are genetically structured and may help prevent further decline in population abundances or extinctions (Manel et al. 2003).

Chinook Salmon have specific requirements for spawning habitat that are characterized by streamflow, water depth, gradient, and other variables (Reiser and White 1988; Bjornn and Reiser 1991; DeVries 1997; Geist and Dauble 1998; Moir and Pasternack 2010). Geomorphic processes that affect spawning habitat are constrained by the hierarchical properties of nested catchments within drainage basins (Miller et al. 2008). Therefore, site-specific variation is associated with the position of habitat in the network and may be a predictor for genetic structure at the reach scale. However, salmonids may undergo different costs that occur en route when traveling within the network (e.g., between spawning and overwintering habitat, or during upstream migration). The cumulative effect of riverscape variables, both site specific and en-route, should be considered in assessments of spatial genetic structure.

Fall-run Chinook Salmon are a genetically distinct phenotype that spawn in river main stems and tributaries, returning from September to December and spawning from late October through January (Moran et al. 2013; Clemento et al. 2014; Davis et al. 2017). The primary objective of this study was to assess whether riverscape indicators (e.g., gradient, elevation, stream width, or stream depth) influenced observed genetic structure in fall-run Chinook Salmon, after accounting for the influence of geographic distance and network structure. The different drivers of within-run genetic structure were assessed by first testing if the geographic distance among tributaries predicted spatial genetic structure. Then, the effect of riverscape features on genetic structure was assessed using both site-based and en-route or effective distance to characterize differences. To our knowledge, this is the first time that an en-route or effective distance approach has been applied to a riverscapes genetics study.

Methods

Study Area

The Siletz River is 120 river kilometers long, with headwaters located in the Central Oregon Coast Range. Its catchment area is 523 km^2 with average annual discharge of 1,500 m^3/s (https://waterdata.usgs.gov/nwis/inventory?). The river drains principally volcanic lithology and is located in a Mediterranean climate where water flow is sourced by winter rains. Six anadromous salmon populations spawn in the Siletz River: spring and fall Chinook Salmon, summer and winter steelhead *Oncorhynchus mykiss*, Coho Salmon *O. kisutch*, and Coastal Cutthroat Trout *O. clarkii clarkii* (Davis et al. 2017). In addition to natural disturbances, historic logging and damming have made substantial and long-lasting alterations to water, sediment and wood flows, and habitat availability throughout the catchment (Miller 2010).

Environmental Layers

Combined utilities in ArcGIS 9.3, ArcMap, and NetMap were used to develop a synthetic stream layer with modeled environmental data layers in vector format of the Siletz River. NetMap is a community-supported geographic analysis platform containing standardized digital catchment data that are commonly used for analysis of freshwater systems (Benda et al. 2007). Relevant riverscape variables were selected for analysis based on review of the ecological literature, knowledge of habitat characteristics most likely to affect gene flow, and dispersal of fall-run Chinook Salmon. After assessing collinearity among variables, four riverscape variables (elevation, gradient, channel depth, and channel width), and two geographic distance variables (the path of the river between sampled reaches [waterway distance] and distance to mouth) were selected for analysis.

Dispersal requires sufficient water velocities and depths during migration to pass obstacles (e.g., waterfalls, log jams). Streamflow velocity and stream depth are two important factors that also influence formation of suitable spawning habitat for Chinook Salmon (Isaak et al. 2007; Hamann et al. 2014). However, because only one flow gauge exists on the Siletz River, this variable was not used for analysis. Channel depth (m) and channel width (m) were modeled as a power function of mean annual flow, drainage area, or precipitation (Benda et al. 2007). Gradient (m/m) was calculated from 10 m digital elevation models (DEMs). Spatial resolution used in NetMap was at a finer resolution than the spatial scale used during genetic sampling. The reach-scale descriptors of physical habitat conditions, stream depth, stream width, and elevation were averaged within each genetically sampled reach for site-based analyses. To facilitate modeling of riverscape resistance (see below), stream gradient (m/m) was first standardized by dividing by the smallest observed gradient and then converted into degree (hereafter, "standardized gradient"), resulting in the lowest gradient having a value of 1.

The National Hydrography Database (www.usgs.gov/core-science-systems/ngp/national-hydrography) stream layer at 1:100,000 scale was clipped to Hydrologic Unit Code 8: 17100204, representing the Siletz and Yaquina catchments. Using the Network Analyst extension in ArcMap, waterway distance was calculated using the downstream boundary for each reach. This measure of distance follows the path of the river and is a more biologically meaningful measure than straight-line Euclidean distance for freshwater organisms.

Spawning Surveys and Genetic Data Collection

Tissue samples were removed from carcasses of fall-run Chinook Salmon by the Oregon Department of Fish and Wildlife Coastal Chinook Research and Monitoring Program (CCRMP) during yearly spawning surveys in 2011 and 2012. The long-term CCRMP survey consisted of 34 adjacent reaches that were distributed throughout the Siletz River and identified by upstream and downstream Global Positioning System coordinates. Although all reaches were sampled, tissue samples were successfully collected from carcasses within only 23 reaches. Furthermore, reaches that had low sample sizes ($N < 9$) were combined with samples from an adjacent reach where appropriate, but samples were not combined if the adjacent reach spanned a confluence. Therefore, a total of 17 reaches were used in the final analysis (Figure 4). Tissue samples were assigned to a reach according to the downstream boundary of the reach where each carcass was collected. Spatial locations for combined reaches were assigned the downstream reach boundary of the upstream reach, representing an approximate midpoint (shared middle boundary) of the two adjacent reaches (Figure 4).

Genetic Analysis and Population Structure

Genotyping and genetic differentiation of Chinook Salmon in the Siletz River was previously described using 11 neutral microsatellite markers (Davis et al. 2017). Neutral microsatellite markers are not known to be associated with phenotypic expression and are considered to be selectively neutral. Therefore, they may be used to infer demographic processes that have shaped population structure (Holderegger et al. 2006; Kirk and Freeland 2011). This riverscape genetics analysis was conducted using the neutral microsatellite marker genotype data set of fall-run Chinook Salmon from Davis et al. (2017) containing 540 fall-run Chinook Salmon.

Characterization of genetic diversity among loci was evaluated using allelic richness (A_r), a measure of allelic number that corrects for unequal sample sizes using a rarefaction method with HP_RARE (Kalinowski 2005). Observed (H_o) and expected (H_e) heterozygosity were calculated in GenAlEx (Peakall and Smouse 2012).

Population structure is popularly estimated using the fixation index (F_{ST}). Interpretation of the index becomes difficult because it is calculated from estimates of heterozygosity (Weir and Cockerham 1984). Therefore, maximum values are strongly constrained to near zero when the number of alleles per locus is high (Meirmans and Hedrick 2011). Several alternate models have been proposed, including Jost's D (D_{est}), that do not rely on heterozygosity (for details see Gerlach et al.

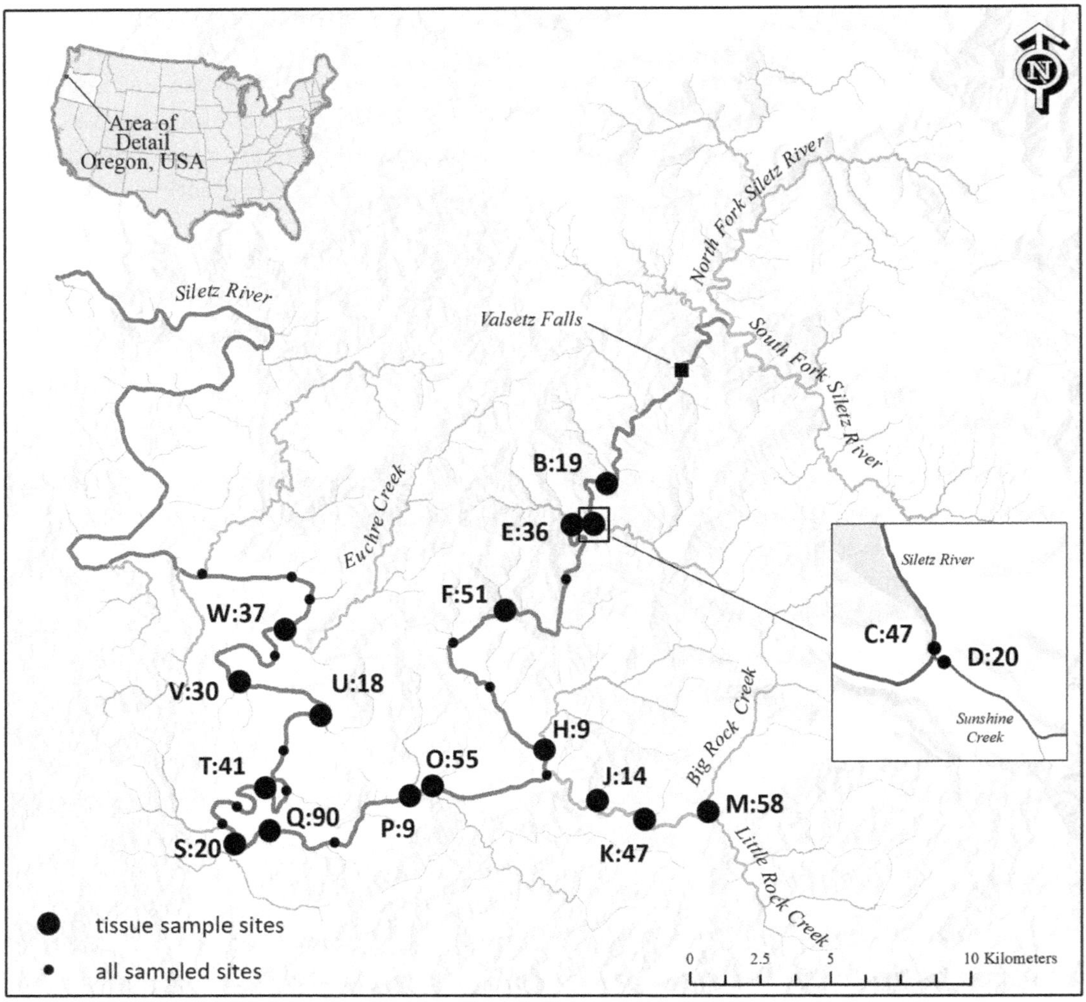

Figure 4. Map of the Siletz River and study reach locations, Oregon. Gray lines represent the main stem and tributaries, black dots represent the downstream boundary of each reach, and numbers represent sample size.

2010; Jost 2008; Meirmans and Hedrick 2011; Whitlock 2011). D_{est} is calculated from the effective number of alleles (Jost 2008), thus allowing a slightly different interpretation of genetic distance. The use of F_{ST} and D_{est} together in this study provides measures of genetic distance that are based on different components of genetic diversity at the spatial resolution of a stream reach. Pairwise genetic distance values were calculated in the program GenAlEx with 1,000 permutations to determine if the values differed significantly from zero, an indication that populations may be genetically distinct (Peakall and Smouse 2012).

Riverscape Genetics of Chinook Salmon

Characterization of genetic diversity was completed in a prior study of Chinook Salmon population genetics in the Siletz River (Davis et al. 2017). The authors found no deviations from Hardy–Weinberg equilibrium, null alleles, or allelic dropout, and further descriptions of microsatellite markers are detailed in that publication. Therefore, we evaluated two hypotheses to explore the relationship between spatial genetic structure and habitat heterogeneity. Each hypothesis evaluated if correlations were best explained through geographic distance or riverscape variation.

Hypothesis 1: isolation by distance.— Within the IBD framework, we predicted that genetic distance would increase with increased waterway distance between reaches, similar to relationships found in larger catchments (Dionne et al. 2008; Petrou et al. 2014). Significance was assessed by Mantel tests (Guillot and Rousset 2013) and multiple regression on distance matrices (MRDM; Lichstein 2006; Oksanen et al. 2015) using the 'ecodist' package in R (Goslee and Urban 2007; R Core Team 2016). Multiple regression on distance matrices was applied with 10,000 permutations to test for significant relationships between a dependent matrix (F_{ST} and D_{est}) and one or multiple predictor matrices (Lichstein 2006). To identify the contribution of each explanatory variable to the overall fit of the model, each dependent matrix was unfolded into vectors that represented pairwise waterway distances. A regression was performed between the response variable and each predictor. Statistical significance was interpreted through permutations. Mantel tests also have been used to compare distance matrices but have received criticism in landscape genetics literature because of elevated risk of type-1 error (Castellano and Balletto 2002; Legendre and Fortin 2010; Diniz-Filho et al. 2013; Guillot and Rousset 2013). Mantel tests were used to provide a comparison among analytical methods, but because results of partial Mantel tests showed similar relationships among models, only MRDM results are presented.

Hypothesis 2: isolation by resistance.— The hypothesis that gene flow is limited by reach-specific spawning habitat conditions was evaluated by examining the differences among reaches in elevation, stream width, and stream depth. Dissimilarity matrices were calculated as the difference between measurements, $x_i - x_j$, where x represents an environmental predictor variable (e.g., stream depth) and sample reaches were represented by i and j. The relationship of genetic distance (i.e., F_{ST} and D_{est}) with waterway distance between reaches was quantified using pairwise measures of genetic distance calculated as linearized (index * [1 – index]–1) values. The correlation between genetic distance and each riverscape variable was quantified individually in three univariate models. Significance among models was assessed as described above. Next, the model with the largest significant R^2 (MRDM) was re-evaluated for significance after accounting for waterway distance between reaches to further evaluate the effect of the predictor while holding waterway distance constant.

In addition to spawning reach conditions, a riverscape genetic approach was used to evaluate the costs associated with traveling between reaches because these costs may also be functionally important components of genetic structure. To do so, we used an integrated network model where edges between reaches represented effective distance (Figure 3). Our approach adapts the effective distance approach from landscape genetics (Figure 1) but is also informed by considerations of current flow in graph theoretic networks as applied in seascape genetics (Figure 2). Increased travel costs were hypothesized to be associated with steeper gradients because steeper gradients maintain faster flows and therefore require more energy to navigate than a river distance with gradual gradients. Model optimization techniques were employed that were similar to approaches used in landscape genetics analysis (Epps et al. 2007; Castillo et al. 2014; Bowlby et al. 2016). The appropriate biologically relevant range or cut-off values for individual variables were determined by testing a wide range of parameters for each model against genetic structure. Therefore, five candidate models were developed to evaluate the hypotheses that steeper or gradual gradients increased resistance to dispersal. A power function modeled as x^y, where x = standardized gradient and y = 0.001, 0.01, 0.1, 1.0, 1.5, was used to produce transformations of gradient. Transformations included a linear increase, an exponentially increasing relationship between resistance and gradient, and a steep initial increase in resistance with increasing gradient followed by a plateau. Transformed gradients were multiplied by the length of a stream unit as it occurs in NetMap and summed along the shortest path to quantify effective distance (Shirk et al. 2010). The shortest path was identified from a river network designed using the Network Analysis and Visualization library and package 'igraph' (R Core Team 2016; Csardi and Nepusz 2006). A pairwise matrix representing cumulative gradient along the shortest path between reaches was compared against pairwise genetic distance. Significance among models was assessed as described above. Univariate tests identified which of the five gradient models were the best predictors of genetic distance. Models were re-evaluated to identify if correlations changed after accounting for the waterway distance between reaches, as described above.

Results

Genetic Analysis

Allelic richness had a mean value of 7.5 that ranged from 7.3 (site J) to 7.8 (site H; Figure 4). Allelic richness was not influenced by distance from the river mouth (Figure 5). Observed heterozygosity had a mean value of 0.85 and ranged between 0.83 (site J) and 0.87 (sites Q and F) but was not significantly influenced by distance from mouth (Figure 5). Pairwise comparisons of genetic distance for both indices ranged between <0.01 and 0.02 ($p < 0.05$; Table 1).

Isolation by Distance and Resistance

Genetic structure was not influenced by waterway distance between reaches (Table 2). However, genetic distance was influenced by elevation (Figure 6). Differences in elevations of sampling locations were correlated with greater genetic differentiation as measured by D_{est} ($R^2 = 0.11$, $P < 0.01$) but not linearized F_{ST} ($R^2 < 0.001$, $P = 0.90$), both as a univariate model and after controlling for waterway distance in a bivariate model (Table 2).

Path-based measures of river characteristics also influenced genetic distance as assessed through a riverscape genetics approach (Figure 7). Model optimization by MRDM showed that for univariate models, steeper channel gradients were more resistant to gene flow as measured by D_{est} (grad$^{1.0}$, $R^2 = 0.11$, $P < 0.001$; grad$^{1.5}$, $R^2 = 0.13$, $P < 0.001$; Table 3). When the gradient model was re-evaluated for correlation with genetic distance after accounting for waterway distance, the multivariate model including grad$^{1.5}$ still displayed significant correlation ($R^2 = 0.14$, $P < 0.01$), indicating that the relationship with gradient was not an artifact of distance. As with results from reach-specific comparisons, linearized F_{ST} did not show significant relationships in any model.

Discussion

This study highlights the importance of a comprehensive RG approach to understand the effects of the environment on genetic variation. Using a more typical analytical framework of differences among reaches, we found that genetic variation was influenced by differences in spawning habitat elevation. However, by incorporating continuous-path analysis in RG, we were able to further identify effects of resistance resulting from the

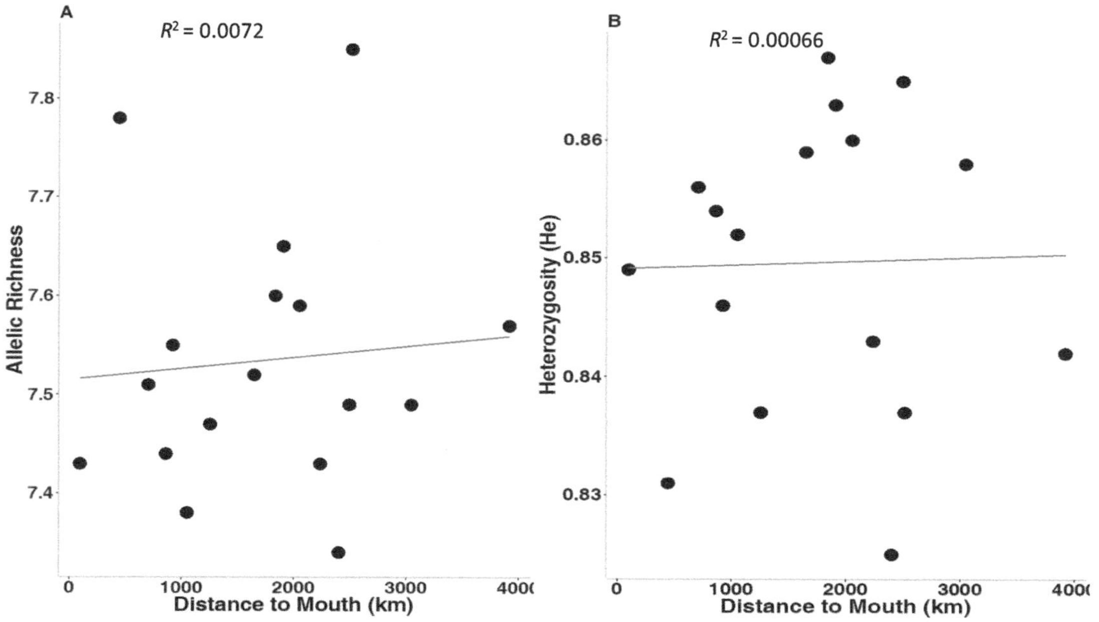

Figure 5. **(A)** Allelic richness and **(B)** expected heterozygosity (H_e) versus distance from Siletz River mouth for each reach.

cumulative effect of gradient incurred en route to spawning habitat. Salmon that traveled through habitat with steeper gradients to spawn in reaches that were at higher elevations were more genetically distinct from those individuals spawning at lower-elevation reaches that required traveling through lower-gradient areas.

Our approach calculated an effective distance based on changes in gradient, and to our knowledge, this is one of the few studies that has begun to explore the contributions of resistance (beyond barriers) on genetic distance that are accrued during upstream migration. However, challenges remain in RG because of the limited statistical tools available for evaluating relationships that capture river flow direction in analyses.

It is worth noting that the interpretation of spatial genetic variation differed between the two indices of genetic diversity that we evaluated. F_{ST} did not resolve spatial genetic variation in fall-run Chinook Salmon, but D_{est} did identify significance resulting from riverscape features. There was an abundance of low D_{est} values that approached zero, which may have affected the interpretation and inference of spatial genetic relationships from a statistical perspective. However, given the simplicity of the Siletz catchment and the connectivity among spawning groups, it is expected that low values would result. Indeed, our approach highlights the power of the RG approach, even when genetic structure is weak. These metrics (F_{ST} and D_{est}) are not interchangeable and describe different components of genetic diversity, highlighting the importance of evaluating the underlying genetic relationships that are used to calculate each index given the hypotheses being tested within each study.

On the Use of Isolation of Distance and Isolation by Resistance Models in Riverscape Genetics

Previous studies that have assessed spatial genetic structure of salmonids within and among catchments using the classical IBD

Table 1. Pairwise genetic distance between fall-run Chinook Salmon sampling locations. Pairwise estimates of linearized Jost's D (D_{est}; above diagonal) and linearized fixation index (F_{ST}; below diagonal) for all population pairs. Significant ($p < 0.05$) D_{est} values are in bold.

	B	C	D	E	F	H	J	K	M	O	P	Q	S	T	U	V	W
B		0.020	<0.001	<0.001	0.001	<0.001	<0.001	<0.001	0.007	<0.001	<0.001	<0.001	<0.001	0.019	0.007	<0.001	<0.001
C	0.011		**0.030**	**0.021**	**0.018**	<0.001	<0.001	0.009	**0.025**	0.007	0.018	**0.019**	0.018	**0.029**	0.005	0.011	**0.018**
D	0.012	0.012		<0.001	<0.001	<0.001	0.013	<0.001	<0.001	0.014	<0.001	0.005	0.000	0.008	0.012	<0.001	0.013
E	0.010	0.008	0.009		0.012	<0.001	<0.001	<0.001	0.010	0.003	<0.001	0.009	0.018	0.030	0.000	0.003	0.004
F	0.009	0.007	0.009	0.007		<0.001	0.006	<0.001	0.004	<0.001	<0.001	<0.001	<0.001	<0.001	<0.001	<0.001	0.008
H	0.018	0.016	0.016	0.015	0.012		<0.001	<0.001	<0.001	<0.001	<0.001	<0.001	<0.001	<0.001	<0.001	<0.001	<0.001
J	0.014	0.010	0.016	0.011	0.012	0.020		<0.001	<0.001	<0.001	<0.001	<0.001	<0.001	<0.001	<0.001	<0.001	<0.001
K	0.009	0.006	0.009	0.006	0.004	0.014	0.009		<0.001	<0.001	<0.001	<0.001	<0.001	<0.001	<0.001	<0.001	<0.001
M	0.009	0.007	0.008	0.007	0.005	0.013	0.011	0.005		0.012	<0.001	0.011	<0.001	0.006	<0.001	<0.001	0.017
O	0.008	0.006	0.010	0.006	0.005	0.013	0.008	0.004	0.006		<0.001	0.001	<0.001	<0.001	<0.001	<0.001	<0.001
P	0.018	0.019	0.019	0.016	0.015	0.019	0.020	0.015	0.016	0.014		<0.001	<0.001	<0.001	<0.001	<0.001	<0.001
Q	0.007	0.006	0.008	0.006	0.004	0.014	0.009	0.003	0.005	0.004	0.012		0.001	<0.001	0.000	<0.001	<0.001
S	0.012	0.010	0.011	0.011	0.009	0.016	0.014	0.009	0.008	0.008	0.018	0.008		<0.001	0.006	<0.001	0.002
T	0.012	0.008	0.010	0.009	0.005	0.015	0.011	0.005	0.006	0.005	0.015	0.004	0.008		0.004	<0.001	0.004
U	0.014	0.010	0.014	0.009	0.009	0.018	0.013	0.010	0.009	0.008	0.017	0.009	0.014	0.011		<0.001	<0.001
V	0.010	0.008	0.010	0.008	0.005	0.014	0.010	0.005	0.006	0.005	0.015	0.005	0.008	0.006	0.011		<0.001
W	0.010	0.008	0.011	0.007	0.007	0.014	0.011	0.006	0.007	0.005	0.015	0.005	0.010	0.007	0.009	0.007	

Table 2. Models of pairwise genetic distance (linearized D_{est} and F_{ST}) as a function of pairwise difference between riverscape variables measured at each reach using multiple regression on distance matrix. Riverscape variables: WD = waterway distance between reaches, elev = elevation, width = stream width, and depth = stream depth. Asterisk (*) represents significant values.

	D_{est}				F_{st}			
	Coefficient	*P*-value	R^2	*P*-value	Coefficient	*P*-value	R^2	*P*-value
Index ~ WD	2.41E-08	0.6514	<0.01	0.6514	<0.00001	0.60	<0.01	0.60
Index ~ depth	3.28E-03	0.3537	<0.01	0.3537	<0.001	0.99	<0.01	0.99
Index ~ width	4.96E-05	0.5833	<0.01	0.5833	−1.04E-07	1.00	<0.01	1.00
Index ~ elev	5.28E-05	<0.01	0.11	<.01*	<0.001	0.90	<0.01	0.90
Index ~ elev(WD)			0.12*	<.01*			0.004	0.90
Int	0.001	0.93			0.009	0.71		
elev	6.10E-05	<0.01			−8.3E-07	0.95		
WD	−5.23E-08	0.3			<0.001	0.62		

framework have established waterway distance as a consistent predictor for observed genetic isolation with genetic distance measured primarily by F_{ST} (Gomez-Uchida et al. 2009; Meeuwig et al. 2010; Petrou et al. 2014; Harris et al. 2015). However, these relationships are commonly identified among groups at spatial extents that spanned thousands of kilometers and included multiple catchments (Castric et al. 2001; Ozerov et al. 2012; Bowlby et al. 2016). Olsen et al. (2010) determined that differences in spatial genetic structure as measured by F_{ST} for Chinook Salmon in the Yukon, Kuskowim, and Nortin River watersheds was influenced by the number of major drainages and flow velocity. The authors also

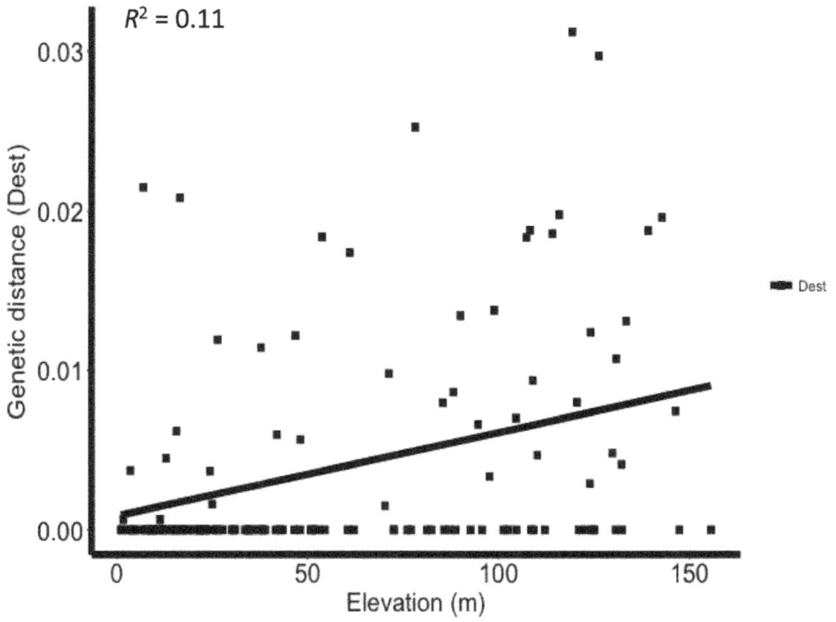

Figure 6. Elevation versus genetic distance between all possible pairs of reaches.

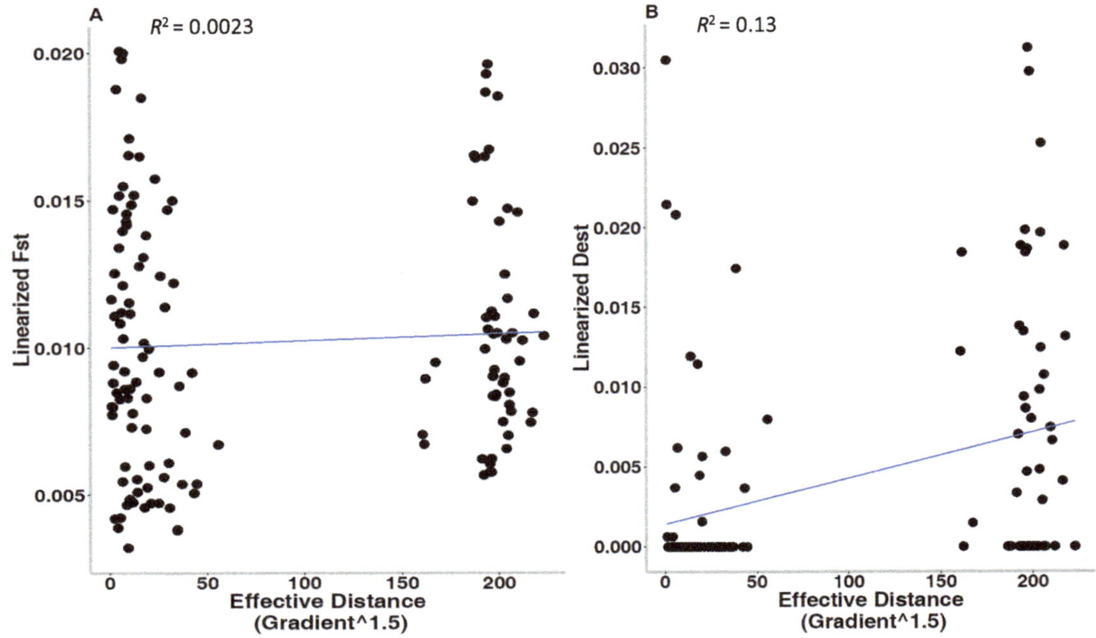

Figure 7. Genetic distances (**[A]** linearized F_{ST}; **[B]** D_{est}) versus effective distances estimated between all possible pairs of reaches. Effective distance, in this case, is a power function transformation of standardized gradient multiplied by the length of stream segment and summed along the shortest path. The gap within each figure results from data transformation where effective distances are not present within the river. Results indicate that effective distance for F_{ST} is relatively flat, whereas effective distance increases for D_{est}.

found that relationships were not maintained at intermediate spatial extents and riverscape indicators did not explain genetic structure better than geographic distance (Olsen et al. 2010). Likewise, similar patterns between isolation by resistance (IBR) and IBD have been identified at large and small spatial extents among Rocky Mountain Sculpin *Cottus* sp. populations in Alberta, Canada. Ruppert et al. (2017) described genetic structure among drainages within the species' eastern range using F_{ST}. The authors found significant IBD was identified across multiple catchments, but at a small spatial extent (i.e., within Lee Creek or St. Mary River). The difference in elevation among sites (IBR) was also a predictor of spatial genetic variation in some, but not all, catchments (Ruppert et al. 2017). The complexity of interactions between IBD and IBR highlights the need for continued investigation of specific riverscape features on spatial genetic variation at multiple spatiotemporal scales and extents. Tools and methods being developed for riverscape genetics analysis will help identify how riverscapes facilitate or impede genetic exchange and resolve patterns of genetic structure (Landguth et al. 2012; Davis et al. 2018).

Incorporating Network and Spatial Extent in Riverscape Genetics

Although the spatial structure of riverine systems influences connectivity and consequently affects dispersal, many of the relationships between the physical network, population distribution, and genetic structure are not well understood (Le Pichon et al. 2006; Leps et al. 2015). Thinking about the riverscape as a branched network of interconnected tributaries is an advancement in riverine ecology

Table 3. Models of pairwise genetic distance (linearized D_{est} and F_{ST}) as a function of cumulative gradient using multiple regression on distance matrix. WD = waterway distance between reaches; grad. = gradient. Asterisk (*) represents significant values.

	D_{est}				F_{st}			
	Coefficient	*P*-value	R^2	*P*-value	Coefficient	*P*-value	R^2	*P*-value
Index ~ grad$^{0.001}$			0.03	0.12			0.01	0.51
Integer	1.70E-03	0.93			0.01	0.26		
gradient	1.03E-07	0.12			<0.001	0.51		
Index ~ grad$^{0.01}$			0.03	0.12			0.01	0.5
Integer	1.73E-03	0.92			0.01	0.26		
gradient	1.02E-07	0.12			<0.001	0.5		
Index ~ grad$^{0.1}$			0.04	0.09			0.01	0.53
Integer	1.57E-03	0.94			0.01	0.27		
gradient	8.52E-08	0.09			<0.001	0.53		
Index ~ grad$^{1.0}$			0.11*	<0.001*			<0.001	0.98
Integer	8.53E-04	0.99			0.01	0.52		
gradient	5.22E-09	< 0.01			<0.001	0.98		
Index ~ grad$^{1.5}$			0.13*	<0.001*			<0.01	0.79
Integer	1.44E-03	0.99			0.01	0.62		
gradient	2.88E-10	<0.01			<0.001	0.79		
Index ~ grad$^{1.5}$ + WD			0.14*	< 0.01*				
Integer	2.50E-03	0.78						
Pwr1.5	3.13E-10	<0.001						
WD	-5.50E-08	0.35						

that is becoming increasingly incorporated into analyses (Wiens 1989; Thorp et al. 2006; Campbell Grant et al. 2007; Altermatt 2013; Borrett et al. 2014). Theoretical literature discussing connectivity in freshwater river systems has provided models and frameworks through which this scale of analysis may be accomplished (Frissell et al. 1986; Montgomery 1999; Benda et al. 2004; Campbell Grant et al. 2007; Wipfli et al. 2007; Borrett et al. 2014).

We demonstrated that IBR is detectable within a high-gene-flow population of fall-run Chinook Salmon in a relatively small catchment. We also showed that incorporating effective distance in a riverscape genetics framework helped expand interpretation of IBR by considering resistance along the path that is traveled to reach spawning habitat. Although genetic variation is weak within fall-run Chinook Salmon at the spatial extent of our analyses, the methods employed in this study highlight the usefulness of incorporating effective distance analysis into riverine research.

Research Gaps

Several options are available for calculating genetic distance. Convention and the need to make comparisons across catchments or among taxa dictate whether one or several metrics are employed. However, agreement on the best choice has not been reached (Jost 2008; Gerlach et al. 2010; Meirmans and Hedrick 2011; Whitlock 2011). Given our results, we recommend ensuring that at least F_{ST} and D_{est} are considered and compared.

There is potential for RG to provide information that would enhance our ability to identify specific catchment and riverscape features that are directly associated with population persistence. Expansion of RG methods to interpret how the environment affects genetic structure throughout the lifetime of individuals (e.g., return migration, spawning, rearing, smolting, and out-migration) would greatly improve our ability to enhance population-scale survival in freshwater habitats.

Differences between terrestrial, marine, and freshwater systems need to be explicitly considered when applying the various analytical methods available (Selkoe et al. 2016). For example, directional water flow is an important component of riverine habitats, whereas in terrestrial systems, there is no force that has a similar effect on flora and fauna (Davis et al. 2018). Directionality has yet to be integrated into existing LG and SG methods, and although it was considered in our study (results not shown), we were unable to identify any relationships. There is great potential for continued growth in the field as these challenges and others are addressed.

Geographic distance was not a predictor of genetic distance within our study, although in larger systems (i.e., greater than 1,000 km), distance is a consistent predictor of spatial variation. Given that smaller catchments may hold life history variation that is important to long-term population persistence, there is a need to understand the relationships that maintain this diversity. One component of maintaining genetic diversity is to protect the components of the watershed that contribute to gene flow. The combined context of reach-specific and effective distance variables indicated the value of a RG approach to better understand relationships among groups within the river. Continued investigation using RG at small spatial extents and incorporating network relationships are needed to reveal a better understanding of the effects of dispersal on population structure. Both are important for the conservation and management of highly migratory species.

Management Implications

Habitat fragmentation and flow changes that alter functional connectivity within riverscapes have caused increased resistance to dispersal and loss of access to suitable spawning habitat. For Pacific salmon, such physical changes have contributed to population decreases, restructuring of source-sink dynamics (Schick and Lindley 2007), and reduced range distributions (Yoshiyama et al. 1998). To combat these effects at state levels, entities like the Pacific Fishery Management Council have developed Pacific Coast Salmon Fishery Management Plans (PFMC 2016). Those plans list management objectives for Endangered Species Act-listed species, escapement goals, and catch limits and describe essential fish habitat for the management of commercial and recreational salmon fisheries. However, management at this spatial extent is based upon the status of larger, regional management units (e.g., the North Oregon Coast Chinook Salmon ESU). The ESUs are combinations of unique spawning groups from multiple, smaller river basins and do not account for finer-scale life history, genetic, and phenotypic variability that is likely present within smaller catchments (Schindler et al. 2010; Davis et al. 2017). Effective management should incorporate such diversity considerations.

Within riverscapes, geographic and hydrologic characteristics are often correlated, and identifying how each or both affect genetic structure is difficult at best. Nonetheless, the ability of conservation and management to ensure long-term viability of salmon populations or establish effective recovery strategies for any threatened species is dependent on understanding how these processes affect dispersal and subsequent gene flow. Riverscape genetics methods that include network relationships would enable more informed interpretation of these fine-scale relationships, thereby ensuring the design of effective recovery strategies for Pacific salmonids in local catchments and contributing to their stability and conservation.

Acknowledgments

Sample collection for this research was supported by the Oregon Department of Fish and Wildlife Coastal CCRMP and staff of the Confederated Tribes of Siletz Indians (CTSI). We thank Kelly Christiansen and Kathryn Ronnenberg for the creation of maps and figures that were used throughout the manuscript. Also, we are grateful to Stan van de Wetering at CTSI for sample collection and insight regarding the distribution of salmonids within the Siletz River. Funding for this research was provided by CTSI project no. Y0434A, the Coastal Oregon Marine Experimental Station (COMES) project no. ASF501, the Living Marine Resource Cooperative Science Center LMRCSC Tab grant no.14-01, and National Oceanic and Atmospheric Administration Educational Partnership Program with Minority Serving Institutions Graduate Research and Training Scholarship Program. Samples were collected and handled under the IAUCUC protocol #4124. Finally, we thank Bob Hughes and three anonymous reviewers who provided constructive feedback on the original manuscript.

References

Altermatt, F. 2013. Diversity in riverine metacommunities: a network perspective. Aquatic Ecology 47:365–377.

Auerbach, D. A., and N. L. Poff. 2011. Spatiotemporal controls of simulated metacommunity dynamics in dendritic networks. Journal of the North American Benthological Society 30:235–251.

Banks, S. C., M. P. Piggott, J. E. Williamson, U. Bové, N. J. Holbrook, and L. B. Beheregaray. 2007. Ocean variability and coastal topography shape genetic structure in a long-dispersing sea urchin. Ecology 88:3055–3064.

Benavente, J. N., L. W. Seeb, J. E. Seeb, I. Arismendi, C. E. Hernandez, G. Gajardo, R. Galleguillos, M. I. Cádiz, S. S. Musleh, and D. Gomez-Uchida. 2015. Temporal genetic variance and propagule-driven genetic structure characterize naturalized Rainbow Trout (*Oncorhynchus mykiss*) from a Patagonian lake impacted by trout farming. PLOS (Public Library of Science) ONE [online serial] 10(11):e0142040.

Benda, L., L. Poff, D. Miller, T. Dunne, G. Reeves, G. Pess, and M. Pollock. 2004. The network dynamics hypothesis: how channel networks structure riverine habitats. BioScience 54:413–427.

Benda, L. D., D. Miller, K. Andras, P. Bigelow, G. Reeves, and D. Michael. 2007. NetMap: a new tool in support of watershed science and resource management. Forest Science 53:206–219.

Bjornn, T. C., and D. W. Reiser. 1991. Habitat requirements of salmonids in streams. Pages 83–138 *in* W. R. Meehan, editor. Influences of forest and rangeland management on salmonid fishes and their habitats. American Fisheries Society, Special Publication 19, Bethesda, Maryland.

Blanchet, F. G., P. Legendre, R. Maranger, D. Monti, and P. Pepin. 2011. Modelling the effect of directional spatial ecological processes at different scales. Oecologia 166:357–368.

Borrett, S. R., J. Moody, and A. Edelmann. 2014. The rise of network ecology: maps of the topic diversity and scientific collaboration. Ecological Modelling 293:111–127.

Bowlby, H. D., I. A. Fleming, and A. J. F. Gibson. 2016. Applying landscape genetics to evaluate threats affecting endangered Atlantic Salmon populations. Conservation Genetics 17:823–838.

Bradbury, I. R., L. C. Hamilton, M. J. Robertson, C. E. Bourgeois, A. Mansour, and J. B. Dempson. 2014. Landscape structure and climatic variation determine Atlantic Salmon genetic connectivity in the northwest Atlantic. Canadian Journal of Fisheries and Aquatic Sciences 71:246–258.

Brenkman, S. J., J. J. Duda, C. E. Torgersen, E. Welty, G. R. Pess, R. Peters, and M. L. McHenry. 2012. A riverscape perspective of Pacific salmonids and aquatic habitats prior to large-scale dam removal in the Elwha River, Washington, USA. Fisheries Management and Ecology 19:36–53.

Campbell Grant, E. H., W. H. Lowe, and W. F. Fagan. 2007. Living in the branches: population dynamics and ecological processes in dendritic networks. Ecological Letters 10:165–175.

Castellano, S., and E. Balletto. 2002. Is the partial mantel test inadequate? Evolution 56:1871–1873.

Castillo, J. A., C. W. Epps, A. R. Davis, and S. A. Cushman. 2014. Landscape effects on gene flow for a climate-sensitive montane species, the American Pika. Molecular Ecology 23:843–856.

Castric, V., F. Bonney, and L. Bernatchez. 2001. Landscape structure and hierarchical genetic diversity in the Brook Charr, *Salvelinus fontinalis*. Evolution 55:1016–1028.

Clemento, A. J., E. D. Crandall, J. C. Garza, and E. C. Anderson. 2014. Evaluation of a single nucleotide polymorphism baseline for genetic stock identification of Chinook Salmon (*Oncorhynchus tshawytscha*) in the California current large marine ecosystem. U.S. National Marine Fisheries Service Fishery Bulletin 112:112–130.

Cote, D., D. G. Kehler, C. Bourne, and Y. F. Wiersma. 2009. A new measure of longitudinal connectivity for stream networks. Landscape Ecology 24:101–113.

Csardi, G., and T. Nepusz. 2006. The igraph software package for complex network research. InterJournal Complex Systems 1695. Available: http://igraph.org.

Davis, C. D., C. W. Epps, R. L. Flitcroft, and M. A. Banks. 2018. Refining and defining riverscape genetics: how rivers influence population genetic structure. Water [online serial] 5(2):e1269.

Davis, C. D., J. C. Garza, and M. Banks. 2017. Identification of multiple genetically distinct populations of Chinook Salmon (*Oncorhynchus tshawytscha*) in a small coastal watershed. Environmental Biology of Fishes 100:923–993.

DeVries, P. 1997. Riverine salmonid egg burial depths: review of published data and implications for scour studies. Canadian Journal of Fisheries and Aquatic Sciences 54:1685–1698.

Diniz-Filho, J. A., T. N. Soares, J. S. Lima, R. Dobrovolski, V. L. Landeiro, M. P. Telles, T. F. Tangel, and L. M. Bini. 2013. Mantel test in population genetics. Genetics and Molecular Biology 36:475–485.

Dionne, M., F. Caron, J. J. Dodson, and L. Bernatchez. 2008. Landscape genetics and hierarchical genetic structure in Atlantic Salmon: the interaction of gene flow and local adaptation. Molecular Ecology 17:2382–2396.

Epps, C. W., J. D. Wehausen, V. C. Bleich, S. G. Torres, and J. S. Brashares. 2007. Optimizing dispersal and corridor models using landscape genetics. Journal of Applied Ecology 44:714–724.

Fausch, K. D., C. E. Torgersen, C. V. Baxter, and H. W. Li. 2002. Landscapes to riverscapes: bridging the gap between research and conservation of stream fishes. BioScience 52:1–16.

Flitcroft, R., K. Burnett, J. Snyder, G. Reeves, and L. Ganio. 2014. Riverscape patterns among years of juvenile Coho Salmon in midcoastal Oregon: implications for conservation. Transactions of the American Fisheries Society 143:26–38.

Fraser, D. J., L. K. Weir, L. Bernatchez, M. M. Hansen, and E. B. Taylor. 2011. Extent and scale of local adaptation in salmonid fishes: review and meta-analysis. Heredity 106:404–420.

Frissell, C. A., W. J. Liss, C. E. Warren, and M. D. Hurley. 1986. A hierarchical framework for stream habitat classification: viewing streams in a watershed context. Environmental Management 10:199–215.

Galbraith, H. S., D. T. Zanatta, and C. C. Wilson. 2015. Comparative analysis of riverscape genetic structure in rare, threatened and common freshwater mussels. Conservation Genetics 16:845–857.

Galindo, H. M., D. B. Olson, and S. R. Palumbi. 2006. Seascape genetics: a coupled oceanographic-genetic model predicts population structure of Caribbean corals. Current Biology 16:1622–1626.

Ganio, L. M., C. E. Torgersen, and R. E. Gresswell. 2005. A geostatistical approach for describing spatial pattern in stream networks. Frontiers in Ecology and Evolution 3:138–144.

Geist, D. R., and D. D. Dauble. 1998. Redd site selection and spawning habitat use by fall Chinook Salmon: the importance of geomorphic features in large rivers. Environmental Management 22:655–669.

Gerlach, G. G., A. Jueterbock, P. Kraemer, J. Deppermann, and P. Harmand. 2010. Calculations of population differentiation based on G_{ST} and D: forget G_{ST} but not all of statistics! Molecular Ecology 19:3845–3852.

Gomez-Uchida, D., T. W. Knight, and D. E. Ruzzante. 2009. Interaction of landscape and life history attributes on genetic diversity, neutral divergence and gene flow in a pristine community of salmonids. Molecular Ecology 18:4854–4869.

Goslee, S. C., and D. L. Urban. 2007. The ecodist package for dissimilarity-based analysis of ecological data. Journal of Statistical Software 22:1–19.

Groot, C., and L. Margolis. 1991. Pacific salmon life histories. University of British Columbia Press, Vancouver.

Guillot, G., and F. Rousset. 2013. Dismantling the Mantel tests. Methods in Ecology and Evolution 4:336–344.

Hall, L. A., and S. R. Beissinger. 2014. A practical toolbox for design and analysis of landscape genetics studies. Landscape Ecology 29:1487–1504.

Hamann, E. J., B. P. Kennedy, D. C. Whited, and J. A. Stanford. 2014. Spatial variability in spawning habitat selection by Chinook Salmon (*Oncorhynchus tshawytscha*) in a wilderness river. River Research and Applications 30:1099–1109.

Hand, B. K., C. C. Muhlfeld, A. A. Wade, R. P. Kovach, D. C. Whited, S. R. Narum, A. P. Matala, M. W. Ackerman, B. A. Garner, J. S. Kimball, J. A. Stanford, and G. Luikart. 2016. Climate variables explain neutral and adaptive variation within salmonid metapopulations: the importance of replication in landscape genetics. Molecular Ecology 25:689–705.

Harris, L. N., R. Bajno, C. P. Gallagher, I. Koizumi, L. K. Johnson, K. L. Howland, E. B. Taylor, and J. D. Reist. 2015. Life-history characteristics and landscape attributes as drivers of genetic variation, gene flow, and fine-scale population structure in northern Dolly Varden (*Salvelinus malma malma*) in Canada. Canadian Journal of Fisheries and Aquatic Sciences 72:1477–1493.

Holderegger, R., U. Kamm, and F. Gugerli. 2006. Adaptive vs. neutral genetic diversity: implications for landscape genetics. Landscape Ecology 21:797–807.

Isaak, D. J., R. F. Thurow, B. E. Rieman, and J. B. Dunham. 2007. Chinook Salmon use of spawning patches: relative roles of habitat quality, size, and connectivity. Ecological Applications 17:352–364.

Johansson, M. L., A. Filipe, D. C. Reed, P. T. Raimondi, N. C. Coelho, M. A. Young, P. T. Drake, C. A. Edwards, K. Cavanaugh, J. Assis, L. B. Ladah, T. W. Bell, J. A. Coyer, D. A. Siegel, E. A. Serrão. 2015. Seascape drivers of *Macrocystis pyfriera* population genetic structure in the northeast Pacific. Molecular Ecology Notes 24:4866–4885.

Jost, L. O. U. 2008. G_{ST} and its relatives do not measure differentiation. Molecular Ecology 17:4015–4026.

Kanno, Y., J. C. Vokoun, and B. H. Letcher. 2011. Fine-scale population structure and riverscape genetics of Brook Trout (*Salvelinus fontinalis*) distributed continuously along headwater channel networks. Molecular Ecology 20:3711–3729.

Kalinowski, S. T. 2005. hp-rare 1.0: a computer program for performing rarefaction on measures of allelic richness. Molecular Ecology Notes 5:187–189.

Kirk, H., and J. R. Freeland. 2011. Applications and implications of neutral versus non-neutral markers in molecular ecology. International Journal of Molecular Science 12:3966–3988.

Landguth, E. L., C. C. Muhlfeld, and G. Luikary. 2012. CDFISH: an individual-based, spatially- explicit, landscape genetics simulator for aquatic species in complex riverscapes. Conservation Genetics Resources 4:133–136.

Le Pichon, C., G. Gorges, P. Boët, J. Baudry, F. Goreaud, and T. Faure. 2006. A spatially explicit resource-based approach for managing stream fishes in riverscapes. Environmental Management 37:322–335.

Legendre, P., and M. J. Fortin. 2010. Comparison of the Mantel test and alternative approaches for detecting complex multivariate relationships in the spatial analysis of genetic data. Molecular Ecology Resources 10:831–844.

Leps, M., J. D. Tonkin, V. Dahm, P. Haase, and A. Sundermann. 2015. Disentangling environmental drivers of benthic invertebrate assemblages: the role of spatial scale and riverscape heterogeneity in a multiple stressor environment. Science of the Total Environment 536:546–556.

Lichstein, J. W. 2006. Multiple regression on distance matrices: a multivariate spatial analysis tool. Plant Ecology 188:117–131.

Liggins, L., E. A. Treml, and C. Riginos. 2013. Taking the plunge: an introduction to understanding seascape genetic studies and using biophysical models. Geography Compass 7:173–196.

Manel, S., M. K. Schwartz, G. Luikart, and P. Taberlet. 2003. Landscape genetics: combining landscape ecology and population genetics. Trends in Ecology and Evolution 18:189–197.

McRae, B. H. 2006. Isolation by resistance. Evolution 60:1551–1561.

Meeuwig, M. H., C. S. Guy, S. T. Kalinowski, and W. A. Fredenberg. 2010. Landscape influences on genetic differentiation among Bull Trout populations in a stream-lake network. Molecular Ecology 19:3620–3633.

Meirmans, P. G., and P. W. Hedrick. 2011. Assessing population structure: F_{ST} and related measures. Molecular Ecology Resources 11:5–18.

Miller, D. J., K. Burnett, and L. Benda. 2008. Factors controlling availability of spawning habitat for salmonids at the basin scale. Pages 103–120 *in* D. Sear and P. DeVries, editors. Salmonid spawning habitat in rivers: physical controls, biological responses, and approaches to remediation. American Fisheries Society, Symposium 65, Bethesda, Maryland.

Miller, R. R. 2010. Is the past present? Historical splash-dam mapping and stream disturbance detection in the Oregon coastal province. Master's thesis. Oregon State University, Corvallis.

Moir, H. J., and G. B. Pasternack. 2010. Substrate requirements of spawning Chinook Salmon (*Oncorhynchus tshawytscha*) are dependent on local channel hydraulics. River Research and Applications 26:456–468.

Montgomery, D. R. 1999. Process domains and the river continuum. Journal of the American Water Resources Association 35:397–410.

Moran, P., D. J. Teel, M. A. Banks, T. D. Beacham, M. R. Bellinger, S. M. Blankenship, J. R. Candy, J. C. Garza, J. E. Hess, S. R. Narum, L. W. Seeb, W. D. Templin, C. G. Wallace, and C. T. Smith. 2013. Divergent life-history races do not represent Chinook Salmon coast-wide: the importance of scale in Quaternary biogeography. Canadian Journal of Fisheries and Aquatic Sciences 70:415–435.

Nehlsen, W., J. E. Williams, and J. A. Lichatowich. 1991. Pacific salmon at the crossroads: stocks at risk from California, Oregon, Idaho, and Washington Fisheries 16(2):4–21.

O'Malley, K. G., and M. A. Banks. 2008. A latitudinal cline in the Chinook Salmon (*Oncorhynchus tshawytscha*) clock gene: evidence for selection on PolyQ length variants. Proceedings of the Royal Society B 275:2813–2821.

O'Malley, K. G., D. P. Jacobson, R. Kurth, A. J. Dill, and M. A. Banks. 2013. Adaptive genetic markers discriminate migratory runs of Chinook Salmon (*Oncorhynchus tshawytscha*) amid continued gene flow. Evolutionary Applications 6:1184–1194.

Oksanen, J., F. Guillaume Blanchet, R. Kindt, P. Legendre, P. R. Minchin, R. B. O'Hara, G. L. Simpson, P. Solymos, M. H. H. Stevens, and H. Wagner. 2015. vegan: community ecology package. Available: https://cran.r-project.org/web/packages/vegan/vegan.pdf.

Olsen, J. B., T. D. Beacham, M. Wetklo, L. W. Seeb, C. T. Smith, B. G. Flannery, and J. K. Wenburg. 2010a. The influence of hydrology and waterway distance on population structure of Chinook Salmon *Oncorhynchus tshawytscha* in a large river. Journal of Fish Biology 76:1128–1148.

Olsen, J. B., P. A. Crane, B. G. Flannery, K. Dunmall, W. D. Templin, and J. K. Wenburg. 2010b. Comparative landscape genetic analysis of three Pacific salmon species from subarctic North America. Conservation Genetics 12:223–241.

Ozerov, M., A. Veselov, J. Lumme, and C. Primmer. 2012. "Riverscape" genetics: river characteristics influence the genetic structure and diversity of anadromous and freshwater Atlantic Salmon (*Salmo salar*) populations in northwest Russia. Canadian Journal of Fisheries and Aquatic Sciences 69:1947–1958.

Peakall, R., and P. Smouse. 2012. GenAlEx 6.5: genetic analysis in Excel. Population genetic software for teaching and research, an update. Bioinformatics 28:2537–2539.

Peterson, E. E., and J. M. Ver Hoef. 2014. STARS: an ArcGIS toolset used to calculate the spatial information needed to fit spatial statistical models to steam network data. Journal of Statistical Software 56:1–17.

Peterson, E. E., J. M. Ver Hoef, D. J. Isaak, J. A. Falke, M. J. Fortin, C. E. Jordan, K. McNyset, P. Monestiez, A. S. Ruesch, A. Sengupta, N. Som, E. A. Steel, D. M. Theobald, C. E. Torgersen, and S. J. Wenge. 2013. Modelling dendritic ecological networks in space: an integrated network perspective. Ecological Letters 16:707–719.

Petrou, E. L., J. E. Seeb, L. Hauser, M. J. Witteveen, W. D. Templin, and L. W. Seeb. 2014. Fine-scale sampling reveals distinct isolation by distance patterns in Chum Salmon (*Oncorhynchus keta*) populations occupying a glacially dynamic environment. Conservation Genetics 15:229–243.

PFMC (Pacific Fisheries Management Council). 2016. Pacific coast salmon fishery management plan for commercial and recreational salmon fisheries off the coasts of Washington, Oregon, and California as amended through amendment 19. PFMC, Portland, Oregon.

Pilger, T. J., K. B. Gido, D. L. Propst, J. E. Whitney, and T. F. Turner. 2015. Comparative conservation genetics of protected endemic fishes in an arid-land riverscape. Conservation Genetics 16:875–888.

R Core Team. 2018. R: a language and environment for statistical computing. R Foundation for Statistical Computing, Vienna, Austria. Available: www.R-project.org.

Reeves, G., L. Benda, K. Burnett, P. A. Bisson, and J. R. Sedell. 1995. A disturbance-based ecosystem approach to maintaining and restoring freshwater habitats of evolutionarily significant units of anadromous salmonids in the Pacific Northwest. Pages 334–349 *in* J. L. Nielsen and D. A. Powers, editors. Evolution and the aquatic ecosystem: defining unique units in population conservation. American Fisheries Society, Symposium 17, Bethesda, Maryland.

Reiser, D. W., and R. G. White. 1988. Effects of two sediment size-classes on survival of steelhead and Chinook Salmon eggs. North American Journal of Fisheries Management 8:432–437.

Riginos, C., and L. Liggins. 2013. Seascape genetics: populations, individuals, and genes marooned and adrift. Geography Compass 7:197–216.

Roux, C., A. Alber, M. Bertrand, L. Vaudor, and H. Piégay. 2015. "Fluvial corridor": a new ArcGIS toolbox package for multiscale riverscape exploration. Geomorphology 242:29–37.

Ruppert, J. L. W., P. M. A. James, E. B. Taylor, T. Rudolfsen, M. Veillard, C. S. Davis, D. Watkinson, and M. S. Poesch. 2017. Riverscape genetic structure of a threatened and dispersal limited freshwater species, the Rocky Mountain Sculpin (*Cottus* sp.). Conservation Genetics 18:925–937.

Schick, R. S., and S. T. Lindley. 2007. Directed connectivity among fish populations in a riverine network. Journal of Applied Ecology 44:1116–1126.

Schindler, D. E., R. Hilborn, B. Chasco, C. P. Boatright, T. P. Quinn, L. A. Rogers, and M. S. Webster. 2010. Population diversity and the portfolio effect in an exploited species. Nature (London) 465:609–612.

Selkoe, K. A., J. R. Watson, C. White, B. Hornin, M. Iacchei, S. Mitarai, D. A. Siegel, D. Gaines, and R. J. Toonen. 2010. Taking the chaos out of genetic patchiness: seascape genetics revels ecological and oceanographic drivers of genetic patterns in three temperate reef species. Molecular Ecology 19:3708–3726.

Selkoe, K. A., K. T. Scribner, and H. M. Galindo. 2016. Waterscape genetics: applications of landscape genetics to rivers, lakes, and seas. Pages 220–246 *in* N. Balkenhol, S. A. Cushman, A. T. Storfer, and L. P. Waits, editors. Landscape genetics: concepts, methods, applications. Wiley, Chichester, UK.

Shirk, A. J., D. O. Wallin, S. A. Cushman, C. G. Rice, and K. I. Warheit. 2010. Inferring landscape effects on gene flow: a new model selection framework. Molecular Ecology 19:3603–3619.

Thorp, J. H., M. C. Thoms, and M. D. Delong. 2006. The riverine ecosystem synthesis: biocomplexity in river networks across space and time. River Research and Applications 22:123–147.

Waples, R. 2001. Characterizing diversity in salmon from the Pacific Northwest. Journal of Fish Biology 59:1–41.

Ward, J. V., and J. A. Stanford. 1983. The serial discontinuity concept of lotic ecosystems. Pages 29–42 *in* T. D. Fontaine and S. M. Bartell, editors. Dynamics of lotic ecosystems. Scientific Publishers, Ann Arbor, Michigan.

Weir, B. S., and C. C. Cockerham. 1984. Estimating *F*-statistics for the analysis of population structure. Evolution 38:1358–1370.

Whitlock, M. C. 2011. G_{ST} and D do not replace F_{ST}. Molecular Ecology 20:1083–1091.

Wiens, J. A. 1989. Spatial scaling in ecology. Functional Ecology 3:385–397.

Wipfli, M. S., J. S. Richardson, and R. J. Naiman. 2007. Ecological linkages between headwaters and downstream ecosystems: transport of organic matter, invertebrates, and wood down headwater channels. Journal of the American Water Resources Association 43:72–85.

Wohl, E., N. Kramer, V. Ruiz-Villanueva, D. N. Scott, F. Comiti, A. M. Gurnell, H. Piégay, K. B. Lininger, K. L. Jaeger, D. M. Walters, and K. D. Fausch. 2019. The natural wood regime in rivers. BioScience 69:259–273.

Yoshiyama, R. M., F. W. Fisher, and P. B. Moyle. 1998. Historical abundance and decline of Chinook Salmon in the Central Valley region of California. North American Journal of Fisheries Management 18:487–521.

Zeller, K. A., K. McGarigal, and A. R. Whiteley. 2012. Estimating landscape resistance to movement: a review. Landscape Ecology 27:777–797.

American Fisheries Society Symposium 90:57–82, 2019
© 2019 by the American Fisheries Society

Climatic Variation Drives Growth Potential of Juvenile Chinook Salmon along a Subarctic Boreal Riverscape

Jeffrey A. Falke*
U.S. Geological Survey, Alaska Cooperative Fish and Wildlife Research Unit
University of Alaska Fairbanks
Post Office Box 757020, Fairbanks, Alaska, 99775, USA

Brock M. Huntsman
Division of Forestry and Natural Resources, West Virginia University
4100 Agricultural Sciences Building
Post Office Box 6108, Morgantown, West Virginia 26506, USA

Erik R. Schoen
Institute of Arctic Biology, University of Alaska Fairbanks
Post Office Box 757000, Fairbanks, Alaska, 99775, USA

Abstract.—Climatic variation is a key driver of freshwater physical processes that in turn control stream fish growth and population dynamics at fine spatial scales and species distributions across broad landscapes. A recent downturn in Chinook Salmon *Oncorhynchus tshawytscha* returns across the Yukon River basin, Alaska, USA and Yukon Territories, Canada has led to hardship among user groups and increased interest in understanding how freshwater processes affect population persistence within this important commercial, recreational, and subsistence fishery. Here, we present results for the Chena River basin, interior Alaska, where we used field observations and riverscape-scale spatially explicit models to assess the influence of stream temperature on juvenile Chinook Salmon growth potential among years (2003–2015) and across 438 stream kilometers. We ran bioenergetic simulations for warm and cool year scenarios and contrasted temperature model precision and growth among different habitat types (small and large tributaries, main stem, and side channels) based on field estimates of growth, size, diet, and measured stream temperatures. Stream temperature regimes predicted from remotely sensed land surface temperatures were precise during the open water season ($R^2 > 0.87$; root-mean-squared error < 1.1°C), although the relationship was weakest in groundwater-mediated tributary habitats. Field observations revealed salmon were 67% larger by mass (g) in September during a warm year versus a cool year from main-stem sites. Bioenergetic simulations predicted that, on average, growth potential was 42% higher in warm years, although growth potential varied across the riverscape as much as 60% between cool upstream and warm downstream habitats. Climate variability is clearly an

* Corresponding author: jfalke4@alaska.edu

important driver of freshwater habitat conditions and has a large role in controlling freshwater growth of juvenile salmon. A better understanding of how climate influences growth conditions in different habitat types and across broad landscapes will be critical for conservation and management of Alaskan Chinook Salmon stocks under an expected warmer and more variable climate.

Introduction

Climatic variation sets the context within which freshwater processes (i.e., flow and thermal regimes) have shaped the evolution of life histories in stream ecosystems (Poff et al. 1997; Rypel 2014; Hand et al. 2016; Kovach et al. 2016). The magnitude of climate variation varies by ecosystems, latitude, and the temporal scale (e.g., interannual, intraannual, monthly, and daily) at which it is examined. For example, cyclical climate patterns such as the Pacific Decadal Oscillation and the El Niño–Southern Oscillation are key drivers of freshwater processes through their effects on air temperature and precipitation patterns (Bieniek et al. 2011). At annual time scales (e.g., within years), climate drives ice-out on lakes and rivers, peak flows from snowmelt timing, summer droughts and floods, and the seasonality of flows. In turn, population dynamics and phenology of freshwater aquatic organisms are coupled with climate variation through effects of flows and stream temperature on vital rates (i.e., birth, death, and immigration and emigration), predator–prey dynamics, and individual growth. Given rapid changes in Earth's climate and because aquatic organisms are particularly susceptible to shifting flow regimes and temperature, a better understanding of the links among climatic variation, freshwater processes, and fish population dynamics is warranted.

Stream temperature is arguably the most important environmental constraint controlling the growth and population dynamics of aquatic ectotherms in stream ecosystems (Caissie 2006; Webb et al. 2008; Steel et al. 2017). Fish growth responds to temperature via a dome-shaped relationship, increasing with temperature up to an optimum and decreasing rapidly with further warming (Brett et al. 1969; Kitchell et al. 1977). These thermal performance curves vary markedly among species (Magnuson et al. 1979) and also within species as a function of body size, feeding rate, and food quality (Beauchamp 2009). As a result, variation in water temperature through time is an important control on fish growth and population dynamics.

Stream temperatures within riverscapes (sensu Fausch et al. 2002) commonly follow a longitudinal pattern of cool headwaters to warm main stems (Poole and Berman 2001). However, hydrologic controls within stream networks can lead to discontinuities in longitudinal patterns in temperature (Ward and Stanford 1983). For example, lateral connectivity (e.g., floodplains), groundwater upwellings, and glacial meltwater can create a mosaic of thermal conditions within the riverscape (Poole and Berman 2001; Snyder et al. 2015; Merriam et al. 2017). At large spatial extents, this complexity may be exacerbated because not all freshwater habitats are equally sensitive to climatic variation (Luce et al. 2014; Snyder et al. 2015; Isaak et al. 2016; Mauger et al. 2016). For example, temperatures in streams with significant groundwater contributions may be more resilient to changing air temperatures than streams dominated by overland flow (Snyder et al. 2015). In combination with other landscape-scale factors that affect stream temperature (e.g., network position, land management; Torgersen et al. 1999; Peterson and Ver Hoef 2010; Steel et al. 2017), knowledge of spatial patterns among habitat types within and among watersheds is valuable information for supporting management of aquatic resources (Isaak et al. 2014).

Statistical techniques that provide spatially explicit predictions of stream temperature across riverscapes have made significant progress in recent years (e.g., Peterson and Ver Hoef 2010; Jones et al. 2014; McNyset et al. 2015; Vatland et al. 2015), yet, to date, such predictions are typically coarse water temperature descriptors for a defined period (e.g., mean August stream temperature, Isaak et al. 2010; 7-d average daily maximum stream temperature, Falke et al. 2016). However, temperature predictions at finer temporal grains (e.g., daily or weekly means) are needed to capture variation in fish growth because fish growth integrates temperature through time. Moreover, spatiotemporal variability in temperature may play an important role in fish growth and is rarely captured in coarse-scale temperature predictions (Armstrong and Schindler 2013; Baldock et al. 2016). For example, juvenile salmon growth is often reduced in systems with greater range and variability of temperatures, as compared to growth in systems with stable temperatures, even when average temperatures over a given period are similar (Geist et al. 2010). Thus, characterization of stream thermal regimes (i.e., temporal patterns) across complex riverscapes is a basic requirement to better understand, predict, and forecast patterns of fish growth across broad spatial extents.

Application of bioenergetics models in fisheries research has observed an exponential increase since the early 1990s (Hartman and Kitchell 2008). Fish bioenergetics applies thermodynamic properties to fishes via temperature-dependent physiological constraints to describe energy budgets (Brandt and Hartman 1993). Useful insight to explain patterns in fish ecology at multiple spatial scales can be achieved through fish bioenergetics with clear value for fisheries research. For example, bioenergetics models have helped to explain fish distributions and habitat selection via optimal foraging theory (Hughes 1998; Petty et al. 2014) and preemptive habitat selection (Petty and Grossman 2010; Ayllón et al. 2013). Both food availability and temperature are significant determinants of a fish's energy budget, and both can strongly vary throughout a watershed. Since somatic growth and persistence are often correlated (e.g., large fish typically have higher survival rates, Quinn and Peterson 1996; Sommer et al. 2001), identifying optimal growth conditions throughout a riverscape would be useful for managers trying to maintain or improve persistence and productivity of fish populations.

Chinook Salmon *Oncorhynchus tshawytscha* is a highly valued species throughout its range, supporting important subsistence, recreational, and commercial fisheries. Significant declines in Alaskan Chinook Salmon populations since 2000 have forced fisheries closures and restrictions, affecting rural economies, food security, and cultural well-being, particularly in the Yukon River basin (ADFG 2013; Schindler et al. 2013). Chinook Salmon in Alaska primarily have a stream-type life history (Healey 1991) where offspring spend approximately 1 year in freshwater. Changes in juvenile growth could influence population productivity because large body size is often associated with greater survival in salmon (Henderson and Cass 1991; Moss et al. 2005; Farley et al. 2007; Duffy and Beauchamp 2011). Not surprisingly, most interannual variation in cohort strength of Yukon River Chinook Salmon is determined during the freshwater rearing period and first summer in the ocean (Murphy et al. 2017), where faster freshwater growth rates are associated with faster growth in the marine environment (Ruggerone et al. 2009). Chinook Salmon may achieve faster growth rates in a warming climate, as has been shown for Sockeye *O. nerka* and Coho Salmon *O. kisutch* in some habitats in Southwest Alaska (Schindler et al.

2005; Rich et al. 2009; Armstrong et al. 2010). Therefore, it is necessary to determine how current Chinook Salmon growth in Alaskan streams varies spatially and temporally to assess how climate change may affect the productivity of such an important ecological, economic, and cultural resource.

Our goal with this research was to use a combined stream temperature and bioenergetics modeling approach to predict growth potential of juvenile Chinook Salmon during years with variable climate conditions and across a heterogeneous subarctic riverscape. Specifically, we (1) collected empirical stream temperature, fish growth, and fish diet composition data in years with contrasting climate conditions; (2) applied a spatiotemporally explicit stream temperature model to predict stream temperatures for different climate scenarios and habitat types; and (3) integrated temperature model predictions with a bioenergetics simulation model to predict juvenile Chinook Salmon growth potential across a subarctic riverscape under two representative current climate scenarios.

Methods

Study Area

Our study area was the Chena River basin, Alaska (Figure 1). Draining 5,130 km^2, the Chena River flows 241 km to its confluence with the Tanana River, a major tributary of the Yukon River, near the city of Fairbanks. The river is located in a subarctic climate and boreal forest ecosystem; annual precipitation

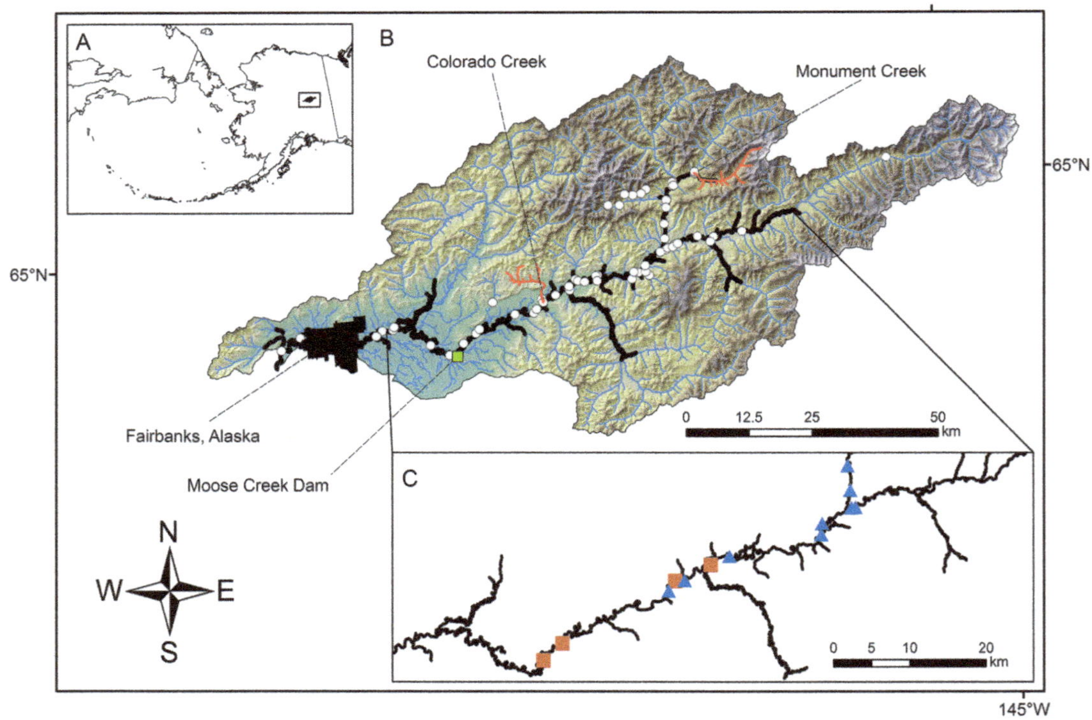

Figure 1. **(A)** Location of the study area in Alaska. **(B)** Chena River basin stream network (blue lines) with juvenile Chinook Salmon distribution (black lines) and location of stream temperature monitoring sites (white dots). Colorado and Monument creeks (red lines) and the location of Moose Creek Dam (green square) are shown. **(C)** Location of sites sampled for prey availability and juvenile Chinook Salmon growth in 2009 (red squares) and 2015 (blue triangles).

averages about 25.4 cm. Flows in the river originate from precipitation, snowmelt, and groundwater from unconfined aquifers. The river is typically ice covered from October through April, and peak flows occur following spring snowmelt and late-summer rain events. Mean annual flow for the Chena River across the period of record (1970–2018) was 38.5 m^3/s, although the river is regulated by the Moose Creek flood-control dam and discharge has been limited to 340 m^3/s since its completion in 1981. Bedrock geology in the basin consists of Precambrian schist and tertiary volcanics and sediments, with unconsolidated sediments comprising most surficial deposits. Vegetation is boreal forest (primarily white and black spruce), grassland, shrubs, bogs, and tundra. Land use above the Moose Creek Dam is almost entirely "general use" (reserves, open space, natural area, remote settlements, and forest land). Below the dam in the city of Fairbanks, land use is a mixture of residential, commercial, industrial, and rural uses.

Of the 2,265 total stream kilometers in the Chena River basin, the distribution of rearing habitats for juvenile Chinook Salmon is estimated to cover approximately 438 km (Figure 1; ADFG 2017; Matter et al. 2018). Rearing habitats range from small- and medium-sized creeks (small tributaries; first to third order) with pool-riffle-run complexes to larger river habitats with large pools (large tributaries and main stem; greater than third order) and numerous sloughs, backwaters, and interconnected ponds (side channels). In addition to Chinook Salmon, the Chena River basin supports a relatively diverse fish assemblage, including Chum Salmon *Oncorhynchus keta*, Arctic Grayling *Thymallus arcticus*, Longnose Sucker *Catostomus catostomus*, Northern Pike *Esox lucius*, Inconnu (also known as Sheefish) *Stenodus leucichthys*, Burbot *Lota lota*, Arctic Lamprey *Lethenteron camtschaticum* (also known as *Lampetra japonica*), Slimy Sculpin *Cottus cognatus*, Lake Chub *Couesius plumbeus*, and four whitefish species.

Stream Temperature Observations

We used a combination of hourly stream temperature observations collected at 21 sites along the Chena River main stem (2006 and 2007; Decker 2010) and from a sensor network of 49 sites deployed throughout the juvenile Chinook Salmon distribution during 2013 to 2015 to quantify stream temperature regimes (Figure 1). Stream temperatures during 2006 and 2007 were collected using Hobo Water Temperature Pro V2 temperature loggers (Onset Computer Corporation, Bourne, Massachusetts) anchored to the substrate near the thalweg with small duckbill anchors and steel cable. Logger data were downloaded periodically during summer months. Stream temperature monitoring during 2013 to 2015 was conducted using Hobo Water Temperature Pro V2 or TidbiT v2 water temperature loggers (Onset Computer Corporation), depending on habitat type. Where boulders or rip rap were present at a site, we used the TidbiT loggers and underwater epoxy protocol of Isaak and Horan (2011). In all other habitats, Pro V2 loggers were placed inside black solar shield polyvinyl chloride housings and anchored to large wood, underwater tree roots, or sand bags filled with cobbles using ultraviolet-resistant plastic zip ties. Data were downloaded yearly during summer months. During 2013 to 2015, pre- and postdeployment calibration of temperature loggers followed guidelines outlined in Dunham et al. (2005).

Riverscape Temperature Model

We used stream temperature observations collected during the ice-free season (April 6 to November 2; Figure 1) to parameterize a spatially and temporally explicit stream temperature model. We used the stream temperature modeling framework described in Falke et al. (2013, 2014) and McNyset et al. (2015) to predict stream temperatures for confluence-to-confluence stream reaches every 8 d during the open water season from 2003 to 2015. Because the model has not been applied to subarctic systems such as the Chena River basin or used in bioenergetics simulations, we describe the model building and diagnostic process here in detail.

We used NetMap (Benda et al. 2007) to derive a synthetic (i.e., digital) stream network for the Chena River basin because the National Hydrography Dataset (USGS 2019) for this region of Alaska has not been updated and is therefore low quality because of coarse-scale, misrepresented flow lines and disconnected and omitted streams. The NetMap model generates a synthetic stream network layer from a digital elevation model based on flow accumulation and channel delineation algorithms (described in Clarke et al. 2008). The result is a network of 50–200-m-long stream reaches linked to the surrounding landscape and attributed with geomorphic characteristics (e.g., drainage area, elevation). We aggregated the 50–200-m stream reaches to confluence-to-confluence reaches (mean = 2.60 km, SD = 2.64) and used this as our spatial grain for all further analyses and prediction.

Based on our NetMap-derived synthetic stream network for the Chena River basin and a 10-m digital elevation model, we used the Spatial Tools for the Analysis of River Systems (STARS) geoprocessing toolbox version 2.0.4 for ArcMap (version 10.2; Environmental Systems Research Institute, Redmond, California) to identify the unique land surface area draining into each reach (reach contributing area [RCA]). The RCAs averaged 3.7 km^2 (SD = 3.2) for the Chena River basin. Elevation (Elev; m) was determined for each RCA by taking the elevation at the outlet node; logger site elevations were assigned based on the RCA within which they occurred.

Eight-day land surface temperature (LST) data were taken from MODIS satellite observations (MOD11A2, 1 km^2 spatial resolution) and downloaded from NASA's Earth Observing System Data and Information System (https://earthdata.nasa.gov) for the years 2003 to 2015. Hierarchical data format files were converted to raster format using the Marine Geospatial Ecology Tools (Roberts et al. 2010), and each 8-d layer (N = 593) was clipped to the extent of the Chena River basin. Because each RCA contained more than one 8-d LST pixel, we took a spatially weighted average of LST values for each RCA and used these values for model development (see below).

We calculated 8-d summaries (rolling means; 8DRMWT) of mean daily stream temperature (°C) for all available data during the open water period (April 6 to November 2) to use as the response in our predictive model. If multiple loggers were located within an RCA, only one was included in the model to avoid pseudoreplication. Based on preliminary analyses and suggestions from McNyset et al. (2015), we developed two half-year fixed-effects generalized linear models fit with data during spring (day of year [DOY] 96–200 = April 6–July 19) and fall (DOY 201–300 = July 20–November 2) of the form

$$\text{8DRMWT} \sim \text{LST} + \text{LST}^2 + \text{DOY} + \text{Elev},$$

where LST is the 8-d spatially weighted average LST for each RCA, LST^2 is a quadratic term, DOY is the midpoint of each 8-d period (e.g., 97, 105…305), and Elev is the elevation (m) at the outlet of each RCA. We used the root-mean-squared error (RMSE) to evaluate model goodness-of-fit and predictive power and Spearman's correlation coefficient to investigate the sensitivity of observed 8DRMWT to LST by habitat type. All temperature model analyses were conducted in R (R Core Team 2018).

We developed climate scenarios based on observed streamflow data from U.S. Geological Survey gaging station number 15514000 near Fairbanks and predicted stream temperature regimes from the LST model during the annual open water period. Although our original intent was to contrast juvenile Chinook Salmon growth potential among low-flow, high-flow, warm, and cool years, preliminary analyses indicated that low-flow years were always warm and high-flow years were always cool. We used median summer flow (m^3/s; annual median of daily mean discharge from April 26 to September 30) as a descriptor of streamflow because lower stream flows have been shown to be related to higher juvenile Chinook Salmon productivity in the Chena River (Neuswanger et al. 2015). Specifically, authors found that low Chinook Salmon population productivity was correlated to years where median summer discharge was above 27.9 m^3/s. We found that median summer flows during the period 2003 to 2015 were highly correlated (Pearson correlation $r = -0.793$, $P < 0.001$; Figure 2) with mean summer stream temperature

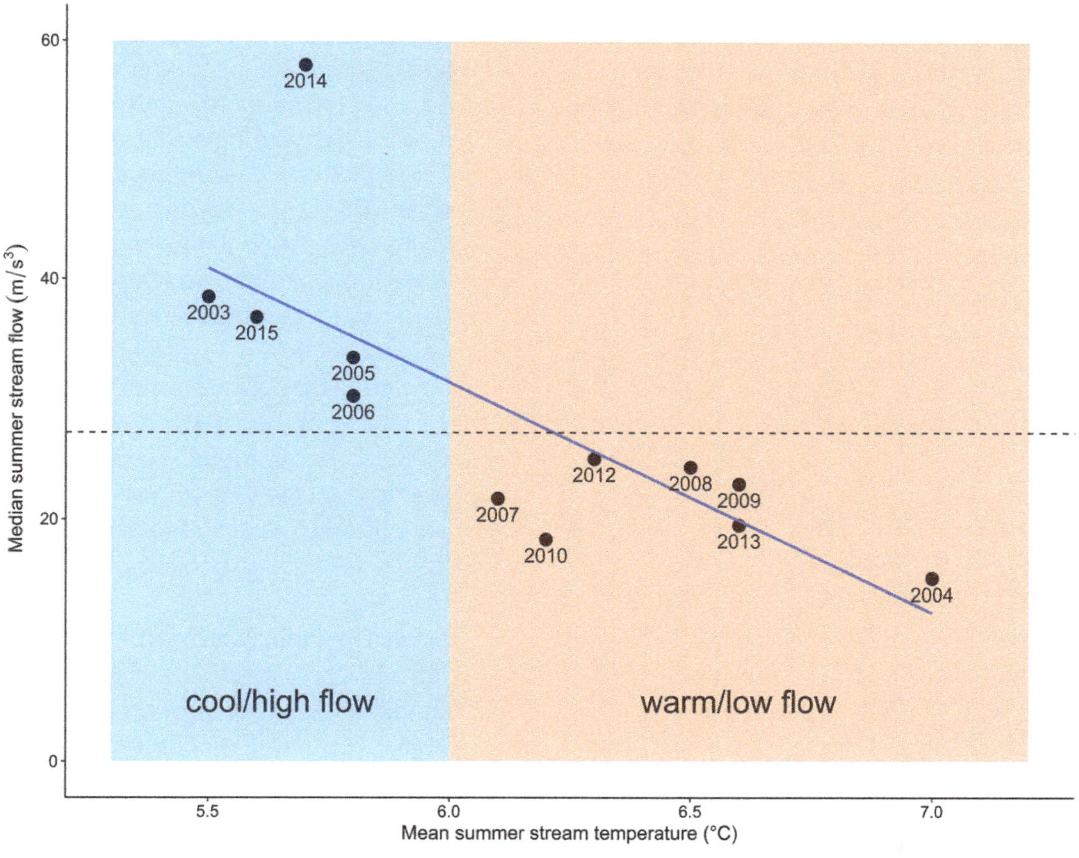

Figure 2. Relationship between mean summer (April 26 to September 30) stream temperature (°C) predicted using the land surface temperature model (see text) and median summer streamflow (m^3/s) from U.S. Geological Survey streamflow gauge 15514000 located on the Chena River near Fairbanks, Alaska, USA for 2003 to 2015. The dashed line indicates the 27.9 m^3/s discharge threshold above which Chinook Salmon productivity has been shown to be lower in this system (Neuswanger et al. 2015). Years were categorized as cool/high flow (<6°C; >27.9 m^3/s) and warm/low flow (>6°C; <27.9 m^3/s) scenarios for bioenergetics simulations.

(°C) predicted from the LST model. Thus, we could not separate low-flow years from warm years or high-flow years from cool years. As a result, we developed two climate scenarios: (1) warm/low flow, and (2) cool/high flow, hereafter referred to as warm and cool. Warm years were defined by measured median summer flow <27.9 m^3/s and mean annual predicted stream temperature >6°C (2004, 2007, 2008, 2009, 2010, 2012, and 2013), and cool years were >27.9 m^3/s and <6°C (2003, 2005, 2006, 2011, 2014, and 2015).

Fish Growth and Diet Composition

We quantified the body size and diet composition of juvenile Chinook Salmon using field data collected during our study and a previous study (Gutierrez 2011). Juvenile (age 0) Chinook Salmon were sampled approximately monthly from June to September during the years 2009 and 2015. In 2009, Gutierrez (2011) sampled fish at four main-stem sites (Figure 1) using dip nets and seines (June to mid-July) and minnow traps (mid-July to mid-September). During each site visit, captured individuals were lightly anesthetized with MS-222 (tricaine methanesulfonate), measured for fork length (mm), and weighed (g). Fish stomach contents were collected using a nonlethal gastric lavage technique, where a 10-mL pipette was inserted into the stomach of each fish and water was pumped into the stomach until all prey items were regurgitated (Meehan and Miller 1978).

In 2015, we sampled juvenile Chinook Salmon at 10 sites (Figure 1); one side-channel habitat and one main-stem habitat were sampled per site (Huntsman and Falke 2019). Fish were sampled using baited minnow traps and depletion seining. Upon capture, individuals were lightly anesthetized using Aqui-S 20E, measured for fork length (mm), and weighed (g). Stomach contents were collected using a lavage technique similar to that used during 2009, where water was pumped into salmon stomachs through a canula with a syringe (Hafs et al. 2011). For samples collected in both years, diet items were preserved in 80% ethanol and juvenile salmon were returned alive to the river following recovery from anesthesia. In the laboratory, diet samples were identified to family for aquatic taxa and order for terrestrial taxa using a dissecting microscope.

Bioenergetics Simulations

We estimated the growth potential (defined here as mean mass [g] in September) of juvenile Chinook Salmon in warm and cool years across their 438-km known distribution (ADFG 2017) in the Chena River basin using a two-step bioenergetics approach. First, we estimated the feeding rates of Chinook Salmon sampled under contrasting thermal conditions using empirical diet composition and growth rate (seasonal body weight) data. Second, we simulated the growth potential of juvenile salmon using modeled stream temperature data (8DRMWT) across the riverscape, assuming they fed at the rate estimated in step 1 and expressed as the proportion of the theoretical maximum daily consumption. Bioenergetics models were run at daily time steps; we summarized each 8-d period from DOY 152 to 274 (approximately June 1 to September 30).

We used a bioenergetics model (Hanson et al. 1997) to estimate the feeding rates of Chinook Salmon in warm (2009) and cool (2015) years based on observed somatic growth rates. For each simulation, the model estimated the feeding rate, expressed as a proportion (P) of the theoretical maximum consumption rate (C_{max}) that would allow a Chinook Salmon to achieve the observed growth rate, given the observed diet composition and temperatures estimated by the LST model. Models were implemented in R using code (A. G. Hansen, Colorado Parks and Wildlife, unpublished data) that replicated the Wisconsin Fish Bioenergetics 3.0 software (Hanson et al. 1997) in a scripted framework that facilitated running repeated simulations. Inputs to estimate *P* for each year included mean (across all sites sampled) fish mass (g) at the beginning (June) and end (September) of summer, predicted 8DRMWT from the LST model (°C), observed diet composition, and literature-based energy densities of prey (J/g; see below).

We developed a simplified prey schedule for input into bioenergetics models based on the observed diet composition of juvenile Chinook Salmon in warm (2009) and cool (2015) years. Invertebrate taxa were categorized as aquatic or terrestrial in origin (sensu McCarthy et al. 2009), and we calculated the proportion of the diet (by mass) that each taxon comprised (averaged across all individual salmon sampled in each month). Diet composition in the bioenergetics simulations were then varied for each month to match the observed diets of juvenile Chinook Salmon in warm (2009) and cool (2015) years (Table 1). We assigned generalized energy density values based on weighted averages of literature-based values (McCarthy et al. 2009) for the prey taxa observed in this study (aquatic = 4,000 J/g; terrestrial = 5,000 J/g).

Bioenergetics simulations were conducted using physiological parameters and a weight-dependent energy density equation for Chinook Salmon (Stewart and Ibarra 1991). We also conducted a set of preliminary simulations using updated parameters developed specifically for juvenile Chinook Salmon in warmer environments (17–25°C; Plumb and Moffitt 2015) to investigate the efficacy of these parameters for subarctic salmon populations. However, preliminary results indicated the updated parameters did not adequately account for the growth rates observed under the cooler temperatures (0.0–18.7°C) observed in our study. The *P*-values ranged from 1.06 to 1.18 with the updated parameters, much higher than typical for wild fish populations (Armstrong and Schindler 2011), and our diet sampling did not suggest that fish were feeding at unusually high ration sizes (e.g., gorging on salmon eggs). Therefore, we used the original Stewart and Ibarra (1991) parameters exclusively for the remainder of the study. We predicted *P* for each RCA within the juvenile Chinook

Table 1. Parameter estimates (±SE) from two linear models (spring = April 6 to July 19; fall = July 20 to November 2) predicting 8-d rolling mean stream temperature (°C) as a function of land surface temperature (LST; °C), day of year (DOY), and elevation (Elev; m). Multiple regression correlation coefficients (R^2) and root-mean-squared errors (RMSE; °C) are shown for both models.

Model	Parameter	Estimate (SE)	*t*	*P*	R^2	RMSE (°C)
Spring	Intercept	–5.6898 ± 0.4422	–12.845	<0.001	0.87	1.14
	LST	–0.0974 ± 0.0410	–2.376	0.017		
	LST^2	0.0104 ± 0.0014	7.391	<0.001		
	DOY	0.0888 ± 0.0037	24.068	<0.001		
	Elev	–0.0123 ± 0.0005	–24.922	<0.001		
Fall	Intercept	20.4244 ± 1.0517	13.420	<0.001	0.93	1.03
	LST	0.0908 ± 0.0118	7.718	<0.001		
	LST^2	0.0038 ± 0.00029	7.782	<0.001		
	DOY	–0.0596 ± 0.0037	–16.060	<0.001		
	Elev	–0.0077 ± 0.0003	–27.145	<0.001		

Salmon distribution ($N = 169$ RCAs) and separately for each year using a permutation approach. For each climate scenario and RCA, we simulated P 100 times for DOY 152–274 based on 8DRMWT drawn randomly from the pool of years within each scenario. For each climate scenario, P was then averaged across all RCAs and used in our riverscape bioenergetics model to estimate juvenile Chinook Salmon growth potential (see below). Thus, P was representative of average conditions across scenario years.

Riverscape Salmon Growth Potential Model

Juvenile Chinook Salmon growth potential (mass [g] in September) was estimated for each RCA and climate scenario (warm and cool years) using a growth rate potential approach. We estimated daily growth (g), based on P, diet composition, prey energy density, and initial fish mass in June, and 8DRMWT. As above, we used representative prey diet composition schedules (i.e., percent aquatic and terrestrial) and associated energy densities for warm (2009) and cool (2015) years. Observed initial juvenile Chinook Salmon mass was similar between the 2 years (see Results), so a value of 0.5 g was used for both warm and cool year simulations. We simulated daily growth increments (g) for the summer growth period (DOY 152–274) 100 times for each RCA based on random draws from the pool of years within each scenario. Growth potential was then expressed as mean September weight (g), the sum of daily growth increments across the summer growth period.

Results

Stream Temperature Observations and Sensitivity

Observed 8DRMWT ranged from 0.0°C to 18.7°C (mean = 3.5; SD = 3.9) based on 3,186 observations across the 70 stream temperature logger sites (Figure 3, top). The distribution of habitat types among stream temperature sites was 22 small tributaries, 6 large tributaries, 33 main stems, and 9 side channels. Rolling means (8DRMWT) varied widely, particularly during midsummer. Based on Spearman's rho (ρ), sensitivity (i.e., correlation) of 8DRMWT to LST was generally high and ranged from 0.59 to 0.96 (all $P < 0.001$; Figure 4). Of the seven relatively less sensitive sites ($\rho < 0.7$), one was a large tributary, two were side channels, and four were small, groundwater-influenced tributaries, indicating that 8DRMWT did not track LST as strongly in those habitats.

Riverscape Temperature Model

Regression diagnostics for models predicting 8DRMWT indicated that LST, DOY, and Elev explained 87% of the variation in 8DRMWT for spring (DOY 96–200) and 93% for fall (DOY 201–306), respectively (Table 1; Figure 3, bottom). On average, the precision of predicted 8DRMWT was approximately ±1°C (spring RMSE = 1.14; fall RMSE = 1.03). Examination of model predictions by habitat type showed that main-stem and off-channel habitats tracked observed 8DRMWT closely (Figure 5). However, predictions for sites influenced by cold (Colorado Creek) and warm (Monument Creek) groundwater were less precise (locations noted in Figure 1), particularly at the cold groundwater site where predictions were biased positively up to 4°C. The LST model underpredicted 8DRMWT during spring at the warm groundwater site (approximately 2°C), though predicted values converged with observed data by midsummer.

Fish Growth and Diet Composition

Monthly diet and size data were collected during June–September from 208 juvenile (age 0) Chinook Salmon in 2009 and 985 in

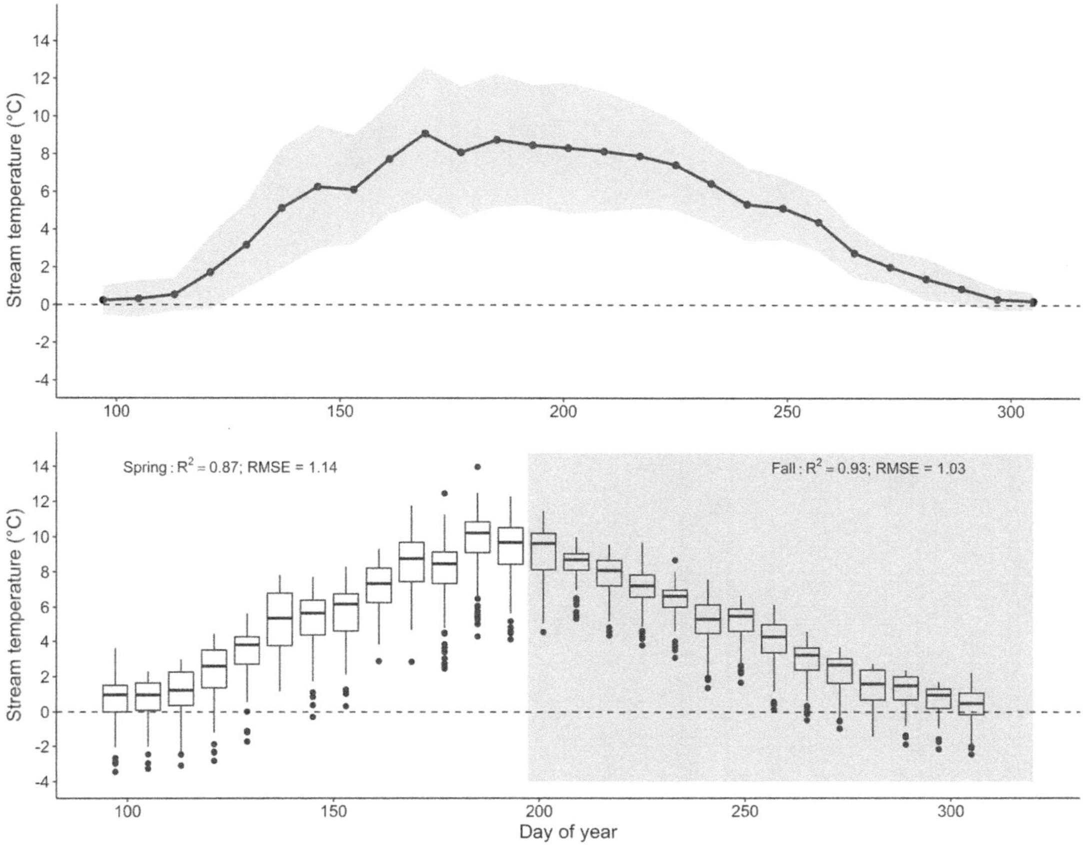

Figure 3. Mean observed (N = 70 sites; top panel) and predicted (lower panel) stream temperatures (°C) from the land surface temperature model for each 8-d period (day of year; x-axis) during the open water season (mid-April to late October) averaged across the Chena River basin, Alaska. Gray shading in top panel represents 95% confidence intervals. In the lower panel, boxplots represent 25th and 75th quartiles, horizontal lines are medians, and dots are outliers. Regression diagnostics (multiple R^2 and root-mean-squared error [RMSE]) are shown for spring and fall land surface temperature models.

2015 (Table 2). Although mean lengths (mm) during summer months were similar between years and mean weights (g) were nearly identical during June for both years (2009; mean = 0.4; SE = 0.0; 2015; mean = 0.5; SE = 0.0), by September, mean weight was 67% higher (mean = 5.5; SE = 0.3) in the warm year (2009) relative to the cool year (2015; mean = 3.3; SE = 0.0; Figure 6).

Juvenile Chinook Salmon diets and the proportion of aquatic- and terrestrial-origin prey items varied between the 2 years (Table 2). In 2009, diets were dominated (≥62% by mass) by aquatic invertebrate prey items (primarily adult Chironomidae) throughout the summer (June–September; Gutierrez 2011). In contrast, during 2015, terrestrial-origin prey items (Diptera) were dominant during early summer (82% of diets), but proportions of the two categories were approximately 1:1 during mid- to late-summer months (July to September).

Bioenergetics Simulations

Feeding rates (P) of juvenile Chinook Salmon were higher in warm (mean = 0.52; SD =

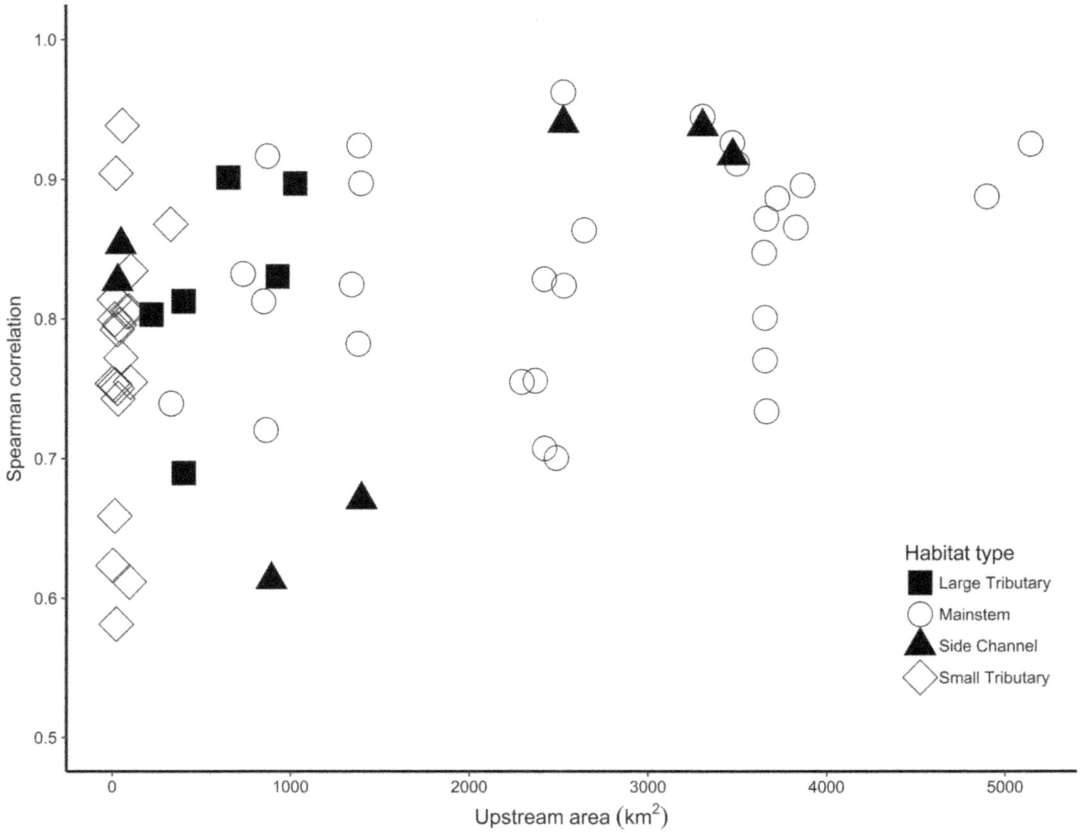

Figure 4. Spearman correlations between observed stream temperature and land surface temperature model predictions by habitat type arranged by the upstream area of each site.

0.05) relative to cool (mean = 0.43; SD = 0.05) years, suggesting that variation in juvenile Chinook Salmon growth was driven directly by temperature effects on metabolism, as well as indirectly by a feeding-related mechanism. Based on the mean *P* and diet proportions for cool and warm years (Table 2), the predicted weights of juvenile Chinook Salmon closely matched observed growth trajectories across the summer months in the cool (2015) and warm (2019) years (Figure 6). Growth potential under observed and predicted 8DRMWT (warm scenario) was similar across summer months for mainstem and off-channel habitats (Figure 7). However, growth potential under observed temperature was higher (1.5 g) than predicted temperatures in September at a warm groundwater site and much lower (2.9 g) at a cold groundwater site.

Riverscape Growth Potential

Riverscape growth potential (mean September weight [g]) varied between scenarios and across the Chena River basin (Figure 8). Growth potential increased from upstream to downstream under both scenarios, with the highest growth potential predicted to occur in the lowermost reaches in the basin. Growth potential was 43% higher in warm years (mean = 4.91 g; SD = 0.99; range = 2.16–6.75) relative to cool years (mean =

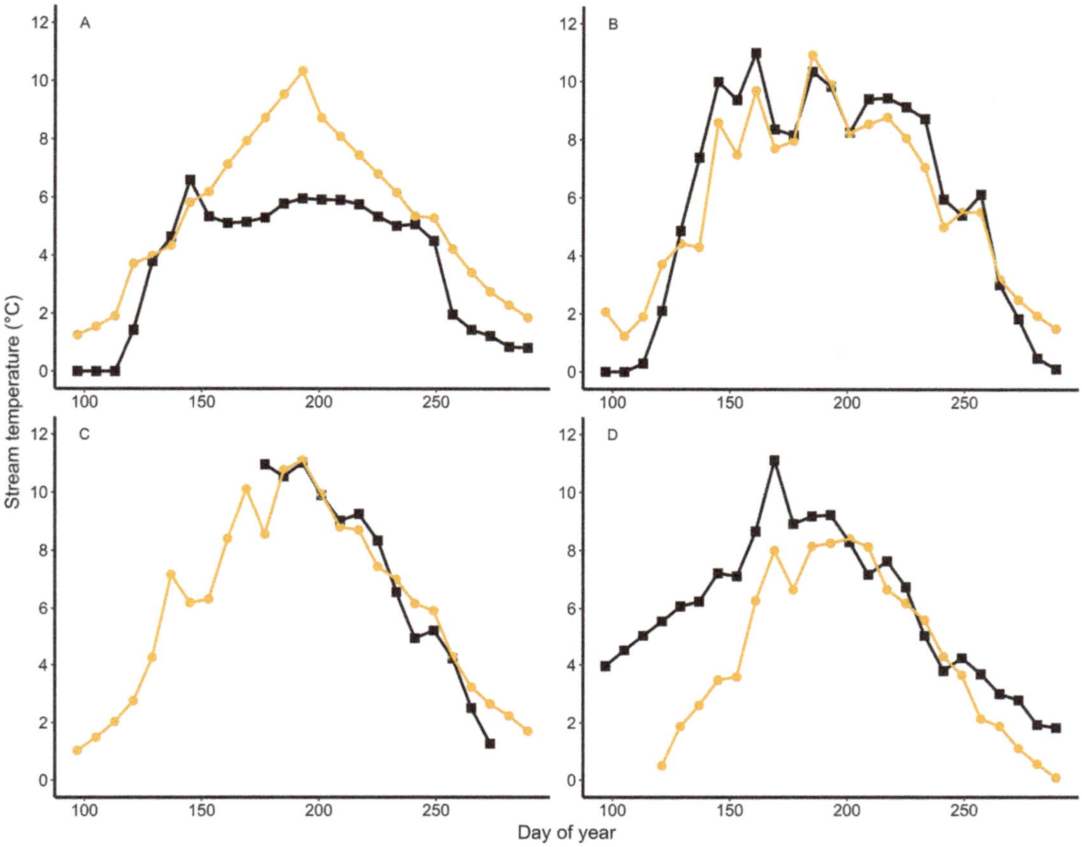

Figure 5. Observed (black squares and lines) and predicted (orange dots and lines) stream temperatures by day of year (April 1 to September 30) for four habitat types: **(A)** cold groundwater, **(B)** main stem, **(C)** off-channel, and **(D)** warm groundwater.

Table 2. Number sampled (*N*), length (mm; mean ± SE), weight (g; mean ± SE), and proportion of aquatic- and terrestrial-origin items in diets of juvenile Chinook Salmon from the Chena River, Alaska during summer months during warm, low-flow (2009) and cool, high-flow (2015) years.

Year	Month	*N*	Length (mm)	Weight (g)	Aquatic	Terrestrial
2009	June	54	38.6 ± 0.7	0.4 ± 0.0	0.86	0.14
	July	65	47.6 ± 1.4	1.4 ± 0.0	0.89	0.11
	August	73	69.2 ± 0.7	3.7 ± 0.1	0.83	0.17
	September	16	77.8 ± 1.4	5.5 ± 0.3	0.62	0.38
2015	June	207	40.7 ± 0.2	0.5 ± 0.0	0.18	0.82
	July	368	55.3 ± 0.4	1.5 ± 0.0	0.88	0.12
	August	132	68.4 ± 0.5	2.7 ± 0.1	0.90	0.10
	September	278	74.9 ± 0.3	3.3 ± 0.0	0.97	0.03

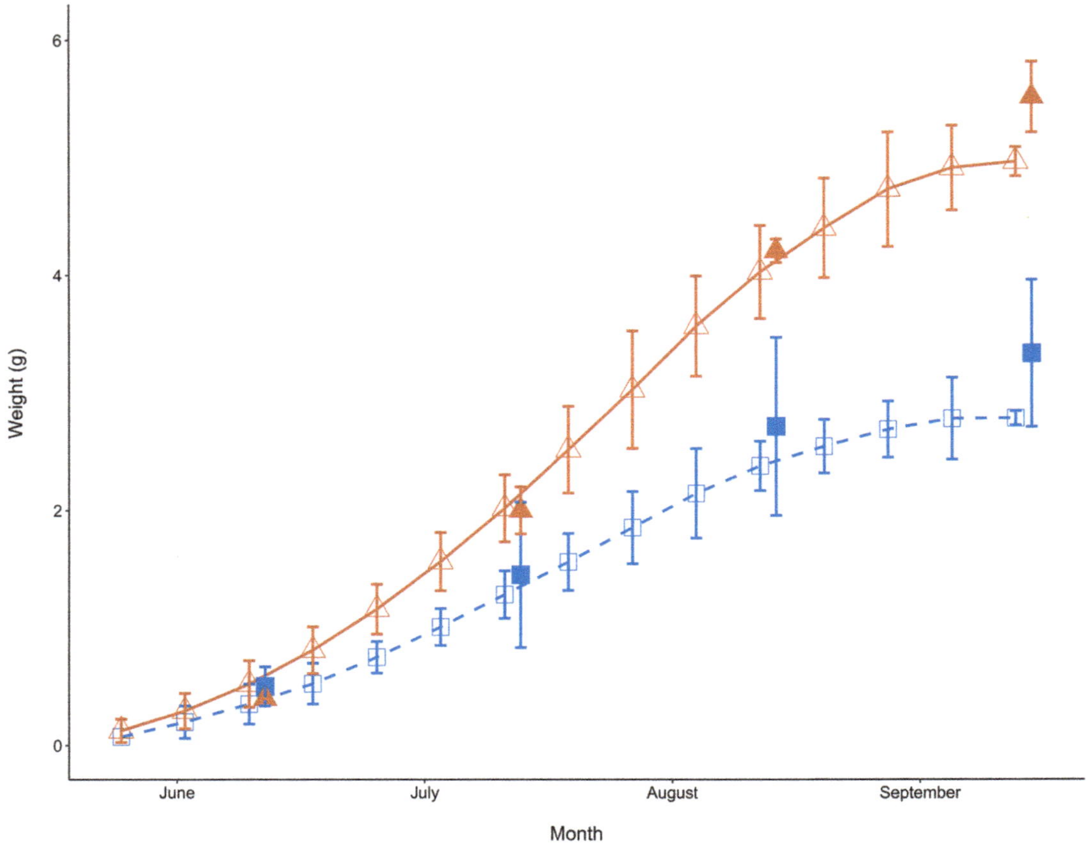

Figure 6. Observed (solid symbols) and predicted (open symbols) juvenile Chinook Salmon weights (g) for 2009 (vermillion triangles; warm year) and 2015 (blue squares; cool year). Predicted weights shown for a reach located in the center of the field sampling sites, modeled using predicted temperatures from all warm and cool years, respectively. Error bars are standard errors and 95% confidence intervals across observed field data and model simulations, respectively.

3.46 g; SD = 0.70; range = 1.53–4.89) across the 169 RCAs, and maximum predicted growth during cool years was near the mean of warm years. Site-specific (i.e., RCA) differences in growth potential between the warm and cool scenarios ranged from 0.62 to 1.97 g (mean = 1.45; SD = 0.29). There was considerable spatial variation in growth potential under both scenarios. Growth potential was predicted to be approximately 3.1 and 3.2 times higher in downstream reaches relative to small tributaries located in the upper basin under the cool and warm scenarios, respectively.

Discussion

By coupling a spatiotemporally explicit stream temperature model with a bioenergetics model and empirical observations of juvenile Chinook Salmon size and diet composition, we were able to simulate growth potential under two contrasting climate scenarios in the Chena River basin, a subarctic boreal ecosystem in Alaska. We found growth conditions to be highly heterogeneous among habitat types (e.g., main stem versus small tributary sites) commonly used by juvenile salmon to rear and also varying longitudinally along

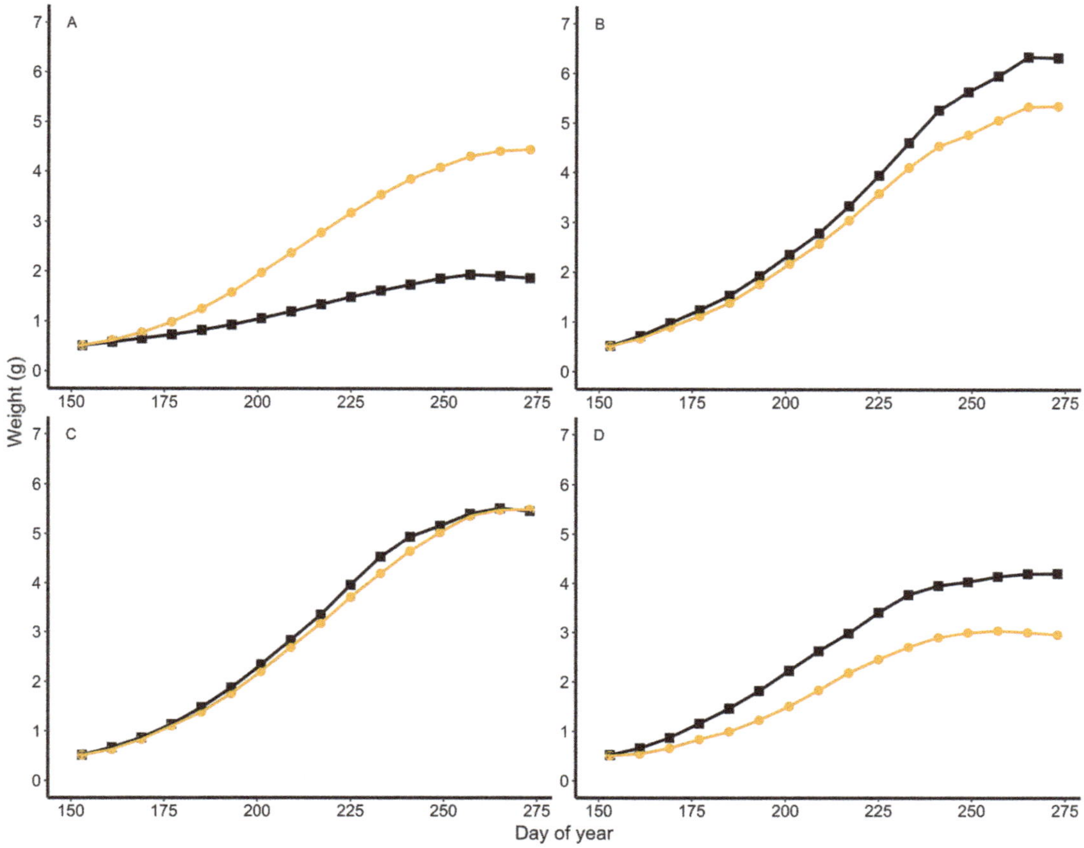

Figure 7. Simulated weight (g) from observed (black squares and lines) and predicted (orange dots and lines) stream temperature regimes in the Chena River basin, Alaska, by habitat type: **(A)** cold groundwater, **(B)** main stem, **(C)** off-channel, and **(D)** warm groundwater.

the riverscape from cold upstream areas to warmer conditions near the river mouth. Climatic variation was a strong driver of growth potential and was markedly different under observed and simulated conditions between contrasting climate years. Growth potential was much higher during warmer, low-flow years relative to cool, high-flow conditions. Because juvenile salmon size entering winter is a critical factor influencing overwinter survival (Cunjak 1988; Beamish and Mahnken 2001), the ability to predict size at summer's end as a function of environmental factors may be a key management and conservation tool (e.g., results could be integrated into stock assessment models). Below, we discuss the implications of strong climate controls on growth and survival of juvenile salmon in heterogeneous environments.

Juvenile Chinook Salmon growth was enhanced under warmer, low-streamflow conditions in the Chena River, located near the northern extent of the species current range. Our results contrast those reported for southern Chinook Salmon populations (Idaho: Crozier and Zabel 2006; California: Sommer et al. 2001), where survival of juveniles was negatively correlated with stream temperature and positively with streamflow. Previous studies in Alaska have documented enhanced growth under warmer conditions for Sockeye Salmon in lakes (Schindler et al. 2005; Rich et

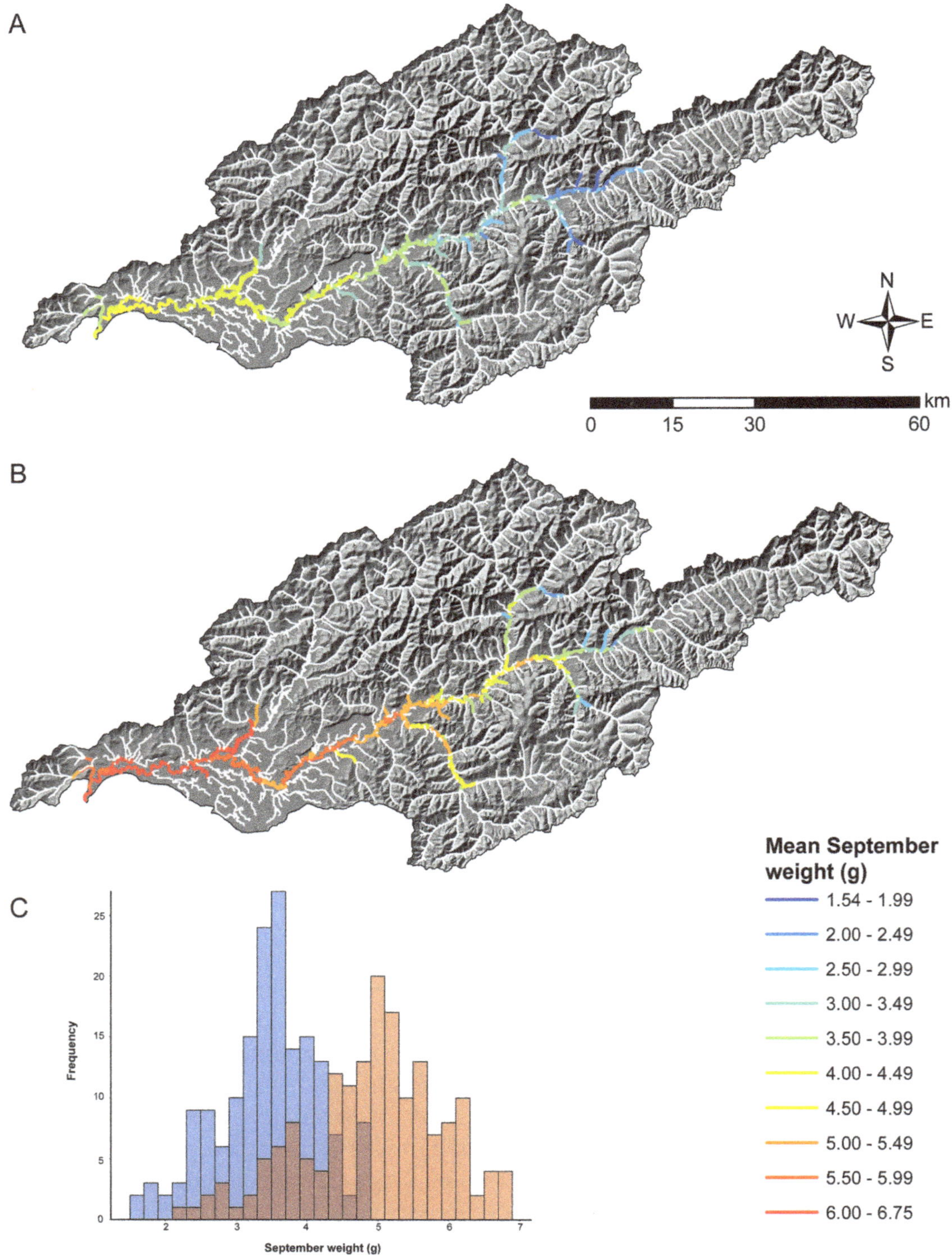

Figure 8. Predictions of juvenile Chinook Salmon growth potential (September weight; g) across the Chena River basin under **(A)** cool and **(B)** warm conditions. Also shown is **(C)** the frequency of predicted juvenile Chinook Salmon growth potential for cool (blue) and warm (vermillion) scenarios. The intermediate color represents the overlap between warm and cool.

al. 2009) and Coho Salmon in small streams (Armstrong and Schindler 2013; Armstrong et al. 2013, 2016). To our knowledge, this is the first study to document such an effect for Chinook Salmon or any salmon species in the Yukon River basin.

Previous work suggested that cold incubation temperatures may limit the upstream distribution of salmon spawning in the Chena River basin (Decker 2010), with lower-quality spawning habitat conditions further limiting the species upstream distribution (Huntsman et al. 2017). However, low overwinter incubation temperatures and lower spawning habitat quality do not appear to translate into poor juvenile rearing conditions, particularly in warm, low-flow years, because juvenile salmon are distributed throughout upstream reaches (ADFG 2017; Matter et al. 2018; J.A.F., unpublished data). Growth was predicted to be moderate to high in those areas under the warm scenario. This is a clear example of habitat complementation (Dunning et al. 1992; Falke et al. 2013), where species occupancy is higher owing to the juxtaposition of nonsubstitutable habitats, even if habitat quality of one type (e.g., spawning) is lower. In our study area, the remote upper portion of the Middle Fork Chena River exhibited the coldest observed and predicted thermal regimes; however, growth and diet data for juvenile Chinook Salmon have yet to be collected in this area. Such data would be useful to refine our riverscape growth potential model.

Is water temperature alone sufficient to explain the observed relationship between high-flow years and low juvenile Chinook Salmon productivity? Our consumption modeling results indicated that the direct metabolic effects of temperature explained more than half of the observed 67% increase in juvenile growth in warm years. This temperature–growth relationship could be an important mechanism helping to explain the relationship between high flows and reduced Chinook Salmon brood year productivity in Yukon River basin streams (Neuswanger et al. 2015; Cunningham et al. 2018) because larger smolts are more likely to survive to adulthood. In addition, the bioenergetics simulations indicated that salmon must have fed at a higher rate during warm, low-flow years than in cool, high-flow years to explain the observed difference in growth. Consistent with this result, high flows in the Chena are associated with low invertebrate drift density and low salmon gut fullness (Gutierrez 2011). Thus, reduced prey density, reduced foraging efficiency, or another feeding-related mechanism likely also contributes to the lower growth rates in cool, high-flow years. Our modeling approach was limited because we were not able to modify consumption rates based on variability in streamflow, although quantifying this relationship would likely provide for an improved model of climate variation impacts on juvenile salmon growth and survival. Such growth-rate potential models (sensu Hughes 1998; Nislow et al. 1999; Kennedy et al. 2008; Weber et al. 2014; Kalb et al. 2018) are a promising approach to disentangle the effects of food and temperature on salmon growth in subarctic river systems.

Our characterization of temperature regimes within the distribution of juvenile Chinook Salmon in the Chena River basin revealed a diverse mosaic of thermal environments. This thermal heterogeneity may confer growth advantages in situations (e.g., cool, high-flow years) when use of main-stem habitats may be disadvantageous. Juvenile Chinook Salmon may behaviorally thermoregulate to achieve digestive rates nearer to optimum by seeking out habitats with relatively warm conditions. For example, Baldock et al. (2016) found that Coho Salmon would thermoregulate by accessing warm off-channel habitats after foraging in cold thalwegs. This supports the hypothesis that juvenile salmon move between highly productive food habitats (e.g., main stems) and refugia habitat (e.g., tributaries) to maximize growth potential (Armstrong et al. 2010, 2013; Baldock et al. 2016; Huntsman et al. 2016). Our current and previous research suggests that juvenile Chinook Salmon within the Chena River may move relatively short distances between adjacent habitats (main stem versus off-channel [Huntsman and Falke 2019] or larger distances (e.g., between main stems and warm groundwater tributaries) to maximize growth and survival. A better understanding of within life stage movements at habitat unit and basin scales are needed to determine how access to diverse habitat mosaics translate into Chena River Chinook Salmon productivity.

Regardless of climate scenario, our bioenergetic simulations indicated that growth potential for juvenile Chinook Salmon was highest in downstream reaches in the Chena River basin. However, river reaches from the Moose Creek Dam (Figure 1) downstream to the confluence with the Tanana River are not core salmon-rearing habitats, likely owing to impacts from the cities of Fairbanks and North Pole, Alaska. These cities are located on the floodplain of the river, much of which now consists of impervious surfaces instead of forests (R. Henszey, U.S. Fish and Wildlife Service, unpublished data), riparian vegetation along the river channel has been removed, and significant bank hardening has occurred by public and private entities (TVWA 2015). Moreover, the natural process of wood transport from the upper to lower Chena River has been severed at Moose Creek Dam as wood is intercepted and removed on an annual basis, particularly following high-flow events when the floodgates are closed. Because logjams are important microhabitats for rearing juvenile Chinook Salmon (Perry 2012; Neuswanger et al. 2015; J.A.F., unpublished data), their absence in the lower river likely contributes to poor rearing habitat quality in these reaches, regardless of water temperature regime. However, because growth potential is high in these reaches, they could be a target for restoration efforts through changes to wood management.

Because of the scarcity of cover in the main stem, as well as greater predator densities, warmer downstream reaches could serve as ecological traps (Jeffres and Moyle 2012), especially in cool years when thermal conditions of lower reaches are attractive for behavioral thermoregulation. Sloughs and other off-channel habitats presently serve as primary velocity refuges to juvenile salmon in the lower Chena River. However, the relative abundance of piscivorous Northern Pike and Burbot is much greater in these sloughs than in the main stem or in sloughs upstream of Moose Creek dam (E.R.S., unpublished data). Juvenile Chinook Salmon are rarely observed in snorkel surveys or minnow trap catches in the lower reaches, but they have been documented in predator stomachs collected in lower river sloughs (Schoen et al. 2018). Clearly tradeoffs exist with respect to management and habitat restoration of juvenile salmon in the lower Chena River, yet our results indicate that improving rearing main-stem habitat conditions may be beneficial in this area.

In this study, we modeled the direct effects of temperature on juvenile salmon growth; however, changes in temperature can also influence salmon indirectly through their prey (Schindler et al. 2005), competitors (Schoen et al. 2015), and predators (Lawrence et al. 2014). Although warmer temperatures appear to provide growth benefits for salmon in the Chena River, warming could also reduce salmon survival rates by allowing warm-adapted predators such as Northern Pike to expand their distribution upstream and enhance their consumption rates. Juvenile Chinook Salmon growth is maximized at roughly 11–14°C (based on the range of body sizes, feeding rates, and prey energy densities used in our simulations), whereas Northern Pike consumption rates peak at roughly 25°C (Bean 2010). We measured stream temperatures between 0°C and 19°C in this study (April to November); temperatures at the upper end of this range may enhance predation risk more than they confer growth benefits to salmon. Projected temperature increases in lakes in Sweden during this century are expected to increase the consumption rates and densities of Northern Pike, allowing them to extirpate Brown Trout *Salmo trutta* in thousands of lakes where these native species now coexist (Hein et al. 2014). It is conceivable that future warming in the Yukon River basin could likewise enhance the impact of native Northern Pike on salmon population dynamics, potentially leading to extirpation of salmon from some habitats. The salmon-focused results of this study can serve as a foundation for future research integrating the direct and indirect effects of temperature on the aquatic food webs supporting salmon.

General circulation models predict a warmer and wetter future for interior Alaska (IPCC 2014). Such changes to the boreal aquatic environment may have significant implications for species such as Chinook Salmon whose population dynamics are closely tied to climate variation. Warmer water temperatures coincident with higher flows would be novel conditions compared to those observed in recent years (e.g., no years are located in upper right quadrant of Figure 2). Ostensibly, warmer conditions may be beneficial in environments such as the Chena River where juvenile salmon may be limited by cold temperatures. However, given the observed negative relationship between Chinook Salmon population productivity and high summer flows in interior Alaska rivers (Neuswanger et al. 2015; Cunningham et al. 2018), changes to flow regimes may limit growth advantages gained by warmer conditions. Moreover, beyond changes to measures of central tendency of flow regimes (e.g., means, medians), the periodicity of seasonal events is also expected to change in the future in the Chena River (Bennett et al. 2015). For example, maximum spring streamflow is increasing in the Chena River basin, indicative of a shift in snowmelt runoff timing, which could lead to higher sustained flows during May and June, critical periods for juvenile salmon growth when individuals are small in size. Thus, the regime shift from snowmelt runoff to rain dominated precipitation is expected to result in lower flows later in summer. Low flows could result in disconnections among important rearing habitats (e.g., main stem and off channel) or reduced access to thermal refuges. To date, predictions of flow regime changes under a changing climate have been conducted at coarse basin scales in interior Alaska (Bennett et al. 2015). Additional fine-scale and habitat-type-specific (e.g., small tributaries, off channel, etc.) predictions of ecologically relevant flow metrics are needed to fully understand the implications for juvenile Chinook Salmon growth potential in subarctic systems.

Management Implications and Future Research

Chinook Salmon are a vital economic, ecological, and cultural resource to Alaska and the Yukon River basin in particular (Schindler et al. 2013; Brown and Godduhn 2015). The large extent, habitat diversity, and remote setting confer many challenges to understanding the effects of climate variation on population vital rates. This is exacerbated by lack of on-the-ground water temperature data, few flow gauges, poor spatial data (e.g., digital elevation models and stream networks), and scarce data on juvenile salmon growth. Fortunately, recent advances in climate modeling, remote sensing, and digital stream network delineation and attribution provided robust spatial data that we integrated with modest empirical stream temperature and fish data. Model frameworks such as those we have developed will be highly useful, particularly in data-poor environments such as the Yukon River basin, and our work in the Chena River basin may serve as an example of what could be done across the greater Yukon River extent, once such data become more available. Because juvenile salmon growth and survival are tightly linked to climate variation and climate change impacts in this region have been and are expected to be serious, a focused effort to quantify the historic and current distribution of thermal and flow regimes and their relationship with juvenile salmon growth will be critical for climate impact assessment. A focus on spatially and temporally explicit, ecologically relevant flow and thermal regime descriptors (Wenger et al. 2010; Neuswanger et al. 2015) will be needed to target management and conservation actions aimed at protecting juvenile Chinook Salmon under a variable climate and across this broad landscape.

Acknowledgments

This work was funded by a grant from the State of Alaska and Alaska Department of Fish and Game (ADFG) to J.A.F., and E.R.S was supported by Alaska EPSCoR National Science Foundation award #OIA-1208927. The authors thank K. McNyset for advice on stream temperature model development, and A. Hansen for providing bioenergetics model code. We also thank S. Decker and the ADFG for providing stream temperature data and J. Adams, S. Corsetti, M. Evenson, J. Fox, and J. Pomeranz for assistance with temperature logger deployment and retrieval. Thanks to M. Perry, L. Gutierrez, and J. Neuswanger for their contributions to the knowledge base on Chinook Salmon in the Chena River from which our modeling and analyses greatly benefitted. This manuscript was greatly improved by comments from D. Beauchamp, L. Jones, R. M. Hughes, and several anonymous reviewers. This research was conducted at the University of Alaska Fairbanks, and the staff and facilities of the Alaska Cooperative Fish and Wildlife Research Unit and the Institute of Arctic Biology were instrumental in the success of this project. Any use of trade, firm, or product names is for descriptive purposes only and does not imply endorsement by the U.S. Government.

References

ADFG (Alaska Department of Fish and Game). 2013. Chinook Salmon stock assessment and research plan, 2013. ADFG, Special Publication No. 13-01, Anchorage.

ADFG (Alaska Department of Fish and Game). 2017. Anadromous waters catalog (AWC). Available: www.adfg.alaska.gov/sf/SARR/AWC/index.cfm?ADFG=main.home. (July 2017).

Armstrong, J. B., and D. E. Schindler. 2011. Excess digestive capacity in predators reflects a life of feast and famine. Nature (London) 476:84–87.

Armstrong, J. B., and D. E. Schindler. 2013. Going with the flow: spatial distributions of juvenile Coho Salmon track an annually shifting mosaic of water temperature. Ecosystems 16:1429–1441.

Armstrong, J. B., D. E. Schindler, K. L. Omori, C. P. Ruff, and T. P. Quinn. 2010. Thermal heterogeneity mediates the effects of pulsed subsidies across a landscape. Ecology 91:1445–1454.

Armstrong, J. B., D. E. Schindler, C. P. Ruff, G. T. Brooks, K. E. Bentley, and C. E. Torgersen. 2013. Diel horizontal migration in streams: juvenile fish exploit spatial heterogeneity in thermal and trophic resources. Ecology 94:2066–75.

Armstrong, J. B., E. J. Ward, D. E. Schindler, and P. J. Lisi. 2016. Adaptive capacity at the northern front: Sockeye Salmon behaviourally thermoregulate during novel exposure to warm temperatures. Conservation Physiology [online serial] 4(1):cow039.

Ayllón, D., G. G. Nicola, I. Parra, B. Elvira, and

A. Almodóvar. 2013. Intercohort density dependence drives Brown Trout habitat selection. Acta Oecologica 46:1–9.

Baldock, J. R., J. B. Armstrong, D. E. Schindler, and J. L. Carter. 2016. Juvenile Coho Salmon track a seasonally shifting thermal mosaic across a river floodplain. Freshwater Biology 61:1454–1465.

Beamish, R. J., and C. Mahnken. 2001. A critical size and period hypothesis to explain natural regulation of salmon abundance and the linkage to climate and climate change. Progress in Oceanography 49:423–437.

Bean, N. J. 2010. An improved bioenergetics model for Northern Pike (*Esox lucius*) of Box Canyon Reservoir, Pend Oreille River, Washington. Master's thesis. Eastern Washington University, Cheney.

Beauchamp, D. A. 2009. Bioenergetic ontogeny: linking climate and mass-specific feeding to life-cycle growth and survival of salmon. Pages 1–19 *in* C. C. Krueger and C. E. Zimmerman, editors. Pacific salmon: ecology and management of western Alaska's populations. American Fisheries Society, Symposium 70, Bethesda, Maryland.

Benda, L., D. Miller, K. Andras, P. Bigelow, G. Reeves, and D. Michael. 2007. Netmap: a new tool in support of watershed science and resource management. Forest Science 53:206–219.

Bennett, K. E., A. J. Cannon, and L. Hinzman. 2015. Historical trends and extremes in boreal Alaska river basins. Journal of Hydrology 527:590–607.

Bieniek, P. A., U. S. Bhatt, L. A. Rundquist, S. D. Lindsey, X. Zhang, and R. L. Thoman. 2011. Large-scale climate controls of interior Alaska river ice breakup. Journal of Climate 24:286–297.

Brandt, S. B., and K. J. Hartman. 1993. Innovative approaches with bioenergetics models: future applications to fish ecology and management. Transactions of the American Fisheries Society 122:731–735.

Brett, J., J. E. Shelbourn, and C. T. Shoop. 1969. Growth rate and body composition of fingerling Sockeye Salmon, *Oncorhynchus nerka*, in relation to temperature and ration size. Journal of the Fisheries Board of Canada 26:2363–2394.

Brown, C. L., and A. Godduhn, editors. 2015. Socioeconomic effects of declining salmon runs on the Yukon River. Alaska Department of Fish and Game, Division of Subsistence, Technical Paper No. 398, Fairbanks.

Caissie, D. 2006. The thermal regime of rivers: a review. Freshwater Biology 51:1389–1406.

Clarke, S. E., K. M. Burnett, and D. J. Miller. 2008. Modeling streams and hydrogeomorphic attributes in Oregon from digital and field data. Journal of the American Water Resources Association 44:459–477.

Crozier, L. G., and R. W. Zabel. 2006. Climate impacts at multiple scales: evidence for differential population responses in juvenile Chinook Salmon. Journal of Animal Ecology 75:1100–1109.

Cunjak, R. A. 1988. Physiological consequences of overwintering in streams: the cost of acclimatization? Canadian Journal of Fisheries and Aquatic Sciences 45:443–452.

Cunningham, C. J., P. A. Westley, and M. D. Adkison. 2018. Signals of large scale climate drivers, hatchery enhancement, and marine factors in Yukon River Chinook Salmon survival revealed with a Bayesian life history model. Global Change Biology 24:4399–4416.

Decker, S. K. S. 2010. Thermal limitations on Chinook Salmon spawning habitat in the northern extent of their range. Master's thesis. University of Alaska Fairbanks, Fairbanks.

Duffy, E. J., and D. A. Beauchamp. 2011. Rapid growth in the early marine period improves the marine survival of Chinook Salmon (*Oncorhynchus tshawytscha*) in Puget Sound, Washington. Canadian Journal of Fisheries and Aquatic Sciences 68:232–240.

Dunham, J., G. Chandler, B. E. Rieman, and D. M. Martin. 2005. Measuring stream temperature with digital data loggers: a user's guide. U.S. Forest Service, RMRS-GTR-150, Fort Collins, Colorado.

Dunning, J. B., B. J. Danielson, and H. R. Pulliam. 1992. Ecological processes that affect populations in complex landscapes. Oikos 65:169–175.

Falke, J. A., J. B. Dunham, D. Hockman-Wert, and R. Pahl. 2016. A simple prioritization tool to diagnose impairment of stream temperature for coldwater fishes in the Great Basin. North American Journal of Fisheries Management 36:147–160.

Falke, J. A., J. B. Dunham, C. E. Jordan, K. M. McNyset, and G. H. Reeves. 2013. Spatial ecological processes and local factors predict the distribution and abundance of spawning by steelhead (*Oncorhynchus mykiss*) across a complex riverscape. PLOS (Public Library of Science) ONE [online serial] 8(11):e79232.

Falke, J. A., R. L. Flitcroft, J. B. Dunham, K. M. McNyset, P. F. Hessburg, and G. H. Reeves. 2014. Climate change and vulnerability of Bull Trout (*Salvelinus confluentus*) in a fire-prone landscape. Canadian Journal of Fisheries and Aquatic Sciences 72:304–318.

Farley, E., J. H. Moss, and R. J. Beamish. 2007. A review of the critical size, critical period hypothesis for juvenile Pacific salmon. North Pacific Anadromous Fish Commission Bulletin 4:311–317.

Fausch, K. D., C. E. Torgersen, C. V. Baxter, and H. W. Li. 2002. Landscapes to riverscapes: bridging the gap between research and conservation of stream fishes. BioScience 52:483–498.

Geist, D. R., Z. Deng, R. P. Mueller, S. R. Brink, and J. A. Chandler. 2010. Survival and growth of juvenile Snake River fall Chinook Salmon exposed to constant and fluctuating temperatures. Transactions of the American Fisheries Society 139:92–107.

Gutierrez, L. 2011. Terrestrial invertebrate prey for juvenile Chinook Salmon: abundance and environmental controls in an interior Alaska river. Master's thesis. University of Alaska Fairbanks, Fairbanks.

Hafs, A. W., J. M. Niles, and K. J. Hartman. 2011. Efficiency of gastric lavage on age-0 Brook Trout and the influence on growth and survival. North American Journal of Fisheries Management 31:530–534.

Hand, B. K., C. C. Muhlfeld, A. A. Wade, R. P. Kovach, D. C. Whited, S. R. Narum, A. P. Matala, M. W. Ackerman, B. A. Garner, and J. S. Kimball. 2016. Climate variables explain neutral and adaptive variation within salmonid metapopulations: the importance of replication in landscape genetics. Molecular Ecology 25:689–705.

Hanson, P., T. Johnson, D. Schindler, and J. Kitchell. 1997. Fish bioenergetics 3.0. Sea Grant Institute, University of Wisconsin, Madison.

Hartman, K. J., and J. F. Kitchell. 2008. Bioenergetics modeling: progress since the 1992 symposium. Transactions of the American Fisheries Society 137:216–223.

Healey, M. 1991. Life history of Chinook Salmon (*Oncorhynchus tshawytscha*). Pages 313–393 *in* C. Groot, and L. Margolis, editors. Pacific salmon life histories. UBC Press, Vancouver.

Hein, C. L., G. Öhlund, and G. Englund. 2014. Fish introductions reveal the temperature dependence of species interactions. Proceedings of the Royal Society of London B 281:20132641.

Henderson, M., and A. Cass. 1991. Effect of smolt size on smolt-to-adult survival for Chilko Lake Sockeye Salmon (*Oncorhynchus nerka*). Canadian Journal of Fisheries and Aquatic Sciences 48:988–994.

Hughes, N. F. 1998. A model of habitat selection by drift-feeding stream salmonids at different scales. Ecology 79:281–294.

Huntsman, B. M., and J. A. Falke. 2019. Main stem and off-channel habitat use by juvenile Chinook Salmon in a sub-Arctic riverscape. Freshwater Biology 64:433–446.

Huntsman, B. M., J. T. Petty, S. Sharma, and E. R. Merriam. 2016. More than a corridor: use of a main stem stream as supplemental foraging habitat by a Brook Trout metapopulation. Oecologia 182:463–473.

Huntsman, B. M., J. A. Falke, J. W. Savereide,

and K. E. Bennett. 2017. The role of density-dependent and -independent processes in spawning habitat selection by salmon in an Arctic riverscape. PLOS (Public Library of Science) ONE [online serial] 12(5):e0177467.

IPCC (International Panel on Climate Change). 2014. Climate change 2014: synthesis report. Contribution of Working Groups I, II and III to the fifth assessment report of the Intergovernmental Panel on Climate Change. R. K. Pachauri, and L. A. Meyers, editors. IPCC, Geneva, Switzerland.

Isaak, D. J., and D. L. Horan. 2011. An evaluation of underwater epoxies to permanently install temperature sensors in mountain streams. North American Journal of Fisheries Management 31:134–137.

Isaak, D. J., C. H. Luce, B. E. Rieman, D. E. Nagel, E. E. Peterson, D. L. Horan, S. Parkes, and G. L. Chandler. 2010. Effects of climate change and wildfire on stream temperatures and salmonid thermal habitat in a mountain river network. Ecological Applications 20:1350–1371.

Isaak, D. J., E. E. Peterson, J. M. Ver Hoef, S. J. Wenger, J. A. Falke, C. E. Torgersen, C. Sowder, E. A. Steel, M. J. Fortin, and C. E. Jordan. 2014. Applications of spatial statistical network models to stream data. Water 1:277–294.

Isaak, D. J., M. K. Young, C. H. Luce, S. W. Hostetler, S. J. Wenger, E. E. Peterson, J. M. Ver Hoef, M. C. Groce, D. L. Horan, and D. E. Nagel. 2016. Slow climate velocities of mountain streams portend their role as refugia for cold-water biodiversity. Proceedings of the National Academy of Sciences of the United States of America 113:4374–4379.

Jeffres, C., and P. Moyle. 2012. When good fish make bad decisions: Coho Salmon in an ecological trap. North American Journal of Fisheries Management 32:87–92.

Jones, L. A., C. C. Muhlfeld, L. A. Marshall, B. L. McGlynn, and J. L. Kershner. 2014. Estimating thermal regimes of Bull Trout and assessing the potential effects of climate warming on critical habitats. River Research and Applications 30:204–216.

Kalb, B. W., B. M. Huntsman, C. A. Caldwell, and M. A. Bozek. 2018. A mechanistic assessment of seasonal microhabitat selection by drift-feeding Rainbow Trout *Oncorhynchus mykiss* in a southwestern headwater stream. Environmental Biology of Fishes 101:257–273.

Kennedy, B. P., K. H. Nislow, and C. L. Folt. 2008. Habitat-mediated foraging limitations drive survival bottlenecks for juvenile salmon. Ecology 89:2529–2541.

Kitchell, J. F., D. J. Stewart, and D. Weininger. 1977. Applications of a bioenergetics model to Yellow Perch (*Perca flavescens*) and Walleye (*Stizostedion vitreum vitreum*). Journal of the Fisheries Board of Canada 34:1922–1935.

Kovach, R. P., C. C. Muhlfeld, R. Al-Chokhachy, J. B. Dunham, B. H. Letcher, and J. L. Kershner. 2016. Impacts of climatic variation on trout: a global synthesis and path forward. Reviews in Fish Biology and Fisheries 26:135–151.

Lawrence, D. J., B. Stewart-Koster, J. D. Olden, A. S. Ruesch, C. E. Torgersen, J. J. Lawler, D. P. Butcher, and J. K. Crown. 2014. The interactive effects of climate change, riparian management, and a nonnative predator on stream-rearing salmon. Ecological Applications 24:895–912.

Luce, C., B. Staab, M. Kramer, S. Wenger, D. Isaak, and C. McConnell. 2014. Sensitivity of summer stream temperatures to climate variability in the Pacific Northwest. Water Resources Research 50:3428–3443.

Magnuson, J. J., L. B. Crowder, and P. A. Medvick. 1979. Temperature as an ecological resource. American Zoologist 19:331–343.

Matter, A. N., J. A. Falke, J. A. López, and J. W. Savereide. 2018. A rapid-assessment method to estimate the distribution of juvenile Chinook Salmon in tributary habitats using eDNA and occupancy estimation. North American Journal of Fisheries Management 38:223–236.

Mauger, S., R. Shaftel, J. C. Leppi, and D. J. Rinella. 2016. Summer temperature regimes in southcentral Alaska streams: watershed drivers of variation and potential implications for Pacific salmon. Canadian Journal of Fisheries and Aquatic Sciences 74:702–715.

McCarthy, S. G., J. J. Duda, J. M. Emlen, G. R. Hodgson, and D. A. Beauchamp. 2009. Linking habitat quality with trophic performance of steelhead along forest gradients in the South Fork Trinity River watershed, California. Transactions of the American Fisheries Society 138:506–521.

McNyset, K. M., C. J. Volk, and C. E. Jordan. 2015. Developing an effective model for predicting spatially and temporally continuous stream temperatures from remotely sensed land surface temperatures. Water 7:6827–6846.

Meehan, W. R., and R. A. Miller. 1978. Stomach flushing: effectiveness and influence on survival and condition of juvenile salmonids. Journal of the Fisheries Board of Canada 35:1359–1363.

Merriam, E. R., R. Fernandez, J. T. Petty, and N. Zegre. 2017. Can Brook Trout survive climate change in large rivers? If it rains. Science of the Total Environment 607:1225–1236.

Moss, J. H., D. A. Beauchamp, A. D. Cross, K. W. Myers, E. V. Farley, Jr., J. M. Murphy, and J. H. Helle. 2005. Evidence for size-selective mortality after the first summer of ocean growth by Pink Salmon. Transactions of the American Fisheries Society 134:1313–1322.

Murphy, J. M., K. G. Howard, J. C. Gann, K. C. Cieciel, W. D. Templin, and C. M. Guthrie, III. 2017. Juvenile Chinook Salmon abundance in the northern Bering Sea: implications for future returns and fisheries in the Yukon River. Deep Sea Research Part II: Topical Studies in Oceanography 135:156–167.

Neuswanger, J. R., M. S. Wipfli, M. J. Evenson, N. F. Hughes, and A. E. Rosenberger. 2015. Low productivity of Chinook Salmon strongly correlates with high summer stream discharge in two Alaskan rivers in the Yukon drainage. Canadian Journal of Fisheries and Aquatic Sciences 72:1125–1137.

Nislow, K. H., C. L. Folt, and D. L. Parrish. 1999. Favorable foraging locations for young Atlantic Salmon: application to habitat and population restoration. Ecological Applications 9:1085–1099.

Perry, M. T. 2012. Growth of juvenile Chinook Salmon (*Oncorhynchus tshawytscha*) as an indicator of density-dependence in the Chena River. Master's thesis. University of Alaska Fairbanks, Fairbanks.

Peterson, E. E., and J. M. Van Hoef. 2010. A mixed-model moving-average approach to geostatistical modeling in stream networks. Ecology 91:644–651.

Petty, J. T., and G. D. Grossman. 2010. Giving-up densities and ideal pre-emptive patch use in a predatory benthic stream fish. Freshwater Biology 55:780–793.

Petty, J. T., D. Thorne, B. M. Huntsman, and P. M. Mazik. 2014. The temperature–productivity squeeze: constraints on Brook Trout growth along an Appalachian river continuum. Hydrobiologia 727:151–166.

Plumb, J. M., and C. M. Moffitt. 2015. Re-estimating temperature-dependent consumption parameters in bioenergetics models for juvenile Chinook Salmon. Transactions of the American Fisheries Society 144:323–330.

Poff, N. L., J. D. Allan, M. B. Bain, J. R. Karr, K. L. Prestegaard, B. D. Richter, R. E. Sparks, and J. C. Stromberg. 1997. The natural flow regime. BioScience 47:769–784.

Poole, G. C., and C. H. Berman. 2001. An ecological perspective on in-stream temperature: natural heat dynamics and mechanisms of human-caused thermal degradation. Environmental Management 27:787–802.

Quinn, T. P., and N. P. Peterson. 1996. The influence of habitat complexity and fish size on over-winter survival and growth of individually marked juvenile Coho Salmon (*Oncorhynchus kisutch*) in Big Beef Creek, Washington. Canadian Journal of Fisheries and Aquatic Sciences 53:1555–1564.

R Core Team 2018. R: a language and environment for statistical computing. R foundation for statistical computing, Vienna, Austria. Available: https://www.R-project.org/. (February 2019).

Rich, H. B., T. P. Quinn, M. D. Scheuerell, and D. E. Schindler. 2009. Climate and intraspecific competition control the growth and life history of juvenile Sockeye Salmon (*Oncorhynchus nerka*) in Iliamna Lake, Alaska. Canadian Journal of Fisheries and Aquatic Sciences 66:238–246.

Roberts, J. J., B. D. Best, D. C. Dunn, E. A. Treml, and P. N. Halpin. 2010. Marine geospatial ecology tools: an integrated framework for ecological geoprocessing with ArcGIS, Python, R, Matlab, and C++. Environmental Modelling and Software 25:1197–1207.

Ruggerone, G. T., J. L. Nielsen, and B. A. Agler. 2009. Climate, growth and population dynamics of Yukon River Chinook Salmon. North Pacific Anadromous Fish Commission Bulletin 5:279–285.

Rypel, A. L. 2014. Do invasive freshwater fish species grow better when they are invasive? Oikos 123:279–289.

Schindler, D., C. Krueger, P. Bisson, M. Bradford, B. Clark, J. Conitz, K. Howard, M. Jones, J. Murphy, and K. Myers. 2013. Arctic-Yukon-Kuskokwim Chinook Salmon research action plan: evidence of decline of Chinook Salmon populations and recommendations for future research. Prepared for the AYK Sustainable Salmon Initiative, Anchorage, Alaska.

Schindler, D. E., D. E. Rogers, M. D. Scheuerell, and C. A. Abrey. 2005. Effects of changing climate on zooplankton and juvenile Sockeye Salmon growth in southwestern Alaska. Ecology 86:198–209.

Schoen, E. R., D. A. Beauchamp, A. R. Buettner, and N. C. Overman. 2015. Temperature and depth mediate resource competition and apparent competition between *Mysis diluviana* and kokanee. Ecological Applications 25:1962–1975.

Schoen, E. R., J. A. López, and B. E. Meyer. 2018. Freshwater predation on Chena River juvenile salmon. Final report prepared for the Yukon River Salmon Research and Management Fund. University of Alaska, Fairbanks.

Snyder, C. D., N. P. Hitt, and J. A. Young. 2015. Accounting for groundwater in stream fish thermal habitat responses to climate change. Ecological Applications 25:1397–1419.

Sommer, T. R., M. L. Nobriga, W. C. Harrell, W. Batham, and W. J. Kimmerer. 2001. Floodplain rearing of juvenile Chinook Salmon: evidence of enhanced growth and survival. Canadian Journal of Fisheries and Aquatic Sciences 58:325–333.

Steel, E. A., T. J. Beechie, C. E. Torgersen, and A. H. Fullerton. 2017. Envisioning, quantifying, and managing thermal regimes on river networks. BioScience 67:506–522.

Stewart, D. J., and M. Ibarra. 1991. Predation and production by salmonine fishes in Lake Michigan, 1978–88. Canadian Journal of Fisheries and Aquatic Sciences 48:909–922.

TVWA (Tanana Valley Watershed Association). 2015. Watershed resource action plan for the Chena River. Available: https://dec.alaska.gov/water/wqsar/pdfs/Reports/Chena-River-2015-ADEC.pdf. (February 2019).

Torgersen, C. E., D. M. Price, H. W. Li, and B. A. McIntosh. 1999. Multiscale thermal refugia and stream habitat associations of Chinook Salmon in northeastern Oregon. Ecological Applications 9:301–319.

USGS (U.S. Geological Survey). 2019. National hydrography dataset. Available: www.usgs.gov/core-science-systems/ngp/national-hydrography. (February 2019).

Vatland, S. J., R. E. Gresswell, and G. C. Poole. 2015. Quantifying stream thermal regimes at multiple scales: combining thermal infrared imagery and stationary stream temperature data in a novel modeling framework. Water Resources Research 51:31–46.

Ward, J. V., and J. A. Stanford. 1983. The serial discontinuity concept of lotic ecosystems. Pages 29–41 *in* T. D. Fontane and S. M. Bartell, editors. Dynamics of lotic ecosystems. Ann Arbor Science, Ann Arbor, Michigan.

Webb, B. W., D. M. Hannah, R. D. Moore, L. E. Brown, and F. Nobilis. 2008. Recent advances in stream and river temperature research. Hydrological Processes 22:902–918.

Weber, N., N. Bouwes, C. E. Jordan, and B. Jonsson. 2014. Estimation of salmonid habitat growth potential through measurements of invertebrate food abundance and temperature. Canadian Journal of Fisheries and Aquatic Sciences 71:1158–1170.

Wenger, S. J., C. H. Luce, A. F. Hamlet, D. J. Isaak, and H. M. Neville. 2010. Macroscale hydrologic modeling of ecologically relevant flow metrics. Water Resources Research 46:1–10.

American Fisheries Society Symposium 90:83–103, 2019
© 2019 by the American Fisheries Society

Characterizing Natural Barriers to Nonnative Stream Fauna in Hawai'i

Brendan Martin and Yin-Phan Tsang*
Department of Natural Resources and Environmental Management
University of Hawai'i at Mānoa
1910 East-West Road, Sherman 242, Honolulu, Hawaii, 96822, USA

Ralph W. Tingley III
Missouri Cooperative Fish and Wildlife Research Unit, School of Natural Resources
University of Missouri
302 Anheuser-Busch Natural Resources Building, Columbia, Missouri 65211, USA

Hannah Clilverd
Department of Natural Resources and Environmental Management
University of Hawai'i at Mānoa
1910 East-West Road, Sherman 242, Honolulu, Hawai'i, 96822, USA
and
Centre for Ecology and Hydrology
Bush Estate, Penicuik, Midlothian, EH26 0QB, UK

Dana M. Infante
Department of Fisheries and Wildlife, Michigan State University
1405 South Harrison Road, Suite 318, East Lansing, Michigan 48823, USA

Abstract.—Waterfalls, caused by the abrupt changes of elevation in streams, are natural barriers that influence the distribution and dispersion of aquatic species. The resulting habitat fragmentation has contributed to species specialization as well as barriers that inhibit passage of nonnative species upstream. In Hawai'i, it is assumed that nonnative species are unable to pass waterfall barriers, yet they are present above some waterfalls, possibly facilitated by human introduction. In this study, we used a landscape approach to identify likely human introductions and examine the ability of nonnative stream fauna to bypass waterfalls. We identified the human activities associated with the high likelihood of species introduction. We found that when a local catchment has a population density >4.24 people/km^2 or road length density is >0.01 km/km^2, the presence of nonnative species in the stream is likely a result of human introduction. After filtering human facilitated introduction, we also assessed the potential waterfall climbing ability of 14 nonnative taxa. We found that 12 out of the 14 taxa were absent upstream of waterfalls, indicative of their inability to traverse waterfalls. Only two species, Tahitian prawn (also known as monkey river shrimp) *Macrobrachium*

* Corresponding author: tsangy@hawaii.edu

lar and American bullfrog *Rana catesbiana*, seem able to pass waterfalls. This study highlights the role that people play in facilitating species introductions in otherwise inaccessible habitats. Without human interference, waterfalls can be considered effective barriers to nonnative species and can be instrumental in supporting nonnative species eradication and control strategies.

Introduction

Stream habitat connectivity is one of the most important hydrological conditions that determines the distribution of freshwater species (Fullerton et al. 2010). Unfragmented free-flowing rivers are essential for species that migrate within rivers and between the ocean and freshwaters to complete their life cycles. Any disruption of hydrologic connectivity could isolate habitats, resulting in changes to assemblages. Most commonly noted are anthropogenic barriers, including weirs and dams that have severely altered the natural flow regime of rivers in many systems worldwide (Poff et al. 1997; Pedroli et al. 2002; Palmer et al. 2005; Liermann et al. 2012; Cooney and Kwak 2013).

Waterfalls and other natural barriers can also restrict migrations of stream fauna (Reznick and Ghalambor 2005), and several studies suggest that they shape native species composition and specialization within both temperate and tropical systems. For example, Robinson and Rand (2005) concluded that waterfalls and cascades limited colonization of fishes along an elevation gradient in a southern Appalachian watershed. In rivers of the Amazon basin, waterfalls influence species distributions (Barbosa et al. 2015) and the compositional dissimilarity of fish assemblages (Torrente-Vilara et al. 2011). Similarly, waterfalls act as natural barriers on the Orinoco River and result in fragmented subbasins that contribute to high neoendemic species richness (i.e., richness of those species arising from in situ cladogenetic speciation; Dias et al. 2013). In tropical island systems of the Pacific, waterfalls also play a substantial role shaping assemblages. For example, native stream assemblages in Hawai'i (Keith 2003; Walter et al. 2012) and New Zealand (McDowall 2010) are structured longitudinally by species-specific abilities to migrate upstream, likely the result of differing abilities to traverse rapids, torrents, or waterfalls.

In addition to shaping native species assemblages, waterfalls can also act as barriers to the upstream movement of nonnative species (Pusey et al. 1998; Hardman et al. 2002). For instance, Kerby et al. (2005) found that red swamp crayfish *Procambarus clarkii*, an invasive species in southern California streams, largely spread downstream and that management efforts to remove invasive organisms should focus on stream segments that are bounded by a downstream barrier. In Australia, managers have eradicated introduced Rainbow Trout *Oncorhynchus mykiss* and protected populations of native Mountain Galaxias *Galaxias olidus* in areas of the Cotter River above a natural barrier (Lintermans 2000). Similar concepts of using natural waterfalls or artificial waterway barriers to prevent dispersal of nonnative taxa are well adapted for restoring freshwater ecosystems (Rahel 2013). In regions where native species of conservation interest are exceptionally adapted climbers, using waterfalls as a conservation tool may be especially effective. Diadromous examples of waterfall-climbing species are found in Australia (Merrick and Schmida 1984; Ebner et al. 2011), New Zealand (McDowall 1995), and tropical island systems (e.g., McDowall 2003; Cooney and Kwak 2013). These species range in their climbing ability, but many are able to occupy high-elevation habitats upstream of steep waterfalls. Not only are these habitats often inaccessible to humans and thus more likely to retain pristine conditions, they also provide refuge from predatory invasive species occupying downstream habitats (Kwak et al. 2007).

Across the Hawaiian Islands, the distribution of nonnative stream species is hypothesized to be limited by natural barriers such as waterfalls (Brasher et al. 2006). However, anthropogenic interference complicates our understanding of nonnative species mobility. Humans often introduce nonnative species for food sources, for mosquito control, or from aquaria (Maciolek 1984; Brown et al. 1999). The presence of nonnative species above a waterfall barrier can occur via two possible modes of dispersal. First, the species may be introduced downstream of the waterfall and pass the waterfall to access upstream reaches. Second, the species may be delivered to reaches upstream of the waterfall by human introductions (Figure 1a). Assessments of the mobility and dispersal history of Hawaiian nonnative species are lacking, in part because documentation of species introductions (e.g., date, location, and number of species) is neither precise nor comprehensive enough to determine if the presence of species above waterfalls resulted from human introduction. Further, observing the mobility of these species in the field is challenging, particularly in remote locations. Whereas laboratory experiments have been used to study the jumping or climbing abilities of stream species in systems with low waterfalls and natural barriers (e.g., Kondratieff and Myrick 2005; Holthe et al. 2005), such approaches are impractical given the difficulties of replicating the extreme heights (>30 m), steep slopes, and substrate roughness of waterfalls in Hawaiʻi. A more practical method is to investigate upstream movement of nonnative species using a landscape approach.

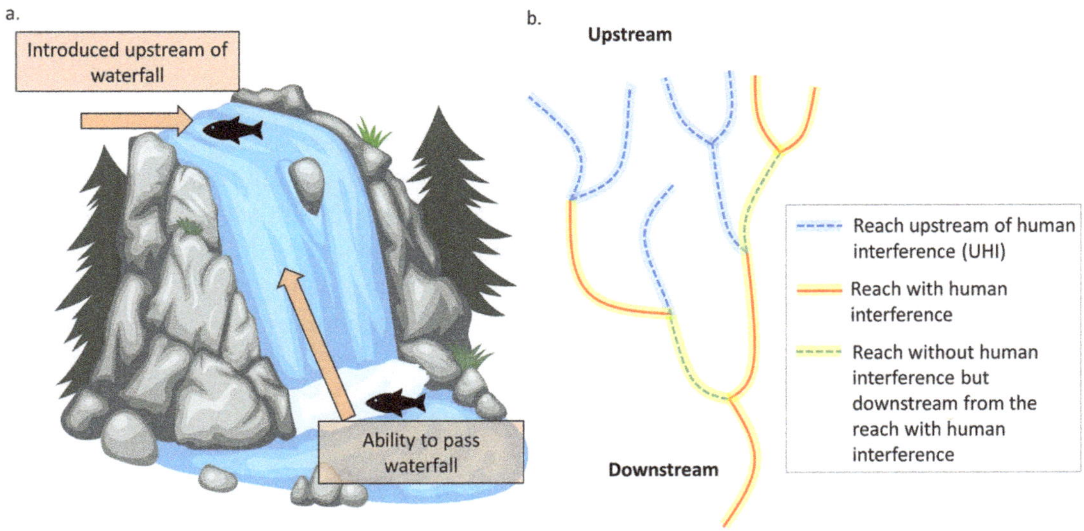

Figure 1. **(a)** Illustration of the possible sources or pathways of introduced species present above waterfalls; **(b)** illustration of the stream reaches upstream of human interference (i.e., UHI reaches) highlighted in bolded blue that were used to evaluate the role of waterfalls as barriers to nonnative species upstream movement in this study. Note that all reaches were initially used to examine the relationships between anthropogenic activities in the local catchment on the presence of nonnative taxa, regardless of the presence of a waterfall.

Our primary objective was to use a landscape assessment to gain insight into the role of anthropogenic interference and natural barriers in influencing the distribution of nonnative stream fauna in Hawai'i. We hypothesized that anthropogenic activities are the main drivers of nonnative species distribution in Hawai'i, and that in the absence of such interference, nonnative species are constrained by waterfalls. We also hypothesized that different species exhibit different abilities to pass waterfalls. To address our hypotheses, we synthesized census information from statewide biological surveys and data sets indicative of human activities in the catchment (i.e., population and road density) to assess the role anthropogenic interferences play in the distribution of nonnative species. After filtering out effects of human interference, we examined the influence of waterfall height on the distribution of nonnative species by identifying waterfall locations via satellite imagery and estimating waterfall height from changes in elevation. The information obtained from this study should advance our knowledge of the ability of nonnative stream species in Hawai'i to traverse natural barriers, providing insight for management and conservation of stream ecosystems.

Methods

Study Area

We investigated the distribution of nonnative freshwater fauna in streams of the five largest Hawaiian Islands (Hawai'i, Maui, Moloka'i, O'ahu, and Kaua'i; Figure 2). These islands, formed by lava flows from large shield volcanoes, increase in age from east (Hawai'i) to west (Kaua'i), range in size from 683 to 10,451 km^2 and vary in elevation from sea level to 4,200 m (Juvik et al. 1998). High relief combined with substantial rainfall exceeding 10 m/year in some areas (Giambelluca et al. 2013) has resulted in intensive weathering and erosion of volcanic rock to form steep slopes and waterfalls (Lau and Mink 2006; Pederson 2010). Because of the young geological age of the island of Hawai'i (eruptions from one of its volcanoes occurred in 2018) and the subsequent high porosity of young basalt, substantial portions of the island have few perennial streams.

Stream Spatial Framework

We first examined the relationship between nonnative species presence and landscape variables indicative of anthropogenic activities across all stream reaches regardless of the presence of natural barriers. We used a spatial framework developed by Tingley et al. (2019) that was built from a modified version of the 1:24,000 National Hydrography Dataset (http://nhd.usgs.gov). Stream reaches (or segments) are the basic spatial unit of this framework and are defined as confluence to confluence portions of the stream, or portions of the stream bounded by a confluence and an ecological boundary (i.e., waterfall; Wieferich et al. 2015). The land area drained directly to a reach is called the local catchment (Wang et al. 2011) and was the spatial unit in which we summarized anthropogenic activities. We chose to summarize anthropogenic activities in the local catchment as opposed to the network catchment (i.e., the land area draining all reaches upstream of a given reach; Wang et al. 2011) because we were interested in understanding the more immediate introduction sources to the stream, which would most likely occur locally.

Biological Data Preparation

Presence–absence data were generated by using a combination of data sets provided by the Hawai'i Division of Aquatic Resources (Hawai'i DAR)—a presence–absence survey data set from 2008 to 2014 and an abundance data set from 1989 to 2010. We selected those periods based on data availability and the

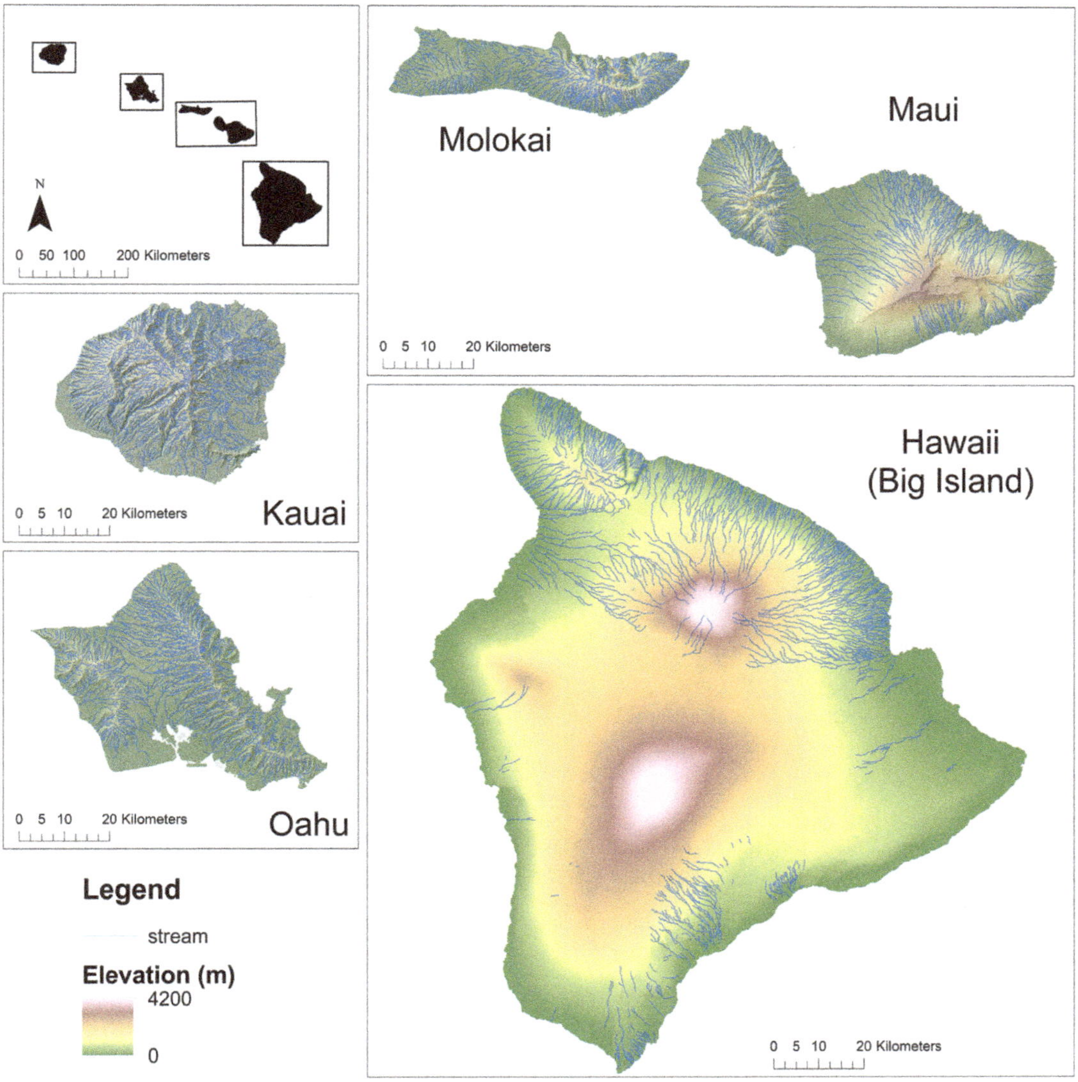

Figure 2. The five largest islands of Hawai'i, with major streams and island topography.

temporal range of the landscape data used in our analyses (see below). We linked biological survey locations to individual stream reaches with ArcMap (ESRI, Redlands, California) following a protocol described in Daniel et al. (2015). We individually assessed survey points greater than 50 m from stream lines for accuracy by matching stream names within the modified National Hydrography Dataset with stream identification information for each survey attributed by the Hawai'i DAR. We omitted cases where locations and stream descriptions were in disagreement and finalized a data set of 8,397 surveys (n = 452 from presence–absence data set; n = 7,945 from abundance data set; Figure 3).

We selected nonnative species for analyses by weighing management concerns and data availability. Management concerns were identified through personal communications with stream biologists in the Hawai'i DAR, U.S. Fish and Wildlife Service, and Pacific Biosciences Research Center, as well previous studies (Maciolek 1984; Hawai'i Cooperative

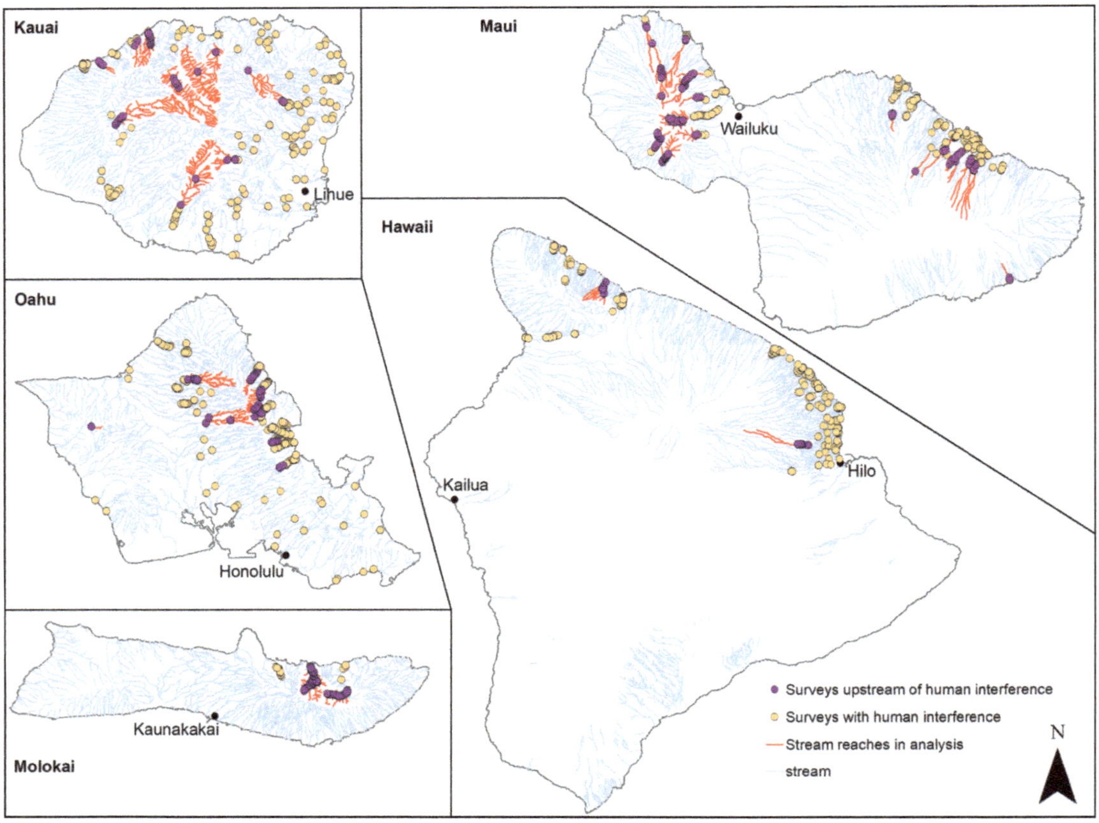

Figure 3. Locations of biological surveys used in this study superimposed on streams across the five main Hawaiian Islands. Stream reaches upstream of human interference are colored in red. Islands are not drawn to scale.

Park Service Unit 1990; Brasher 2003; Snow and Witmer 2010). Focal species included all fishes in the families Poeciliidae, Cichlidae, Centrarchidae, and Salmonidae and the order Siluriformes, as well as Tahitian prawn (also known as monkey river shrimp) *Macrobrachium lar*, red swamp crayfish *Procambarus clarkia,* American bullfrog *Rana catesbiana*, and Pond Loach (also known as Oriental Weatherfish) *Misgurnus anguillicaudatus*. Three tilapia species (Blackchin Tilapia *Sarotherodon melanotheron*, Mozambique Tilapia *Oreochromis mossambicus*, and Red Belly Tilapia *Tilapia zillii*) were aggregated into one taxon (*Tilapia*) prior to analysis. As in some surveys, tilapia were not identified to species because of poor visibility or species hybridization. We further determined species for final analyses based on data availability. Species or taxon groups were considered present if they were found in at least one survey in a given reach, and considered absent otherwise (following Steen et al. 2008; Elith et al. 2011). Additionally, only species or taxon groups that occurred in three or more reaches were included, resulting in a final set of 14 nonnative species or taxa (hereafter, referred to as "taxa"; Table 1) in a total of 129 watersheds (i.e., entire stream network draining to the ocean).

Summary of Human Activity Data Sets

Human population density and road length density were used to characterize human activities that may have facilitated establishment

Table 1. Selected nonnative stream taxa and their places of origin. Much of the information was gathered from the U.S. Geological Survey Nonindigenous Aquatic Species Program (https://nas.er.usgs.gov).

Common name	Scientific name	Place of origin
Tahitian prawn	*Macrobrachium lar*	Indo-West Pacific: East Africa to the Ryukyu Islands and the Marquesas (Holthuis 1980). Australia, French Polynesia (Marquesas), Guam, Indonesia, Japan, Kenya, Madagascar, Malaysia, Mauritius, Mozambique, New Caledonia, Northern Mariana Islands, Papua New Guinea, Philippines, Taiwan, China, Tanzania (De Grave 2013).
American bullfrog	*Rana catesbiana*	Eastern United States (McKercher and Gregoire 2019)
Green Swordtail	*Xiphophorus helleri*	Middle America from Rio Nautla (=Rio Nantla), Veracruz, Mexico to northwestern Honduras (Rosen 1960, 1979; Page and Burr 1991; Greenfield and Thomerson 1997; Miller et al. 2005).
Pond Loach	*Misgurnus anguillicaudatus*	Eastern Asia from Siberia south to northern Vietnam, including Japan (Berg 1948-1949; Masuda et al. 1984; Talwar and Jhingran 1992).
Western Mosquitofish	*Gambusia affinis*	Gulf Slope drainages from Florida panhandle to Mexico; Mississippi River basin from central Indiana and Illinois south to the Gulf of Mexico (Rauchenberger 1989; Page and Burr 1991).
Smallmouth Bass	*Micropterus dolomieu*	St. Lawrence and Great Lakes, Hudson Bay (Red River), and Mississippi River basins from southern Quebec to North Dakota and south to northern Alabama and eastern Oklahoma; Atlantic and Gulf Slope drainages from Virginia to central Texas (Page and Burr 1991).
Red swamp crayfish	*Procambarus clarkii*	Gulf coastal plain from the Florida panhandle to Mexico; southern Mississippi River drainage to Illinois (Hobbs 1989; Taylor et al. 2007).
Largemouth Bass	*Micropterus salmoides*	St. Lawrence and Great Lakes, Hudson Bay (Red River), and Mississippi River basins from southern Quebec to Minnesota and south to the Gulf of Mexico; Atlantic Slope drainages from North Carolina to Florida; Gulf Slope drainages from southern Florida into northern Mexico (Page and Burr 2011).
Guppy	*Poecilia reticulata*	Lesser Antilles and northern South America from western Venezuela to Guyana (Rosen and Bailey 1963; Page and Burr 1991).
Common Molly	*Poecilia sphenops*	Atlantic Slope of eastern Mexico from north of Veracruz City to Guatemala (Brett and Turner 1983; Miller 1983).
Cuban Limia	*Limia vittata*	Streams, lakes, estuaries, coastal lagoons, and mangrove swamps throughout Cuba and Isle of Pines (Lee et al. 1980).

Table 1. Continued.

Common name	Scientific name	Place of origin
Tilapia	*Tilapiini* (Cichlidae) spp.	Tropical and subtropical Africa
Convict Cichlid	*Archocentrus nigrofasciatus* (also *Cichlasoma nigrofasciatum*)	Central America, including Pacific Slope drainages of Guatemala, from Río Suchiate to Río Grande de Taracoles in northwestern Costa Rica. Atlantic Slope drainages in Central America, including Honduras, from Río Aguan to Río Guarumo in Panama (Conkel 1993). Distribution maps in Bussing (1998) and Conkel (1993).
Bristlenosed Catfish	*Ancistrus* cf. *temminckii*	Tropical South America (Armbruster 2004).

of nonnative species. Both data sets are available statewide as geographic information system (GIS) layers in the 2010 TIGER census (U.S. Census Bureau 2012). Summaries of population density (people/km^2) and road length density (km/km^2) were generated for all stream reaches within their local catchments (following McKinney 2002). We also gathered information on human-made waterways that intersect stream reaches, such as canals and agricultural ditches (Hawai'i DAR 2004), to examine alternative pathways for nonnative species colonization.

Assessing Effects of Human Activities on Nonnative Species Presence–Absence

We performed an indicator species analysis (ISA) for the 14 nonnative taxa. First proposed by Dufrêne and Legendre (1997), an ISA associates differences in numbers of species with differences in an environmental factor. We used the program TITAN (Threshold Indicator Taxa Analysis; Baker and King 2010) to implement the ISA; TITAN repeatedly divides sites into two groups along an environmental gradient and calculates an indicator value (IndVal) for the high end and for the low end of the environmental factor. A high IndVal for one group (and corresponding low indicator values for the other group) suggests a high species occurrence or abundance that is associated with the environmental factor. TITAN then examines the IndVal along the environmental gradient to determine a "change point," a value of the environmental factor that is associated with the increasing occurrence of a species. In this study, we used TITAN to associate the taxa with the two anthropogenic factors and determine the change point that results in increasing occurrence of each taxon. We used a 0.95 purity and 0.95 reliability level to determine the significance of the change point and used the lowest significant value among all 14 taxa as the critical ecosystem load, indicating human interference for each anthropogenic factor. When anthropogenic factors in a stream reach had a value greater than or equal to the identified critical "ecosystem-load" values, we considered the reach to be affected by human interference and a location where nonnative taxa presence could potentially result from human-facilitated introduction.

Identification of Waterfalls and Estimation of Heights

We created a comprehensive waterfall GIS layer using a three-step process. First, we gathered point locations of waterfalls throughout the state from three data sources: the National Hydrography Dataset (http://nhd.usgs.gov/), the Hawaiʻi Statewide GIS Program (http://planning.hawaii.gov/gis/), and the World Waterfall Database (www.worldwaterfalldatabase.com/). Second, we used Google Earth Pro (Google Inc., Mountain View, California) to inspect the locations in the existing records and identify additional waterfalls that were not previously documented within the stream network of the 129 watersheds with nonnative taxa data. This included inspecting all main channel reaches downstream of the most upstream biological surveys and reaches immediately upstream of the most upstream biological surveys. To better identify waterfalls, we assessed historical satellite imagery from 1984 to 2016, which accounted for different temporal records that included a range of image quality because of differing levels of cloud cover, streamflow magnitude, and vegetative coverage. Finally, we estimated waterfall height, defined as the change of elevation from the top of the waterfall to the bottom of the waterfall, using the Path Measure Tool in Google Earth Pro (WGS84 EGM96 Geoid). We flagged questionable waterfall identifications and excluded them from analysis. We then grouped waterfalls into one of five height classes: <5.0 m, 5.1–10.0 m, 10.1–20.0 m, 20.1–30.0 m, and >30.0 m.

Assessment of Waterfalls as Barriers to Nonnative Species

We examined the ability of the 14 taxa to pass waterfalls without human interference using a multistep approach. First, we eliminated all reaches with values of population density or road length density above the critical ecosystem load in the local catchment, as well as any reach downstream (illustrated as the reaches highlighted in yellow in Figure 1b). This step mitigated the potential for the remaining reaches in our analysis to have a nonnative species presence because of an introduction anywhere in the upstream network. We deemed those reaches as upstream of human interference (hereafter, "UHI reaches"; the reaches highlighted in blue in Figures 1b and 3). We then assessed the presence of nonnative taxa in biological surveys on UHI reaches in relation to the presence and height of waterfalls. Last, for those taxa that were sampled in UHI reaches above waterfalls, we determined whether canals or agricultural ditches that intersect the reaches were possible sources of nonnative taxa presence.

Results

Association of Nonnative Species with Anthropogenic Activities

TITAN revealed that both population density and road length density were associated with presence of nonnative taxa. The presence of 11 of the 14 nonnative taxa increased with increasing population density, six with purity and reliability values >0.95 (Table 2). Among the six taxa with significant change points, Green Swordtail exhibited the lowest value (4.24 people/km^2). This indicates that when local population density of a given reach exceeds 4.24 people/km^2, occurrence of at least one nonnative taxon is likely. The presence of 13 of the 14 nonnative taxa increased with increasing local road density and for three taxa purity and reliability values was >0.95 (Table 3). Of the three taxa with significant change points, Western Mosquitofish exhibited the lowest value (0.01 km/km^2). This suggests that when the local road density exceeds 0.01 km/km^2, at least one taxon or species group is likely to occur. Using these two change-point values, we identified a total of 1,208 UHI reaches for use in assessing the effects of natural barriers on species movement. The number of

Table 2. Associations between the 14 nonnative taxa and local catchment population density (people/km^2) generated using TITAN (Threshold Indicator Taxa Analysis). Change points (cp, people/km^2; 500 bootstrap iterations) are reported. Frequency (freq) indicates the number of occurrences per taxon. Maximum group (maxgrp) indicates the response direction, with 1 = negative (–) and 2 = positive (z+). Indicator value is the IndVal statistic (Dufrêne and Legendre 1997), scaled 0–100%. Observation probability (obsiv.prob) indicates the probability of obtaining an equal or larger IndVal score from random data. *Z* score is the corresponding z score for the indicator value. Purity is the mean proportion of correct response direction (z–) or (z+) assignments; reliability is the mean proportion of *p*-values ≤0.05 among 500 iterations. The median z score (z.median) indicates the median z score magnitude across all bootstrap replicates. Bolded taxa and cp highlight the taxa with significant change points (i.e., met both a 0.95 purity and 0.95 reliability). (*) Indicates the lowest significant change point (i.e., critical ecosystem load).

Taxa	cp	Freq	Max grp	IndVal	Obsiv.prob	*Z* score	Purity	Reliability	z.median
Green Swordtail	***4.24**	87	2	16.47	0.004	11.18	0.998	1	11.15
Guppy	**5.63**	77	2	18.71	0.004	14.83	1	1	15.84
Common Molly	**121.50**	11	2	6.89	0.004	7.93	0.978	0.996	9.84
Cuban Limia	64.50	4	2	2.48	0.004	6.06	0.854	0.854	6.91
Western Mosquitofish	**4.90**	65	2	13.94	0.004	11.71	1	1	12.49
Tilapia	**265.81**	27	2	37.96	0.004	25.25	1	1	26.67
Convict Cichlid	**64.50**	7	2	5.13	0.004	9.22	0.996	0.986	10.51
Largemouth Bass	14.76	4	2	1.29	0.076	2.09	0.522	0.696	4.61
Smallmouth Bass	0.00	4	1	3.14	0.032	4.81	0.754	0.774	6.39
Pond Loach	0.03	7	1	1.63	0.252	0.73	0.534	0.48	2.46
Bristlenosed Catfish	87.66	4	2	2.98	0.004	7.26	0.986	0.904	7.27
American bullfrog	23.37	26	2	5.71	0.004	4.51	0.804	0.968	6.22
Tahitian prawn	0.00	246	1	46.59	0.104	2.19	0.756	0.892	3.26
Red swamp catfish	960.26	20	2	10.16	0.076	3.02	0.716	0.822	4.98

Table 3. Associations between the 14 taxa and local catchment road length density (km/km^2) generated using TITAN (Threshold Indicator Taxa Analysis). Change points (cp, km/km^2; 500 bootstrap iterations) are reported. Frequency (freq) indicates the number of occurrences per taxon. Maximum group (maxgrp) indicates the response direction, with 1 = negative (z–) and 2 = positive (z+). Indicator value is the IndVal statistic (Dufrêne and Legendre 1997), scaled 0–100%. Observation probability (obsiv.prob) indicates the probability of obtaining an equal or larger IndVal score from random data. Z score is the corresponding z score for the Indicator value. Purity is the mean proportion of correct response direction (z–) or (z+) assignments, reliability is the mean proportion of *p*-values ≤0.05 among 500 iterations. The median z score (z.median) indicates the median z score magnitude across all bootstrap replicates. Bolded taxa and cp highlight the taxa with significant change points (i.e., met the both a 0.95 purity and 0.95 reliability). (*) Indicates the lowest significant change point (i.e., the critical ecosystem load).

Taxa	cp	Freq	Max grp	IndVal	Obsiv.prob	*Z* score	Purity	Reliability	z.median
Green Swordtail	0.85	87	2	10.60	0.004	3.55	0.826	0.926	4.10
Guppy	**0.49**	77	2	15.17	0.004	11.43	1	1	10.89
Common Molly	1.09	11	2	3.18	0.004	4.82	0.974	0.946	6.59
Cuban Limia	0.44	4	2	1.24	0.084	2.43	0.958	0.578	2.85
Western Mosquitofish	***0.01**	65	2	11.76	0.004	7.65	0.998	1	9.33
Tilapia	**7.39**	27	2	25.87	0.004	18.44	1	1	20.15
Convict Cichlid	0.00	7	2	2.03	0.012	3.89	0.996	0.9	4.87
Largemouth Bass	3.24	4	2	1.38	0.068	2.46	0.972	0.604	3.51
Smallmouth Bass	0.47	4	2	1.24	0.06	2.23	0.96	0.632	3.21
Pond Loach	0.00	7	2	1.46	0.148	1.23	0.452	0.432	2.34
Bristlenosed Catfish	0.00	4	2	0.92	0.26	1.05	0.546	0.298	2.10
American bullfrog	6.99	26	2	7.77	0.024	3.43	0.918	0.88	5.07
Tahitian prawn	0.00	246	1	49.69	0.168	1.83	0.514	0.896	3.16
Red swamp crayfish	0.00	20	2	4.03	0.004	3.87	0.936	0.9	4.68

biological surveys located on UHI reaches totaled 343 on Kaua'i, 408 on O'ahu, 612 on Moloka'i, 662 on Maui, and 141 on Hawai'i.

Assessment of Waterfalls and Their Role as Barriers to Nonnative Species

We identified a total of 1,099 waterfalls across the five main islands of Hawai'i (Figure 4). The compiled waterfall data set generated for this analysis is publicly available (https://doi.org/10.5066/P9EA3GCD). Among them, 340 waterfalls were located on UHI reaches (Figure 4). These waterfalls were evaluated with the biological survey locations. Among surveys on UHI reaches, most were located upstream of waterfalls <5.0 m high. Only on Maui were surveys conducted upstream of waterfalls >30.0 m high. See Table 4 for a summary of the biological surveys by island within the five classes of estimated waterfall height.

Twelve of the 14 taxa evaluated, with the exception of Tahitian prawn and American bullfrog, were absent from UHI reaches above waterfalls (Table 5). Tahitian prawn occurred in 390 surveys located on UHI reaches. Of these surveys, 27 occurred upstream of falls less than 5.0 m high (two reaches on Moloka'i and one reach on Maui), 3 occurred upstream of 10.1–20.0 m falls (one reach on Maui), and 4 occurred upstream of 20.1–30.0 m falls (one reach on Kaua'i). American Bullfrog occurred in seven UHI reaches; three occurred upstream of 10.0–20.1 m falls (all in one stream on the island of Hawai'i). Agricultural ditches or ca-

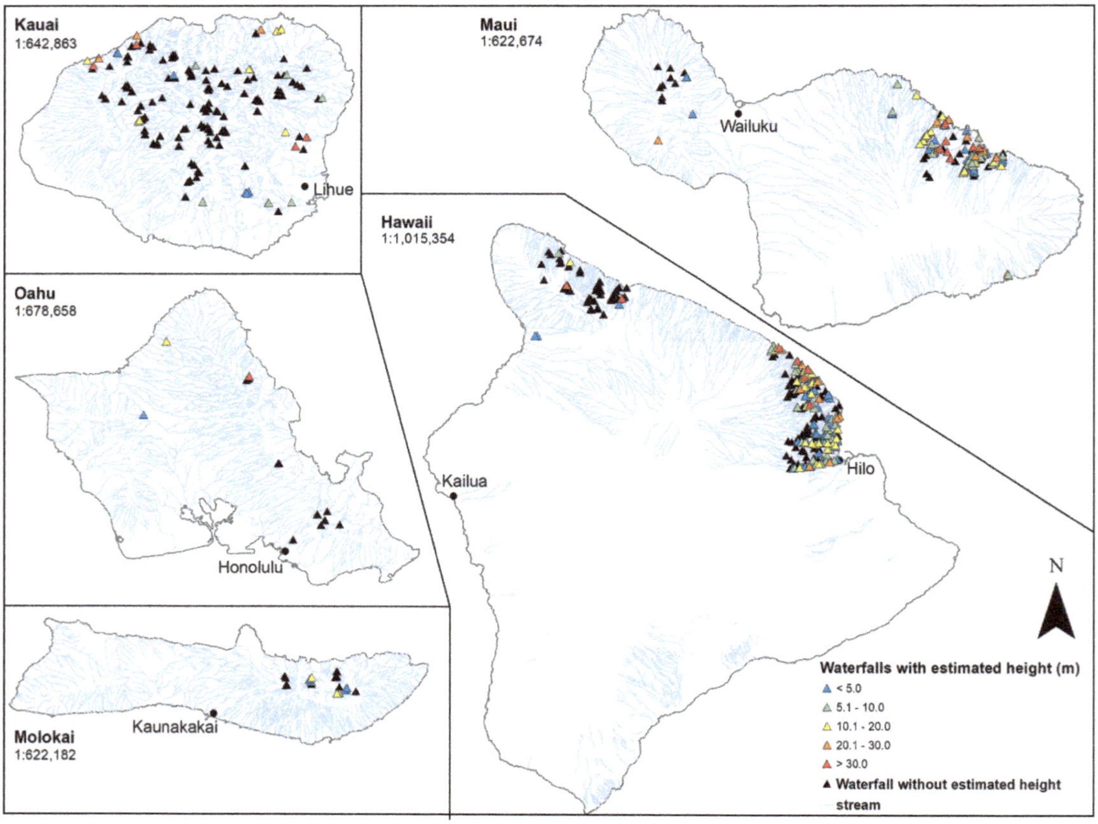

Figure 4. Identified waterfalls in the five major islands of Hawai'i (*n* = 1,099, data are available via https://doi.org/10.5066/P9EA3GCD). Islands are not drawn to scale.

Table 4. Number of surveys that occurred upstream of human interference (UHI) by island. Surveys on UHI reaches were grouped by occurrence above various waterfall height classes (<5.0 m, 5.1–10.0 m, 10.1–20.0 m, 20.1–30.0 m, and >30.0 m).

Island	# surveys	# surveys UHI	Number of UHI surveys upstream of waterfalls height (m) <5.0	5.1–10.0	10.1–20.0	20.1–30.0	>30.0
Kauaʻi	1,662	345	29	1	0	34	0
Oʻahu	2,049	408	0	0	0	0	0
Molokaʻi	904	612	113	13	67	0	0
Maui	1,999	662	98	49	41	48	4
Hawaiʻi	2,539	141	16	0	41	0	0

nals did not intersect any of the UHI reaches that contained American bullfrog or Tahitian prawn, reducing the likelihood of these waterways serving as introduction paths in the sampled reaches.

Discussion

Overview

We used a landscape approach to examine the role of anthropogenic activities on the dispersal of nonnative stream fauna in Hawaiʻi and to examine whether waterfalls serve as natural barriers. We found that the presence of nonnative taxa increased with increasing human activities, a strong indication that human introductions play a role in the proliferation of nonnative species. We were able to identify reaches below levels of human activities (i.e., 4.24 people/km^2 or 0.01 road km/km^2) in reaches that would result in human introduction, and only assessed

Table 5. Summary of the presence of taxa that were upstream of human interference (UHI). Taxa were summarized within waterfall height classes (<5.0 m, 5.1–10.0 m, 10.1–20.0 m, 20.1–30.0 m, and >30.0 m) based on the occurrence upstream of a given waterfall.

Taxa	# surveys	# surveys UHI	Number of species/taxon group presences by waterfall class (m) <5.0	5.1–10.0	10.1–20.0	20.1–30.0	>30.0
Tahitian prawn	1,527	390	27	0	3	4	0
American bullfrog	43	7	0	0	3	0	0
Green Swordtail	544	48	0	0	0	0	0
Pond Loach	22	7	0	0	0	0	0
Western Mosquitofish	143	6	0	0	0	0	0
Smallmouth Bass	28	4	0	0	0	0	0
Red swamp crayfish	118	2	0	0	0	0	0
Largemouth Bass	11	1	0	0	0	0	0
Guppy	310	0	–	–	–	–	–
Common Molly	62	0	–	–	–	–	–
Cuban Limia	5	0	–	–	–	–	–
Tilapia	57	0	–	–	–	–	–
Convict Cichlid	12	0	–	–	–	–	–
Bristlenosed Catfish	30	0	–	–	–	–	–

nonnative taxa presence–absence upstream of these reaches. We found that most nonnative taxa (12 of 14) were absent upstream of waterfalls regardless of waterfall height, with the two exceptions being Tahitian prawn and the American bullfrog. Tahitian prawn has an amphidromous life cycle similar to some native Hawaiian stream species, using their tarsal claws to migrate upstream and traverse barriers (Benbow et al. 2002). Also, American bullfrogs can temporarily leave the water and move around barriers through the terrestrial environment (Gahl et al. 2009). Our data support the hypothesis that waterfalls indeed serve as natural barriers to most of the nonnative stream fauna and were found to limit the upstream movement of most nonnative taxa in Hawai'i. Our study highlights the role that people play in facilitating species introductions in otherwise inaccessible habitats, which is one of the biggest challenges in nonnative species control. Our results also highlight the need to consider migratory behavior and different life stages of the species prior to the design and implementation of management strategies, and that the presence of a waterfall may be an effective conservation tool in some Hawaiian streams.

Upstream Movement of Nonnative Species

Our results indicate that Tahitian prawn and American bullfrog can occur above waterfalls in areas with limited human interference. This finding provides evidence that supports previous accounts of the suspected climbing ability of Tahitian prawn (Yamamoto and Tagawa 2000). We were unable to determine a limiting waterfall height for either species, but Tahitian prawn were present above many different waterfall height categories in our assessment, which suggested that they are more than capable of surpassing smaller waterfalls less than 5.0 m but that larger waterfalls may act as partial barriers (10.0–30.0 m). No adequate conclusion was drawn for movement limitations of the American bullfrog, but previous studies have shown that the species dispersal abilities were enhanced in wetter areas, as moisture is required for substantial overland dispersal (Smith and Green 2005; Gahl et al. 2009). It is therefore likely that given adequate humidity (i.e., areas that receive substantial rainfall or remain wet), waterfall height may not be a limiting factor for the upstream movement of the American bullfrog.

Major agricultural ditches have been suspected as a potential means for the movement of introduced species between watersheds in Hawai'i (Glenn Higashi, Hawai'i Department of Land and Natural Resources, personal communication). Some of these ditches were established more than 100 years ago to meet agricultural demand (Wilcox 1996). Because most were built before species introductions, we considered ditches as potential pathways for introduction or dispersion. However, our results did not indicate that Tahitian prawn or American bullfrog were dependent on agricultural ditches for establishment in UHI reaches. It is important to note that these waterways may still operate as modes of movement for these two nonnative species. It is possible that in lower reaches where agricultural ditches are more common, species movement is likely or at least plausible, but the high correlations with other human activities limits our ability to explore this relationship. More surveys on UHI reaches may provide further insight.

Additional Research and Management Needs

This assessment encompassed a statewide investigation to evaluate waterfalls with nonnative species without human interference, which are mostly located in headwater streams. There are notable differences in the number and types of waterfalls across islands. For instance, Hawai'i Island and Maui have the greatest number of waterfalls and waterfalls that terminate in the ocean were mostly observed on these two islands. Terminal waterfalls, in particular, may inhibit the dispersal of nonnative species that are able to disperse through marine environments (Nico and Walsh 2011), such as the Tahitian prawn (Yamamoto and Tagawa 2000) and Blackchin Tilapia (Brown et al. 1999). Further examination of the relationship between terminal waterfalls and presence of Tahitian prawn and Blackchin Tilapia could validate this suspicion. Because terminal waterfalls may act as barriers to the mobility of nonnative species at the river mouth, they can potentially provide pristine habitats free from nonnative species that could become priority areas for the conservation of some native climbing stream fauna.

Although not assessed in this study, suckermouth catfishes (family Loricariidae, including Bristlenosed Catfish [also known as Bushymouth Catfish) and Armored Catfish [also known as Suckermouth Catfish] *Hypostomus plecostomus*) have been reported to pass small (1.0–2.0 m) waterfalls and move upstream moderate distances. For example, a waterway restoration project in Waihee Stream on O'ahu that resulted in removal of an overhung edge and replacement with a 1.0–2.0-m near-vertical surface greatly improved passage for fishes. Native fishes as well as Bristlenosed Catfish were observed upstream of the structure where they were previously absent (Glenn Higashi, Hawai'i Department of Land and Resources, and Gordon Smith, Hawai'i Fish Habitat Partnership, personal communications). Additionally, Bristlenosed Catfish have been observed in the upper reaches of Helemano Stream on O'ahu at approximately 2,100 m elevation (Cory Yap, Pacific Biosciences Research Center, personal communication). Collectively, these observations support the idea that Bristlenosed Catfish and Armored Catfish are capable of traversing small obstructions to colonize upstream reaches. Bristlenosed and Armored catfishes feed on benthic algae, invertebrates, and detritus, which overlaps with some of the niche requirements of native fishes (Kinzie 1988). However, in contrast to native fish, these catfishes exhibit facultative air breathing, which may allow populations to become established in streams with low water quality (Brown et al. 1999). Therefore, they are considered habitat and resource competitors to Hawaiian native species. More data are needed, including studies of their ability to climb waterfalls, to quantify their impact on native species.

Effects of Barriers on Introduced Species and Stream Ecosystems

Previous studies have examined the effect of barriers on the distribution of native or nonnative aquatic taxa (e.g., Robinson and Rand 2005; Boizard et al. 2009). Several studies have examined the effect of dams on freshwater fishes, and others have shown the effects of downstream channel slope (e.g., Adams et al. 2000). Using census data, Han et al. (2008) investigated 76 species within 200 drainages in Japan and found that the occurrence of 20 species, mostly diadromous taxa, were negatively affected by downstream dams, and the occurrence of 12 largely nonnative species were positively associated with the presence of dams. Jellyman and Harding (2012) used a fish database and 30 (out of 2,000) dams in New Zealand to examine the assemblage composition above and below dammed and undammed sites. They concluded that dams restricted the movement of diadromous species, resulting in low percent composition above dams, and that dams created artificial lentic environments that promoted the establishment of nonnative species. Liermann et al. (2012) examined the impact of dam obstruction on freshwater species richness, endemism, or diadromy at a global scale and concluded that lampreys *Lampetra* spp., eels *Anguilla* spp., and shads *Alosa* spp. are at particular risk of species loss. Our study, based on 14 species and 129 watersheds with natural waterfall barriers on five tropical islands, supports these findings and indicates the importance of climbing abilities to the dispersal and distribution of stream fauna.

Further, several studies have revealed the roles of nonnative species within streams with waterfall barriers. Tilzey (1976) found that the presence of nonnative Brown Trout *Salmo trutta* or Rainbow Trout was the sole cause of the disappearance of Mountain Galaxias below the barriers in New South Wales, Australia. Similarly, Townsend (1996) showed that native galaxiid fishes were restricted to headwaters above large waterfalls that prevented the upstream movement of Brown Trout in New Zealand. Little has been studied on the presence or co-occurrence of native and nonnative species in Hawai'i. Further work is needed to better understand the native and nonnative interactions in Hawaiian stream ecosystems.

Challenges and Management Implications

Hawai'i is regarded for the biodiversity of its endemic species; however, many are now listed as threatened or endangered because of nonnative invasive species, urbanization, and climate change, which have dramatically altered assemblage structure and ecosystem processes (Maciolek 1984; Vitousek et al. 1987; Brasher 2003; Vorsino et al. 2014; Gon et al. 2018). Although Hawai'i currently prohibits bringing any living plant or organism into the state, many of the nonnative stream taxa were historically brought in for food, for biological control, for sportfishing, or as aquarium pets. Further, there is no regulation in place to prevent or prohibit nonnative stream fauna, such as aquarium fishes, from being disposed or released into streams. We argue that the knowledge and awareness of potential impacts of nonnative stream species is still lacking. Communicating the severity of impacts that nonnative species pose for Hawaiian streams should be a goal for environmental public awareness campaigns to stop introductions of nonnative species. Specifically, developing curricula on native and nonnative stream species and promoting early education in classrooms may be an effective strategy, with some local efforts in Hawai'i seeing successes (e.g., Nā Wai 'Ekolu, www.nawaiekolu.org). Hosting public outreach events and engaging the local community to participate in conservation of streams could also have an immediate and strong effect (e.g., Waipā Foundation, www.waipafoundation.org).

Biological stream sampling is currently a low priority within the state of Hawai'i, but increasing sampling effort can have numerous benefits. Additional biological surveys in areas previously surveyed as well as areas underrepresented in the current data set (e.g., areas upstream of human influences, agricultural ditches, reservoirs, urban areas, and estuarine areas) would bolster the ability to understand habitat requirements and potential barriers to nonnative species. Continued monitoring can also facilitate early detection of newly introduced species, which is critical for containment and eradication. In addition, updated information on nonnative species could indicate the severity of the species invasion and suggest locations to prioritize management action.

With adequate geographic information, landscape assessments have the potential to provide additional knowledge to aid management and policy. In instances where funding is limited and laboratory studies are difficult or impossible, perceived understanding of species ecology is often used to make management decisions. Our study shows that using a landscape approach can provide evidence for local knowledge and understanding, which can bolster support for management actions. In addition, the geographic and landscape information generated through this study can be used in prioritization and planning initiatives when we know where species are likely, or unlikely, to occur. We can now use this information to both prioritize areas for protection and identify connected reaches that, if restored, are unlikely to be recolonized by nonnative species. Proactive management, such as public awareness initiatives, fencing, and limiting access, could be implemented to enhance and ensure the effectiveness of nonnative species control. In places like Hawaiʻi where many streams and ecosystems are heavily and extensively degraded, it is important to take these steps to protect unique species.

Acknowledgments

We thank Glenn Higashi and the Division of Aquatic Resources, Hawai ʻi State Department of Natural and Land Resources for providing the biological survey data and sharing their knowledge of Hawaiian aquatic ecosystems and their field expertise and observations. We also thank Dan Polhemus and Gordon Smith (U.S. Fish and Wildlife Service) as well as Cory Yap (Pacific Biosciences Research Center) for their guidance and knowledge of introduced species in Hawaiʻi. We also thank the Michigan State University Aquatic Landscape Ecology Laboratory and Kyle Herreman for analytical support. In addition, we thank Robert Hughes, Kai Chen, and two anonymous reviewers for their thoughtful comments that improved the clarity of this chapter. The College of Tropical Agriculture and Human Resources, University of Hawaii at Mānoa provided funding support for the first two authors.

References

Adams, S. B., C. A. Frissell, and B. E. Rieman. 2000. Movements of nonnative Brook Trout in relation to stream channel slope. Transactions of the American Fisheries Society 129:623–638.

Armbruster, J. W. 2004. Phylogenetic relationships of the suckermouth armoured catfishes (Loricariidae) with emphasis on the Hypostominae and the Ancistrinae. Zoological Journal of the Linnean Society 141:1–80.

Baker, M. E., and R. S. King. 2010. A new method for detecting and interpreting biodiversity and ecological community thresholds. Methods in Ecology and Evolution 1:25–37.

Barbosa, T. A. P., N. L. Benone, T. O. R. Begot, A. Gonçalves, L. Sousa, T. Giarrizzo, L. Juen, and L. Montag. 2015. Effect of waterfalls and the flood pulse on the structure of fish assemblages of the middle Xingu River in the eastern Amazon basin. Brazilian Journal of Biology 75(3):78–94.

Benbow, M. E., L. L. Orzetti, M. D. McIntosh, and A. J. Burky. 2002. A note on cascade climbing of migrating goby and shrimp postlarvae in two Maui streams. Micronesica 34:243–248.

Berg, L. S. 1948–1949. Freshwater fishes of the USSR and adjacent countries. Israel Program for Scientific Translations, Jerusalem, Israel.

Boizard, J., P. Magnan, and B. Angers. 2009. Effects of dynamic landscape elements on fish dispersal: the example of Creek Chub (*Semotilus atromaculatus*). Molecular Ecology 18:430–441.

Brasher, A. M. D. 2003. Impacts of human disturbances on biotic communities in Hawaiian streams. BioScience 53:1052–1060.

Brasher, A. M. D., C. D. Luton, S. L. Goodbred, and R. H. Wolff. 2006. Invasion patterns along elevation and urbanization gradients in Hawaiian streams. Transactions of the American Fisheries Society 135:1109–1129.

Brett, B. L. H., and B. J. Turner. 1983. Genetic divergence in the Poecilia sphenops complex in Middle America. Biochemical Systematics and Ecology 11:127–137.

Brown, L. R., A. M. Brasher, B. C. Harvey, and M. Mathew. 1999. Success and failure of nonindigenous aquatic species in stream system: case studies from California and Hawaii. Pages 415–430 *in* B. Raton, editor. Success and failure of nonindigenous aquatic species in stream system: case studies from California and Hawaii. Lewis Publishers, Boca Raton, Florida.

Bussing, W. A. 1998. Peces de las aguas continentales de Costa Rica [Freshwater fishes of Costa Rica.] Editorial de la Universidad de Costa Rica, San José, Costa Rica.

Conkel, D. 1993. Cichlids of North and Central America. Tropical Fish Hobbyist Publications, Neptune City, New Jersey.

Cooney, P. B., and T. J. Kwak. 2013. Spatial extent and dynamics of dam impacts on tropical island freshwater fish assemblages. BioScience 63(3):176–190.

Daniel, W. M., D. M. Infante, R. M. Hughes, Y.-P. Tsang, P. C. Esselman, D. Wieferich, K. Herreman, A. R. Cooper, L. Wang, and W. W. Taylor. 2015. Characterizing coal and mineral mines as a regional source of stress to stream fish assemblages. Ecological Indicators 50:50–61.

De Grave, S. 2013. Macrobrachium lar. The IUCN Red List of Threatened Species 2013:e. T197866A2503299. Available: http://dx.doi.org/10.2305/IUCN.UK.2013-1.RLTS.T197866A2503299.en. (March 1 2019).

Dias, M. S., J.-F. Cornu, T. Oberdorff, C. A. Lasso, and P. A. Tedesco. 2013. Natural fragmentation in river networks as a driver of speciation for freshwater fishes. Ecography 36:683–689.

Dufrêne, M., and P. Legendre. 1997. Species assemblages and indicator species: the need for a flexible asymmetrical approach. Ecological Monographs 67:345–366.

Ebner, B. C., P. A. Thuesen, H. K. Larson, and P. Keith. 2011. A review of distribution, field observations and precautionary conservation requirements for sicydiine gobies in Australia. Cybium 35:397–414.

Elith, J., S. J. Phillips, T. Hastie, M. Dudík, Y. E. Chee, and C. J. Yates. 2011. A statistical explanation of MaxEnt for ecologists. Diversity and Distributions 17:43–57.

Fullerton, A. H., K. M. Burnett, E. A. Steel, R. L. Flitcroft, G. R. Pess, B. E. Feist, C. E. Torgersen, D. J. Miller, and B. L. Sanderson. 2010. Hydrological connectivity for riverine fish: measurement challenges and research opportunities. Freshwater Biology 55:2215–2237.

Gahl, M. K., A. J. K. Calhoun, and R. Graves. 2009. Facultative use of seasonal pools by American bullfrogs (*Rana catesbeiana*). Wetlands 29:697–703.

Giambelluca, T. W., Q. Chen, A. G. Frazier, J. P. Price, Y.-L. Chen, P.-S. Chu, J. K. Eischeid, and D. M. Delparte. 2013. Online rainfall atlas of Hawai'i. Bulletin of the American Meteorological Society 94:313–316.

Gon, S. M., III, S. L. Tom, and U. Woodside. 2018. 'Āina momona, honua au loli—productive lands, changing world: using the Hawaiian footprint to inform biocultural restoration and future sustainability in Hawai'i. Sustainability [online serial] 10(10):article 3420.

Greenfield, D. W., and J. E. Thomerson. 1997. Fishes of the continental waters of Belize. University Press of Florida, Gainesville.

Han, M., M. Fukushima, S. Kameyama, T. Fukushima, and B. Matsushita. 2008. How do dams affect freshwater fish distributions in Japan? Statistical analysis of native and nonnative species with various life histories. Ecological Research 23:735–743.

Hardman, M., L. M. Page, M. H. Sabaj, J. W. Armbruster, and J. H. Knouft. 2002. A comparison of fish surveys made in 1908 and 1998 of the Potaro, Essequibo, Demerara, and coastal river drainages of Guyana Ichthyological Exploration of Freshwaters 13:225–238.

Hawai'i Cooperative Park Service Unit. 1990. Hawaii stream assessment: a preliminary appraisal of Hawaii's stream resources: state of Hawaii. Hawai'i Department of Land and Natural Resources, Report R84, Honolulu.

Hawai'i DAR (Hawai'i Division of Aquatic Resource). 2004. Ditches [data file]. Available: http://planning.hawaii.gov/gis/download-gis-data/. (June 2010).

Hobbs, H. H., Jr. 1989. An illustrated checklist of the American crayfishes (Decapoda, Astacidae, Cambaridae, Parastacidae). Smithsonian Contributions to Zoology 480.

Holthe, E., E. Lund, B. Finstad, E. B. Thorstad, and R. S. McKinley. 2005. A fish selective obstacle to prevent dispersion of an unwanted fish species, based on leaping capabilities. Fisheries Management and Ecology 12(2):143–147.

Holthuis, L. B. 1980. Shrimps and prawns of the world. An annotated catalogue of species of interest to fisheries. FAO (Food and Agriculture Organization of the United Nations) Fisheries Synopsis 125.

Jellyman, P. G., and J. S. Harding. 2012. The role of dams in altering freshwater fish communities in New Zealand. New Zealand Journal of Marine and Freshwater Research 46:475–489.

Juvik, S. P., J. O. Juvik, and T. R. Paradise. 1998. Atlas of Hawaiʻi. University of Hawaiʻi Press, Honolulu.

Keith, P. 2003. Biology and ecology of amphidromous Gobiidae of the Indo-Pacific and the Caribbean regions. Journal of Fish Biology 63:831–847.

Kerby, J. L., S. P. D. Riley, L. B. Kats, and P. Wilson. 2005. Barriers and flow as limiting factors in the spread of an invasive crayfish (*Procambarus clarkii*) in Southern California streams. Biological Conservation 126:402–409.

Kinzie, R. A., III. 1988. Habitat utilization by Hawaiian stream fishes with reference to community structure in oceanic island streams. Environmental Biology of Fishes 22(3):179–192.

Kondratieff, M. C., and C. A. Myrick. 2005. Two adjustable waterfalls for evaluating fish jumping performance. Transactions of the American Fisheries Society 134:503–508.

Kwak, T. J., P. B. Cooney, and C. H. Brown. 2007. Fishery population and habitat assessment in Puerto Rico streams. Puerto Rico Department of Natural and Environmental Resources, Phase 1 Final Report Federal Aid in Sport Fish Restoration Project F-50, San Juan.

Lau, L. S., and J. F. Mink. 2006. Hydrology of the Hawaiian Islands. University of Hawai'i Press, Honolulu.

Lee, D. S., C. R. Gilbert, C. H. Hocutt, R. E. Jenkins, D. E. McAllister, and J. R. Stauffer. 1980. Atlas of North American freshwater fishes. North Carolina State Museum of Natural History, Raleigh.

Liermann, C. R., C. Nilsson, J. Robertson, and R. Y. Ng. 2012. Implications of dam obstruction for global freshwater fish diversity. BioScience 62:539–548.

Lintermans, M. 2000. Recolonization by the Mountain Galaxias *Galaxias olidus* of a montane stream after the eradication of Rainbow Trout *Oncorhynchus mykiss*. Marine and Freshwater Research 51:799–804.

Maciolek, J. A. 1984. Exotic fishes in Hawaii and other islands of Oceania. Pages 131–136 *in* W. R. Courtenay and J. R. Stauffer, Jr., editors. Distribution, biology, and management of exotic fishes. Johns Hopkins University Press, Baltimore, Maryland.

Masuda, H., K. Amaoka, C. Araga, T. Uyeno, and T. Yoshino, editors. 1984. The fishes of the Japanese archipelago. Tokai University Press, Tokyo.

McDowall, R. M. 1995. Seasonal pulses in migrations of New Zealand diadromous fish and the potential impacts of river mouth closure. New Zealand Journal of Marine and Freshwater Research 29:517–526.

McDowall, R. M. 2003. Hawaiian biogeography and the islands' freshwater fish fauna. Journal of Biogeography 30:703–710.

McDowall, R. M. 2010. New Zealand freshwater fishes: an historical and ecological biogeography. Springer, Fish & Fisheries Series 32, New York.

McKercher, L., and D. R. Gregoire. 2019. Lithobates catesbeianus (Shaw, 1802). U.S. Geological Survey, Nonindigenous Aquatic Species Database, Gainesville, Florida. Available: https://nas.er.usgs.gov/queries/factsheet.aspx?speciesid=71. (March 2019).

McKinney, M. L. 2002. Influence of settlement time, human population, park shape and age, visitation and roads on the number of alien plant species in protected areas in the USA. Diversity and Distributions 8:311–318.

Merrick, J. R., and G. E. Schmida. 1984. Australian freshwater fishes: biology and management. Griffen Press, Adelaide, South Australia.

Miller, R. R. 1983. Checklist and key to the mollies of Mexico (Pisces: Poeciliidae: *Poecilia*, subgenus *Mollienesia*). Copeia 1983:817–822.

Miller, R. R., W. L. Minckley, and S. M. Norris. 2005. Freshwater fishes of México. University of Chicago Press, Chicago.

Nico, L. G., and S. J. Walsh. 2011. Non-indigenous freshwater fishes on tropical Pacific islands: a review of eradication efforts. Pages 97–107 *in* C. R. Veitch, M. N. Clout, and D. R. Towns, editors. Island invasives: eradication and management. International Union of Conservation and Nature, Gland, Switzerland.

Page, L.M., and B. M. Burr. 1991. A field guide to freshwater fishes: North America north of Mexico. Houghton Mifflin, Boston.

Page, L. M., and B. M. Burr. 2011. Peterson field guide to freshwater fishes of North America north of Mexico. Houghton Mifflin Harcourt, Boston.

Palmer, M. A., E. S. Bernhardt, J. D. Allan, P. S. Lake, G. Alexander, S. Brooks, J. Carr, S. Clayton, C. N. Dahm, J. Follstad Shah, D. L. Galat, S. G. Loss, P. Goodwin, D. D. Hart, B. Hassett, R. Jenkinson, G. M. Kondolf, R. Lave, J. L. Meyer, T. K. O'Donnell, L. Pagano, and E. Sudduth. 2005. Standards for ecologically successful river restoration. Journal of Applied Ecology 42:208–217.

Pederson, D. T. 2010. The waterfalls of Kauai, Hawaii. The Hydrogeologist 71:9–12.

Pedroli, B., G. de Blust, K. van Looy, and S. van Rooij. 2002. Setting targets in strategies for river restoration. Landscape Ecology 17:5–18.

Poff, N. L., J. D. Allan, M. B. Bain, J. R. Karr, K. L. Prestegaard, B. D. Richter, R. E. Sparks, and J. C. Stromberg. 1997. The natural flow regime: a paradigm for river conservation and restoration. BioScience 47:769–784.

Pusey, B. J., A. H. Arthington, and M. G. Read. 1998. Freshwater fishes of the Burdekin River, Australia: biogeography, history and spatial variation in community structure. Environmental Biology of Fishes 53:303–318.

Rahel, F. J. 2013. Intentional fragmentation as a management strategy in aquatic systems. BioScience 63:362–372.

Rauchenberger, M. 1989. Systematics and biogeography of the genus *Gambusia* (Cyprinodontiformes, Poecilidae). American Museum Novitates 2951.

Reznick, D. N., and C. K. Ghalambor. 2005. Selection in nature: experimental manipulations of natural populations. Integrative and Comparative Biology 45:456–462.

Robinson, J. L., and P. S. Rand. 2005. Discontinuity in fish assemblages across an elevation gradient in a southern Appalachian watershed, USA. Ecology of Freshwater Fish 14:14–23.

Rosen, D. E. 1960. Middle American poecillid fishes of the genus *Xiphophorus*. Bulletin of the Florida State Museum, Biological Sciences 5:57–242.

Rosen, D. E. 1979. Fishes from the uplands and intermontane basins of Guatemala: revisionary studies and comparative geography. Bulletin of the American Museum of Natural History 162:267–376.

Rosen, D. E., and R. M. Bailey. 1963. The poeciliid fishes (Cyprinodontiformes): their structure, zoogeography, and systematics. Bulletin of the American Museum of Natural History 126.

Smith, M. A., and D. M. Green. 2005. Dispersal and the metapopulation paradigm in amphibian ecology and conservation: are all amphibian populations metapopulations? Ecography 28:110–128.

Snow, N. P., and G. Witmer. 2010. American bullfrogs as invasive species: a review of the introduction, subsequent problems, management options, and future directions. U.S. Department of Agriculture, Animal and Plant Health Inspection Service, National Wildlife Research Center, Fort Collins, Colorado.

Steen, P. J., T. G. Zorn, P. W. Seelbach, and J. S. Schaeffer. 2008. Classification tree models for predicting distributions of Michigan stream fish from landscape variables. Transactions of the American Fisheries Society 137:976–996.

Talwar, P. K., and A. G. Jhingran. 1992. Inland fishes of India and adjacent countries, volume 2. A. A. Balkema, Rotterdam, Netherlands.

Taylor, C. A., G. A. Schuster, J. E. Cooper, R. J. DiStefano, A. G. Eversole, P. Hamr, H. H. Hobbs, III, H. W. Robison, C. E. Skelton, and R. F. Thoma. 2007. A reassessment of the conservation status of crayfishes of the United States and Canada after 10+ years of increased awareness. Fisheries 32:372–389.

Tilzey, R. 1976. Observations on interactions between indigenous Galaxiidae and introduced salmonidae in the Lake Eucumbene catchment, New South Wales. Marine and Freshwater Research 27:551–564.

Tingley, R. W., III, D. M. Infante, R. A. MacKenzie, A. R. Cooper, and Y.-P. Tsang. 2019. Identifying natural catchment landscape influences on tropical stream organisms: classifying stream reaches of the Hawaiian Islands. Hydrobiologia 826:67–83.

Torrente-Vilara, G., J. Zuanon, F. Leprieur, T. Oberdorff, and P. A. Tedesco. 2011. Effects of natural rapids and waterfalls on fish assemblage structure in the Madeira River (Amazon basin). Ecology of Freshwater Fish 20:588–597.

Townsend, C. R. 1996. Invasion biology and ecological impacts of Brown Trout *Salmo trutta* in New Zealand. Biological Conservation 78(1–2):13–22.

U.S. Census Bureau. 2012. 2010 Census TIGER/Line Shapefiles [machine-readable data files]. Available: www.census.gov/geo/maps-data/data/tiger-data.html. (May 2015).

Vitousek, P. M., L. L. Loope, and C. P. Stone. 1987. Introduced species in Hawaii: biological effects and opportunities for ecological research. Trends in Ecology and Evolution 2:224–227.

Vorsino, A. E., L. B. Fortini, F. A. Amidon, S. E. Miller, J. D. Jacobi, J. P. Price, S. O. Gon, III, and G. A. Koob. 2014. Modeling Hawaiian ecosystem degradation due to invasive plants under current and future climates. PLOS (Public Library of Science) ONE [online serial] 9(5):e95427.

Walter, R. P., J. D. Hogan, M. J. Blum, R. B. Gagne, E. F. Hain, J. F. Gilliam, and P. B. McIntyre. 2012. Climate change and conservation of endemic amphidromous fishes in Hawaiian streams. Endangered Species Research 16:261–272.

Wang, L., D. Infante, P. Esselman, A. Cooper, D. Wu, W. Taylor, D. Beard, G. Whelan, and A. Ostroff. 2011. A hierarchical spatial framework and database for the National River Fish Habitat Condition Assessment. Fisheries 36:436–449.

Wieferich, D. J., W. M. Daniel, and D. M. Infante. 2015. Enhancing the utility of the NHDPlus river coverage: characterizing ecological river reaches for improved management and summary of information. Fisheries 40:562–564.

Wilcox, C. 1996. Sugar water: Hawaii's plantation ditches. University of Hawai'i Press, Honolulu.

Yamamoto, M. N., and A. W. Tagawa. 2000. Hawaii's native and exotic freshwater animals. Mutual Publishing, Honolulu, Hawai'i.

American Fisheries Society Symposium 90:105–123, 2019
© 2019 by the American Fisheries Society

Landscape Approaches to Understanding Invasions in Inland Lakes: Rainbow Smelt as a Case Study

Karen M. Alofs*
School for Environment and Sustainability, University of Michigan
440 Church Street, Ann Arbor, Michigan 48109, USA

Abstract.—Lakes, like islands, have been model systems for testing important concepts in landscape ecology. Lake assemblage and community composition, generally, and the occurrence of invaders, specifically, are controlled by a range of factors across scales. Here, I use the example of Rainbow Smelt *Osmerus mordax* invasions in inland lakes to illustrate common problems in both predictive and explanatory models of invasive species distributions across landscapes. Using variables related to dispersal and regional- and lake-scale environment, I fitted a series of boosted regression tree models to examine the factors that explain Rainbow Smelt invasion success. These models illustrate the potential effects of extrapolation and nonequilibrium conditions, the role of human activities, and the difficulty of understanding the importance of biotic interactions in the spread of invasives. Understanding the factors controlling invasions should inform management and conservation of inland lake ecosystems. For this to be effective, a mechanistic framework is needed to untie correlations in potential driving factors. Emerging data sets with fine spatial grain and broad spatial extent will support the transition from correlative models to mechanistic understanding of aquatic invasions.

Introduction

Landscape tools are powerful for studying invasions, and as invasions spread across landscapes, they are useful for testing ecological theory on biogeography, dispersal, and ecological assembly. Since MacArthur and Wilson described the theory of island biogeography (MacArthur and Wilson 1967), islands have been key systems for understanding landscape-scale biodiversity patterns and the processes that underlie them. Lakes, as patches of aquatic habitat in a larger terrestrially dominated matrix, are similarly key systems for understanding landscape patterns (Tonn and Magnuson 1982). Lakes have also been heralded as sentinels of environmental change because of their sensitivity and the manner in which they integrate changes from their surrounding landscapes (Adrian et al. 2009; Allerton et al. 2019, this volume; Jacobson et al. 2019, this volume). As clearly defined repeatable units with varying abiotic and biotic habitat characteristics and quantifiable propagule pressure, lakes are also model systems for testing invasion theory (MacIsaac et al. 2007).

Here, I focus on understanding species invasions in inland lakes where invasions have become widespread, often with negative impacts on native populations, assemblages, communities, and ecosystems (Ricciardi and MacIsaac 2011). In North American lentic

* Corresponding author: kmalofs@umich.edu

ecosystems, invasive species range from fish and dreissenid mussels to invertebrates like spiny water flea *Bythotrephes longimanus* and aquatic plants like Eurasian watermilfoil *Myriophyllum spicatum*. Overall, studies of invasions in tropical lakes are less common than in temperate systems. In part, this may be because globally the greatest concentration and surface area of lakes occurs at high northern, glaciated latitudes (Verpoorter et al. 2014).

Ecologists sometimes conceptualize the factors that control which species occur in a particular location, or lake, as a series of filters through which a regional species pool passes (Tonn et al. 1990). The first of these filters encompasses broad regional factors that include historical biogeography, dispersal barriers, climate, and geomorphology. Later filters include both abiotic conditions and biotic interactions that affect resource availability, physiology, and behavior. In applying the environmental filter framework to invasive species, MacIsaac et al. (2007) highlighted that human introductions often allow invasive species to bypass broad biogeographic filters obfuscating historical patterns. In this chapter, I categorize factors that influence the distribution of invasive fishes across landscapes as three filters (Figure 1). First, dispersal-related factors include catchment connectivity and human transport. Second, regional factors include climate, geology, and land use. And finally, lake-scale factors include lake morphology, water quality and chemistry, and the richness and composition of the resident biotic community. Factors related to each of these filters are indicated in Figure 1 and elsewhere in the chapter, with purple, orange, and green representing dispersal, regional, and lake factors, respectively.

Understanding the factors influencing the distribution of invasive species across a landscape can help inform predictions and guide targeted management actions. Identifying factors controlling dispersal may help limit spread, for example, by informing decisions related to catchment connectivity (Allen and Ramcharan 2001; Fausch et al. 2009) or altering policy regarding the use of live bait fish (Drake et al. 2010; Drake and Mandrak 2014a, 2014b). Alternatively, regional and lake-scale factors can be used to identify areas to target prevention efforts (Vander Zanden and Olden 2008) or obligate resources that may control the growth of invasive populations (Lurgi et al. 2016).

Species distribution models (SDMs), sometimes called ecological niche models, correlate occurrence patterns with environmental factors across landscapes. Species distribution models have been used for both prediction of which sites might be vulnerable to species invasions and explanation of which factors may contribute to invasion success (e.g., Wang and Jackson 2014; Gallardo et al. 2015; Romero-Alvarez et al. 2017). Species distribution models have also become increasingly common tools for understanding the influence of changing environmental conditions on fish and fisheries (e.g., Thuiller 2003; Buisson and Grenouillet 2009; Brodie et al. 2015; Moore et el. 2016; Rose et al. 2016; McGarvey et al. 2018). Species distribution models must be used with awareness that they are based on observational and correlative relationships. Additionally, SDMs assume that the relationship between species occurrences and environments reflect equilibrium conditions, an assumption that is often violated as invasive species spread across landscapes (Gallien et al. 2012). The factors defining a species realized niche in one range may not accurately predict its potential niche when introduced to novel conditions or species interactions (Broennimann et al. 2007; Zimmermann et al. 2010; Gallien et al. 2012). Predictive models are, therefore, often criticized for extrapolating the relationships defined in a species native range to new regions. Further,

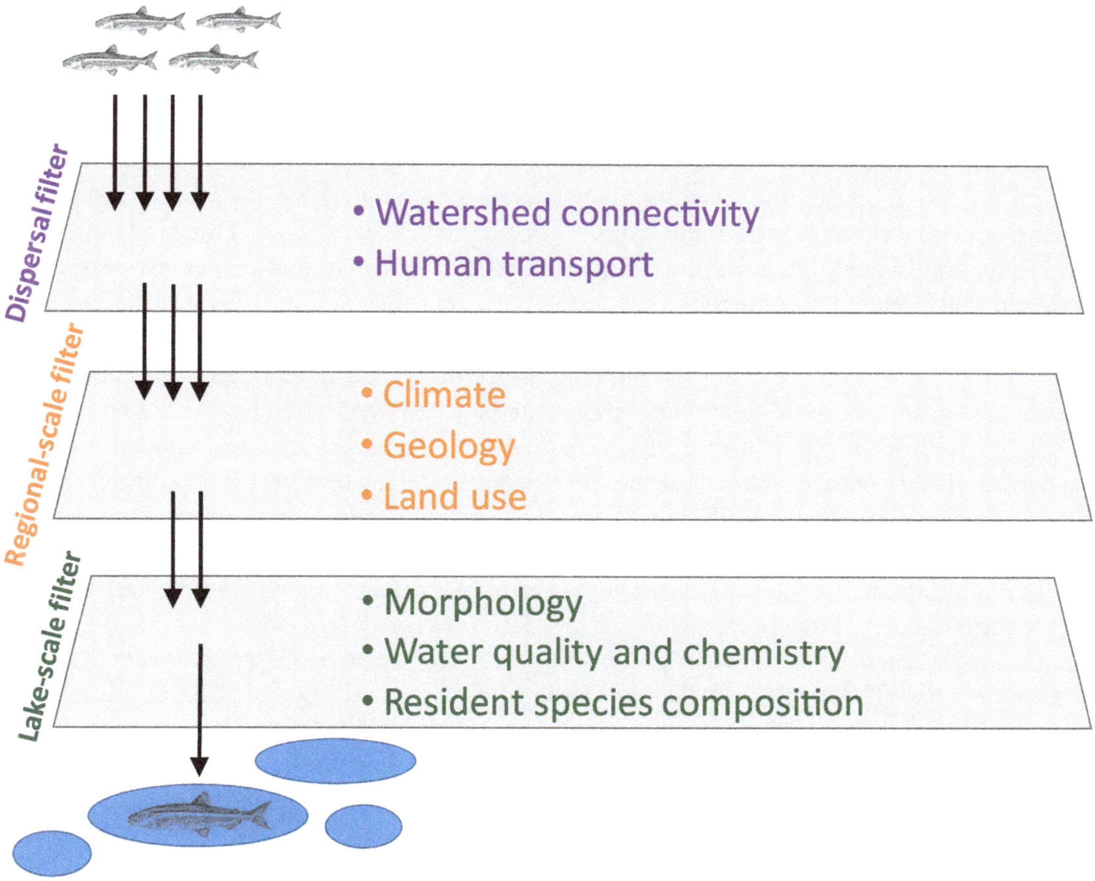

Figure 1. Factors influencing the distribution of invasive species across a landscape. Here, Rainbow Smelt *Osmerus mordax* in lakes across Ontario can be conceptualized as a series of filters, including dispersal, regional-scale and lake-scale filters. Factors related to dispersal, regional, and lake filters are indicated in purple, orange, and green, respectively.

when using presence–absence data, one must consider whether absences reflect unsuitable conditions or potentially suitable conditions in areas where an invasive species has not yet spread (Elith and Leathwick 2009).

Species distribution models rely upon the availability of landscape-scale georeferenced habitat data to map the potential distribution of invasive species. The availability of regional climate data has advanced our understanding but may have also biased our impression, of the importance of climate in determining species distributions (Elith 2017). In contrast, tools for classifying the connectivity of lakes within catchments and regional databases of such are only recently available (Fergus et al. 2017; Allerton et al. 2019). Additionally, few regions have quantitative measures of human-related dispersal of invasive species across landscapes. Several proxy variables, however, have been used to estimate relative propagule pressure among lakes, including distance to or along roads, boat launches or traffic patterns, and socioeconomic indicators (Leung et al. 2006; Vander Zanden and Olden 2008; Muirhead and MacIsaac 2011; Gallardo and Aldridge 2013; Drake and Mandrak 2014a).

Using observational correlative data, it is difficult to separate the influence of local-scale abiotic and biotic factors on invasions. Early hypotheses on the relationship between diversity and invasibility suggested that more diverse communities may resist invasions by limiting the availability of ecological niches (Elton 1958). Experimental studies have demonstrated biotic resistance to invasions in freshwater and suggest that it may be stronger in lentic than lotic environments (Alofs and Jackson 2014). Across landscapes in observational studies, however, sites with the highest species richness have also been those most vulnerable to invasions (Stohlgren et al. 2003). Literature on this paradox illustrates the difficulty of separating abiotic and biotic effects on species invasions across landscape extents. Positive correlations between species richness and invasion success have been attributed to abiotic factors that favor both native and invasive species alike (Naeem et al. 2000; Shea and Chesson 2002). Overall, environmental factors and species interactions are statistically confounded, which makes it difficult to attribute relative effects when SDMs are jointly fitted to abiotic and biotic predictors (Guisan and Thuiller 2005). As a consequence of this confounding, and because abiotic data are often more readily available at landscape extents, SDMs are often built on abiotic predictors alone (Elith and Leathwick 2009).

As a case-study highlighting landscape approaches to understanding invasions in lakes, I focus on Rainbow Smelt. Rainbow Smelt is a fish that has invaded inland lakes across eastern North America, the Great Lakes basin, and beyond. It often establishes large populations that affect native species by competition and predation (Evans and Loftus 1987). Rainbow Smelt is native to both Atlantic and Pacific coastal drainages in North America and to inland lakes in southern Maine. Four inland lakes populations in Ontario are believed to be the result of inundation by the Champlain Sea, roughly 11,000 years ago (Evans and Loftus 1987).

As is the case with most aquatic invaders, Rainbow Smelt can spread to new locations by natural dispersal through connected waterways or by introductions related to human activities. Downstream spread through connected waterways has been repeatedly observed in lakes across Ontario (Evans and Loftus 1987). Human-related dispersal can have long-distance and overland components because Rainbow Smelt are introduced to lakes both intentionally and unintentionally as bait and fertilized eggs (Evans and Loftus 1987). Moreover, spatially explicit population modeling indicates that Rainbow Smelt dispersal by human transportation outweighs dispersal through stream networks (Hrabik and Magnuson 1999). Studies examining Rainbow Smelt invasions illustrate the different types, scales, and applications of SDMs.

Drake and Lodge (2006) developed broadscale predictions of Rainbow Smelt distribution. Their model is an example of predictions built from presence-only data. They used known Rainbow Smelt presences from various sources to develop a genetic algorithm for rule-set production (GARP) model of suitable habitat across North America based on six regional climate variables, along with elevation and slope. Their model illustrated potential Rainbow Smelt habitat across the continent without lake-specific predictions. Broadscale predictive models of invasions often focus on the relationship between species occurrences and climate variables. In part, this reflects the strength of the influence of climate on species distributions, but likely also the availability of globally coherent climate data (Elith 2017).

In contrast to the broadscale model, Mercado-Silva et al. (2006) focused on lake-specific variables. They developed classification trees with data from lakes in Rainbow Smelt's native range in southern Maine. In their trees, the occurrence of Rainbow Smelt was related to maximum lake depth, lake area, and Secchi depth. Classification trees illustrate interactive effects of environmental variables on species distributions through a branching structure, where the importance of one factor may depend on the state of a second factor. For example, Mercado-Silva et al. (2006) found that Secchi depth was only important for predicting Rainbow Smelt occurrence in lakes with maximum depth between 8.9 and 12.3 m that were less than 102.8 ha in area. Shoreline perimeter and pH were excluded as predictors, although pH has been related to Rainbow Smelt occurrence (Evans and Loftus 1987). Models parametrized in their native range were used to predict Rainbow Smelt invasions in northern Maine, Ontario, and Wisconsin lakes. Mercado-Silva et al. (2006) additionally considered whether lakes with suitable habitat also held fish species likely to be affected by Rainbow Smelt invasions.

To further evaluate vulnerability to invasions across landscapes, Vander Zanden and Olden (2008) presented a framework combining data representative of introduction, establishment, and impact stages of species invasions. They incorporated data on boat landings, where introductions were likely; classification models of suitable habitat; and the locations of vulnerable species to clarify which lakes in Wisconsin should be targeted for prevention efforts.

Here, I use a comprehensive data set of lake environmental variables across a large region to develop SDMs of Rainbow Smelt invasions. I fit boosted regression tree models (BRTs), an approach that combines the advantages of two methods. First, regression trees relate responses (in this case, the presence or absence of Rainbow Smelt) to binary splits in predictors (such as lake characteristics or regional climate). And second, boosting as an ensemble method blends models to improve prediction. There are many advantages to using BRTs. They are flexible to nonlinearities and interactions, handle different types of predictors, accommodate missing data, do not require data transformation or elimination of outliers, and need as few as 100–150 observations for stable results (Elith et al. 2008). I examine the importance of dispersal and regional and lake-level factors in explaining the distribution of Rainbow Smelt in Ontario lakes. I also demonstrate the effects of (1) absence from potentially inaccessible lakes, (2) recently available estimates of propagule pressure from human activities, and (3) potentially confounded abiotic and biotic predictors on our understanding of the factors controlling Rainbow Smelt invasions across landscapes.

Methods

Data Collection

Ontario is a Canadian province with around 250,000 lakes distributed across a land-use and climate gradient with the warmest and most populated and developed area in the southeast and the much colder and less developed region in the north. In 2008, the Ministry of Natural Resources and Forestry began a Broad-scale Monitoring (BsM) Program surveying lakes across the province. This program includes coordinated fish surveys using large- and small-mesh gill nets coupled with data on lake bathymetry, water quality, and aerial surveys of angling activity in summer and winter. Details of survey methods and abiotic variables measured and calculated can be found in Sandstrom et al. (2013) and Chu et al. (2016). I used BsM data to refine the SDM for Rainbow Smelt and better understand the factors controlling the invasion in a series of four models (described in Species Distribution Models below).

In model selection, I included variables from each of the three, above-described filters (Figure 1). I used only variables that have previously been related to Rainbow Smelt distributions, fish assemblage grouping, or invasion processes. Measures of lake connectivity within the catchment were elevation and the Strahler and Shreve stream order of the outflow stream (as in Martin and Soranno 2006). As proxies of human-related dispersal, I included measures of angling pressure and an index of live-bait angling effort. Angling pressure was derived from aerial surveys of angler activities in summer (vessels and shore anglers) and winter (ice huts). The index of live-bait angler effort is derived from a gravity model that incorporated angler behavior and movement from a social survey and least-cost routing throughout a provincial road network (Drake and Mandrak 2010). If bycatch of a given species is equal to effort, then the index should represent propagule pressure. The index of live-bait angling effort differs from measures of angling pressure by directly incorporating propensity to transport live bait such as Rainbow Smelt.

Regional-scale variables comprised climate, geology, and anthropogenic pressures. Climate variables included normals of degree-days above 5°C and total precipitation calculated over 1981–2010 (Chu et al. 2016). Total dissolved solids concentration, measured in each lake, was used as a surrogate broadscale surficial geology indicator in Ontario (Baldwin et al. 2000). A catchment stress index summarized the intensity of a range of human activities related to development, including farming, forestry, manufacturing, waste management, dwellings, air discharge, and road densities (calculated as a "watershed stress index" in Chu et al. 2015).

Lake-scale morphological variables included surface area, maximum depth, and the proportion of habitat in the littoral zone (<4.6 m). Lake-scale water-quality variables included sodium (mg/L), dissolved organic carbon (DOC; mg/L), and Secchi depth measured in spring. Several abiotic variables, mostly lake-scale, were not included in model selection because they were highly correlated (Spearman rho >0.70) with other variables, including maximum annual air temperature, maximum surface water temperature, mean depth, summer Secchi depth, pH, conductivity, alkalinity, calcium, magnesium, chloride, and dissolved inorganic carbon.

Species Distribution Models

I implemented models using the gbm.step function in the 'dismo' package (Hijmans et al. 2017) in R (R Core Team 2017). After testing different model parameters, all models were fitted with 10-fold cross validation using a tree complexity of two (which allowed for two-way interactions), learning rate = 0.005, and bag fraction = 0.75. Cross validation is often used in cases where there are a limited number of observations within a data set and independent data are not available for model validation. In building BRTs, cross validation is used to validate the optimal number of trees in the final model (Elith et al. 2008). Tenfold cross-validation divides the data into 10 subsets and then combines 9 subsets in different combinations to create 10 training sets, where each omits a different subset that is used to test predictive performance. Final models included between 1,400 and 2,400 trees.

I developed four models. The first model used data for 714 lakes across the province, and Rainbow Smelt was recorded as present in 70 of those lakes (Figure 2A). This model was extrapolated over areas of the province where Rainbow Smelt does not yet occur (spatial extrapolation model). The second model excluded lakes that fell in "tertiary watersheds," with no Rainbow Smelt presence recorded for any lakes within the tertiary watershed (spatial restriction model). Excluded lakes likely included absences where habitat suitability has not yet been tested. Tertiary watersheds comprise 145 subdivisions of secondary watersheds, which are 28 subdivisions of three primary Ontario watersheds in the federal Water Resources Index Inventory Fil-

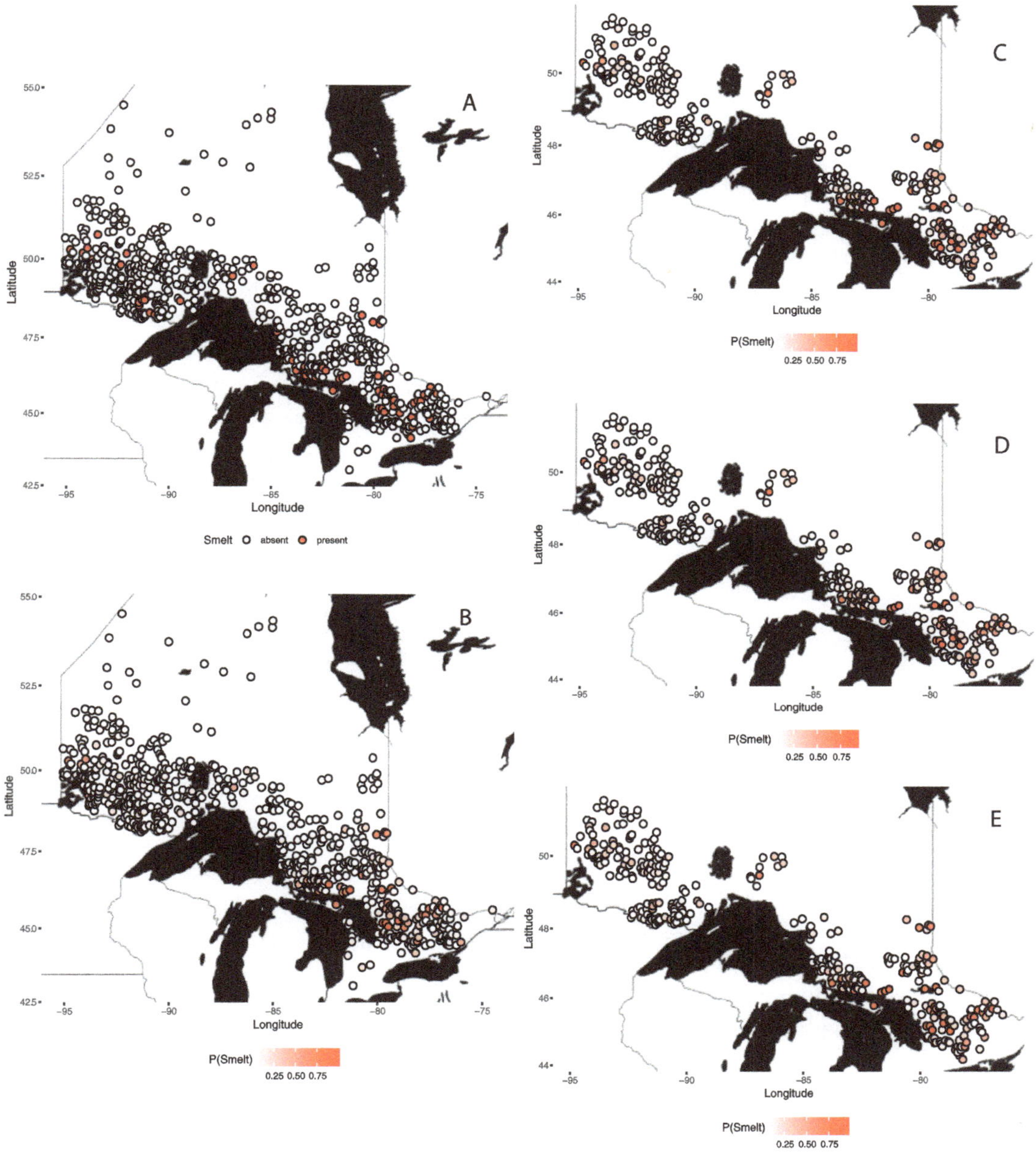

Figure 2. **(A)** The distribution of Rainbow Smelt presences (red circles) and absences (open circles) across Ontario in data from the Ontario Ministry of Natural Resources and Forestry Broad-scale Monitoring Program. Also, the predicted probability of Rainbow Smelt occurrence (higher probabilities of occurrence in darker red) in models built **(B)** with the complete provincial data set, **(C)** with only those data from tertiary catchments where Rainbow Smelt have been introduced, **(D)** without human-related predictors, and **(E)** with both abiotic and biotic (species richness) predictors.

ing System. The tertiary watersheds range in size from 700 to 31,000 km^2 (OMNRF 2015). All three levels of "watersheds" are simply administrative geographic polygons—not true catchments or watersheds (Omernik et al. 2017). For the second, and remaining, models, selection was based on a set of 354 lakes, again including 70 lakes with Rainbow Smelt present. The third model illustrated the value of human-related predictors by excluding angling pressure and bait angling estimates, as well as the catchment stress index (no human factors model). The final model included potential biotic interactions, specifically fish species richness (species richness model). Species richness is higher in northwestern and southern Ontario and is correlated with several abiotic factors (Chu et al. 2016). Across the landscape, richness has been related to patterns of postglacial recolonization (Mandrak 1995) and also to lake area, mean depth, pH, and elevation (Jackson and Harvey 1989; Minns 1989). Using each model, I predicted the probability of Rainbow Smelt occurrence in the lakes dataset and mapped these, using the 'ggmap' package (Kahle and Wickham 2013), to compare across model predictions to the observed distribution of Rainbow Smelt.

I evaluated model fit by several metrics. Threshold independent metrics included examining the area under the receiver operating characteristic curve (AUC) and explained deviance. The area under the receiver operating characteristic curve has been criticized because it weighs omission and commission errors equally and is a poor evaluation metric to assess the performance of presence-only SDMs (Lobo et al. 2008; Peterson et al. 2008). Explained deviance in BRT models was calculated as mean total deviance*(1 – mean residual deviance/mean total deviance), which can range from zero to one. I also calculated class-specific accuracy, including both sensitivity, the proportion of actual presences correctly predicted, and specificity, the proportion of actual absences that were correctly predicted. These measures were calculated using a threshold that maximized the sum of sensitivity and specificity, which maximizes the correct prediction rate across both presences and absences and is well suited for species with low prevalence in a data set (Liu et al. 2013).

I also examined the relative importance of explanatory factors from each of the three filters. Relative importance was calculated as the number of times a variable was selected for splitting in trees, weighted by the squared improvement to the model as a result of each split, and averaged over all trees (Friedman and Meulman 2003). This represents the influence of each predictor variable and is scaled so that the sum of relative importance across all of the predictors in a model adds to 100. I used partial dependence plots to show the effect of individual predictors on Rainbow Smelt occurrence after accounting for the average effects of all other predictors in the model. I similarly plotted important interactions between predictors in the BRT models (further described in Elith et al. 2008).

Results

Model performance differed most in explained deviance and sensitivity (Table 1). When evaluated by AUC, there was excellent fit to training data (by Swets 1988 criteria); all model AUCs >0.90, and good fit in 10-fold cross validation (for all models, $0.8 \leq AUC < 0.9$). The area under the receiver operating characteristic curve was highest in the spatial extrapolation model and lowest in the model excluding human-related predictors. Explained deviance and sensitivity showed similar patterns but with greater differences between models. There was little difference between spatial restriction and species richness models by these measures. There was also little change in specificity across models. Mapping predictions from each of the models (Figure 2) demonstrated, for example, that the model extrapolating across the province underpredicted Rainbow Smelt occurrence in northwestern Ontario.

Table 1. Evaluation of four models. AUC = area under the receiver operating characteristic curve, cv AUC = cross-validation AUC, TPR = true positive rate, and TNR = true negative rate.

	Extrapolation	Spatial restriction	No human factors	Species richness
AUC	0.96	0.95	0.93	0.96
cv AUC	0.89	0.84	0.83	0.85
Explained deviance	0.70	0.49	0.42	0.51
Threshold (max TPR + TNR)	0.12	0.20	0.30	0.21
Sensitivity (TPR)	0.91	0.79	0.63	0.79
Specificity (TNR)	0.91	0.95	0.98	0.95

The four consistently most important variables included in model selection were proportion of littoral habitat, live bait, maximum depth, and sodium concentration (Table 2). Additional important variables across models included precipitation and angling pressure. Partial dependence plots (Figure 3) demonstrate that Rainbow Smelt occur in lakes with high live-bait angling effort, proportionally smaller littoral habitat, deeper waters, and higher sodium concentrations. Whereas some effects increased or declined

Table 2. Variables used and relative importance of each across four models. DOC = dissolved organic carbon; TDS = total dissolved solids. Factors related to dispersal, regional, and lake filters are indicated in purple, orange, and green, respectively.

Extrapolation		Spatial restriction		No human factors		Species richness	
Bait	19.32	Proportion littoral	18.47	Proportion littoral	18.73	Fish species richness	17.94
Maximum depth	18.97	Bait	14.36	Sodium	17.44	Proportion littoral	17.71
Proportion littoral	10.03	Maximum depth	13.83	Elevation	16.6	Bait	10.61
Precipitation	8.72	Sodium	12.1	Precipitation	12.04	Sodium	9.69
Angling pressure	6.5	Elevation	9.97	Maximum depth	11.53	Maximum depth	8.77
Sodium	6.45	Angling pressure	8.34	Area	9.59	Angling pressure	8.51
Stress index	5.89	Precipitation	5.55	DOC	5.93	Precipitation	8.12
Shreve	5.82	Area	5.12	Shreve	3.67	Elevation	4.77
DOC	5.55	DOC	4.22	Strahler	2.18	DOC	4.04
Elevation	5.1	Stress index	2.26	Degree-days	1.2	Area	3.89
Degree-days	3.08	Strahler	2.03	TDS	0.73	Stress index	2.07
Area	1.67	Shreve	1.68	Spring Secchi	0.37	Degree-days	1.43
Strahler	1.28	Degree-days	0.87			Shreve	1.02
TDS	0.89	Spring Secchi	0.67			Strahler	0.69
Spring Secchi	0.72	TDS	0.51			TDS	0.59
						Spring Secchi	0.14

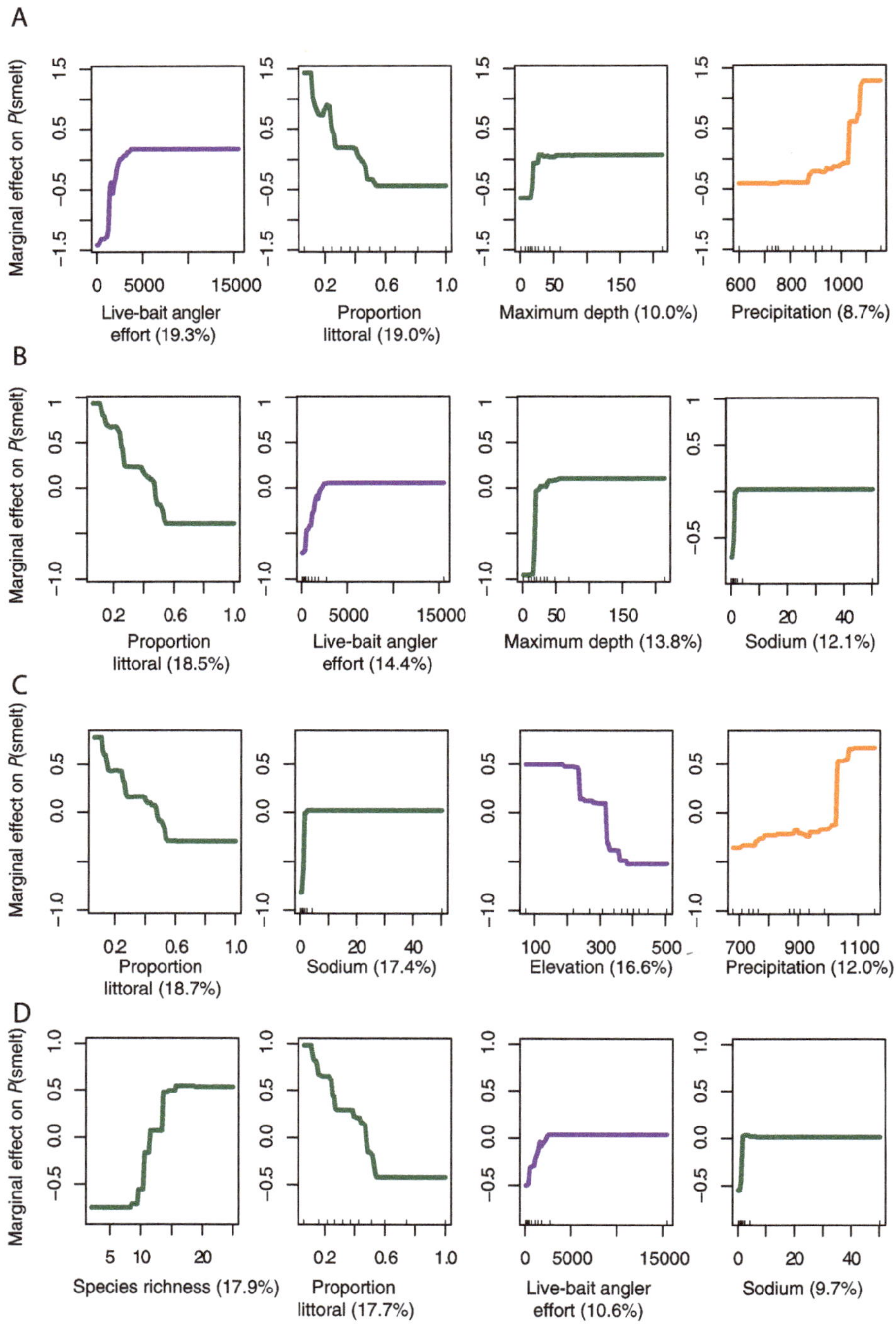

Figure 3. Partial dependence plots show the effect of predictors on Rainbow Smelt occurrence after accounting for the average effects of all other predictors in the model. Percentages are relative importance of each factor in a given model. Here, these are plotted for the four most important predictors in models built **(A)** using the complete provincial data set, **(B)** with data only from tertiary catchments where Rainbow Smelt have been introduced, **(C)** without human-related predictors, and **(D)** with both abiotic and biotic (species richness) predictors. Factors related to dispersal, regional, and lake filters are indicated in purple, orange, and green, respectively.

with predictors, many demonstrated some threshold. For example, Rainbow Smelt was unlikely to occur in lakes with maximum depths less than 25 m or sodium concentrations less than 3 mg/L. Fish species richness had the highest variable importance when both biotic and abiotic factors were included.

Predictors with threshold effects often had strong interactions with other predictors. As illustrated in Figure 4, the effect of bait angling is nearly nil in lakes less than 25 m deep, and elevation is likewise not important in lakes with large littoral habitats. The strongest two-way interactions across models (those with interaction strength greater than five) included live-bait angler effort*maximum depth, precipitation*Shreve order, and proportion littoral*elevation (interaction strength = 8.50, 8.27, and 6.75, respectively) in the spatial extrapolation model. When human-related predictors were excluded from the model, proportion littoral*elevation and precipitation*elevation strongly interacted (interaction strength = 8.20 and 5.46, respectively). Species richness*precipitation was a strong interaction in the model with both biotic and abiotic factors included (interaction strength = 6.92).

Discussion

As a case study, Rainbow Smelt invasions allow us to better understand the challenges in explaining and predicting the distribution of invasive species across regional landscapes. Previous SDMs for Rainbow Smelt have focused on regional (Drake and Lodge 2006) or local (Mercado-Silva et al. 2006) factors. Using BRTs, I was able to evaluate the relative importance of dispersal, regional, and local factors in explaining the occurrence of the invasive fish. The series of models presented here illustrate how methodological choices and data availability may bias insights drawn from observational studies and correlative data.

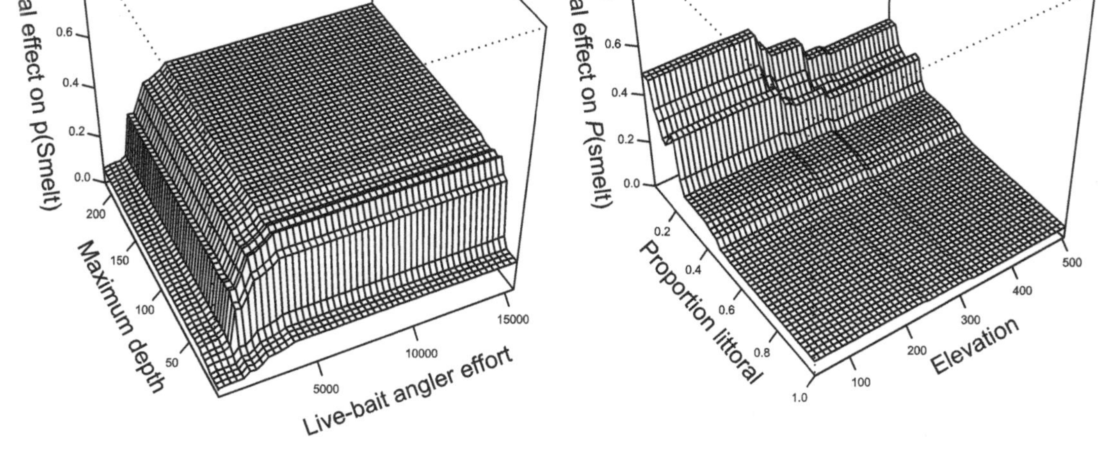

Figure 4. Three-dimensional joint partial dependence plots depicting the most important interactions in the complete provincial model. As examples, **(A)** the interaction between maximum depth and live-bait angling and **(B)** the interaction between the proportion of littoral habitat and elevation.

Overall, integrating across environmental filters suggests that regional factors are less important than human related-dispersal or local lake factors in explaining the pattern of Rainbow Smelt distribution in Ontario. Similar to previous models (Mercado-Silva et al. 2006), depth was an important predictor of Rainbow Smelt occurrence. In contrast, I failed to identify a strong influence of water clarity (Secchi depth). This could be related to differences in modeling methods, the spatial locations and variables included in the models, or shifts in the realized species niche with invasions into previously untested or novel lake habitats. Sodium was also a relatively important predictor on BRT models, with very low concentrations having negative impacts on Rainbow Smelt occurrence. This may reflect salinity tolerance and adaptations related to the anadromous life cycle in coastal environments (Ayer et al. 2005).

One challenge in using SDMs to study the pattern of invasions across landscapes is that these patterns reflect nonequilibrium conditions. Here, I examined the potential influence of considering observed absences from untested habitats in catchments where the invasion has not spread. Excluding lakes in these uninvaded areas from model selection greatly reduced explained deviance and sensitivity. The model built on the full provincial data set, however, may misrepresent the factors limiting Rainbow Smelt distribution. Many of the uninvaded catchments are in less populated and wetter regions in northern and western Ontario. Including these absences may have inflated the importance of angling with bait, human-related stress, and precipitation. Excluding these absences, all of these factors dropped in importance. Precipitation has been linked to catch and size distributions of fishes in Ontario lakes, using BRT models (Chu et al. 2016). But the potential mechanisms driving correlations between precipitation and fish assemblages remain unclear (Chu et al. 2016). In Ontario, understanding the importance of climate- and human-related factors is challenging because variables often covary across the province (Chu et al. 2016).

Human-related factors are increasingly influencing the distribution of species across landscapes. This is emphasized by the low sensitivity and true presence rate in the model with no human-related factors. In aquatic systems, humans facilitate the dispersal of invasive species, but also range expansions by fishes (Alofs et al. 2014). Fish dispersal is often by illegal or unintended introductions and therefore is difficult to quantify (Litvak and Mandrak 1993; Kerr 2006). Removing baitfish angling effort, angling pressure, and the stress index from our models reduced explained deviance and sensitivity (Table 1). The estimate of bait fish angling effort, derived from angler surveys and road networks (Drake et al. 2010), was one of the most important predictors of Rainbow Smelt distribution when they were included in model selection. Drake and Mandrak (2014b) estimated the probability of Rainbow Smelt bycatch being sold as bait, carried to an uninvaded lake, and released as 0.0008 or approximately 1 in 1,200 trips. Whereas this is a very low per-trip probability, there are an estimated 4.2 million yearly angling trips distributed broadly across the province that use live bait fish (Drake and Mandrak 2014a, 2014b). Given this, it is likely that the relationship between Rainbow Smelt occurrence and bait use identified in the BRT model is not merely correlative, but that bait-release underlies many Rainbow Smelt invasions. Data estimating human impacts across landscape scales, particularly dispersal-related data, are difficult to derive but are likely important for understanding the spread of many invasive species.

Potential confounding between biotic and abiotic factors can limit the improvement in explanatory power related to incorporating biotic predictors in addition to more readily available and commonly used abiotic predictors (Elith and Leathwick 2009; Elith 2017). Early tests, however, found surprisingly little overlap in the variation in species occurrence explained by abiotic and biotic predictors (Zimmermann et al. 2010). When included with abiotic predictors, fish species richness was the strongest predictor of Rainbow Smelt occurrence, but adding richness improved model fit very little and the explained deviance increased very little with no change in sensitivity (Table 1). Rainbow Smelt was more likely to occur in lakes with more fish species. The positive relationship between richness and invasibility is a pattern that can reflect the influence of abiotic variables covarying with species richness (Naeem et al. 2000). Covariance between abiotic and biotic factors supports the common use of abiotic variables as proxies that are assumed to capture the predictive power of biotic drivers (Elith 2017). Native and invasive fish distributions are often controlled by similar abiotic factors, making it difficult to understand the importance of species interactions independent of environmental variation. In this case study, fish species richness, like Rainbow Smelt occurrence, is highest in deep lakes with relatively little littoral habitat in regions of Ontario where angling efforts and baitfish usage are high. Because of the correlative nature of the data, we cannot determine whether these abiotic factors alone drive Rainbow Smelt occurrence or whether richness contributes to invasibility. I am not aware of evidence of facilitation or mutualism between Rainbow Smelt and native fish species that would produce the positive effect observed in the model. I would be more inclined to attribute a negative relationship between richness and invasibility to biotic interactions, including competition and predation.

Challenges and Policy Implications

Using Rainbow Smelt as a case study, I have highlighted several challenges for understanding species invasions across landscapes. First, fitting SDMs assumes equilibrium between occurrences and environmental gradients. Species invasions are dynamic and their patterns likely do not represent equilibrium conditions over much of their invaded range. Second, dispersal clearly plays a key role in understanding the patterns of invasions. Both natural dispersal through river–lake networks (Melles et al. 2015; Allerton et al. 2019) and long-distance dispersal driven by human behaviors (Drake et al. 2010, 2014) are difficult to quantify and require large amounts of data. Third, the influence of species interactions in controlling invasions remains unclear because abiotic and biotic variables are often confounded across large regional scales. Finally, invasions and many ecological processes are controlled by multivariate interacting drivers, and thresholds, nonlinearities, and interactions make understanding their influences at landscape scales challenging (Turner and Gardner 2015).

Developing predictions of future species distributions is further complicated by uncertainty in quantifying future environmental conditions (Porfirio et al. 2014). The difficulty in untangling which factors control the spread of invasive species across landscapes with changing and interactive environmental conditions makes using correlative models to guide policy decision challenging. The models presented here, however, strongly suggest that the use of live bait fish is contributing to the spread of invasive Rainbow Smelt into inland lakes across Ontario. Drake and Mandrak (2014b) outlined recommendations for limiting the impacts of this dispersal pathway, including harvest, movement, and release restrictions. Rainbow Smelt have been implicated in population declines and local extirpations of species, including Walleye *Sander vitreus*, Yellow Perch *Perca flavescens*, and Cisco *Coregonus artedi* (Hrabik et al. 1998; Mercado-Silva et al. 2007). Further, Rainbow Smelt have now been observed in nearly 10% of lakes surveyed recently across the province, underlying the need to enact policy limiting their spread.

Additional Research and Management Needs

Several avenues should be pursued to make species distribution modeling more useful in the management of invasive species. First, given that these models have been applied for more than a decade, we need to better test their accuracy by validating their predictions across independent data sets, which is rarely done (MacIsaac et al. 2007). Further, SDMs should more commonly be combined with data on potential impact (as in Vander Zanden and Olden 2008). Finally, I join others in suggesting the next important step in understanding the factors controlling invasions in lakes, and other ecosystems, is transitioning from correlative to mechanistic models, including models incorporating physiology, life history, phenology, functional traits, spread dynamics, and spatially explicit dispersal (Buckley et al. 2010; Gallien et al. 2010; Elith 2017; Lustig et al. 2017). Applying mechanistic models to make predictions across landscapes is data-intensive and this has limited their use and development, to date (Elith 2017).

The transition to mechanistic models will be facilitated by new and emerging data that will allow us to merge broad spatial extent and fine spatial grain. The dearth of wide coverage in hydrological data sets has limited our ability to predict suitable aquatic environments for invasives (MacIsaac et al. 2007). Since the turn of the century, the availability of spatial data has been exponentially increasing including downscaled general circulation models, Landsat imagery, social media, and community-sourced data (Turner and Gardner 2015). Recently available lake data sets include LAGOS-NE (Soranno et al. 2017) and LakeCat (Hill et al. 2018). LAGOS-NE includes land-use, geologic, climatic, and hydrologic data for more than 51,000 lakes in the northeastern United States. For a subset of these lakes, there is also data on in situ water quality (e.g., phosphorus, chlorophyll, and Secchi depth). LakeCat includes landscape features for more than 378,000 lakes across the conterminous United States. In many regions, however, we still lack detailed information at landscape scales for many of the human-related, local lake-scale, and biological variables we found predictive for Rainbow Smelt. Such data are key in developing landscape scale tools for managing the spread and impact of invasive species and thus worthy of investment.

In an era of multiple interacting stressors affecting freshwater ecosystems, more research is needed to understand the current and future impacts of aquatic invasive species. The impacts of Rainbow Smelt on coldwater adapted Cisco populations in Wisconsin lakes is expected to be less important than the predicted impacts of climate change (Sharma et al. 2011). Additional investigations on the relative influence of climate, invasions, and other stressors are needed for fish populations, biotic assemblages and communities, and aquatic ecosystem services.

Acknowledgments

I thank Andrew Drake for providing estimated live-bait angler effort data. I am grateful to the Ontario Ministry of Natural Resources and Forestry for collecting and sharing lake and fish data. I thank Brian Shuter, Nigel Lester, Henrique Giacomini, Cindy Chu, and Dak de Kerckhove for their perspectives on and assistance with this data and their modeling. I received helpful manuscript reviews from Lizhu Wang, Robert Hughes, and two anonymous reviewers. Finally, thank you to Don Jackson and his University of Toronto laboratory for their support and assistance during the development of this project.

References

Adrian, R., C. M. O'Reilly, H. Zagarese, S. B. Baines, D. O. Hessen, W. Keller, D. M. Livingstone, R. Sommaruga, D. Straile, E. Van Donk, G. a Weyhenmeyer, and M. Winder. 2009. Lakes as sentinels of climate change. Limnology and Oceanography 54:2283–2297.

Allen, Y. C., and C. W. Ramcharan. 2001. Dreissena distribution in commercial waterways of the US: using failed invasions to identify limiting factors. Canadian Journal of Fisheries and Aquatic Sciences 58:898–907.

Allerton, M. L., N. E. Jones, S. J. Melles, and B. J. Schmidt. 2019. Ecologically relevant structural metrics for complex stream–lake networks. Pages 125–139 *in* R. M. Hughes, D. Infante, L. Wang, K. Chen, and B. F. Terra, editors. Advances in understanding landscape influences on freshwater habitats and biological assemblages. American Fisheries Society, Symposium 90, Bethesda, Maryland.

Alofs, K., and D. Jackson. 2014. Meta-analysis suggests biotic resistance in freshwater environments is driven by consumption rather than competition. Ecology 95:3259–3270.

Alofs, K. M., D. A. Jackson, and N. P. Lester. 2014. Ontario freshwater fishes demonstrate differing range-boundary shifts in a warming climate. Diversity and Distributions 20:123–136.

Ayer, M. H., C. Benton, W. V King, J. Kneebone, S. Elzey, M. Toran, K. Grange, and D. L. Berlinsky. 2005. Development of practical culture methods for Rainbow Smelt larvae. North American Journal of Aquaculture 67:202–209.

Baldwin, D. J. B., J. R. Desloges, and L. E. Band. 2000. Physical geography of Ontario. Chapter 2 *in* A. H. Perera, D. L. Euler, and I. D. Thompson, editors. Ecology of a managed terrestrial landscape: patterns and processes of forest landscapes in Ontario. University of British Columbia Press, Vancouver.

Brodie, S., A. J. Hobday, J. A. Smith, J. D. Everett, M. D. Taylor, C. A. Gray, and I. M. Suthers. 2015. Modelling the oceanic habitats of two pelagic species using recreational fisheries data. Fisheries Oceanography 24:463–477.

Broennimann, O., U. A. Treier, H. Müller-Schärer, W. Thuiller, A. T. Peterson, and A. Guisan. 2007. Evidence of climatic niche shift during biological invasion. Ecology Letters 10:701–709.

Buckley, L. B., M. C. Urban, M. J. Angilletta, L. G. Crozier, L. J. Rissler, and M. W. Sears. 2010. Can mechanism inform species' distribution models? Ecology Letters 13:1041–1054.

Buisson, L., and G. Grenouillet. 2009. Contrasted impacts of climate change on stream fish assemblages along an environmental gradient. Diversity and Distributions 15:613–626.

Chu, C., N. P. Lester, H. C. Giacomini, B. J. Shuter, and D. A. Jackson. 2016. Catch-per-unit-effort and size spectra of lake fish assemblages reflect underlying patterns in ecological conditions and anthropogenic activities across regional and local scales. Canadian Journal of Fisheries and Aquatic Sciences 73:535–546.

Chu, C., C. K. Minns, N. P. Lester, N. E. Mandrak, and J. Rosenfeld. 2015. An updated assessment of human activities, the environment, and freshwater fish biodiversity in Canada. Canadian Journal of Fisheries and Aquatic Sciences 72:135–148.

Drake, D. A. R., H. H. Harvey, and N. E. Mandrak. 2010. Least cost transportation networks predict spatial interaction of invasion vectors. Ecological Applications 20:2286–2299.

Drake, D. A. R., and N. E. Mandrak. 2014a. Bycatch, bait, anglers, and roads: quantifying vector activity and propagule introduction risk across lake ecosystems. Ecological Applications 24:877–894.

Drake, D. A. R., and N. E. Mandrak. 2014b. Ecological risk of live bait fisheries: a new angle on selective fishing. Fisheries 39:201–211.

Drake, D. A. R., R. Mercader, T. Dobson, and N. E. Mandrak. 2014. Can we predict risky human behaviour involving invasive species? A case study of the release of fishes to the wild. Biological Invasions 17:309–326.

Drake, J. M., and D. M. Lodge. 2006. Forecasting potential distributions on nonindigenous species with a genetic algorithm. Fisheries 31:9–16.

Elith, J. 2017. Predicting distributions of invasive species. Pages 93–129 *in* A. P. Robinson, T. Walshe, M. A. Burgman, and M. Nunn, editors. Invasive species: risk assessment and management. Cambridge University Press, Cambridge, UK.

Elith, J., and J. R. Leathwick. 2009. Species distribution models: ecological explanation and prediction across space and time. Annual Review of Ecology, Evolution, and Systematics 40:677–697.

Elith, J., J. R. Leathwick, T. Hastie, and J. R. Leathwick. 2008. A working guide to boosted regression trees. Journal of Animal Ecology 77:802–813.

Elton, C. 1958. The ecology of invasions by animals and plants. University of Chicago Press, Chicago.

Evans, D. O., and D. H. Loftus. 1987. Colonization of inland lakes in the Great Lakes region by Rainbow Smelt, *Osmerus mordax*: their freshwater niche and effects on indigenous fishes. Canadian Journal of Fisheries and Aquatic Sciences 44(S2):s249–s266.

Fausch, K. D., B. E. Rieman, J. B. Dunham, M. K. Young, and D. P. Peterson. 2009. Invasion versus isolation: trade-offs in managing native salmonids with barriers to upstream movement. Conservation Biology 23:859–870.

Fergus, C. E., J. F. Lapierre, S. K. Oliver, N. K. Skaff, K. S. Cheruvelil, K. Webster, C. Scott, and P. Soranno. 2017. The freshwater landscape: lake, wetland, and stream abundance and connectivity at macroscales. Ecosphere [online serial] 8(8):e01911.

Friedman, J. H., and J. J. Meulman. 2003. Multiple additive regression trees with application in epidemiology. Statistics in Medicine 22:1365–1381.

Gallardo, B., and D. C. Aldridge. 2013. The "dirty dozen": socio-economic factors amplify the invasion potential of 12 high-risk aquatic invasive species in Great Britain and Ireland. Journal of Applied Ecology 50:757–766.

Gallardo, B., A. Zieritz, and D. C. Aldridge. 2015. The importance of the human footprint in shaping the global distribution of terrestrial, freshwater and marine invaders. PLOS (Public Library of Science) ONE [online serial] 10(5):e0125801.

Gallien, L., R. Douzet, S. Pratte, N. E. Zimmermann, and W. Thuiller. 2012. Invasive species distribution models: how violating the equilibrium assumption can create new insights. Global Ecology and Biogeography 21:1126–1136.

Gallien, L., T. Münkemüller, C. H. Albert, I. Boulangeat, and W. Thuiller. 2010. Predicting potential distributions of invasive species: where to go from here? Diversity and Distributions 16:331–342.

Guisan, A., and W. Thuiller. 2005. Predicting species distribution: offering more than simple habitat models. Ecology Letters 8:993–1009.

Hijmans, R. J., S. Phillips, J. Leathwick, and J. Elith. 2017. Dismo: species distribution modeling. R package version 1.0–12. Available: http://CRAN.R-project.org/package=dismo.

Hill, R. A., M. H. Weber, R. M. Debbout, S. G. Leibowitz, and A. R. Olsen. 2018. The lake-catchment (LakeCat) dataset: characterizing landscape features for lake basins within the conterminous USA. Freshwater Science 37:208–221.

Hrabik, T., J. Magnuson, and A. S. McLain. 1998. Predicting the effects of Rainbow Smelt on native fishes in small lakes: evidence from long-term research on two lakes. Canadian Journal of Fisheries and Aquatic Sciences 55:1364–1371.

Hrabik, T. R., and J. Magnuson. 1999. Simulated dispersal of exotic Rainbow Smelt (*Osmerus mordax*) in a northern Wisconsin lake district and implications for management. Canadian Journal of Fisheries and Aquatic Sciences 56(S1):35–42.

Jackson, D. A., and H. H. Harvey. 1989. Biogeographic associations in fish assemblages: local versus regional processes. Ecology 70:1472–1484.

Jacobson, P. C., G. J. A. Hansen, L. G. Olmanson, K. E. Wehrly, C. L. Hein, and L. B. Johnson. 2019. Loss of coldwater fish habitat in glaciated lakes of the Midwestern United States after a century of land use and climate change. Pages 141–157 *in* R. M. Hughes, D. Infante, L. Wang, K. Chen, and B. de F. Terra, editors. Advances in understanding landscape influences on freshwater habitats and biological assemblages. American Fisheries Society, Symposium 90, Bethesda, Maryland.

Kahle, D., and H. Wickham. 2013. ggmap: spatial visualization with ggplot2. The R Journal 5:144–161.

Kerr, S. J. 2006. An historical review of fish culture, stocking and fish transfers in Ontario, 1865–2004. Ontario Ministry of Natural Resources, Peterborough.

Leung, B., J. M. Bossenbroek, and D. M. Lodge. 2006. Boats, pathways, and aquatic biological invasions: estimating dispersal potential with gravity models. Biological Invasions 8:241–254.

Litvak, M. K., and N. E. Mandrak. 1993. Ecology of freshwater baitfish use in Canada and the United States. Fisheries 18(12):6–13.

Liu, C., M. White, and G. Newell. 2013. Selecting thresholds for the prediction of species occurrence with presence-only data. Journal of Biogeography 40:788–789.

Lobo, J. M., A. Jiménez-Valverde, and R. Real. 2008. AUC: a misleading measure of the performance of predictive distribution models. Global Ecology and Biogeography 17:145–151.

Lurgi, M., K. Wells, M. Kennedy, S. Campbell, and D. A. Fordham. 2016. A landscape approach to invasive species management. PLOS (Public Library of Science) ONE [online serial] 11(7):e0160417.

Lustig, A., S. P. Worner, J. P. W. Pitt, C. Doscher, D. B. Stouffer, and S. D. Senay. 2017. A modeling framework for the establishment and spread of invasive species in heterogeneous environments. Ecology and Evolution 7:8338–8348.

MacArthur, R. H., and E. O. Wilson. 1967. The theory of island biogeography. Princeton University Press, Princeton, New Jersey.

MacIsaac, H. J., L.-M. Herborg, and J. R. Muirhead. 2007. Modeling biological invasions of inland waters. Pages 347–368 *in* F. Gherardi, editor. Biological invaders in inland waters: profiles, distribution, and threats. Springer, New York.

Mandrak, N. E. 1995. Biogeographic patterns of fish species richness in Ontario lakes in relation to historical and environmental factors. Canadian Journal of Fisheries and Aquatic Sciences 52:1462–1474.

Martin, S. L., and P. A. Soranno. 2006. Lake landscape position: relationships to hydrologic connectivity and landscape features. Limnology and Oceanography 51:801–814.

McGarvey, D. J., M. Menon, T. Taylor, S. Tassone, J. Reese, M. Vergamini, and E. Kellogg. 2018. On the use of climate covariates in aquatic species distribution models: are we at risk of throwing out the baby with the bath water? Ecography 41:695–712.

Melles, S. J., C. Chu, K. M. Alofs, and D. A. Jackson. 2015. Potential spread of Great Lakes fishes given climate change and proposed dams: an approach using circuit theory to evaluate invasion risk. Landscape Ecology 30:919–935.

Mercado-Silva, N., J. D. Olden, J. T. Maxted, T. R. Hrabik, M. J. Vander Zanden, and N. P. Street. 2006. Forecasting the spread of invasive Rainbow Smelt in the Laurentian Great

Lakes region of North America. Conservation Biology 20:1740–1749.

Mercado-Silva, N., G. G. Sass, B. M. Roth, S. Gilbert, and M. J. Vander Zanden. 2007. Impact of Rainbow Smelt (*Osmerus mordax*) invasion on Walleye (*Sander vitreus*) recruitment in Wisconsin lakes. Canadian Journal of Fisheries and Aquatic Sciences 64:1543–1550.

Minns, C. K. 1989. Factors affecting fish species richness in Ontario lakes. Transactions of the American Fisheries Society 118:533–545.

Moore, C., J. C. Drazen, B. T. Radford, C. Kelley, and S. J. Newman. 2016. Improving essential fish habitat designation to support sustainable ecosystem-based fisheries management. Marine Policy 69:32–41.

Muirhead, J. R., and H. J. MacIsaac. 2011. Evaluation of stochastic gravity model selection for use in estimating non-indigenous species dispersal and establishment. Biological Invasions 13:2445–2458.

Naeem, S., J. M. H. Knops, D. Tilman, K. M. Howe, T. Kennedy, and S. Gale. 2000. Plant diversity increases resistance to invasion in the absence of covarying extrinsic factors. Oikos 91:97–108.

Omernik, J. M., G. E. Griffith, R.M. Hughes, M.H. Weber, and J. B. Glover. 2017. How misapplication of the hydrologic unit framework diminishes the meaning of watersheds. Environmental Management 60:1–11.

OMNRF (Ontario Ministry of Natural Resources and Forestry) 2015. User guide for watershed. Available: www.sse.gov.on.ca/sites/MNR-PublicDocs/EN/CMID/Watershed%20-%20User%20Guide.pdf. (March 2019).

Peterson, A. T., M. Papeş, and J. Soberón. 2008. Rethinking receiver operating characteristic analysis applications in ecological niche modeling. Ecological Modelling 213:63–72.

Porfirio, L. L., R. M. B. Harris, E. C. Lefroy, S. Hugh, S. F. Gould, G. Lee, N. L. Bindoff, and B. Mackey. 2014. Improving the use of species distribution models in conservation planning and management under climate change. PLOS (Public Library of Science) ONE [online serial] 9(11):e113749.

R Core Team. 2017. R: a language and environment for statistical computing. R Foundation for Statistical Computing, Vienna, Austria.

Ricciardi, A., and H. J. MacIsaac. 2011. Impacts of biological invasions on freshwater ecosystems. Pages 211–224 *in* D. M. Richardson, editor. Fifty years of invasion ecology: the legacy of Charles Elton. Blackwell Publishing, Hoboken, New Jersey.

Romero-Alvarez, D., L. E. Escobar, S. Varela, D. J. Larkin, and N. B. D. Phelps. 2017. Forecasting distributions of an aquatic invasive species (*Nitellopsis obtusa*) under future climate scenarios. PLOS (Public Library of Science) ONE [online serial] 12(7):e0180930.

Rose, P. M., M. J. Kennard, D. B. Moffatt, F. Sheldon, and G. L. Butler. 2016. Testing three species distribution modelling strategies to define fish assemblage reference conditions for stream bioassessment and related applications. PLOS (Public Library of Science) ONE [online serial] 11(1): e0146728.

Sandstrom, S., M. Rawson, and N. Lester. 2013. Manual of instructions for broad-scale fish community monitoring using North American (NA1) and Ontario small mesh (ON2) gillnets. Queen's Printer for Ontario, Peterborough, Ontario.

Sharma, S., M. J. Vander Zanden, J. J. Magnuson, and J. Lyons. 2011. Comparing climate change and species invasions as drivers of coldwater fish population extirpations. PLOS (Public Library of Science) ONE [online serial] 6(8):e22906.

Shea, K., and P. Chesson. 2002. Community ecology theory as a framework for biological invasions. Trends in Ecology and Evolution 17(4):170–176.

Soranno, P. A., L. C. Bacon, M. Beauchene, K. E. Bednar, E. G. Bissell, C. K. Boudreau, M. G. Boyer, M. T. Bremigan, S. R. Carpenter, J.

W. Carr, K. S. Cheruvelil, S. T. Christel, M. Claucherty, S. M. Collins, J. D. Conroy, J. A. Downing, J. Dukett, C. E. Fergus, C. T. Filstrup, C. Funk, M. J. Gonzalez, L. T. Green, C. Gries, J. D. Halfman, S. K. Hamilton, P. C. Hanson, E. N. Henry, E. M. Herron, C. Hockings, J. R. Jackson, K. Jacobson-Hedin, L. L. Janus, W. W. Jones, J. R. Jones, C. M. Keson, K. B. S. King, S. A. Kishbaugh, J. F. Lapierre, B. Lathrop, J. A. Latimore, Y. Lee, N. R. Lottig, J. A. Lynch, L. J. Matthews, W. H. McDowell, K. E. B. Moore, B. P. Neff, S. J. Nelson, S. K. Oliver, M. L. Pace, D. C. Pierson, A. C. Poisson, A. I. Pollard, D. M. Post, P. O. Reyes, D. O. Rosenberry, K. M. Roy, L. G. Rudstam, O. Sarnelle, N. J. Schuldt, C. E. Scott, N. K. Skaff, N. J. Smith, N. R. Spinelli, J. J. Stachelek, E. H. Stanley, J. L. Stoddard, S. B. Stopyak, C. A. Stow, J. M. Tallant, P. N. Tan, A. P. Thorpe, M. J. Vanni, T. Wagner, G. Watkins, K. C. Weathers, K. E. Webster, J. D. White, M. K. Wilmes, and S. Yuan. 2017. LAGOS-NE: a multi-scaled geospatial and temporal database of lake ecological context and water quality for thousands of US lakes. GigaScience [online serial] 6(12):giw014.

Stohlgren, T. J., D. T. Barnett, and J. T. Kartesz. 2003. The rich get richer: patterns of plant invasions in the United States. Frontiers in Ecology and the Environment 1(1):11–14.

Swets, K. 1988. Measuring the accuracy of diagnostic systems. Science 240:1285–1293.

Thuiller, W. 2003. BIOMOD—optimizing predictions of species distributions and projecting potential future shifts under global change. Global Change Biology 9:1353–1362.

Tonn, W. M., and J. J. Magnuson. 1982. Patterns in the species composition and richness of fish assemblages in northern Wisconsin lakes. Ecology 63:1149–1166.

Tonn, W. M., J. J. Magnuson, M. Rask, and J. Toivonen. 1990. Intercontinental comparison of small-lake fish assemblages: the balance between local and regional processes. The American Naturalist 136:345–375.

Turner, M. G., and R. H. Gardner. 2015. Landscape ecology in theory and practice. Springer, New York.

Vander Zanden, M. J., and J. D. Olden. 2008. A management framework for preventing the secondary spread of aquatic invasive species. Canadian Journal of Fisheries and Aquatic Sciences 65:1512–1522.

Verpoorter, C., T. Kutser, D. A. Seekell, and L. J. Tranvik. 2014. A global inventory of lakes based on high-resolution satellite imagery. Geophysical Research Letters 41:6396–6402.

Wang, L., and D. L. Jackson. 2014. Shaping up model transferability and generality of species distribution modeling for predicting invasions: implications from a study on Bythotrephes longimanus. Biological Invasions 16:2079–2103.

Zimmermann, N. E., T. C. Edwards, C. H. Graham, P. B. Pearman, and J. C. Svenning. 2010. New trends in species distribution modelling. Ecography 33:985–989.

American Fisheries Society Symposium 90:125–139, 2019
© 2019 by the American Fisheries Society

Ecologically Relevant Structural Metrics for Complex Stream–Lake Networks

Meghan L. Allerton*
Environmental and Life Sciences Graduate Program, Trent University
1600 West Bank Drive, Peterborough, Ontario, K9J 7B8, Canada

Nicholas E. Jones
River and Stream Ecology Laboratory, Aquatic Research and Monitoring Section
Ontario Ministry of Natural Resources and Forestry, Trent University
DNA Building, 2140 East Bank Drive, Peterborough, Ontario, K9J 7B8, Canada

Stephanie J. Melles
Chemistry and Biology Department, Ryerson University, Ryerson Urban Water
350 Victoria Street, Toronto, Ontario, M5B 2K3, Canada

Bastian J. Schmidt
River and Stream Ecology Laboratory, Aquatic Research and Monitoring Section
Ontario Ministry of Natural Resources and Forestry, Trent University
DNA Building, 2140 East Bank Drive, Peterborough, Ontario, K9J 7B8, Canada

Abstract.—Lakes are common features on the landscape in many regions of the world and have important impacts on the ecological function and habitat conditions of connected rivers and streams. Yet lakes are rarely incorporated in studies of river and stream network structure. Using stream and lake spatial data of catchments in Ontario, metrics can be developed that quantify the spatial structure of complex stream–lake networks. This case study represents the initial step in a larger project to develop a complete, parsimonious set of metrics that would make the link between stream–lake network structure and aquatic ecosystem function at the landscape scale. We present three new metrics that capture essential information about the spatial distribution and size of lakes in aquatic networks. These metrics can be used as descriptive measures to understand differences among catchments and as explanatory variables in predictive modeling. Developing stream–lake network measures is an important first step towards integrating lakes and streams for aquatic assessments, improving our ability to effectively manage and conserve aquatic ecosystems in many parts of the world.

Introduction

Stream–lake networks define the aquatic landscape of regions where lakes are abundant. This includes large areas of Russia, Sweden, Finland, and Norway, as well as the Canadian Shield region of North America, which spans an area of about 4.6 million km^2 (Jones et al. 2014). Despite their broad extent, these lake-network systems are understudied.

* Corresponding author: meghanallerton@trentu.ca

This volume and the preceding one (Hughes et al. 2006) are evidence that significant progress has been made in understanding landscape-scale influences on both stream and lake ecosystems in recent decades. However, little attention has been given to streams and lakes as integrated networks, resulting in a lack of "system understanding" (Wurtsbaugh et al. 2015).

Individual lakes have moderating effects on outflowing streams (lake outlets), and these effects attenuate downstream, altering longitudinal gradients from headwaters to outlet and modifying the ecological conditions described by the river continuum concept (Vannote et al. 1980). Lakes increase the hydrologic, wood, and sediment storage of aquatic networks, altering flow and substrate regimes that define the physical structure of streams (e.g., bed material composition and channel profile; Wohl et al. 2019). Diel fluctuations in water temperature are reduced, and flow variability is moderated by the presence of upstream lakes, resulting in more stable conditions (Jones 2010). In addition, seasonal export of warm lake-surface water and increased nutrient concentrations, along with the delivery of organic matter, phytoplankton, and zooplankton contributes to greater productivity in lake outflows, as is evident by the high density of benthic invertebrates typically observed at these areas (Richardson and Mackay 1991; Jones 2010; Baker et al. 2016). These effects are temporally variable (e.g., Leach et al. 2017) and are observed at varying magnitude and spatial extent below individual lakes (Ellis and Jones 2013). For example, seston flowing into outlet streams is typically removed from the water rapidly by filter-feeding insects (Richardson and Mackay 1991), whereas water temperature may take several kilometers to reach an equilibrium temperature. The relative influence of lakes also depends on where they are positioned in a network. Headwater streams are likely more heavily influenced by lakes than large rivers, as conditions in large rivers are already more similar to lake conditions.

Lake effects can be viewed within the context of a stream network or drainage area. Depending on the location of a drainage "pour point," which defines the delineation of a drainage area, there may be several lakes located within an upstream catchment. We refer to this as the "network scale" as it includes the land area contributing to a stream along with the set of interconnected surface water bodies and watercourses. Consistent with other chapters in this volume, we use "catchment" to refer to relatively small drainages and "basin" for larger drainages. The network scale spans catchment and basin scales.

The combined influence of multiple lakes makes the ecology of stream–lake networks distinct at the landscape level. Each lake changes the ecology of the stream (e.g., flow, temperature, sediment, wood, and nutrients), and these changes cumulatively combine across stream networks. Collectively, lakes can alter the physical and chemical characteristics of entire stream–lake networks and, in turn, influence the abundance and distribution of biota and their life history. There is evidence, for example, that fish species such as Brook Trout *Salvelinus fontinalis* and Arctic Grayling *Thymallus arcticus* are highly mobile within stream–lake systems, using lentic and lotic habitats at different times and for different purposes, including foraging, spawning, and migration (Addison and Wilson 2010; Heim et al. 2018). Greater overall habitat complexity in stream–lake networks may enhance the resiliency of fish populations by supporting a range of age-classes and behavioral traits (Heim et al. 2018) and by providing refuges that allow for recolonization of the network following disturbance (Pépino et al. 2016). In general, the diversity of habitats in stream–lake networks may be expected to support more diverse communities than would be found in systems without lakes (Jones 2010). Jones et al. (2014) concluded that lake size and position within the network must receive more attention before predictive models of flow regime can be developed.

Such findings suggest clear ecological differences between stream networks with and without connected lakes. However, our ability to explain and predict these differences is limited because of poor understanding of how the unique structure of each stream–lake system influences its ecology, including life history characteristics, community composition and resilience, and recovery from disturbance. We suggest that understanding and quantifying stream–lake network structure is essential to understanding ecosystem function and will provide necessary tools for the effective management and conservation of these systems.

A growing body of research aims to quantify the structure of stream networks in order to relate physical patterns of stream habitat occurrence and biodiversity. For example, the network dynamics hypothesis (Benda et al. 2004) provided a framework for predicting how network structure interacts with disturbance to produce spatiotemporal heterogeneity of habitat types. Metrics of network structure included basin size and shape, confluence symmetry ratio, drainage density, confluence density, network pattern, and local network geometry (Benda et al. 2004). Recent advances have been made through the application of network theory and related metrics to quantify the structure of river networks, with an emphasis on relating measures of network/habitat connectivity to population dynamics and dispersal (e.g., Van Looy et al. 2014; Chaput-Bardy et al. 2017; Sarhad et al. 2017), biodiversity (Altermatt 2013; Carrara et al. 2014; Tornwall et al. 2017), and ecological impacts of impoundments (e.g., Cote et al. 2009).

Network-scale lake effects are influenced both by stream network structure (e.g., tributaries can dilute lake effects) and drainage-area characteristics such as slope and riparian cover (Jones 2010). Therefore, the development of an ecologically relevant set of stream–lake network measures would build on previous studies to include metrics of three categories: (1) lake characteristics; (2) stream network pattern, configuration, and connectivity; and (3) catchment characteristics (see Table 1). Quantifying relevant lake attributes through the development of network-scale metrics of lake influence is the subject of this chapter.

Because lake effects at the network scale represent the combined influence of all connected lakes, research on the downstream influence of individual lakes and man-made reservoirs provides insight into the relevant measures of lake characteristics. Notably, Ward and Stanford (1983) recognized that downstream impacts of reservoirs depend on position within a catchment (i.e., headwaters or lower reaches). Similarly, both the magnitude and downstream extent of effects of natural lakes are related to position: as stream size increases from headwaters to river mouth, lake influence theoretically decreases, whereas the distance of attenuation (e.g., biotic community changes, physical changes) increases (Jones 2010). However, lake influence on flow regime is also a function of lake size relative to stream size (Spence 2006), and this relative relationship likely affects distance of attenuation for other physical and biological changes (Jones 2010).

Metrics of stream–lake network structure must therefore include measures that summarize information about the position and size of each lake in the network. When multiple connected lakes are present, the extent of lake influence through a network also depends on the number of lakes and distances between lakes. If the downstream zone of influence

Table 1. Classes of stream–lake network metrics.

Type of metric	Description	Examples
Measures of lake characteristics	Network-scale metrics that capture or summarize information about lake size, shape, number, position, and other related attributes	Percent cover of connected lakes; density of connected lakes; percent of river kilometers flowing through lakes (Jones 2010); lake count balance point; lake area balance point
Measures of network pattern, configuration, and connectivity	Network-scale metrics of the physical structure of the stream network, including measures developed based on graph theory approaches	Drainage density; confluence density (Benda et al. 2004); dendritic connectivity index (Cote et al. 2009); integral index of connectivity (Pascual-Hortal and Saura 2006); probability of connectivity (Chaput-Bardy et al. 2017)
Measures of drainage basin/catchment characteristics	Metrics of drainage area size and shape, land cover, geology, topography, and climate	Drainage basin area; basin modified form factor (Benda et al. 2004); skewness index (Mineau et al. 2015); land cover and climate indices

is less than the distance between lakes, then the entire stream–lake network will show a greater diversity of habitat types: stream segments influenced by lakes and those that are not (Jones 2010).

Ideally, a set of metrics would provide sufficient information to allow one to reconstruct the relevant physical characteristics of a system. In addition to basic catchment and network attributes such as drainage area size and shape and drainage density, simple metrics of lake number and surface cover serve as a starting point. More information is needed, however, to describe where lake area is distributed and how lake effects interact with network configuration and catchment attributes to determine physical habitat characteristics and biological responses.

This is particularly clear in the context of modeling and classifying flow regimes in lake-rich catchments (Jones et al. 2014). Flow regime is a defining feature of river systems, and it can be altered dramatically by the presence of lakes. By increasing the hydrologic storage of watersheds, lakes act as buffers of hydrologic variability by providing base flow and dissipating flood pulses that result from heavy precipitation (Jones et al. 2014; Woo et al., in press). In Canadian Shield systems, the abundance of lakes creates flows that are less variable and more predictable than systems in neighboring regions (Jones et al. 2014). Broadscale flow regime classifications based on empirical data may capture this effect, but where data are lacking, flow regimes cannot be accurately modeled using deductive approaches without incorporating the effect of lakes in the upstream catchment area. Furthermore, simple metrics of lake influence (e.g., lake area as a percentage of catchment area, which has been used in catchment characterizations or in linking catchment attributes to hydrology; e.g., Detenbeck et al. 2005) do not provide enough information to explain or accurately predict flow and other key processes in lake-rich catchments. A set of metrics that captures all relevant aspects of stream–lake network structure will fill this gap and greatly enhance predictive modeling capabilities for these systems.

This chapter presents the results of a preliminary analysis focused on the development of new stream–lake network metrics that capture information about the number, size, and position of lakes. Metrics are calculated for a sample drainage area, the Petawawa River basin in Ontario and several of its tributary catchments, to explore the usefulness/effectiveness of these new metrics in capturing essential information about network structure across landscape scales. The work represents the foundation for a larger project to develop a complete, parsimonious set of stream–lake and catchment-wide metrics that will quantify the relevant structural elements of these systems, including lake attributes, network configuration and connectivity, and drainage basin characteristics.

Methods

Study Area

The Petawawa River is located in eastern Ontario, roughly 250 km northeast of Toronto, and drains into the Ottawa River, which forms the border between the provinces of Ontario and Quebec (outlet at 45.909° N, 77.259° W). Situated on the Canadian Shield, the course of the river and its tributaries is largely dictated by geology and shows a "deranged" drainage pattern (Trenhaile 2013) with hundreds of connected lakes, making it an ideal study area. Almost 90% of the 4,150 km^2 drainage basin is within Algonquin Provincial Park. Although there are legacy impacts of logging operations of the past two centuries, the landscape remains relatively undisturbed and reforested, and is mainly used for recreational purposes. The Petawawa basin and seven tributary catchments are illustrated in Figure 1.

Data

Stream and lake spatial data layers were produced by the Ontario Ministry of Natural Resources and Forestry and originally accessed through Land Information Ontario. Lake layers were produced from the Ontario Hydrographic Network - Waterbody data set, downloaded January 2014 and processed in ArcMap. A stream network layer was produced using Arc Hydro and source data from the Ontario Integrated Hydrology package, which includes raster and vector (watercourse) data. Several stages of data processing were undertaken to preprocess stream and lake layers, integrate lakes into the stream network, and assign Strahler and Shreve orders to stream reaches. Petawawa River and tributary drainage areas were delineated using the Ontario Flow Assessment Tool (OMNRF 2017), which is provided by the Ontario Ministry of Natural Resources and Forestry and uses the Ontario Integrated Hydrology spatial data. Surface area of water bodies in the Petawawa basin ranges from 5.5×10^{-5} to 27 km^2.

Analysis

Using ArcMap and R, and the spatial data layers described above, standard summary metrics were calculated for the Petawawa River basin and its catchments (Table 2). The same data layers were then used to calculate three new metrics of lake characteristics: lake count and area balance points and the count-area balance point ratio (lake count balance point divided by lake area balance point). Balance points are the locations on the main-stem path at which half of the lakes (lake count balance point) or half of the lake area (lake area balance point) is positioned upstream, and half downstream. These points were quantified as percent of main-stem length, where 0%

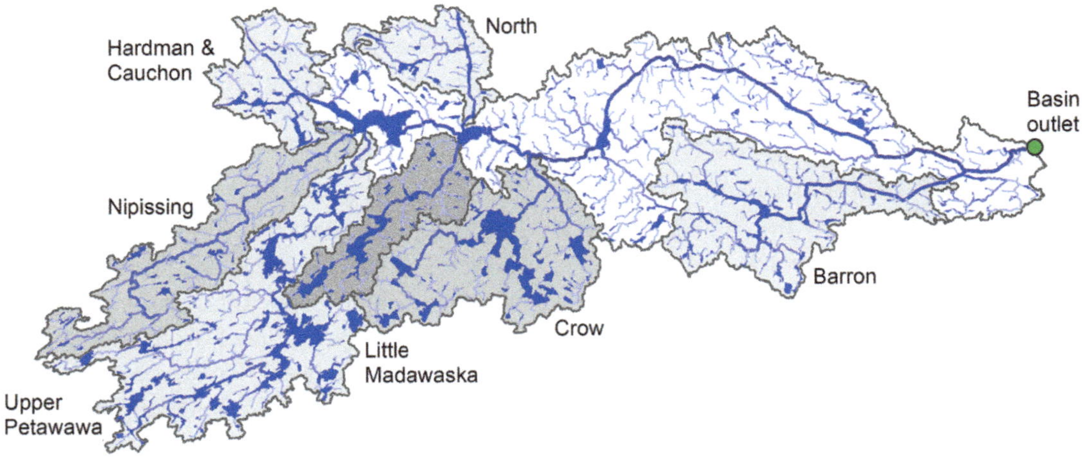

Figure 1. Petawawa River basin and seven catchments.

is the headwater source and 100% is the river mouth. The main stem of the Petawawa River and each of its tributaries was defined as the longest flow path. The total area of upstream lakes and count of upstream lakes were calculated for each stream segment within the main stem of each catchment and used to calculate balance point metrics. Stream segments were defined as sections of stream between two confluences or where streams intersect lakes. Lake orders were assigned to all connected lakes based on the Strahler order of the outflowing stream. The percentages of the total number of lakes and total area of

Table 2. Standard summary metrics. Catchments: 1—Crow, 2—upper Petawawa, 3—Nipissing, 4—Hardman and Cauchon, 5—Little Madawaska, 6—North, and 7—Barron.

		Catchments						
	Petawawa	1	2	3	4	5	6	7
Drainage basin area (km^2)	4,150	502	741	411	202	245	174	505
Main-stem length (km)	232.5	65.4	102.1	99.8	24.4	45.2	38.0	66.5
Basin modified form factor	0.08	0.12	0.07	0.04	0.34	0.12	0.12	0.11
Max Strahler order	6	5	4	4	5	4	4	5
Max Shreve order	1,071	127	204	123	57	69	47	124
Total number of lakes	3,972	433	674	388	174	180	273	502
Number of connected lakes	1,307	150	182	102	57	60	84	206
Percent of lakes connected	32.9	34.6	27.0	26.3	32.8	33.3	30.8	41.0
Density of connected lakes (count/km^2)	0.3	0.3	0.2	0.2	0.3	0.2	0.5	0.4
Total area of connected lakes (km^2)	329	64	100	11	15	27	10	25
Percent cover of connected lakes	7.9	12.8	13.6	2.6	7.4	11.0	5.5	5.0
Total length of channels (km)	2,280	225	328	312	95	121	100	270
Drainage density (m/km^2)	549	449	442	758	471	495	574	536

lakes in the Petawawa basin that belong to each lake order were then determined.

Results

Standard summary metrics for the Petawawa basin and its catchments are displayed in Table 2. Basin and network metrics provide basic information about the size and shape of the drainage basin and size and density of the stream network. Other metrics could be added, but these are a starting point in reconstructing the network. Simple lake metrics add information about the surface cover, number, and density of lakes. With this information, we pictured the Petawawa as a medium-sized system with an elongated basin, moderate drainage density, and high density of connected lakes, which cover ~8% of the basin area. The catchment results reveal variation in these characteristics between catchments. For example, surface cover of connected lakes ranges from 2.6% in the Nipissing River catchment to 13.6% in the adjacent upper Petawawa catchment. Because these two catchments have the same lake density, lakes in the upper Petawawa also must be larger on average. Whereas these simple lake metrics facilitate basic comparisons between catchments, there is little we can glean about the environment and ecology of each catchment without further information on where the lake area is distributed.

Figure 2 illustrates how total area and count of upstream (connected) lakes increase with distance along the main-stem path from headwaters to mouth for the Petawawa basin and two of its catchments. Thresholds are readily apparent where lake area or count increase markedly, and these jumps represent positions in the basin or catchment where large tributaries join the main stem. Thresholds in lake area may also represent large lakes located on the main-stem path. Visual inspection of area plots reveals how lakes are positioned in each catchment, and comparison with count plots provides insight into whether increases in area are the result of a few large lakes or many small lakes. For example, most of the lake area in the Crow River catchment is positioned in the lower segments, and the large increase in area at ~52 km along the main stem is the result of a small number of large lakes.

Lake count and area balance points represent one way of quantifying this type of information in catchment-scale metrics. Results of balance point calculations are shown in Table 3. The lake area balance point adds to our picture of each catchment by providing essential information about the distribution of lake area. For example, in the Crow River catchment, lakes cover ~13% of the catchment, and half of that area is positioned on the lower 20% of the main stem and its tributaries.

The relative positions of the lake count and lake area balance points, expressed as their ratio (Table 3), provide further information about the distribution and position of lake sizes. When the ratio is 1, the balance points coincide and half the lake area is divided among half the lakes in both upstream and downstream directions. A ratio greater than 1 indicates that the lake count balance point is downstream of the lake area balance point, and therefore, upstream lakes are larger on average. A ratio of less than 1 occurs when the lake count balance point is upstream of the lake area balance point and downstream lakes are larger on average. The Crow River has a ratio of 0.91, which indicates that lakes on the lower 20% of the drainage system are larger on average than those on the upper 80%. In contrast, the Little Madawaska River has a ratio of 1.33, indicating that the average size of lakes in the upper portion of the catchment is greater than the lower portion. The upstream lake area balance point for the Little Madawaska River is 49%, indicating that lake area is well balanced within the catchment (along the main stem). Although the percent

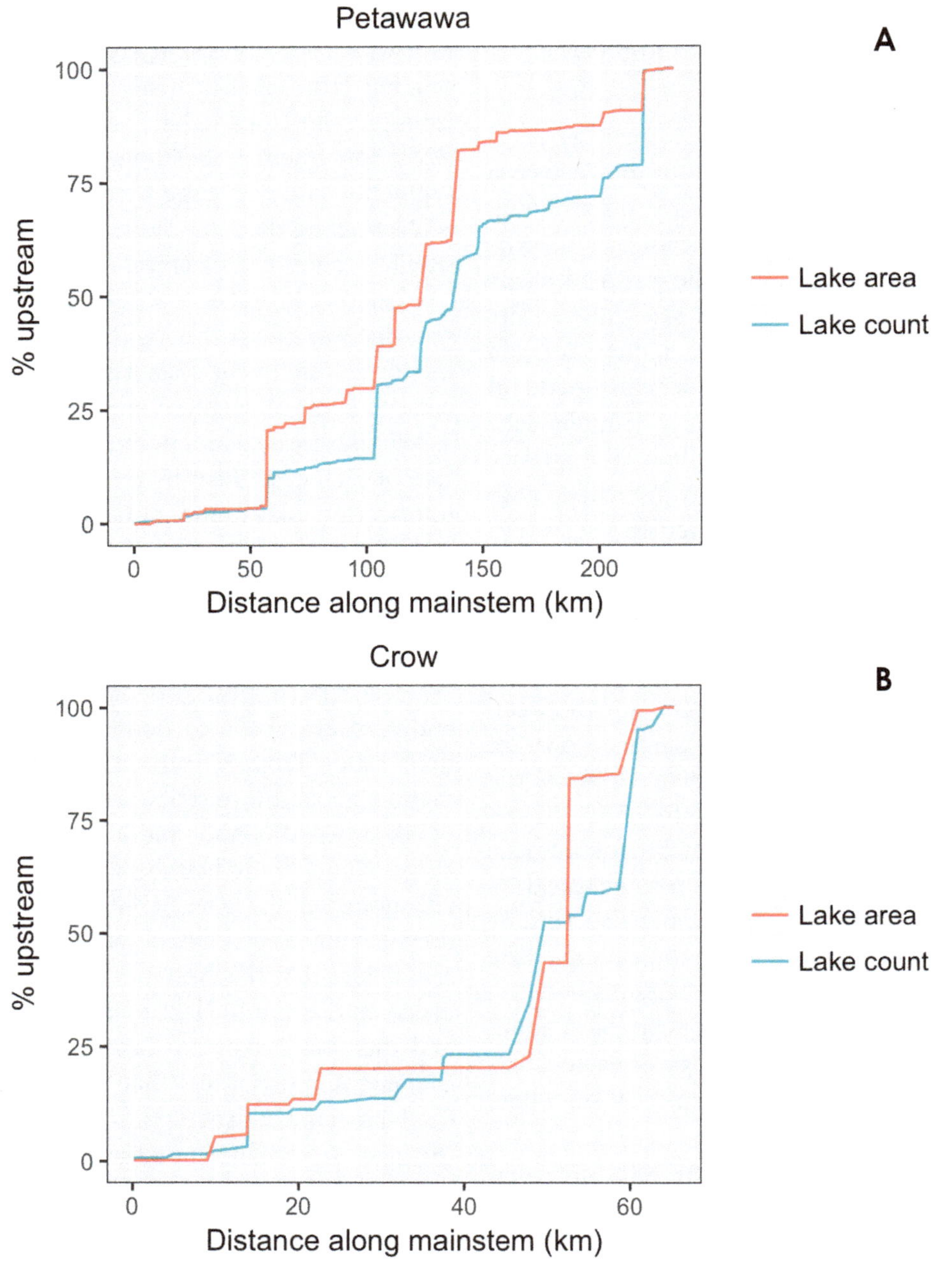

Figure 2. Percent of total (connected) lake area and count located upstream with distance along the main-stem path from headwaters to mouth for **(A)** the entire Petawawa River basin, **(B)** the Crow River catchment, and **(C)** the Little Madawaska catchment.

cover of lakes in these two catchments (Crow and Little Madawaska) may have suggested greater overall lake influence in the Crow River, this additional information indicates that the opposite is likely true.

Discussion

At regional extents, factors such as geology and glacial history determine the general pattern of natural lakes (Jeffries and Mills 1990), and simple metrics like percent cover of lakes

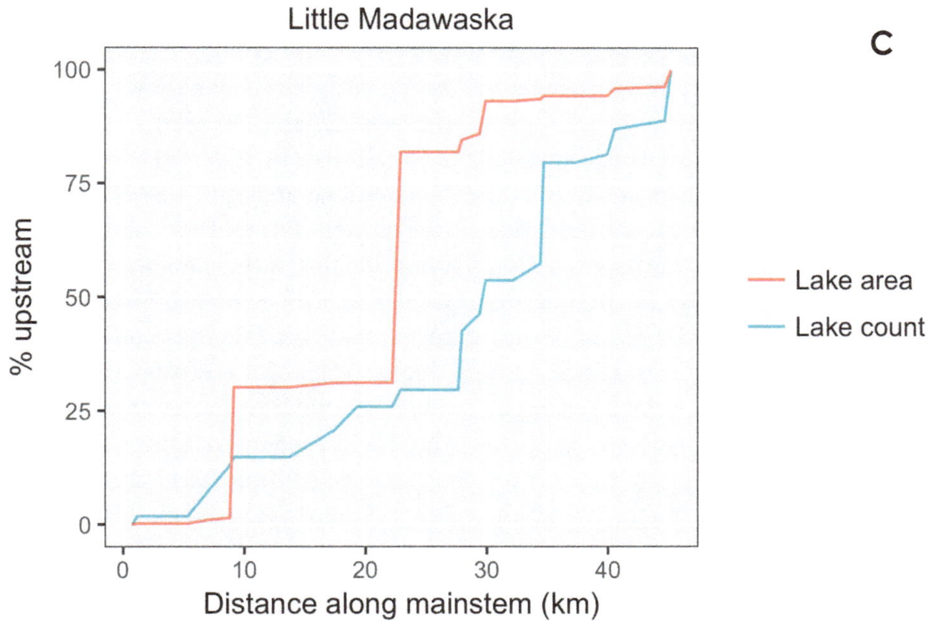

Figure 2. Continued.

capture regional differences. However, our results show that there is significant variation among catchments, both in percent cover of lakes and in distribution of lake area. On the Canadian Shield, the surficial geology is such that water flow and lake formation appears deranged or random (Trenhaile 2013). Standard summary metrics are not adequate to capture this variation or to assess the relative degree of lake influence at the catchment scale, which was identified as the most appropriate scale for linking landscape characteristics to hydrology (McManamay et al. 2012) and for the managing species that show high levels of movement within catchments, such as Brook Trout (Addison and Wilson 2010).

This case study shows that lake count and area balance points (and their ratio) give us a clearer picture of the structure of the Petawawa system. Total area of connected lakes is distributed almost equally between the upper and lower halves of the entire Petawawa network, and average lake size is greater in the upper portion of the basin. Catchment comparisons show that our selected metrics capture local differences between catchments

Table 3. Mainstem balance points. Catchments: 1—Crow, 2—upper Petawawa, 3—Nipissing, 4—Hardman and Cauchon, 5—Little Madawaska, 6—North, and 7—Barron.

		Catchments						
	Petawawa	1	2	3	4	5	6	7
Lake count balance point (% of main-stem length)	58.7	73.2	55.6	35.7	81.3	65.1	57.6	48.6
Lake area balance point (% of main-stem length)	52.9	80.1	55.6	40.1	94.1	49.1	57.6	48.6
Count-area balance point ratio	1.11	0.91	1.00	0.89	0.86	1.32	1.00	1.00

and provide essential information about the spatial distribution of lake area that is missing from standard summary metrics such as percent cover of lakes.

We recognize that although our metrics begin to paint a picture of how lakes may modify stream characteristics, the depiction is far from complete. Furthermore, metric usefulness is likely related to catchment size because as catchment size increases, so too does the range of stream sizes to which lakes may be connected. For example, lakes in the upper portion of the Petawawa basin are connected to streams ranging from Strahler order one through five, so at the extent of the Petawawa basin, more information is clearly needed to describe lake size to stream size relationships.

Another approach would make use of lake order to summarize information about the position of lakes. Lake order can be used to group lakes based on approximate size of the connected stream, which is correlated with hydrologic position (Martin and Soranno 2006). Percentages of total lake area and total count of lakes within each lake order for the entire Petawawa basin are shown in Figure 3. This additional information about how total area and count of lakes is divided between each lake order allows us to determine the average size of lakes of each order. For example, fourth-order lakes account for 33% of total lake area and just 3% of lake count, indicating that in this basin, fourth-order lakes are few, large, and well connected. Although stream order is an imperfect stand-in for stream size (Hughes and Omernik 1983; Hughes et al. 2011), this information gives us an idea of the lake size to stream size relationship within the basin. Other lake characteristics can be summarized based on lake order; however, further research is needed to develop network-scale metrics that capture this information and would have greater usefulness in modeling applications that link network structure to ecological function.

Next Steps

There is a need to develop and test ecologically relevant metrics of the structure of stream–lake networks at multiple scales (subcatchment, catchment, basin, etc.). The data presented here are the result of a preliminary exploration meant to highlight the types of

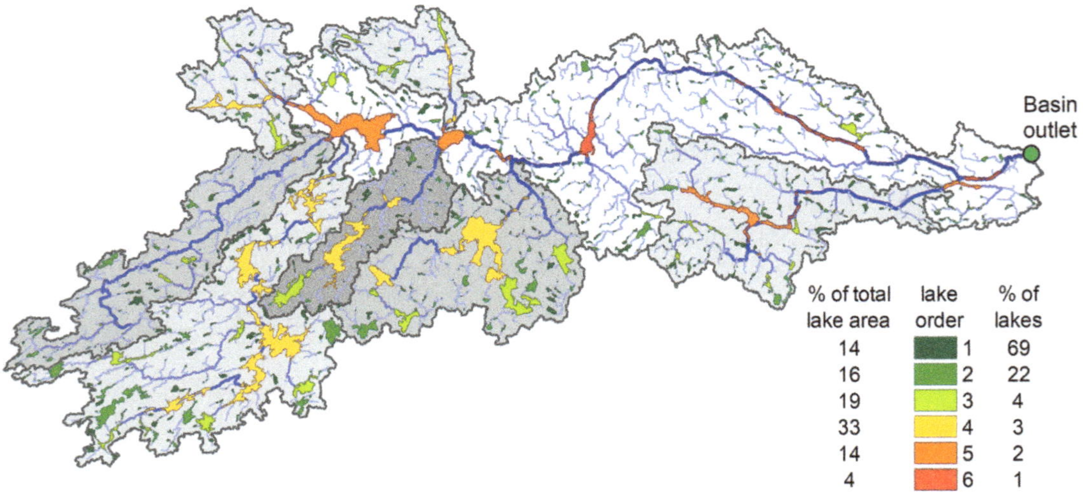

Figure 3. Percent of total lake area and percent of total count of lakes within each lake order for the entire Petawawa River basin.

information needed, some of the challenges in their development, and considerations going forward. The next step will be generation of a larger set of metrics that capture structural aspects, including the number, size, and position of lakes, as well as lake shape and volume, inlet–outlet configuration, lake-to-lake distances, and network configuration. Concurrent research on the downstream distance of attenuation of individual lake effects also may allow for the development of a cumulative lake-effect measure (e.g., percent of system influenced by lakes; see Jones 2010) derived from the physical and biological zones of influence of each lake in a network. The complete set of metrics will be computed for a large number of catchments and drainage basins in Ontario, representing a diversity of sizes and characteristics. It will then be possible to analyze the actual and theoretical range and distribution of metric values to determine their usefulness in clearly distinguishing a variety of network patterns. Principal component analysis will be used to examine intercorrelation among metrics and minimize redundancy (Olden and Poff 2003). Although we focus our analysis within Ontario, we anticipate results that will be widely applicable across the Canadian Shield region and elsewhere.

The outcome of this work will be a standard set of metrics for characterizing stream–lake network pattern, which will also serve as predictors in landscape-based (deductive) hydrology models (see Olden et al. 2011) and other modeling applications that link ecosystem structure and function in these systems. When combined with measures of relevant catchment attributes, including climate, topography, groundwater input, land cover, and riparian cover, these metrics will be the basis for more accurate predictive modeling of essential components of the physical habitat of these systems, particularly flow, temperature, and sediment regimes. Defining relevant metrics depends on understanding the important relationships between these processes, network structure, and catchment characteristics.

Applications for Conservation and Management

The ecology of river systems is strongly influenced by the spatial and temporal patterns of flow and temperature. Environmental conditions of lake outlets are predictable: typically, they have a more stable flow regime, warmer water temperatures in summer and fall, and large amounts of seston fueling filter-feeding insects and higher productivity. Overall, the conditions are more benign (or less harsh) in stream–lake networks than in lakeless systems (Jones et al. 2014). The pattern of these conditions varies over space and time, and this physical habitat template largely determines the ecology of streams, including the relative importance of biotic and abiotic interactions, dominant life-history characteristics observed, and the resiliency and response of species and communities to disturbance (Poff and Ward 1990). Accordingly, classification of flow regimes, temperature regimes, and other physical conditions can be used to establish class-ecology relationships (Melles et al. 2014). For example, flow–ecology relationships are developed based on relationships between fish traits and hydrologic metrics (Bruckerhoff et al. 2019), and fish and macroinvertebrate assemblage composition and modeled flow metrics (Patrick and Yuan 2017).

The presence and spatial pattern of network lakes also can be directly related to the distribution and abundance of aquatic species. For example, Pépino et al. (2016) found that the relative abundance of Brook Trout in headwater streams increases with the presence of lakes. Furthermore, organisms like amphibians and invertebrates may move between structurally isolated lakes and stream–lake networks, and isolated lakes can support landscape-level biodiversity (Scheffer et al. 2006). Quantifying the spatial configuration and connectivity of stream and lake habitats will improve our ability to model landscape-level patterns of biodiversity, as well as the distribution and abundance of species of interest. In addition to species of high economic or conservation value, this may include invasive species, which are often introduced into lakes and whose rate of dispersal can depend on network connections with other lakes (Lamberti et al. 2010). Knowledge of network connectivity can be used to assess vulnerability to invasive species and target efforts to control their spread (Melles et al. 2015; Perry et al. 2017).

A province-wide aquatic ecosystem classification is currently being developed for Ontario's streams and rivers (Melles et al. 2014), and given the abundance of lakes in this jurisdiction, we expect that lakes will play a significant role in shaping the classification. Relevant metrics of upstream lake characteristics computed for every stream segment can be used along with segment-scale attributes to group contiguous segments into ecologically similar stream units, and then to cluster these units into classes (Melles et al. 2014). The integration of lakes is expected to significantly improve the accuracy of the aquatic ecosystem classification, enhancing our capacity to predict habitat characteristics and ecosystem responses to disturbance in areas where local data are lacking. Stream–lake integration will enable more appropriate selection of representative reference sites for monitoring programs and stream ecosystem research (Melles et al. 2014), and sustained monitoring of representative sites will be key to the effective management of these ecosystems.

Given the many threats to the long-term ecological integrity and health of freshwater systems, conservation planning is more important now than ever. Effective policies and strategies for the design and management of conservation areas need to incorporate measures that ensure both the representation and persistence of biodiversity (Nel et al. 2010). The applications mentioned above contribute to both. Flow classes can be used as biodiversity surrogates (Olden et al. 2011), especially where local data are lacking, aiding in the selection of representative communities targeted for conservation efforts. Persistence of biodiversity is incorporated in conservation planning by allowing for adequate rates of dispersal and access to critical habitats required by a species, which is accomplished through the maintenance of habitat connectivity (Nel et al. 2010). This includes not only longitudinal connectivity through a network, but also lateral connectivity between the network and other aquatic habitats, including lakes and wetlands (Hermoso et al. 2012). Understanding and quantifying the spatial pattern of lakes is therefore essential to catchment-scale conservation planning in lake-rich regions.

Acknowledgments

The authors thank Jim Buttle for his input in the initial stages of this project and Wayne Robinson, Lizhu Wang, and an anonymous reviewer for their helpful comments. We acknowledge the support of both Trent University and the Ontario Ministry of Natural Resources and Forestry.

References

Addison, P. A., and C. C. Wilson. 2010. Matching management to biological scale: connectivity among lacustrine Brook Trout populations. North American Journal of Fisheries Management 30:1132–1142.

Altermatt, F. 2013. Diversity in riverine metacommunities: a network perspective. Aquatic Ecology 47:365–377.

Baker, M. A., C. D. Arp, K. J. Goodman, A. M. Marcarelli, and W. A. Wurtsbaugh. 2016. Stream–lake interaction: understanding coupled hydro-ecological systems. Pages 321–348 *in* J. B. Jones and E. H. Stanley, editors. Stream ecosystems in a changing environment. Academic Press, London.

Benda, L., N. L. Poff, D. Miller, T. Dunne, G. Reeves, G. Pess, and M. Pollock. 2004. The network dynamics hypothesis: how channel networks structure riverine habitats. BioScience 54:413–427.

Bruckerhoff, L. A., D. R. Leasure, and D. D. Magoulick. 2019. Flow–ecology relationships are spatially structured and differ among flow regimes. Journal of Applied Ecology 56:398–412.

Carrara, F., A. Rinaldo, A. Giometto, and F. Altermatt. 2014. Complex interaction of dendritic connectivity and hierarchical patch size on biodiversity in river-like landscapes. The American Naturalist 183:13–25.

Chaput-Bardy, A., N. Alcala, J. Secondi, and S. Vuilleumier. 2017. Network analysis for species management in rivers networks: application to the Loire River. Biological Conservation 210:26–36.

Cote, D., D. G. Kehler, C. Bourne, and Y. F. Wiersma. 2009. A new measure of longitudinal connectivity for stream networks. Landscape Ecology 24:101–113.

Detenbeck, N. E., V. J. Brady, D. L. Taylor, V. M. Snarski, and S. L. Batterman. 2005. Relationship of stream flow regime in the western Lake Superior basin to watershed type characteristics. Journal of Hydrology 309:258–276.

Ellis, L. E., and N. E. Jones. 2013. Longitudinal trends in regulated rivers: a review and synthesis within the context of the serial discontinuity concept. Environmental Reviews 21:136–148.

Heim, K. C., C. D. Arp, M. S. Whitman, and M. S. Wipfli. 2018. The complementary role of lentic and lotic habitats for Arctic Grayling in a complex stream–lake network in Arctic Alaska. Ecology of Freshwater Fish 1–13.

Hermoso, V., M. J. Kennard, and S. Linke. 2012. Integrating multidirectional connectivity requirements in systematic conservation planning for freshwater systems. Diversity and Distributions 18:448–458.

Hughes, R. M., and J. M. Omernik. 1983. An alternative for characterizing stream size. Pages 87–102 *in* T. D. Fontaine, III and S. M. Bartell, editors. Dynamics of lotic ecosystems. Ann Arbor Press, Ann Arbor, Michigan.

Hughes, R. M., P. R. Kaufmann, and M. H. Weber. 2011. Strahler order versus stream size. Journal of the North American Benthological Society 30:103–121.

Hughes, R. M., L. Wang, and P. W. Seelbach, editors. 2006. Landscape influences on stream habitats and biological assemblages. American Fisheries Society, Symposium 48, Bethesda, Maryland.

Jeffries, M., and D. Mills. 1990. Freshwater ecology: principles and applications. Belhaven Press, London.

Jones, N. E. 2010. Incorporating lakes within the river discontinuum: longitudinal changes in ecological characteristics in stream–lake networks. Canadian Journal of Fisheries and Aquatic Sciences 67:1350–1362.

Jones, N. E., B. J. Schmidt, and S. J. Melles. 2014. Characteristics and distribution of natural flow regimes in Canada: a habitat template approach. Canadian Journal of Fisheries and Aquatic Sciences 71:1616–1624.

Lamberti, G. A., D. T. Chaloner, and A. E. Hershey. 2010. Linkages among aquatic ecosystems. Journal of the North American Benthological Society 29:245–263.

Leach, J. A., W. Lidberg, L. Kuglerova, A. Peralta-Tapia, A. Agren, and H. Laudon. 2017. Evaluating topography-based predictions of shallow lateral groundwater discharge zones for a boreal lake–stream system. Water Resources Research 53:5420–5437.

Martin, S. L., and P. A. Soranno. 2006. Lake landscape position: relationships to hydrologic connectivity and landscape features. Limnology and Oceanography 51:801–814.

McManamay, R. A., D. J. Orth, C. A. Dolloff, and E. A. Frimpong. 2012. Regional frameworks applied to hydrology: can landscape-based frameworks capture the hydrologic variability? River Research and Applications 28:1325–1339.

Melles, S. J., N. E. Jones, and B. J. Schmidt. 2014. Aquatic ecosystem classification for Ontario: a technical proposal. Ontario Ministry of Natural Resources, Aquatic Research Series 2013-05, Ottawa.

Melles, S. J., C. Chu, K. M. Alofs, and D. A. Jackson. 2015. Potential spread of Great Lakes fishes given climate change and proposed dams: an approach using circuit theory to evaluate invasion risk. Landscape Ecology 30:919–935.

Mineau, M. M., W. M. Wollheim, and R. J. Stewart. 2015. An index to characterize the spatial distribution of land use within watersheds and implications for river network nutrient removal and export. Geophysical Research Letters 42:6688–6695.

Nel, J. L., B. Reyers, D. J. Roux, N. D. Impson, and R. M. Cowling. 2010. Designing a conservation area network that supports the representation and persistence of freshwater biodiversity. Freshwater Biology 56:106–124.

Olden, J. D., and N. L. Poff. 2003. Redundancy and the choice of hydrologic indices for characterizing streamflow regimes. River Research and Applications 19:101–121.

Olden, J. D., M. J. Kennard, and B. J. Pusey. 2011. A framework for hydrologic classification with a review of methodologies and applications in ecohydrology. Ecohydrology 5:503–518.

OMNRF (Ontario Ministry of Natural Resources and Forestry). 2017. User guide for Ontario flow assessment tool (OFAT). OMNRF, Peterborough.

Pascual-Hortal, L., and S. Saura. 2006. Comparison and development of new graph-based landscape connectivity indices: towards the prioritization of habitat patches and corridors for conservation. Landscape Ecology 21:959–967.

Patrick, C. J., and L. L. Yuan. 2017. Modeled hydrologic metrics show links between hydrology and the functional composition of stream assemblages. Ecological Applications 27:1605–1617.

Pépino, M., M. A. Rodríguez, and P. Magnan. 2016. Incorporating lakes in stream fish habitat models: are we missing a key landscape attribute? Canadian Journal of Fisheries and Aquatic Sciences 635:629–635.

Perry, G. L. W., K. A. Moloney, and T. R. Etherington. 2017. Using network connectivity to prioritise sites for the control of invasive species. Journal of Applied Ecology 54:1238–1250.

Poff, N. L., and J. V. Ward. 1990. Physical habitat template of lotic systems: recovery in the context of historical pattern of spatiotemporal heterogeneity. Environmental Management 14:629–645.

Richardson, J. S., and R. J. Mackay. 1991. Lake outlets and the distribution of filter feeders: an assessment of hypotheses. Oikos 62:370–380.

Sarhad, J., S. Manifold, and K. E. Anderson. 2017. Geometric indicators of population persistence in branching continuous-space networks. Journal of Mathematical Biology 74:981–1009.

Scheffer, M., G. J. Van Geest, K. Zimmer, E. Jeppesen, M. Sondergaard, M. G. Butler, M. A. Hanson, S. Declerck, and L. De Meester. 2006. Small habitat size and isolation can promote species richness: second-order effects on biodiversity in shallow lakes and ponds. Oikos 112:227–231.

Spence, C. 2006. Hydrological processes and streamflow in a lake dominated watercourse. Hydrological Processes 20:3665–3681.

Tornwall, B. M., C. M. Swan, and B. L. Brown. 2017. Manipulation of local environment produces different diversity outcomes depending on location within a river network. Oecologia 184:663–674.

Trenhaile, A. S. 2013. Geomorphology: a Canadian perspective, 5th edition. Oxford University Press Canada, Don Mills, Ontario.

Van Looy, K., J. Piffady, C. Cavillon, T. Tormos, P. Landry, and Y. Souchon. 2014. Integrated modelling of functional and structural connectivity of river corridors for European otter recovery. Ecological Modelling 273:228–235.

Vannote, R. L., G. W. Minshall, K. W. Cummins, J. R. Sedell, and C. E. Cushing. 1980. The river continuum concept. Canadian Journal of Fisheries and Aquatic Sciences 37:130–137.

Ward, J. V., and J. A. Stanford. 1983. The serial discontinuity concept of lotic ecosystems. Pages 29–42 *in* T. D. Fontaine and S. M. Bartell, editors. Dynamics of lotic ecosystems. Ann Arbor Science, Ann Arbor, Michigan.

Wohl, E., N. Kramer, V. Ruiz-Villanueva, D. N. Scott, F. Comiti, A. M. Gurnell, H. Piégay, K. B. Lininger, K. L. Jaeger, D. M. Walters, and K. D. Fausch. 2019. The natural wood regime in rivers. BioScience 69:259–273.

Woo, M., R. Thorne, and L. Brown. In press. Comparison of runoff and river flow in two large northern basins. Hydrology Research, doi: 10.2166/nh.2018.199.

Wurtsbaugh, W. A., N. A. Heredia, B. G. Laub, C. S. Meredith, H. E. Mohn, S. E. Null, D. A. Pluth, B. B. Roper, W. C. Saunders, D. K. Stevens, R. H. Walker, and K. Wheeler. 2015. Approaches for studying fish production: do river and lake researchers have different perspectives? Canadian Journal of Fisheries and Aquatic Sciences 72:149–160.

American Fisheries Society Symposium 90:141–157, 2019
© 2019 by the American Fisheries Society

Loss of Coldwater Fish Habitat in Glaciated Lakes of the Midwestern United States after a Century of Land Use and Climate Change

Peter C. Jacobson*
Minnesota Department of Natural Resources
603 1st Street West, Park Rapids, Minnesota 56470, USA

Gretchen J. A. Hansen
Department of Fisheries, Wildlife, and Conservation Biology, University of Minnesota
135 Skok Hall, 2003 Upper Buford Circle, St. Paul, Minnesota 55108, USA

Leif G. Olmanson
Remote Sensing and Geospatial Analysis Laboratory
Department of Forest Resources, University of Minnesota
115 Green Hall, 1530 Cleveland Avenue North, St. Paul, Minnesota 55108, USA

Kevin E. Wehrly
Institute for Fisheries Research
Michigan Department of Natural Resources and University of Michigan
400 North Ingalls G250, Ann Arbor, Michigan 48109, USA

Catherine L. Hein
Wisconsin Department of Natural Resources
101 South Webster Street, Madison, Wisconsin 53707, USA

Lucinda B. Johnson
Natural Resources Research Institute, University of Minnesota Duluth
5013 Miller Trunk Highway, Duluth, Minnesota 55811, USA

Abstract.—Freshwater resources are threatened by multiple stressors, and identifying the relative roles of each is necessary for effective management. The relative contributions of eutrophication and climate change to the loss of coldwater fish habitat was estimated for 5,220 stratified lakes in Michigan, Minnesota, and Wisconsin. A hypolimnetic oxygen model was coupled with a landscape disturbance model to estimate change in oxythermal habitat since European settlement. The general additive model predicted late-summer oxythermal habitat conditions as a function of remotely sensed Secchi depth (mean of 1995–2014 values), geometry ratio (area$^{0.25}$/maximum depth), and mean July air temperature (1995–2014). The use of the remotely sensed water-clarity variable allowed modeling of lakes that do not have in situ water-quality data. The landscape dis-

* Corresponding author: peter.jacobson@state.mn.us

turbance model predicted remotely sensed water clarity as a function of levels of catchment land-use disturbance, geometry ratio, and proportion of glacial outwash soils in each catchment. Historic coldwater habitat was estimated for undisturbed landscape conditions and an earlier period of climate and then compared to contemporary conditions. Eutrophication and climate change substantially reduced coldwater fish habitat over the past century in many stratified lakes in Minnesota, Wisconsin, and Michigan. The greatest loss of coldwater habitat occurred in lakes with substantial land-use changes in their catchments, primarily in the Great Plains and Eastern Temperate Forests ecoregions. Oxythermal habitat in many other lakes in the Northern Forests ecoregion remained intact, with only modest changes primarily because of warming climate. To maintain coldwater habitat, deep, clear lakes in the forested ecoregions of Minnesota, Wisconsin, and Michigan should receive high priority for catchment protection efforts. Protective catchment land-use measures will be needed for coldwater fish to survive further climate warming.

Introduction

Glaciated landscapes are characterized by the presence of deep, thermally stratified lakes that support diverse fish assemblages with different habitat and thermal tolerances. Many cold-adapted fish species, such as Lake Trout *Salvelinus namaycush*, Lake Whitefish *Coregonus clupeaformis*, Cisco *C. artedi*, and Burbot *Lota lota*, seek summer refuge in the cold hypolimnia of these stratified lakes if oxygen concentrations are sufficient (Dillon et al. 2003; Lyons et al. 2017). However, summer hypolimnetic oxygen concentrations are declining in some lakes because of eutrophication (Evans et al. 1996; Scavia et al. 2014) and climate change (Stefan et al. 1996; Jacobson et al. 2010). Decomposition of organic material substantially intensifies hypolimnetic oxygen depletion rates in lakes undergoing cultural eutrophication resulting from increased primary production. And where lakes are experiencing earlier springs and later falls, longer stratification periods can also reduce critical late-summer hypolimnetic oxygen concentrations (Foley et al. 2012). Coldwater fish suffer substantial summer mortality events when hypolimnetic oxygen concentrations are inadequate and epilimnetic temperatures are lethally warm (Jacobson et al. 2008). Therefore, there is a critical need to understand and quantify how eutrophication and climate change have changed coldwater oxythermal habitat in stratified lakes.

Many lakes in the Midwestern United States have experienced the effects of eutrophication for more than a century (Umbanhowar et al. 2011; Heathcote et al. 2013). Native prairies were extensively converted to agriculture starting in the 1850s, and land conversions were 95% complete by 1910 (Heathcote et al. 2013). A sharp increase in the use of synthetic fertilizers began in the 1950s as modern agronomic methods rapidly developed (Tilman et al. 2002). Corn and soybean fields now dominate the region where native prairies once existed (Lark et al. 2015). The removal of natural, perennial vegetation, and intensification of agricultural practices accelerated sediment and nutrient loading to Midwestern lakes (Dietz et al. 2015), despite recent soil conservation efforts (Heathcote et al. 2013). Other lands were converted to urban uses, where increased impervious surfaces, disturbed soils, and wastewater discharges increased nutrient loading rates as well (Carpenter et al. 1998; Ramstack et al. 2004). Many lakes in the intensive agricultural and urban areas of Minnesota are considered severely impaired by eutrophication (Heiskary et al. 1987), and their hypolimnia experience significant periods of hypoxia (Jacobson et al. 2010).

Globally, lake temperatures have increased (O'Reilly et al. 2015), stratification has intensified (Kraemer et al. 2015), and ice cover durations have decreased (Magnuson et al. 2000) with climate warming. However, lake responses to climate change vary considerably with lake morphometry, location, and water clarity (Kraemer et al. 2015; Winslow et al. 2015; Rose et al. 2016), with deep, seasonally ice-covered lakes having the fastest warming rates (O'Reilly et al. 2015). These changes in lake thermal conditions have altered lake fish assemblages. Notably, populations of an important coldwater fish, Cisco, have declined in Indiana (Honsey et al. 2016), Michigan (Latta 1995), Minnesota (Jacobson et al. 2012), and Wisconsin (Lyons et al. 2017) in recent years.

Disentangling the effects of multiple ecological stressors changing concurrently, such as climate warming and eutrophication, can be challenging. Considerable scientific attention has resulted in the development of novel approaches to the issue (Moorhouse et al. 2014). For example, landscape models that directly incorporate specific ecological stress variables allow for the exploration of individual effects on important biological responses (Jacobson et al. 2017). Many of these models have the potential to approximate predisturbance or reference conditions (Wang et al. 2008) by estimating the set of parameters associated with no anthropogenic stress. The effects of individual ecological stressors can be estimated by comparing contemporary, anthropogenically stressed conditions to historic predisturbance levels.

Therefore, the goal of this study was to estimate the relative contributions of eutrophication and climate change to loss of coldwater fish habitat in Michigan, Minnesota, and Wisconsin lakes. A hypolimnetic oxygen model was coupled with a landscape disturbance model to estimate change in oxythermal habitat since European settlement. We analyzed spatial and stressor-specific patterns of coldwater habitat changes and then suggest possible management strategies to mitigate those changes.

Methods

Study Area

Lakes in the glaciated Midwestern states of Michigan, Minnesota, and Wisconsin formed the basis for the study. These lakes are located across three major ecoregions (Omernik Level 1: Omernik 1987) that are distinct in their environmental conditions and water quality (Omernik et al. 1988; Olmanson et al. 2013). The Northern Forests ecoregion is represented by heavily forested areas in the northern portions of all three states and is interspersed with many lakes, ponds, wetlands, and bogs. Although a few lakes are plagued by algal blooms (Binding et al. 2011), most have good water quality (Cheruvelil et al. 2013; Cross and Jacobson 2013). The Eastern Temperate Forests ecoregion is a transition zone between forested areas to the north and agricultural areas to the south and includes the major metropolitan areas of Detroit, Chicago, Milwaukee, and Minneapolis–St. Paul. It has the highest proportion of urban and suburban land among the ecoregions in this region. In addition, there are a number of agricultural lands (both cultivated crops and livestock production) within the Eastern Temperate Forest ecoregion, and the lakes have a wide range of water qualities (Cheruvelil et al. 2013). The Great Plains ecoregion lies primarily in southwestern Minnesota and is heavily cultivated with row crops (primarily corn and soybeans), with few native prairies remaining. Lake water-quality problems in this ecoregion include high nutrient and suspended sediment loads and intense algal blooms (Heiskary et al. 1987; Read et al. 2015).

Modeling Approach

A hypolimnetic oxygen model was coupled with a water quality model to estimate change in oxythermal habitat since European settlement (Figure 1). The water quality model first related the effects of land-use disturbance, lake depth, and catchment soils on lake water clarity. Then, the hypolimnetic oxygen model was related to the effects of eutrophication and climate on late-summer oxythermal habitat (Figure 2). Finally, we estimated historic conditions of climate and land use as predisturbance values in the lake water quality and hypolimnetic oxygen models.

We used the lake water quality model developed by Cross and Jacobson (2013), which predicted lake-specific epilimnetic total phosphorus concentrations as a function of lake morphometry, catchment soil characteristics, and catchment disturbance. The model is structured so that the catchment land-use disturbance variable can be set to zero to estimate predisturbance conditions (assumed to represent conditions prior to European settlement). This model was revised to use water clarity, rather than epilimnetic total phosphorus concentration. Remotely sensed water clarity (Secchi depth) data were available for several thousand Minnesota, Wisconsin, and Michigan lakes, which is considerably more than total phosphorus data that require in situ sampling. The revised model predicted Secchi depth as a function of lake morphometry (geometry ratio), catchment soil characteristics (proportion of glacial outwash soils), and catchment disturbance (proportion of agriculture, urban, and mining land uses in catchment). Geometry ratio (lake area$^{0.25}$/maximum depth with units of $m^{-0.5}$) is a measure of depth, relative to lake size and has been useful for classifying stratification status of lakes (Fang et al. 2012).

The hypolimnetic oxygen model was based on the model of Jacobson et al. (2010), which uses the oxythermal parameter TDO_3 as a measure of coldwater habitat quality. TDO_3 is the temperature at 3 mg/L of dissolved oxygen determined by interpolation of observed temperature/dissolved oxygen profiles. The original Jacobson et al. (2010) model predicted late-summer TDO_3 as a function of lake morphometry (geometry ratio), climate (mean July air temperature), and lake productivity (epilimnetic total phosphorus con-

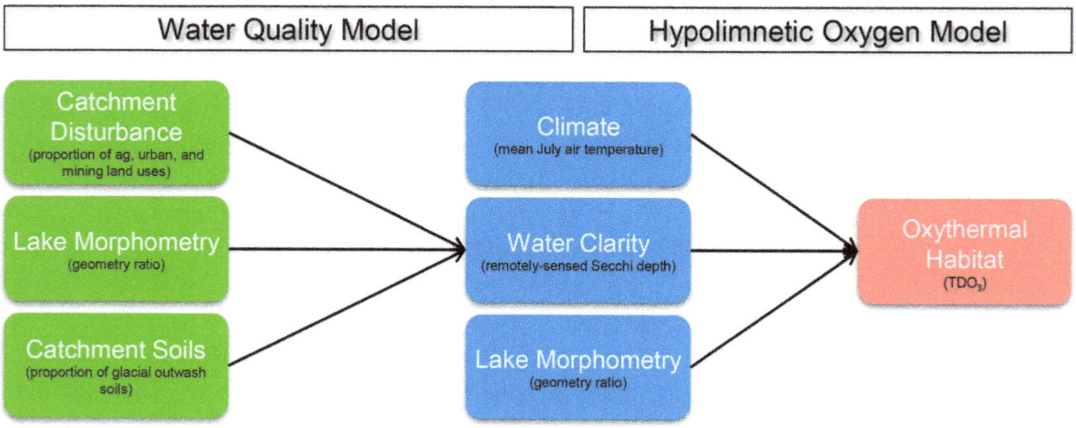

Figure 1. Conceptual diagram of the coupled water quality and hypolimnetic models that predict oxythermal habitat (with actual variables in parenthesis).

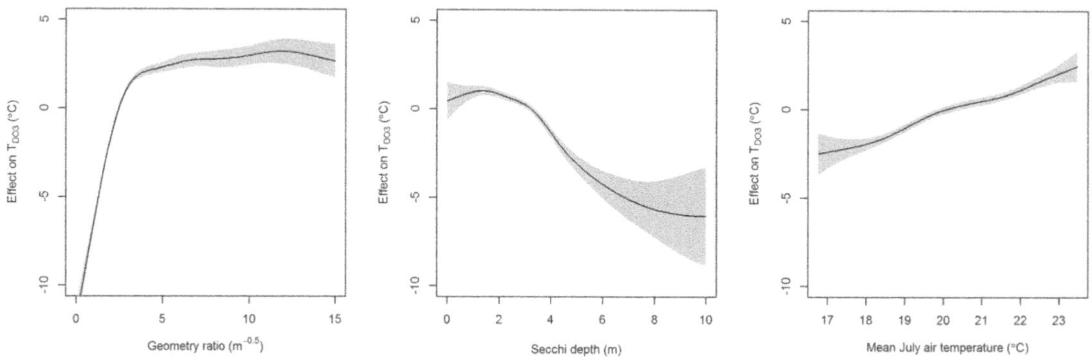

Figure 2. Generalized additive model response effects of variables that predicted August TDO_3 (°C) in 2,087 lakes across Michigan, Minnesota, and Wisconsin. Shaded area represents ±2 standard errors).

centration). We revised the original model to include water clarity as the lake productivity variable, rather than epilimnetic total phosphorus concentration, for compatibility with the water quality model.

Historic climate conditions were estimated using existing air temperature data collected at the earliest dates available in the PRISM database (1895–1914; PRISM Climate Group 2018). Mean July air temperatures were calculated for the periods from 1895 through 1914 and 1991 through 2010 from PRISM data to estimate predisturbance and contemporary climate conditions. We assigned the 4-km-gridded PRISM data to each geographic lake center using the 'geoknife' package in R (Read et al. 2016). Historic Secchi depths were estimated by setting the catchment land-use disturbance variable to zero in the lake water quality model. Since land-use disturbance likely occurred in parts of the Midwestern United States before 1895, when air temperature data became available for the entire region, the predisturbance period could not be defined precisely. So, we assumed that predisturbance conditions generally corresponded to the climate that was present from 1895 through 1914 and a period when land-use disturbance was minimal. Rahmstorf et al. (2017) noted that 1907 was the start of a significant change point in global temperatures, with relatively cool periods in the decades before that year, which would provide support for a generalized period of predisturbance. Late-summer TDO_3 values were then estimated for the historic and contemporary periods to explore stressor-specific oxythermal habitat changes in 714 Cisco lakes and 4,596 other stratified lakes in the region.

Generalized additive models (Hastie and Tibshirani 1990) were used for both the hypolimnetic oxygen and water clarity models. Since response variables are described by the sum of additive predictor variables, generalized additive models are especially amenable for disentangling the effects of multiple ecological stressors (Jacobson et al. 2017). The 'mgcv' (multiple generalized cross validation) package in R was used for the development of the generalized additive models (Wood 2006). The 'mgcv' implementation fitted a series of penalized regression splines as smoothing functions for each predictor variable and suggested degrees of freedom for smooth terms by minimizing the generalized cross-validation score. Percent deviance explained was used as a measure of model fit. In this analysis, additive effects in climate and water clarity from the hypolimnetic oxygen generalized additive model were used to partition changes in late summer TDO_3 for each stressor. Changes in oxythermal habitat were also compared across ecoregions.

Data Sources

Lake morphometry data were assembled from state agencies of Michigan, Minnesota, and Wisconsin and from a regional assessment of lake habitats by the Midwest Glacial Lakes Fish Habitat Partnership (www.midwestglaciallakes.org). Proportion of 2001 National Land Cover Database (Homer et al. 2007) disturbed land uses (agriculture, urban, and mining categories) within lake catchments was used to quantify catchment disturbance. Individual catchments were delineated by the Midwest Glacial Lakes Fish Habitat Partnership, and total lake catchment extents were calculated by accumulating all individual catchments upstream of each lake. Proportions of coarse soils in the catchment of each lake were calculated from the U.S. Geological Survey 1 km lithology GIS data set (Soller and Packard 1998).

Remote-sensed water clarity data derived from the optical properties of Landsat satellite images for Michigan, Minnesota, and Wisconsin for the period from 1991 through 2010 were retrieved from state, federal, and university sources (Chipman et al. 2004; Fuller et al. 2004, 2011; Olmanson et al. 2008). Complete sets of remote-sensed and catchment data were available for 8,977 lakes in the region: Michigan (1,600 lakes), Minnesota (3,483 lakes), and Wisconsin (3,894 lakes).

Mean late-summer TDO_3 values were calculated from August dissolved oxygen/temperature profiles measured from 1991 through 2010 retrieved from state-agency water quality data sources for 2,087 lakes in the region: Michigan (264 lakes), Minnesota (1,316 lakes), and Wisconsin (507 lakes). Measured TDO_3 values were generated by linear interpolation of dissolved oxygen and temperature profile measurements. Only profiles from the deepest basin in a lake were used for TDO_3 calculations. Presence of Cisco lakes within the region were identified by state fisheries management agencies.

Results

Model Development

Oxythermal habitat, as measured by August TDO_3, was modeled for 2,087 lakes in the region (Figure 2) based on geometry ratio, remote-sensed Secchi depth, and mean July air temperature (significance of all variables $P < 0.001$ and deviance explained = 45.7%). Geometry ratio had the largest effect on August TDO_3, with a strong positive relationship (loss of oxythermal habitat) for lakes with geometry ratios less than 4.0 $m^{-0.5}$. The influence of geometry ratio lessened at values greater than 4.0 $m^{-0.5}$, after which lakes were unstratified and had no hypolimnetic cold water. Secchi depth negatively influenced August TDO_3 in a nonlinear fashion and had the largest effect at water clarity greater than 2.0 m. Mean July air temperatures significantly increased August TDO_3 throughout the range observed in the region.

The lake water quality model variables of geometry ratio, catchment outwash soils, and land-use disturbance were significant (all variables $P < 0.001$ and deviance explained = 65.5%) in explaining remote-sensed Secchi depth for 8,977 lakes in the region (Figure 3). The effect of geometry ratio on water clarity was especially strong, with a sharp negative response below 4.0 $m^{-0.5}$. Catchment land-use disturbance also strongly influenced remote-sensed Secchi depth, where agricultural and urban land uses were related to decreased water clarity (Olmanson et al. 2013). Although statistically significant, the magnitude of the effect of outwash soils on water clarity was considerably less than geometry ratio and catchment disturbance.

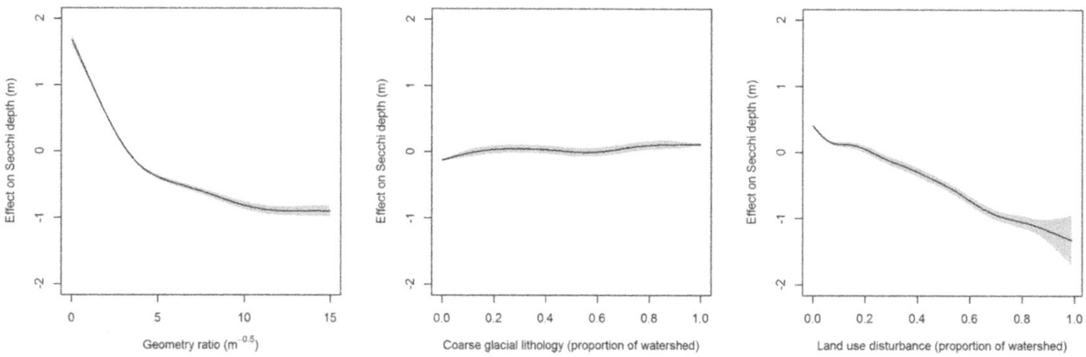

Figure 3. Generalized additive model response effects of variables that predicted Secchi depth (m) in 8,977 lakes across Michigan, Minnesota, and Wisconsin. Shaded area represents ±2 standard errors).

Changes from Historic to Contemporary Conditions

Application of the lake water quality model to predisturbance conditions demonstrated a wide range of estimated historic water clarity in 5,310 stratified lakes throughout the region (Figure 4). The greatest water clarity existed in the deepest lakes and in areas with the greatest amount of sandy outwash soils. Michigan had a large proportion of lakes with good water clarity. Contemporary water clarity was considerably reduced, especially in areas where lands had been converted to agricultural and urban uses. Estimated changes in Secchi depths, from historic to contemporary periods, revealed a substantial decrease in water clarity in the southwestern portion of the region.

Historic mean July air temperatures were the lowest in northeast Minnesota, northern Wisconsin, and the Upper Peninsula of Michigan (Figure 4). Contemporary climates demonstrated a similar geographical pattern. However, changes in mean July air temperatures revealed that the greatest increases occurred in central and southern Minnesota, western Wisconsin, and the eastern portion of the Upper Peninsula of Michigan.

Application of the historic climate and predisturbance land-use conditions to the oxythermal habitat model revealed that a wide range of historic August TDO_3 values existed throughout the region (Figure 4). The geographic distribution followed patterns in mean July air temperatures but were strongly mediated by lake geometry ratio. Variation in geometry ratios across the region produced an extensive heterogeneous spatial pattern in historic August TDO_3 values. Contemporary August TDO_3 values illustrated a similar degree of spatial heterogeneity. However, changes in August TDO_3 values from historic to contemporary periods revealed a striking loss of oxythermal habitat in distinct parts of the region. Lakes in west-central Minnesota and southern Michigan experienced the greatest losses of oxythermal habitat (as much as 4°C). Increases in August TDO_3 values of between 0°C and 2°C occurred in most other lakes in the region.

Stressor-Specific and Ecoregional Losses of Oxythermal Habitat

The greatest losses of oxythermal habitat from historic to contemporary conditions occurred in the Great Plains and Eastern Temperate Forests ecoregions (mean decreases in August TDO_3 of approximately 1°C). Losses of coldwater fish habitat in the Northern Forests ecoregion were markedly less (mean

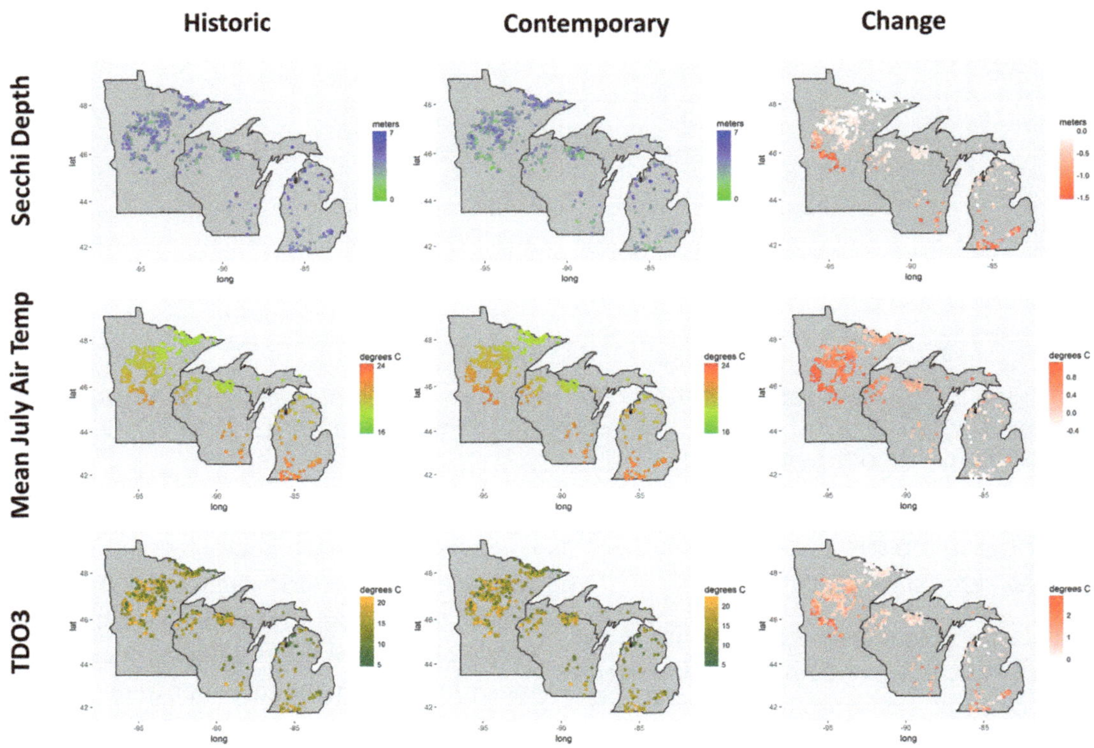

Figure 4. Estimated historic and contemporary values of Secchi depth (m), mean July air temperature (°C), and August TDO_3 (°C) for 5,310 stratified lakes across Michigan, Minnesota, and Wisconsin. Change represents contemporary minus historic values.

decreases in August TDO_3 of less than 1°C). Although most changes in August TDO_3 were less than 2°C for all ecoregions, some lake-specific values exceeded 3°C and 4°C (Figure 5). The effects of decreased water clarity on oxythermal habitat loss were generally greater than the effects of increased mean July air temperatures (Figure 5). Lake-specific increases in August TDO_3 from loss of water clarity were occasionally as large as 4°C, primarily in lakes with disturbed catchments. In contrast, lake-specific increases in August TDO_3 due to increased mean July air temperatures were nearly always less than 1°C.

Loss of coldwater habitat in Cisco lakes (Figure 6) was slightly greater than that in other stratified lakes. Increases in August TDO_3 resulting from changing climate were usually less than 0.5°C, as were increases due to changing water clarity. Total increases in August TDO_3 from both stressors were generally less than 1°C of TDO_3. However, much greater losses occurred in lakes where land-use change significantly decreased water clarity.

Discussion

Eutrophication and climate change have substantially diminished coldwater fish habitat over the past century in many lakes of the Midwestern United States. The greatest losses of coldwater habitat occurred in lakes where extensive land-use changes have occurred, primarily in the Great Plains and Eastern Temperate Forests ecoregions. Prairies and forests around many of these lakes have been converted to agricultural and urban land uses, with important consequences for water

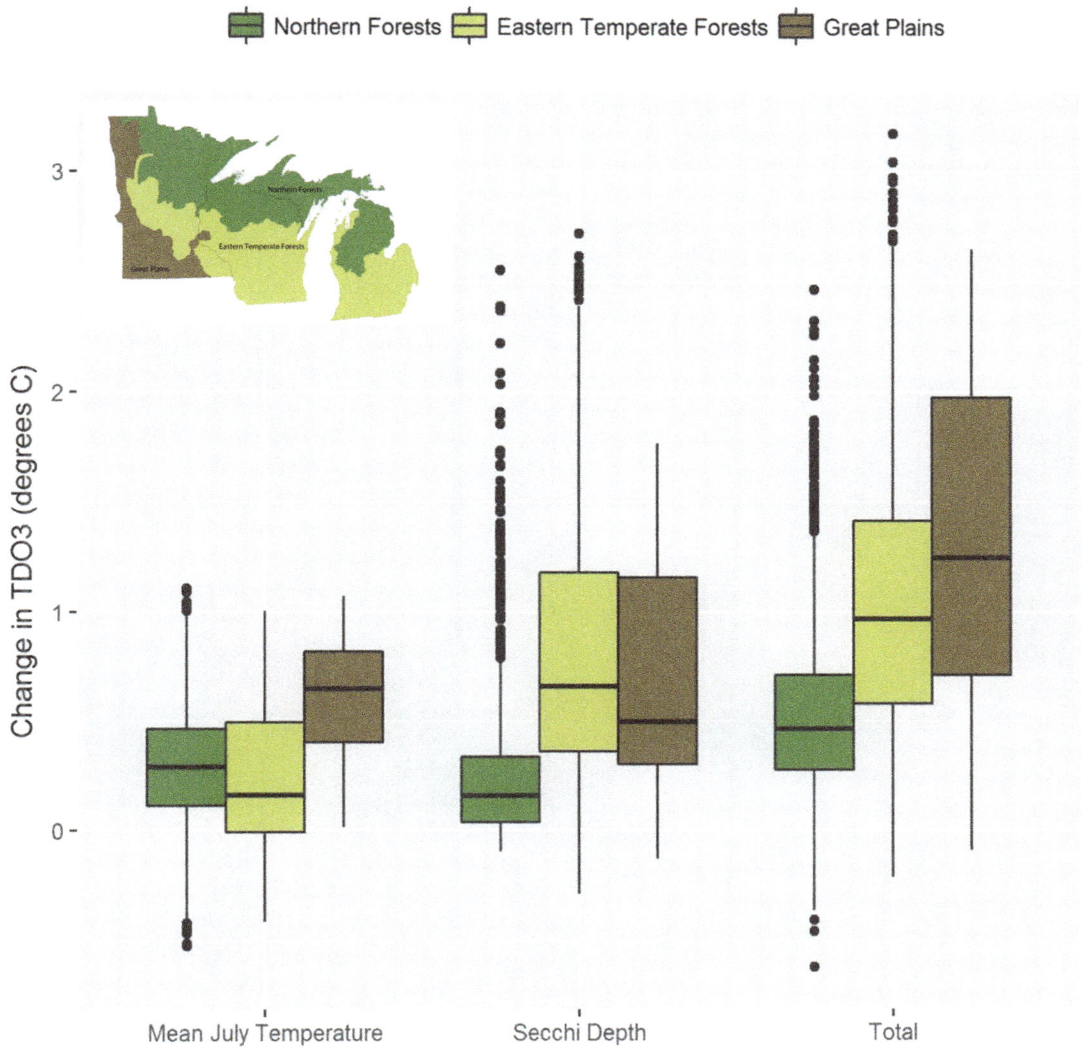

Figure 5. Boxplots of estimated change in August TDO_3 (°C) values from historic to contemporary conditions for 5,310 stratified lakes across Michigan, Minnesota, and Wisconsin by Omernik (1987) Level 1 ecoregions. Bold horizontal lines represent medians, boxes represent interquartile ranges (25th–75th percentiles), and extended dotted lines represent 1.5 interquartile ranges.

quality and coldwater fish habitat. Increased primary productivity from greater nutrient loading rates likely increased hypolimnetic oxygen depletion rates and reduced critical coldwater habitat at the end of summer. In contrast, coldwater habitats in many lakes of the Northern Forests ecoregion have largely remained intact, with only modest changes resulting from a warming climate. Nutrient loading rates have likely remained low from catchments that have stayed forested through contemporary periods.

In this study, eutrophication diminished coldwater fish habitat more than climate warming in the region. The dominance of the eutrophication effect likely resulted from the scale of land-use transformations that have occurred. Productive soils and suitable

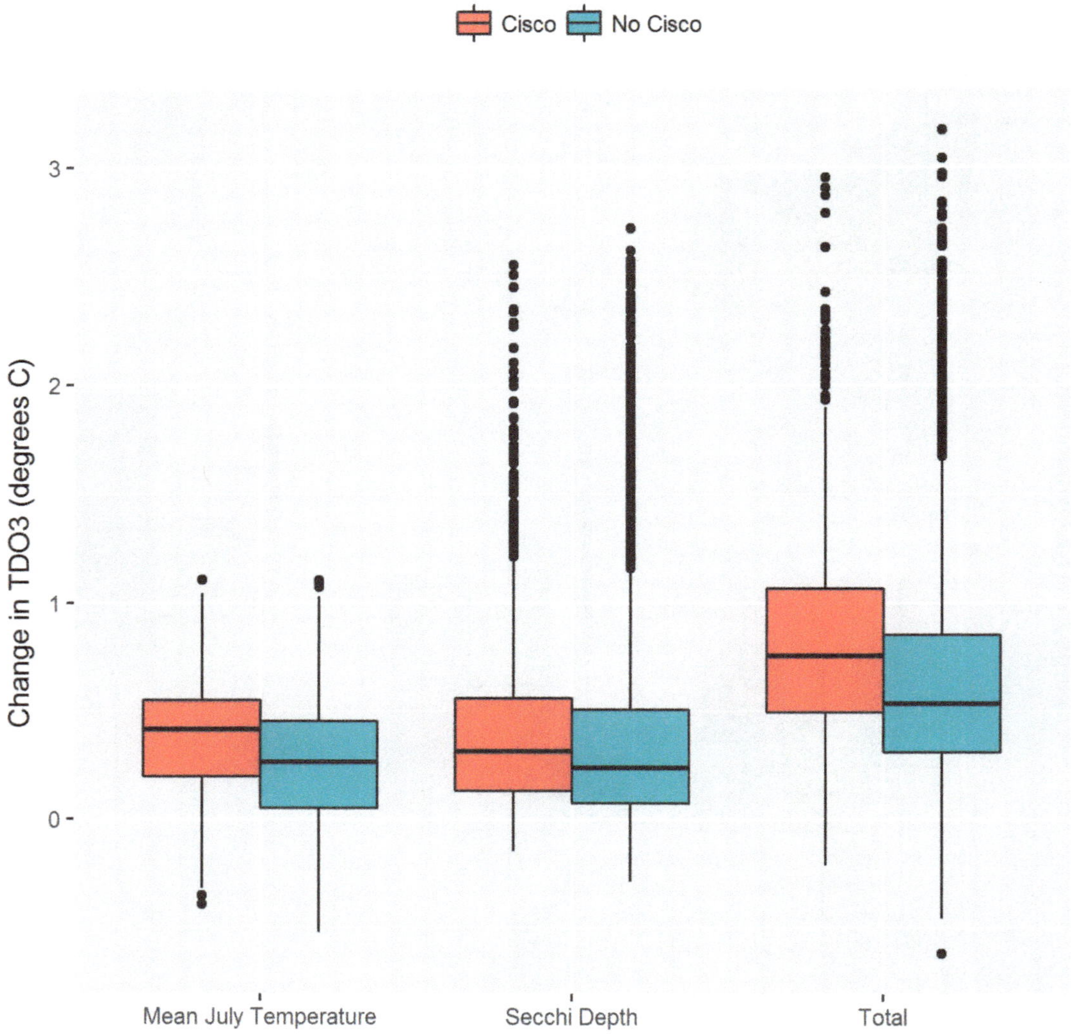

Figure 6. Boxplots of estimated change in August TDO_3 (°C) values from historic to contemporary conditions for 5,310 stratified lakes across Michigan, Minnesota, and Wisconsin by Cisco presence/absence and by ecological stressor. Bold horizontal lines represent medians, boxes represent interquartile ranges (25th–75th percentiles), and extended dotted lines represent 1.5 interquartile ranges.

climates drove a rapid expansion of intensive agriculture throughout the Midwestern United States over the period (Heathcote et al. 2013; Lark et al. 2015). Nutrient loading rates and lake phosphorus concentrations in the most agriculturally intensive portion of the region are greatly elevated over natural background levels (Heiskary et al. 1987; Cross and Jacobson 2013). Organic carbon burial rates in Minnesota lakes also increased markedly after lands were converted to agriculture (Dietz et al. 2015), which may have concurrently affected hypolimnetic oxygen depletion rates. A similar analysis in Minnesota also indicated that eutrophication drove changes in fish assemblages more than climate warming (Jacobson et al. 2017). In contrast, model predictions by Herb et al. (2014)

suggested that future climate warming may be a greater threat than further changes in land use. However, future land-use changes in the Midwest are challenging to predict. Considerably more work is required to understand the roles of both climate warming and land-use change in the future. Furthermore, atmospheric phosphorus deposition is increasing nationally (Stoddard et al. 2016), which may be accelerating eutrophication throughout the Midwest, especially in oligotrophic systems.

Although climate-driven losses of coldwater habitat were generally smaller than those from eutrophication, portions of the region have experienced substantial increases in July air temperatures during the past century. Lakes in west-central Minnesota experienced July air temperature increases as large as 1.5°C. Unfortunately, those same lakes also experienced significant conversions of land to agricultural uses and late-summer TDO_3 values increased by as much as 3°C as a result. Oxythermal habitat losses of that magnitude reduce coldwater fish habitat resources because there is a sharp gradient of habitat quality as a function of TDO_3 (Jacobson et al. 2010). A 3°C shift in TDO_3 could make some historic Lake Trout lakes uninhabitable for Lake Trout. Similarly, a 3°C shift in TDO_3 would result in a decline of 10% in probability of occurrence of Cisco in the lakes modeled by Jacobson et al. (2010). The combined effects of eutrophication and climate warming on oxythermal habitat have likely played a role in declining Cisco populations throughout the region (Jacobson et al. 2012; Honsey et al. 2016; Lyons et al. 2017).

The comparatively smaller change of oxythermal habitats in the Northern Forests ecoregion is an important finding. Coldwater habitat remains excellent in many lakes of northern Michigan, Minnesota, and Wisconsin. Paleo-reconstructions of hypolimnetic oxygen concentrations also indicated few changes in south-central Ontario lakes since the 19th century (Quinlan and Smol 2002). Similarly, paleolimnological reconstructions of total phosphorus concentrations in Minnesota lakes of the Northern Forests ecoregion remained relatively unchanged since European settlement times (Ramstack et al. 2004). Although many of these lakes have extensive cabin development, intact forested catchments beyond the shoreline have apparently been sufficient to protect water quality. Cross and Jacobson (2013) noted that including all lands within a catchment was a better predictor of lake phosphorus concentrations than only considering near-lake land uses. Predictions of excellent future oxythermal habitats in northern Minnesota formed the basis for a Cisco "refuge" lake program (Fang et al. 2012; Jacobson et al. 2013). Protecting catchments of these deep, clear lakes will ensure that they have the water quality necessary to provide suitable coldwater habitat in a climate-warmed future.

Although a climate warming effect was successfully captured with a mean July air temperature variable, future modeling efforts would benefit from exploring other metrics. For example, mean July air temperature only partially captured the effects of duration of stratification. As spring and ice-out dates arrive earlier, increased durations of stratification will likely have profound effects on critical end-of-summer oxythermal habitats. Metrics that directly incorporate duration of stratification would likely improve the accuracy of oxythermal habitat models. Increased nutrient and sediment loads from increased precipitation events also need to be investigated. Nutrient loading rates accelerate significantly after large storm runoff events and subsequent effects on hypolimnetic depletion rates are likely large. Climate can also affect rates of lake productivity and eutrophication (Umbanhowar et al. 2011). The introduction of a climate/eutrophication interaction term into a hypolimnetic oxygen depletion model could improve accuracy and relevance. As these complex processes of land use and climate change effects on hypolimnetic oxygen dynamics are better understood, hindcasting (and forecasting) models of coldwater fish habitat will also improve.

The incorporation of remote-sensed Secchi depth as a variable in the lake water quality model prominently expanded the number of modeled lakes and increased the power of analyses. The ability to capture this key water quality parameter over a wide region allowed the effects of massive landscape change on hypolimnetic oxygen resources to be quantified for a large number of lakes. Recent advancements in remote sensing allow the mapping of more water quality variables on a regional basis. The factors that control water clarity (chlorophyll, colored dissolved organic material, and suspended sediments) can now be mapped for lakes using Landsat 8 and Sentinel 2 data (Olmanson et al. 2015). Cloud computing and supercomputers are opening the potential for near real-time water quality monitoring and creation of large water quality databases that could be used to improve future fish habitat modeling efforts.

The generalized additive modeling approach was able to disentangle the effects of land-use change from climate change on coldwater fish habitat loss. The ability to visualize the magnitude and shape of each stressor response was especially valuable. Apportioning loss to each stressor will allow fisheries and land-use managers to understand which stressors are operating and then to apply targeted conservation measures for individual lakes. Although the sequence of land-use change differed throughout the region, the capacity of the lake water quality model to estimate a predisturbance condition based on a zero land-use disturbance value was also valuable. A limitation of the approach is that changes in hypolimnetic oxygen conditions were assumed to follow existing gradients of climate and water clarity conditions. Lake-specific trajectories of change were almost certainly more complex. Although water quality in most Northern Forest ecoregion lakes has likely recovered from early, widespread logging, changes in hypolimnetic oxygen depletion rates were likely not linear. Paleolimnological reconstructions of hypolimnetic oxygen concentrations (similar to Quinlan and Smol 2002; Brodersen and Quinlan 2006) in Midwestern lakes would provide interesting and relevant comparisons to these modeling results.

Management Implications

The powerful economic forces that profoundly transformed the landscape in the Midwestern United States are still operating today (Heathcote et al. 2013; Lark et al. 2015). Agriculture on the most conducive soils and suitable climates has intensified in recent years (Wright and Wimberly 2013) with growing fertilizer application rates (Good and Beatty 2011). Despite significant agronomic improvements, agricultural yields may still not increase rapidly enough to satisfy future global food demand, resulting in unrelenting demand for more agricultural land (Ray et al. 2013). Accelerated use of Midwestern corn for biofuels has increased pressure to further convert lands to agriculture in recent years (Lark et al. 2015). In addition, advances in agronomic methods and technology for farming the previously less-desirable sandy soils of glacial outwash plains have allowed profitable agriculture to develop when significant groundwater resources are available for irrigation. Conversions of pine forests to irrigated agriculture have been accelerating in recent years, producing large yields of highly valued crops (e.g., potatoes and corn). These recent conversions of forest land in Minnesota were noted as an agricultural expansion hotspot in a national analysis of land-use change (Lark et al. 2015). Pressures to convert forest lands to agriculture will likely increase in the future as global demand for these commodities continues to increase (Ray et al. 2013). Many lake catchments in previously lightly disturbed areas of the region will face unprecedented risks for water-quality degradation as land conversions accelerate. Losses of oxythermal habitat and coldwater fish resources will subsequently accelerate.

Curiously, hypoxic zones in the Gulf of Mexico have received the greatest attention for diminishing aquatic oxygen resources resulting from Midwestern U.S. land-use change (Rabalais et al. 2002). Yet, the same excessive nutrient loads that move down the Mississippi River to the Gulf of Mexico, also create large hypoxic zones in Midwestern lakes (Jacobson et al. 2010). The total volume of hypoxic water in the predominantly mesotrophic and eutrophic lakes throughout the region in late summer is likely large. Perhaps recognition of the severity of ecological losses in both the Gulf of Mexico and important Midwestern lakes will spur action to reduce the excessive nutrient loads.

The continued existence of many lakes in the Northern Forests ecoregion that sustain excellent coldwater habitat is a promising finding. Maintaining intact forests within the catchments of these deep, clear lakes will be critically important for preserving that coldwater habitat as the climate warms. The Cisco refuge lake program in Minnesota (Fang et al. 2012; Jacobson et al. 2013) should be extended to other states. Deep, clear lakes with excellent oxythermal habitat resources should receive high priority for catchment protection efforts (Lyons et al. 2017). Protecting forests with fee-title acquisition and private land conservation easements should be the primary focus when extensive forests are intact within lake catchments. When some agriculture is present, rigorous best management practices such as vegetated buffer strips, tiling prohibition, and cover crops will be required to reduce sediment and nutrient loss from fields. Similarly, rigorous urban best management practices (e.g., stormwater capture) will be required for any development that occurs in their catchments. Protective catchment land-use measures such as these will be needed for coldwater fish to survive further climate warming in the region.

Acknowledgments

We thank the many agency staff who collected water quality data throughout the three states. We also thank Joe Nohner, Lizhu Wang, and an anonymous reviewer for their helpful reviews of the manuscript.

References

Binding, C. E., T. A. Greenberg, and R. P. Bukata. 2011. Time series of algal blooms in Lake of the Woods using the MERIS maximum chlorophyll index. Journal of Plankton Research 33:1847–1852.

Brodersen, K. P., and R. Quinlan. 2006. Midges as palaeoindicators of lake productivity, eutrophication and hypolimnetic oxygen. Quaternary Science Reviews 25:1995–2012.

Carpenter, S. R., N. F. Caraco, D. L. Correll, R. W. Howarth, A. N. Sharpley, and V. H. Smith. 1998. Nonpoint pollution of surface waters with phosphorus and nitrogen. Ecological Applications 8:559–568.

Cheruvelil, K. S., P. A. Soranno, K.E. Webster, and M. T. Bremigan. 2013. Multi-scaled drivers of ecosystem state: quantifying the importance of the regional spatial scale. Ecological Applications 23:1603–1618.

Chipman, J. W., T. M. Lillesand, J.E. Schmaltz, J.E. Leale, and M. J. Nordheim. 2004. Mapping lake water clarity with Landsat images in Wisconsin, USA. Canadian Journal of Remote Sensing 30:1–7.

Cross, T. K., and P. C. Jacobson. 2013. Landscape factors influencing lake phosphorus concentrations across Minnesota. Lake and Reservoir Management 29:1–12.

Dietz, R. D., D. R. Engstrom, and N. J. Anderson. 2015. Patterns and drivers of change in organic carbon burial across a diverse landscape: insights from 116 Minnesota lakes. Global Biogeochemical Cycles 29:708–727.

Dillon, P. J., B. J. Clark, L. A. Molot, and H. E. Evans. 2003. Predicting the location of optimal habitat boundaries for Lake Trout (*Salvelinus namaycush*) in Canadian Shield lakes. Canadian Journal of Fisheries and Aquatic Sciences 60:959–970.

Evans, D. O., K. H. Nicholls, Y. C. Allen, and M. J. McMurtry. 1996. Historical land use, phosphorus loading, and loss of fish habitat in Lake Simcoe, Canada. Canadian Journal of Fisheries and Aquatic Sciences 53(Supplement 1):194–218.

Fang, X., L. Jiang, P. C. Jacobson, H. G. Stefan, S. R. Alam, and D. L. Pereira. 2012. Identifying Cisco refuge lakes in Minnesota under future climate scenarios. Transactions of the American Fisheries Society 141:1608–1621.

Foley, B., I. D. Jones, S. C. Maberly, and B. Rippey. 2012. Long-term changes in oxygen depletion in a small temperate lake: effects of climate change and eutrophication. Freshwater Biology 57:278–289.

Fuller, L. M., S. S. Aichele, and R. J. Minnerick. 2004. Predicting water quality by relating Secchi-disk transparency and chlorophyll-a measurements to satellite imagery for Michigan inland lakes, August 2002: U.S. Geological Survey, Scientific Investigations Report 2004-5086, Reston, Virginia.

Fuller, L. M., R. S. Jodoin, and R. J. Minnerick. 2011. Predicting lake trophic state by relating Secchi-disk transparency measurements to Landsat-satellite imagery for Michigan inland lakes, 2003-05 and 2007-08. U.S. Geological Survey, Scientific Investigations Report 2011-5007, Reston, Virginia.

Good, A. G., and P. H. Beatty. 2011. Fertilizing nature: a tragedy of excess in the commons. PLOS (Public Library of Science) Biology [online serial] 9(8):e1001124.

Hastie, T. J., and R. J. Tibshirani 1990. Generalized additive models. Chapman and Hall, London.

Heathcote, A. J., C. T. Filstrup, and J. A. Downing. 2013. Watershed sediment losses to lakes accelerating despite agricultural soil conservation efforts. PLOS (Public Library of Science) ONE [online serial] 8(1):e53554.

Heiskary, S. A., C. B. Wilson, and D. P Larsen. 1987. Analysis of regional patterns in lake water quality: using ecoregions for lake management in Minnesota. Lake and Reservoir Management 3:337–344.

Herb, W. R., L. B. Johnson, P. C. Jacobson, and H. G. Stefan. 2014. Projecting cold-water fish habitat in lakes of the glacial lakes region under changing land use and climate regimes. Canadian Journal of Fisheries and Aquatic Sciences 71:1334–1348.

Homer, D., J. Dewitz, J. Fry, M. Coan, N. Hossain, C. Larson, N. Herold, A. McKerrow, J. N. VanDriel, and J. Wickham. 2007. Completion of the 2001 national land cover database for the conterminous United States. Photogrammetric Engineering and Remote Sensing 73:337–341.

Honsey, A. E., S. B. Donabauer, and T. O. Höök. 2016. An analysis of lake morphometric and land-use characteristics that promote persistence of Cisco in Indiana. Transactions of the American Fisheries Society 145:363–373.

Jacobson, P. C., T. S. Jones, P. Rivers, and D. L. Pereira. 2008. Field estimation of a lethal oxythermal niche boundary for adult Cisco in Minnesota Lakes. Transactions of the American Fisheries Society 137:1464–1474.

Jacobson, P. C., H. G. Stefan, and D. L. Pereira. 2010. Coldwater fish oxythermal habitat in Minnesota lakes: influence of total phosphorus, July air temperature, and relative depth. Canadian Journal of Fisheries and Aquatic Sciences 67:2002–2013.

Jacobson, P. C., T. K. Cross, J. Zandlo, B. N. Carlson, and D. L. Pereira. 2012. The effects of climate change and eutrophication on Cisco *Coregonus artedi* abundance in Minnesota lakes. Advances in Limnology 63:417–427.

Jacobson, P. C., X. Fang, H. G. Stefan, and D. L. Pereira. 2013. Protecting Cisco (*Coregonus artedi* Lesueur) oxythermal habitat from climate change: building resilience in deep lakes using a landscape approach. Advances in Limnology 64:323–332.

Jacobson, P. C., G. J. A. Hansen, B. J. Bethke, and T. K. Cross. 2017. Disentangling the effects of a century of eutrophication and climate warming on freshwater lake fish assemblages. PLOS (Public Library of Science) ONE [online serial] 12(8):e0182667.

Kraemer, B. M., O. Anneville, S. Chandra, M. Dix, E. Kuusisto, D. M. Livingstone, A. Rimmer, S. G. Schladow, E. Silow, E., L. M. Sitoki, R. Tamatamah, Y. Vadeboncoeur, and P. B. McIntyre. 2015. Morphometry and average temperature affect lake stratification responses to climate change. Geophysical Research Letters 42:4981–4988.

Lark, T. J., J. M. Salmon, and H. K. Gibbs. 2015. Cropland expansion outpaces agricultural and biofuel policies in the United States. Environmental Research Letters [online serial] 10:044003.

Latta, W. C. 1995. Distribution and abundance of the Lake Herring (*Coregonus artedi*) in Michigan. Michigan Department of Natural Resources, Fisheries Division Research Report 2014, Lansing.

Lyons, J., T. P. Parks, K. L. Minahan, and A. S. Ruesch. 2017. Evaluation of oxythermal metrics and benchmarks for the protection of Cisco (*Coregonus artedi*) habitat quality and quantity in Wisconsin lakes. Canadian Journal of Fisheries and Aquatic Sciences 75:600–608.

Magnuson, J. J., D. M. Robertson, B. J. Benson, R. H. Wynne, D. M. Livingstone, T. Arai, R. A. Assel, R. G. Barry, V. Card, and E. Kuusisto. 2000. Historical trends in lake and river ice cover in the Northern Hemisphere. Science 289:1743–1746.

Moorhouse, H. L., S. McGowan, M. D. Jones, P. Barker, P. R. Leavitt, S. A. Brayshaw, and E. Y. Haworth. 2014. Contrasting effects of nutrients and climate on algal communities in two lakes in the Windermere catchment since the late 19th century. Freshwater Biology 59:2605–2620.

Olmanson, L. G., M. E. Bauer, and P. L. Brezonik. 2008. A 20-year Landsat water clarity census of Minnesota's 10,000 lakes. Remote Sensing of Environment 112:4086–4097.

Olmanson, L. G., P. L. Brezonik, and M. E. Bauer. 2013. Geospatial and temporal analysis of a 20-year record of Landsat-based water clarity in Minnesota's 10,000 lakes. Journal of the American Water Resources Association 50:48–761.

Olmanson, L. G., P. L. Brezonik, and M. E. Bauer. 2015. Remote sensing for regional lake water quality assessment: capabilities and limitations of current and upcoming satellite systems, Pages 111–140 *in* T. Younos and T. E. Parece, editors. Advances in watershed science and assessment. Springer, The Handbook of Environmental Chemistry 33, Berlin.

Omernik, J. M. 1987. Ecoregions of the conterminous United States. Annals of the Association of American Geographers 77:118–125.

Omernik, J. M., D. P. Larsen, C. M. Rohm, and S. E. Clarke. 1988. Summer total phosphorus in lakes: a map of Minnesota, Wisconsin, and Michigan, USA. Environmental Management 12:815–825.

O'Reilly, C. M., S. Sharma, D. K. Gray, S. E. Hampton, J. S. Read, R. J. Rowley, P. Schneider, J. D. Lenters, P. B. McIntyre, B. M. Kraemer, and 54 others. 2015. Rapid and highly variable warming of lake surface waters around the globe. Geophysical Research Letters 42:773–781.

PRISM Climate Group. 2018. PRISM climate data [online database]. PRISM Climate Group, Oregon State University. Available: http:///prism.oregonstate.edu.

Quinlan, R., and J. P. Smol. 2002. Regional assessment of long-term hypolimnetic oxygen changes in Ontario (Canada) shield lakes using subfossil chironomids. Journal of Paleolimnology 27:249.

Rabalais, N. N., R. E. Turner, and W. J. Wiseman. 2002. Gulf of Mexico hypoxia, a.k.a. "The Dead Zone." Annual Review of Ecology, Evolution, and Systematics 33:235–63.

Ramstack, J. M., S. C. Fritz, and D. R. Engstrom. 2004. Twentieth century water quality trends in Minnesota lakes compared with presettlement variability. Canadian Journal of Fisheries and Aquatic Sciences 61:561–576.

Rahmstorf, S., G. Foster, and N. Cahill. 2017. Global temperature evolution: recent trends and some pitfalls. Environmental Research Letters [online serial] 12:054001.

Ray, D. K., N. D. Mueller, P.C. West, and J. A. Foley. 2013. Yield trends are insufficient to double global crop production by 2050. PLOS (Public Library of Science) ONE [online serial] 8(6):e66428.

Read, E. K., V. P. Patil, S. K. Oliver, A. L. Hetherington, J. A. Brentrup, J. A., Zwart, K. M. Winters, J. R. Corman, E. R. Nodine, R. J. Woolway, H. A. Dugan, A. Jaimes, A. B. Santoso, G. S. Hong, L. A. Winslow, P. C. Hanson, and K. C. Weathers. 2015. The importance of lake-specific characteristics for water quality across the continental United States. Ecological Applications 25:943–955.

Read, J. S., J. I. Walker, A. P. Appling, D. L. Blodgett, E. K. Read, and L. A. Winslow. 2016. geoknife: reproducible web-processing of large gridded datasets. Ecography 39:354–360.

Rose, K. C., L. A. Winslow, J. S. Read, and G. J. A. Hansen. 2016. Climate-induced warming of lakes can be either amplified or suppressed by trends in water clarity. Limnology and Oceanography Letters 1:44–53.

Scavia, D., J. D. Allan, K. K. Arend, S. Bartell, D. Beletsky, N. S. Bosch, S. B. Brandt, R. D. Briland, I. Daloglu, J. V. DePinto, D. M. Dolan, M. A. Evans, T. M Farmer, D. Goto, H. Han, T. O. Hook, R. Knight, S. A. Ludsin, D. Mason, A. M. Michalak, R. P. Richards, J. J Roberts, D. K Rucinski, E. Rutherford, D. J. Schwab, T .M. Sesterhenn, H. Zhang, and Y. Zhou. 2014. Assessing and addressing the re-eutrophication of Lake Erie: Central Basin hypoxia. Journal of Great Lakes Research 40:226–246.

Soller, D. R., and P. H. Packard. 1998. Map showing the thickness and character of Quaternary sediments in the glaciated United States east of the Rocky Mountains: surficial Quaternary sediments. United States Geological Survey, Reston, Virginia.

Stefan, H. G., M. Hondzo, X. Fang, J. G. Eaton, and J. H. McCormick. 1996. Simulated long-term temperature and dissolved oxygen characteristics of lakes in the northcentral United States and associated fish habitat limits. Limnology and Oceanography 41:1124–1135.

Stoddard, J. L., J. Van Sickle, A.T. Herlihy, J. Brahney, S. Paulsen, D. V. Peck, R. Mitchell, and A. I. Pollard. 2016. Continental-scale increase in lake and stream phosphorus: are oligotrophic systems disappearing in the United States? Environmental Science and Technology 50:3409–3415.

Tilman, D., K. G. Cassman, P. A. Matson, P. A. Naylor, and S. Polasky. 2002. Agricultural sustainability and intensive production practices. Nature (London) 418:671–677.

Umbanhowar, C. E., P. Camill, and J. A. Dorale. 2011. Regional heterogeneity and the effects of land use and climate on 20 lakes in the big woods region of Minnesota. Journal of Paleolimnology 45:151–166.

Wang, L., T. Brendan, P. Seelbach, A. Cooper, D. Allan, R. Clark, and M. Wiley. 2008. Landscape based identification of human disturbance gradients and reference conditions for Michigan streams. Environmental Monitoring and Assessment 141:1–17.

Winslow, L. A., J. S. Read, G.J.A. Hansen, and P. C. Hanson. 2015. Small lakes show muted climate change signal in deepwater temperatures. Geophysical Research Letters 42:355–361.

Wood, S. 2006. Generalized additive models: an introduction with R. Chapman and Hall/CRC Boca Raton, Florida.

Wright, C. K., and M. C. Wimberly. 2013. Recent land use change in the Western Corn Belt threatens grasslands and wetlands. Proceedings of the National Academy of Sciences of the United States of America 110:4134–4139.

American Fisheries Society Symposium 90:159–177, 2019
© 2019 by the American Fisheries Society

Assessing Site-Selection Strategies for Modeling the Influence of Landscape Factors on Stream Fish Assemblages

Lindsey A. Bruckerhoff* and Keith B. Gido
Kansas State University, Division of Biology
116 Ackert Hall, Manhattan, Kansas 66506, USA

Abstract.—Linking landscape features, both natural and human-altered, to aquatic ecosystem structure and function is a fundamental objective in landscape ecology and freshwater science, but this process is data- and resource-intensive. Quantifying how landscape stressors influence aquatic communities requires balancing logistic and financial constraints with effectively sampling the landscape to capture gradients of interest. There are a variety of ways to balance these constraints, such as using existing data, handpicked site selection, or a statistical site-selection scheme. Poor sampling design reduces statistical power; however, we do not know how differences in site-selection designs influence our ability to measure ecological responses to landscape gradients. We quantified how the distribution of sample sites across landscape gradients affected the measured responses of stream fish assemblages to these gradients at different sample sizes. Specifically, we used randomization tests to compare the variability in the responses of fish assemblage structure (species richness and composition) to catchment area and land use (agricultural land) with manipulated distributions (random, highly skewed, and uniform) of sites across these landscape gradients. Assemblage composition was more sensitive than species richness to sampling design, and we observed less variability in the detected response of assemblage composition when samples were distributed uniformly across landscape gradients, especially when sample sizes were small. Although strong responses to environmental gradients, such as species richness to catchment area, are robust to sampling distributions, large sample size and a uniform distribution of samples might help elucidate more subtle responses to environmental gradients.

Introduction

Lotic communities face multiple stressors as humans continue to modify landscapes (Roth et al. 1996; Allan et al. 1997; Allan 2004). Land use change is one of the primary drivers of biodiversity loss in freshwater systems (Sala et al. 2000), so determining how landscapes influence the structure and function of riverine systems is a top goal in freshwater ecology and management. Landscape alterations influence freshwater systems through a variety of mechanisms at multiple spatial scales (Schlosser 1991; Townsend et al. 2003; Allan 2004). For example, land use changes alter physiochemical attributes of stream habitats through sedimentation, nutrient en-

* Corresponding author: lbrucke@ksu.edu

richment, contaminant pollution, hydrologic alteration, riparian modification, and loss of woody debris (summarized in Allan 2004). These impacts may be the result of landscape change at multiple spatial scales, including changes in areas directly connected to the river channel, areas within the floodplain and riparian zones, and areas with indirect contact through surface water, groundwater, and sediment flows into the river channel (Wang et al. 2006). Landscape alterations interact with natural characteristics across these spatial scales, as well as with stochastic (such as precipitation or temperature) and historic (biogeography) factors, to produce ecological patterns across river networks.

Because multiple environmental gradients interact at multiple spatial extents to drive patterns of lotic community structure, quantifying ecological responses to even a single landscape factor is a data-intensive task. Capturing coarse-scale gradients of environmental features requires sampling large areas and collecting many samples. In addition to the intensive sampling required to capture terrestrial landscape features, researchers must also consider the hierarchical structure of riverine systems. Broadscale landscape features may interact with properties of the stream network (e.g., stream size or location in the network). For example, in one study, insects unique to headwaters were at greater risk of extirpation from urbanization than species common throughout the stream network (Smith and Lamp 2008). To capture these types of interactive effects, stream network features must also be adequately sampled. As a result, researchers are often constrained in their ability to sample enough sites to capture complex responses across multiple environmental gradients, and it is critical to develop efficient sampling strategies to optimize allocation of effort.

There are many biomonitoring data sets collected by a variety of institutions (e.g., local, state, and federal governments; citizen science; universities; and museums) available to study impacts of landscape change on various ecological responses (see Buss et al. 2015 for summary of national data sets). To save time and money, many studies have used these existing data sets to assess how landscape patterns influence stream systems. We performed a cursory review of papers written since 2002 assessing the effects of land use on stream fish assemblages. Our review included the first 50 papers found from major freshwater science and landscape journals. From this review, we found that 18 out of 50 papers used pre-existing data sets (Table 1). Although few people may criticize scientists for being resourceful, there is potential bias associated with using existing landscape-stream data sets because data were likely collected with different objectives, sampling designs, and sampling methods (Deweber et al. 2014; Maas-Hebner et al. 2015). For example, the U.S. Environmental Protection Agency collects biological, chemical, and physical data at stream sites throughout the country through the National Aquatic Resource Surveys (NARS). Stream sites in this program are randomly selected based on stream size, spatial distribution (interspersed across the United States), and accessibility (USEPA 2012). Thus, they may not be ideal for studying, for example, the influence of impervious surfaces on aquatic communities within a state because they were not collected with this intention. The NARS data likely cover a full gradient of stream sizes in the study area but may not capture the full gradient of impervious surfaces. In addition to using pre-existing data, many studies also used what we classified as handpicked sites (Table 1). In those studies, authors provided little to no rationale for choosing sample sites. For example, an author investigating the influence of impervious surfaces on fish assemblages may indicate that sites were sampled along a gradient of urban-

Table 1. Site-selection methods used in 50 papers investigating the influence of landscape factors on fish assemblage data. These papers were compiled by conducting a cursory review of papers written since 2002 assessing the effects of land use on stream fish assemblages and included the first 50 papers found from major freshwater science and landscape journals. Hand-picked designs were those with vague description indicating that sites were distributed across environmental gradients, but no formal method for the allocation of sample units across these gradients was provided. Studies were classified as using existing data if they indicated the use of data from a publicly available data set from a public or private institution or used data collected in another study for a different purpose. Designed studies were those in which a clear description of the sample unit allocation (random, systematic, etc.) was provided. Species distribution models (SDMs) and historic occurrences represents samples that were allocated based on species distribution models or historic occurrence data.

Site-selction method	Number of papers
Handpicked	20
Existing data	18
Not provided	6
Designed	5
SDM and historic occurrences	1

ization intensity but not provide any information about how that gradient was quantified or how sites were chosen.

Our observation that many broadscale analyses quantifying impacts of landscape features on stream biota used existing data or provided little rationale for site selection led us to ask how sample-site selection across environmental gradients influences the measured response of ecological metrics. Specifically, we tested how the measured relationships between stream fish assemblage structure (species richness and composition) and landscape factors (both natural and human land use changes) varied with different sampling designs (random, highly skewed, or uniformly distributed) of sample sites (stream segments) across landscape gradients. We also compared variation in measured responses across different sample sizes to evaluate the importance of sample designs at different levels of effort. This study provides insight into the importance of study design when conducting landscape studies and may serve as a resource to those designing studies to quantify the relationship between landscape change and stream biotic responses.

Methods

General Framework

We combined two data sets to quantify how the distribution of sample sites across environmental gradients influenced measured ecological responses. Our two response variables were fish species richness (hereafter, richness) and assemblage composition, and the predictors were catchment area and the proportion of agricultural lands in the catchment of each site. We calculated catchment areas based on data included in the National Hydrography Dataset (USGS 2016) using Arc Hydro Tools in ArcMap 10.4 (ESRI 2011). Use of the term "catchment" in this paper refers to true upstream catchments delineated from the downstream point of all sites (Omernik et al. 2017). We then calculated the proportion of agricultural land within each catchment using Landsat raster data from the Kansas Satellite Image Database (KARS 2005). Agricultural land in this data set was classified as cropland planted with corn, soybeans, sorghum, winter wheat, or alfalfa, or land used as fallow or planted with multiple crops, but did not include land classified as pasture (Peterson et al. 2010). We defined a site as a stream segment, or the portion of stream between two confluences. Our general framework consisted of grouping similar sites based on each predictor (e.g., small catchment area, low proportion of agriculture), randomly drawing sites from each group to fit different sample designs (random, right-skewed, and uniform), and modeling the relationship between the ecological responses (richness or composition) and gradients of interest (e.g., linear regression of richness versus catchment area). We repeated draws of sites for each relationship at varying sample sizes (25, 50, 75, and 100) 1,000 times (without replacement), developed models for each draw, and assessed the distribution of estimated coefficients across those 1,000 draws. We were interested in the variability in our ability to detect a response (the proportion of times we detected a nonzero slope or significant relationship). All analyses were completed in R 3.4.2 (R Core Team 2017).

Fish Assemblage Data Sets

We combined fish assemblage data for streams in the Flint Hills Ecoregion of Kansas (Figure 1) using existing records from the Kansas Department of Wildlife, Parks, and Tourism (KDWPT) and Kansas State University (KSU). The Kansas Department of Wildlife, Parks, and Tourism conducts stream surveys to monitor distributions of native fish throughout different parts of Kansas each year; this data set contains data from different sites collected between 1995 and 2012. The KDWPT sites are selected using a variety of methods depending on the goals of sampling for each particular year. The KSU data were

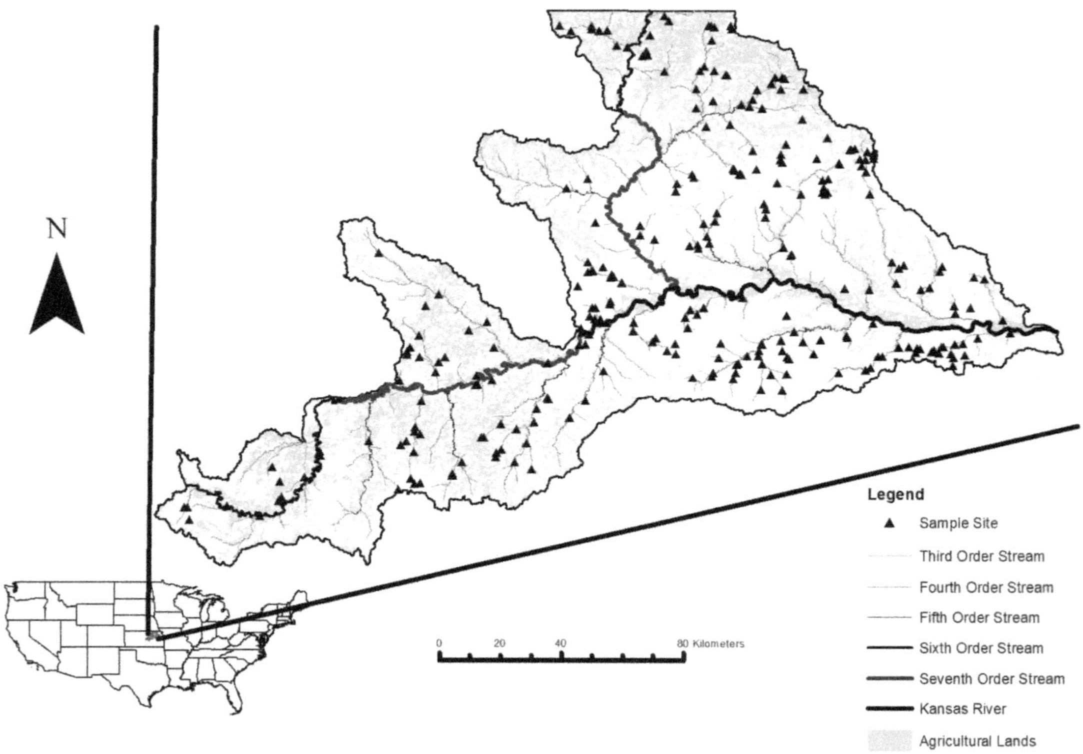

Figure 1. Locations of sites included in the combined fish assemblage data set sampled by the Kansas Department of Wildlife, Parks, and Tourism and Kansas State University. All sites were located in the Kansas River basin and the Flint Hills Ecoregion of Kansas, USA.

collected in the summer of 2017 to determine the influence of multiple stressors on stream fish assemblages. These sites were selected using a random stratified design across a range of stream sizes and gradient of agricultural land use. All sites in the KSU data set were wadeable and sampled using backpack electrofishing and seining. We combined these two data sets to increase the pool of sites used to create manipulated sampling distributions. All results in this study are constrained by the ecological responses apparent in the global pool of sites. In other words, the largest effect sizes produced by combinations of sample sizes and sampling distributions were constrained by the effect sizes observed in the global data set. In addition, our random samples actually represent random samples of the sites in this combined data set, not random samples of the environmental gradients. For the sake of this study, we consider patterns observed in the global pool of sites to represent the true relationships. We then compared how our ability to detect these relationships changed with different combinations of sample sizes and site-selection designs.

Sample Distributions

We chose to test fish assemblage responses across environmental gradients with random, right-skewed, and uniform distributions (Figure 2). We chose these distributions because they are (based on the authors' experience) commonly encountered by researchers conducting landscape studies. Completely random and random-stratified designs are commonly used to designate samples across space (Manly 2001; Quinn and Keough 2002). Because completely random sampling does not account for probabilities of encountering different environmental gradients, more samples are likely needed to capture a full range of variability across gradients. Random-stratified designs distribute sites across strata representing different levels of gradients of interest, but samples are randomly designated within each stratum. If samples are allocated equally across strata, (i.e., equal-random-stratified), this design gives rise to uniform distributions of predictor variables, whereas purely random designs will produce distributions that mirror the proportional distribution of gradients across the landscape. Because samples are allocated equally across strata, fewer samples are needed to capture variability of environmental gradients of interest. However, the choice of strata may be inappropriate for other studies or other inferences. Highly skewed distributions may arise from a combination of the natural distribution of the gradient across the landscape and bias associated with sampling. For example, within a drainage basin, stream networks are dominated by small streams (regardless of how you measure stream size). Within a given area, a histogram of stream sizes will likely be right-skewed, so a randomly sampled stream network will produce right-skewed distributions of stream size. In addition, stream-size bias may also arise from sampling bias. Smaller streams are typically easier to access and sample relative to larger streams, potentially adding further bias toward a right-skewed distribution of sample sites. We therefore distributed our samples randomly, right-skewed, and uniformly (using an equal-random-stratified design) to capture those common sampling distributions.

Landscape Predictors and Ecological Responses

We chose catchment area and the proportion of agricultural lands in upstream catchments as predictor variables because we expected to see differences in our ability to detect a response between those variables. The influence of catchment area on fish assemblage structure is well documented for stream fishes (Matthews 1998). Although this pattern may not be considered an ecological response

Gradient Distribution	Frequency Distribution	Produced By	Strengths	Weaknesses
Random	Mirrors landscape composition	Completely random sampling	Sample units independent Do not need *a priori* knowledge about the system	May not capture gradients of interest Wasted effort in common habitats
Highly Skewed	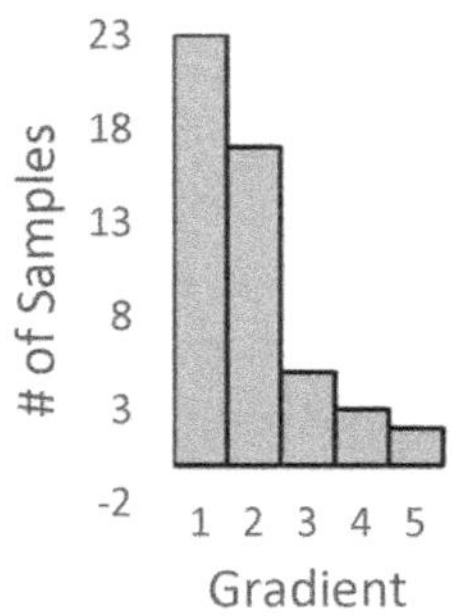	Biased sampling	Convenient, may be able to achieve high sample size	Not enough samples across entire gradient Wasted effort around some value of gradient Sample units may not be independent
Uniform	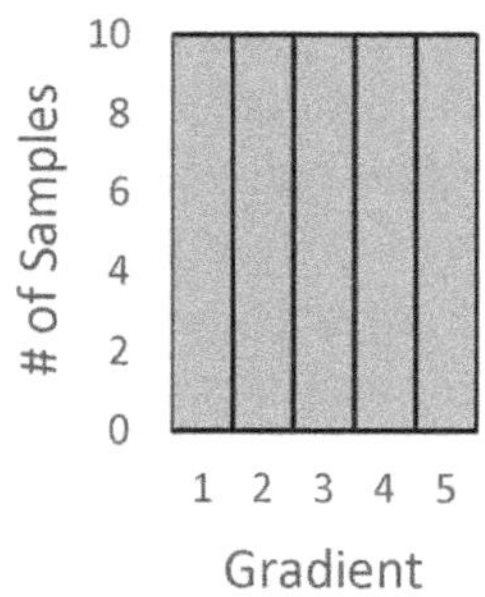	Equal-random-stratified sampling	No wasted effort Entire gradient sampled	Need *a priori* knowledge about the systems Potential spatial clumping

Figure 2. Random, highly skewed, and uniform gradient distributions are produced by different sampling distributions, and there are different strengths and weaknesses associated with each distribution.

in the same way assemblages respond to land use change, this known pattern allowed us to confidently explore the influence of sample design. Because we expected a strong effect, we predicted the influence of sample design on detecting a relationship between catchment area and fish assemblages would be less important. Land use effects on fish assemblages, however, were expected to be harder to detect because of multiple indirect influences on instream physical and chemical conditions that influence fish (Berkman et al. 1986; Schlosser 1991; Hering et al. 2006); therefore, we expected the importance of study design to be more conspicuous when assessing the influence of agricultural lands.

We chose richness and assemblage composition (Jaccard dissimilarity) as response variables because they are basic descriptors of assemblage structure. Jaccard dissimilarity was used rather than other distance matrices because it is based on presence–absence data, and abundance data from these data sets were unreliable due to the variety of methods and range of years included. Species richness and Jaccard dissimilarity capture both univariate (richness) and multivariate (composition) responses so that we could compare the influence of sample design on response variables with different statistical properties.

Randomization Tests

To quantify the influence of sample design on the response of richness to landscape factors, we developed ordinary least-squares linear regression models for 1,000 random draws of sites with varying sample sizes (N = 25, 50, 75, and 100) across the three sample distributions (random, right-skewed, and uniform) of catchment area and the proportion of catchment agricultural lands. Before drawing, we removed sites with catchment areas greater than 1,000 km^2 or less than 2.7 km^2 because those few outlier sites influenced the classification scheme used to draw samples. From this final data set, we classified sites into five strata (Table 2) for each predictor (catchment area and proportion of agricultural lands) using the Jenks natural breaks classification method (Jenks 1977). The Jenks algorithm finds natural groups inherent in the data set and maximizes differences between classes. We used these groups to draw samples with uniform (equal number of samples in each group) and highly right-skewed (dominated by low values) distributions for both landscape predictors. Catchment area was log-transformed in regression models to improve linearity. Across each sample size and distribution combination, we calculated the mean slope and 95% confidence intervals in the predicted slopes across the 1,000 draws. We also calculated the proportion of draws that produced slopes of zero (indicating no relationship). We tested for positive slopes for the relationship between richness and catchment area because we expected richness to increase with increasing stream size. We also tested for negative relationships between the proportion of agricultural lands and richness because we expected decreases in richness with increasing upstream area converted for agricultural use.

To evaluate the relationship between fish assemblage composition and landscape features, we used Mantel tests to quantify the concordance of fish assemblage dissimilarity and landscape features across sites. We started with the same global pool of sites used in richness regressions (outlier sites based on catchment area removed) and then removed rare species (those not present across at least

Table 2. Predictor variables were classified into five strata using the natural Jenks algorithm to produce distributions with right-skewed and uniform distribution.

	Range				
Variable	Stratum 1	Stratum 2	Stratum 3	Stratum 4	Stratum 5
Proportion of agricultural land	0.00–10.20	10.30–25.40	25.50–48.20	48.30–67.60	67.70–89.00
Catchment area (km^2)	2.85–10.27	10.28–27.67	27.68–79.64	79.65–246.23	246.24–864.19

5% of the sites). We used the same strata used in regression models to draw sites exhibiting the three sampling distributions. For each draw, a Jaccard dissimilarity index representing the proportion of unique taxa between samples was computed for each pair of sites. We also calculated the differences between catchment areas or proportion of agricultural land between these same pairs of sites. Two matrices, one of Jaccard dissimilarity and one of differences in environmental gradients, were generated to represent all pairwise comparisons for each set of samples. We then completed a permutational (N = 1,000) Mantel test to determine if the correlation between assemblage distance and difference in catchment area or proportion of agricultural lands between all pairs of sites was significant. We calculated the proportion of tests with nonsignificant Mantel tests and assessed the distribution of P-values across all draws and sample distributions.

We tested for correlations between the distance between sites, differences in richness, and assemblage dissimilarity to test for potential issues regarding spatial autocorrelation. Richness was not correlated with distance between sites (Pearson's r = 0.04) and assemblage dissimilarity was only moderately correlated with distance between sites (Pearson's r = 0.32), so we did not correct for spatial proximity in our analysis. We also tested for correlations between the two predictor variables, catchment area and proportion of agriculture, which were also not correlated (Pearson's r = 0.08, Figure 3).

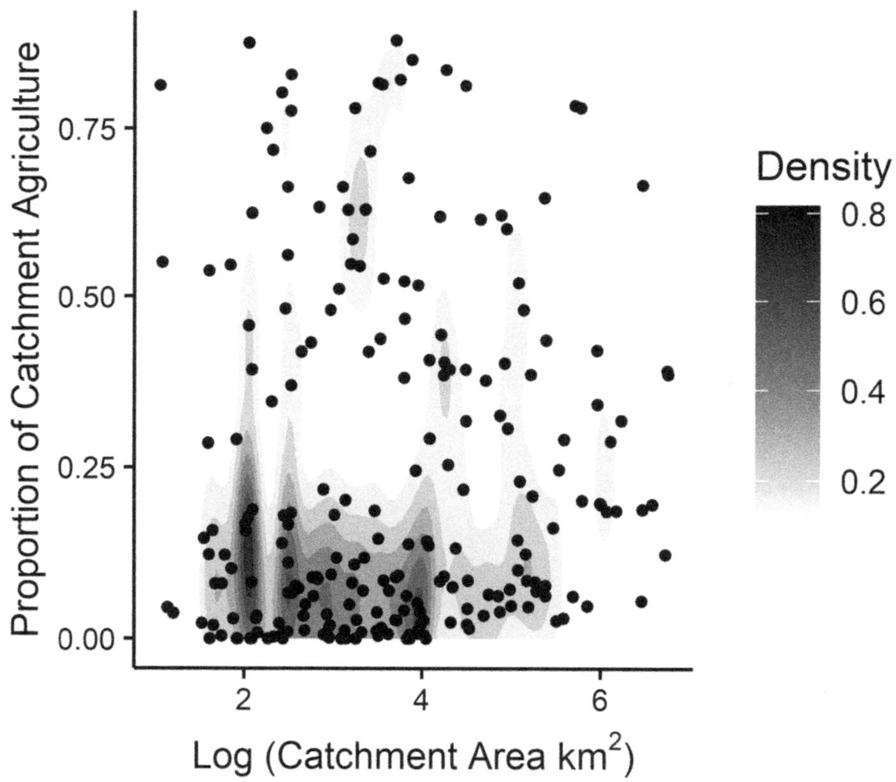

Figure 3. The density of sites across log-transformed catchment area (km^2) and the proportion of catchment agriculture. Shading represent kernel density estimates of sites, with darker values representing a higher density of sites.

Results

Our combined fish assemblage data set had 279 sites after removing outliers. For richness regressions, 72 species were included in the global species pool. After removing rare species, there were 33 in the global pool for composition comparisons. Strata classifications allowed us to produce a gradient of skewness in the distribution of sites across landscape gradients (Table 3). Sites in the combined data set were evenly distributed across the gradient of catchment size, but most sites had low-catchment agriculture regardless of catchment size (Figure 3).

Fish Species Richness

Richness had a significant, positive relationship ($P < 0.001$, $m = 2.35$, adjusted $R^2 = 0.29$) with catchment area in the global data set (Figure 4a). This positive relationship was detected across all combinations of sample sizes and sampling distributions (Figure 5a).

Table 3. Mean skewness values of the distribution of agricultural land across draws with different sample sizes and random, right-skewed, and uniform distributions. Distributions of catchment area across the three distributions exhibited similar skewness values as these.

Gradient distribution	Sample size	Skewness
Random	25	0.91
	50	0.91
	75	0.91
	100	0.92
Right-skewed	25	1.27
	50	1.24
	75	1.16
	100	1.17
Uniform	25	0.16
	50	0.16
	75	0.16
	100	0.16

Richness also had a significant, negative relationship ($P = 0.005$, $m = -3.93$, adjusted $R^2 = 0.03$) with the proportion of agricultural land in the catchment of each site (Figure 4b). Mean slopes derived from randomization tests varied between −5.02 and −3.48 across all sample sizes and sampling distributions, with lower, steeper slope means for uniform distributions across all sample sizes (Figure 5b). Confidence intervals did not overlap zero for uniform distributions at samples sizes of 50, 75, or 100 (Figure 5b), suggesting robust analyses at those levels of effort. The proportion of draws that produced nonnegative slopes for the relationship between richness and the proportion of agricultural lands decreased with increased sample size and was consistently lowest for uniform sampling distributions (Figure 6).

Assemblage Composition

Jaccard dissimilarity was significantly correlated with difference in catchment area ($P < 0.001$) and the proportion of agricultural lands ($P < 0.001$) in the global data set. Mantel tests evaluating the correlation between Jaccard dissimilarity and catchment area were significant across all sampling distributions at sample sizes greater than 50 and for all sample sizes with a uniform distribution (Figure 7a). When the sample size was 25, 6% and 4% of the draws produced nonsignificant Mantel test for random and right-skewed distributions (Figure 7a).

There was more variability, however, in the distribution of P-values when testing for associations between assemblage composition and catchment agriculture (Figure 7b). There were at least some nonsignificant Mantel tests across all combinations of sample sizes and sampling distributions, except uniform distributions with sample sizes of 100 (Figure 7b). The proportion of nonsignificant Mantel tests decreased with sample size and was consistently the lowest

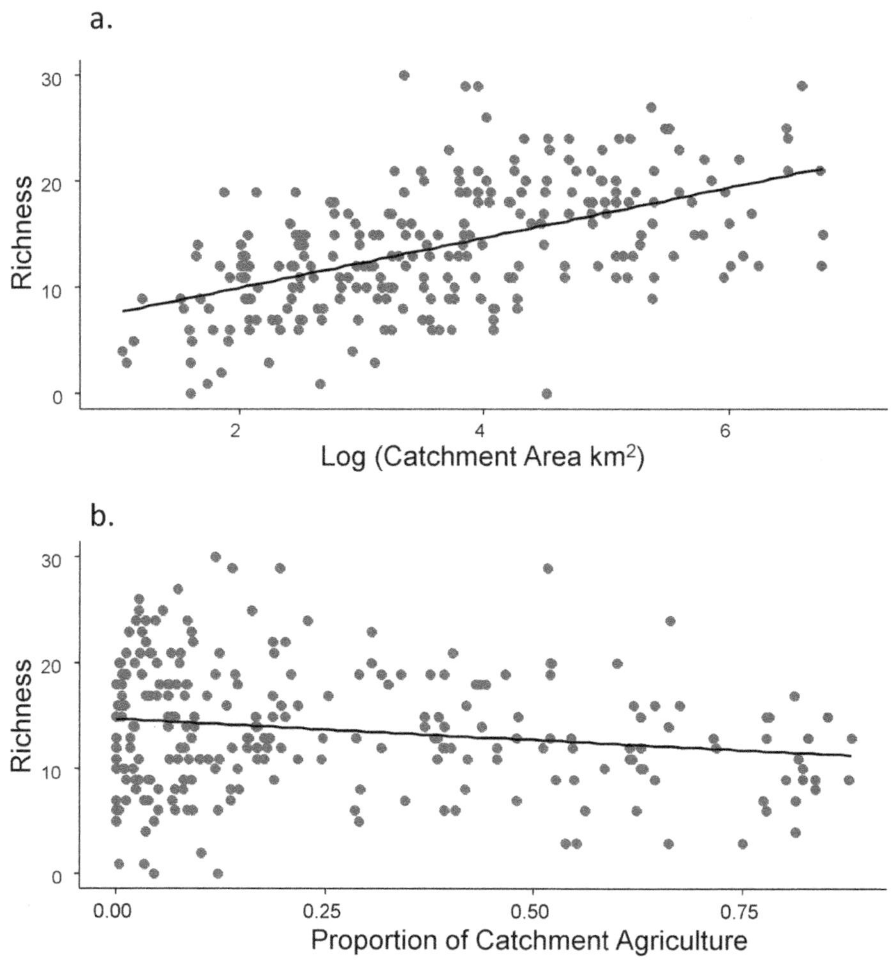

Figure 4. Fish species richness **(a)** increased ($P < 0.001$, $m = 2.35$, adjusted $R^2 = 0.29$) with increased catchment area and **(b)** decreased ($P = 0.005$, $m = -3.93$, Adjusted $R^2 = 0.03$) with increased proportion of agricultural land in each catchment across the global pool of sites.

for uniform distributions (Figure 8). More than 50% of draws produced nonsignificant Mantel tests at sample sizes of 25 with random and right-skewed distributions, whereas less than 2% of draws were not significant with uniform distributions and sample sizes greater than 25 (Figure 8).

Discussion

Study design, as indicated by differing predictor variable distributions, influenced the variability in our ability to detect fish assemblage responses. We observed less variability in the response of both fish species richness and assemblage composition to catchment area versus the proportion of catchment agriculture. This fit our prediction that land use effects on fish assemblages would be weaker than the influence of catchment area. Similarly, we observed more variation in our ability to detect these relationships using Jaccard dissimilarity than species richness. However, we consistently observed less variation with uniform distributions and large sample sizes for both Jaccard index and species richness responses

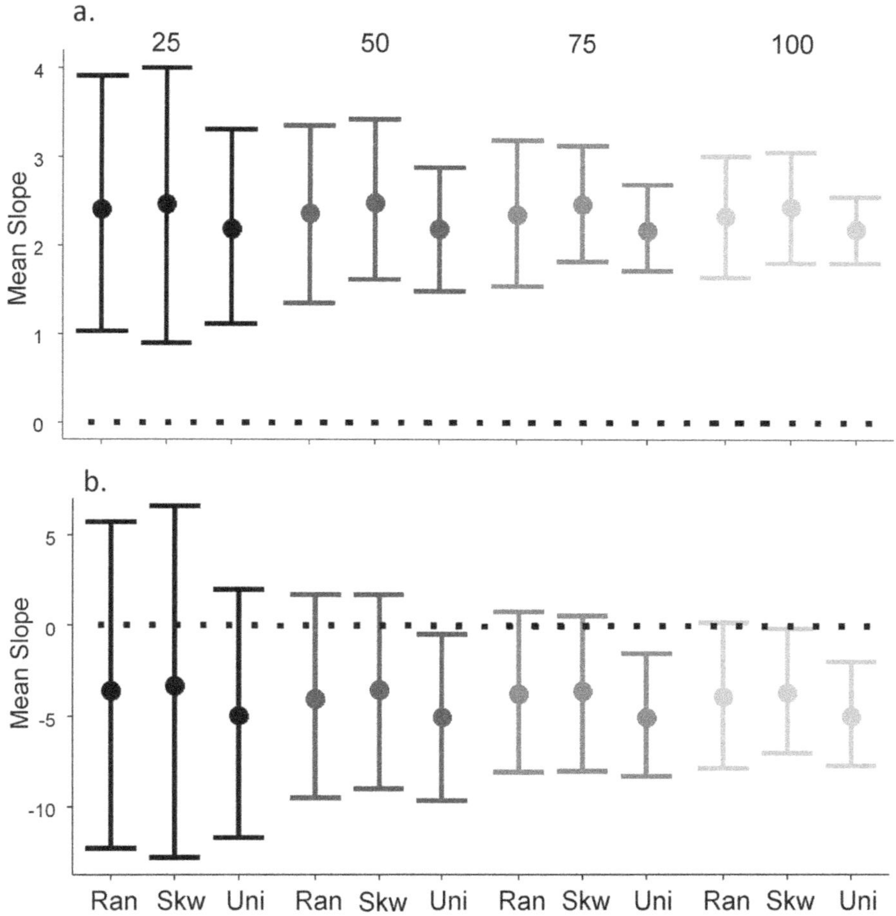

Figure 5. Mean slope and confidence intervals derived from 1,000 iterations of regressions of stream fish species richness against **(a)** catchment area and **(b)** the proportion of agricultural land in the catchment for N = 25, 50, 75, and 100 samples. Samples were drawn randomly (Ran), right-skewed (Skw), and uniformly (Uni) from an existing data set. The dotted line at mean slope = 0 indicates when no relationship was observed between richness and catchment area or proportion catchment agriculture.

to catchment agriculture. This indicates that sample design is especially important to consider when investigating relatively weak or multivariate ecological relationships. We expect this result to be consistent across different regions, although the strength of the relationship between fish assemblage structure (species richness and composition) and environmental gradients may vary.

We observed more variability in assemblage responses when samples sizes were small, but sample design could offset those differences. For example, our ability to detect correlations between assemblage dissimilarity and differences in catchment area or catchment agriculture declined with fewer samples but was improved when those samples were from a uniform variable distribution. This is an important consideration because factors such as study objectives, timelines, funding, personnel, and other logistic constraints (e.g., access, distance between sites) limit the number of sample sites included in a study (Hughes and Peck 2008). Although statisti-

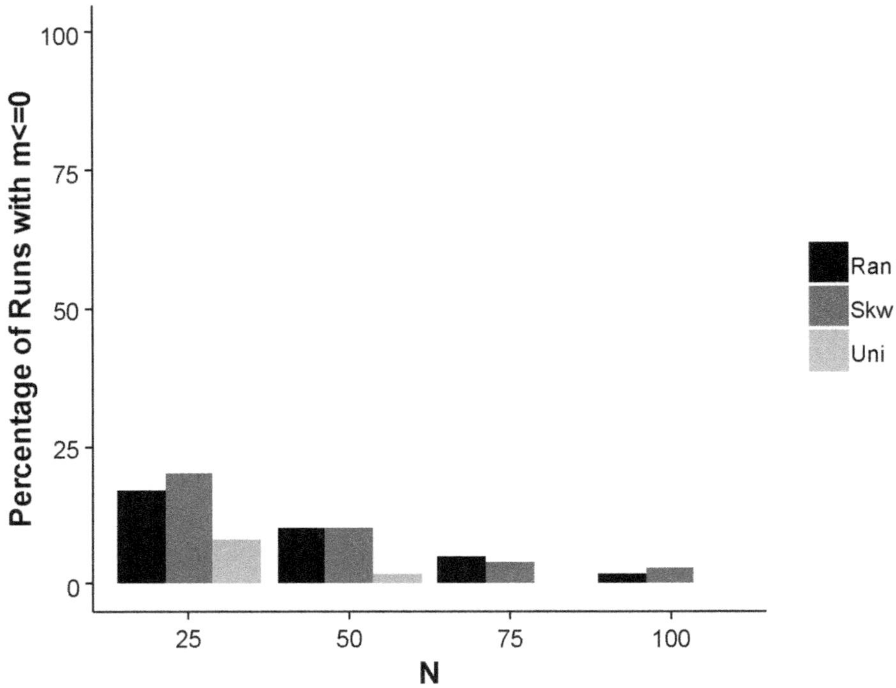

Figure 6. The proportion of runs (N = 1,000) with slopes less than or equal to zero, indicating no relationship between catchment land use and fish species richness for sample sizes of 25, 50, 75, and 100 for random (Ran), right-skewed (Skw), and uniform (Uni) distributions.

cal power is always a consideration when determining how many sites to sample (Toft and Shea 1983; Fairweather 1992; Quinn and Keough 2002), our study and others (Albert et al. 2010) suggest that poor sample design can reduce statistical power by incorporating unnecessary variability. Further, if a response is tested across multiple environmental gradients, power will further be limited, as more samples are needed to capture multiple gradients. Thus, efficient sample design across multiple gradients may help optimize the designation of sample units to improve statistical power (see Additional Research and Management Needs).

We observed less variation in our ability to detect relationships between fish assemblage structure and environmental gradients when sites were drawn across uniform gradients. Uniform designation of sites incorporates prior knowledge about the system by requiring identification of important environmental gradients before sampling. These designations are achieved using equal-random-stratified sampling designs (Figure 2), in which an equal number of samples are allocated to each stratum but samples are randomly allocated within strata. Equal-random-stratified designs are ideal as gradients can be sampled efficiently without wasting effort oversampling any part of the gradient (Olsen and Peck 2008, Herlihy et al. 2019, this volume; Robinson et al. 2019, this volume). Similar to the current study, Hirzel and Guisan (2002) observed that equal-stratified designs were the most accurate and robust when predicting habitat suitability and the presence–absence of virtual species in simulated terrestrial landscapes relative to random, regular, or proportional-stratified designs. However, other studies comparing ecological responses across different sample designs produced

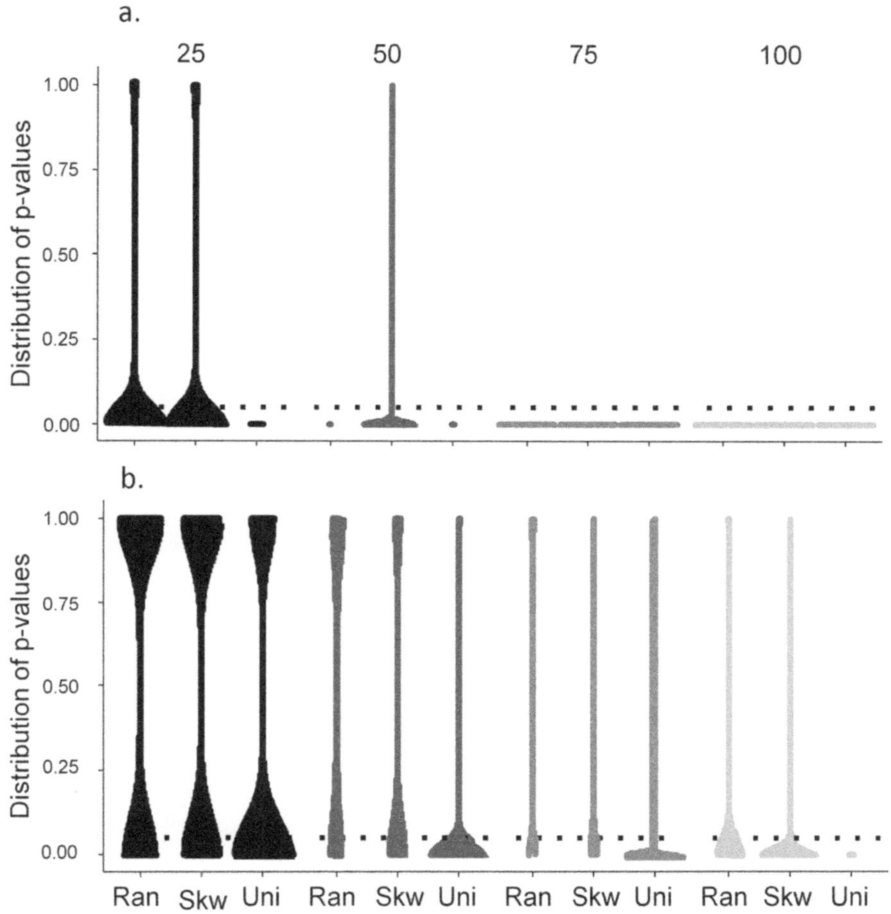

Figure 7. Distribution of *P*-values obtained from Mantel tests of difference in **(a)** catchment area and **(b)** the proportion of agricultural land in the catchment versus fish assemblage Jaccard dissimilarity for sample sizes of 25, 50, 75, and 100. Samples were drawn randomly (Ran), highly right-skewed (Skw), and uniformly (Uni) from an existing data set.

mixed results and do not always suggest the use of uniform gradient distributions. For example, Albert et al. (2010) used methods similar to those used in this study and discovered that the efficiency of different sample designs was dependent on the types of parameters being estimated. Mean values of biological variables were best estimated by simple-random sampling designs, whereas optima were best predicted using model-based designs (iterative learning methods that incorporate prior knowledge; Albert et al. 2010). Mohler (1983) used simulations to recommend sampling designs that favor the extremes of gradients to best quantify abundance patterns of multiple species. Alternatively, sample design did not influence the prediction accuracy of species distribution models (SDMs) because the extent of individual species distributions had more influence on SDM predictions than sample design (Tessarolo et al. 2014). These studies suggest that sample design is (usually) important and influences the ability to detect a response, but different designs may be better suited for different response variables. Similarly, the importance of sample design may also vary with different modeling approaches, with some statistical models being

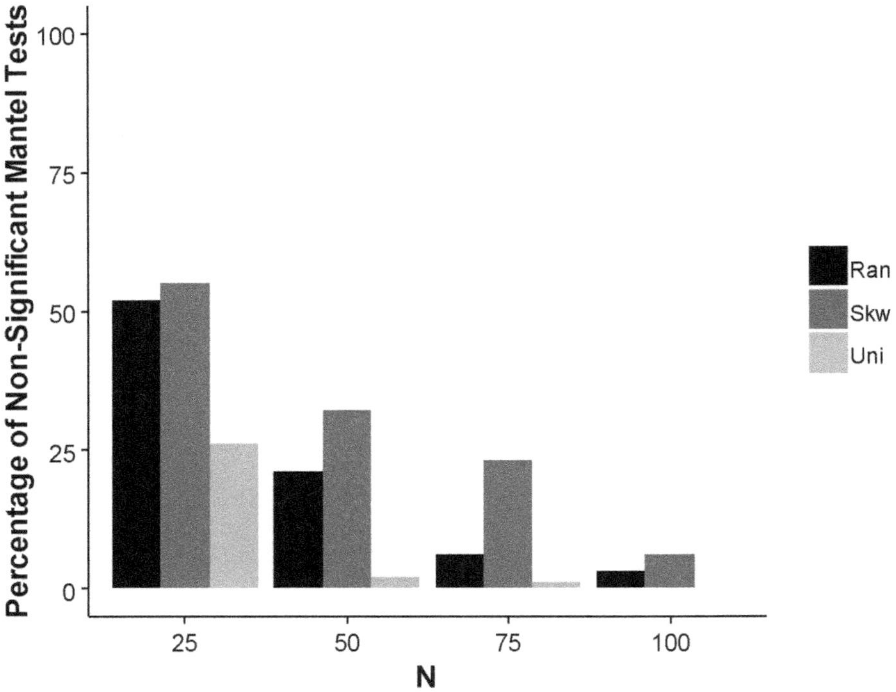

Figure 8. The proportion of runs (N = 1,000) with nonsignificant Mantel tests indicating no relationship between upstream land use and fish assemblage composition for sample sizes of 25, 50, 75, and 100 for random (Ran), right-skewed (Skw), and uniform (Uni) distributions.

more robust to sampling distributions. More work is needed to determine which type of designs are ideal for different ecological responses (presence–absence, abundance, diversity metrics, disturbance indices, etc.) and different types of statistical models.

Comparisons of sample designs are primarily studied using either real or simulated data. We drew samples from existing data to maintain some realism in our ecological relationships. Our data set included some of the biases that are apparent in true sample schemes (such as issues with accessibility or sampling close to roads) and provide realistic estimates of variability that would be detected in the study system. Albert et al. (2010) called this approach a "semivirtual simulation." Whereas the decision to use a semivirtual approach allowed us to capture realism, it did not allow us to account for some potential bias. For example, we did not account for differences in detection probabilities of species across environmental gradients. In addition, our observed ecological responses may also be biased due to the different methods of site selection used by our two data sources. Although we potentially could have avoided some of this bias by using a larger data set, such as one of the publicly available data sets mentioned in the Introduction, we would have had to alter the geographic extent of our study. Alternatively, we could have simulated data to compare different gradient distributions. Simulated data sets are useful because the true correlation structure between the response and predictor is known (Olden et al. 2004), making them useful in assessing the performance of different statistical tests used in ecology (Jackson 1993; Olden and Jackson 2000; Olden et al. 2004). Although we did not use simulated data in this study, other comparisons of sample designs that did use simulated data (Hirzel and Guisan 2002) produced similar results.

Additional Research and Management Needs

Our study is a first step in assessing the influence of sample design on the ability to detect ecological responses in stream systems. More work is needed to determine how unique properties of stream systems can be incorporated into sampling designs. The structure and function of streams is driven by heterogeneity at multiple spatial scales (hierarchical filters; Poff 1997), and stochastic events (disturbance; Townsend 1989). Although these processes have been integrated and studied in terrestrial systems using classic landscape principles like hierarchical patch dynamics (Wu and Loucks 1995), unique properties of dendritic networks challenge direct application of these ideas in streams (Benda et al. 2004). For example, confluences have unique geomorphological properties (Poole 2002), and the placement and number of confluences within a network may create uneven patterns of heterogeneity within and between networks (Benda et al. 2004). These unique properties of confluences and the spatial position of sites within stream networks are known to influence relationships between fish assemblage metrics and local environmental conditions (Hitt and Angermeier 2008). The age of geomorphic landforms might also vary from headwaters to main stems (Benda et al. 2004), and other attributes of stream ecosystems (e.g., sediment size, functional trait composition, primary production, and allochthonous inputs) vary longitudinally in stream networks (Vannote et al. 1980). Heterogeneity is also inherent in studies with extents that include multiple ecoregions or cross biogeographic boundaries (Pinto et al. 2009). Those sources of heterogeneity are common in broadscale studies across stream networks and should be considered when designing studies to measure effects of landscape change of stream ecosystems. Further work is needed to assess the effect of these attributes (location in the network, proximity to a confluence, differences in drainage density between networks, ecoregion, etc.) on the ability to detect ecological responses to landscape changes. One promising method proposed to account for spatially structured patterns of heterogeneity are spatial stream network models that capture spatial autocorrelation in stream networks resulting from confluences, network connectivity, and the direction of stream flow (Ver Hoef et al. 2006; Ver Hoef and Peterson 2010).

Controlling for these sources of variability while simultaneously comparing the effects of multiple landscape stressors requires many samples and it is likely difficult to efficiently capture multiple landscape gradients (both natural and altered). Although we did not investigate multiple gradients in the current study, we know statistical power will decline as more strata are included in a statistical model and more samples are needed to capture multiple gradients. Obtaining a representative sample of multiple stressors across stream networks is complicated by both difficulties in calculating a priori landscape metrics for streams and the lack of sampling schemes based on multiple gradients. Typically, variables describing upstream catchments (e.g., catchment area, proportion of catchment land-cover classes) are calculated post hoc for sample sites (reach or segments). This makes stratifying across these types of variables a priori difficult, as one would have to compute catchment-based metrics for every possible site within a study area. Estimates of catchment-based metrics can be made at coarser spatial scales (e.g., using hydrologic units or ecoregions), but these estimates may not accurately reflect the true upstream conditions for an individual sample unit, complicating the use of stratified designs for catchment-based predictors. In addition, current sampling strategies are largely based on single-predictor variables. Stratified sampling requires classifying gradients into multilevel strata, and capturing multiple gradients requires sampling all factorial combinations of strata levels. This is difficult to do when the natural gradient distribution is skewed because all factorial combinations of strata may not exist. For example, in the current study, it would be difficult to find sites with both intermediate-sized-catchment areas and intermediate levels of agriculture or both large catchment areas and high-catchment agriculture, but there are many sites with low-catchment agriculture at a variety of catchment areas (Figure 3). Response surface methodology, a sequential process involving several designed experiments (Box and Wilson 1951), has been suggested as a method to incorporate multiple predictor variables when trying to estimate the optimal value of a response (e.g., peak abundance of species), and its application to ecology has been explored in terrestrial systems (Albert et al. 2010). Other options for capturing multiple gradients may include compressing variability in multidimensional environmental space using ordination techniques (such as principal coordinate analysis) and stratifying samples in lower dimensional space (across a few axes; Stevens and Olsen 2004; (Olsen and Peck 2008)). Overall, a methodological framework for capturing multiple strata in stream networks is needed.

Policy Implications

Understanding the effect of study design on our ability to detect ecological responses can help inform the management of riverine systems by improving the interpretation of studies and by helping managers effectively allocate sampling effort to detect ecological relationships. Inefficient study designs may run the risk of producing type II errors (false negative), and researchers may conclude that environmental gradients are not important when in reality they are. For example, if policymakers were interested in the impact of agriculture on fish assemblage structure, they should be aware that a lack of pattern might result from low statistical power, as observed in our results. Thus, management recommendations should consider the nature of the data, including the spatial distribution of sample sites, prior to any decision-making process. Thorough descriptions of study design and the use of post hoc power analysis may help provide clarity on how much confidence can be placed in results so policymakers and managers can make decisions based on studies with robust designs. Beyond robust study designs, the biological significance is also an important consideration when translating science into policy and management decisions. Although this paper was focused on the influence of study design on detecting statistical relationships, detected statistical relationships may not always translate into biologically significant responses. Detailed investigation into individual species responses, for example, might be more relevant to specific management goals.

Conclusions

Overall, sample design influences our ability to detect ecological responses to environmental gradients, especially when sample sizes are low. This point has already been developed as a main component of study design theory, but we emphasize that sample design is particularly important when assessing subtle or complex ecological phenomena. The unique attributes of stream networks and the interactions between terrestrial and aquatic environments complicate understanding many landscape–stream ecology relationships. Stream ecologists should strive to develop new ways to overcome these inconveniences to maximize the probability of detecting complex but ecologically important relationships. Implementation of improved study designs is needed to help us uncover environment–ecological relationships, especially as humans continue to modify many landscape and stream attributes.

Acknowledgments

We thank Mark VanScoyoc and Ryan Waters from the Kansas Department of Wildlife, Parks, and Tourism for providing fish assemblage data; Michael Estey, Pamela Moore, and the Habitat and Population Evaluation Team of the U.S. Fish and Wildlife Service for financial support for data collection and motivation to complete this project; and Garrett Hopper for providing useful comments of earlier versions of this paper. We also thank three peer reviewers and Bob Hughes for suggestions for improving this manuscript.

References

Albert, C. H., N. G. Yoccoz, T. C. Edwards, C. H. Graham, N. E. Zimmermann, and W. Thuiller. 2010. Sampling in ecology and evolution: bridging the gap between theory and practice. Ecography 33:1028–1037.

Allan, J. D. 2004. Influence of land use and landscape setting on the ecological status of rivers. Limnetica 23:187–198.

Allan, J. D., D. L. Erickson, and J. Fay. 1997. The influence of catchment land use on stream integrity across multiple spatial scales. Freshwater Biology 37:149–161.

Benda, L., N. L. Poff, D. Miller, T. Dunne, G. Reeves, G. Pess, and M. Pollock. 2004. The network dynamics hypothesis: how channel networks structure riverine habitats. BioScience 54:413.

Berkman, H. E., C. F. Rabeni, and T. P. Boyle. 1986. Biomonitors of stream quality in agricultural areas: fish versus invertebrates. Environmental Management 10:413–419.

Box, G. E. P., and K. B. Wilson. 1951. On the experimental attainment of optimum conditions. Journal of the Royal Statistical Society 13:1–45.

Buss, D. F., D. M. Carlisle, T. S. Chon, J. Culp, J. S. Harding, H. E. Keizer-Vlek, W. A. Robinson, S. Strachan, C. Thirion, and R. M. Hughes. 2015. Stream biomonitoring using macroinvertebrates around the globe: a comparison of large-scale programs. Environmental Monitoring and Assessment 187.

Deweber, J. T., Y.-P. Tsang, D. M. Krueger, J. B. Whittier, T. Wagner, D. M. Infante, and G. Whelan. 2014. Importance of understanding landscape biases in USGS gage locations: implications and solutions for managers. Fisheries 39:155–163.

ESRI (Environmental Systems Research Institute). 2011. ArcGIS desktop. ESRI, Redlands, California.

Fairweather, P. G. 1992. Statistical power and design requirements for environmental monitoring. Marine and Freshwater Research 42:55–67.

Hering, D., R. K. Johnson, S. Kramm, S. Schmutz, K. Szoszkiewicz, and P. F. M. Verdonschot. 2006. Assessment of European streams with diatoms, macrophytes, macroinvertebrates and fish: a comparative metric-based analysis of organism response to stress. Freshwater Biology 51:1757–1785.

Herlihy, A. T., J. C. Sifneos, R. M. Hughes, D. V. Peck, and R. M. Mitchell. 2019. Lotic fish assemblage clusters across the conterminous United States and their associations with environmental variables. Pages 385–408 *in* R. M. Hughes, D. M. Infante, L. Wang, K. Chen, and B. F. Terra, editors. Advances in understanding landscape influences on freshwater habitats and biological assemblages. American Fisheries Society, Symposium 90, Bethesda, Maryland.

Hirzel, A. H., and A. Guisan. 2002. Which is the optimal sampling strategy for habitat suitability modelling. Ecological Modelling 157:331–341.

Hitt, N. P., and P. L. Angermeier. 2008. River-stream connectivity affects fish bioassessment performance. Environmental Management 42:132–150.

Hughes, R. M., and D. V. Peck. 2008. Acquiring data for large aquatic resource surveys: the art of compromise among science, logistics, and reality. Journal of the North American Benthological Society 27:837–859.

Jackson, D. A. 1993. Stopping rules in principal components analysis: a comparison of heuristical and statistical approaches. Ecology 74:2204–2214.

Jenks, G. F. 1977. Optimal data classification for choropleth maps. University of Kansas, Department of Geography, Occasional Paper No. 2, Lawrence.

KARS (Kansas Applied Remote Sensing). 2005. Kansas satellite image database [online database]. Available: https://kars.ku.edu/geodata/.

Maas-Hebner, K. G., M. J. Harte, N. Molina, R. M. Hughes, C. Schreck, and J. A. Yeakley. 2015. Combining and aggregating environmental data for status and trend assessments: challenges and approaches. Environmental Monitoring and Assessment 187:1–16.

Manly, B. 2001. Statistics for environmental science and management. Chapman and Hall/CRC, Boca Raton, Florida.

Matthews, W. J. 1998. Patterns in freshwater fish ecology. Springer US, Boston, Massachusetts.

Mohler, C. L. 1983. Effect of sampling pattern on estimation of species distributions along gradients. Vegetatio 54:97–102.

Olden, J. D., and D. A. Jackson. 2000. Torturing data for the sake of generality: how valid are our regression models? Ecoscience 7:501–510.

Olden, J. D., M. K. Joy, and R. G. Death. 2004. An accurate comparison of methods for quantifying variable importance in artificial neural networks using simulated data. Ecological Modelling 178:389–397.

Olsen, A. R., and D. V. Peck. 2008. Survey design and extent estimates for the Wadeable Stream Assessment. Journal of the North American Benthological Society 27:822–836.

Omernik, J. M., G. E. Griffith, R. M. Hughes, J. B. Glover, and M. H. Weber. 2017. How misapplication of the hydrologic unit framework diminishes the meaning of watersheds. Environmental Management 60:1–11.

Peterson, D. L., S. L. Egbert, and E. A. Martinko. 2010. 2005 Kansas land cover patterns: phase II final report. KARS (Kansas Applied Remote Sensing), Lawrence, Kansas.

Pinto, B. C. T., F. G. Araujo, V. D. Rodrigues, and R. M. Hughes. 2009. Local and ecoregion effects on fish assemblage structure in tributaries of the Rio Paraíba do Sul, Brazil. Freshwater Biology 54:2600–2615.

Poff, N. L. 1997. Landscape filters and species traits: towards mechanistic understanding and prediction in stream ecology. Journal of the North American Benthological Society 16:391–409.

Poole, G. C. 2002. Fluvial landscape ecology: Addressing uniqueness within the river discontinuum. Freshwater Biology 47:641–660.

Quinn, G. P., and M. J. Keough. 2002. Experimental design and data analysis for biologists. Cambridge University Press, Cambridge, UK.

R Core Team. 2017. R: a language and environment for statistical computing. R Foundation for Statistical Computing, Vienna, Austria.

Robinson, W. A., M. Lintermans, J. H. Harris,and F. Guarino. 2019. A landscape-scale electrofishing monitoring program can evaluate fish responses to climatic conditions in the Murray–Darling River system, Australia. Pages 179–201 *in* R. M. Hughes, D. M. Infante, L. Wang, K. Chen, and B. F. Terra, editors. Advances in understanding landscape influences on freshwater habitats and biological assemblages. American Fisheries Society, Symposium 90, Bethesda, Maryland.

Roth, N. E., J. D. Allan, and D. L. Erickson. 1996. Landscape influences on steam biotic integrity assessed at multiple spatial scales. Landscape Ecology 11:141–156.

Sala, O. E., F. S. Chapin Iii, J. J. Armesto, E. Berlow, J. Bloomfield, R. Dirzo, E. Huber-Sanwald, L. F. Huenneke, R. B. Jackson, A. Kinzig, R. Leemans, D. M. Lodge, H. A. Mooney, M. Oesterheld, N. L. Poff, M. T. Sykes, B. H. Walker, M. Walker, D. H. Wall, F. S. Chapin, M. Oesterheld, and D. H. Wall16. 2000. Global biodiversity scenarios for the year 2100. Science 287:1770–1774.

Schlosser, I. J. 1991. Stream fish ecology: a landscape perspective. BioScience 41:704–712.

Smith, R. F., and W. O. Lamp. 2008. Comparison of insect communities between adjacent headwater and main-stem streams in urban and rural watersheds. Journal of the North American Benthological Society 27:161–175.

Stevens, D. L., and A. R. Olsen. 2004. Spatially balanced sampling of natural resources. Journal of the American Statistical Association 99:262–278.

Tessarolo, G., T. F. Rangel, M. B. Araújo, and J. Hortal. 2014. Uncertainty associated with survey design in species distribution models. Diversity and Distributions 20:1258–1269.

Toft, C. A., and P. J. Shea. 1983. Detecting community-wide patterns: estimating power strengthens statistical inference. The American Naturalist 122:618–625.

Townsend, C. R. 1989. The patch dynamics concept of stream community ecology. Journal of North American Benthological Society 8:36–50.

Townsend, C. R., S. Dolédec, R. Norris, K. Peacock, and C. Arbuckle. 2003. The influence of scale and geography on relationships between stream community composition and landscape variables: description and prediction. Freshwater Biology 48:768–785.

USEPA (U.S. Environmental Protection Agency). 2012. National rivers and streams assessment 2013–2014: site evaluation guidelines. USEPA, Office of Water, EPA-841- B-12-008, Washington, D.C.

USGS (U.S. Geological Survey. 2016. National hydrography dataset. United State Geological Survey, Reston, Virginia.

Vannote, R. L., W. G. Minshall, K. W. Cummins, J. R. Sedell, and C. E. Cushing. 1980. The river continuum concept. Canadian Journal of Fisheries and Aquatic Sciences 37:130–137.

Ver Hoef, J. M., E. Peterson, and D. Theobald. 2006. Spatial statistical models that use flow and stream distance. Environmental and Ecological Statistics 13:449–464.

Ver Hoef, J. M., and E. E. Peterson. 2010. A moving average approach for spatial statistical models of stream networks. Journal of the American Statistical Association 105:6–18.

Wang, L., P. Seelbach, and R. Hughes. 2006. Introduction to landscape influences on stream habitats and biological assemblages. Pages 1–23 *in* R. M. Hughes, L. Wang, and P. W. Seelbach, editors. Landscape influences on stream habitats and biological assemblages. American Fisheries Society, Symposium 48, Bethesda, Maryland.

Wu, J., and O. L. Loucks. 1995. From balance of nature to hierarchical patch dynamics: a paradigm shift in ecology. The Quarterly Review of Biology 70:439–466.

American Fisheries Society Symposium 90:179–201, 2019
© 2019 by the American Fisheries Society

A Landscape-Scale Electrofishing Monitoring Program Can Evaluate Fish Responses to Climatic Conditions in the Murray–Darling River System, Australia

WAYNE A. ROBINSON*
Murray Darling Basin Authority
General Post Office Box 409, Canberra 2601, Australia
and
Charles Sturt University
Old Sydney Road, Thurgoona, New South Wales 2640, Australia

MARK LINTERMANS
Murray Darling Basin Authority
General Post Office Box 409 Canberra 2601, Australia
and
Institute for Applied Ecology, University of Canberra
Canberra, ACT 2601, Australia

JOHN H. HARRIS
Cooperative Research Centre for Freshwater Ecology
Canberra, ACT 2601, Australia

FIORENZO GUARINO
Murray Darling Basin Authority
General Post Office Box 409 Canberra, 2601, Australia
and
Institute for Applied Ecology, University of Canberra
Canberra, ACT 2601, Australia

Abstract.—We document a simple electrofishing-only monitoring program for assessing fish assemblages across large spatial extents. First, we demonstrate the justification for using only electrofishing for the monitoring. Second, we demonstrate the usefulness of having a well-designed surveillance-monitoring program in place to demonstrate the effect of landscape disturbances. Implementing electrofishing alone means that multiple sites can be sampled in a single day and there is no need to return to clear nets or traps within a sampling site. Whereas electrofishing alone does not return full species lists within sampled sites, we demonstrate that when data are aggregated up to the watershed or catch-

* Corresponding author: wayne.robinson@csu.edu.au

ment extent, more than 90% of species are included. Analyses that do not require a census of species, such as bioassessment of river health can be readily carried out using electrofishing data. The Murray–Darling basin, Australia, was sampled with the recommended large-extent electrofishing program between 2004 and 2012, a period that saw the region subjected to large-scale variations in river flow levels spatially and temporally. We fit generalized additive models to the electrofishing data in conjunction with river flow data to document large-extent relationships between fish species occurrence and relative flow levels for the previous 3 d, 3 months, or 3 years. We found that several small-bodied species, Eastern Mosquitofish *Gambusia holbrooki*, Flathead Gudgeon *Philypnodon grandiceps*, and Australian Smelt *Retropinna semoni*, were more likely to be collected when conditions were drier in the past 3 d to 3 months, whereas common medium and large-bodied species were less likely to be collected when flow was lower over the previous 3 months to 3 years.

Introduction

Freshwater ecosystems are among the most imperiled in the world with a range of well-documented threats, including overallocation of water for human needs, habitat loss and destruction, degraded flow regimes and water quality, overharvesting of fish species, and invasion by alien species (Malmqvist and Rundle 2002; Dudgeon et al. 2006; Helfman 2007). Overlying all of these existing threats is the prospect of increasing temperature and altered rainfall and streamflow patterns under climate change (Ficke et al. 2007; Morrongiello et al. 2011), with regulated river basins predicted to be affected more than unregulated systems (Palmer et al. 2008; Balcombe et al. 2011). Declines in the condition and functioning of freshwater ecosystems worldwide have stimulated increased management attention, but there are few large-extent assessment programs to measure the cumulative effectiveness of combined management interventions (Davies et al. 2010).

Australia is the driest inhabited continent and includes rivers with particularly variable flow regimes (Puckridge et al. 1998; Arthington and Pusey 2003). Australian rivers have very low runoff and, on average, only 12% of rainfall is collected in rivers (NLWRA 2002). Annual evaporation frequently exceeds annual rainfall, stimulating significant development of irrigation and water storage in some regions. The Murray–Darling basin (MDB) in southeastern Australia (Figure 1) covers more than a million square kilometers, or about 14% of Australia's total area, provides about 40% of Australia's total agricultural production, and accounts for 50% of the nation's irrigated agricultural water use (MDBA 2010). It was recognized in 2000 that millions of dollars were being spent on initiatives to improve river health in the MDB, but there was no overarching monitoring program that could be used to assess the effectiveness of those investments. To address this deficiency, the concept and framework for an ongoing assessment of river health were prepared (Whittington et al. 2001), with the assessment to be known as the MDB Sustainable Rivers Audit (SRA). The SRA was a surveillance monitoring program to report on the status and long-term trend in ecological health of rivers at the river valley (catchment) extent.

The SRA aimed to provide a statistically robust assessment of river health in the MDB by reporting on the condition of the hydrology, fish, macroinvertebrates, and physical form for each of the 23 component river valleys (Whittington et al. 2001). We acknowledge that river health is a construct, open to

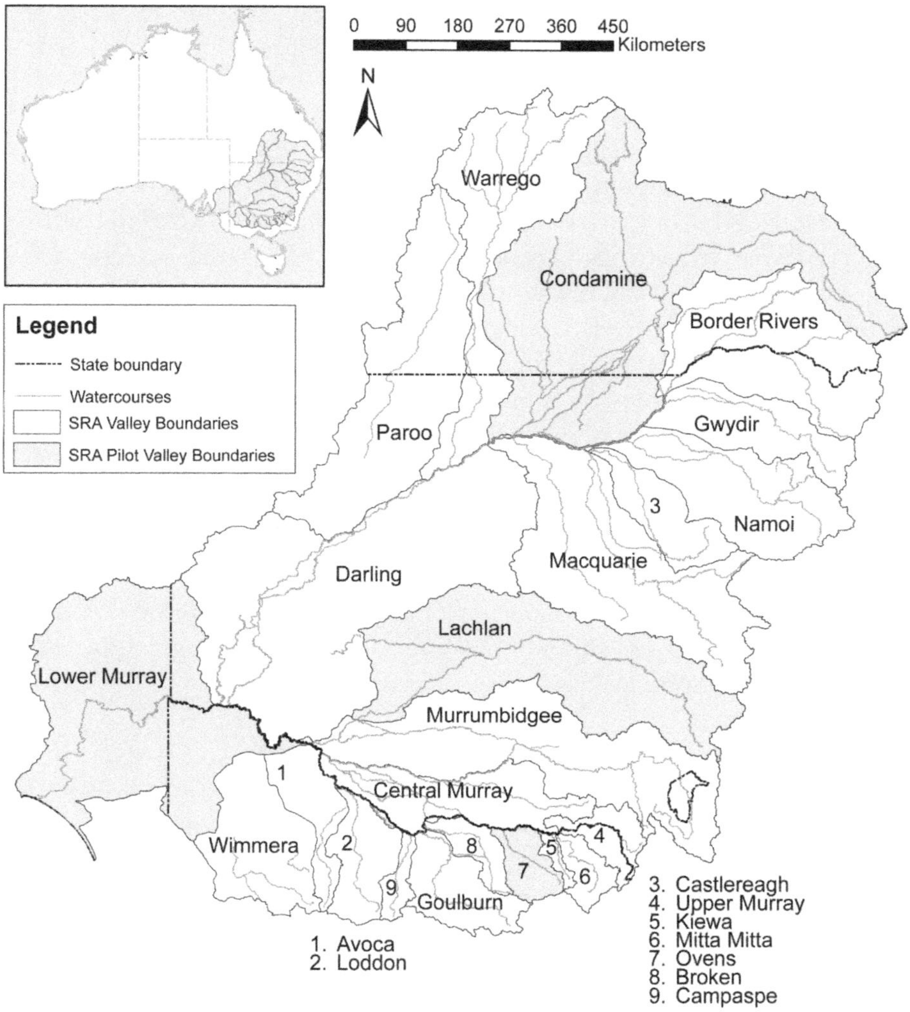

Figure 1. Catchments and hydrologic units in the Murray–Darling basin, southeastern Australia. Shaded areas were sampled as part of the pilot Sustainable Rivers Audit in 2003 and all catchments or hydrologic units were sampled three times between 2004 and 2012.

numerous interpretations (e.g., Norris and Thoms 1999), and focus this paper on the influence of sampling types on estimates of entire fish assemblages within MDB rivers. The use of these data in the river health context is not considered here. The MDB crosses five state and territory borders, and numerous jurisdictions and agencies are involved in resource management, including fisheries (Koehn and Lintermans 2012). Prior to the SRA, there was no standardized sampling procedure that was consistent among research programs or agencies, with many different combinations of active and passive fishing gears, such as electrofishing, nets, and traps, in use. Electrofishing (Anon 1997) had only missed one freshwater species in the New South Wales Rivers Survey (NSWRS; Harris and Gehrke 1997), highlighting the potential for that method to be used alone for monitoring fish assemblages, without the more time-consuming passive sampling methods, ultimately leading to significant improvements in the rapidity of assessments. Unlike nets and traps, electrofishing is a quick, active sampling technique enabling multiple sites to be sampled in a day. If electrofishing alone provided data suitable for assessing fish assemblage condition, it could potentially allow more cost-efficient use of sampling resources.

The SRA can thus be viewed as an early warning "surveillance" monitoring program, as it made no attempt to associate assessments to individual stressors or management actions. Nine years after the SRA program was implemented, we had access to a medium-term, large-extent data set designed with a probabilistic sampling methodology. During the same period, the MDB had undergone a one-in-a-thousand-year drought (van Dijk et al. 2013) and been subject to extreme flooding events. Consequently, an opportunity existed that allowed investigation of the influences of external factors, such as climate on MDB fish assemblages.

Surveillance monitoring programs such as the SRA are often maligned (e.g., Nichols and Williams 2006) because they do not target specific management actions and are frequently treated only as an early warning system (e.g., Vos et al. 2000). The current paper recounts the pilot the SRA carried out in 2003 with the aim of establishing a cost-effective, standard sampling methodology enabling assessment of fish assemblage condition across the MDB. Data were collected on fish individuals, populations, and assemblages through use of a combination of electrofishing and passive sampling methods in four river valleys. We explain the initial justification for the final SRA program adopting an electrofishing-only program after analyzing those pilot data by documenting differences in fish assemblage similarity between using electrofishing alone or the combination of electrofishing and passive sampling methods at the river valley and subcatchment extents. We then substantiate the added value gained from adopting a large-extent surveillance monitoring program such as the SRA by demonstrating how the data collected were used to answer landscape-extent responses by fish to environmental factors or climatic conditions through use of statistical modeling. For our example, we used generalized additive models (GAMs) to relate fish species occurrence to spatial and temporal variations in river levels in the MDB during the period of 2004 to 2012, which included the flow extremes mentioned above. We asked the question "Are native and alien fish in the MDB more or less likely to be collected when the catchment has experienced wet or dry conditions"? We used 3-d, 3-month, and 3-year river flow levels as our measure of wetness or dryness and identified fish species that were more or less likely to be collected given recent (e.g., 3 d) or longer term (3 months or 3 years) flow levels.

Study Location

The MDB is one of the largest basins in the world (the river system extends some 3,750 km from its headwaters to the sea) and is subject to six different legislative jurisdictions. The three major rivers of the basin are the longest rivers in Australia, including the Murray (2,530 km length), the Darling (2,740 km), and the Murrumbidgee (1,690 km) rivers, and there are a further 21 major rivers in the basin (Figure 1). More than 50% of fish habitats and streams are lowlands (below 200 m altitude). Lowland rivers are generally meandering, slow-flowing, and turbid, often surrounded by extensive floodplains containing billabongs (oxbow lakes), swamps, and river red gum *Eucalyptus camaldulensis* forests. In the arid west and north of the basin, rivers may dry to a series of waterholes in low-rainfall years. The upland streams are often rocky-bottomed, swift-flowing, and clear with limited or no floodplains or billabongs. Boulders and bedrock may replace fallen wood as major fish habitats, and the streams often contain a series of alternating riffles, pools, and runs in contrast to the predominantly pool habitat of lowland rivers (Lintermans 2007).

River flows are highly variable from one year to another, but river regulation has reduced this natural variability, with small floods now much less frequent than previously. The rivers have low runoff, on average, with the majority of the basin's runoff coming from a relatively small proportion of the total catchment. For example, the Darling River catchment, while comprising 10% of the basin's area, contributes only 0.4% of its mean annual runoff. By contrast, the upper Murray River catchment occupies only 1.4% of the basin's area but contributes 17.3% of mean annual runoff. River regulation, in general, has seen the replacement of fluvial habitats by extensive lentic conditions in weir pools (Mallen-Cooper and Zampatti 2018).

Climatic Conditions 2004 to 2012

Between 1997 and 2010, the MDB experienced the Millennium Drought (van Dijk et al. 2013), so called because of a one-in-a-thousand-year severity, causing reductions in streamflow and habitat loss that exacerbated ongoing threats to native fishes. The drought broke in 2010 with the flooding of long-dry floodplains in 2010 and 2012 (Thiem et al. 2017). During the 2010–2011 flood, flows in the Murray River increased from stable low flows of 5,000–10,000 ML/d to bank-full at ~35,000 ML d in late November 2010 and then peaked at ~93,000 ML/d in early February 2011 (Leigh and Zampatti 2013).

Methods

The Pilot Sustainable Rivers Audit 2003

The fish component of the pilot SRA was conducted in four river valleys selected to represent environmental conditions and river types found in the MDB: the Condamine-Culgoa in Queensland, the Lachlan in New South Wales, the Ovens in Victoria, and the lower Murray in South Australia (Figure 1). The primary reporting unit for fish assemblage condition in the SRA is the river valley; however, it was also considered plausible prior to the pilot study that fish assemblages could be expected to show affinities within process zones based on geomorphic characteristics (source, transportational, and depositional zones for sediments). Three of the pilot river valleys were stratified into three valley process zones (VPZs). In the lower Murray River, the whole river is geomorphologically classified as depositional, but three broad zones of differing landscape features and associated fish assemblages were established for reporting. These were the floodplain and anabranch section from Wentworth to Overland Corner (surrogate for source VPZ), the strongly confined channel section from Overland Corner to Mannum (surrogate for transport VPZ) and the estuarine-influenced section from Mannum to the Murray mouth (surrogate for depositional VPZ). The number of sampling sites within each river valley was allocated proportionally according to the area of each VPZ, whereas the location of each site within each VPZ was random within the stream network. To enable calculation of basic summary statistics in small zones, a minimum of three sites were allocated to the Condamine River VPZs and five to the Lachlan River VPZs. After rounding the proportionally allocated sample sizes, 92 sites across the four pilot valleys were sampled (Table 1). A site was a 1-km-long river reach, and electrofishing, fyke nets, and bait traps were used to sample fish within each site. The pilot study design sample size was based on the standard deviation of the estimates of proportion of native fish by biomass (PNB) between sites in the same valley in the NSWRS (Harris and Gehrke 1997) and power analyses following Zar (1984). The PNB index was chosen as it was

Table 1. Numbers of sites sampled for fish in the four river valleys and their valley process zones in the pilot Sustainable Rivers Audit in 2003. For the lower Murray, the numbers in parentheses show the sites in the surrogate valley process zones.

River valley	Valley process zone			Valley total
	Source	Transportational	Depositional	
Condamine	3	6	12	21
Lachlan	5	5	16	26
Lower Murray	(12)	(8)	(4)	24
Ovens	7	7	7	21

considered less likely to be affected by differences in sampling effort between gear types and durations in the NSWRS. The sampling effort used within each site in the pilot study was based on an oversampling principle agreed to by a technical panel consisting of members from each state agency. Full details of sampling techniques are given by the Murray–Darling Basin Commission (MDBC 2004), with summaries provided here.

Electrofishing

Wadeable streams were sampled using backpack electrofishing, and larger streams and rivers were sampled using boat electrofishing. In seven of the sites, geomorphic variations in the stream bed required a mix of backpack and boat electrofishing.

Electrofishing boats (large: >4 m long, with 7.5 or 9 kW generators; small: <4 m long, with 2.5–3.5 KW generators) were used according to stream size and access constraints, including river width, river depth, length of habitat available to sample, and accessibility for boats. Electrofisher output settings were not standardized. Operator judgment was used instead to select settings that nondestructively maximized efficiency at each site. Electrofishing dip nets were standardized across all sampling teams, with a knotless stretched mesh size of 6 mm. Sampling was carried out in 2-min shots (elapsed time). Immobilized fish were dipnetted from the water into a live well to recover prior to identification, measurement, and release. Wherever possible, 15 shots were made at each site. If a site was considered suitable for a large boat, a minimum of eight shots were made. Shots were distributed so that all habitats were sampled in approximate proportion to their occurrence at that site. Shots were as independent as possible and incorporated bank and mid-channel habitats. Shots of 2-min elapsed time generally resulted in 60–70 s of power-on time.

Backpack electrofishers (Smith Root 24V, model 12) with standard square-wave settings were used, with a 30-cm-diameter anode ring without an anode net. Collected fish were allowed to recover before release. Electrofisher output settings were not standardized; operator judgment was used instead to choose settings that nondestructively maximized efficiency at each site (Anon 1997). Sampling was carried out in 5-min (elapsed time) shots, with eight shots at each site. While slowly walking upstream, intermittent power-on time was used during shots to avoid herding fish. The dip nets used for backpack electrofishing had a maximum mesh size of 6-mm stretch mesh. As for the boat electrofishing, habitat was sampled in approximate proportion to its occurrence at each site. In deep areas (i.e., pools), the wadeable habitat was sampled with backpack electrofishing, and unwadeable areas were sampled using fyke nets, bait traps, or boat electrofishing if possible. Elapsed time shots of 5 min generally resulted in 130–140 s of power-on time. All electrofishing was carried out in daylight using pulsed direct current, and in accordance with the provisions of the Australian Code of Electrofishing Practice (Anon 1997).

Nets and Traps

Fyke nets and concertina bait traps were intended to be used in all sites, with nets and traps set 2 h before sunset and retrieved 2 h after sunrise. Fyke nets and concertina bait traps were placed independent of the electrofishing areas. Four fyke nets were set at each site. Each had two 5-m-long wings and three funnels and a 60-cm-high net entrance D-ring. The fyke nets and wings were constructed from 4-ply twisted nylon, with 10-mm stretched mesh diameter. Nets were generally set with the mouth of the net facing downstream and the cod end tied above the water level. Each net had a polystyrene float in the cod end to avoid mortality of air-breathing vertebrates such as platypus *Ornithorhyncus anatinus* or turtles, in case of rising water levels.

Ten commercially available concertina bait traps (500 μm mesh, 22 × 22 × 40 cm, 5-cm opening) were deployed at each site. Traps were not baited but fitted with yellow 12-h light sticks (Cyalume). Traps were set in the best available habitat at each site, generally slow-flowing or backwater areas with a maximum depth of approximately 1 m. At two sites in the lower Murray River, bait traps and fyke nets were not used, and at two sites in the Ovens River, fyke nets were not used.

Considerations influencing timing of sampling included flow conditions (low-flow, high-flow, peak-irrigation periods) and seasonal effects on fish life cycles and water temperature. For all states except South Australia, sampling was conducted between March and early June in 2002. High irrigation flows in the Murray River affect the sampling availability of sites in the lower Murray River with 18 sites sampled in April–May 2002 and 6 in September 2002. Sampling was discontinued if a flood occurred during or immediately prior to sampling, and sites were resampled as soon as possible after flows receded.

The primary variables for fish measured at each site were species identity, number of each species caught, fish lengths, and health and condition of individuals (the incidence of external parasites, lesions, diseases, and abnormalities was recorded for all measured fish). All individual fish larger than 15 mm length were counted and identified to species, with the exception of carp gudgeons *Hypseleotris* spp., whose taxonomy is confused with a number of species and hybrids present in the MDB (Bertozzi et al. 2000). Individuals smaller than 15 mm length were not included in the sample because of concerns about the inefficiency of the gear types used for fish of this size. Exclusion of small fish also removed the need to employ specialized collecting apparatus for larval fish and the need for routine laboratory identification of samples, thus reducing the time and cost associated with sampling each site.

The primary sampling unit was the site, rather than the individual shots, nets, or traps taken within each site. For example, a small headwater stream site may have consisted of eight individual electrofishing shots from a backpack unit, whereas a deep-channel lowland stream site may have consisted of 15 individual electrofishing shots from a large electrofishing boat. The assumption for the analysis was that the sampling effort per site used in the pilot SRA was deemed to return unbiased data for that ≤1 km site, even though the number of shots may have varied. This was justified by considering that there was no attempt to assess raw abundance or other variables such as biomass; all analyses considered relative occurrences. Further, there was no pretense that all fish or all species were collected within a site. The aim of the analyses was to estimate how many species may have been missed within a river valley if only electrofishing was used. Because the analysis treated the site as the primary sampling unit, all shots/nets within a method type were pooled within the site and the data are comparable across sites.

We used two simple measures to compare how the addition of the passive gear (nets and traps) to the electrofishing-only data could affect the SRA assessments. First, we compared the assemblage returned by each of the different sampling methods with results from the combined methods for each VPZ, each valley, and all four valleys combined using Jaccard similarity. Second, we compared the proportion of biomass from native fish, which we deemed independent of the species assemblage comparison. Jaccard similarity is an ideal measure because it is easily interpreted as proportions of shared species, whereas proportion of native fish species biomass is useful in the SRA bioassessment context because it is a useful measure of fish assemblage condition. We performed a mixed model to compare proportion of native biomass estimates from electrofishing only with estimates from all methods combined. Sites were treated as the subject nested within VPZs, which were a fixed effect crossed with rivers.

The Sustainable Rivers Audit Methodology 2004 to 2012

The full SRA sampling strategy (Davies et al. 2008) was based on the following:

- Each catchment in the basin was sampled once every 3 years.
- Each catchment had a minimum of 20 sites sampled and a minimum of 7 sites sampled in each (redefined) VPZ.
- Catchments with less than three VPZs had more sites per VPZ; catchments with more than three VPZs had more than the minimum of 20 sites per catchment.
- The basic randomized site-selection strategy employed was a mixed longitudinal rotating panel design (Lesser and Kalsbreek 1997).
- 50% of the sites were resampled in repeat visits to each catchment.
- Each site usually had either 8 backpack or 12 boat electrofishing shots taken.
- Some sites with a mix of habitat types had a mix of boat and backpack shots taken.
- Unbaited bait traps were also placed at every SRA site with the intention of improving within-site species lists, but the data are not used at the valley scale and are not presented here.

Initial rarefaction analysis (W. A. Robinson, unpublished data) found that 174 of 180 sites sampled in the SRA in 2004 returned the full species list within each sampling site and that 95% of VPZs had all species collected (0.05 chance of a zone not having all species). The probabilistic design of the SRA—different site selection probabilities for sites in the same catchment but in different zones—precluded a valley scale cumulative species list assessment. However, the wet/dry analysis that follows in this chapter proceeds under the assumption that valley species lists have minimal missing species. This is reasonable, given that each valley is made up of at least two and usually three VPZs and the probability of an individual species being missing in all three VPZ lists is therefore $(1 - 0.95)^2 = 0.0025$ or $(1 - 0.95)^3 = 0.000125$. In the case where a species is not well sampled by electrofishing in general, then it should be missing from the list altogether and did not contribute to the 0.05 error rate. Therefore, it is not relevant to the current climatic analyses anyway.

Large-Scale Fish Responses to Climatic Conditions

Hydrology.—Hydrological conditions experienced at sampled SRA sites were assessed based on river height gauging station records. Gauging stations were chosen from the current commissioned streamflow gauge stations listed in the Australian Water Resource Information System (BOM 2011). Gauge station metadata such as geographic location, date sampling began, and parameters measured were used to evaluate whether the site would be suitable for associating with a sampled SRA site (at least 25 years of daily flow data were required). At the time of this analysis, the source data table held 6,741 possible candidate gauge stations associated with SRA sample sites.

A base map showing the river network, the 3,726 SRA sampled sites, and the 6,741 gauge stations was constructed. This was supplemented with the Geofabric Surface Network (BOM 2011), which indicated all the complex flow paths (with directions) that existed in the landscape. We then applied a decision-based nearest neighbor tool in ArcGIS (ESRI 2011) to snap SRA sample sites to the nearest hydrologically relevant gauge station.

Proximity within the same stream or subcatchment unit was the primary rule for associating SRA sites to gauge stations, with each SRA site being inspected via desktop analysis and allocated to the nearest most representative gauge. Where nearby gauge stations did not exist, the association was made by decision rules which included (1) the closest downstream gauge, considering stream order, or (2) if nearest gauge station was greater than 50 km, then a suitable neighboring stream of similar order with a gauge was chosen, or (3) in the large distributary systems, the nearest upstream/downstream gauge was chosen regardless of distance. The analysis resulted in a total of 547 gauge stations being associated to 3,726 SRA sites across all themes, with 1,570 fish (site × date) samples having hydrological data available.

For every gauging station selected, we constructed flow duration curves via the HYFLOW function using Hydstra (Kisters 2014), and values were interpolated to obtain the desired quantiles. The percentile (otherwise known as the percentage of flow events that equaled or exceeded the nominated value) was calculated by inputting a nominated flow value into the distributional expression and extracting the percentile, and the extracted data were used to calculate dryness indices.

Dryness indices.—The flow level in each station for the past 3 d, 3 months, and 3 years was converted to a percentile given the long-term flow records (as described above) and then to an 8-point scale (Table 2). The 3-year limit on the scales was used to align with the 3 yearly return visits by the SRA fish program to sites, preserving temporal independence between fish samples.

Analyses of Fish and Climatic Conditions

Sustainable Rivers Audit fish data.—We analyzed the fish data collected by the SRA surveys between 2004 and 2012. We hypothesized that the harsher environmental condi-

Table 2. Conversion of flow level percentiles to an 8-point dryness scale (higher score is drier). The percentile is the percentage of flow events that equalled or exceeded the nominated value. For example, a percentile of 1% means that it is an extremely high flow value, as 99% of historical flows are lower.

Flow percentile	<1%	1–5%	5–10%	10–30%	30–70%	70–90%	90–99%	99%+
Dryness score	1	2	3	4	5	6	7	8

tions would favor native fish over alien fish species. Consequently, the objective of the analyses was to determine if there were any relationships between the likelihood of collecting each fish species and water level in the sites, a surrogate measure of dryness.

The historical distributions for all fish species mapped by Lintermans (2007) were used to generate the potential list of valleys in which each species was expected to occur. Lintermans (2007) used state agency databases, museum collections, and data sets of individual researchers. Species that occurred in the SRA database beyond the Lintermans (2007) maps had their expected ranges extended to include the additional valleys in addition to the Lintermans map for this analyses. Only species that occurred in at least 5% of sites across the entire study are included in the analyses. Site selection probabilities vary between the SRA VPZs, and individual site probabilities were used to weight occurrences to estimate the final proportion of site occurrences for each species in each valley. Each dryness index was normalized prior to calculations, and a single score (weighted by stream length) was calculated for the 3-d, 3-month, and 3-year dryness scores for each valley.

We investigated the relationships between each dryness index, and the proportion of sites each species occurred in using GAMs (Hastie and Tibshirani 1990), which are suitable when the dependent variable (proportion of site occurrences) is not normally distributed and the relationship is not expected to be linear. For each GAM, we modeled the proportional occurrence of each fish species in the valleys as a binomial response variable and fit a smoothing spline with three degrees of freedom using the dryness scores as a predictor, one index at a time. We included year as a fixed effect and fit a separate model to each fish species and dryness index combination. We present the results in an exploratory context using graphs and highlight only models where the relationship was statistically significant (analysis of deviance $P > \chi^2 < 0.05$). To simplify the graphs for interpretation, we present the relationships as a locally weighted polynomial regression that smooths the relationship across the 9 years of sampling.

Results

Sustainable Rivers Audit Pilot Data

A total of 13,952 fish from 27 species (20 native, 7 alien) were caught using all methods combined in the 92 sites sampled. Twenty-four species and a total of 6,900 fish were captured by electrofishing. Most individuals were captured in the Condamine valley, followed by the lower Murray, Lachlan, and Ovens, respectively. Seven species had fewer than 20 individuals captured in total. Bony Herring *Nematalosa erebi* comprised 42% and Eastern Mosquitofish *Gambusia holbrooki* comprised approximately 9% of the total electrofishing catch.

The most widespread native species from the electrofishing catch was Bony Herring, recorded at 47 sites, with Golden Perch *Macquaria ambigua* and Australian Smelt *Retropinna semoni* recorded at approximately 40 sites (Table 3). Three native species, Olive Perchlet, Southern Purple-spotted Gudgeon, and Hyrtl's Tandan, were only recorded from single sites. At the site where Hyrtl's Tandan was caught, electrofishing produced only one individual, whereas the other methods yielded 17 additional individuals.

Species detected in only one site included Silver Perch (one individual), Macquarie Perch (two individuals), and Southern Pygmy Perch (two individuals), and none of these were collected by electrofishing (Table 3). On the other hand, only one Oriental Weatherfish was recorded and it was collected by electrofishing, whereas 83 of 88 individuals of Southern Purple-spotted

Table 3. Number of sites from which each fish species was recorded using electrofishing only (and using all methods) in the pilot SRA in 2003. $N = 92$.

Species	No. sites	Species	No. sites	Species	No. sites
Olive Perchlet *Ambassis agassizii*	1 (5)	Carp gudgeons	22 (49)	Southern Pygmy Perch *Nannoperca australis*	0 (1)
Silver Perch	0 (1)	Spangled Perch *Leiopotherapon unicolor*	13 (15)	Bony Herring	47 (47)
Goldfish *Carassius auratus*	35 (45)	Golden Perch *Macquaria ambigua*	39 (43)	Hyrtl's Tandan	1 (1)
Tawny Hardyhead *Craterocephalus fulvus*	16 (21)	Macquarie Perch	0 (1)	Rainbow Trout *Oncorhynchus mykiss*	9 (9)
Common Carp *Cyprinus carpio*	63 (68)	Trout-cod *Maccullochella macquariensis*	4 (4)	Redfin Perch *Perca fluviatilis*	13 (13)
Two-spined Blackfish *Gadopsis bispinosus*	14 (14)	Murray Cod *Maccullochella peelii*	5 (5)	Flathead Gudgeon *Philypnodon grandiceps*	3 (10)
River Blackfish *Gadopsis marmoratus*	3 (3)	Murray River Rainbowfish *Melanotaenia fluviatilis*	28 (28)	Australian Smelt	43 (47)
Mountain Galaxias *Galaxias olidus*	12 (15)	Oriental Weatherfish *Misgurnus anguillicaudatus*	1 (1)	Brown Trout *Salmo trutta*	10 (10)
Eastern Mosquitofish	26 (29)	Southern Purple-spotted Gudgeon *Mogurnda adspersa*	1 (1)	Dewfish *Tandanus tandanus*	6 (8)

Gudgeon were collected by electrofishing. Electrofishing missed at least one species in all VPZs except Zone 3 of the lower Murray and the Lachlan source and depositional zones. At the whole valley scale, Silver Perch in the Condamine and Golden Perch and Southern Pygmy Perch in the Ovens River would have been missed if electrofishing had been used alone. Those three species were only recorded as one or two small individuals in those valleys.

Fish species lists generated using electrofishing alone averaged 86% of the total fish species collected within a VPZ (Figure 2). The Lachlan transportational zone, which only included five sites, only had 61% of species re-

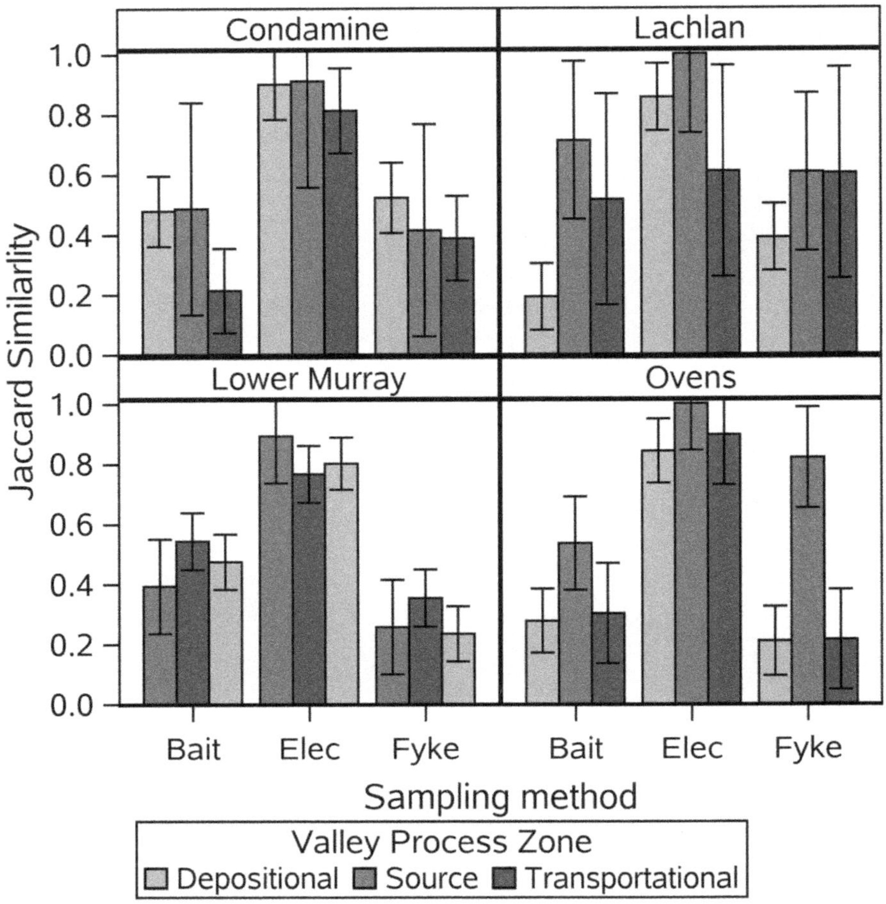

Figure 2. Valley process zone average Jaccard similarity (±95% confidence interval) between species lists generated in each site using three different sampling methods compared to the pooled methods species list.

turned by electrofishing alone (Figure 2). On the other hand, either fyke nets or concertina bait traps alone only averaged 42% similarity in species composition to all methods combined lists (Figure 2). The three zones with the largest number of sites (Lachlan, lower Murray, and Condamine depositional zones) all had 89% or more of species returned by electrofishing alone.

At the river-valley scale, electrofishing alone returned more than 80–91% of species (Figure 3), and across the entire data set had 94% similarity to the complete pooled species list. Fyke nets and bait traps alone always provided fewer than 60% of species than electrofishing (Figure 3).

Proportion Native Biomass

Across all sites in the four valleys, 880 kg of fish were caught during the pilot SRA, and electrofishing accounted for 799.5 kg (90.9%) of this. Bait traps only contributed 5.2 kg and fyke nets 75.2 kg of total biomass. Native fish averaged 35% of total biomass across all methods and all sites and 34% of electrofishing biomass. There was no difference between proportion native biomass estimates using just electrofishing or all meth-

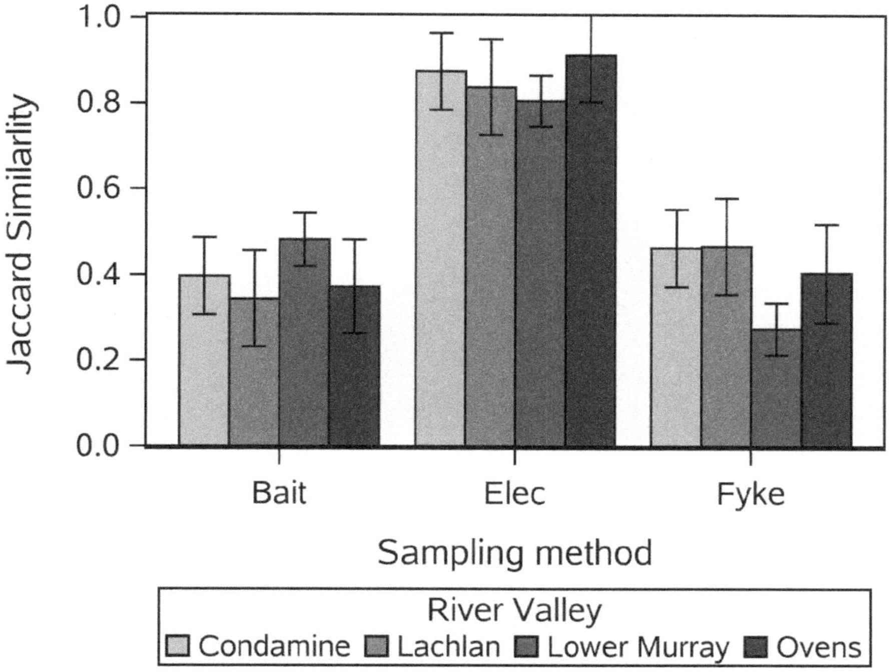

Figure 3. Valley average Jaccard similarity (±95% confidence interval) between species lists generated using three different sampling methods compared to the pooled methods species list.

ods combined ($F = 0.32$, df = 1, 382, $p = 0.57$; Figure 4).

Fish Responses to Climatic Conditions

Thirteen species occurred in 5% or more of sites, and of these, only *Hypseleotris* spp. showed no relationship with any of the dryness scores. Only three species (Redfin Perch [also known as Eurasian Perch], Flathead Gudgeon, and Brown Trout) showed relationships with the 3-d dryness score, and these were all weak positive, as indicated by the almost horizontal slope of the relationship (Figure 5). Of these three species, Redfin Perch also showed a positive relationship with the 3-month dryness, but not 3-year dryness, suggesting that it is more likely to be collected when recent conditions within the sites that are drier than average (Figure 5). Three small-bodied species, Eastern Mosquitofish, Flathead Gudgeon, and Australian Smelt, all showed a tendency to be collected when conditions were drier (Figure 5). On the other hand, two native common medium- and large-bodied species (Bony Herring and Golden Perch) and two alien large-bodied species (Goldfish and Common Carp) showed a negative relationship with long-term dryness. The iconic large-bodied Murray Cod showed no relationship with 3-d or 3-month dryness and only a weak relationship with 3-year dryness.

Discussion

Electrofishing Only

In this study, electrofishing (active) alone returned fish species lists for VPZs with higher similarity to those returned by all methods (active and passive), demonstrating its usefulness in obtaining a representative sample, albeit not a census of fish species diversity (see Ebner et al. 2008). This was more than

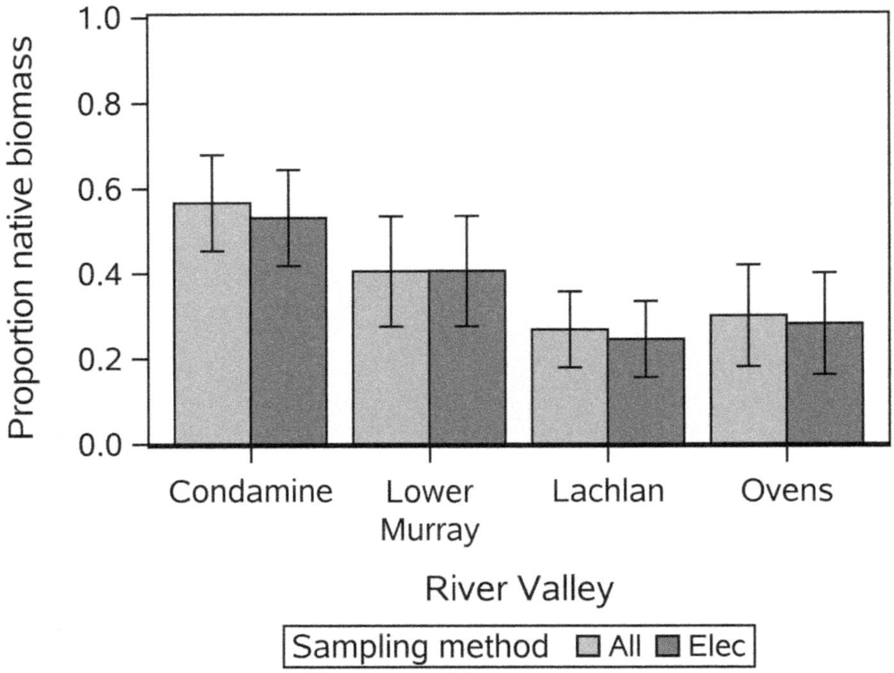

Figure 4. Average proportion of native fish biomass (±95% confidence interval), comparing data from all gear types and electrofishing alone in river valleys in the pilot Sustainable Rivers Audit.

adequate for the purpose of the SRA because research on reef fish assemblages in the Philippines suggest that detection of trend or change in condition assessment through time can be made with as few as 75% of species represented (Molloy et al. 2010). After sampling 92 sites in four valleys, the electrofishing protocol only failed to capture three rare, smaller species (or individuals) that were collected by other methods (Table 3), and simple fish condition metrics were not altered (Figure 4). This emphasizes that the SRA fish sampling protocol, although not intended to provide an inventory of fish species in sites and zones of the MDB, provided enough data to assess fish assemblage condition and to detect changes in condition through time at the catchment scale. The practical benefits of being able to sample more than one site per day with electrofishing far outweigh the cost of losing some information on rare species, without affecting integrated fish assemblage condition assessments. Combined with the consistency in valley integrated fish condition scores (not presented here), we found that there also is marginal benefit for integrated fish condition assessment in using multiple sampling methods.

Electrofishing is commonly used as the sole collection method for assessing river health in international programs, including Europe (FAME Consortium 2004), South Africa (CSIR 2007), and the United States (USEPA 2016; Herlihy et al. 2019, this volume). The results for the biomass metric in our study were also very consistent at the valley scale, returning equivalent scores when using only electrofishing data or when adding the passive sampling data. This was expected because electrofishing collected the greater majority (90.9%) of total biomass. On the other hand, electrofishing alone returned only 49.5% of total fish abundance because

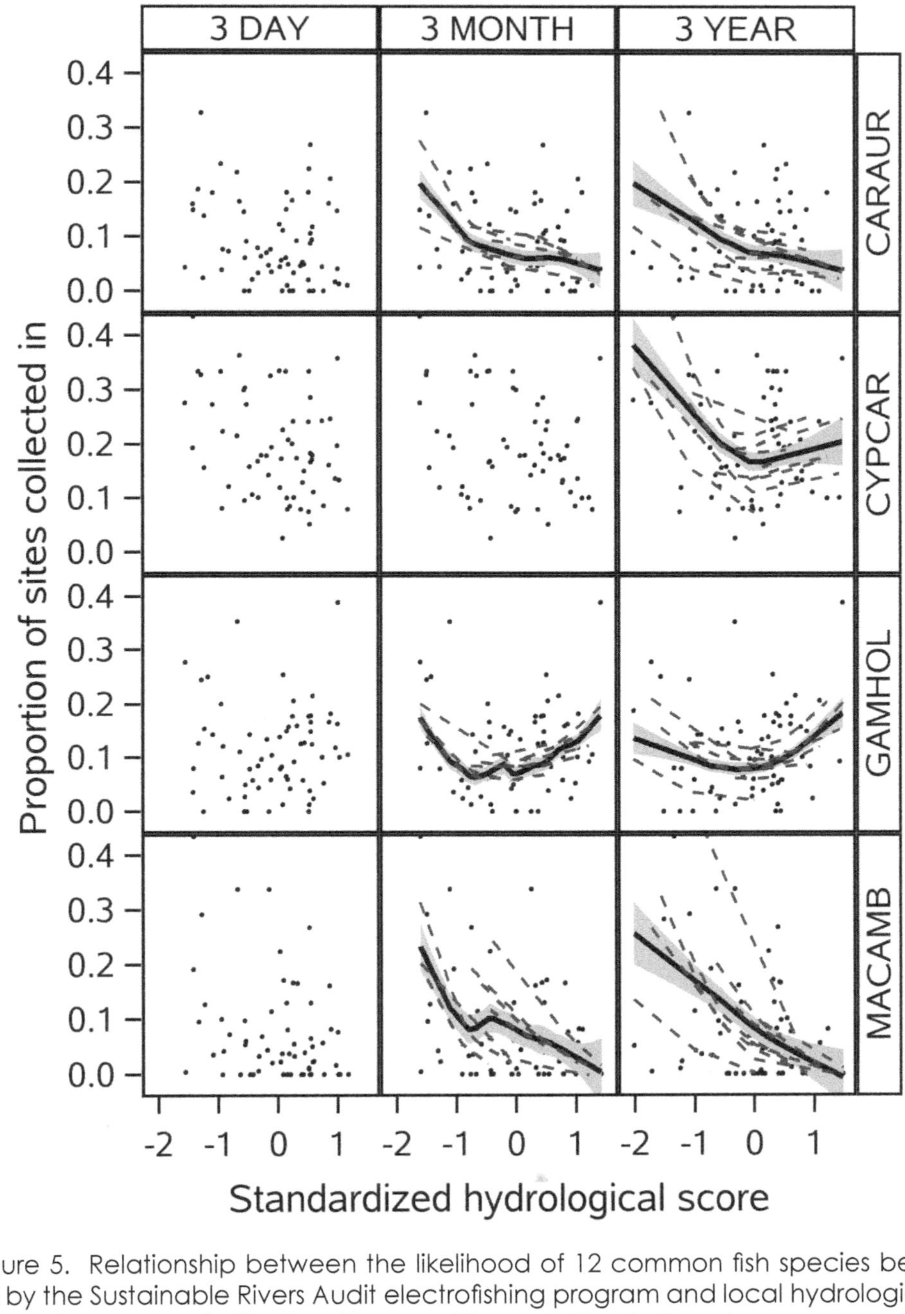

Figure 5. Relationship between the likelihood of 12 common fish species being collected by the Sustainable Rivers Audit electrofishing program and local hydrological conditions (a higher score indicates drier conditions relative to long-term averages). Separate locally weighted polynomial regression lines for each sampling year are fit as dashed lines while the solid line is the average overall relationship (±95% confidence interval). Lines are only on plots where the relationship was significant using a GAM analysis ($p < 0.05$). CARAUR = *Carassius auratus*, CYPCAR = *Cyprinus carpio*, GAMHOL = *Gambusia holbrooki*, MACAMB = *Macquaria ambigua*, MACPEE = *Maccullochella peelii*, MELFLU = *Melanotaenia fluviatilis*, NEMERE = *Nematalosa erebi*, PERFLU = *Perca fluviatilis*, PHIGRA = *Philypnodon grandiceps*, SALTRU = *Salmo trutta*, LEIUNI = *Leiopotherapon unicolor*, RETSEM = *Retropinna semoni*.

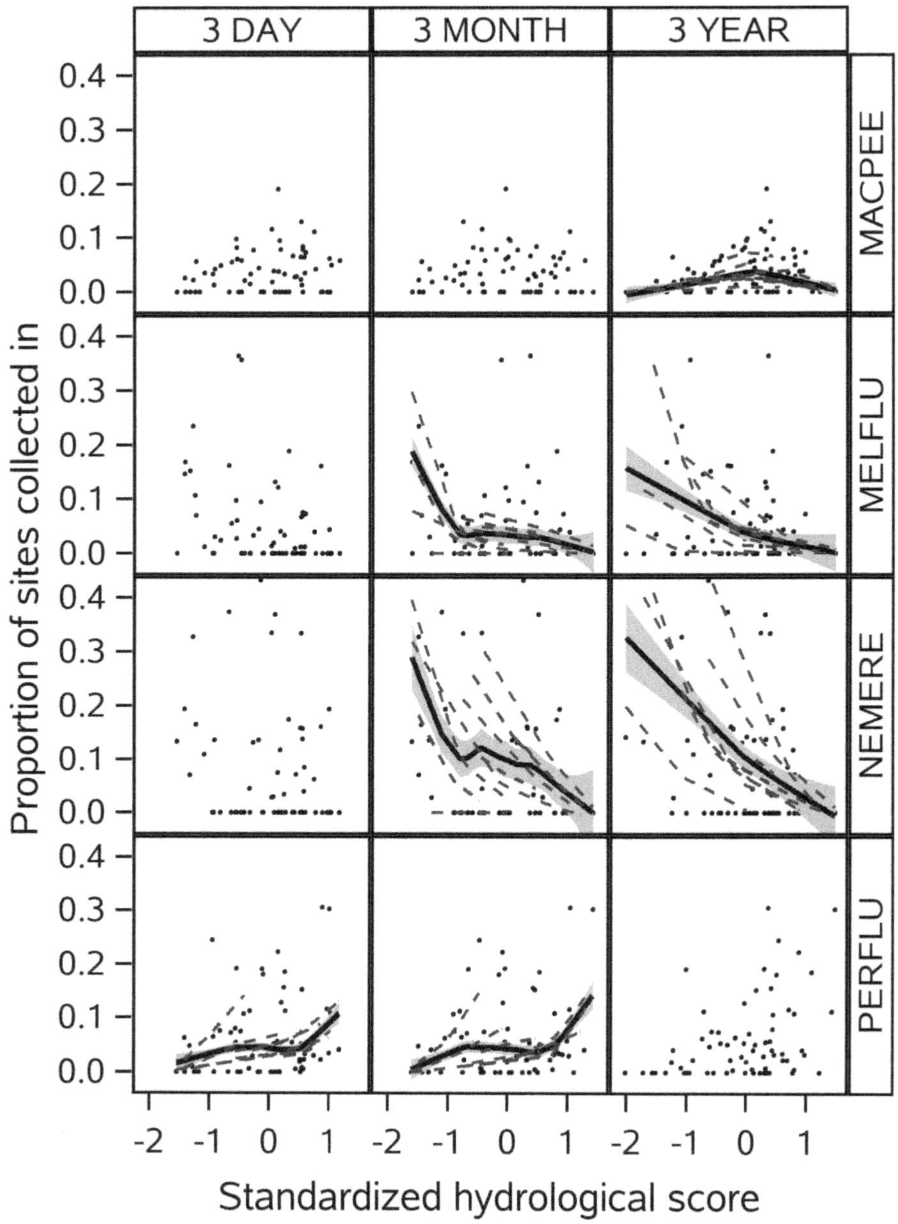

Figure 5. Continued.

the majority of individuals and species captured in bait traps were small-bodied species or individuals. Thus, there is a potential bias for assemblage condition scores because if only electrofishing data are used, small species or individuals may be underrepresented in some metrics. The large-bodied species (Silver Perch and Golden Perch) that were missed by electrofishing in the pilot study in the Condamine and Ovens rivers were also small individuals. Thus, the use of electrofishing-only data are unlikely to be sensitive when monitoring requires information about recruitment or juvenile fish. In such cases, supplementary and or targeted sampling should be considered (Lintermans 2016).

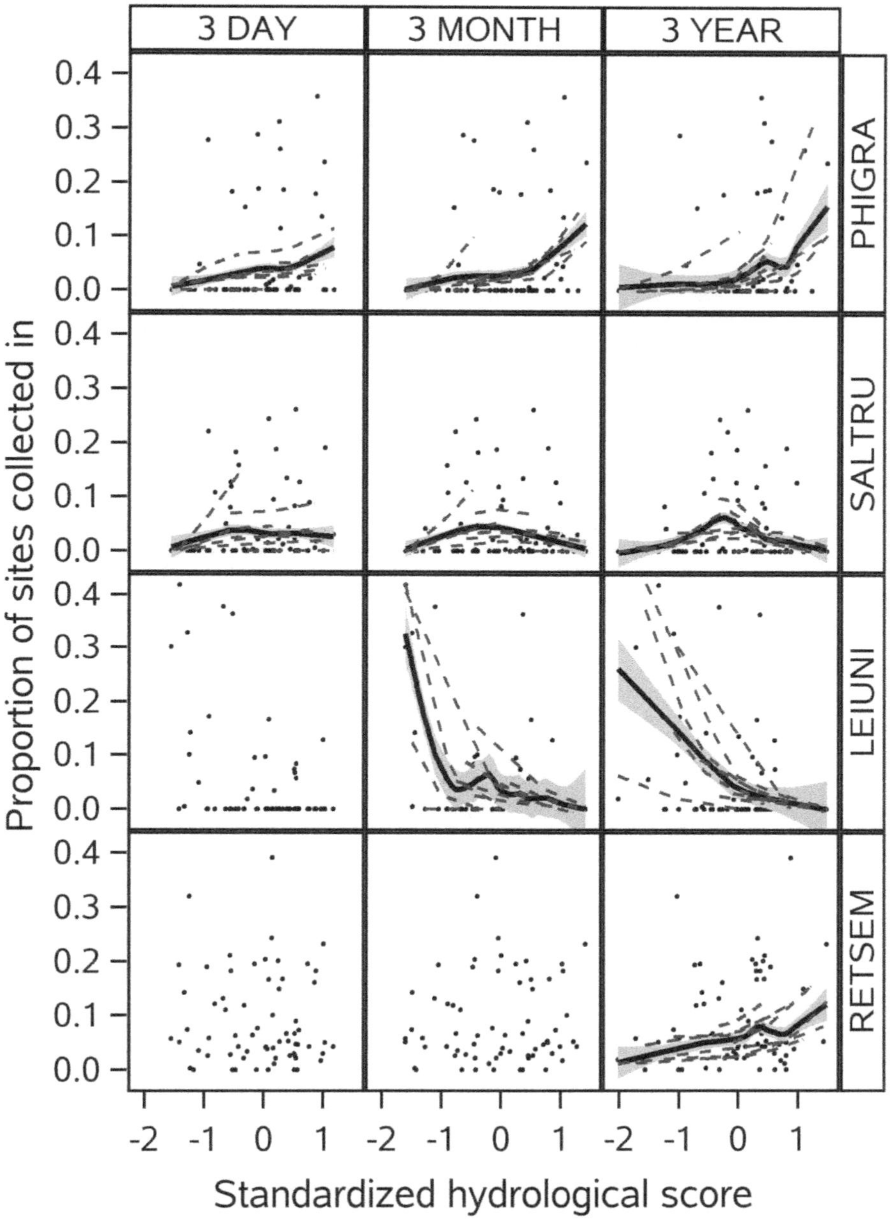

Figure 5. Continued.

Even though the fish assemblage condition assessment of river valleys in the MDB can be adequate using electrofishing alone, augmentation could potentially add value for future projects. For example, some agencies or researchers may wish to use SRA data as baseline or reference data in future research or programs that may not necessarily be monitoring fish assemblage condition. For example, threatened species management, recreational fisheries issues, monitoring responses to environmental flows, or other specific management interventions may require more complete species inventories, or target particular species or life stages (e.g., juveniles; see Lintermans 2016; Lintermans and

Robinson 2018). Therefore, to expand species lists in primary data, we suggest that adding a sampling method that could rapidly sample the small-bodied fish species and individuals underrepresented by standard levels of electrofishing effort (Neebling and Quist 2011; Pompeu et al. 2019, this volume) could be of benefit. Subsequently, bait traps were incorporated into the full SRA sampling program, but their potential to improve the representation of small-bodied fish is yet to be assessed, and their data were not included in this paper. The combined sampling approach (electrofishing plus bait traps) developed during the pilot SRA is now widely accepted in fish assessment projects and represents a long-overdue standard fish sampling methodology in southeastern Australia. The collaborating jurisdictions are able to further supplement sampling to meet any additional needs, as well as satisfying SRA requirements.

The pilot SRA found that using electrofishing and a stratified random site selection protocol returned primary data that could be used to derive the proportion of native fish biomass, and an integrated assessment of fish assemblage condition that were comparable to assessments using additional passive sampling. However, the findings are scale-dependent. The SRA treats sites as the sampling unit and river valleys (or VPZs) as the reporting unit. Projects intending to answer questions at smaller extents could treat the site as the reporting extent and the individual electrofishing shots as the sampling units (e.g., Ebner et al. 2008). The SRA electrofishing sampling protocol did not attempt to place confidence on site-extent scores. The SRA methodology and data focus on fish of rivers and streams, not wetlands, oxbow lakes, or floodplains. If future assessments are intended to represent these off-channel habitats, then a different sampling protocol will be needed. For example, the detectability of some species in deepwater environments using electrofishing needs further investigation and its effects on individual metrics needs to be documented (Neebling and Quist 2011). Electrofishing is also inefficient and returns poor representativeness in high conductivity environments (>1,000 μS/cm), and in lowland rivers, individual fish vary in susceptibility to capture according to their length and local turbidity (Lyon et al. 2014).

Comparisons between the three lower Murray VPZs need to be treated cautiously because they are actually all in a depositional zone but had the VPZ tags applied for the sake of the pilot SRA sampling strategy. Considering the other valleys, the source VPZs in the Lachlan and Ovens rivers returned identical assemblages whether using electrofishing or all gear types in almost every site (Figure 2), and this reflects the depauperate nature of fish assemblages in these VPZs in MDB rivers. There are usually four or fewer species in these headwater sites in the MDB, and all are well sampled using electrofishing when present. On the other hand, the depositional VPZs in all valleys (including all three Lower Murray VPZs) all returned average Jaccard similarity of 80–90% between electrofishing-only and all-methods species lists. This consistency reflects the overall similarity of the type and number of fish species in these typically slow-flowing lowland river sites. The whole of valley similarities (Figure 4), while reasonably similar across all four pilot valleys, do reflect the same trend. That is, the Ovens River, which has the highest proportion of source zone (48%), has the highest average similarity in fish assemblages across all sites, whereas the Lower Murray, which is all depositional zone, had the lowest average similarity in fish assemblages. Biomass in all rivers (Figure 4) was dominated by Common Carp (55% to 75% total biomass), but the average proportion of native biomass was influenced by differences in the assemblages of native species in each river. The native large-bodied species Golden Perch made up an average of 14% and

17% of total biomass in the Condamine and lower Murray rivers, both of which are dominated by transportational or depositional zones (93% and 100%, respectively; Figure 1). These two rivers also featured 10% and 14%, respectively, of biomass from the native Bony Bream *Nematalosa erebi*. Latitude was also related to landscape-scale differences as the Condamine River, in the northern MDB, had none of the large-bodied Murray Cod collected. Murray Cod, however, contributed an average of 10% of total biomass in the Lachlan and lower Murray River sites and 21% of total biomass in the Ovens River sites.

Relationship with Climatic Conditions

We speculate that small-bodied species may be more sampleable by electrofishing in drier locations because turbidity lessens with low/no flow or fish occupy more habitats as flow lessens. Both these possibilities could be further investigated by drilling down into the data further but are beyond the scope of the exercise here. Short-term dryness may reduce habitat size and lead to displacement movement and aggregations of fish into remnant areas, whereas long-term dryness will likely affect recruitment, migration, and population sizes. Fish crowding in reduced habitats areas would certainly influence catchability, especially if small fish are pushed to fringe areas by predators or if the water clarity is greater.

At the large extent of assessment provided by the SRA, there is considerable redundancy in our analyses. For example, modeling large lowland and nonnative species (e.g., Common Carp) in this paper included upland sites in the analyses. Understanding distributions of species within catchments (relationships of species occurrence with altitude) would no doubt reduce the unexplained variation to some extent. To this end, all of the modeling in this paper was also performed using VPZs rather than valleys as the reporting unit. However, this did not change the results noticeably, as all zones in a catchment tended to have the same dryness characteristics, and the plots and interpretation were more difficult because of the larger number of data points. A future analysis could fit a spatial or altitudinal covariate to help explain some of the geographical variation before attempting to look for climate-related variation.

We used proportion of occurrence for simplicity in this paper; however, there are a multitude of similar and simple analyses that we also performed on this data set. Generalized additive models are a highly flexible option (e.g., Zuur et al. 2009) that can be used to model fish abundance (catch per unit effort), recruitment, fish size distributions, any of the SRA metrics (including proportion native biomass) and their relationships to any number of hydrological variables, including river levels, or other readily available variables such as rainfall.

Research Gaps and Challenges

In our opinion, commitment to long-term funding is the one of the best assets any monitoring program can have. Competition for the monitoring dollar means that surveillance type monitoring programs must return value for money to maintain funding. Thus, a major challenge for surveillance-type monitoring is access to ongoing funding in the current climate, which demands quick return on investment and favors using intervention-based monitoring strategies. These types of strategies can be very efficient at returning answers quickly to specific interventions (for example, increased environmental flows for fish migration). But on the other hand, large-extent responses to integrated management regimes cannot be so readily assessed. For example, a native fish strategy (Hames et al. 2014) used local reach based interventions involving restoration of habitat, flows, water quality, conservation polic and more, but the overall objective for the program was at the whole of MDB extent. Clearly, monitoring that can report at the whole basin extent is required, so the challenge remains to integrate surveillance and intervention monitoring programs that can be adapted to return optimal return for investment at landscape extents.

Management and Policy Implications

The pilot SRA provided a significant service to landscape extent assessment of fish assemblages and river health in Australia. The SRA protocol within sites has been adopted as standard best practice by all the MDB states and is used in statewide programs that extend far beyond the basin itself. When this monitoring is employed across large areas, we have monitoring programs that use standardized effort and probabilistic site-selection strategies where appropriate. This extended footprint means that data collected from various programs at differing spatial and temporal scales can be aggregated (e.g., Larsen et al. 2007; Pompeu et al. 2019) and value added to existing programs. When the monitoring is for single sites, or local-extent responses to interventions, the within-site electrofishing-only protocol can easily be adapted by adding sampling methods, such as netting.

Summary

The SRA is surveillance monitoring set up to assess aquatic ecological condition and to detect changes in fish assemblage condition in rivers at large spatial extents, but it does not attempt to infer any cause and effect associated with the changes. However, if analyses are carefully chosen and relationships between outputs and causal processes are well understood, surveillance monitoring like the SRA can provide standardized data collected in a manner that can guide and inform adaptive management (e.g., Bunn et al. 2010).

The scope of aquatic rehabilitation programs grows (Palmer et al. 2007), but many interventions are not specifically monitored and the requirement for cost-effective surveillance assessment approaches will continue to expand (Lindenmayer and Likens 2009, 2010). The use of active sampling methods that permit sites to be assessed more rapidly will receive increased scrutiny, and we and others (Herlihy et al. 2019; Pommpeu et al. 2019) have demonstrated that such approaches can still return valid, robust assessments of fish assemblage condition. Further, we have been able to show that surveillance monitoring programs can add value simply by being in place when unplanned changes in natural conditions, such as droughts or floods occur.

Acknowledgments

Thanks to all of the SRA pilot team, including all of the field teams who sampled fish, and to members of the SRA Taskforce and other agency and nonagency experts who guided the development of the pilot SRA. We are indebted to Michael Wilson, who managed the SRA fish program during the climatic conditions study period. We also thank Bianca de Freitas Terra and three anonymous reviewers for their comments on an earlier manuscript.

References

Anon. 1997. Australian code of electrofishing practice. New South Wales Fisheries, Publication No. 1, Plymont.

Arthington, A. H., and B. J. Pusey. 2003. Flow restoration and protection in Australian rivers. River Research and Applications 19:377–395.

Balcombe, S. R., F. Sheldon, S. J. Capon, N. R. Bond, W. L. Hadwen, N. Marsh, and S. J. Bernays. 2011. Climate-change threats to native fish in degraded rivers and floodplains of the Murray–Darling basin, Australia. Marine and Freshwater Research 62:1099–1114.

Bertozzi, T., M. Adams, and K. F. Walker. 2000. Species boundaries in carp gudgeons (Eleotridae: *Hypseleotris*) from the River Murray, South Australia: evidence for multiple species and extensive hybridization. Marine and Freshwater Research 51:805–815.

BOM (Bureau of Meteorology). 2011. Geofabric surface network - V2.1. Bioregional assessment source dataset. Available: http://data.bioregionalassessments.gov.au/dataset/ed3acf9b-888c-4d53-b376-ecab89781651. (December 2018).

Bunn, S. E., E. G. Abal, M. J. Smith, S. C. Choy, C. S. Fellows, B. D. Harch, M. J. Kennard, and F. Sheldon. 2010. Integration of science and monitoring of river ecosystem health to guide investments in catchment protection and rehabilitation. Freshwater Biology 55(Supplement 1): 223–240.

Davies, P. E., J. H. Harris, T. Hillman, and K. F. Walker. 2008. SRA Report 1: a report on the ecological health of rivers in the Murray–Darling basin, 2004–2007. Murray–Darling Basin Commission, Canberra, Australia. Available: www.mdba.gov.au/sites/default/files/archived/mdbc-sra-reports/1-SRA_Report_1_tech_full.pdf. (December 2018).

Davies, P. E., J. H. Harris, T. Hillman, and K. F. Walker. 2010. The Sustainable Rivers Audit: assessing river ecosystem health in the Murray–Darling basin, Australia. Marine and Freshwater Research 61:764–777.

Dudgeon, D., A. H. Arthington, M. O. Gessner, Z.-I. Kawabata, D. J. Knowler, C. Leveque, R. J. Naiman, A.-H. Prieur-Richard, D. Soto, M. L. J. Stiassny, and C. A. Sullivan. 2006. Freshwater biodiversity: importance, threats, status and conservation challenges. Biological Reviews 81:163–182.

Ebner, B. C., J. D. Thiem, D. M. Gilligan, M. Lintermans, I. J. Wooden, and S. Linke. 2008. Estimating species richness and per unit effort from boat electro-fishing in a lowland river in temperate Australia. Austral Ecology 33:891–901.

ESRI (Environmental Systems Research Institute). 2011. ArcGIS desktop: release 10. ESRI, Redlands, California.

FAME Consortium. 2004. Manual for application of the European fish index (EFI). A fish-based method to assess the ecological status of European rivers in support of the Water Framework Directive. Version 1.1. Available: https://fame.boku.ac.at/downloads/manual_Version_Februar2005.pdf. (April 2019).

Ficke, A., C. Myrick, and L. Hansen. 2007. Potential impacts of global climate change on freshwater fisheries. Reviews in Fish Biology and Fisheries 17:581–613.

Hames, F., A. Townsend, G. Ringwood, P. Clunie, and J. McPhail. 2014. Effective engagement of the Native Fish Strategy is delivered by coordinated and contextual effort. Ecological Management and Restoration 15:13–27.

Harris, J. H., and P. C. Gehrke. 1997. Fish and rivers in stress: the NSW rivers survey. New South Wales Fisheries Office of Conservation and the Cooperative Research Centre for Freshwater Ecology, in association with the New South Wales Resource and Conservation Assessment Council, Cronulla, New South Wales.

Hastie, T. J., and R. J. Tibshirani. 1990. Generalized additive models. Chapman and Hall, New York.

Helfman, S. 2007. Fish conservation: a guide to understanding and restoring global aquatic biodiversity and fishery resources. Island Press, Washington, D.C.

Herlihy, A. T., J. C. Sifneos, R. M. Hughes, D. V. Peck, and R. M. Mitchell. 2019. Lotic fish assemblage clusters across the conterminous United States and their associations with environmental variables. Pages 385–408 *in* R. M. Hughes, D. M. Infante, L. Wang, K. Chen, and B. F. Terra, editors. Advances in understanding landscape influences on freshwater habitats and biological assemblages. American Fisheries Society, Symposium 90, Bethesda, Maryland.

Kisters. 2014. Hydstra/TS time series data management V7.27.2. Available: www.kisters.com.au. (August 2015).

Koehn, J. D., and M. Lintermans. 2012. A strategy to rehabilitate fishes of the Murray–Darling basin, south-eastern Australia. Endangered Species Research 16:165–181.

Larsen, D. P., A. R. Olsen, S. H. Lanigan, C. Moyer, K. K. Jones, T.M. Kincaid, T. M. 2007. Sound survey designs can facilitate integrating stream monitoring data across multiple programs. Journal of the American Water Resources Association 43:384–397.

Lesser, V. M., and W. D. Kalsbreek. 1997. A comparison of periodic survey designs employing multi-stage sampling. Environmental and Ecological Statistics 4:117–130.

Lindenmayer, D. B., and G. E. Likens. 2009. Adaptive monitoring: a new paradigm for long-term research and monitoring. Trends in Ecology and Evolution 24:482–486.

Lindenmayer, D. B., and G. E. Likens. 2010. The science and application of ecological monitoring. Biological Conservation 143:1317–1328.

Lintermans, M. 2007. Fishes of the Murray–Darling basin: an introductory guide. Murray–Darling Basin Commission, Canberra, Australia.

Lintermans, M. 2016. Finding the needle in the haystack: comparing sampling methods for detecting an endangered freshwater fish. Marine and Freshwater Research 67:1740–1749.

Lintermans, M., and W. Robinson. 2018. The extent and adequacy of monitoring of Australian threatened freshwater fish. Pages 85–99 *in* S. Legge, D. Lindenmayer, N. Robinson, B. Scheele, D. Southwell, and B. Wintle, editors. Monitoring threatened species and ecological communities. CSIRO Publishing, Collingwood, Victoria, Australia.

Lyon, J. P., T. Bird, S. Nicol, J. Kearns, J. O'mahony, C. R. Todd, I. G. Cowx, and C. J. A. Bradshaw. 2014. Efficiency of electrofishing in turbid lowland rivers: implications for measuring temporal change in fish populations. Canadian Journal of Fisheries and Aquatic Sciences 71:878–886.

Mallen-Cooper, M., and B. Pl Zampatti. 2018. History, hydrology and hydraulics: rethinking the ecological management of large rivers. Ecohydrology [online serial] 11(5):e1965.

Malmqvist, B., and S. Rundle. 2002. Threats to the running water ecosystems of the world. Environmental Conservation 29(2):134–153.

MDBA (Murray–Darling Basin Authority). 2010. Guide to the proposed basin plan: overview. MDBA, Canberra, Australia. Available: www.mdba.gov.au/publications/archived-information/basin-plan-archives/guide-proposed-basin-plan. (April 2019).

MDBC (Murray–Darling Basin Commission). 2004. Fish theme pilot audit technical report – Sustainable Rivers Audit. MDBC, Canberra, Australia. Available: www.mdba.gov.au/sites/default/files/archived/mdbc-sra-reports/5_Fish_Tech_report_may_2004.pdf. (April 2019).

Molloy, P. P., J. A. Anticamara, J. L. Rist, and A. C. J. Vincent. 2010. Frugal conservation: what does it take to detect changes in fish populations? Biological Conservation 143:2532–2542.

Morrongiello, J. R., S. Beatty, J. Bennett, D. A. Crook, D. N. E. N. Ikedife, M. Kennard, A. Kerezsy, M. Lintermans, D. G. McNeil, B. Pusey, and T. Rayner. 2011. Climate change and its implications for Australia's freshwater fish. Marine and Freshwater Research 62:1082–1098.

Neebling, T. E., and M. C. Quist. 2011. Comparison of boat electrofishing, trawling, and seining for sampling fish assemblages in Iowa's nonwadeable rivers. North American Journal of Fisheries Management 31:390–402.

Nichols, J. D., and B. K. Williams. 2006. Monitoring for conservation. Trends in Ecology and Evolution 21668–673.

NLWRA (National Land and Water Resources Audit). 2002. Australian Water Resources Assessment 2000. Land and Water Australia, Canberra. Available: http://lwa.gov.au/programs/national-land-and-water-resources-audit. (April 2019).

Norris, R. H., and M. C. Thoms. 1999. What is river health? Freshwater Biology 41:197–209.

Palmer, M., J. D. Allan, J. Meyer, and E. S. Bernhardt. 2007. River restoration in the twenty-first century: data and experiential future efforts. Restoration Ecology 15:472–481.

Palmer, M. A., C. A. Reidy Liermann, C. Nilsson, M. Florke, J. Alcamo, P. S. Lake, and N. Bond. 2008. Climate change and the world's river basins: anticipating management options. Frontiers in Ecology and the Environment 6:81–89.

Pompeu, P. S., C. G. Leal, D. R. Carvalho, N. T. Junqueira, M. A. Castro, and R. M. Hughes. 2019. Effects of catchment land use on stream fish assemblages in the Brazilian savanna. Pages 303–320 *in* R. M. Hughes, D. M. Infante, L. Wang, K. Chen, and B. F. Terra, editors. Advances in understanding landscape influences on freshwater habitats and biological assemblages. American Fisheries Society, Symposium 90, Bethesda, Maryland.

Puckridge, J. T., F. Sheldon, K. F. Walker, and A. Boulton. 1998. Flow variability and the ecology of large rivers. Marine and Freshwater Research 49:55–72.

Thiem, J. D., I. J. Wooden, L. J. Baumgartner, G. L. Butler, J. P. L. Forbes, and J. Conallin. 2017. Recovery from a fish kill in a semi-arid Australian river: can stocking augment natural recruitment processes? Austral Ecology 42:218–226.

USEPA (U. S. Environmental Protection Agency). 2016. National Rivers and Stream Assessment 2008–2009: a collaborative survey. USEPA, Office of Water and Office of Research and Development, EPA/841/R-16/007, Washington, D.C.

van Dijk, A. I. J. M., H. E. Beck, R. S. Crosbie, R. A. M. de Jeu, Y. Y. Liu, G. M. Podger, B. Timbal, and N. R. Viney. 2013. The Millennium Drought in southeast Australia (2001–2009): natural and human causes and implications for water resources, ecosystems, economy, and society. Water Resources Research 49:1040–1047.

Vos, P., E. Meelis, and W. J. Ter Keurs. 2000. A framework for the design of ecological monitoring programs as a tool for environmental and nature management. Environmental Monitoring and Assessment 61:317–344.

Whittington, J., J. Coysh, P. Davies, F. Dyer, B. Gawne, I. Lawrence, P. Liston, R. Norris, W. Robinson, and M. Thoms. 2001. Development of a framework for the Sustainable Rivers Audit. Cooperative Research Centre for Freshwater Ecology, Technical Report #7/2001, Canberra, Australia.

Zar, J. H. 1984. Biostatistical analysis. Prentice-Hall, Upper Saddle River, New Jersey.

Zuur A.F., E. N. Ieno, N. J. Walker, A. A. Saveliev, and G. M. Smith. 2009. GLM and GAM for count data. Pages 209–244 *in* Mixed effects models and extensions in ecology with R. Springer, New York.

American Fisheries Society Symposium 90:203–227, 2019
© 2019 by the American Fisheries Society

Influences of Local and Landscape Characteristics, Including Past Timber Harvest Practices, on Large Wood in Streams Draining Young-Growth Southeast Alaska Catchments

Jared A. Ross* and Dana M. Infante
Department of Fisheries and Wildlife, Michigan State University
1405 South Harrison Road, Suite 318, East Lansing, Michigan 48823, USA

Douglas J. Martin
Martin Environmental
2103 North 62nd Street, Seattle, Washington 98103, USA

Arthur R. Cooper and Kyle Herreman
Department of Fisheries and Wildlife, Michigan State University
1405 South Harrison Road, Suite 318, East Lansing, Michigan 48823, USA

Abstract.—Currently, much is known about influences of landscape attributes, including timber harvest practices, on large wood dynamics in streams. Comparatively, much less is known about influences of catchment attributes on Southeast Alaska streams, in part because of a historical lack of consistent catchment-scale data available for the region. As in forested regions elsewhere, large wood is an important resource to stream habitats and fishes in Southeast Alaska. Therefore, the purpose of this study was to characterize catchment-scale influences, including various timber harvest practices, on large wood in Southeast Alaska stream sites. We delineated local catchment boundaries for all stream reaches in the region and summarized landscape influences, including natural and harvest-related attributes in local and network catchments. Relative amounts of variation in four large wood habitat variables explained by natural versus harvest practice-related landscape attributes were evaluated to compare different influences on 28 randomly selected study sites. We used those results to predict variation in large wood variables from our sites through use of both natural and harvest practice-related catchment attributes to identify those that may be most influential to large wood. Natural characteristics, including catchment area, deciduous forests, forested wetlands, and catchment slope, all had significant influences on large wood variables, as did various measures of contemporary and historical timber harvest practices. We found that large wood length was positively related to conventional harvest after 1990, suggesting the potential effectiveness of contemporary logging regulations in protecting large wood characteristics. In our study, both natural and timber harvest practice-related attributes had mea-

* Corresponding author: rossjare@msu.edu

sureable influences on stream habitat, underscoring the importance of considering catchment-scale attributes, including riparian management schemes, for managing Southeast Alaska streams.

Introduction

In Southeast Alaska streams, large wood (LW) is a critical habitat factor for supporting multiple species of Pacific salmon *Oncorhynchus* spp., many of which are culturally and socioeconomically important and a primary focus of management actions in the region. Large wood in Southeast Alaska streams benefits fishes in multiple ways. It promotes channel complexity, forms pools, and contributes to the storage and processing of sediments and organic matter (Robison and Beschta 1990; Martin and Shelly 2017; Wohl et al. 2019). In low to moderate gradient streams of the region, scour pools formed by LW are common (Martin 2001). Compared to LW from deciduous species in the region, LW originating from coniferous trees has a greater size potential and decays more slowly in streams (Harmon et al. 1986; Bilby et al. 1999). Additionally, LW from coniferous trees may have a greater capacity to create pools within stream channels (Ross et al. 2019). Pools provide essential habitat for juvenile salmonids (Bryant et al. 2005; Bryant et al. 2007), are used by adult fishes for resting purposes during migration, and provide refuge for juvenile anadromous species and resident fishes (e.g., *Oncorhynchus* spp. and *Salvelinus* spp.) during periods of flooding, as well as cover from predators (Walker et al. 2018). Finally, LW effectively retains coarse substrates important for lithophilic spawning fishes, including anadromous Pacific salmon (Martin and Shelly 2017).

The amount, type, and movement of LW to and through stream channels is influenced by properties of the landscapes they drain. Many studies conducted outside of Alaska describe how landscape characteristics (e.g., land use and land cover) affect streams and fishes, with attributes summarized in catchments as well as in riparian buffers and related to response variables of interest (e.g., Wang et al. 2001, 2006; Elosegi and Johnson 2003; Burnett et al. 2006; Johnson et al. 2006; Hemstad and Newman 2006; Keeton et al. 2017). In Alaska, much is known about influences of landscape factors in riparian zones on LW in streams. For example, in a subset of Southeast Alaska streams having old growth riparian zones, 99% of recruited LW was found to come from within 30 m of the channel, and recruitment processes varied by differences in stream size and channel type, characteristics that vary within and across catchments (Murphy and Koski 1989). In another study of Alaska catchments harvested under contemporary logging practices (which require retention of streamside riparian buffers), 95% of recruited LW came from within 20 m of the stream channel. In addition, the occurrence of wood recruited by windthrow increased in the outer edge of buffers left after logging (Martin and Grotefendt 2007). Aquatic productivity is also enhanced through allochthonous input of litter and insects (Allan et al. 2003; Wipfli et al. 2007), and wood-formed pools and sediment retention depend on quantities and types of riparian vegetation (Gomi et al. 2006).

Comparatively much less is known about catchment-scale influences on stream characteristics, including LW, in Alaska. One study in the region found that the longevity of debris jams decreases with catchment area, and large streams move logs farther and more often than smaller streams (Martin and Benda 2001). Additionally, bank erosion, an important LW input process, is driven in part by catchment size (Martin and Benda 2001), as well as catchment geomorphology (Murphy and Koski 1989).

The lack of information regarding catchment-scale influences on LW is due in part to the inability to summarize landscape characteristics specific to ecologically relevant drainage boundaries in Alaska.

In Southeast Alaska, large-scale timber harvests represent the greatest potential anthropogenic landscape disturbance to streams of the region (Crawford et al. 2016; Cooper et al. 2019, this volume). Timber harvest policies in Southeast Alaska have changed over time, with no rules surrounding practices prior to 1980. The initial Alaska Forest Resources and Practices Act (FRPA 1978) restricted how timber could be harvested, felled, and transported near streams supporting anadromous fishes, but still allowed logging up to stream banks when valuable timber was available. Amendments in 1990 required the retention of 20-m-wide riparian buffers on private timberlands along streams supporting anadromous fishes (FRPA 1978).

Substantial research has occurred to identify influences of timber harvest practices on stream ecosystems prior to 1980 (e.g., Johnson et al. 1986; Murphy 1995) and after 1990 (e.g., Martin and Grotefendt 2007; Martin and Shelly 2017), though little research has focused on the period of initial Alaska Forest Resources and Practices Act regulations in place from approximately 1980 to 1990 (but see Ross et al. 2019). Nonetheless, restoration schemes are currently being implemented across the region's extensive young growth forests (USFS 2014). Understanding how harvest practices within catchments affect stream habitat would provide useful information for rehabilitating habitats in catchments that have undergone historic and contemporary timber harvest practices. Therefore, the goal of this study was to characterize how catchment landscape attributes, including both contemporary and historical timber harvest practices, influence LW in Southeast Alaska streams. We accomplished this by completing three objectives. First, we used a geographic information system (GIS) to delineate catchments for all Southeast Alaska stream reaches and summarized a variety of landscape attributes within catchments throughout the region. Next, we broadly assessed the influences of natural versus harvest practice-related landscape attributes on the region's streams by comparing relative amounts of variation in four LW variables. Finally, we used natural and timber harvest practice-related landscape attributes to predict multiple measures of LW in a subset of Southeast Alaska streams. The collective findings yield new insights for management.

Methods

Study Region

The Southeast Alaska region contains more than one-quarter of the world's coastal temperate rainforests (Olson et al. 2001) and is part of the 1,100-island Alexander Archipelago. The island-dominated region is drained by relatively short stream networks that flow through glacially carved, high-relief mountainous terrain from headwaters to river mouths. The majority of the region (~80%) is part of the largest national forest in the United States, the Tongass National Forest. Native Alaskan corporations manage just under 3% of the land in Southeast Alaska, as granted by the Native Claims Settlement Act (1971). Coniferous Sitka spruce *Picea sichensis*, western red-cedar *Thuja plicata*, and western hemlock *Tsuga heterophylla* dominate the region's old growth forests. Deciduous red alder *Alnus rubra* is common in floodplains and on other disturbed areas such as landslides or areas with past fire and logging activity (Rot et al. 2000; Villarin et al. 2009). Multiple species of anadromous fishes are supported by streams in the region, including recreationally and commercially important Pacific salmon and char *Salvelinus* spp.

Catchment Delineation

Streams were identified from the 1:63,360 scale National Hydrography Dataset (USGS 2012a), which maps locations of nearly 200,000 stream reaches in the region that extend from headwaters to confluences, confluences to confluences, or confluences to stream mouths. For each reach, we delineated its local catchment boundaries based on a 60-m digital elevation model (DEM; USGS 2012b), which was the finest resolution DEM available for the region at the time this work was conducted. Local catchment delineations were generated using a script developed in the Python programming language. Local catchments are defined as the land area draining directly to the downstream end of any stream reach (Figure 1; Wang et al. 2011). Additionally, we defined network catchments as the land area within all upstream local catchments draining to the downstream end of any given reach, including its local catchment (Figure 1; Wang et al. 2011).

Study Sites

All potential study sites were located on native Alaskan-owned timberlands throughout the region, and all timber was managed by Sealaska Timber Corporation. The sites had experienced a single harvest of old-growth trees. Nonetheless, pre-commercial thinning occurred adjacent to most potential study sites between 1993 and 2009. Thinning reduces the density of very small, naturally regenerated trees (<10 cm diameter at breast height [dbh]), thereby increasing growth of desirable, unthinned trees. We initially identified all potential study sites using GIS data and aerial imagery to ensure that they met the following criteria. They had to have at least one bank harvested between 1980 and 1990 (Ross et al. 2019) and were identified as

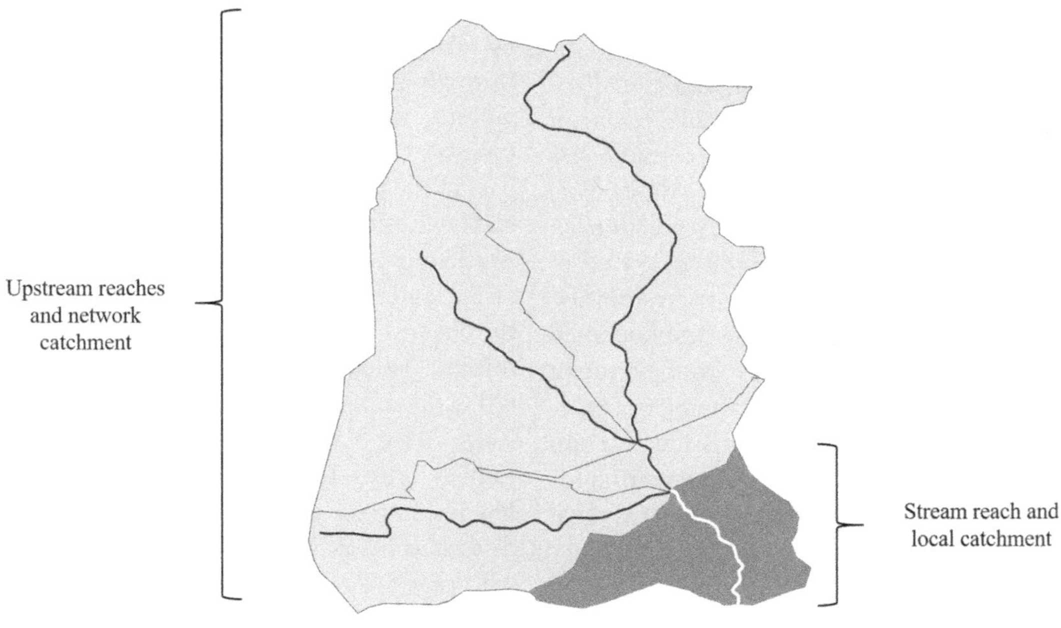

Figure 1. Example of local and network catchments for a stream reach of interest. The local catchment includes the land area draining to the reach of interest and between its defining tributaries. The network catchment includes the local catchment as well as the land area draining to all upstream reaches.

supporting anadromous fishes based on multiple data sets for the region (Paustian et al. 1992; Johnson and Blossom 2018). The state of Alaska manages and updates the Anadromous Waters Catalog, which uses anadromous fish collections to identify streams that provide migration, spawning, or rearing habitat for anadromous fishes (Johnson and Blossom 2018). We used the 2011 version of this data set to help identify potential sample sites for this study. Potential sites were further reduced to being classified as either floodplain or moderate gradient, mixed control stream channels following Paustian et al. (1992). Those two stream types were selected in part because of their capacity to retain abundant and functioning LW (Paustian et al. 1992), their capacity to support the suite of anadromous fishes in the region, and their potential sensitivity to disturbances resulting from timber harvest practices (Woodsmith et al. 2005). Other channel types in the region, including those with high gradients (>6%) and/or confined channels, have a reduced capacity to retain LW or support anadromous fishes compared to floodplain and moderate gradient, mixed control stream types (Paustian et al. 1992). After the target population of potential sites was identified, actual sample sites were selected randomly using a generalized random-tessellation stratified design (Stevens and Olsen 2004) to ensure that sites were spatially distributed across the entire target population of potential sites. In June and July of 2011, measurements of multiple LW characteristics were taken from 28 sample sites draining young-growth watersheds on Prince of Wales and Chichagof islands (Figure 2).

Variables Attributed to Local and Network Catchments of All Stream Reaches in the Region

A variety of natural landscape characteristics were summarized within local and network catchments for nearly 200,000 stream reaches in the region, and we summarized landscape attributes within network catchments using a process described in Tsang et al. (2014). These characteristics included catchment area, with catchment boundaries derived from the 60-m DEM (described above). Local catchment slope was also created from the 60-m DEM by calculating the maximum percent change in elevation of a single cell with all of its neighboring cells, averaged across the local catchment. Although slope is sensitive to properties of the DEM used to generate this information, this was the finest resolution data set available for all of Southeast Alaska at the time this work was conducted. The National Wetlands Inventory (USFWS 2013) was acquired for Alaska, and we summarized the percentage of differing types of wetlands in local and network catchments for the entire region. The National Land Cover Database for Alaska (NLCD 2001; Homer et al. 2007) was also acquired, and attributes from this database were summarized as percentages of total land cover in local and network catchments (Table 1). Land cover included characterizations of forested land cover such as deciduous, evergreen, or mixed (i.e., deciduous and evergreen) forests, as well has human-dominated land covers such as urban or agricultural land uses. Those variables were attributed to every local and network catchment in the study region as potential predictor variables in subsequent analyses.

Landscape Variables Attributed to Local and Network Catchments of Study Sites

Timber harvest practices and logging road data were obtained from managers at Sealaska Timber Corporation and from the U.S. Forest Service. Timber harvest policies have changed multiple times in Southeast Alaska, and data were summarized into three time periods corresponding to those regulatory

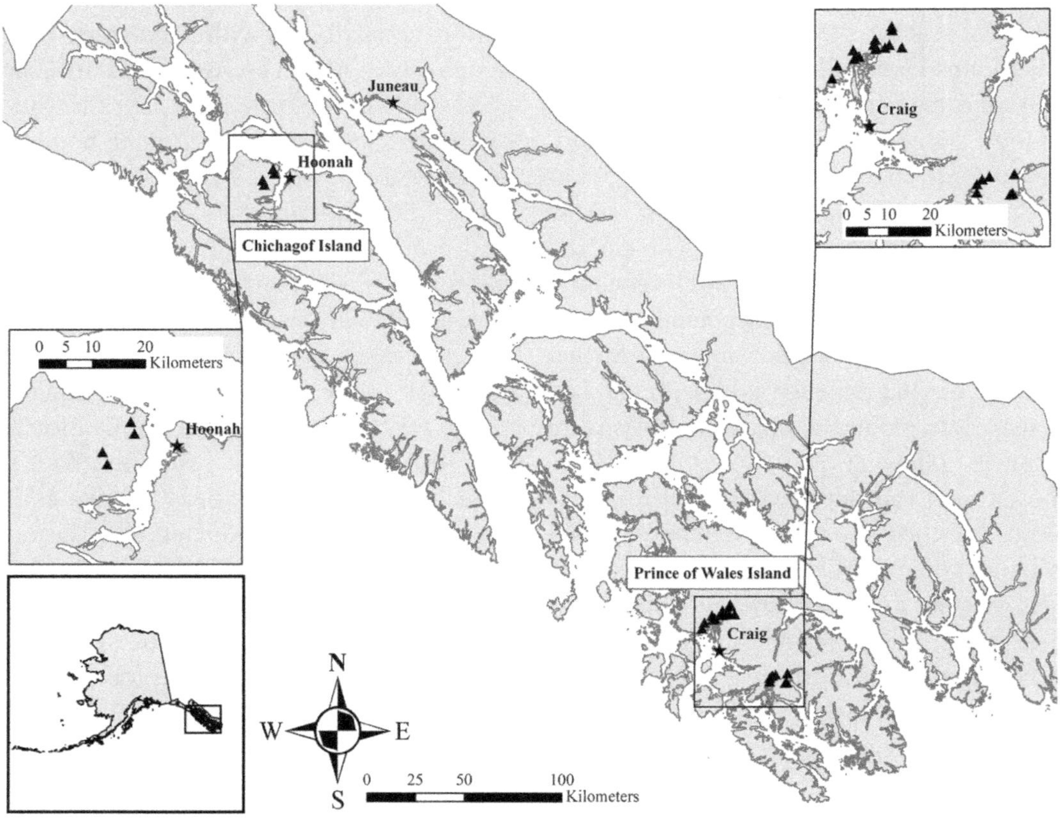

Figure 2. Locations of 28 study sites where large wood habitat data were collected on Chichagof and Prince of Wales islands. Black triangles represent sample site locations and stars are nearby towns.

changes. The first time period included ruleless harvest that occurred before 1980; during this time, no standards existed regarding how timber could be harvested, felled, and/or transported near streams. Logging could occur within the riparian zone, and trees could be harvested directly up to the stream bank. Additionally, there were no rules surrounding removal of large wood from channels (i.e., stream clearing), a historically common logging practice shown to have long-term, negative effects on stream habitat (Murphy 1995; Mellina and Hinch 2009; Miller 2010). In 1978, the initial Forest Resources and Practices Act was adopted for Alaska, and these FRPA regulations were in place between roughly 1980 and 1990. During this time, high-value trees were typically still logged directly up to stream banks (riparian clear-cut). Removal of large wood from channels did not occur, and many streams with riparian timber harvest during this period have similar loads of wood and pool habitats 30–40 years postharvest (Ross et al. 2019). In 1990 and after, amendments to the Alaska FRPA required leaving 20-m uncut riparian buffers along all streams on private timberlands supporting anadromous fishes (contemporary clear-cut). Besides characterizing differences in time, we considered two differences in timber harvest practices. Conventional clear-cut logging

Table 1. Summary statistics of landscape attributes for local and network catchments from 28 sampled stream sites. T1 refers to local catchment variables while T2 refers to network catchment variables. Logging before 1990 allowed riparian clearcutting; logging before 1980 was unrestricted. Local catchment variables used in the analyses are indicated by (*) whereas (**) indicates that the network catchment variable was used.

		Local (T1)				Network (T2)			
	Code	Mean	Min	Max	SD	Mean	Min	Max	SD
Timber harvest-related landscape variables									
Conventional pre-1980 (%)[a]	CONpre80	1.76	0.00	27.27	6.49	0.46	0.00	6.43	1.44
Conventional 1980–1990 (%)*[a]	CON8090	41.58	5.78	80.23	21.03	30.29	3.64	77.55	22.43
Conventional post 1990 (%)**[a]	CONpo90	10.30	0.00	37.81	11.88	14.24	0.00	49.10	12.69
Other 1980–1990 (%)[a]	OT8090	0.25	0.00	3.46	0.79	0.18	0.00	2.12	0.55
Other post-1990 (%)*[a]	OTpo90	10.34	0.00	31.61	9.81	14.07	0.00	41.91	11.93
Logging road/stream crossing density (#/km^2)**[a]	LogXDens	5.58	0.00	75.73	14.41	3.00	0.00	8.92	2.53
Natural landscape variables									
Area (km^2)**[b]	areasqkm	0.71	0.02	2.83	0.75	4.24	0.12	19.75	5.63
Max slope (%)*[b]	slpMAX	3.52	1.21	7.13	1.97	3.82	1.47	6.54	1.38
Deciduous forest (%)*[c]	decid	6.59	0.00	37.93	9.02	3.95	0.00	37.93	7.01
Evergreen forest (%)[c]	evgrn	36.64	0.00	65.62	16.91	42.03	0.00	66.98	19.77
Mixed forest (%)[c]	mixd	27.40	2.74	57.14	15.29	19.17	2.75	44.63	12.65
Forested wetlands (%)**[d]	WETfor	33.17	0.00	100.00	30.77	25.69	0.00	93.16	22.65

[a] U.S. Forest Service and Sealaska Timber Corporation.
[b] Digital elevation model (DEM; USGS 2012b).
[c] National Land Cover Dataset (2001).
[d] National Wetlands Inventory (2013).

aims to harvest all trees in a given harvest unit, while other (i.e., selective) harvesting practices target specific, high-value trees for harvest and is less invasive than clear-cut logging. Harvest data (ruleless, riparian clear-cut, contemporary clear-cut, other) were summarized for the three time periods for both local and network catchments from all 28 sampled sites (Table 1). Densities of logging road/stream crossings (#/km^2) were also calculated and summarized for local and network catchments.

Large Wood Data

Large wood data were collected at the 28 study sites, which extended for a length of at least 20 times the wetted stream width (Harrelson et al. 1994). For this study, we define LW as any log at least 10 cm in diameter and 2 m in length within the bankfull channel, following Robison and Beschta (1990) and Martin (2001). The diameter of the center of each log occurring within the site was measured and assigned to the following size categories: small (10–30 cm), medium (30–60 cm), large (60–90 cm) and extra-large (>90 cm). For subsequent analyses, we combined the large and extra-large categories into a single variable. The length of each log was measured to the nearest 3 m.

All logs for which the source could be identified were considered a LW recruit and classified by its recruitment process, with possible recruitment processes including bank erosion, windthrow, broken stem (i.e., assumed windthrow), mortality, landslide, cut, beaver, or unknown. Information on whether a recruit was in or above the bankfull channel was also recorded. Recruits found in the bankfull channel typically interacted with the channel bed to create pools and/or hold substrates, LW, or other organic material. Recruits located above the bankfull channel were generally at or above bankfull, spanning across the bankfull channel but not typically associated with physical stream characteristics at lower flows. Large wood recruits were additionally classified as either deciduous or conifer species.

From the site data, we calculated a variety of variables and selected a subset for analysis (Ross 2013). These included average LW length (m) per site, large and extra-large sized LW density (#/km >60 cm dbh), medium-sized LW density (#/km 30–60 cm dbh), and the ratio of the number of recruits found in the bankfull channel to those found above the bankfull channel (Table 2). The length of LW is important because longer versus shorter pieces provide a larger volume of LW given the positive relationship between tree height and dbh (Arcangeli et al. 2013), and larger pieces of wood have a greater capacity to function within the channel. Similarly, the width of LW is important, as wider pieces proportionately function more than narrower pieces (Ross et al. 2019) and are also more resistant to transport downstream, thus providing greater stability.

Table 2. Large wood variables used in analysis. INABOVE is the ratio of recruits in versus above the channel. MEDLWKM is the density (#/km) of medium-sized pieces of large wood. LXLWKM is the density (#/km) of large- and extra-large-sized pieces of large wood (>60 cm dbh). LWLENGTH is the average length of large wood pieces (m).

Variable	Mean	Min	Max	SD
INABOVE	1.73	0.33	5.60	1.33
MEDLWKM (#/km)	106.10	20.83	197.53	52.17
LXLWKM (#/km)	57.61	10.42	160.49	37.70
LWLENGTH (m)	4.70	3.35	6.84	1.03

Because of this, the density of the largest pieces (i.e., >60 cm) was also selected. Medium-sized LW density was also selected because it is functionally important to streams in a manner similar to larger pieces of wood, and the medium size-class encompasses the average size that red alder tends to achieve (Worthington et al. 1962). Furthermore, the maximum diameter for the medium size-class (i.e., 60 cm) is approximately the largest that Sitka spruce will grow in 80 years in Oregon, where trees are on average larger than in Southeast Alaska (Smith et al. 1984), so 30–60 cm dbh (i.e., medium size-class) is the maximum size regenerated trees would be able to achieve in these previously harvested stands. Finally, the ratio of LW recruits that are located within the bankfull channel as opposed to positioned above the bankfull channel was selected to characterize the functioning potential of LW recruited to the set of study sites.

Data Preparation

All predictor and response variables were transformed to maximize linearity and normality prior to running subsequent analyses. Continuous data were transformed by using natural log, count data were transformed by using square root, and proportions were transformed by using arcsin square root.

Comparing Relative Amounts of Variation in Large Wood Variables Explained by Natural and Timber Harvest Practice-Related Catchment Variables

A constrained ordination analysis was run in CANOCO to assess the relative importance of timber harvest practices versus natural landscape factors on LW in the study sites. To determine whether a canonical correspondence analysis (CCA) or redundancy analysis (RDA) was more appropriate given the data distribution, a detrended correspondence analysis (DCA) was first performed to assess the gradient lengths in the response data set. This analysis suggested that the data had shorter gradient lengths and followed a linear distribution, making RDA a more appropriate technique (ter Braak and Prentice 1988). Redundancy analysis is a direct gradient analysis and can be used to quantify variation in a set of response variables by constraining their ordination to a linear combination of independent predictors (i.e., landscape variables, ter Braak and Prentice 1988). An equal number of harvest-related and natural landscape variables were chosen so as not to over- or underestimate variance explained as a result of differences in numbers of predictor variables.

Predicting Large Wood Variables from Natural and Timber Harvest Practice-Related Landscape Attributes

Two sets of regression models were developed to relate specific LW variables to timber harvest practice-related and natural landscape attributes. Because of the known influences of stream size (i.e., network catchment area) on LW recruitment by bank erosion and fluvial transport (e.g., Martin and Benda 2001), each LW variable was initially predicted using only network catchment area to better compare the performance of subsequent models that included additional variables (following Burnett et al. 2006). Models having an Akaike information criterion (AIC) near or lower than the model with catchment area alone were considered to perform as well or better than the network catchment area only model. Next, we performed a best subsets regression (R 'leaps' package) on each LW variable with the suite of natural and timber harvest practice attributes to identify the five best models, characterized by those having the largest adjusted R^2. We considered only significant models that contained no more than four explanatory variables to avoid model overfitting and only those within two AIC units of the "best" model. While rules of thumb exist to guide consideration of the maximum number of predictors to sites, the ratio used here is similar to those used in previous studies (e.g., Johnston et al. 1990; Burnett et al. 2006). Further, a greater sample size per independent variable is more critical when developing equations to predict response variables at unsampled sites compared to questions addressing significant associations between variables (Maxwell 2000; Knofczynski and Mundfrom 2008). Collinearity was assessed based on a variance inflation factor threshold of three (VIF < 3).

Results

Natural Landscape Characteristics of Study Site Catchments

Wide ranges of values were observed in natural landscape characteristics within catchments of the sampled study sites. Catchment slope ranged from 1.21% to 7.13% and averaged 3.52% (Table 1). Deciduous forests averaged about 7% of local catchment area. Network catchment area averaged 4.24 km^2 and ranged from less than 1 km^2 to just under 20 km^2. The percent of network catchment area comprised of forested wetlands ranged from zero to a maximum greater than 90% but averaged slightly more than 25% across all site catchments.

Timber Harvest Practice-Related Characteristics of Study Site Catchments

Timber harvest practices occurring from 1980 to 1990 (CON8090, riparian clear-cutting) was the most pervasive practice across study sites; it averaged more than 40% across local catchments and exceeded 80% for one site (Table 1). Other harvest after 1990 (OTpo90) averaged 10% of local catchment area and ranged from 0% to slightly greater than 30% for the study sites. Contemporary clear-cutting occurring in network catchments after 1990 (CONpo90) averaged about 14% across sites with a maximum of 49%. Logging roads were present in all network catchments, and logging road/stream crossing densities (LogXDens) averaged 3/km^2 across all sites. Riparian clear-cutting (ruleless) before 1980 (CONpre80) was uncommon within network catchments and ranged from 0% to more than 6% but averaged less than 1% across all study sites.

Comparing Variation Explained by Natural versus Timber Harvest Practice-Related Landscape Attributes

Variance partitions from the CCA analysis suggest that the four natural and four harvest-related landscape factors evaluated in this study explained 44% of the variation in the set of LW variables tested. Of the explained variation, timber harvest practice-related factors explained about 54% and natural factors explained about 28%, while about 18% was explained by the interactions among harvest-related and natural landscape factors (Figure 3). Plotting landscape and habitat factors based on CCA axes scores (Figure 4) supported findings shown by best subsets regression models (described below). Catchment area, local deciduous forest, and local catchment riparian clear-cutting from 1980 to 1990 were positively associated with all four LW variables along axis 1, whereas local other harvest after 1990 was negatively associated with those variables. Along axis 2, contemporary clear-cutting occurring after 1990 was positively associated with length of wood and medium-sized LW density (Figure 4).

Predicting Large Wood Variables Using Natural and Timber Harvest Practice-Related Landscape Attributes

A range of variation in average LW length (Table 3), ratio of recruits located within the bankfull channel versus recruits positioned above the bankfull channel (Table 4), and

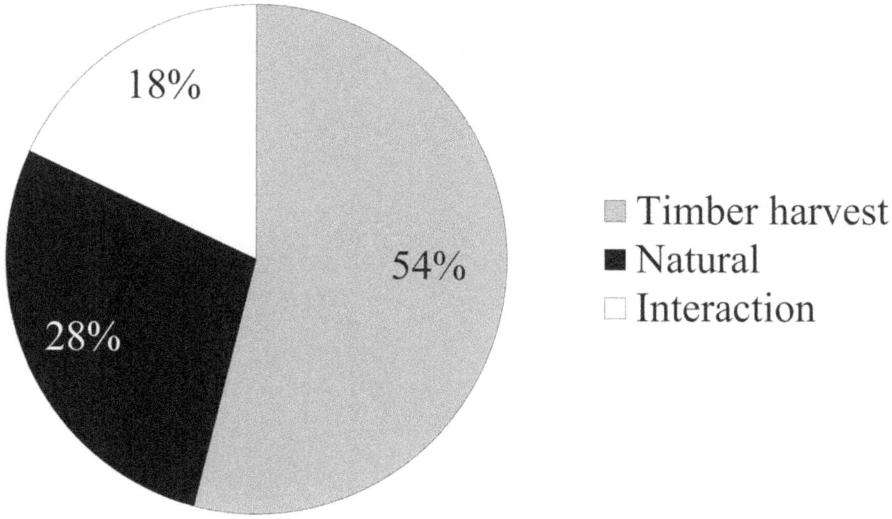

Figure 3. Results of variance partitioning from the redundancy analysis assessing the relative importance of various types of landscape factors on large wood habitat factors. A total of 44% of the variation in large wood habitat factors is partitioned into timber harvest-related factors in gray, natural factors in black, and the interaction in white.

medium-sized LW density (Table 5) was explained by using a best subsets regression approach that derived multiple significant (all P-values < 0.02) linear models from both natural and harvest-related landscape variables. The procedure only derived one significant model (P-value < 0.05) predicting large and extra-large size (dbh > 60 cm) LW density, and this included only timber harvest practice-related variables (Table 6).

Large wood length (LWLENGTH)

The average length of LW was significantly (P-values < 0.0001) related to natural and harvest-related landscape variables in five linear regression models. The amount of variation explained (adjusted R^2) by these models ranged from 0.57 to 0.59 (Table 3), and all five included network catchment area (T2areasqkm), local catchment deciduous forest (T1decid), and network catchment contemporary clear-cutting after 1990 (T2CONpo90), with the best model based on adjusted R^2 containing only these three variables (Table 3). In all models, average length of LW in the study sites was positively associated with these landscape variables. Additional models included network-catchment logging road/stream crossing density (T2LogXDens), network-catchment forested wetlands (T2WETfor), local catchment riparian clear-cutting from 1980 to 1990 (T1CON8090), and local catchment other harvest after 1990 (T1OTpo90). Network catchment area (T2areasqkm) had the greatest contribution in all models predicting LW length, followed by local deciduous forest (T1decid) and network catchment contemporary clear-cutting after 1990 (T2CONpo90) based on standardized beta coefficients.

Position of large wood recruits (INABOVE)

The ratio of recruits that were in versus above the bankfull channel was best predicted by a model incorporating one harvest-related and three natural landscape variables (P-value < 0.01). The amount of variation explained in this model had an adjusted R^2 = 0.34 (Table 4). Riparian clear-cutting occurring from 1980 to 1990 in the local catchment explained the

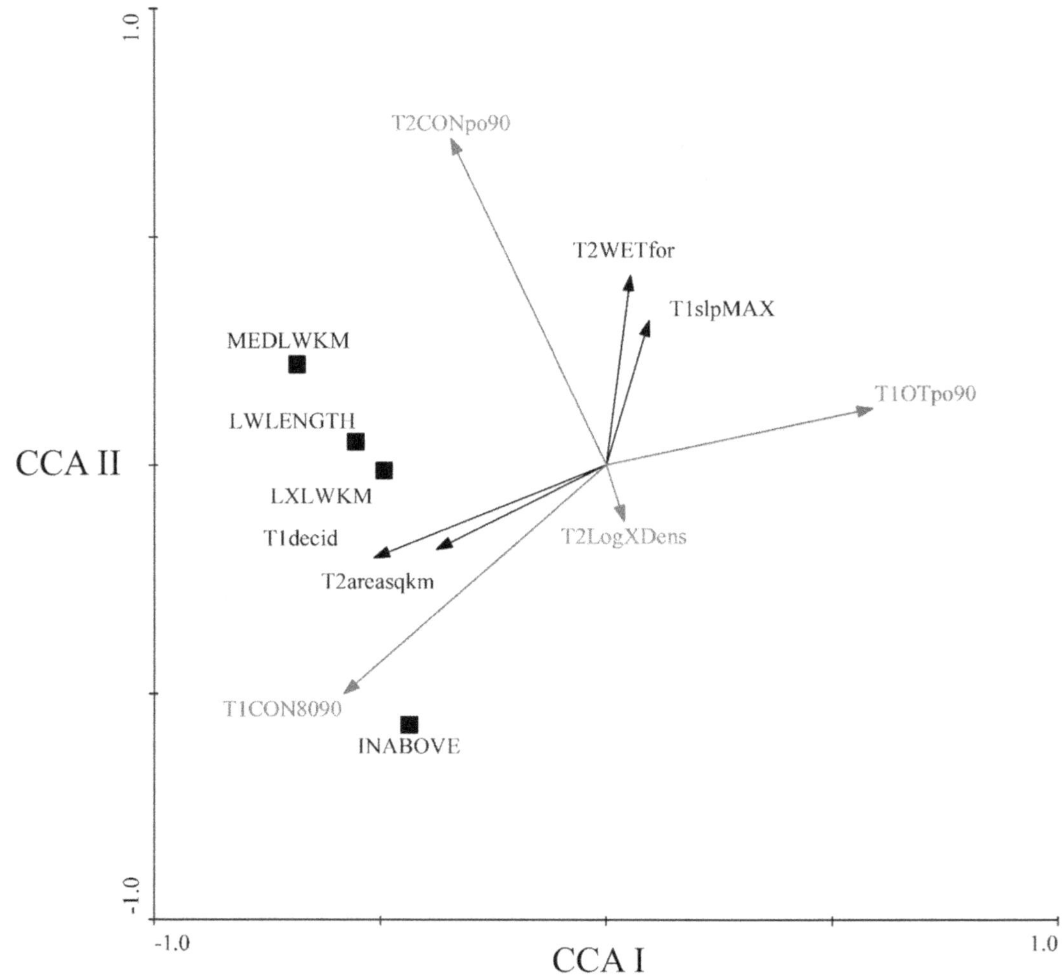

Figure 4. Canonical correspondence analysis (CCA) axis 1 versus CCA axis 2. Arrows represent natural (black text) and timber-harvest related (gray text) factors, black squares represent large wood (LW) response variables. The length of the arrows indicates the strength of relationship between the landscape attributes with LW variables. Abbreviations are defined in Tables 1 and 2.

most variation and was positively associated with this habitat variable in all models. Local deciduous forest, local maximum slope, and network catchment area were additionally included in this model, with maximum slope having the only negative relationship with the habitat variable. Four subsequent models that included both local catchment riparian clear-cutting occurring from 1980 to 1990 and local maximum slope also included local deciduous forest as having a positive association in two of the remaining four models. Local other harvest after 1990 and network catchment density of logging road/stream crossings were negatively related to the LW habitat factor and included in two of the five models.

Medium-Sized Large Wood Density (MEDLWKM)

Variation (adjusted $R^2 = 0.37$, Table 5) was explained in the density of medium-sized LW by two measures of timber harvest practice

Table 3. Results from best subsets regression for average large wood length. Models are in order of most variance explained, predictor variables are ordered by their influence on the model based on standardized beta coefficients. Variable codes are in Table 1, T1 represents local catchment variables while T2 represents network catchment variables, and ΔAIC represents the difference in Akaike information criterion when compared to a model containing network catchment area alone. VIF refers to variance inflation factor.

Model	Landscape variable	Adjusted R^2	Standard beta	F	Model significance	VIF	ΔAIC
1	T2areasqkm	0.59	0.66	13.70	0.000021	1.02	−6.89
	T1decid		0.32			1.01	
	T2CONpo90		0.28			1.02	
2	T2areasqkm	0.58	0.68	10.36	0.00006	1.04	−5.79
	T1decid		0.34			1.04	
	T2CONpo90		0.28			1.02	
	T2LogXDens		0.11			1.05	
3	T2areasqkm	0.57	0.68	10.03	0.000075	1.09	−5.22
	T1decid		0.34			1.08	
	T2CONpo90		0.28			1.03	
	T2WETfor		0.07			1.15	
4	T2areasqkm	0.57	0.65	9.89	0.000083	1.07	−4.96
	T1decid		0.31			1.09	
	T2CONpo90		0.28			1.08	
	T1CON8090		0.03			1.17	
5	T2areasqkm	0.57	0.66	9.86	0.000085	1.07	−4.92
	T1decid		0.32			1.05	
	T2CONpo90		0.28			1.02	
	T1OTpo90		−0.02			1.10	

and two natural landscape variables (P-value < 0.01). Variables included in this model (network catchment contemporary clear-cutting after 1990, local deciduous forest, network catchment forested wetlands, and local riparian clear-cutting from 1980 to 1990) all had positive relationships with medium-sized large wood density. Three subsequent models significantly (P-values < 0.02) explained variation (adjusted $R^2 = 0.32–0.33$) in this LW variable. All three included network catchment contemporary clear-cutting after 1990 and network catchment forested wetlands, whereas local catchment deciduous and local catchment other harvest after 1990 were included in two of the remaining three models. Other harvest after 1990 in the local catchment had the only negative association with medium-sized LW density in any model. Network catchment area was related to medium-sized LW density in only one of the four regression models and was the weakest contributing landscape variable included for that model.

Large-Sized and Extra-Large-Sized (>60 cm dbh) Large Wood Density (LXLWKM)

Large- and extra-large-sized LW density was significantly (P-value < 0.05) related to only timber harvest practice factors. The amount

Table 4. Results from best subsets regression for the ratio of large wood recruits that are in or above the bankfull channel. Models are in order of most variance explained, predictor variables are ordered by their influence on the model based on standardized beta coefficients. Variable codes are in Table 1. T1 represents local catchment variables while T2 represents network catchment variables, and ΔAIC represents the difference when compared to a model containing network catchment area alone. VIF refers to variance inflation factor.

Model	Landscape variable	Adjusted R^2	Standard beta	F	Model significance	VIF	ΔAIC
1	T1CON8090	0.34	0.40	4.50	0.008	1.11	−8.20
	T1decid		0.28			1.10	
	T1slpMAX		−0.26			1.01	
	T2areasqkm		0.23			1.03	
2	T1CON8090	0.33	0.43	4.38	0.009	1.08	−7.88
	T1slpMAX		−0.24			1.03	
	T1decid		0.21			1.12	
	T1OTpo90		−0.21			1.06	
3	T1CON8090	0.33	0.53	4.33	0.009	1.07	−7.74
	T1slpMAX		−0.36			1.31	
	T2LogXDens		−0.23			1.37	
	T1OTpo90		−0.20			1.10	
4	T1CON8090	0.32	0.50	4.23	0.01	1.18	−7.47
	T1slpMAX		−0.36			1.30	
	T2LogXDens		−0.22			1.42	
	T1decid		0.19			1.20	
5	T1CON8090		0.56	5.22	0.006	1.05	−8.08
	T1slpMAX	0.32	−0.41			1.23	
	T2LogXDens		−0.28			1.28	

of variation explained for this model was relatively low (adjusted R^2 = 0.16, Table 6). Harvest-related landscape variables explaining this LW variable include local catchment riparian clear-cutting occurring from 1980 to 1990, which had a positive association, whereas a negative association existed between this habitat variable and local other harvest after 1990.

Discussion

This study highlights differences in catchment characteristics occurring throughout Southeast Alaska and identifies important natural and human influences resulting from timber harvest practices on LW in a subset of the region's streams. Results from the CCA support associations between this set of LW variables with major landscape drivers and show that 28% of the variation in the LW variables can be attributed to natural characteristics, whereas 54% of that variation is explained by timber harvest practice-related variables. Best subsets regression predictions of LW variables from landscape variables indicate that LW is significantly associated with multiple characteristics summarized in catchments of study sites. This includes both natural and timber harvest practice-related attributes. To our knowledge,

Table 5. Results from best subsets regression for medium-sized large wood density. Models are in order of most variance explained, predictor variables are ordered by their influence on the model based on standardized beta coefficients. Variable codes are in Table 1. T1 represents local catchment variables while T2 represents network catchment variables, and ΔAIC represents the difference when compared to a model containing network catchment area alone. VIF refers to variance inflation factor.

Model	Landscape variable	Adjusted R^2	Standard beta	F	Model significance	VIF	ΔAIC
1	T2CONpo90	0.37	0.50	4.96	0.005	1.08	−10.37
	T1decid		0.39			1.11	
	T2WETfor		0.37			1.15	
	T1CON8090		0.34			1.20	
2	T1decid	0.33	0.40	4.37	0.009	1.11	−8.79
	T2CONpo90		0.40			1.02	
	T1OTpo90		−0.26			1.05	
	T2WETfor		0.26			1.08	
3	T1decid	0.32	0.49	4.24	0.01	1.08	−8.44
	T2CONpo90		0.39			1.03	
	T2WETfor		0.35			1.15	
	T2areasqkm		0.25			1.09	
4	T2CONpo90	0.32	0.47	4.15	0.011	1.09	−8.19
	T2CON8090		0.39			1.16	
	T1OTpo90		−0.31			1.02	
	T2WETfor		0.28			1.11	

this is the first study comparing relationships between multiple catchment variables and LW in Southeast Alaska stream sites.

Natural Landscape Characteristics

Network catchment area (T2areasqkm) was included in all five models produced for average LW length, and more than half of the variance in this variable was explained by network catchment area (Table 3). An increase in LW length with increasing network catchment area follows previous findings from the Pacific Northwest and Alaska. Wood frequency and volume is greater in

Table 6. Results from best subsets regression for large- and extra-large-sized large wood density. Predictor variables are ordered by their influence on the model based on standardized beta coefficients. Variable codes are in Table 1. T1 represents local catchment variables while T2 represents network catchment variables, and ΔAIC represents the difference when compared to a model containing network catchment area alone. VIF refers to variance inflation factor.

Model	Landscape variable	Adjusted R^2	Standard beta	F	Model significance	VIF	ΔAIC
1	T1CON8090	0.16	0.36	3.50	0.046	1.01	−4.11
	T1OTpo90		−0.28			1.01	

wider streams (Hough-Snee et al. 2014) and is, on average, longer in larger streams, as LW is generally stored when wood length is longer than the stream width (Martin and Benda 2001). Wider channels tend to be a characteristic of higher-order streams that have larger drainage areas (Vannote et al. 1980). Therefore, larger streams typically have longer pieces of wood found in the bankfull channel compared to smaller streams. When stream channels are wider than the tallest tree found in the riparian zone, they may have a greater capacity to transport wood (Benda et al. 2002). The results also concur with Bilby and Ward (1989), who explained more than 75% of variance in LW diameter, length, and volume through 22 sampled reaches of western Washington sites through use of stream width alone (as an indicator of stream size).

Maximum slope in the local catchment (T1slpMAX) was negatively associated with one LW variable, the ratio of recruits located within versus positioned above the bankfull channel (Table 4). It was included in all five models, although it was never the most influential landscape factor based on the beta coefficient. We found that there were fewer recruits located within the bankfull channel than spanning above the bankfull channel in catchments with steeper gradients. Narrower and more confined stream sites may restrict the capacity for LW to intersect stream channels (Liquori et al. 2008; Kraft et al. 2011). May and Gresswell (2003) found that 33% of locally recruited wood pieces were spanning the full width or suspended over the channel of headwater streams in a coastal Oregon catchment. Furthermore, high-gradient constrained channels have low LW storage because of greater stream power and subsequent capacity to transport instream LW. These channel types also meander less than low gradient and floodplain reaches, resulting in overall low LW recruitment in the absence of wood-delivering landslides (May and Gresswell 2003).

Deciduous forest land cover in local catchments (T1decid) was positively associated with average LW length in all five models predicting this factor (Table 3), although its influence was always second to that of network catchment area (T2areasqkm). This positive association indicates that streams with more deciduous forests in local catchments have, on average, longer pieces of instream LW across our study sites. Deciduous red alder is an early successional tree species that commonly establishes in disturbed landscapes, including harvested areas (Rot et al. 2000; Villarin et al. 2009). Aggressively harvested lands theoretically would have had large trees and subsequently long pieces of wood recruited from these riparian zones prior to logging. However, remnant large pieces of wood originating from old-growth forests prior to harvest coupled with large proportions of deciduous trees regenerating following logging could explain this association. In a separate analysis of data from these study sites, Ross et al. (2019) found similar loads of wood in stream sites with riparian timber harvest under the initial FRPA (i.e., 1980–1990) compared to sites with no harvest and sites with riparian buffers retained during harvest. They additionally found upwards of 73% deciduous riparian trees along their study sites. Another potential factor may be related to growth rates because red alder in Southeast Alaska grows rapidly during early stages of succession (Deal et al. 2004). Higher densities and death rates occur for alder versus conifers (Minore and Weatherly 1994; Hibbs and Giordano 1996), initially providing more pieces of deciduous wood than conifer stands because of quickly growing deciduous juvenile trees (Andrus et al. 1988; Gomi et al. 2006). Hanley and Hoel (1996) also found that 40 years after timber harvest in Southeast Alaska, riparian stands became dominated by red alder trees large enough to be considered LW (i.e., 15–25 cm dbh). This may alter size distributions and functioning capacity of LW in streams having harvested catchments over time because deciduous trees are smaller and decompose faster than conifers in streams (Bilby et al. 1999), and it may take upwards of 140 years for large and extra-large conifer trees (>60 cm dbh) to re-establish following harvest of old growth forests (Murphy and Koski 1989).

Our regression results also indicate that greater densities of medium-sized LW can be found in catchments with greater proportions of forested wetlands throughout network catchments (T2WETfor; Table 5). Wetlands have a significant influence on catchment hydrology, and there are many examples of wetlands reducing stream flooding and reducing high flows by storing large amounts of precipitation (e.g., Maltby 1991; Hey and Philippi 1995). This could potentially limit LW transport out of stream reaches draining catchments with wetlands having high water storing capacity, encouraging retention of wood as a result of lack of transport. In contrast, three out of four models also included a positive influence of deciduous forests in local catchments on medium-sized LW density. So, forested wetlands in catchments of our study sites may be dominated by naturally regenerated red alder that would have had 30–40 years after harvest to reach their maximum diameter potential, which falls within the medium LW size category (Worthington et al. 1962).

Timber Harvest Practice-Related Landscape Characteristics

Timber harvests occurring from 1980 to 1990 in Southeast Alaska did not require retention of riparian vegetation along streams. Timber harvest practices occurring during this period were related to all four LW variables tested (Tables 3–6), although it was most strongly associated with LW recruit position (Table 4). We found that with more riparian clear-cutting from 1980 to 1990 in local catchments (T1CON8090), more LW recruits were located within bankfull channels versus above the bankfull channel. Whereas this may result from LW being left within streams after harvest (Gomi et al. 2001; Benda et al. 2002), only 4% of LW recruitment was attributed to timber harvest in our study. As LW ages and decays over time, it breaks down and will be located within the channel more often than more recently recruited LW (Powell et al. 2009). The higher proportion of LW recruits positioned within these streams may be legacy pieces that were recruited to the stream before timber harvest practices occurred, whereas the lower proportion of LW located above the channel in these streams may have been recruited more recently and likely originated from the young growth stands. In fact, Martin and Shelly (2017) found a peak in functioning wood (which tends to be positioned within versus above the bankfull channel) approximately three decades after recruitment in similar Southeast Alaska stream sites. Consequently, catchments having more conventional harvest may produce relatively more LW recruits located within the bankfull channel because the loss of riparian trees would eliminate new LW recruitment for decades. This would skew the ratio toward relatively more instream recruits occurring as remnant pieces prior to harvest decay, which were eventually repositioned from above the channel into the channel (Powell et al. 2009).

Riparian clear-cutting occurring from 1980 to 1990 in local (T1CON8090; Table 5, model 1) and network (T2CON8090; Table 5, model 4) catchments was also positively associated with the density of medium-sized LW. In Southeast Alaska, local disturbances to riparian zones, such as timber harvest, landslides, and bank erosion, can remove patches of old-growth conifers, which can promote quickly growing red alder to colonize (Rot et al. 2000; Villarin et al. 2009). Because red alder grow more rapidly than conifers (Deal et al. 2004), they also die more quickly and potentially enter the stream as LW, and the medium size-class is the largest these trees tend to grow (Worthington et al. 1962). For this study, only LW recruits were determined to be deciduous or conifer, so we lacked sufficient data to determine the type of all logs measured.

Forested buffers along streams supporting anadromous fishes were required after 1990. Our landscape variable characterizing this practice, contemporary clear-cutting after 1990 in network catchments (T2CONpo90), was positively related to average LW length in all models, although it was typically less influential than natural attributes (i.e., network catchment area and local catchment deciduous forest). In a separate analysis of the data from these sites, Ross et al. (2019) found relatively more large- and extra-large trees in riparian zones with one bank harvested between 1980 and 1990 compared to sites with two banks harvested during this period. Although they did not present data on harvested stream sites that had riparian buffers retained, the lack of timber harvest in buffers would theoretically have greater proportions of their riparian zones comprised of large and extra-large (>60 cm dbh) trees compared to those harvested prior to buffer requirements. This is because larger trees are more valuable for timber and would not have been left in areas experiencing riparian clear-cutting. This may result in taller trees being available as LW to streams with riparian buffers retained after timber harvest. A similar relationship is present with medium-sized LW density. More medium-sized LW was observed in stream catchments with more contemporary clear-cutting after 1990 in network catchments (T2CONpo90). Timber harvest practices can increase LW input in Southeast Alaska, not only from harvest related debris being left over (Gomi et al. 2001), but also from increased windthrow of standing riparian buffers. The increased windthrow after harvest generally occurs near the outer edge of the buffer (Martin and Grotefendt 2007; Martin and Shelly 2017), so the tree boles reaching the stream from such an influence would likely be within the medium size-class.

Other types of harvest after 1990 in local catchments (T1OTpo90) were negatively related to the density of large and extra-large LW (dbh > 60 cm; Table 6). There was only one model significantly predicting the largest wood in streams, and it included two measures of harvest; however, it was a comparatively weak model and explained little variation. Recruitment rates of large- and extra- large trees (dbh > 60 cm) to streams is generally low in the region (Martin and Benda 2001), and the negative relationship between other harvest after 1990 and the density of the largest pieces of LW may be a product of selective harvest methods targeting very large trees from the stream buffer—a practice allowed under the post 1990 FRPA (Martin et al. 1998).

Considerations for Future Studies

Our study showed that both local and network catchment variables characterizing natural land cover and timber harvest practices explained variation in a suite of ecologically important LW variables. Although links between the riparian zone and instream LW have been demonstrated in Alaska, our findings re-enforce the idea that influences on LW characteristics can occur at extents larger than the riparian zone alone. Additional opportunities exist to better understand influences of catchment-extent attributes on stream characteristics, including LW. This may include collecting data from streams having similar ruleless, riparian clear-cutting and contemporary clear-cutting harvest intensities within their catchments to better discriminate impacts from differing timber harvest practices on LW or other response variables of interest.

Natural landscape characteristics, contemporary and historic land use, and LW characteristics differ throughout catchments of Southeast Alaska, so it is important to clarify that these results were from streams that have undergone timber harvest under the initial, and least restrictive, timber harvest policies. Relationships between landscape characteristics and large wood variables likely differ in streams draining landscapes derived of extensively old growth forests or those that have been harvested under more-restrictive contemporary logging regulations. Furthermore, thinning of previously harvested areas, a factor not accounted for in this study, has been and continues to be a common practice throughout the region (USFS 2014). So, understanding how these patterns in LW factors play out at multiple scales, under different land uses, or under different policies regulating timber harvest is important when characterizing the current condition and developing management prescriptions to restore and protect stream habitat in the region.

We acknowledge that use of a 60-m DEM limits the accuracy of catchment boundaries delineated through this work. Regardless, it was the finest-scale resolution coverage available consistently for the entire region at the time this work was conducted. Further, we feel that the 60-m DEM can sufficiently characterize broad differences in elevation because of the relatively short stream networks and high relief from headwaters to river mouths and is broadly applicable to the natural resource managers interested in having comparable information summarized across all of Southeast Alaska.

Finally, expanding this work to delineate local and network catchments for the entire state of Alaska would greatly improve the ability to consistently assess the condition of aquatic resources across the state within ecologically relevant spatial units that correspond to the ecology and nature of flowing waters (Omernik et al. 2017). The paucity of consistently summarized spatial information available to characterize not only these bodies of water, but the landscapes draining to them restricts efforts to assess their condition or to plan and prioritize management actions. Such efforts have generally been forced to use spatial units much coarser than catchments (e.g., HUC 12s; Crawford et al. 2016; Sloat et al. 2018), which diminishes the ability to discriminate landscape influences on stream habitats and biota when compared to local catchments (e.g., Omernik et al. 2017; Cooper et al. 2019).

Management Implications

Patterns in the amount, size, and input of LW in streams having old-growth, forested catchments have been shown to vary by stream size, channel type, and riparian condition (Murphy and Koski 1989; Bilby and Ward 1991; Keeton et al. 2007). Critical fish habitat characteristics including amounts and sizes of pools and channel substrates are additionally influenced by large wood and differ considerably in streams having unrestricted harvest practices (e.g., stream clearing) compared to those in old growth catchments (Heifetz et al. 1986). Input of LW is generally low in Southeast Alaska (Martin and Benda 2001), though decay can take multiple centuries depending on LW size (Murphy and Koski 1989). Decades after harvest, the LW loads and size distributions from our sites are similar to those having old-growth riparian zones (Ross et al. 2019). However, as remnant pieces of LW in these channels continue to decay and be transported, depleted loads and skewed size distributions toward smaller pieces may result over time as harvested riparian zones can take well over a century to recover to old-growth levels in Southeast Alaska (Taylor 1934; Murphy and Koski 1989).

The state of Alaska includes roughly 20% of all U.S. river length, and 40% of U.S.

fresh surface waters are located there (USGS 1993, 2012a). With relatively little research conducted to study effects of catchment landscape attributes on LW characteristics in the region, the results of our study can aid managers of these vast resources. By understanding the degree to which human versus natural landscape factors affect stream habitat, we can better understand the degree to which humans have altered stream systems and indirectly fish assemblages. This has proven useful in identifying the landscape-scale attributes that are most influential to LW variables in the region, thus offering managers better insights for assessing threats to streams. Additionally, the development of consistently derived catchment boundaries throughout Alaska will allow decision makers and natural resource managers to make meaningful regional, statewide, or national comparisons of landscape influences on stream systems in support of achieving their objectives.

Acknowledgments

Funding for this research was provided by Sealaska Corporation as well as the U.S. Fish and Wildlife Service through the National Fish Habitat Partnership (F13AC00565 and F11AC00709). Support also came from a scholarship from the Schrems West Michigan Chapter of Trout Unlimited. Thanks to Emil Tucker from the U.S. Forest Service for helping to provide timber harvest activities information. I. Cadiente and A. Hughes were very helpful in collecting LW data. R. Grotefendt deserves thanks for help in study design and identifying suitable sample sites. Comments from R. Hughes and G. Reeves as well as two anonymous reviewers improved previous versions of this chapter.

References

Allan, J. D., M. S. Wipfli, J. P. Caouette, A. Prussian, and J. Rogers. 2003. Influence of streamside vegetation on inputs of terrestrial invertebrates to salmonid food webs. Canadian Journal of Fisheries and Aquatic Sciences 60:309–320.

Andrus, C. W., B. A. Long, and H. A. Froehlich. 1988. Woody debris and its contribution to pool formation in a coastal stream 50 years after logging. Canadian Journal of Fisheries and Aquatic Sciences 45:2080–2086.

Arcangeli, C., M. Klopf, S. E. Hale, T. A. R. Jenkins, and H. Hasenauer. 2013. The uniform height curve method for height-diameter modelling: an application to Sitka spruce in Britain. Forestry: An International Journal of Forest Research 87:177–186.

Benda, L. E., P. Bigelow, and T. M. Worsley. 2002. Recruitment of wood to streams in old- growth and second-growth redwood forests, northern California, USA. Canadian Journal of Forest Research 32:1460–1477.

Bilby, R. E., J. T. Heffner, B. R. Fransen, J. W. Ward, and P. A. Bisson. 1999. Effects of immersion in water on deterioration of wood from five species of trees used for habitat enhancement projects. North American Journal of Fisheries Management 19:687–695.

Bilby, R. E., and J. W. Ward. 1989. Changes in characteristics and functions of woody debris with increasing size of streams in western Washington. Transactions of the American Fisheries Society 118:368–378.

Bilby, R. E., and J. W. Ward. 1991. Characteristics and function of large woody debris in streams draining old-growth, clear-cut, and second-growth forests in southwestern Washington. Canadian Journal of Fisheries and Aquatic Sciences 48:2499–2508.

Bryant, M. D., R. T. Edwards, and R. D. Woodsmith. 2005. An approach to effectiveness monitoring of floodplain channel aquatic habitat: salmonid relationships. Landscape and Urban Planning 72:157–176.

Bryant, M. D., T. Gomi, and J. J. Piccolo. 2007. Structures linking physical and biological processes in headwater streams of the Maybeso watershed, Southeast Alaska. Forest Science 53:371–383.

Burnett, K. M., G. H. Reeves, S. E. Clarke, and K. R. Christiansen. 2006. Comparing riparian and catchment influences on stream habitat in a forested, montane landscape. Pages 175–197 *in* R. M. Hughes, L. Wang, and P. W. Seelbach, editors. Landscape influences on stream habitats and biological assemblages. American Fisheries Society, Symposium 48, Bethesda, Maryland.

Cooper, A. R., D. M. Infante, J. A. Ross, W. M. Daniel, K. Herreman, C. C. Estes, and G. E. Whelan. 2019. Contemporary and future landscape-based anthropogenic disturbances to streams in Alaska: implications for anadromous fish habitat conservation. Pages 229–253 *in* R. M. Hughes, D. M. Infante, L. Wang, K. Chen, and B. de F. Terra, editors. Advances in understanding landscape influences on freshwater habitats and biological assemblages. American Fisheries Society, Symposium 90, Bethesda, Maryland.

Crawford, S., Whelan, G., Infante, D. M., Blackhart, K., Daniel, W. M., Fuller, P. L., Birdsong, T., Wieferich, D. J., McClees-Funinan, R., Stedman, S. M., Herreman, K., and P. Ruhl. 2016. Through a fish's eye: the status of fish habitats in the United States 2015. National Fish Habitat Partnership. Available: http://assessment.fishhabitat.org/. (March 2019).

Deal, R. L., P. E. Hennon, E. H. Orlikowska, and D. V. D'Amore. 2004. Stand dynamics of mixed red alder–conifer forests of Southeast Alaska. Canadian Journal of Forest Research 34:969–980.

Elosegi, A., and L. B. Johnson. 2003. Wood in rivers in developed landscapes. Pages 337–354 *in* S. V. Gregory, K. B. Staley, and A. Gurnell, editors. The ecology and management of wood in world rivers. American Fisheries Society, Symposium 37, Bethesda, Maryland.

FRPA (Alaska Forest Resources and Practices Act). 1978. AS 41.17.

Gomi, T., A. C. Johnson, R. L. Deal, P. E. Hennon, E. H. Orlikowska, and M. S. Wipfli. 2006. Factors affecting distribution of wood, detritus, and sediment in headwater streams draining managed young growth red alder–conifer forests in Southeast Alaska. Canadian Journal of Forest Research 36:725–737.

Gomi, T., R. C. Sidle, M. D. Bryant, and R. D. Woodsmith. 2001. The characteristics of woody debris and sediment distribution in headwater streams, southeastern Alaska. Canadian Journal of Forest Research 31:1386–1399.

Hanley, T. A., and T. Hoel. 1996. Species composition of old-growth and riparian Sitka spruce–western hemlock forests in southeastern Alaska. Canadian Journal of Forest Research 26:1703–1708.

Harmon, M. E., J. F. Franklin, F. J. Swanson, P. Sollins, S. V. Gregory, J. D. Lattin, N. H. Anderson, S. P. Cline, N. G. Aumen, J. R. Sedell, G. W. Lienkaemper, K. Cromack, and K. W. Cummins. 1986. Ecology of coarse woody debris in temperate ecosystems. Advances in Ecological Research 15:133–302.

Harrelson, C. C., C. L. Rawlins, and J. P. Potyondy. 1994. Stream channel reference sites: an illustrated guide to field technique. U.S. Forest Service, General Technical Report RM-245, Fort Collins, Colorado.

Heifetz, J., M. L. Murphy, and K. V. Koski. 1986. Effects of logging on winter habitat of juvenile salmonids in Alaskan streams. North American Journal of Fisheries Management 6:52–58.

Hemstad, N. A., and R. M. Newman. 2006. Local and landscape effects of past forest harvest on stream habitat and fish assemblages. Pages 413–427 *in* R. M. Hughes, L. Wang, and P. W. Seelbach, editors. Landscape influences on stream habitats and biological assemblages. American Fisheries Society, Symposium 48, Bethesda, Maryland

Hey, D. L., and N. S. Philippi. 1995. Flood reduction through wetland restoration: the upper Mississippi River basin as a case history. Restoration Ecology 3:4–17.

Hibbs, D. E., and P. A. Giordano. 1996. Vegetation characteristics of alder-dominated riparian buffer strips in Oregon Coast Range streams. Northwest Science 70:213–222.

Homer, C., J. Dewitz, J. Fry, M. Coan, N. Hossain, C. Larson, N. Herold, A. McKerrow, J. N. VanDriel, and J. Wickham. 2007. Completion of the 2001 National Land Cover Database for the conterminous United States. Photogrammetric Engineering and Remote Sensing 73:337–341.

Hough-Snee, N., A. Kasprak, B. B. Roper, and C. S. Meredith. 2014. Direct and indirect drivers of instream wood in the interior Pacific Northwest, USA: decoupling climate, vegetation, disturbance, and geomorphic setting. Riparian Ecology and Conservation 2:14–34.

Johnson, S. W., J. Heifetz, and K. V. Koski. 1986. Effects of logging on the abundance and seasonal distribution of juvenile steelhead in some southeastern Alaska streams. North American Journal of Fisheries Management 6:532–537.

Johnson, L., G. E. Host, J. Olker, and C. Richards. 2006. Landscape-and reach-scale predictors of large wood abundance in low-gradient streams. Pages 151–173 *in* R. M. Hughes L. Wang, and P. W. Seelbach editors, Influences of landscapes on stream habitats and biological assemblages. American Fisheries Society, Symposium 48, Bethesda, Maryland.

Johnson, J., and B. Blossom. 2018. Catalog of waters important for spawning, rearing, or migration of anadromous fishes – Southeastern Region, effective June 1, 2018. Alaska Department of Fish and Game, Special Publication No. 18-05, Anchorage.

Johnston, C. A., N. E. Detenbeck, and G. J. Niemi. 1990. The cumulative effect of wetlands on stream water quality and quantity: a landscape approach. Biogeochemistry 10:105–141.

Keeton, W. S., E. M. Copeland, S. Mazeika, P. Sullivan, and M. C. Watzin. 2017. Riparian forest structure and stream geomorphic condition: implications for flood resilience. Canadian Journal of Forest Research 47:476–487.

Keeton, W. S., C. E. Kraft, and D. R. Warren. 2007. Mature and old-growth riparian forests: structure, dynamics, and effects on Adirondack stream habitats. Ecological Applications 17:852–868.

Knofczynski, G. T., and D. Mundfrom. 2008. Sample sizes when using multiple linear regression for prediction. Educational and Psychological Measurement 68:431–442.

Kraft, C. E., D. R. Warren, and W. S. Keeton. 2011. Identifying the spatial pattern of wood distribution in northeastern North American streams. Geomorphology 135:1–7.

Liquori, M. D. Martin, L. Benda, R. Coats, and D. Ganz. 2008. Scientific literature review of forest management effects on riparian functions for anadromous salmonids. Report of Sound Watershed Consulting to the California State Board of Forestry and Fire Protection, Sacramento.

Maltby, E. 1991. Wetland management goals: wise use and conservation. Landscape Urban Planning 20:9–18.

Martin, D. J. 2001. The influence of geomorphic factors and geographic region on large woody debris loading and fish habitat in Alaska coastal streams. North American Journal of Fisheries Management 21:429–440.

Martin, D. J., and L. E. Benda. 2001. Patterns of instream wood recruitment and transport at the watershed scale. Transactions of the American Fisheries Society 130:940–958.

Martin, D. J., and R. A. Grotefendt. 2007. Stand mortality in buffer strips and the supply of woody debris to streams in Southeast Alaska. Canadian Journal of Forest Research 37:36–49.

Martin, D. J., M. E. Robinson, and R. A. Grotefendt. 1998. The effectiveness of riparian buffer zones for protection of salmonid habitat in Alaska coastal streams. Prepared for Sealaska Corporation, Juneau, Alaska and Alaska Forest Association, Ketchikan.

Martin, D. J., and A. Shelly. 2017. Temporal trends in stream habitat on managed forestlands in coastal Southeast Alaska. North American Journal of Fisheries Management 37:882–902.

Maxwell, S. E. 2000. Sample size and multiple regression analysis. Psychological Methods 5:434–458.

May, C. L., and R. E. Gresswell. 2003. Spatial and temporal patterns of debris-flow deposition in the Oregon Coast Range, USA. Geomorphology 57:135–149.

Mellina, E., and S. G. Hinch. 2009. Influences of riparian logging and in-stream large wood removal on pool habitat and salmonid density and biomass: a meta-analysis. Canadian Journal of Forest Research 39:1280–1301.

Miller, R. R. 2010. Is the past present? Historical splash-dam mapping and stream disturbance detection in the Oregon Coastal Province. Master's thesis. Oregon State University, Corvallis.

Minore, D., and H. G. Weatherly. 1994. Riparian trees, shrubs, and forest regeneration in the coastal mountains of Oregon. New Forests 8:249–263.

Murphy, M. L., and K. V. Koski. 1989. Input and depletion of woody debris in Alaska streams and implications for streamside management. North American Journal of Fisheries Management 9:427–436.

Murphy, M. L. 1995. Forestry impacts on freshwater habitat of anadromous salmonids in the Pacific Northwest and Alaska: requirements for protection and restoration. National Oceanic and Atmospheric Administration, Coastal Ocean Office, Coastal Ocean Program Decision Analysis Series No. 7, Silver Spring, Maryland.

Olson, D. M., E. Dinerstein, E. D. Wikramanayake, N. D. Burgess, G. V. N. Powell, E. C. Underwood, J. A. D'Amico, I. Itoua, H. E. Strand, J. C. Morrison, C. J. Loucks, T. F. Allnutt, T. H. Ricketts, Y. Kura, J. F. Lamoreaux, W. W. Wettengel, P. Hedao, and K. R. Kassem. 2001. Terrestrial ecoregions of the world: a new map of life on Earth. BioScience 51:933–938.

Omernik, J. M., G. E. Griffith, R. M. Hughes, J. B. Glover, and M. H. Weber. 2017. How misapplication of the hydrologic unit framework diminishes the meaning of watersheds. Environmental Management 60:1–11.

Paustian, S. J., K. Anderson, D. Blanchet, S. Brady, M. Cropley, J. Edgington, J. Fryxell, G. Johnejack, D. Kelliher, M. Kuehn, S. Maki, R. Olsen, J. Seesz, and M. Wolanek. 1992. A channel type user's guide for the Tongass National Forest, Southeast Alaska. U.S. Forest Service, Technical Paper 26, Juneau, Alaska.

Powell, S. R., L. D. Daniels, and T. A. Jones. 2009. Temporal dynamics of large woody debris in small streams of the Alberta foothills, Canada. Canadian Journal of Forest Research 39:1159–1170.

Robison, E. G., and R. L. Beschta. 1990. Identifying trees in riparian areas that can provide coarse woody debris to streams. Forest Science 36:790–801.

Ross, J. A. 2013. Influences of timber management and natural landscape factors on anadromous streams of southeastern Alaska: relating local and catchment factors to aquatic habitat. Master's thesis. Michigan State University, East Lansing.

Ross, J. A., D. M., Infante, D. J. Martin, and M. Rey. 2019. Effects of riparian timber harvest on Southeast Alaska stream habitat after 30–40 years: insights for management. North American Journal of Fisheries Management 39:328–342.

Rot, B. W., R. J. Naiman, and R. E. Bilby. 2000. Stream channel configuration, landform, and riparian forest structure in the Cascade Mountains, Washington. Canadian Journal of Fisheries and Aquatic Sciences 57:699–707.

Sloat, M. R., G. H. Reeves, and K. R. Christiansen. 2018. Stream network geomorphology mediates predicted vulnerability of anadromous fish habitat to hydrologic change in southeast Alaska. Global Change Biology 23:604–620.

Smith, S. H., J. F. Bell, F. R. Herman, and T. See. 1984. Growth and yield of Sitka spruce and western hemlock at Cascade Head Experimental Forest, Oregon. U.S. Forest Service, Pacific Northwest Research Station, Research Paper PNW-325, Portland, Oregon.

Stevens, D. L., and A. R. Olsen. 2004. Spatially balanced sampling of natural resources. Journal of the Statistical Association 99:262–278.

Taylor, R. F. 1934. Yield of second-growth western hemlock-Sitka spruce stands in southeastern Alaska. U.S. Department of Agriculture, Technical Bulletin 412, Washington D.C.

ter Braak, C. J. F., and I. C. Prentice. 1988. A theory of gradient analysis. Advanced Ecological Research 18:271–313.

Tsang, Y.-P, D. Wieferich, K. Fang, D. M. Infante, and A. R. Cooper. 2014. An approach for aggregating upstream catchment information to support research and management of fluvial systems across large landscapes. SpringerPlus [online serial] 3:589.

USFS (U.S. Forest Service). 2014. Tongass young growth management strategy 2014. Available: www.fs.usda.gov/Internet/FSE_DOCUMENTS/fseprd559973.pdf. (March 2019).

USFWS (U.S. Fish and Wildlife Service). 2013. National Wetlands Inventory [online database]. U.S. Fish and Wildlife Service, Washington, D.C. Available: www.fws.gov/wetlands/. (March 2019).

USGS (U.S. Geological Survey). 1993. National water summary 1990–1991: hydrologic events and stream water quality. U.S. Geological Survey, Water Supply Paper 2400, Reston, Virginia.

USGS (U.S. Geological Survey). 2012a. National Hydrography Dataset for Alaska. Available: http://prd-tnm.s3-website-us-west-2.amazonaws.com/?prefix=StagedProducts/Hydrography/NHD/State/HighResolution/GDB/. (March 2019).

USGS (U.S. Geological Survey). 2012b. Alaska 2 Arc-second digital elevation models (DEMs) - USGS national map 3DEP downloadable data collection. Available: https://catalog.data.gov/dataset/national-elevation-dataset-ned-alaska-2-arc-second-downloadable-data-collection-national-geosp. (March 2019).

Vannote, R. L., G. W. Minshall, K. W. Cummins, J. R. Sedell, and C. E. Cushing. 1980. The river continuum concept. Canadian Journal of Fisheries and Aquatic Sciences 37:130–137.

Villarin, L. A., D. M. Chapin, and J. E. Jones. 2009. Riparian forest structure and succession in second growth stands of the central Cascade Mountains, Washington, USA. Forest Ecology and Management 257:1375–1385.

Walker, R. H., B. M. Maitland, T. N. LaSharr, M. N. Rosing, and M. Ben-David. 2018. Fate of juvenile salmonids stranded in off-channel pools: implications for nutrient transfers. Aquatic Sciences 80:10.

Wang, L., J. Lyons, P. Kanehl, and R. Bannerman. 2001. Impacts of urbanization on stream habitat and fish across multiple spatial scales. Environmental Management 28:255–266.

Wang, L., D. Infante, P. Esselman, A. Cooper, D. Wu, W. Taylor, D. Beard, G. Whelan, and A. Ostroff. 2011. A hierarchical spatial framework and database for the National River Fish Habitat Condition Assessment. Fisheries 36:436–449.

Wang, L., P. W. Seelbach, and J. Lyons. 2006. Effects of levels of human disturbance on the influence of catchment, riparian, and reach-scale factors on fish assemblages. Pages 199–219 *in* R. M. Hughes, L. Wang, and P. W. Seelbach, editors. Landscape influences on stream habitats and biological assemblages. American Fisheries Society, Symposium 48, Bethesda, Maryland.

Wipfli, M. S., J. S. Richardson, and R. J. Naiman. 2007. Ecological linkages between headwaters and downstream ecosystems: transport of organic matter, invertebrates, and wood down headwater channels. Journal of the American Water Resources Association 43:72–85.

Wohl, E., N. Kramer, V. Ruiz-Villanueva, D. N. Scott, F. Comiti, A. M. Gurnell, H. Piegay, K. B. Lininger, K. L. Jaeger, D. M. Walters, and K. D. Fausch. 2019. The natural wood regime in rivers. BioScience 69:259–273.

Woodsmith, R. D., J. R. Noel, and M. L. Dilger. 2005. An approach to effectiveness monitoring of floodplain channel aquatic habitat: channel condition assessment. Landscape and Urban Planning 72:177–204.

Worthington, N. P., R. H. Ruth, and E. E. Matson. 1962. Red alder, its management and utilization. U.S. Forest Service, Miscellaneous Publication 881, Washington, D.C.

American Fisheries Society Symposium 90:229–253, 2019
© 2019 by the American Fisheries Society

Contemporary and Future Landscape-Based Anthropogenic Disturbances to Alaska Streams: Implications for Anadromous Fish Habitat Conservation

ARTHUR R. COOPER*, DANA M. INFANTE, JARED A. ROSS, WESLEY M. DANIEL, AND KYLE HERREMAN
Department of Fisheries and Wildlife, Michigan State University
1405 South Harrison Road, Suite 318, East Lansing, Michigan 48823, USA

CHRISTOPHER C. ESTES
Chalk Board Enterprises, LLC
Post Office Box 221322, Anchorage, Alaska 99522, USA

GARY E. WHELAN
Michigan Department of Natural Resources, Fisheries Division
Post Office Box 30446, Lansing, Michigan 48909, USA

Abstract.—Fluvial threat assessments characterize the potential for fluvial habitat conditions to be degraded by differing types and intensities of anthropogenic activities occurring on the landscape, ultimately affecting stream biota. We present a threat assessment for fluvial habitats in Alaska based on six anthropogenic disturbance indices representing urbanization, agriculture/timber harvest, stream fragmentation, point-source pollution, infrastructure, and mines. These indices were combined to develop an overall index of contemporary threat of habitat alteration throughout Alaska using the 12-digit U.S. Geological Survey hydrologic unit code (HUC-12) framework (n = 13,997) and at a finer spatial resolution using local and network catchments for individual stream reaches within Southeast Alaska (n = 207,092). Overall, contemporary threat of habitat alteration indices showed that ~96% of fluvial habitats both statewide and for Southeast Alaska were at low or very low levels; however, anadromous fish habitats were under greater human pressure with nearly double the amount of moderately to severely disturbed habitats when compared to all fluvial habitats. We further evaluated potential future threats to fluvial habitats from mineral mining activities and climate change. More than 86% of existing mine claims statewide and 99% of claims in Southeast Alaska occur in areas of low and very low contemporary threat to fluvial habitats for anadromous fishes. Under climate change, July air temperatures are projected to increase ~1.9°C, on average, by mid-century within HUC-12s containing anadromous fish streams, indicating immense potential to warm streams with anadromous fishes within the state. This fluvial threat assessment demonstrates that overall threats from contemporary anthro-

* Corresponding author: coopera6@msu.edu

pogenic disturbance factors are generally low with localized areas of high intensity. However, future threats from mining and climate change have considerable potential to alter fluvial habitats for anadromous fishes in Alaska, particularly those currently unaltered by anthropogenic disturbances.

Introduction

Aquatic habitats are under enormous pressure from human activities on the landscape, causing freshwater organisms to be one of the most imperiled groups of organisms on the planet (Dudgeon et al. 2006; Strayer and Dudgeon 2010). To effectively mitigate the effects of human landscape pressures on freshwater habitats, fisheries and watershed managers require the ability to assess the types and intensities of stressors throughout regions of interest to understand their influences on freshwaters (Ormerod et al. 2010; Schinegger et al. 2016). Characterizing anthropogenic pressures consistently over large spatial extents and at multiple spatial scales can provide decision makers with important information with which to effectively prioritize where management actions are needed for conservation (protection, rehabilitation, or improvement) of aquatic habitats (Fausch et al. 2002; Falcone et al. 2010; Wang et al. 2011). Mapping the spatial extent and frequency of those pressures throughout aquatic landscapes (e.g., river basins) facilitates spatially explicit identification of potentially altered aquatic habitats. Such information improves our ability to investigate how those habitats may respond to multiple anthropogenic disturbances (e.g., McCluney et al. 2014) and to establish more effective mitigation goals and actions.

One approach for characterizing stressor-response relationships of anthropogenic disturbances on aquatic habitats is the empirical evaluation of these relationships. That approach requires fish assemblage, habitat, and landscape data. However, consistent, comparable fish assemblage and stream habitat data are often limiting over large regions (e.g., Esselman et al. 2011). In the absence of such data, an alternative is a threat-based approach, which relies on existing knowledge of effects of anthropogenic disturbances on aquatic habitats and/or the biotic assemblages they support (e.g., Danz et al. 2007; Mattson and Angermeier 2007; Esselman et al. 2011; Paukert et al. 2011). For fluvial systems, threat-based approaches often involve summarizing types and intensities of multiple pressures (e.g., urban land use, dams, etc.) that could affect habitat, typically culminating in development of an overall threat index that integrates combined influences of multiple stressors (Mattson and Angermeier 2007; Brown and Froemke 2012; Esselman et al. 2011; Paukert et al. 2011). Such cumulative, threat-based indices are used to infer the likely ecological condition of fluvial systems for conservation planning purposes (Wang et al. 2008; Theobald 2013; Thornbrugh et al. 2018). Threat-based assessments are particularly important in data-limited regions that have a paucity of information available to calibrate indices, including habitat or biological data (Falcone et al. 2010). They are also useful in areas where information on ecologically relevant assessment units, such as individual stream reaches and catchments, may be unavailable (Thieme et al. 2007). In such instances, threat-based approaches and associated threat mapping can offer managers useful information for decision making regarding the relative threats of differing contemporary sources of anthropogenic disturbance and for anticipating where future threats may be especially problematic.

The state of Alaska has vast freshwater resources with 687,966 km^2 of wetlands, 3 million lakes, and 1.3 million km of river, representing approximately 20% of river length and 40% of freshwater surface water of the entire United States (USGS 1993, 2012a). Many of these freshwater habitats support anadromous fishes that use fluvial habitats for migration, spawning, and feeding. These species are culturally and socioeconomically important statewide and globally. Despite low human population density throughout most of the state, anthropogenic pressures nonetheless are a known source of degradation to fluvial habitats and, thus, have the potential to negatively affect aquatic organisms, including anadromous fishes. Some of the most pervasive human activities in Alaska include those associated with natural resource extraction activities such as timber harvest or mining. Negative effects of timber harvest on streams can include reduced amounts of pool habitat and large wood (e.g., Mellina and Hinch 2009), increased sedimentation (Murphy 1995), and hydrologic alteration, which is often observed immediately following timber harvest (Surfleet and Skaugset 2013). Roads supporting access to logging and mining activities lead to culverts at stream crossings that block migrating fish (Davis and Davis 2011; MacPherson et al. 2012), with increased potential for landslides from poorly placed logging roads (Murphy 1995). Mining alters stream hydrology (Chambers et al. 2012; USEPA 2012) and is a source of contaminants that are harmful to aquatic organisms (Gray et al. 2000; Bernhardt and Palmer 2011). Whereas urban and agricultural land uses comprise only a small percentage of Alaska land area (~0.1%; MRLC 2015), locally intense areas of urbanization and agriculture degrade aquatic habitats (e.g., Allan 2004). Similarly, infrastructure associated with human settlements and transport of goods and resources, such as roads, pipelines, waterways, and railways, often alter stream geomorphology and floodplain dynamics through their construction in stream valleys (Blanton and Marcus 2013; 2014). Dams disrupt connectivity in stream networks, blocking or reducing fish migrations, which is particularly problematic for migratory fishes that must travel great distances throughout river networks to access habitats required to complete spawning, incubation, and rearing life history stages (Mattocks et al. 2017). Despite these numerous potential sources of fluvial habitat alteration, little is known about the spatial configuration and relative intensities of anthropogenic disturbances among fluvial habitats throughout Alaska. Such information would provide a basis for evaluating contemporary fluvial habitat conditions and assessing future threats to fluvial habitats and the biota they support.

In addition to the wide-ranging threats from contemporary anthropogenic disturbances, Alaska faces enormous potential future threats to fluvial habitats. Rapid warming associated with climate change poses a significant threat to northern ecosystems, including those in Alaska (Hinzman et al. 2005; Milner et al. 2009; Wassmann et al. 2011). Not only does climate change threaten infrastructure such as roads, buildings, and pipelines in Alaska (Melvin et al. 2017), it also threatens stream habitats and the organisms they support (e.g., Shanley and Albert 2014; Sloat et al. 2017), including warming stream temperatures that can result from warmer air temperatures and altered stream hydrology resulting from changes in precipitation and melting of snow, ice, and permafrost (Bryant 2009; Lisi et al. 2015; Valentin et al. 2018; Winfree et al. 2018). Stream temperature is a major factor influencing fishes, especially the coldwater species of Alaska streams whose life history strategies have evolved across specific stocks throughout their wide-ranging native distributions (Bryant 2009; Schindler et al. 2010). Climate change has the potential to alter timing of salmon migration (Kovach et al. 2015) , increase or decrease salmon growth rates (Falke et al. 2019, this volume), and influence salmon life history phenology (Sparks et al. 2019). Although contemporary and historic threats from mining are numerous, future mining has the potential to alter fluvial habitats that are otherwise relatively undisturbed from other human activities on the landscape, particularly in Alaska where many landscapes are considered intact. Further, mineral contaminants from some mining developments can negatively affect salmon (Hughes et al. 2016), including impairing chemosensory functions and increased vulnerability to predation of juvenile salmon (McIntyre et al. 2008; 2012). Because of the far-reaching and long-lasting effects of mines on aquatic ecosystems both during and after mining operations (Hughes et al. 2016), improving knowledge of these threats by mapping potential future mining locations provides an opportunity to account for this future threat. By incorporating information on future threats, in addition to current threats, resource managers can evaluate and mitigate threats from both contemporary and future anthropogenic pressures (e.g., Mantyka-Pringle et al. 2016).

Therefore, the goal of our study was to evaluate contemporary pressures as well as two key future threats, climate change and mineral mining, to Alaskan fluvial habitats. Specifically, we summarize contemporary threats represented by six anthropogenic disturbance categories, including urbanization, timber harvest and agriculture, point-source pollution, mines, infrastructure, and stream fragmentation. In addition, we evaluate future threats to Alaskan fluvial habitats from both mineral mining claims and projected climate change. Finally, we consider these contemporary and future threats in reference to anadromous fish habitats, providing new information to aid in current and future conservation of anadromous fishes in Alaska.

Methods

Spatial Framework

We used 12-digit U.S. Geological Survey hydrologic units (HUC-12s; n = 13,997, mean = 123.42 km^2, SD = 321.31 km^2) as the base spatial units for assessing statewide landscape-based threats to stream habitats (i.e., statewide Alaska, Figure 1). Hydrologic units represent subdivisions of drainage basins that are somewhat similarly sized, and they differ from true catchments, which encompass entire landscape areas draining to a given stream point (Omernik et al. 2017; Figure 1). HUC-12s were used throughout Alaska because they were the finest-resolution spatial unit available to discriminate landscape influences on stream conditions. In Southeast Alaska, however, local catchments delineated for stream reaches from the 1:63,360 National Hydrography Dataset (USGS 2012a) were the finest resolution spatial unit available (Ross et al. 2019, this volume). Stream reaches are defined as sections of streams occurring between confluences, headwaters and confluences, and confluences and lake/stream outlets, providing a finer-scale view of anthropogenic disturbance in the region than would be available with HUC-12s (Figure 1). A local catchment is defined as the land area draining directly to a stream reach. Because coastal drainage areas situated between fluvial stream valleys could contain streams supporting fish habitat not represented at the 1:63,360 scale of the stream data set used in this study, we also delineated and included local catchments from those areas in this study. Local catchments were delineated from a 60-m digital elevation model (USGS 2012b; see Ross et al. 2019 for an overview). In total, 207,092 local catchments (mean = 0.49 km^2, SD = 6.94 km^2) were delineated for Southeast Alaska. In Southeast Alaska, we also delineated network catchments for each stream reach,

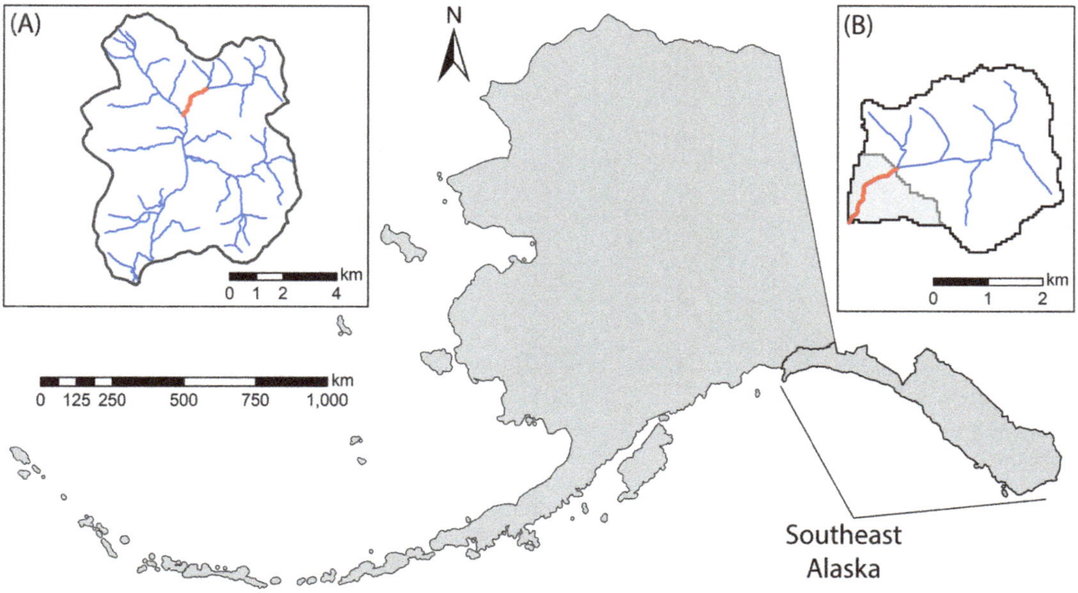

Figure 1. Spatial units used for threat assessment, including USGS HUC-12s throughout Alaska (example in **A**) and local and network catchments for stream reaches in Southeast Alaska (examples in **B**). Streams from the NHD are shown in blue. A target stream reach is shown in red in (A) as one reach out of many within a HUC-12 and in (B), which also shows its local catchment (gray area) and network catchment (black boundary).

representing the entire upstream drainage area for a given stream reach, including the reach's local catchment (Figure 1; Wang et al. 2011), with an approach described in Tsang et al. (2014).

Summarizing Contemporary Anthropogenic Landscape Pressures

Multiple anthropogenic landscape pressure variables were assembled to characterize the contemporary condition of Alaska stream habitats. Variables were suggested and approved by members and affiliates of five partnerships representing local, regional, and national interests in conserving and restoring fish habitat throughout Alaska, including the Southeast Alaska Fish Habitat Partnership, the Matanuska-Susitna Basin Salmon Habitat Partnership, the Kenai Peninsula Fish Habitat Partnership, the Southwest Alaska Salmon Habitat Partnership, and the Western Native Trout Initiative (www.fishhabitat.org/the-partnerships). In total, 20 anthropogenic landscape pressure data sets were identified and summarized as variables (Table 1). Because of the differing spatial extents and availability of data sets, the number and type of pressure variables differed between Southeast Alaska (n = 17) and statewide for Alaska (n = 18). In particular, spatial data sets representing timber harvest and forestry roads were available in Southeast Alaska but not statewide, whereas row crop and pasture land cover was not present in Southeast Alaska but occurred in other regions of the state. Variables were attributed to HUC-12s and local catchments following Wang et al. (2011), with variables summarized within spatial units as percentages of total areas comprised by different land uses and forest harvest or a density for point-based (#/km^2; e.g., culverts, point-source pollution inputs) or line-based (km/km^2; e.g., pipelines, roads) data sets.

Table 1. Anthropogenic landscape pressure data sets and threat subindices used in evaluating contemporary threats to fluvial habitats in Alaska. MRLC = Multi-Resolution Land Cover Characteristic Consortium, USFS = U.S. Forest Service, ADFG = Alaska Department of Fish and Game, NABD = National Anthropogenic Barrier Database, EPA = U.S. Environmental Protection Agency, ADEC = Alaska Department of Environmental Conservation, TIGER = Topologically Integrated Geographic Encoding and Referencing, and ASGDC = Alaska State Geo-Spatial Data Clearinghouse.

Threat index/variable	Units	Year	Source	Statewide Alaska	Southwest Alaska
Urbanization					
Population density	#/km^2	2010	U.S. Census	x	x
Developed open space	%	2011	MRLC	x	x
Developed low intensity	%	2011	MRLC	x	x
Developed medium intensity	%	2011	MRLC	x	x
Developed high intensity	%	2011	MRLC	x	x
Agriculture/timber harvest					
Pasture/hay	%	2011	MRLC	x	
Cultivated crops	%	2011	MRLC	x	
Conventional forest harvest	%	2012	USFS and Sealaska		x
Fragmentation					
Culvert density	#/km^2	2014	Tongass Forest Service and ADFG	x	x
Dam density	#/km^2	2012	NABD	x	x
Point-source pollution					
Toxic Release Inventory site density	#/km^2	2013	EPA	x	x
National Pollution Discharge Elimination System site density	#/km^2	2013	EPA	x	x
Contaminated site density	#/km^2	2015	ADEC	x	x
303d impaired waters	%	2010	EPA	x	x
Infrastructure					
Road length density	km/km^2	2014	TIGER	x	
Railroad length density	km/km^2	2006	ASGDC	x	x
Pipeline length density	km/km^2	2006	ASGDC	x	x
Road length density (including forestry roads)	km/km^2	2012	Southeast Alaska GIS Library		x
Airport/landing strips	#/km^2	2006	ASGDC	x	x
Mines					
Active and prospect mines	#/km^2	2008	ASGDC	x	x

Cumulative Threat to Stream Habitat from Contemporary Pressures

In developing an overall cumulative threat score, we first grouped individual anthropogenic pressure variables (e.g., Danz et al. 2007; Falcone et al. 2010) into six categories statewide and for Southeast Alaska. Those pressures represent specific types of disturbances, including urbanization, fragmentation, point-source pollution, infrastructure, agriculture and timber harvest, and mines. Each subindex was developed from one to five anthropogenic disturbance variables (Table 1), and agriculture and timber harvest were combined to create a single subindex based on the request of partnerships described earlier. This step was justified because agriculture is an uncommon land-use activity in Alaska, and its potential effects on fish habitat are expected to be more similar to that of timber harvest than effects from other types of disturbances. Subindices were developed by rescaling variables from zero to one (e.g., Theobald 2013) using variable minimum/maximum values, then summing all values in each subindex together and further rescaling resulting subindex scores from zero to one (Crawford et al. 2016). The resulting normalized scores for each disturbance subindex within each spatial unit allowed for comparison of scores among all spatial units (Crawford et al. 2016). To create a cumulative threat index of habitat alteration for HUC-12s statewide, all six subindices of disturbances were summed, resulting in a theoretical maximum value of 6, with higher scores representing greater potential threat. For Southeast Alaska, each of the six subindices were scored for each spatial scale, including local and network catchments, and were summed independently to create two habitat condition indices for each stream reach. Both local and network catchment indices had a theoretical maximum value of 6 and were averaged, giving each an equal weighting to each catchment scale in the cumulative threat of habitat alteration score. To make a continuous statewide map of Alaska at the HUC-12 level, cumulative habitat alteration threat scores were scaled up to HUC-12s using area-weighted averages as assigned to local catchments.

We classified the cumulative threat of habitat alteration scores to characterize the varying degrees of threat to fluvial habitats by subdividing scores into five condition classes: very low, low, moderate, high, and very high (Crawford et al. 2016). First, spatial units with an overall score of zero were assigned to the "very low" threat class, which indicates that no anthropogenic disturbances were detected within these spatial units. Next, the remaining four classes (low, moderate, high, and very high) were developed using Jenks natural breaks (Jenks 1967) on the nonzero scores. The Jenks method assigned reaches into four groups based on an optimization process that minimizes variation within and maximizes variation between groups based on cumulative threat of habitat alteration scores. We analyzed threat scores by mapping and calculating the percentage of spatial units that were identified within the five threat classes, facilitating assessment of patterns of both individual subindices and the overall cumulative index. These relationships were examined for all HUC-12s statewide and all local and network catchments in Southeast Alaska, and separately for HUC-12s and local/network catchments containing anadromous fish streams (described below).

Summarizing Future Landscape Pressures

We evaluated two potential future threats to fluvial habitats in Alaska: mineral mining and climate change. Mineral mine claim data representing both state (ADNR 2018a) and federal (ADNR 2018b) mine claim areas were obtained to evaluate the spatial relationships between fluvial spatial units and mine claim polygons. Because claims can have overlapping areas, we dissolved polygons in the state and federal mine claim data sets using a geographic information system (GIS; ArcMap 10.4; ESRI, Redlands, California) to avoid double-counting of overlapping areas in each data set. We viewed current mine claims as a measure of potential future mining activities within our spatial units (i.e., HUC-12s and local catchments), with greater claim area (measured as a percentage of claim area within spatial units) indicating a higher potential for future mining activity and subsequently a greater threat of altering fluvial habitats. We summarized federal claims, state claims, and combined total claims to determine the amount of area and percentage of claims in the five habitat alteration threat classes, facilitating comparison of potential future mining threats versus contemporary threats of habitat alteration among spatial units.

To assess future climate change threats on fluvial habitats, we evaluated projected air temperature increases with respect to historical baselines for areas containing anadromous fishes. Warming may facilitate range expansions in arctic streams for anadromous salmon and decrease them elsewhere (Rinne et al. 2005; Winfield et al. 2016; Falke et al. 2019). For this study, we assumed that warming air temperatures are a threat to stream habitats and anadromous fishes because air temperature increases could encourage snow and ice melting, leading to substantially altered flow regimes in streams draining landscapes with prevalent snow and ice cover (however, see Falke et al. 2019). Additionally, air temperature increases warm stream temperatures. Whereas stream temperature measures and models are currently not available for much of the state, air temperature has been shown to be highly correlated with stream temperatures in many regions of Alaska (Kovach et al. 2013, 2015; Fellman et al. 2014; Shanley and Albert 2014; Lisi et al. 2015).

Historical and projected air temperature data developed by Scenarios Network for Alaska + Arctic Planning (SNAP; International Arctic Research Center at the University of Alaska Fairbanks) were used in this study. For historical data, we used July air temperature from 1901 to 2006 (SNAP 2018a). We obtained climate change projections representing an average of five general circulation models (NCAR-CCSM4, GFDL-CM3, GISS-E2-R, IPSL-CM5A-LR, and MRI-CGCM3) under the CMIP5 ensemble (SNAP 2018b), models known to be robust for the Alaska region (Walsh et al. 2018). Although there are differing representative concentration pathways (RCPs) representing different climate-forcing scenarios (Van Vuuren et al. 2011), we chose models in RCP 6.0 representing a single mid-range scenario to use in our analysis. Both projected and historical data sets have been downscaled to a 771 m resolution by SNAP through use of PRISM (PRISM Climate Group, Oregon State University) climate data (Walsh et al. 2018). We averaged historical 1901–2006 July air temperature to create a historical baseline for Alaska's warmest month and averaged climate change data from 2040 to 2059 to create a mid-century July mean air temperature projection. The fluvial habitat threats from climate change were measured as the difference in future versus historical air temperature projections (e.g., Melvin et al. 2017). We only summarized potential threats from climate change for HUC-12s throughout Alaska, and not for local catchments in Southeast Alaska because of the coarse resolution of the climate data.

Threats to Anadromous Fish

To assess threats to anadromous fish fluvial habitats, we identified HUC-12s (n = 6,212) statewide and local/network catchments in

Southeast Alaska (n = 30,010) containing streams that support anadromous fishes as indicated by Alaska's Anadromous Waters Catalog (AWC; Johnson and Blossom 2018a–f). First developed in 2008, the AWC is updated annually through ongoing collection and observation of 21 anadromous fish species from three families (Table 2) and includes habitats used for migration, spawning, and rearing. The AWC represents a legally defined data set, identifying locations of anadromous waters that are afforded special legal protection status within the state; however, many streams within the state have yet to be surveyed for the presence of anadromous fishes (e.g., Woody and O'Neal 2016). To evaluate relationships between contemporary and future threat factors for HUC-12s with anadromous fishes, we conducted a principal component analysis (PCA) using a correlation matrix approach because of the differing units among factors. Prior to running the PCA, we arcsine-square-root-transformed contemporary threat factors (ranging from 0 to 1) and future mining claim areas (expressed as a proportion) and log-transformed future versus historical air temperature projections. We implemented the PCA in R (version 3.4.2; R Core Team, Vienna, Austria) with the 'prcomp' function in the 'stats' package.

Results

Contemporary Threat of Habitat Alteration for Alaska

The frequency of particular anthropogenic disturbances varied considerably both state-

Table 2. List of anadromous fish species from the Alaska Department of Fish and Game's Anadromous Waters Catalog.

Family	Scientific name	Common name
Osmeridae	*Hypomesus olidus*	Pond Smelt
	Osmerus mordax	Rainbow Smelt
	Spirinchus thaleichthys	Longfin Smelt
	Thaleichthys pacificus	Eulachon
Petromyzontidae	*Lampetra ayresii*	River Lamprey
	Lethenteron camtschaticum	Arctic Lamprey
	Entosphenus tridentatus	Pacific Lamprey
Salmonidae	*Coregonus laurettae*	Bering Cisco
	Coregonus nasus	Broad Whitefish
	Coregonus pidschian	Humpback Whitefish
	Coregonus sardinella	Least Cisco
	Oncorhynchus clarkii	Cutthroat Trout
	Oncorhynchus gorbuscha	Pink Salmon
	Oncorhynchus keta	Chum Salmon
	Oncorhynchus kisutch	Coho Salmon
	Oncorhynchus mykiss	Steelhead
	Oncorhynchus nerka	Sockeye Salmon
	Oncorhynchus tshawytscha	Chinook Salmon
	Salvelinus alpinus	Arctic Char
	Salvelinus malma	Dolly Varden
	Stenodus leucichthys	Inconnu

wide and for Southeast Alaska. Among HUC-12s in statewide Alaska, mine density and road length density were the most prevalent disturbances found in 18.9% of HUC-12s, followed by human population density (10.2%) and low-intensity urban land use (9.9%) (Table 3). Variables representing Toxic Release Inventory site density, Section 303d of the Federal Clean Water Act (impaired waters), and airport/landing strip density were the least frequent disturbances, generally occurring in fewer than 0.25% of HUC-12s. In Southeast Alaska, timber harvest, road length density, and human population density were the most prevalent anthropogenic disturbances in both local and network catchments, ranging from 5.9% to 8.4% of local catchments and 10.7–12.2% of network catchments. Anthropogenic disturbance variables representing point-source pollution site densities, airport/landing strip density, and dam density were infrequent for Southeast Alaska, occurring in very few local and network catchments. Overall, the prevalence of many anthropogenic disturbances is higher in HUC-12s compared to local and network catchments in Southeast Alaska; however, intensity of the anthropogenic disturbances as measured by mean and maximum values is greater in Southeast Alaska because of the finer spatial resolution in spatial units.

Cumulative threat of fluvial habitat alteration scores ranged from 0 to 2.64 for HUC-12s for statewide Alaska. When broken into the five habitat alteration threat classes, a large majority of land surface area was considered very low (59.3%) and low (36.3%) threat, whereas only 3.2% of area was considered moderate, 1.0% high, and 0.2% very high threat. In comparison, scores for HUC-12s containing anadromous streams had a smaller percentage of land area in the very low class (44.1%) and a greater combined amount of moderate to very high alteration threat at 7.4% compared to 4.4% for all HUC-12s statewide (Figure 2A). When mapped, the cumulative threat of habitat alteration in Alaska showed great spatial variability (Figure 2A). In particular, threat of habitat alteration was driven largely by spatial patterns in subindices representing urbanization, such as the municipality of Anchorage and city of Fairbanks, and associated interconnecting infrastructure and fragmentation among human settlements, whereas active mines resulted in more diffuse areas of higher threat (Figure 2B). Areas of higher intensity threat in Southeast Alaska were driven by multiple factors, including timber harvest and urbanization.

For the Southeast Alaska assessment based on local and network catchments, cumulative habitat alteration threat scores ranged from 0 to 1.79. A high percentage of land area occurred in the very low threat class (78.5%) relative to the low threat class (17.4%). Compared to HUC-12s throughout Alaska, Southeast Alaska had similar percentages in the moderate (2.9%), high (1.0%), and very high (0.2%) threat classes. Also similar to HUC-12 results, local and network catchments with anadromous streams had less area considered very low threat of habitat alteration at 64.6% and a greater amount of land area with moderate to very high threat (9.4%) compared to all local and network catchments in Southeast Alaska (4.1%) (Figure 3A). Areas with higher threat intensities included areas of extensive timber harvest and forestry roads, such as the northern Chichagof, Kupreanof, and Prince of Wales islands, as well as developed areas around Juneau (Figure 3A). When compared to HUC-12 results constrained to Southeast Alaska, local and network catchment-level results produce enhanced detail of threats (Figure 3B). Specifically, results for HUC-12s for Southeast Alaska indicate higher levels of moderate to very high levels of threat (~27%) compared to only 4.1% when evaluated among all local and network catchments.

Table 3. Percent, mean, and maximum of nonzero cases of anthropogenic disturbance variables for HUC-12s in statewide Alaska and local and network catchments for Southeast Alaska.

Threat index/variable	Units	Statewide Alaska % nonzero	Statewide Alaska Mean	Statewide Alaska Max	Southeast Alaska (local) % nonzero	Southeast Alaska (local) Mean	Southeast Alaska (local) Max	Southeast Alaska (network) % nonzero	Southeast Alaska (network) Mean	Southeast Alaska (network) Max
Urbanization										
Population density	#/km^2	10.15	4.877	1115.415	5.92	18.642	5132.605	11.39	1.999	1236.397
Developed open space	%	3.28	0.720	26.810	0.16	12.094	100.000	0.145	1.394	34.783
Developed low intensity	%	9.88	0.634	24.578	2.72	9.872	100.000	4.163	2.529	67.241
Developed medium intensity	%	3.31	0.239	12.197	0.19	12.175	100.000	0.213	0.997	17.073
Developed high intensity	%	1.46	0.165	3.481	0.12	11.016	100.000	0.138	1.066	31.690
Agriculture/Timber harvest										
Pasture/hay	%	0.99	0.267	4.348	–	–	–	–	–	–
Cultivated crops	%	1.11	1.653	33.779	–	–	–	–	–	–
Conventional forest harvest	%	–	–	–	7.27	28.984	100.000	10.74	19.680	100.000
Fragmentation										
Culvert density	#/km^2	4.32	0.062	0.726	0.64	8.318	208.247	2.646	0.996	69.416
Dam density	#/km^2	0.34	0.013	0.050	0.01	4.075	12.621	0.076	0.184	10.158
Point Source Pollution										
Toxic Release Inventory site density	#/km^2	0.05	0.007	0.014	0	1.285	1.285	0.008	0.026	0.043
National Pollution Discharge Elimination System site density	#/km^2	3.19	0.022	0.350	0.07	9.968	138.831	0.165	0.375	29.750
Contaminated site density	#/km^2	3.26	0.029	0.652	0.05	8.144	138.831	0.137	0.462	24.500
303d impaired waters	%	0.23	0.201	0.515	0.04	100.000	100.000	0.07	15.245	100.000
Infrastructure										
Road length density	km/km^2	18.90	0.179	6.540	–	–	–	–	–	–
Railroad length density	km/km^2	1.12	0.073	0.421	0.05	0.816	4.529	0.18	0.218	1.628
Pipeline length density	km/km^2	2.32	0.071	0.338	0.06	1.916	17.481	0.13	0.307	3.304
Road length density (including forestry roads)	km/km^2	–	–	–	8.38	2.354	33.341	12.15	1.347	28.597
Airport/landing strips	#/km^2	0.26	0.009	0.021	0.01	12.231	141.394	0.121	1.593	141.394
Mines										
Active and prospect mines	#/km^2	18.94	0.028	1.513	0.24	3.893	138.831	1.651	0.267	27.766

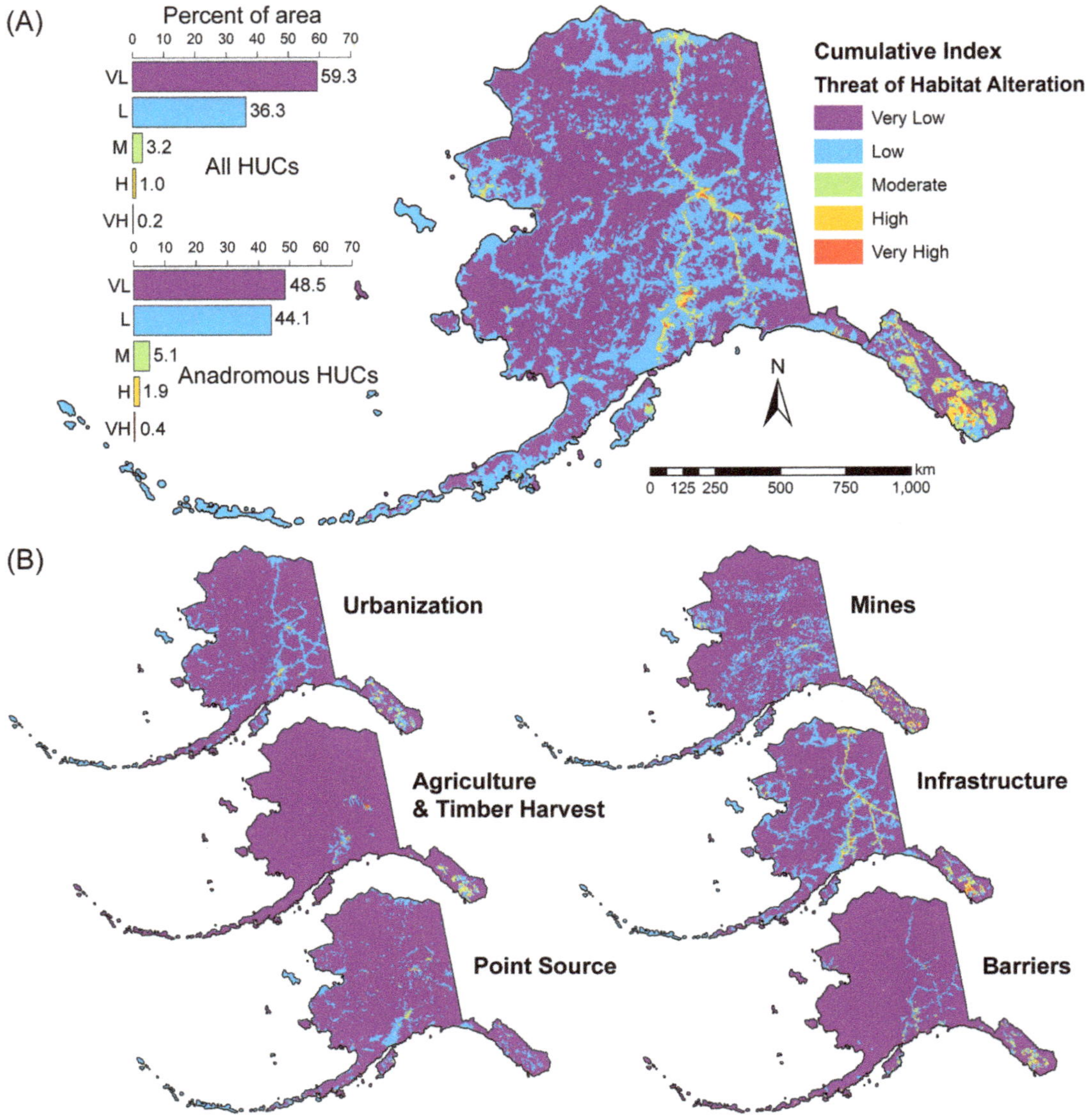

Figure 2. **(A)** Contemporary threat of habitat alteration scores for HUC-12s throughout Alaska (n = 13,997) and **(B)** individual maps of the subindices used to derive the cumulative index. Map classes reflect Jenks natural breaks of the cumulative and subindex scores. Bar plots indicate percent of area within very low (VL), low (L), moderate (M), high (H), and very high (VH) threat classes for all HUCs statewide and for those with anadromous fishes.

Future Threat of Habitat Alteration for Alaska

Mineral mine claim areas total 12,798 km^2 statewide (~0.7% of state's land area; Table 4), with the vast majority being state claims (96%) versus federal claims (4%). In Southeast Alaska, this trend was reversed; federal mine claim area was greater than state mine claims, 66% compared to 34%, based on a total claim area of 369.5 km^2 (~0.3% of Southeast Alaska area). The vast majority of mine claim areas occurred in spatial units with lower overall threat of habitat alteration (Figure 4).

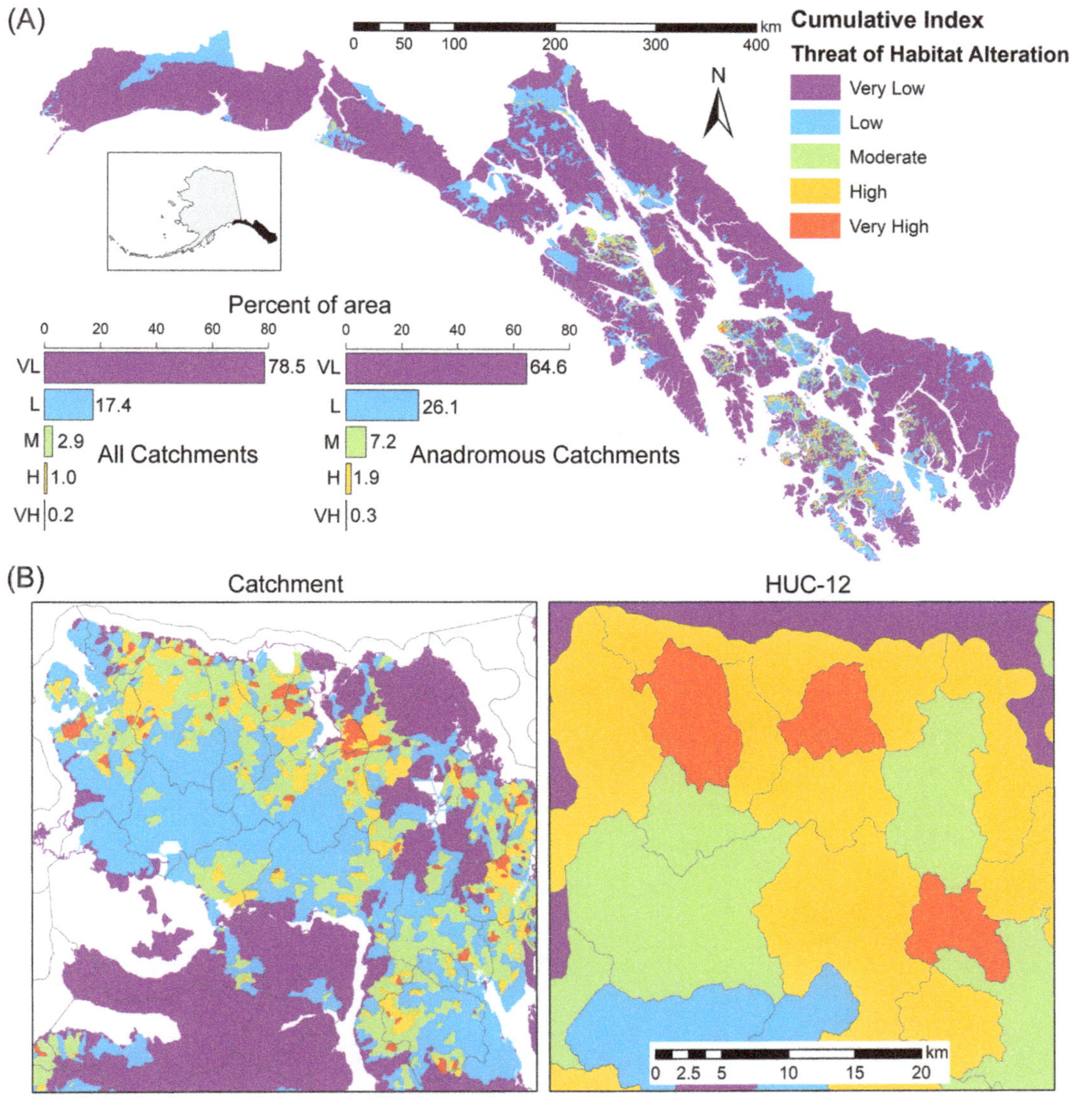

Figure 3. **(A)** Contemporary threat of habitat alteration scores for local and network catchments in Southeast Alaska (n = 207,092) and **(B)** comparison of index scores based on local and network catchments (left; mapped to local catchments) and HUC-12s (right) for an example region. Bar plots indicate percent of area within very low (VL), low (L), moderate (M), high (H), and very high (VH) threat classes for all local and network catchments in Southeast Alaska and for those with anadromous fishes.

For example, more than 86% of existing mine claims occur within HUC-12s with low and very low contemporary threat for statewide Alaska, whereas 99% of claim areas occur in local catchments with low and very low threat for Southeast Alaska. Results for spatial units with anadromous streams matched those for all fluvial spatial units, with 86% of HUC-12s and 99% of local catchment claim areas with low and very low contemporary threat.

Regarding climate change, mid-century July air temperatures are projected to increase for all HUC-12s statewide, ranging from 1.14°C to 2.50°C (mean = 1.81°C, SD =

Table 4. Area and percent of state, federal, and overall mineral mine claims by contemporary threat class for HUC-12s in statewide Alaska and local catchments in Southeast Alaska.

		State		Federal		Overall	
Region	Threat class	Area (km^2)	Percent	Area (km^2)	Percent	Area (km^2)	Percent
Statewide Alaska	Very low	2541.7	20.7	43.8	8.6	2585.5	20.2
	Low	8346.3	67.9	181.5	35.8	8527.8	66.6
	Moderate	1085.2	8.8	245.8	48.4	1331.0	10.4
	High	295.8	2.4	29.8	5.9	325.6	2.5
	Very high	21.1	0.2	6.7	1.3	27.8	0.2
Southeast Alaska	Very low	74.8	59.2	148.8	61.2	223.6	60.5
	Low	50.4	39.9	91.9	37.8	142.3	38.5
	Moderate	1.1	0.9	2.2	0.9	3.3	0.9
	High	0.1	0.0	0.1	0.0	0.2	0.0
	Very high	0.0	0.0	0.0	0.0	0.0	0.0

0.25°C). Similarly, for HUC-12s containing streams with anadromous fishes, projected temperature increases ranged from 1.15°C to 2.50°C (mean = 1.86°C, SD = 0.22°C). Many HUC-12s with anadromous streams and historically colder July air temperatures (e.g., <12°C) exhibit the greatest potential increase in future July air temperatures (Figure 5A). Areas in west-central interior Alaska exhibited the least change (Figure 5B), whereas areas in Southeast Alaska, northwest coast, the Aleutian Islands, and areas near Anchorage

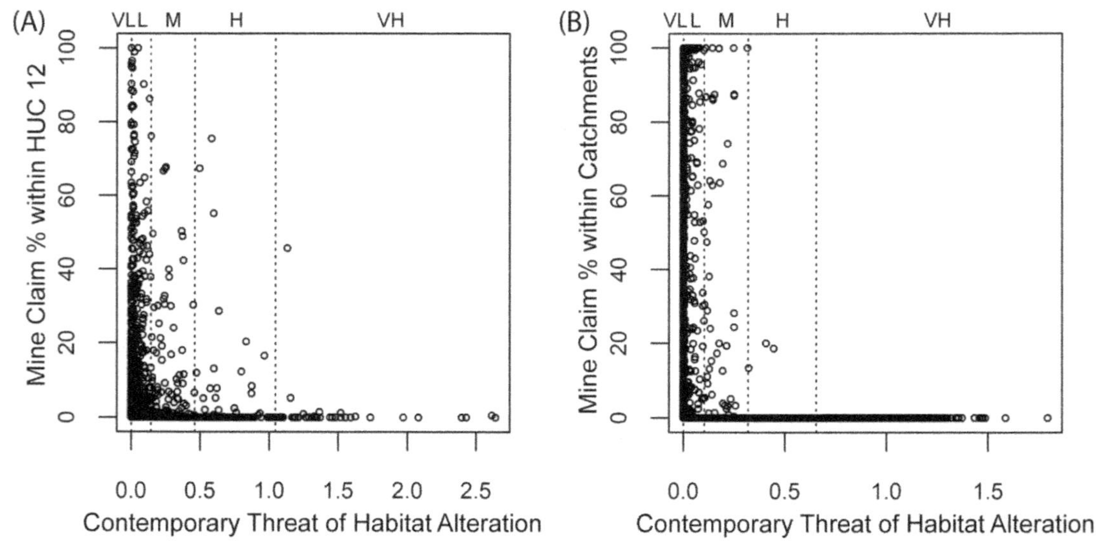

Figure 4. Comparison of future threats from mineral mining and contemporary threat of habitat alteration scores for **(A)** statewide Alaska and **(B)** Southeast Alaska. Vertical dashed lines represent breaks for contemporary threat classes, including very low (VL), low (L), moderate (M), high (H), and very high (VH).

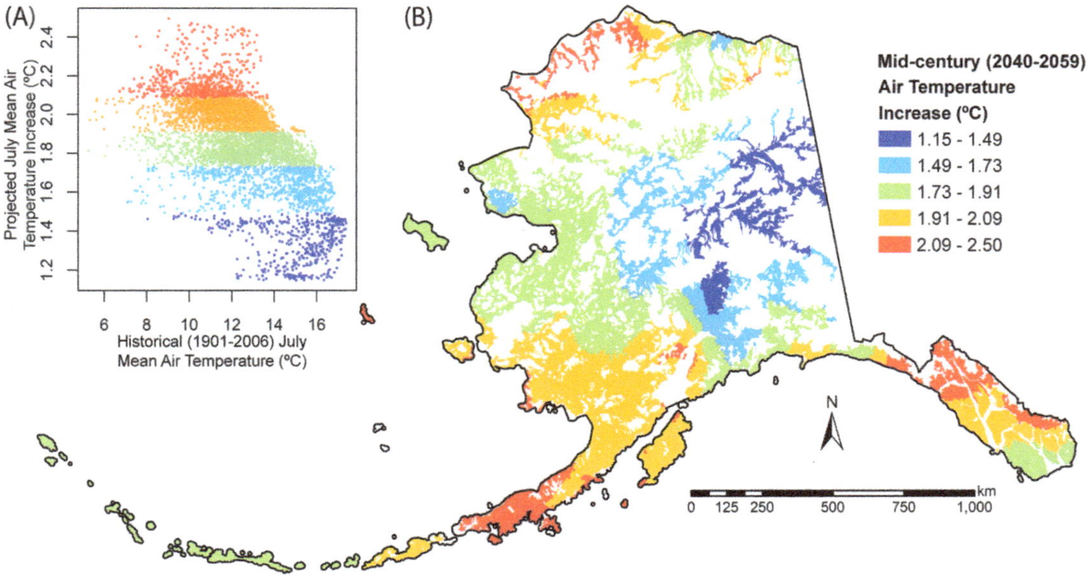

Figure 5. **(B)** Projected mid-century (2040–2059) July air temperature increases from an ensemble model of five general circulation models (NCAR-CCSM4, GFDL-CM3, GISS-E2-R, IPSL-CM5A-LR, and MRI-CGCM3) under representative concentration pathway 6.0 for HUC-12s containing streams with anadromous fishes (A; n = 6,212) for Alaska as compared to **(A)** a historical (1901–2006) baseline. Jenks natural breaks were used to create the temperature categories depicted on the map and graph.

showed the greatest change in air temperature for anadromous HUC-12s.

Principal component analysis of contemporary and future threat factors for HUC-12s with anadromous fishes resulted in threats loading on three separate principal components that collectively explain 78.4% of total variance (Table 5). Contemporary threats from urbanization, agriculture, infrastructure, point source pollution, and fragmentation grouped on the first principal component, representing 41.2% of total variance. Threats from

Table 5. Principal component analysis (PCA) results for contemporary and future threat factors for HUC-12s containing streams with anadromous fishes (n = 6,212). Dominant factors for individual axes are shown in bold.

Subindex category	PC1	PC2	PC3
Urbanization	**0.331**	−0.071	0.033
Agriculture/timber harvest	**0.368**	−0.078	−0.113
Fragmentation	0.195	−0.057	−0.038
Point source	0.143	0.002	0.032
Infrastructure	**0.821**	−0.053	−0.084
Mines (current)	0.096	−0.076	**0.977**
Mines (future)	0.012	0.026	0.140
Climate change	−0.115	**−0.988**	−0.058
Variance explained (%)	41.2	24.9	12.3

climate change loaded strongly on the second principal component axis (24.9% of total variance), whereas both current and future mining loaded onto the third principal component axis (12.3%). These results suggest that future mining and climate change have the potential to influence fluvial habitats that are largely uninfluenced by contemporary threats from urbanization, agriculture, infrastructure, point-source pollution, and fragmentation.

Discussion

This study provides a first look at landscape-based anthropogenic disturbance conditions for rivers throughout the state of Alaska, with emphasis on waters that are important migration, spawning, and rearing habitat for socioeconomically and culturally important anadromous fishes. This information provides a starting point for assessing fluvial conservation opportunities using both contemporary and future threat factors by identifying both the spatial distribution and frequency of underlying sources of stress. These threat-based assessments provide an additional source of information in guiding broadscale decision making for fluvial habitat conservation and management, particularly in regions lacking data for empirically driven analyses.

Comparison of Threats in Alaska Versus the Conterminous United States

As would be expected, Alaska differs from the conterminous United States in that the frequency and intensity of anthropogenic pressures are much lower, and in some cases, their configuration across the landscape is far more diffuse (e.g., mines and timber harvest), resulting in fewer highly altered aquatic systems across the state as compared to the conterminous United States (Crawford et al. 2016; USEPA 2016). For example, urban and agriculture combine to make up ~28% of land area in the conterminous United States compared to ~0.1% in Alaska (MRLC 2015). In fact, the population of Alaska is approximately 740,000 people, with roughly 40% of residents living in the municipality of Anchorage (U.S. Census Bureau 2017). Given that Alaska is approximately one-fifth the area of the conterminous 48 states, vast areas throughout the state are essentially void of continuous human-induced stressors that are prominent in the conterminous United States. Whereas the current assessment demonstrates that parts of the state have localized areas of high-intensity anthropogenic disturbance, contemporary threats to fluvial habitats are generally low throughout the state, suggesting that these largely intact aquatic habitats, which are a globally rare resource (Vörösmarty et al. 2010; Hughes 2019), warrant special protection. These conditions provide the opportunity to chart a more measured path in developing resources and balancing socioeconomic benefits to minimize negative affects to intact habitats and ecosystems in Alaska.

Contemporary Threats to Anadromous Fish Habitats

Despite the low levels of anthropogenic disturbance to fluvial habitats throughout much of Alaska, our study indicates that threats to anadromous fluvial habitats are comparatively higher than when all fluvial habitats are considered. In particular, the amount of "very low" threat class, representing areas where no anthropogenic disturbances were detected, was much lower for anadromous fishes at 44.0% compared to 59.0% statewide for all HUC-12s, and 65.6% for anadromous local and network catchments compared to 78.6% for all local and network catchments in Southeast Alaska. These results suggest that anadromous habitats are under greater anthropogenic pressures relative to all fluvial habitats within the state, underscoring the importance of understanding patterns in pressures for guiding anadromous fish habitat conservation and management actions. For instance, areas of human population growth and urban development will need to be monitored for effects on stream habitats and organisms, as urbanized areas of the state such as Anchorage have known impairments to water quality, physical habitat, and biotic communities in streams (Milner and Oswood 2000; Ourso and Frenzel 2003). Anadromous streams with barriers such as dams and culverts present opportunities for direct management intervention to rehabilitate systems by increasing habitat connectivity to important spawning and rearing habitats. Prime opportunities involve reconstructing poorly constructed culverts associated with forestry roads, which block both spawning adult salmon as well as the movement of juvenile salmon among seasonal rearing habitats (Murphy 1995) along with greatly increasing inputs of roadway origin sediments to these sensitive habitats. For example, restoration of a weir and culverts for Chester Creek in Anchorage resulted in a 300% increase in Coho Salmon *Oncorhynchus kisutch* runs (Myers and Nieraeth 2016), demonstrating barrier mitigation in an urbanized stream. Use of threat-based assessments can help pinpoint both location and likely sources of stress related to altered habitats across broad geographic regions and help guide both impairment inventory and habitat rehabilitation work.

Future Threats to Anadromous Fish Habitats

Results of this study indicate that future threats from mining and climate change have considerable potential to alter fluvial habitats throughout Alaska. Air temperatures are projected to increase 1.15–2.50°C for HUC-12s containing streams with anadromous fishes by mid-century, suggesting an enormous potential to warm anadromous streams. Recent studies have evaluated specific climate change influences on anadromous salmon species within the state. Specifically, Kovach et al. (2015) studied the timing of salmon migrations in Southeast Alaska and concluded that future climatic variation could amplify existing interannual variability in salmon migration timing. Analysis of thermal regimes in 80 streams in Southwest Alaska by Lisi et al. (2015) highlighted the influence of topography on stream temperatures, with rain-dominated streams situated in catchments with lower elevations and slopes having greater sensitivity to changes in summer air temperature compared to steeper, higher elevation snowmelt-driven catchments. Similarly, Winfree et al. (2018) examined the thermal regimes of 47 salmon-bearing streams in Southeast Alaska and found that lower-gradient streams with greater lentic influences have greater thermal sensitivity to changes in summer air temperatures than higher gradient streams. Evaluating egg hatching and emergence in 25 Sockeye Salmon *Oncorhynchus nerka* populations in western Alaska, Sparks et al. (2019) concluded that climate change would have minor influences on life history phenology and that the inherent variability in timing among populations may help mitigate climate change influences. Although not addressed in the current study, changes in future precipitation patterns could also result in differential effects to fluvial habitats and biota in Alaska. For instance, evaluation of multiple salmon species by Sloat et al. (2017) suggested that differences in habitat use among species could drive climate change threats based on projected flows among confined versus unconfined stream channels. Damages to infrastructure from climate change (Melvin et al. 2017) could result in future threats to fluvial habitats from damage to pipelines, culverts, and roads. Although the focus of this study has been on anadromous fishes, potamodromous fishes in Alaska, such as Rainbow Trout *Oncorhynchus mykiss*, Arctic Grayling *Thymallus arcticus*, and Inconnu (also known as Sheefish) *Stenodus leucichthys* are equally important to consider in future threat-based studies in the region, given their mobility, fishery value, and use of multiple habitats (e.g., Fraley et al. 2016).

Despite encompassing a relatively small land area (0.7% of statewide area with mine claims and 0.3% in Southeast Alaska), the literature suggests that mineral mining can have widespread effects on fluvial habitats and aquatic organisms. A single, upstream mine has the potential to alter downstream fish assemblages (Daniel et al. 2015), with effects on aquatic taxa persisting well after mining operations have ended and reclamation efforts have been conducted (Gray et al. 2000; Giam et al. 2018). Similarly, a mine that impedes fish passage can have both upstream and downstream influences on connectivity depending on the mining location within the catchment. We documented extensive potential habitat alteration for anadromous fishes to be associated with mineral mines, particularly those currently unaltered by anthropogenic disturbances. In addition to mineral mines, large oil, gas, coal, and sand and gravel deposits exist within the state, and these future threats were not addressed in our study. For instance, in evaluating potential conservation planning areas for five salmon species in the Matanuska-Susitna River basin, Witt and Hammill (2018) assessed threats from both urban development and fossil fuel extraction (including coal, oil, and gas deposits) as potential threat factors. They found that conservation planning scenarios that included all threat factors achieved more species conservation targets and were ultimately more cost-effective. We recommend that additional analyses on the potential future threats from all forms of mining be conducted for all of Alaska's fluvial habitats. This includes not only the evaluation of individual mines, but also the broadscale impact of all mining-related activities, including infrastructure and urban development, at catchment, basin, and statewide levels.

Data Limitations

In our study, use of local and network catchments in Southeast Alaska compared to coarser HUC-12s greatly improved our ability to assess threats that can manifest at finer spatial resolutions (e.g., dams, point-source pollution sites, urbanization, etc.), whereas HUC-12s may adequately represent certain threats occurring over broader spatial scales (e.g., climate change). In particular, HUC-12s represent subdivisions of the stream network and can encompass many stream reaches, losing the ability to discriminate localized contributions to individual stream reaches that can be accomplished with use of local catchments. Further, use of network catchments and their corresponding aggregated upstream attributes facilitates summarization of landscape-scale anthropogenic disturbances to stream reaches, whereas additional processing is required to enforce topological relationships among HUC-12s to account for upstream contributing areas (Omernik et al. 2017; Witt and Hammill 2018). For fluvial habitat assessments, the HUC-based framework has the additional limitation of including areas contributing to nonfluvial habitats such as lakes and coastal areas (Omernik et al. 2017). By delineating catchments and establishing topological relationships among stream reaches, development of the high-resolution National Hydrography Dataset Plus (NHDPlus) in Alaska will provide a consistent spatial framework to assess anthropogenic disturbance factors at both local and network catchment scales throughout the state.

Important threats to fluvial habitats could not be incorporated into the assessment because of data availability limitations, including current and historic timber harvest across the entire state, coal mining, sand and gravel mining, oil/gas extraction and various water withdrawal, diversion, and reservation of water uses. Further, this study does not evaluate the effects of past anthropogenic disturbances such as the large-scale placer and dredge mining for gold, including logging that occurred to support these and other disturbances stemming from natural resource extraction. Based on those absences, threat scores developed in this study underestimate anthropogenic threats to stream habitats in certain regions (Crawford et al. 2016). This and similar future assessments also serve to inform gap analyses that will result in establishing more meaningful and cost-effective future data collection priorities. Filling significant data gaps more strategically will ultimately contribute to improvements in the precision and accuracy of this and other similar natural resource assessments. As consistently assembled statewide data sets representing these additional historic and contemporary anthropogenic disturbances become available, this would represent an opportunity to refine fluvial threat-based assessments in Alaska. Further, including data sets representing natural landscape features such as topography and areas of snow and ice cover would aid in interpreting threats to fluvial habitats.

Policy Implications

Fluvial threat assessments accounting for ongoing and future threats can identify habitats that have low levels of contemporary anthropogenic disturbances but also may be more resilient to future threats from climate change. These habitats are prime candidates for conservation actions and ensuring that any human activities are compatible with these intact systems. Conservation strategies informed by these assessments that provide key information on the potential of the system can be implemented, such as promoting best management and mitigation practices for human activities such as housing developments and resource extraction in proximity to these fluvial habitats. For instance, many areas of Alaska have been extensively logged, with most future logging occurring in second-growth forests, and management plans focused on how to best meet timber resource and ecological goals for secondary forests have been developed (USFS 2014). Further, regulations surrounding logging of old-growth forests have evolved over time to reduce negative effects of historical unregulated activities by requiring best management practices to be implemented during harvest (Ice et al. 2010). Implementation of these best management practices have mitigated many historically negative impacts of logging (Cristan et al. 2016; but see Rhodes et al. 1994; Frissell et al. 2014), though our current assessment does not account for variation in rules surrounding how logging has occurred. Incorporating this type of information into future assessment efforts and policy decisions may improve our ability to more accurately depict the potential threat of fluvial habitat alteration from timber harvest under varying degrees of regulations surrounding logging rules.

Conclusions

Alaska represents one of the few regions globally where a majority of fluvial habitats still have sufficient natural flow regimes, water levels, and water quality for sustaining indigenous aquatic ecosystems. Maintaining those fluvial habitat conditions is paramount to the continuation of the largest self-sustaining anadromous fish populations in the world, particularly in the face of future threats from natural resource extraction and climate change. Given the cultural and socioeconomic importance of fisheries to the Alaskan citizens and the world, continued assessment of landscape-based anthropogenic disturbances in the state and incorporating their effects on biota will be critical to taking proactive management actions ensuring these globally unique systems continue their remarkable productivity.

Acknowledgments

Funding for this study was provided by the U.S. Fish and Wildlife Service and the National Fish Habitat Partnership Board. We are grateful to the Southeast Alaska Fish Habitat Partnership, the Matanuska-Susitna Basin Salmon Habitat Partnership, the Kenai Peninsula Fish Habitat Partnership, the Southwest Alaska Salmon Habitat Partnership, and the Western Native Trout Initiative for feedback and contributions to this study. We thank Erin Tracy, Robert Hughes, and two anonymous reviewers for reviewing an earlier version of this manuscript.

References

ADNR (Alaska Department of Natural Resources) 2018a. Alaska DNR state mining claims. Available: http://asgdc.alaska.gov/#137. (September 2018)

ADNR (Alaska Department of Natural Resources) 2018b. Federal mining claims in Alaska. Available: Available: www.asgdc.state.ak.us/#76. (September 2018).

Allan, J. D. 2004. Landscapes and riverscapes: the influence of land use on stream ecosystems. Annual Review of Ecology, Evolution, and Systematics 35:257–284.

Bernhardt, E. S., and M. A. Palmer. 2011. The environmental costs of mountaintop mining valley fill operations for aquatic ecosystems of the central Appalachians. Annals of the New York Academy of Sciences 1223:39–57.

Blanton, P., and W. A. Marcus. 2013. Transportation infrastructure, river confinement, and impacts on floodplain and channel habitat, Yakima and Chehalis rivers, Washington, USA. Geomorphology 189:55–65.

Blanton, P., and W. A. Marcus. 2014. Roads, railroads, and floodplain fragmentation due to transportation infrastructure along rivers. Annals of the Association of American Geographers 104:413–431.

Brown, T. C., and P. Froemke. 2012. Nationwide assessment of nonpoint source threats to water quality. BioScience 62:136–146.

Bryant, M. D. 2009. Global climate change and potential effects on Pacific salmonids in freshwater ecosystems of Southeast Alaska. Climatic Change 95:169–193.

Chambers, D., R. Moran, L. Trasky, S. Bryce, L. Danielson, L. Fulkerson, J. Goin, R. M. Hughes, J. Konigsberg, R. Spies, G. Thomas, M. Trenholm, and T. Wigington. 2012. Bristol Bay's wild salmon ecosystems and the Pebble Mine: key considerations for a large-scale mine proposal. Wild Salmon Center and Trout Unlimited, Portland, Oregon.

Crawford, S., G. Whelan, D. M. Infante, K. Blackhart, W. M. Daniel, P. L. Fuller, T. Birdsong, D. J. Wieferich, R. McClees-Funinan, S. M. Stedman, K. Herreman, and P. Ruhl. 2016. Through a fish's eye: the status of fish habitats in the United States 2015. Available: http://assessment.fishhabitat.org/. (March 2019).

Cristan, R. L., M. W. Aust, M. C. Bolding, S. M. Barrett, J. F. Munsell, and E. Schilling. 2016. Effectiveness of forestry best management practices in the United States: literature review. Forest Ecology and Management 360:133–151.

Daniel, W. M., D. M. Infante, R. M. Hughes, Y. P. Tsang, P. C. Esselman, D. Wieferich, K. Herreman, A. R. Cooper, L. Wang, and W. W. Taylor. 2015. Characterizing coal and mineral mines as a regional source of stress to stream fish assemblages. Ecological Indicators 50:50–61.

Danz, N. P., G. J. Niemi, R. R. Regal, T. Hollenhorst, L. B. Johnson, J. M. Hanowski, R. P. Axler, J. J. Ciborowski, T. Hrabik, V. J. Brady, J. R. Kelly, J. A. Morrice, J. C. Brazner, R. W. Howe, C. A. Johnston, and G. E. Host. 2007. Integrated measures of anthropogenic stress in the US Great Lakes basin. Environmental Management 39:631–647.

Davis, J. C., and G. A. Davis. 2011. The influence of stream-crossing structures on the distribution of rearing juvenile Pacific salmon. Journal of the North American Benthological Society 30:1117–1128.

Dudgeon, D., A. H. Arthington, M. O. Gessner, Z. I. Kawabata, D. J. Knowler, C. Lévêque, R. J. Naiman, A. H. Prieur-Richard, D. Soto, M. L. J. Stiassny, and C. A. Sullivan. 2006. Freshwater biodiversity: importance, threats, status and conservation challenges. Biological Reviews of the Cambridge Philosophical Society 81:163–182.

Esselman, P. C., D. M. Infante, L. Wang, D. Wu, A. R. Cooper, and W. W. Taylor. 2011. An index of cumulative disturbance to river fish habitats of the conterminous United States from landscape anthropogenic activities. Ecological Restoration 29:133–151.

Falcone, J. A., D. M. Carlisle, and L. C. Weber. 2010. Quantifying human disturbance in watersheds: variable selection and performance of a GIS-based disturbance index for predicting the biological condition of perennial streams. Ecological Indicators 10:264–273.

Falke, J. A., B. M. Huntsman, and E. R. Schoen. 2019. Climatic variation drives growth potential of juvenile Chinook Salmon along a subarctic boreal riverscape. Pages 57–82 *in* R. M. Hughes, D. M. Infante, L. Wang, K. Chen, and B. F. Terra, editors. Advances in understanding landscape influences on freshwater habitats and biological assemblages. American Fisheries Society, Symposium 90, Bethesda, Maryland.

Fausch, K. D., C. E. Torgersen, C. V. Baxter, and H. W. Li. 2002. Landscapes to riverscapes: bridging the gap between research and conservation of stream fishes: a continuous view of the river is needed to understand how processes interacting among scales set the context for stream fishes and their habitat. BioScience 52:483–498.

Fellman, J. B., S. Nagorski, S. Pyare, A. W. Vermilyea, D. Scott, and E. Hood. 2014. Stream temperature response to variable glacier coverage in coastal watersheds of Southeast Alaska. Hydrological Processes 28:2062–2073.

Fraley, K. M., J. A. Falke, R. Yanusz, and S. Ivey. 2016. Seasonal movements and habitat use of potamodromous Rainbow Trout across a complex Alaska riverscape. Transactions of the American Fisheries Society 145:1077–1092.

Frissell, C. A., R. J. Baker, D. DellaSala, R. M. Hughes, J. R. Karr, D. A. McCullough, R. K. Nawa, J. Rhodes, M. C. Scurlock, and R. C. Wissmar. 2014. Conservation of aquatic and fishery resources in the Pacific Northwest: implications of new science for the aquatic conservation strategy of the Northwest Forest Plan. Coast Range Association, Corvallis, Oregon.

Giam, X., J. D. Olden, and D. Simberloff. 2018. Impact of coal mining on stream biodiversity in the US and its regulatory implications. Nature Sustainability 1:176.

Gray, J. E., P. M. Theodorakos, E. A. Bailey, and R. R. Turner. 2000. Distribution, speciation, and transport of mercury in stream-sediment, stream-water, and fish collected near abandoned mercury mines in southwestern Alaska, USA. Science of the Total Environment 260:21–33.

Hinzman, L. D., N. D. Bettez, W. R. Bolton, F. S. Chapin, M. B. Dyurgerov, C. L. Fastie, B. Griffith, R. D. Hollister, A. Hope, H. P. Huntington, A. M. Jensen, G. J. Jia, T. Jorgenson, D. L. Kane, D. R. Klein, G. Kofinas, A. H. Lynch, A. H. Lloyd, A. D. McGuire, F. E. Nelson, W. C. Oechel, T. E. Osterkamp, C. H. Racine, V. E. Romanovsky, R. S. Stone, D. A. Stow, M. Sturm, C. E. Tweedie, G. L. Vourlitis, M. D. Walker, D. A. Walker, P. J. Webber, J. M. Welker, K. S. Winker, and K. Yoshikawa. 2005. Evidence and implications of recent climate change in northern Alaska and other arctic regions. Climatic Change 72:251–298.

Hughes, R. M. 2019. Ecological integrity: conceptual foundations and applications. In E. Wohl, editor. Oxford bibliographies in environmental science. Oxford University Press, New York.

Hughes, R. M., F. Amezcua, D. M. Chambers, W. M. Daniel, J. S. Franks, W. Franzin, D. MacDonald, E. Merriam, G. Neall, P. dos Santos Pompeu, L. Reynolds, and C. A. Woody. 2016. AFS position paper and policy on mining and fossil fuel extraction. Fisheries 41:12–15.

Ice, G. G., E. Shilling, and J. Vowell. 2010. Trends for forestry best management practices implementation. Journal of Forestry 108:267–273.

Jenks, G. F. 1967. The data model concept in statistical mapping. International Yearbook of Cartography 7:186–190.

Johnson, J., and B. Blossom. 2018a. Catalog of waters important for spawning, rearing, or migration of anadromous fishes - Arctic region, effective June 1, 2018. Alaska Department of Fish and Game, Special Publication No. 18-02, Anchorage.

Johnson, J., and B. Blossom. 2018b. Catalog of waters important for spawning, rearing, or migration of anadromous fishes - Interior region, effective June 1, 2018. Alaska Department of Fish and Game, Special Publication No. 18-03, Anchorage.

Johnson, J., and B. Blossom. 2018c. Catalog of waters important for spawning, rearing, or migration of anadromous fishes - Southcentral region, effective June 1, 2018. Alaska Department of Fish and Game, Special Publication No. 18-04, Anchorage.

Johnson, J., and B. Blossom. 2018d. Catalog of waters important for spawning, rearing, or migration of anadromous fishes - Southeastern region, effective June 1, 2018. Alaska Department of Fish and Game, Special Publication No. 18-05, Anchorage.

Johnson, J., and B. Blossom. 2018e. Catalog of waters important for spawning, rearing, or migration of anadromous fishes - Southwestern region, effective June 1, 2018. Alaska Department of Fish and Game, Special Publication No. 18-06, Anchorage.

Johnson, J., and B. Blossom. 2018f. Catalog of waters important for spawning, rearing, or migration of anadromous fishes - Western region, effective June 1, 2018. Alaska Department of Fish and Game, Special Publication No. 18-07, Anchorage.

Kovach, R. P., J. E. Joyce, J. D. Echave, M. S. Lindberg, and D. A. Tallmon. 2013. Earlier migration timing, decreasing phenotypic variation, and biocomplexity in multiple salmonid species. PLOS (Public Library of Science) ONE [online serial] 8(1):e53807.

Kovach, R. P., S. C. Ellison, S. Pyare, and D. A. Tallmon. 2015. Temporal patterns in adult salmon migration timing across Southeast Alaska. Global Change Biology 21:1821–1833.

Lisi, P. J., D. E. Schindler, T. J. Cline, M. D. Scheuerell, and P. B. Walsh. 2015. Watershed geomorphology and snowmelt control stream thermal sensitivity to air temperature. Geophysical Research Letters 42:3380–3388.

MacPherson, L. M., M. G. Sullivan, A. L. Foote, and C. E. Stevens. 2012. Effects of culverts on stream fish assemblages in the Alberta foothills. North American Journal of Fisheries Management 32:480–490.

Mantyka-Pringle, C. S., T. G. Martin, D. B. Moffatt, J. Udy, J. Olley, N. Saxton, F. Sheldon, S. E. Bunn, and J. R. Rhodes. 2016. Prioritizing management actions for the conservation of freshwater biodiversity under changing climate and land-cover. Biological Conservation 197:80–89.

Mattocks, S., C. J. Hall, and A. Jordaan. 2017. Damming, lost connectivity, and the historical role of anadromous fish in freshwater ecosystem dynamics. BioScience 67:713–728.

Mattson, K. M., and P. L. Angermeier. 2007. Integrating human impacts and ecological integrity into a risk-based protocol for conservation planning. Environmental Management 39:125–138.

McCluney, K. E., N. L. Poff, M. A. Palmer, J. H. Thorp, G. C. Poole, B. S. Williams, M. R. Williams, and J. S. Baron. 2014. Riverine macrosystems ecology: sensitivity, resistance, and resilience of whole river basins with human alterations. Frontiers in Ecology and the Environment 12:48–58.

McIntyre, J. K., D. H. Baldwin, D. A. Beauchamp, and N. L. Scholz. 2012. Low-level copper exposures increase visibility and vulnerability of juvenile Coho Salmon to Cutthroat Trout predators. Ecological Applications 22:1460–1471.

McIntyre, J. K., D. H. Baldwin, J. P. Meador, and N. L. Scholz. 2008. Chemosensory deprivation in juvenile Coho Salmon exposed to dissolved copper under varying water chemistry conditions. Environmental Science and Technology 42:1352–1358.

Mellina, E., and S. G. Hinch. 2009. Influences of riparian logging and in-stream large wood removal on pool habitat and salmonid density and biomass: a meta-analysis. Canadian Journal of Forest Research 39:1280–1301.

Melvin, A. M., P. Larsen, B. Boehlert, J. E. Neumann, P. Chinowsky, X. Espinet, J. Martinich, M. S. Baumann, L. Rennels, A. Bothner, D. J. Nicolsky, and S. S. Marchenko. 2017. Climate change damages to Alaska public infrastructure and the economics of proactive adaptation. Proceedings of the National Academy of Sciences of the United States of America 114:E122–E131.

Milner, A. M., and M. W. Oswood. 2000. Urbanization gradients in streams of Anchorage, Alaska: a comparison of multivariate and multimetric approaches to classification. Hydrobiologia 422:209–223.

Milner, A. M., L. E. Brown, and D. M. Hannah. 2009. Hydroecological response of river systems to shrinking glaciers. Hydrological Processes 23:62–77.

MRLC (Multi-Resolution Land Cover Characteristics Consortium). 2015. NLCD 2011 land cover (Alaska). Available: www.mrlc.gov/data/nlcd-2011-land-cover-alaska. (March 2019).

Murphy, M. L. 1995. Forestry impacts on freshwater habitat of anadromous salmonids in the Pacific Northwest and Alaska: requirements for protection and restoration. National Oceanic and Atmospheric Administration, Coastal Ocean Program Decision Analysis Series 7, Silver Spring, Maryland.

Myers, R. L., and S. Nieraeth. 2016. Restoration and Coho Salmon carrying capacity of Chester Creek, an urban stream in anchorage, Alaska. Northwest Science 90:146–158.

Omernik, J. M., G. E. Griffith, R. M. Hughes, J. B. Glover, and M. H. Weber. 2017. How misapplication of the hydrologic unit framework diminishes the meaning of watersheds. Environmental Management 60:1–11.

Ormerod, S. J., M. Dobson, A. G. Hildrew, and C. R. Townsend. 2010. Multiple stressors in freshwater ecosystems. Freshwater Biology [online serial] 55(s1):1–4.

Ourso, R. T., and S. A. Frenzel. 2003. Identification of linear and threshold responses in streams along a gradient of urbanization in Anchorage, Alaska. Hydrobiologia 501:117–131.

Paukert, C. P., K. L. Pitts, J. B. Whittier, and J. D. Olden. 2011. Development and assessment of a landscape-scale ecological threat index for the lower Colorado River basin. Ecological Indicators 11:304–310.

Rhodes, J. J., D. A. McCullough, and F. A. Espinosa. 1994. A coarse screening process for evaluation of the effects of land management activities on salmon spawning and rearing habitat in ESA consultations. Columbia River Inter-Tribal Fish Commission, Report #94-04, Portland, Oregon.

Rinne, J. N., R. M. Hughes, and B. Calamusso, editors. 2005. Historical changes in large river fish assemblages of the Americas. American Fisheries Society, Symposium 45, Bethesda, Maryland.

Ross, J. A., D. M. Infante, D. J. Martin, A. Cooper, and K. Herreman. 2019. Influences of local and landscape variables, including past timber harvest practices, on large wood in streams draining young-growth Southeast Alaska catchments. Pages 203–227 *in* R. M. Hughes, D. M. Infante, L. Wang, K. Chen, and B. F. Terra, editors. Advances in understanding landscape influences on freshwater habitats and biological assemblages. American Fisheries Society, Symposium 90, Bethesda, Maryland.

Schindler, D. E., R. Hilborn, B. Chasco, C. P. Boatright, T. P. Quinn, L. A. Rogers, and M. S. Webster. 2010. Population diversity and the portfolio effect in an exploited species. Nature (London) 465:609–612.

Schinegger, R., M. Palt, P. Segurado, and S. Schmutz. 2016. Untangling the effects of multiple human stressors and their impacts on fish assemblages in European running waters. Science of the Total Environment 573:1079–1088.

Shanley, C. S., and D. M. Albert. 2014. Climate change sensitivity index for Pacific Salmon habitat in southeast Alaska. PLOS (Public Library of Science) ONE [online serial] 9:e104799.

Sloat, M. R., G. H. Reeves, and K. R. Christiansen. 2017. Stream network geomorphology mediates predicted vulnerability of anadromous fish habitat to hydrologic change in Southeast Alaska. Global Change Biology 23:604–620.

SNAP (Scenarios Network for Alaska + Arctic Planning). 2018a. Projected monthly and derived temperature products - 771m CMIP5/AR5. Available: http://ckan.snap.uaf.edu/dataset/projected-monthly-and-derived-temperature-products-771m-cmip5-ar5. (September 2018).

SNAP (Scenarios Network for Alaska + Arctic Planning). 2018b. Historical monthly and derived temperature products - 771m CRU TS. Available: http://ckan.snap.uaf.edu/dataset/historical-monthly-and-derived-temperature-products-771m-cru-ts. (September 2018).

Sparks, M. M., J. A. Falke, T. P. Quinn, M. D. Adkison, D. E. Schindler, K. Bartz, D. Young, and P. A. H. Westley. 2019. Influences of spawning timing, water temperature, and climatic warming on early life history phenology in western Alaska Sockeye Salmon. Canadian Journal of Fisheries and Aquatic Sciences 76:123–135.

Strayer, D. L., and D. Dudgeon. 2010. Freshwater biodiversity conservation: recent progress and future challenges. Journal of the North American Benthological Society 29:344–358.

Surfleet, C. G., and A. E. Skaugset. 2013. The effect of timber harvest on summer low flows, Hinkle Creek, Oregon. Western Journal of Applied Forestry 28:13–21.

Theobald, D. M. 2013. A general model to quantify ecological integrity for landscape assessments and US application. Landscape Ecology 28:1859–1874.

Thieme, M., B. Lehner, R. Abell, S. K. Hamilton, J. Kellndorfer, G. Powell, and J. C. Riveros. 2007. Freshwater conservation planning in data-poor areas: an example from a remote Amazonian basin (Madre de Dios River, Peru and Bolivia). Biological Conservation 135:484–501.

Thornbrugh, D. J., S. G. Leibowitz, R. A. Hill, M. H. Weber, Z. C. Johnson, A. R. Olsen, J. E. Flotemersch, J. L. Stoddard, and D. V. Peck. 2018. Mapping watershed integrity for the conterminous United States. Ecological Indicators 85:1133–1148.

Tsang, Y. P., D. Wieferich, K. Fung, D. M. Infante, and A. R. Cooper. 2014. An approach for aggregating upstream catchment information to support research and management of fluvial systems across large landscapes. SpringerPlus [online serial] 3:589.

U.S. Census Bureau. 2017. Alaska population estimates July 1, 2017. Available: www.census.gov/quickfacts/ak. (January 2017).

USEPA (U.S. Environmental Protection Agency). 2012. An assessment of potential mining impacts on salmon ecosystems of Bristol Bay, Alaska. USEPA, EPA910-R-12-004a, Washington, D.C.

USEPA (U.S. Environmental Protection Agency). 2016. National rivers and streams assessment 2008–2009: a collaborative survey. USEPA, EPA/841/R-16/007, Washington, D.C.

USFS (U.S. Forest Service). 2014. Tongass young growth management strategy 2014. Available: www.fs.usda.gov/Internet/FSE_DOCUMENTS/fseprd559973.pdf. (February 2019).

USGS (U.S. Geological Survey). 1993. National water summary 1990–1991: hydrologic events and stream water quality. U.S. Geological Survey, Water Supply Paper 2400, Reston, Virginia.

USGS (U.S. Geological Survey). 2012a. National Hydrography Dataset [for Alaska]. Available: www.usgs.gov/core-science-systems/ngp/national-hydrography. (March 2019).

USGS (U.S. Geological Survey). 2012b. Alaska 2 Arc-second Digital Elevation Models (DEMs) - USGS National Map 3DEP downloadable data collection. Available: https://catalog.data.gov/dataset/national-elevation-dataset-ned-alaska-2-arc-second-downloadable-data-collection-national-geosp. (March 2019).

Valentin, M. M., R. J. Viger, A. E. Van Beusekom, L. E. Hay, T. S. Hogue, and N. L. Foks. 2018. Enhancement of a parsimonious water balance model to simulate surface hydrology in a glacierized watershed. Journal of Geophysical Research: Earth Surface 123:1116–1132.

Van Vuuren, D. P., J. Edmonds, M. Kainuma, K. Riahi, A. Thomson, K. Hibbard, G. C. Hurtt, T. Kram, V. Krey, J.-F. Lamarque, T. Masui, M. Meinshausen, N. Nakicenovic, S. J. Smith, and S. K. Rose. 2011. The representative concentration pathways: an overview. Climatic Change 109:5–31.

Vörösmarty, C. J., P. B. McIntyre, M. O. Gessner, D. Dudgeon, A. Prusevich, P. Green, S. Glidden, S. E. Bunn, C. A. Sullivan, C. Reidy Liermann, and P. M. Davies. 2010. Global threats to human water security and river biodiversity. Nature (London) 467:555.

Walsh, J. E., U. S. Bhatt, J. S. Littell, M. Leonawicz, M. Lindgren, T. A. Kurkowski, P. A. Bieniek, R. Thoman, S. Gray, and T. S. Rupp. 2018. Downscaling of climate model output for Alaskan stakeholders. Environmental Modelling and Software 110:38–51.

Wang, L., T. Brenden, P. Seelbach, A. Cooper, D. Allan, R. Clark, and M. Wiley. 2008. Landscape based identification of human disturbance gradients and reference conditions for Michigan streams. Environmental Monitoring and Assessment 141:1–17.

Wang, L., D. Infante, P. Esselman, A. Cooper, D. Wu, W. Taylor, D. Beard, G. Whelan, and A. Ostroff. 2011. A hierarchical spatial framework and database for the national river fish habitat condition assessment. Fisheries 36:436–449.

Wassmann, P., C. M. Duarte, S. Agusti, and M. K. Sejr. 2011. Footprints of climate change in the Arctic marine ecosystem. Global Change Biology 17:1235–1249.

Winfield, I. J., C. Baigun, P. A. Balykin, B. Becker, Y. Chen, A. F. Filipe, Y. V. Gerasimov, A. L. Godinho, R. M. Hughes, J. D. Koehn, D. N. Kutsyn, V. Mendoza-Portillo, T. Oberdorff, A. M. Orlov, A. P. Pedchenko, F. Pletterbauer, I. G. Prado, R. Rösch, and S. J. Vatland. 2016. International perspectives on the effects of climate change on inland fisheries. Fisheries 41:399–405.

Winfree, M. M., E. Hood, S. L. Stuefer, D. E. Schindler, T. J. Cline, C. D. Arp, and S. Pyare. 2018. Landcover and geomorphology influence streamwater temperature sensitivity in salmon bearing watersheds in Southeast Alaska. Environmental Research Letters [online serial] 13(6):064034.

Witt, A., and E. Hammill. 2018. Using systematic conservation planning to establish management priorities for freshwater salmon conservation, Matanuska-Susitna basin, AK, USA. Aquatic Conservation: Marine and Freshwater Ecosystems 28:994–1003.

Woody, C. A. and S. L. O'Neal. 2016. Fish surveys in headwater streams of the Nushagak and Kvichak River drainages, Bristol Bay, Alaska, 2008–2010. Fisheries Research and Consulting, Anchorage, Alaska. Available: www.nature.org/media/alaska/awc_dec_2010.pdf. (March 2019).

American Fisheries Society Symposium 90:255–274, 2019
© 2019 by the American Fisheries Society

Does the Proportion of Crop Fields in a Small Catchment Predict Coupled Benthic-Hyporheic Invertebrate Responses?

Junjiro N. Negishi*
Faculty of Environmental Earth Science, Hokkaido University
N10, W5, Sapporo, Hokkaido 060-0810, Japan

Abstract.—The applicability of catchment indicators for predicting aquatic invertebrate responses in both vertical and lateral dimensions of river ecosystems is poorly understood. Therefore, our objective was to determine whether landscape indicators can predict coupled benthic-hyporheic responses in river and riparian environments. To do so, we assessed the relationships between the proportion of crop fields, instream habitat conditions, and abundance of larval and adult Ephemeroptera, Plecoptera, and Trichoptera (EPT) taxa in (and from) benthic and hyporheic zones of a 15-km segment of the Satsunai River (catchment area: 725 km^2), eastern Hokkaido, northern Japan. Invertebrates were collected using colonization traps (hyporheic taxa), Surber sampler (benthic taxa), and Malaise traps (adults). We first identified insect taxa belonging to each of the benthic dwellers (EPT, excluding chloroperlid Plectoptera) and hyporheic dwellers (chloroperlid Plectoptera) based on the relative unit-volume abundances in benthic zone (7-cm surficial part of riverbed) and hyporheic zone (30 and 50 cm deep). Nitrate concentration and total chlorophyll, the abundance of larval insects in both zones, and adult insects from the benthic zone increased with an increasing proportion of crop fields. In contrast, the abundance of adult invertebrates originating from the hyporheic zone maximized in the area with an intermediate proportion of crop fields. We attributed this disparity to the spatial variability in total availability of functional hyporheic habitat in a vertical dimension, which could not be explained by the abundance estimates on a unit-volume basis. Overall, the proportion of crop fields in a catchment can be used to predict numerically similar (coupled) abundance responses of EPT larvae in river benthic and hyporheic zones. Furthermore, the use of landscape indicators in predicting coupled benthic-hyporheic responses in the riparian zone can be improved with information on the vertical extent of functional hyporheic habitat.

Introduction

Catchment characteristics such as surficial geology types, climate conditions, and land use determine runoff patterns and flow regime as well as channel morphology, thereby affecting habitat conditions of aquatic organisms (Richards et al. 1996; Poff et al. 1997; Tague and Grant 2004). Modifications of catchment land-surface characteristics commonly result in environmental changes

* Corresponding author: negishi@ees.hokudai.ac.jp

in receiving rivers. There is an ample evidence that natural events as well as anthropogenic activities in catchments, such as wildfires, urban development and agriculture, affect runoff pattern and water quality (Hollis 1975; Moody and Martin 2001; Sliva and Williams 2001; Poor and McDonnell 2007). Benthic macroinvertebrates have been extensively studied in relation to catchment human activities, and their assemblage structures often exhibit predictive relationships with landscape indicators (Roy et al. 2003; Hodkinson and Jackson 2005; Beyene et al. 2009). Consequently, catchment landscape indicators are useful in predicting the status of riverine biotic assemblages (Gergel et al. 2002). Structure, function, and diversity of riverine fauna are maintained through interactions among surface water, riparian zone, and groundwater (Ward 1989). Understanding the applicability of catchment landscape indicators in the context of multidimensionally connected ecosystem components contributes to efficiently assessing and maintaining or improving the ecological condition of rivers.

The hyporheic zone forms a dynamic ecotone where surface water and shallow groundwater mix through porous sediments under control of spatial and temporal variations in channel characteristics (streambed pressure, bed mobility, alluvial volume, and hydraulic conductivity) (Tonina and Buffington 2009; Robertson and Wood 2010). This zone is potentially extensive in alluvial gravel-bed rivers because of relatively porous sediment and abundant deposits of alluvium (Stanford and Ward 1993). Crucial ecological functions of hyporheic zone include temperature regulation, nutrient cycling, aquatic metabolism, and habitat provision for rich fauna (Valett et al. 1994; Bo et al. 2006; Boulton et al. 2010; Bengtsson et al. 2014; DelVecchia et al. 2016). Invertebrates with relatively small and/or elongated bodies (e.g., Crustacea, Oligochaeta, and some Plecoptera species), which are morphologically preadapted to space-limited life within sediments, dominate the hyporheic zone (Stanford and Gaufin 1974; Bo et al. 2006; Omesová et al. 2008; Käser 2010), whereas most Ephemeroptera, Plecoptera, and Trichoptera (EPT) dominate the surficial zones of riverbeds (Lencioni and Spitale 2015; Negishi et al., in press). Assemblage structures of both hyporheic and benthic invertebrates reflect the surrounding environment, such as food resource availability and water quality (Quinn and Hickey 1990; Strayer et al. 1997; Marmonier et al. 2012). Relatively few studies have examined the effects of catchment anthropogenic activities on hyporheic invertebrates (Boulton et al. 1997; Moldovan et al. 2013), whereas none rigorously examined the use of landscape indicators in predicting the assemblage structure of hyporheic invertebrates.

Ecological interactions between rivers and riparian areas through various types of matter exchanges affect consumers and trophic interactions in recipient ecosystems (Sabo and Power 2002; Baxter et al. 2005; Richardson et al. 2010). The emergence of aquatic insects from rivers after their metamorphosis into adults is the major vector of subsidy from rivers to riparian zones. Because of such interactions, changes in rivers can cause far-reaching impacts across ecosystem boundaries by altering the quality and quantity of subsidy (Walters et al. 2008; Abbott et al. 2009; Terui et al. 2018). Those previous studies did not focus on the detailed processes concerning how changes in larval insects in rivers were reflected in the abundance of emergent aquatic insects. Instead, they examined assemblage composition, abundance, and/or biomass of benthic macroinvertebrates or adult insects as indicators of spatial variability in the quality and quantity of subsidy.

Some hyporheic invertebrates are insects, such as some Plecoptera (Stanford and Ward 1988; Lencioni and Spitale 2015; DelVecchia et al. 2016); thus, the response of adult insects to environmental changes in rivers may be better explained by understanding the responses of hyporheic and epigean insect taxa separately. Furthermore, little is known about the use of landscape indicators in predicting the variations in spatial subsidy from rivers to riparian zones (Krell et al. 2015).

Water pollution associated with overenrichment of P and N is a widespread concern for the ecosystem health of surface waters (Carpenter et al. 1998). Diffuse N pollution associated with agriculture is a leading contributor to water quality impairment (Ahearn et al. 2005; Hayakawa et al. 2006; USEPA 2016). In moderately polluted rivers, an increase in nutrients compared to background intact conditions can result in stimulated primary productivity and positive cascading effects on secondary productivity, along with assemblage structure changes in benthic zones (Welch et al. 1992; Johnson et al. 2013; Terui et al. 2018). However, when the supply reaches a certain threshold, nutrient supply may yield negative effects, stressing the system (Niyogi et al. 2007). Such negative stresses may be associated with sediment inputs from agricultural fields, which could also reduce abundance or taxonomic richness of invertebrates (Sueyoshi et al. 2016). The wealth of knowledge concerning the effects of nutrient pollution has largely focused on benthic assemblages. Changes in hyporheic assemblage structure in response to surface water nutrient pollution have been reported in fewer cases (Scarsbrook and Fenwick 2003; Pacioglu and Moldovan 2016). Negishi et al. (in press) demonstrated positive numerical increases of both benthic and hyporheic invertebrates through bottom-up processes (i.e., coupled benthic-hyporheic responses) that were triggered by increased primary production partially resulting from nutrient inputs such as nitrate from catchment agricultural fields.

Therefore, our objective was to determine if catchment-scale landscape indicators can predict coupled benthic-hyporheic responses in river and riparian environments. To do so, we report the relationships between the proportion of crop fields, instream habitat conditions (nitrate concentration and total chlorophyll of periphytic biofilm), and abundance of both larval and adult EPT in (or from) benthic and hyporheic zones. We focused on EPT because they have terrestrial adults and are biotic vectors linking surface water, hyporheic, and riparian zones. This study builds on our previous determination that some plecopteran species (family: Chloroperlidae) were hyporheic dwellers at 50 cm depth within the riverbed and that the number of benthic and hyporheic invertebrates responded positively to increased nitrate concentrations (Negishi et al., in press). The prior identification of hyporheic taxa allowed us to distinguish between adult aquatic insects emerging from the two zones. We sought to answer four questions: (1) Does the proportion of crop fields in catchments explain habitat condition? (2) Are chloroperlid larvae generally hyporheic taxa at shallower depths within riverbeds? (3) Does the proportion of crop fields explain the spatial variability in abundance of benthic as well as hyporheic insect larvae? (4) Does the proportion of crop fields explain the spatial variability in the abundance of adult insects from benthic as well as hyporheic zones? We hypothesized that the proportion of crop fields explains spatial variations in abundance of both EPT larval and adult phases because of similar bottom-up responses to increased nitrate levels. We predicted that chloroperlids occur more abundantly at 30 cm depths than in the benthic zone. We also predicted that the proportion of crop fields is positively correlated with nitrate concentration and total chlorophyll, thereby leading to higher abundances of larval and adult EPT taxa in both river and riparian environments.

Methods

Study Site

The field study was conducted between October 2015 and August 2017 at 16 sites within a 15-km segment of the Satsunai River, eastern Hokkaido, Japan. The Satsunai River originates from Mount Satsunai (1,895 m above sea level), has a catchment area of 725 km^2, and is a tributary of the Tokachi River (Figure 1). The regional climate is characterized by low air temperature in winter and relatively low precipitation. The hydrograph in typical years was generally characterized by relatively high sustained flow rates for the period between late March to early June and occasional floods in late summer and autumn. Mean (±SD) daily flow rate in 1999–2016 was 11.1 (±17.6) m^3/s at the Kamisatsunai gauging station of the Ministry of Land, Infrastructure, Transport and Tourism (MLIT) located within the study segment. The segment has flood levees on both sides, with the inter-levee width of the channel extending approximately 350 m. The thickness of the deposited sediment was commonly greater than 20 m. More details of the study site are provided in Negishi et al. (2019a). All sites had active low-flow channels comprised of multiple river channels interspersed with islands and bank-connected exposed gravel bars; the area between the active channel and the levees consisted of riparian forest dominated by *Salix* spp. A wastewater treatment plant located between L4 and L5 discharged effluent with a high amount of nutrients through a small tributary, causing changes in water quality, whereas the effects of diffuse source nutrients associated with the surrounding land use was also suggested because of gradual changes in water quality along the channel. During a dry spell in August 2017, the surface water in the section between immediately downstream of site L9_2 and L8 was reduced and eventually completely became subsurface flow with the flow present in the upstream and downstream of this section.

Catchment Land Use

We examined the proportions of different land use types in catchments contributing to each site using a geographic information system (ArcMap, version 10.4.3, ESRI, Redlands, CA, USA). We used the land use data for the year 2000 (100 × 100 m resolution) and a digital elevation model (10 × 10 m resolution) provided by MLIT. We focused on two types of land use (i.e., forest and crop fields) because they comprised a total of more than 85% land use in the study area and were most relevant to the study hypotheses. We estimated the relative percent of each type by using the land use data delineated by catchment boundaries that were generated by the digital elevation model. The along-channel distance of each site from the most upstream site (A3) was measured using aerial photos and 1:25,000 topographic maps.

Instream Habitat Conditions

We measured nitrate (NO_3^-) concentration in surface as well as hyporheic water and total chlorophyll on the riverbed as potential indicators that reflect invertebrate habitat changes through bottom-up nutrient processes. Surface water was collected in clean 250-mL polyethylene bottles two or three times at each of the 10 sites (twice at L1, L2, L3, L4, L5, L6, L7, L8, L9, and L10 in October 2015 and June 2016; additionally once at L1, L2, L7, L8, L9, and L9_2 in July 2017), once at two sites (H1 and H4 in July 2016), and twice at three sites (A1, A2, and A3 in early and mid-July 2017). At five sites (H1, H4, A1, A2, and A3), hyporheic water was also collected

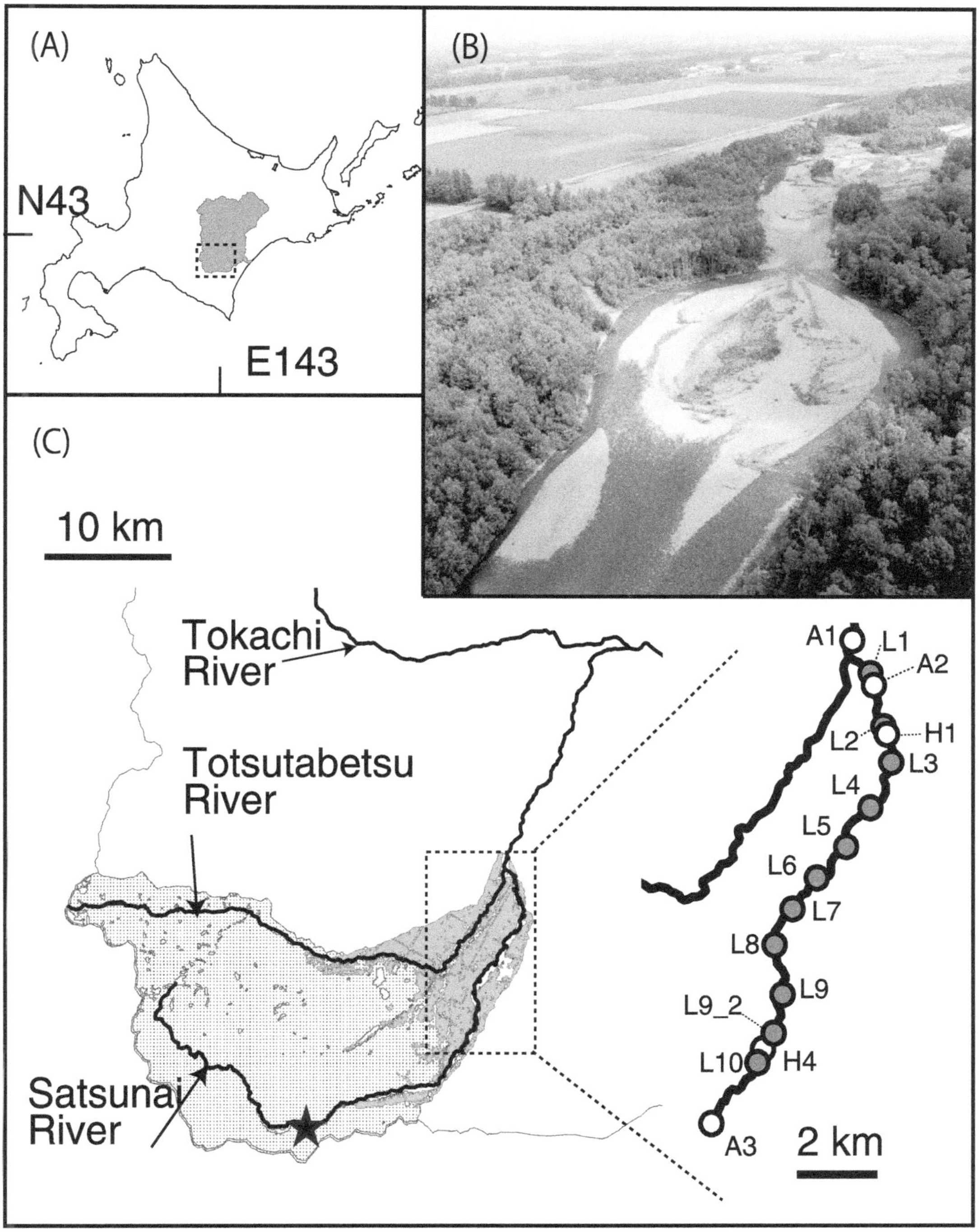

Figure 1. **(A)** The location of the study area in Hokkaido, **(B)** an aerial view of typical landscape consisting of a river channel surrounded by crop fields, and **(C)** major land use types (gray: crop fields; dotted: forest) of the catchment for site A1 and locations of sampling sites in the Satsunai River. A star denotes the Satsunai River Dam, and open and filled circles indicate sites used for collections of hyporheic invertebrates and adult aquatic invertebrates, respectively.

from 50 cm deep in the riverbed through polyvinyl chloride (PVC) wells simultaneously with the collection of corresponding surface water (see Invertebrate Collection below). Total chlorophyll was measured for periphytic biofilm collected twice (October 2014 and June 2015) as a slurry by brushing off the surfaces of two to four cobbles collected randomly in riffles (10 × 10 cm^2 area[s]). Both types of samples were immediately brought back to the laboratory in an ice chest. In the laboratory, water samples were filtered through glass-fiber filters (0.6-μm pore size; GS-25, Advantec Inc., Tokyo, Japan) and kept refrigerated until analyses were conducted within a week from the collection through use of ion chromatography (IA-300, TOA-DKK Co., Tokyo, Japan). The biofilm slurry was filtered through GS-25 filters and the filters were kept frozen until analyses within 3 months. We measured thawed filters for total chlorophyll with the United Nations Educational, Scientific and Cultural Organization method (UNESCO 1966) by using an ultraviolet-visible spectrophotometer (UVmini-1240, Shimadzu Co., Kyoto, Japan).

Invertebrate Collection

We collected benthic invertebrate larvae by using a Surber sampler (0.0625 m^2, 324-μm mesh) to a depth of approximately 7 cm (a range of 6–8 cm; an approximate volume of 4,375 cm^3) and hyporheic larvae using colonization traps (3,800 cm^3 volume; Figure 2).

Colonization traps were made by tightly enclosing sediments, which were collected in situ and consisted of a mixture of particles with diameters ranging from 4 to 64 mm and median particle size (D_{50}) of approximately 15 mm in 4 × 3 mm^2 mesh nylon netting several days prior to their installation. Traps were installed on June 10, 2016 and June 9, 2017 into pits that were excavated in the riverbed by using a backhoe (Figure 2). In 2016, we placed two traps in each of six pits at both H1 and H4 (a total of 24 traps) 50 cm deep in the riverbed (Figure 1). In 2017, we placed two traps at two depths (30 and 50 cm in the riverbed) in A1 (four pits), A2 (four pits), and A3 (six pits) (a total of 56 traps; Figure 1). A well, which was made of a 1-m-long PVC-pipe (an internal diameter of 25 mm), having perforations (12 drilled 8-mm holes) at the lower 10-cm end was installed between two traps at the depth of 50 cm in two randomly chosen pits of each site in 2016 and two (A1 and A2) and three (A3) pits in 2017. Further details of the procedure involved in installations of the traps within the riverbed is provided by Negishi et al. (in press).

Those traps were retrieved manually approximately 1 month after their installations (three randomly selected pits on both July 9 and 16, 2016; two or three randomly selected pits on July 14 and 30 and August 2017). Hyporheic traps were located by nylon strings that were tied to the traps and extended to the surface water and manually excavated by gently removing the deposited sediment above the traps. After reaching the end of the marker strings, traps were swiftly removed from the sediment and transferred into a D-frame net (375-μm mesh), which was held submersed immediately downstream of the trap. Because the sediment within the trap was tightly wrapped in the mesh net (Figure 2), quickly pulling the marker strings allowed for the collection of samples while maintaining their structural integrity and minimizing content loss.

Surber samples were collected on the riverbed vertically above the colonization traps in each pit to represent the corresponding benthic assemblage immediately after trap installation in 2016 and immediately before trap retrieval in both 2016 and 2017. No benthic invertebrates were found in samples collected immediately after the

Figure 2. Photographs of **(A)** colonization traps before the installation, **(B)** the process of installation of traps using a backhoe, and **(C)** a colonization trap immediately after the retrieval.

installation in 2016, and thus, we assumed that both traps and benthic samples had comparable colonization periods starting from the installation days. The invertebrate samples were rinsed off from sediments and were passed through a 500-μm stainless-steel sieve; the organic materials retained on the sieve were preserved in situ with either of 5% formaldehyde solution (2016) or 75% ethanol (2017).

Flying adult aquatic insects were collected at six sites (L1, L2, L7, L8, L9, and L9_2; Figure 1) through use of Malaise traps (ez-Malaise Trap; 165 cm length × 180 cm width × 180 cm height; BugDorm, Megaview Science Co., Ltd., Taichung, Taiwan). Malaise traps were set at the edge of the riparian forest facing the river channel (within 3 m from the edge of the forest) so that the long axis and mesh panel of the traps were aligned parallel to the channel with their lower edges attached to the ground. These traps were set between early April and October in 2017. The insect-collecting bottles of the traps were filled with 75% ethanol and were replaced at intervals of 5–31 d, 16 times. In this study, we focused on data from mid-June to the end of July when the peak of aquatic insect emergence occurred (Terui et al. 2018).

The invertebrate samples were processed within 5 months from collection. To characterize assemblage structure of EPT larvae, Surber and colonization-trap samples were passed through a stack of two sieves of different mesh sizes (1 mm and 500 μm). Ephemeroptera, Plecoptera, and Trichoptera larvae retained on the 1-mm sieve and all EPT adults were identified and enumerated under a stereo microscope to order, family, or species based on available keys (e.g., Maruyama and Takai 2000; Kawai and Tanida 2005).

Statistical Analyses

Our analyses involved four major steps. First, we examined the relationships between along-channel distance and land use types (forest and crop fields) using generalized linear models (GLMs) with distance as a main factor and the relative proportion of each land use type as response variables (Gaussian error distribution). Then, we tested if the proportion of crop fields predicted nitrate concentration and total chlorophyll using generalized linear mixed models (GLMMs) with the land proportion as main factor and time of collection as a random factor (Gaussian error distribution). The differences in nitrate concentration between surface water and hyporheic water was tested by developing a GLMM with sample type (surface water versus hyporheic water) as a main factor with the year of collection and pit identify nested within each site as random factors (Gaussian error distribution). Only the data collected at five sites (H1, H4, A1, A2, and A3) were used.

Second, we identified the EPT taxa that were disproportionately more abundant in each of benthic and hyporheic zones based on the percentages of each taxon in the hyporheic zone to the total abundance in both zones (Negishi et al., in press). Those percentages were calculated for each colonization-trap sample by comparing it with the benthic sample collected at the same pit on the same occasion. We corrected for the volume difference between two sample types (benthic sample was 4,375 cm^3/3,800 cm^3 = 1.15 times larger than the colonization-trap sample). For the 2017 data, we calculated those percentages separately for the samples from the two depths. Percentages close to 0% or 100% implied benthic or hyporheic taxa, respectively. We also tested if the abundance of hyporheic invertebrates (inclusive of non-EPT taxa) in 2017 differed between two depths by developing a GLMM with depth as a main factor with sampling occasion, taxon identity, and colonization-trap identity nested within pit identity as random factors (negative binomial error distribution).

Third, we determined if EPT larval abundance variation was predictable by the landscape indicator using GLMMs. In the models for benthic and hyporheic invertebrates, we included the total abundance of benthic or hyporheic taxa as the response variables, with the proportion of crop fields as a main factor and date of sampling nested within sampling year as a random factor (negative binomial error distribution).

Lastly, we examined if the abundance variation of invertebrate adults was predictable by the landscape indicator using GLMMs. We separately developed models for benthic and hyporheic taxa with the proportion of crop fields as a main factor and date of sampling as a random factor, with the duration of sampling as an offset term (negative binomial error distribution). We included a quadratic term of the main factor in the full model and identified the best model in relation to other reduced models using Akaike's information criterion (AIC). We assigned EPT adult origins for benthic and hyporheic taxa based on their larval stage information.

All analyses were conducted using R (R Core Team 2016) with packages 'glmmADMB' and 'MuMIn.' The significance of the main factor was tested by Wald tests with a statistical significance at $p = 0.05$.

Results

Habitat Condition

Catchment land-use characteristics changed longitudinally with an increasing proportion of crop fields and decreasing proportion of forest in a downstream direction (Figure 3A; the effects of the main factor in GLMs with $p < 0.001$). Nitrate concentration as well as total chlorophyll showed positive linear relationships with the increase in the proportion of crop fields (Figure 3B, 3C; the effects of the main factor in GLMMs with $p < 0.01$). The nitrate concentration at 50 cm depth did not differ from that of the corresponding surface water ($p = 0.56$).

Chloroperlid Larvae as Hyporheic Taxa

In total, 1,846 and 1,277 individuals were caught in the benthic zone in 2016 and 2017, respectively, and a total of 1,706 and 2,858 (1,872 in 30 cm and 986 in 50 cm) individuals were caught in the hyporheic zone in 2016 and 2017, respectively. We focused on eight numerically dominant taxa that comprised more than 99.7% of the total abundances in each zone in each year (Table 1). The benthic zone was dominated by four insect taxa (EPT and Chironomidae), whereas the Oligochaeta and Chironomidae dominated abundance in the hyporheic zone.

The proportion of individuals found in the hyporheic zone differed among taxa (Figure 4). In both years, Chloroperlidae were consistently found mostly in the hyporheic zone, with Oligochaeta and Crustacea also exhibiting greater occurrences there. Ephemeroptera, Plecoptera (except Chloroperlidae), and Trichoptera were largely found in the benthic zone. Chironomidae were abundant in both habitats. Among the EPT taxa, therefore, we classified Chloroperlidae as hyporheic taxa and the other EPT as benthic taxa. These patterns remained similar in both 30 and 50 cm depths in 2017. The abundance of invertebrates found in the hyporheic zone was greater at the 30 cm depth than the 50 cm depth ($p < 0.05$; Table 1).

Spatial Variability in the Abundance of Benthic and Hyporheic Taxa

The abundances of both the benthic and hyporheic EPT larvae were positively related to the proportion of crop fields (Figure 5; $p < 0.001$). The abundances of adult aquatic insects from the benthic zone increased positively with an increasing proportion of crop fields (Figure 6A). In the model selection, the

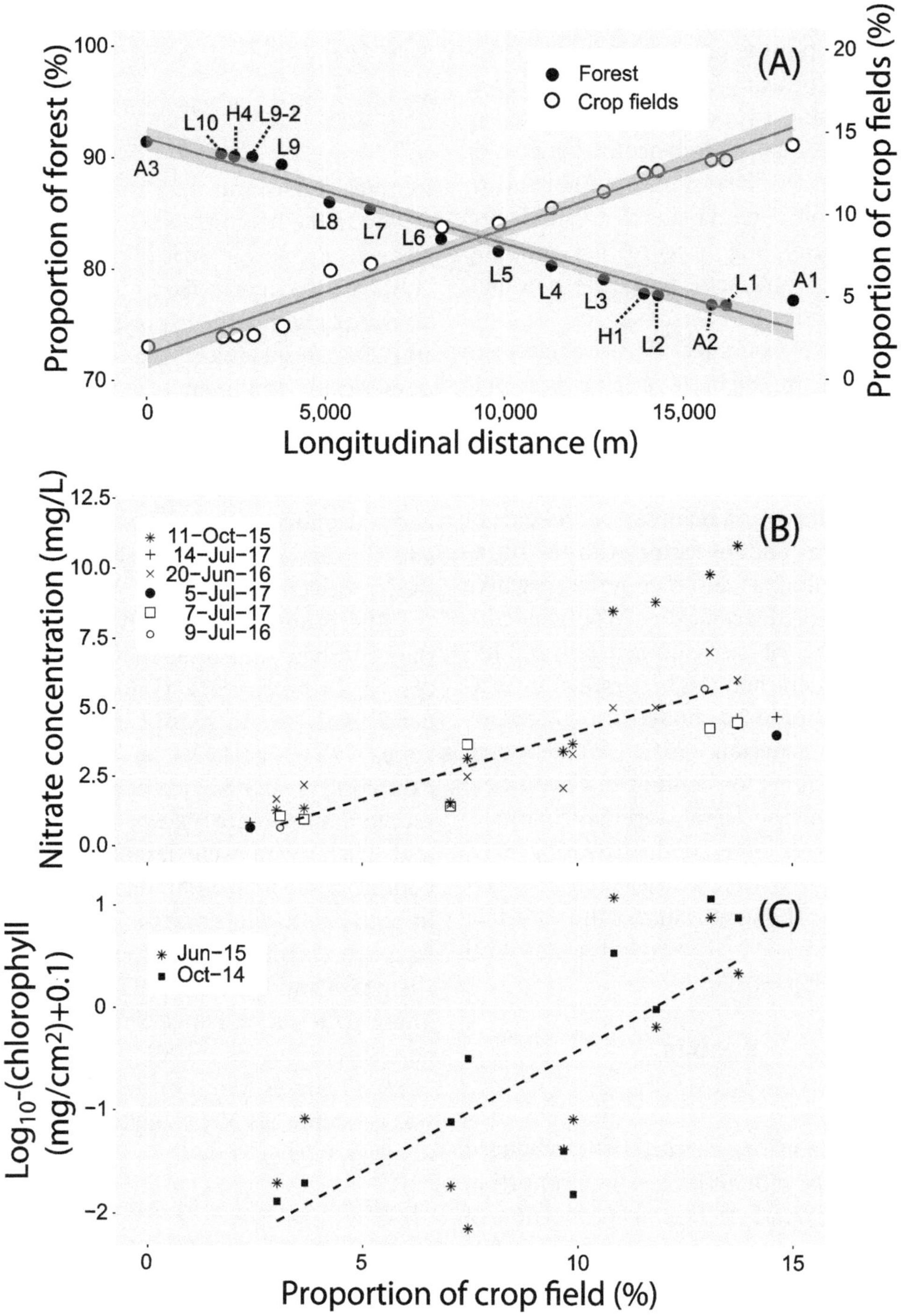

Figure 3. **(A)** The changes in proportion of two major land use types in catchments for sampling sites with an increasing distance from the most upstream site A3; the changes in **(B)** nitrate and **(C)** total chlorophyll with an increasing proportion of crop fields in contributing catchments. Solid lines and gray zones denote regression lines using generalized linear models and 95% confidence intervals, respectively, whereas dashed lines in (B) and (C) denote regression lines using generalized linear mixed models.

Table 1. Mean (±SD) abundance of invertebrate taxa collected in benthic and hyporheic samples in 2016 and 2017. Ephem. = Ephemeroptera; Pleco._others = Plecoptera other than Chloroperlidae; Pleco._Chlor. = Chloroperlidae; Tricho. = Tricoptera; Chiron. = Chironomidae; Tipul. = Tipulidae; Oligo. = Oligochaeta; Crus. = Crustacea.

	2016		2017		
	Benthic	Hyporheic	Benthic	Hyporheic (30 cm)	Hyporheic (50 cm)
Ephem.	66.8 ± 41.8	1.4 ± 1.3	36.6 ± 33.3	3.3 ± 4.0	1.4 ± 2.7
Pleco._others	3.1 ± 4.5	0.1 ± 0.3	11.0 ± 9.3	4.1 ± 8.3	1.2 ± 1.1
Pleco._Chlor.	0.8 ± 1.4	6.4 ± 10.6	0.6 ± 1.1	4.7 ± 7.3	2.5 ± 3.4
Tricho.	29.4 ± 47.1	2.4 ± 3.3	15.9 ± 23.4	2.5 ± 3.3	3.5 ± 8.6
Chiron.	44.2 ± 27.6	38.3 ± 29.7	102.4 ± 110.8	113.5 ± 131.1	51.1 ± 50.3
Tipul.	6.0 ± 6.6	0.8 ± 1.3	13.9 ± 12.8	1.3 ± 1.8	0.9 ± 1.1
Oligo.	2.8 ± 3.6	19.3 ± 31.9	1.6 ± 2.4	25.2 ± 21.4	20.9 ± 21.4
Crus.	0.8 ± 1.6	2.3 ± 3.9	0.1 ± 0.4	1.5 ± 2.2	0.6 ± 1.2

model without a second-order term was the best model with the smallest AIC (Table 2A). The quadratic term was included in models with small differences of AIC compared to the best model, and their coefficients were all positive, indicating support for a concave relationship. On the other hand, the abundance of adult aquatic insects from the hyporheic zone did not progressively increase with an increasing proportion of crop fields. Instead, highest abundances occurred in sites having intermediate proportions of crop fields in their catchments (Figure 6B). In the model selection, the model with the second-order quadratic term was chosen as the best model (Table 2B). The quadratic terms in the two best models had negative coefficients, indicating a convex relationship.

Discussion

Our results were consistent with our predictions. Chloroperlid larvae at 30 cm depths were more abundant than in the benthic zone. Furthermore, water quality (nitrate concentration) and total chlorophyll, as well as the abundance of EPT larvae, were positively correlated with the proportion of crop fields. The proportion of crop fields explained the spatial variations of abundance of larval EPT. In contrast, the abundance of adult chloroperlids originating from the hyporheic zone responded to the proportion of crop fields in a divergent way compared with adult insects from the benthic zone. Numerous studies have shown that catchment indicators, such as proportion of agricultural or urban areas or the amount of vegetation buffer along river channels, have a strong control on in-stream processes and biological assemblages, including aquatic invertebrates (Gergel et al. 2002). To the best of our knowledge, the present study is the first to report the responses of aquatic invertebrates to catchment indicators by considering both vertical and lateral dimensions of a gravel-bed riverscape.

Catchment land use often serves as an effective predictor of surface water quality (Baker 2003). In agriculture-dominated landscapes, residues from fertilizer are one of the common causes of water impairments (Carpenter et al. 1998; Buck et al. 2004; Hayakawa et al. 2006; USEPA 2016). A clear longitudinal increase of nitrate in surface water with an increasing proportion of crop fields in catchments was attributed to the effects from agricultural land use in the upstream area.

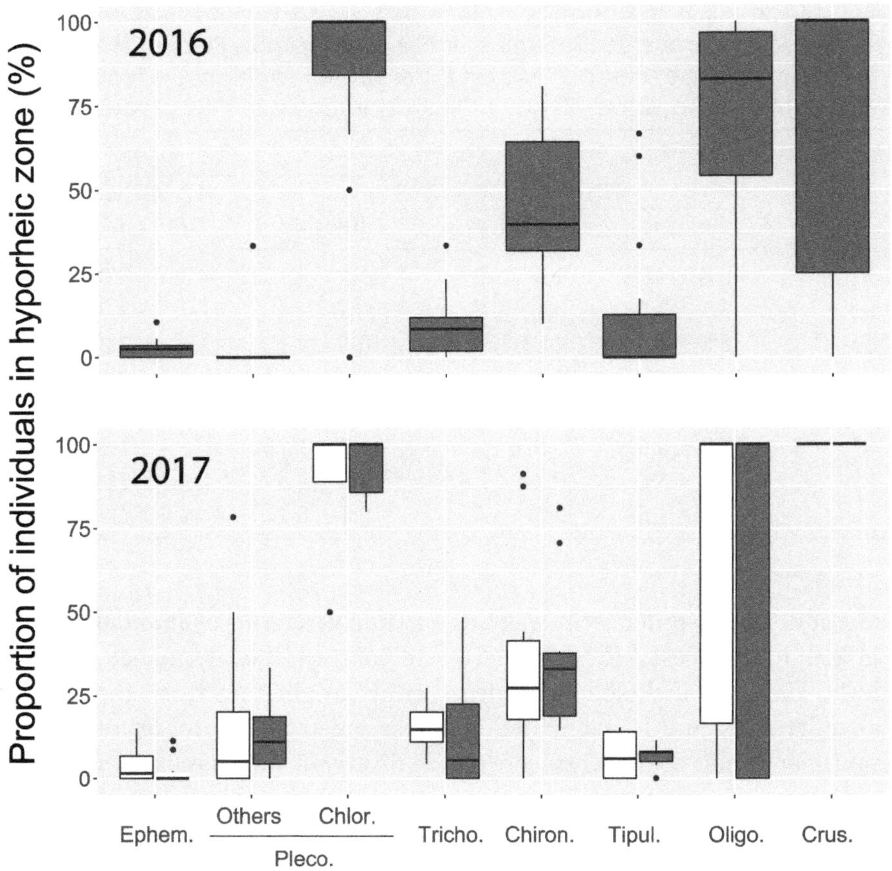

Figure 4. The proportion of individuals found in the hyporheic zone relative to the total number of individuals found in both the hyporheic and benthic zones after correcting for sample volume differences in 2016 and 2017. In 2017, the proportions were estimated for both at the depth of 30 cm (open box) and 50 cm (filled box). Ephem.: Ephemeroptera; Others: Plecoptera other than Chloroperlidae; Pleco._Chlor.: Chloroperlidae; Tricho.: Tricoptera; Chiron.: Chironomidae; Tipul.: Tipulidae; Oligo.: Oligochaeta; Crus.: Crustacea. Top (bottom) edges of box are 75th (25th) percentiles; center line in box is median; the upper (lower) whisker extends from the box edge to the largest (smallest) value no further than 1.5× interquartile ranges of the edge; data beyond the end of the whiskers are outliers and are plotted individually.

This increase of nutrient likely stimulated primary productivity and thus benthic invertebrate abundance through bottom-up effects as seen in other studies (Niyogi et al. 2007). Our previous studies using carbon and nitrogen stable isotope ratios on the same samples collected in 2016 have demonstrated that the changes in primary productivity and benthic invertebrate abundance co-occurred with the increase of nitrogen, with an anthropogenic origin for which land use effects were speculated (Negishi et al. 2019a, in press). Our current study provides quantitative supporting evidence that catchment land use indirectly affected benthic EPT invertebrates by generating a longitudinal gradient of water quality and primary productivity.

Crustaceans, Chloroperlidae, and Oligochaeta were disproportionately more abundant in the hyporheic zone in both years,

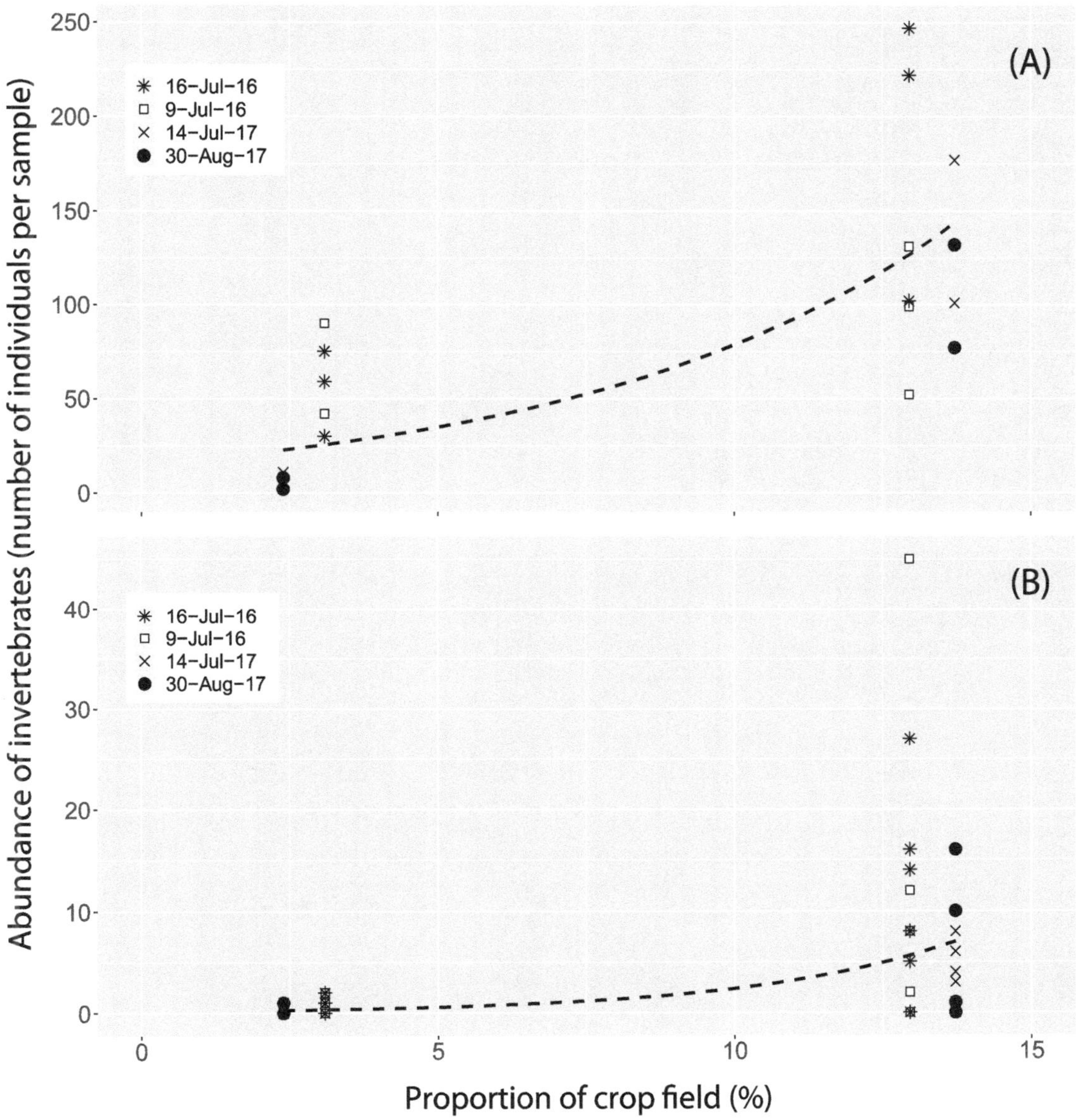

Figure 5. The changes in the abundance of **(A)** benthic and **(B)** hyporheic invertebrates with an increasing proportion of crop fields in catchments. Benthic invertebrates were found in Surber samples (A, only Ephemeroptera, Trichoptera, and Plecoptera, excluding Chloroperlidae, were included as benthic inhabitants), whereas hyporheic invertebrates were collected in hyporheic traps (B, only Chloroperlidae is included as hyporheic inhabitants). Dashed lines in (A) and (B) denote regression lines using generalized linear mixed models.

which was consistent with and strengthened the generality of the findings of the 2016 data (Negishi and al, in press). Previous studies have reported vertical changes in assemblage structure of hyporheic invertebrates, typically with decreasing abundance in deeper zones (e.g., Fraser and Williams 1997; Bo et al. 2006). We also found that the abundance of hyporheic invertebrates was slightly greater at the shallower depth,

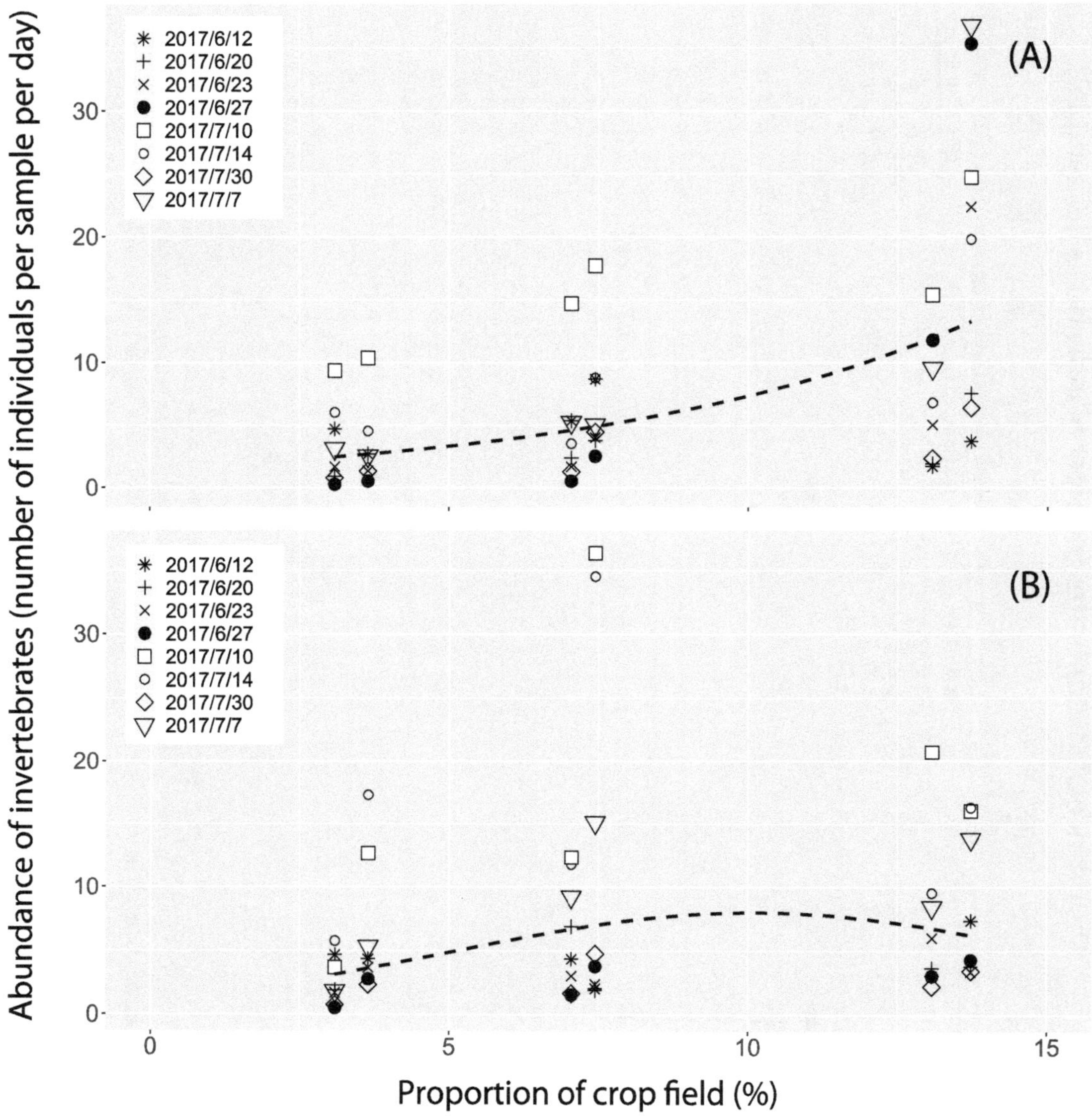

Figure 6. The changes in the abundance of flying adults of **(A)** benthic and **(B)** hyporheic invertebrates with an increasing proportion of crop fields in catchments. Both invertebrates were caught in Malaise traps set at the river–riparian interface (A, only Ephemeroptera, Trichoptera, and Plecoptera, excluding Chloroperlidae, were included as benthic inhabitants), whereas hyporheic invertebrates were collected in hyporheic traps (B, only Chloroperlidae was included as hyporheic inhabitants). Dashed lines in (A) and (B) denote regression lines using generalized linear mixed models.

but the difference from the deeper area was relatively small. This suggests that the zone below the benthic zone (with a depth of approximately 7 cm in our case) down to a depth of 50 cm was similarly functional as habitat for hyporheic invertebrates. Nitrate concentration was similar between surface and hyporheic water consistently in both years, suggesting a high level of hydrological exchange at the surface water and groundwa-

Table 2. Results of generalized linear mixed models on the abundance of adult flying insects from the **(A)** benthic and **(B)** hyporheic zones with the proportion of crop fields and a quadratic term of the proportion of crop fields as main factors and sampling day nested within sampling year as a random factor (negative binomial error distribution). Prop. = the proportion of crop fields; (Prop.) 2 = a quadratic term of the proportion of crop fields; duration = offset term to correct for the effects of variable sampling duration; Null = null model including only intercept. Delta Akaike information criterion (ΔAIC) denotes the difference of AIC between the best model (smallest AIC) and each model.

(A)

Model	Log-likelihood	Number of parameters (*K*)	AIC	ΔAIC	Akaike weight
Prop., duration	−206.294	6	420.6	0	0.269
(Prop.) 2, duration	−206.478	7	421	0.37	0.223
Prop.	−206.635	2	421.3	0.68	0.191
(Prop.) 2	−206.954	3	421.9	1.32	0.139
Prop., (Prop.) 2, duration	−206.227	8	422.5	1.87	0.106
Prop., (Prop.) 2	−206.603	4	423.2	2.62	0.073
duration	−220.452	5	446.9	26.32	0
Null	−221.435	1	448.9	28.28	0

(B)

Model	Log-likelihood	Number of parameters (*K*)	AIC	ΔAIC	Akaike weight
Prop., (Prop.) 2, duration	−198.927	8	407.9	0	0.648
Prop., (Prop.) 2	−199.731	4	409.5	1.61	0.29
Prop., duration	−203.014	6	414	6.17	0.03
Prop.	−203.904	2	415.8	7.95	0.012
(Prop.) 2, duration	−204.096	7	416.2	8.34	0.01
Duration	−206.012	5	418	10.17	0.004
(Prop.) 2	−205.013	3	418	10.17	0.004
Null	−206.976	1	420	12.1	0.002

ter interface. This is also consistent with the highly synchronized temporal fluctuation of water temperature in the two zones (Negishi et al., in press) and likely an indication that the hyporheic zone has high porosity and hydraulic conductivity. Comparable habitat suitability for invertebrates at least to a depth of 50 cm helps explain the presence of similar hyporheic assemblages regardless of depth differences. This probably led to bottom-up positive effects on hyporheic invertebrates, including chloroperlid Plecoptera, in the polluted downstream sites, supporting coupled numerically positive responses of benthic and hyporheic invertebrates.

The vertical dimension of the hyporheic zone has an important implication to the extent of the potential source area of adult aquatic insects emerging from the river and moving laterally to riparian zones. On a unit-volume basis, Chloroperlidae as hyporheic dwellers had an average abundance an order of magnitude less than those of numerically dominant benthic dwellers, such as Ephemeroptera and Trichoptera, in both years. However, this contrasting difference in abundances of EPT from two zones was not reflected in the adult abundance data collected by Malaise traps that practically integrated the spatial extent of the two zones. We can reconcile this disparity by considering the total volume availability of habitat for the two zones. The availability of benthic habitat can be estimated as a product of the superficial zone of the riverbed and the horizontal extent of the riverbed, which is largely two-dimensional relative to hyporheic zone. In contrast, hyporheic zone habitat availability can be estimated in a three-dimensional space as a product of the horizontal extent of the riverbed and the vertical extent of the functional hyporheic zone, which potentially has a greater total volume than benthic habitat alone. If we assume that the hyporheic habitat encompasses the zone ranging between 7 and 50 cm deep, the abundance of hyporheic inhabitants for a unit-area riverbed is roughly estimated as $(50-7)/7 = 6.14$ times the abundance on a unit-volume basis. This is probably a conservative estimate because hyporheic invertebrate abundance is unlikely to decrease acutely as the depth goes beyond 50 cm. Our estimates of the range of abundance of emergent adults of chloroperlids from the hyporheic zone, which considered the vertical extent of hyporheic habitat, became comparable to that from the benthic zone.

The changes in quantity and quality of subsidy from rivers to riparian zones in response to changes in river condition are relatively well known (Walters et al. 2008; Terui et al. 2018). We initially predicted positive numerical responses to water pollution of adult EPT with both benthic and hyporheic origins for larvae. Surprisingly, the predicted response was observed only for benthic taxa. We suspect this was also related to spatial variations of vertical extent of functional habitat for hyporheic invertebrates. There is an indication that the area with intermediate levels of crop fields has greater hyporheic habitat availability, but we observed drying of the river section for this area. Nakao et al. (1987) examined groundwater pathways along with detailed surveys of alluvial deposits and demonstrated that reduced surface flow in this section traveled through a relatively thick layer of porous alluvial deposits and resurfaced downstream. Therefore, it is likely that the relatively low abundance of larval hyporheic taxa on a unit-volume in this section was transformed into a relatively high abundance of their adults because of relatively higher availability of suitable habitat. Therefore, the responses of adult EPT to crop fields from benthic and hyporheic zones were decoupled in this section.

In conclusion, we showed that the proportion of crop fields in catchments had an ecologically predictable relationship with in-stream habitat, larval abundance of EPT in benthic and hyporheic zones, and emergent adult benthic EPT. The adults of hyporheic taxa were better explained by considering the total availability of functional habitat in a vertical dimension. Thus, the proportion of crop fields in catchments can be used to predict assemblage structures of invertebrates in hyporheic zones, as commonly done for benthic invertebrates. Another highlighted implication is that the use of landscape indictors in predicting coupled benthic-hyporheic responses can be improved with information on the vertical extent of functional habitat. Recent developments in modeling groundwater hydrology can be a promising tool (e.g., Poole et al. 2006).

We acknowledge three major limitations in this study as challenges for future studies. First, the study site catchments had relatively low levels of crop fields (up to 15%), which were associated only with numerically positive effects on invertebrates. Second, the study sites were characterized by high hydrological exchanges at the surface–subsurface interface. Therefore, the generality of the findings needs to be tested in other settings, including sites within larger catchments and lower hydrological exchange rates. Third, we omitted Chironomidae from our target taxa because their adults could not be assigned correctly to the habitat of origins because of their occurrence in both zones, incomplete taxonomy, and the presence of semiaquatic species. Because Chironomidae were among the most abundant taxa in all samples, the total abundance of both insect larvae and adults were probably underestimated. Future research linking physical structure of functional habitat, unit-area habitat condition for hyporheic fauna, and subsidy exports and the responses of riparian consumers in the context of human interventions would provide an avenue to substantially advance our understanding of catchment indicators with implications to vertically and laterally connected riverscapes.

Similar to the research gaps, management implications include the need to assess both hyporheic and riparian effects of land uses—not only benthic or water column effects. In addition, our research indicates that even low levels of catchment agriculture can affect benthic and hyporheic invertebrate abundances. Those changes in abundances may cloud undesirable changes in sensitive invertebrate or vertebrate taxa.

Acknowledgments

We are grateful to A. Agui, K. Alam, T. Rahman, Y. Gao, J. Tolod, and Obihiro Regional Office of Hokkaido Development Bureau, Ministry of Land, Infrastructure, Transport and Tourism (MLIT) for their laboratory and field assistance. This study was partly supported by the research fund for the Tokachi River provided by the Ministry of Land, Infrastructure, Transport, and Tourism of Japan as well as JSPS KAKENHI granted to J.N.N. (18H03408 and 18H03407). We thank Kai Chen and three anonymous reviewers for constructive comments on an earlier manuscript.

References

Abbott, I. M., D. P. Sleeman, and S. Harrison. 2009. Bat activity affected by sewage effluent in Irish rivers. Biological conservation 142:2904–2914.

Ahearn, D. S., R. W. Sheibley, R. A. Dahlgren, M. Anderson, J. Johnson, and K. W. Tate. 2005. Land use and land cover influence on water quality in the last free-flowing river draining the western Sierra Nevada, California. Journal of hydrology 313:234–247.

Baker, A. 2003. Land use and water quality. Hydrological processes 17:2499–2501.

Baxter, C. V., K. D. Fausch, and W. C. Saunders. 2005. Tangled webs: reciprocal flows of invertebrate prey link streams and riparian zones. Freshwater Biology 50:201–220.

Bengtsson, M. M., K. Wagner, N. R. Burns, E. R. Herberg, W. Wanek, L. A. Kaplan, and T. J. Battin. 2014. No evidence of aquatic priming effects in hyporheic zone microcosms. Scientific reports 4:5187.

Beyene, A., T. Addis, D. Kifle, W. Legesse, H. Kloos, and L. Triest. 2009. Comparative study of diatoms and macroinvertebrates as indicators of severe water pollution: case study of the Kebena and Akaki rivers in Addis Ababa, Ethiopia. Ecological indicators 9:381–392.

Bo, T., M. Cucco, S. Fenoglio, and G. Malacarne. 2006. Colonisation patterns and vertical movements of stream invertebrates in the interstitial zone: a case study in the Apennines, NW Italy. Hydrobiologia 568:67–78.

Boulton, A. J., T. Datry, T. Kasahara, M. Mutz, and J. A. Stanford. 2010. Ecology and management of the hyporheic zone: stream–groundwater interactions of running waters and their floodplains. Journal of the North American Benthological Society 29:26–40.

Boulton, A. J., M. R. Scarsbrook, J. M. Quinn, and G. P. Burrell. 1997. Land-use effects on the hyporheic ecology of five small streams near Hamilton, New Zealand. New Zealand Journal of Marine and Freshwater Research 31:609–622.

Buck, O., D. K. Niyogi, and C. R. Townsend. 2004. Scale-dependence of land use effects on water quality of streams in agricultural catchments. Environmental Pollution 130:287–299.

Carpenter, S. R., N. F. Caraco, D. L. Correll, R. W. Howarth, A. N. Sharpley, and V. H. Smith. 1998. Nonpoint pollution of surface waters with phosphorus and nitrogen. Ecological Applications 8:559–568.

DelVecchia, A. G., J. A. Stanford, and X. Xu. 2016. Ancient and methane-derived carbon subsidizes contemporary food webs. Nature communications 7:13163.

Fraser, B. G., and D. D. Williams. 1997. Accuracy and precision in sampling hyporheic fauna. Canadian Journal of Fisheries and Aquatic Sciences 54:1135–1141.

Gergel, S. E., M. G. Turner, J. R. Miller, J. M. Melack, and E. H. Stanley. 2002. Landscape indicators of human impacts to riverine systems. Aquatic Sciences 64:118–128.

Hayakawa, A., M. Shimizu, K. P. Woli, K. Kuramochi, and R. Hatano. 2006. Evaluating stream water quality through land use analysis in two grassland catchments. Journal of Environmental Quality 35:617–627.

Hodkinson, I. D., and J. K. Jackson. 2005. Terrestrial and aquatic invertebrates as bioindicators for environmental monitoring, with particular reference to mountain ecosystems. Environmental Management 35:649–666.

Hollis, G. E. 1975. The effect of urbanization on floods of different recurrence interval. Water Resources Research 11:431–435.

Johnson, R. C., H. S. Jin, M. M. Carreiro, and J. D. Jack. 2013. Macroinvertebrate community structure, secondary production and trophic-level dynamics in urban streams affected by non-point-source pollution. Freshwater Biology 58:843–857.

Käser, D. H. 2010. A new habitat of subsurface waters: the hyporheic biotope, by Traian Orghidan (1959). Fundamental and Applied Limnology 176:291–302.

Kawai, T., and K. Tanida editors. 2005. Aquatic insects of Japan: keys to families, genera and species. Tokai University Press, Hadano, Kanagawa, Japan. (In Japanese.)

Krell, B., N. Röder, M. Link, R. Gergs, M. H. Entling, and Schäfer, R. B. 2015. Aquatic prey subsidies to riparian spiders in a stream with different land use types. Limnologica 51:1–7.

Lencioni, V., and D. Spitale. 2015. Diversity and distribution of benthic and hyporheic fauna in different stream types on an alpine glacial floodplain. Hydrobiologia 751:73–87.

Marmonier, P., G. Archambaud, N. Belaidi, N. Bougon, P. Breil, E. Chauvet, C. Claret, J. Cornut, T. Datry, M.-J. Dle-Olivier, B. Dumont, N. Flipo, A. Foulquier, M. Gérino, A. Guilpart, F. Julien, C. Maazouzi, D. Martin, F. Mermillod-Blondin, B. Montuelle, Ph. Namour, S. Navel, D. Ombredane, T. Petite, C. Piscart, M. Pusch, S. Stroffek, A. Robertson, J.-M. Sanchez-Pérez, S. Sauvage, A. Taleb, M. Wantzen, and Ph. Vervier. 2012. The role of organisms in hyporheic processes: gaps in current knowledge, needs for future research and applications. International Journal of Limnology 48:253–266.

Maruyama, H., and M. Takai. 2000. Kawamushi Zukan. Zenkoku Noson Kyoiku Kyokai, Tokyo. (In Japanese.)

Moldovan, O. T., I. N. Meleg, E. Levei, and M. Terente. 2013. A simple method for assessing biotic indicators and predicting biodiversity in the hyporheic zone of a river polluted with metals. Ecological Indicators 24:412–420.

Moody, J. A., and D. A. Martin. 2001. Initial hydrologic and geomorphic response following a wildfire in the Colorado Front Range. Earth Surface Processes and Landforms: The Journal of the British Geomorphological Research Group 26:1049–1070.

Nakao, K., Y. Ishu, and S. Kodomari. 1987.

Mechanism of Wadi Occurrence in Satsunai River. Geophysical Bulletin of Hokkaido University 49:139–150.

Negishi, J. N., A. Hibino, K. Miura, R. Kawanishi, N. Watanabe, and K. Toyoda. In press. Coupled benthic–hyporheic responses of macroinvertebrates to surface water pollution in a gravel-bed river. Freshwater Science 38(3), doi: 10.1086/705000.

Negishi, J. N., A. Terui, N. Badrun, K. Miura, T. Oiso, K. Sumitomo, T. Kyuka, M. Yonemoto, and F. Nakamura. 2019a. High resilience of aquatic community to a 100-year flood in a gravel-bed river. Landscape and Ecological Engineering 15:143–154.

Niyogi, D. K., M. Koren, C. J. Arbuckle, and C. R. Townsend. 2007. Stream communities along a catchment land-use gradient: subsidy-stress responses to pastoral development. Environmental management 39:213–225.

Omesová, M., M. Horsák, and J. Helešic. 2008. Nested patterns in hyporheic meta-communities: the role of body morphology and penetrability of sediment. Naturwissenschaften 95:917–926.

Pacioglu, O., and O. T. Moldovan. 2016. Response of invertebrates from the hyporheic zone of chalk rivers to eutrophication and land use. Environmental Science and Pollution Research 23:4729–4740.

Poff, N. L., J. D. Allan, M. B. Bain, J. R. Karr, K. L. Prestegaard, B. D. Richter, R. E. Sparks, and J. C. Stromberg. 1997. The natural flow regime. BioScience 47:769–784.

Poole, G. C., J. A. Stanford, S. W. Running, and C. A. Frissell. 2006. Multiscale geomorphic drivers of groundwater flow paths: subsurface hydrologic dynamics and hyporheic habitat diversity. Journal of the North American Benthological Society 25:288–303.

Poor, C. J., and J. J. McDonnell. 2007. The effects of land use on stream nitrate dynamics. Journal of Hydrology 332:54–68.

Quinn, J. M., and C. W. Hickey. 1990. Characterisation and classification of benthic invertebrate communities in 88 New Zealand rivers in relation to environmental factors. New Zealand Journal of Marine and Freshwater Research 24:387–409.

R Core Team. 2016. R: a language and environment for statistical computing. R Foundation for Statistical Computing, Vienna, Austria. Available: www.R-project.org/. (April 2019).

Richards, C., L. B. Johnson, and G. E. Host. 1996. Landscape-scale influences on stream habitats and biota. Canadian Journal of Fisheries and Aquatic Sciences 53(S1):295–311.

Richardson, J. S., Y. Zhang, and L. B. Marczak. 2010. Resource subsidies across the land–freshwater interface and responses in recipient communities. River Research and Applications 26:55–66.

Robertson, A. L., and P. J. Wood. 2010. Ecology of the hyporheic zone: origins, current knowledge and future directions. Fundamental and Applied Limnology 176:279–289.

Roy, A. H., A. D. Rosemond, M. J. Paul, D. S. Leigh, and J. B. Wallace. 2003. Stream macroinvertebrate response to catchment urbanisation (Georgia, USA). Freshwater Biology 48:329–346.

Sabo, J. L., and M. E. Power. 2002. River–catchment exchange: effects of riverine subsidies on riparian lizards and their terrestrial prey. Ecology 83:1860–1869.

Scarsbrook, M. R., and G. D. Fenwick. 2003. Preliminary assessment of crustacean distribution patterns in New Zealand groundwater aquifers. New Zealand Journal of Marine and Freshwater Research 37:405–413.

Sliva, L., and D. D. Williams. 2001. Buffer zone versus whole catchment approaches to studying land use impact on river water quality. Water Research 35:3462–3472.

Stanford, J. A., and A. R. Gaufin. 1974. Hyporheic communities of two Montana rivers. Science 185:700–702.

Stanford, J. A., and J. V. Ward. 1988. The hyporheic habitat of river ecosystems. Nature (London) 335:64.

Stanford, J. A., and J. V. Ward. 1993. An ecosystem perspective of alluvial rivers: connectivity and the hyporheic corridor. Journal of the North American Benthological Society 12:48–60.

Strayer, D. L., S. E. May, P. Nielsen, W. Wollheim, and S. Hausam. 1997. Oxygen, organic matter, and sediment granulometry as controls on hyporheic animal communities. Archiv für Hydrobiologie 140:131–144.

Sueyoshi, M., N. Ishiyama, and F. Nakamura. 2016. β-diversity decline of aquatic insects at the microhabitat scale associated with agricultural land use. Landscape and ecological engineering 12:187–196.

Tague, C., and G. E. Grant. 2004. A geological framework for interpreting the low-flow regimes of Cascade streams, Willamette River basin, Oregon. Water Resources Research [online serial] 40(4).

Terui, A., J. N. Negishi, N. Watanabe, and F. Nakamura. 2018. Stream resource gradients drive consumption rates of supplemental prey in the adjacent riparian zone. Ecosystems 21:772–781.

Tonina, D., and J. M. Buffington. 2009. Hyporheic exchange in mountain rivers I: mechanics and environmental effects. Geography Compass 3:1063–1086.

UNESCO (United Nations Educational, Scientific and Cultural Organization). 1966. Determination of photosynthetic pigments in seawater. UNESCO, Monographs on Oceanographic Methodology 1, Paris.

USEPA (U. S. Environmental Protection Agency). 2016. National rivers and streams assessment: a collaborative survey. USEPA, Office of Water and Office of Research and Development, EPA/841/R-16/007, Washington, D.C.

Valett, H. M., S. G. Fisher, N. B. Grimm, and P. Camill. 1994. Vertical hydrologic exchange and ecological stability of a desert stream ecosystem. Ecology 75:548–560.

Walters, D. M., K. M. Fritz, and R. R. Otter. 2008. The dark side of subsidies: adult stream insects export organic contaminants to riparian predators. Ecological Applications 18:1835–1841.

Ward, J. V. 1989. The four-dimensional nature of lotic ecosystems. Journal of the North American Benthological Society 8:2–8.

Welch, E. B., J. M. Quinn, and C. W. Hickey. 1992. Periphyton biomass related to point-source nutrient enrichment in seven New Zealand streams. Water Research 26:669–675.

American Fisheries Society Symposium 90:275–302, 2019
© 2019 by the American Fisheries Society

Multistatus and Multispatial Scale Assessment of Landscape Effects on Benthic Macroinvertebrates in the Neotropical Savanna

Marcos Callisto*
*Laboratório de Ecologia de Bentos, Departamento de Genética, Ecologia E. Evolução
Instituto de Ciências Biológicas, Universidade Federal de Minas Gerais
Avenida Antônio Carlos 6627, Belo Horizonte, Minas Gerais 31270, Brazil*

Diego R. Macedo
*Departamento de Geografia, Instituto de Geociências
Universidade Federal de Mina Gerais
Avenida Antônio Carlos 6627, Belo Horizonte, Minas Gerais 31270, Brazil*

Marden S. Linares
*Laboratório de Ecologia de Bentos, Departamento de Biologia Geral
Instituto de Ciências Biológicas, Universidade Federal de Minas Gerais
Avenida Antônio Carlos 6627, Belo Horizonte, Minas Gerais 31270, Brazil*

Robert M. Hughes
*Amnis Opes Institute, and Department of Fisheries and Wildlife, Oregon State University
2895 SE Glenn, Corvallis, Oregon 97333, USA*

Abstract.—Despite its importance as a global biodiversity hotspot, the Neotropical savanna is threatened by rampant agricultural, hydropower, and mining development. This chapter describes the influence of landscape patterns and land uses on the taxonomic composition and structure of benthic macroinvertebrate assemblages in wadeable streams and hydropower reservoirs in the Neotropical savanna, southeastern Brazil. We used the following approaches: (1) an environmental fragility (erodibility) index, (2) an integrated disturbance index, (3) a hemeroby index of natural vegetation change, (4) the spatial distribution of benthic macroinvertebrate assemblages, (5) macroinvertebrate multimetric indices, and (6) a simplified macroinvertebrate tolerance index for urban streams. We found that land use and anthropogenic disturbances at the catchment scale had significant effects on the structure and functioning of lotic ecosystems, thereby reducing their ability to deliver ecosystem services. Our results also showed that citizen science projects can successfully apply simple, inexpensive methodologies and open an important dialogue between academia and the society at large. This chapter is a synthesis of multistatus and multispatial scale assessment of landscape effects on benthic macroinvertebrates living in headwaters and hydropower dam reservoirs in the Neotropical savanna. Fu-

* Corresponding author: callistrom@ufmg.br

ture challenges include incorporating novel ecological methodologies in ecological syntheses (e.g., eco-bioinformatics), functional trait-based indices and holistic thermodynamic indices, and standardized assessment methodologies. Doing so will further our understanding of the many-layered ecological effects of land use and other anthropogenic disturbances on aquatic biota at landscape scales.

Introduction

The Brazilian Neotropical savanna (Cerrado) biome has high biodiversity but has been historically neglected by the Brazilian government, leading to its devastation by intense land-use alterations (Strassburg et al. 2017). The Cerrado is the second most-threatened South American biodiversity biome after the Amazon and covers about 2 million km^2 or 21% of Brazil. The climate is humid tropical, with a May to September dry season (Hunke et al. 2015). Mean annual precipitation ranges from 800 to 2,000 mm, and mean annual temperature ranges from 18°C to 28°C (Ratter et al. 1997). The vegetation is mostly forest, shrub, and seasonally wet grasslands (Fernandes et al. 2018). It supports many endemic species, a large number of them threatened or extinct, but only around 20% of its natural area is undisturbed (Strassburg et al. 2017). Furthermore, less than 3% of this area occurs in formal environmental protected areas (Ministério do Meio Ambiente, www.mma.gov.br).

Aquatic ecosystems in the region are widely threatened by human land use and climate change (Callisto et al. 2012; Linares et al. 2018a; Macedo et al. 2018). Recent studies have indicated that the most diverse lotic ecosystems are found in streams surrounded by natural vegetation with intact riparian vegetation, high-quality physical habitat structure, and good water quality (e.g., Macedo et al. 2016; Firmiano et al. 2017; Silva et al. 2017). Other recent studies have shown that land use affects benthic organism foods, niche sizes, and organic matter assimilation (Castro et al. 2016). Furthermore, anthropogenic pressures negatively affected streams through loading of excess fine sediments that impair environmental condition (Macedo et al. 2018; Silva et al. 2018).

Dams are one of the most common sources of anthropogenic impacts in the Neotropical savanna (Linares et al. 2018b). Recently, because of the rising demand for renewable energy sources, hydropower dams are an increasingly common sight (Anderson et al. 2015), representing more than 64% of Brazilian electricity generation (EPE 2016). Hydropower dams have substantial negative ecological impacts on lotic ecosystems. Recent studies showed that even small, run-of-river hydropower dams have significant but spatially limited effects on the biological structure and function of the river in which they occur, including the increased presence of invasive nonnative mollusks (Linares et al. 2018b). The presence of such mollusks can also significantly perturb benthic macroinvertebrate assemblages in reservoirs, altering their taxonomic structure and thermodynamics by changing physical habitat structure and opening new energetic pathways (Linares et al. 2017). In heavily disturbed urban reservoirs, local-scale factors such as substrate characteristics are the main physical feature determining benthic assemblage condition (e.g., Molozzi et al. 2013a). In the littoral region of hydropower reservoirs, greater substrate diversity is associated with more microhabitats, food sources, and protection against predators (de Morais et al. 2017). Brazilian fish assemblages are also altered by dams and reservoirs (Pompeu et al. 2012; Terra et al. 2013; Becker et al. 2016; Sanches et al. 2016). Generally, increased siltation and reduced connectivity result in lower benthic macroinvertebrate and native fish species richness and more nonnative invasive mollusk and fish species and individuals, thereby altering ecosystem structure and function.

Habitat fragmentation, siltation, flow regulation, water pollution, and biological invasions are among the main reasons for aquatic biodiversity losses (Dudgeon 2010; Sánchez-Bayo and Wyckhuys 2019; Reid et al. 2019). Landscape changes from agriculture, livestock, and urbanization have dramatically reduced water quality and are the direct cause for many species extinctions (Gardner et al. 2009). The creation of protected areas and riparian reserves have not been fully effective for protecting freshwater species in Brazil (Oliveira-Junior et al. 2017; Leal et al. 2018; Azevedo-Santos et al. 2019). However, headwater streams in national parks and other protected areas remain important for ecological assessments of river basins (Martins et al. 2018; Agra et al. 2019). In addition, conserving high-quality aquatic ecosystems is a strategic goal for maintaining human water supplies in the Neotropical savanna.

This chapter synthesizes applications of multiple ecological indicators applied in the Brazilian Neotropical savanna, including (1) an environmental fragility index for estimating landscape erodibility, (2) an integrated disturbance index for assessing headwater stream and hydropower reservoir disturbance gradients, (3) a hemeroby index (based on the degree of natural vegetation versus human land uses) for determining degree of land cover change from the natural state, (4) spatial distribution of benthic macroinvertebrate assemblages in wadeable streams and hydropower reservoirs, (5) macroinvertebrate multimetric indices for assessing headwater stream condition, and (6) use of a simplified macroinvertebrate tolerance index for evaluating the ecological status of urban streams by public-school teachers and students. The main objective of this chapter is to synthesize those multistatus and multispatial scale assessments of landscape effects on benthic macroinvertebrates occupying headwaters and reservoirs in the Neotropical savanna.

Methods

Stream Sampling Design

We sampled 230 wadeable stream sites (first to fifth order sensu Strahler 1957). Sites were defined at 1:100,000 scale and selected through the use of a spatially dispersed random survey design (Stevens and Olsen 2004; Olsen and Peck 2008). In addition to the 183 randomly selected sites, we sampled 47 hand-picked sites to ensure that both minimally disturbed and highly disturbed sites were represented. The sites were located in the Pandiros River basin and in four different hydrological units (drainage areas within 35 km upstream of each of fourmajor hydropower reservoirs): São Simão, Nova Ponte, Volta Grande, and Três Marias (Figure 1), comprising a total geographic area of 49,100 km^2. Sampling was conducted in September from 2009 to 2016, ensuring that samples were all taken in the low flow, and least naturally variable, season (Santos et al. 2017; Silva et al. 2017).

In each stream site, we measured physical habitat following Peck et al. (2006). The length of each stream site sampled was 40 times its mean wetted width, with a minimum length of 150 m. Each site was divided into 11 equally spaced transects. At each transect, we quantified channel dimensions (e.g., wetted width, depth, and bank-full width), bank angle, riparian vegetation condition (e.g., tree canopy, understory, and ground cover), presence of instream fish cover (e.g., undercut banks, overhanging shrubs, filamentous algae, and macrophytes), and presence of human activities (e.g., pasture, agriculture, trash, and pipes). Between transects, we determined channel slope (with a clinometer) and sinuosity (with a compass), and at every 1.5 m, we recorded flow habitat type (e.g., riffles, pools, glides, etc.) and thal-

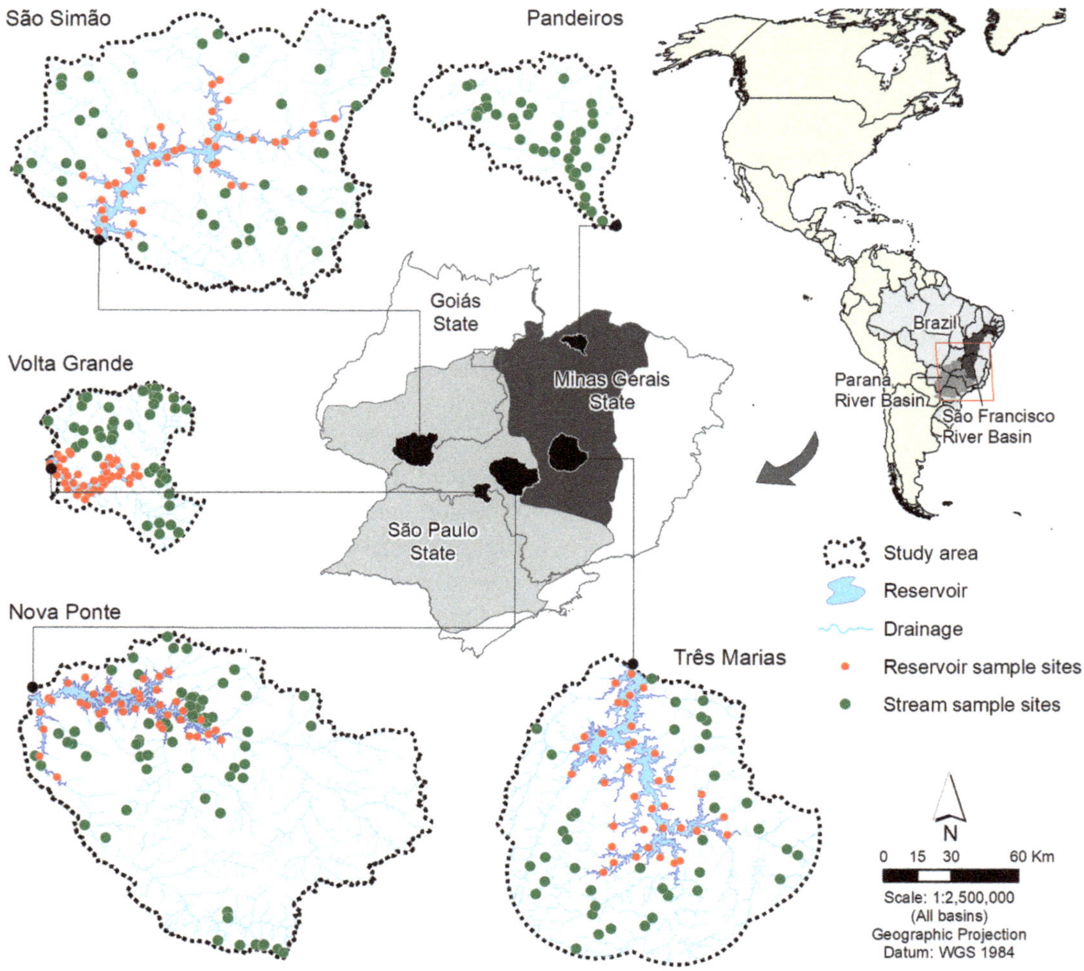

Figure 1. Sampling sites in streams and hydropower reservoirs in the Neotropical savanna.

weg depth. Substrate size was sampled by visually classifying the diameter class (e.g., sand, gravel, or boulder) of a total of 105 individual particles in five systematic points distributed across 21 cross sections of the wetted channel to ensure stable and precise substrate estimates (Kaufmann et al. 1999). We measured instantaneous discharge at a cross section having nonturbulent or near-laminar flow in or near the site (Peck et al. 2006) and calculated physical habitat metrics as described in Kaufmann et al. (1999), but with relative bed stability, calculated as recommended by Kaufmann et al. (2009).

In both stream and reservoir sites, we measured temperature, electrical conductivity, pH, and total dissolved solids in situ with a model YSI 6600 multiprobe meter. In the laboratory, dissolved oxygen, turbidity, total alkalinity, total nitrogen, and total phosphorus were analyzed following APHA (2005).

Reservoir Sampling Design

We selected 40 reservoir sites in each of the four large reservoirs (São Simão, Nova Ponte, Volta Grande, and Três Marias) through use of 1:100,000 scale maps for a total of 160 sites. Pandeiros reservoir is a small run-of-river impoundment in which only two sites were sampled. The four large reservoirs ranged in volume from 2.2×10^6 m^3 to 5.5×10^6 m^3 and in area from 222 to 1,040 km^2 (Braga and Gomieiro 1997; Pinto-Coelho 2013; Callisto et al. 2014). Sites were in the littoral zone and adjacent shoreline riparian areas. The sample design was based on random selection of an initial sampling site; the subsequent 39 other large reservoir sites were marked systematically along the mapped margin at equal intervals of one-fortieth of the total reservoir perimeter. Sampling was conducted in April–May 2010 to 2015, assuming that during this period water levels, habitat complexity, and taxa diversity would be greatest (Becker et al. 2016; Sanches et al. 2016).

We applied the protocol used by the U.S. Environmental Protection Agency (USEPA 2011, 2016) and de Morais et al. (2017) along 150 m of the margin, in 10 consecutive transects at each of the 40 large reservoir sites, producing a total of 400 physical habitat measurements for each large reservoir. Each site was composed of continuous sections of littoral (15×10 m), riparian (15×15 m), and drawdown ($15 \times Y$ m) zones, where Y represents the distance of the drawdown zone, which varied among transects because of differing margin slopes at the habitat transects. The drawdown zone extended a variable distance landward, depending on the amount of reservoir level drop compared with typical high-water levels (modified from USEPA 2011). At each transect, human disturbances in the riparian zone were recorded. Subsequently, these data were used to calculate an index of the nearshore anthropogenic disturbance intensity. The calculated metrics and indices followed Kaufmann et al. (2014).

Landscape Data Acquisition

Landscape characteristics were determined using geographic information system (GIS) procedures. We used Shuttle Radar Topography Mission topographic data (USGS 2005, 2015) to determine catchment boundaries and terrain-derived features (e.g., elevation, slope). Climatic data were provided by the Brazilian Water Agency (ANA 2017b), Brazilian National Institute of Meteorology (INMET 2017), or global geostatistical interpolated data (Hijmans et al. 2005). We calculated the proportions of various geologic units after extracting data from Brazil (2004). For more details about geodynamic data procedures, see Macedo et al. (2014a, 2014b) and Firmiano et al. (2017). We assessed catchment land use and cover for each stream site catchment and for a buffer of 500 m radius from each reservoir site by interpreting multispectral images from the Landsat Thematic Mapper in conjunction with GoogleEarth fine-resolution images (0.6–5 m spatial resolution; Macedo et al. 2014a, 2014b). The hemeroby index, based on the degree of natural vegetation versus human land uses, was calculated for each site, as suggested by Walz and Stein (2014), through use of Mapbiomas (2018) project data. Hemeroby index scores are distributed along a numeric scale, with lower scores corresponding to near natural vegetation and higher scores to severely degraded vegetation (Walz and Stein 2014).

Environmental Fragility Index

An environmental fragility index was determined based on the interaction between vulnerability (the intrinsic or natural potential of catchments to yield fine sediment when disturbed) and anthropogenic influences that act on that vulnerability to augment or moderate erosion rates. We assessed vulnerability using natural landscape data from GIS coverages and assessed influences that act on that vulnerability by vegetation cover and anthropogenic pressures as also indicated by GIS-derived data (Macedo et al. 2018).

Integrated Disturbance Index

An integrated disturbance index (Ligeiro et al. 2013) was calculated for each site based on a buffer disturbance index and a local disturbance index for reservoirs and catchment (CDI) and local disturbance indices (LDI) for streams. For quantifying the LDI, we used the W1_HALL index calculated as described in Kaufmann et al. (1999). This index summarizes the amount of evidence observed in the channel and in the riparian zone for 11 types of disturbances (e.g., buildings, channel revetment, pavement, roads, pipes, trash and landfill, parks and lawns, row crop agriculture, pasture, logging, and mining) along the 11 transects demarked at the stream site. The CDI was based on the human land uses in the catchments and was calculated according to the formula

$$\text{CDI} = 4 \times \%\ \text{urban areas} + 2 \times \%\ \text{agricultural areas} + \%\ \text{pasture areas}.$$

To evaluate anthropogenic impacts on the reservoirs at macro-, meso-, and microscales after Ligeiro et al. (2013), we constructed three disturbance indices: buffer disturbance index (250 m wide from the site), local disturbance index (reservoir modified W1_HALL + drawdown exposure), and micro disturbance index (trophic state index at the sampling station) as described in de Morais et al. (2017).

Benthic Macroinvertebrate Sampling

Benthic organisms were sampled in streams using a D-frame kick net (30 cm aperture, 500 μm mesh). Eleven subsamples were taken at each site following a systematic zig-zag trajectory (Peck et al. 2006; Firmiano et al. 2017). In the reservoirs, two benthic samples were collected at each site by using an Eckman-Birge grab (0.0225 m^2) in the littoral zone and a D-frame kick net along the wetted margin. Laboratory taxonomic identifications of benthic organisms were based on Mugnai et al. (2010) and Hamada et al. (2014, 2018). Individuals were identified mostly to family level, except for Annelida (subclass), Ephemeroptera, Plecoptera, and Trichoptera in streams (genus), Chironomidae in reservoirs (genus), and invasive species (species). We assessed benthic assemblage metrics, including taxonomic richness (rarefied), and a multimetric index (Silva et al. 2017) for stream sites.

Citizen Science

We conducted a 5-year study at 46 urban stream sites that was implemented by 155 teachers and 1,810 students aged from 9 to 18 years and trained in the use of simplified scientific approaches to assess environmental conditions (França et al. 2019). This methodology used a simplified habitat index (Callisto et al. 2002), colorimetric kits to assess water quality, and a simplified biological monitoring working party index ([citizen science biological monitoring working party [CS-BMWP]; Junqueira and Campos 1998; França et al. 2019). The simplified habitat index scored sites qualitatively based on riparian vegetation condition, erosion extent, streambed substrate composition, and pollution evidence. The CS-BMWP index scored sites based on the relative proportions of sensitive, tolerant, and resistant macroinvertebrate taxa (order or class). We then compared these results to three landscape metrics: LD1 (qualitative index of garbage, odor, and sewage), buffer disturbance index (BDI; land use within a 1-km upstream buffer), and integrated disturbance index (IDI; LDI + BDI) (Ligeiro et al. 2013). The physical habitat index and CS-BMWP results were compared with the disturbance indices through the use of a Pearson correlation test. Water quality correlation with the disturbance indices was tested by using a generalized linear model, and the model's significance was tested by an analysis of deviance (*F*-test). In addition, water quality variables were compared with the Brazilian environmental legislation (CONAMA Resolution 357/2005; Brasil 2005).

Results

Landscape Conditions of Reservoirs and Hydrologic Units

The hydrologic units differed in annual rainfall, annual mean air temperature, elevation, geology, and land use and cover (Figure 2). The geology is mostly igneous and sedimentary in São Simão and Três Marias, sedimentary in Pandeiros, metamorphic in Nova Ponte, and igneous in Volta Grande. In terms of elevation, Nova Ponte is higher than the other hydrologic units and São Simão is the lowest. Even though the hydrologic units are not distant from each other, the annual mean air temperature among them varies widely, with warmer sites in São Simão and cooler sites in Nova Ponte. Regarding annual precipitation, a typical subtropical climate occurs in São Simão, Nova Ponte, and Volta Grande, a semiarid climate in Pandeiros, and a mixed climate in Três Marias. Land use and cover was the most natural in Nova Ponte and the least natural in Três Marias, whereas Pandeiros had the greatest percent of natural woodland/park savanna cover.

The hydrologic units showed similar patterns in the environmental fragility and integrated disturbance indices. The Pandeiros River basin, a legally protected area in the Cerrado biome had a slight to moderate degree of fragility; however, São Simão, Nova Ponte, and especially Volta Grande had extensive agriculture areas and high to extreme fragility areas (Figure 3). Stream disturbance level assessments showed that the Pandeiros basin was the least-disturbed hydrologic unit and Volta Grande the most disturbed; Pandeiros, Nova Ponte, and Três Marias possessed the most least-disturbed reference sites, in that order (Figure 4A). Volta Grande and São Simão had no least-disturbed sites. The site-scale IDI scores in hydropower reservoirs showed that São Simão and Três Marias had the most least-disturbed and intermediate sites, whereas Volta Grande was the most disturbed reservoir, and the upper sites in Nova Ponte were less disturbed than the sites near the dam (Figure 4B). High hemeroby scores occurred only for Volta Grande Reservoir, indicating substantial deviation from natural vegetation at the site scale (Figure 5).

Biological Condition of Reservoirs and Hydrologic Units

Benthic assemblage conditions in streams (Figure 6A) and large reservoirs (Figure 6B) considered the number of taxa (mainly Ephemeroptera, Plecoptera, and Trichoptera and Chironomidae genera and families of other aquatic insects), as well as multimetric index scores for streams (Figure 7). Higher taxa richness occurred in the Nova Ponte hydrologic unit and Pandeiros basin, whereas the lowest values occurred in the São Simão and Volta Grande hydrologic units. A spatial zonation was found in most of the reservoirs. Riverine sites farther from dams tended to resemble natural lotic ecosystems in taxa richness; sites closer to dams tended to be more modified. This spatial zonation was found in Pandeiros (not shown), Nova Ponte, and São Simão reservoirs but was not observed in Volta Grande or Três Marias. Using >50% taxa loss as an indicator of poor condition and the top quintile as indicating good condition, we scored all four large reservoirs but São Simão as being in poor condition (Table 1). Using the same taxa richness indicators for streams, we found that the entire stream lengths of the São Simão, Três Marias, and Volta Grande hydrologic units were in poor condition. The Pandeiros and Nova Ponte hydrologic units registered 45% and 75.5% of their stream lengths in poor condition and 15% and 6.5% in good condition, respectively.

The stream sites in Nova Ponte and Pandeiros were dominated by those in good bio-

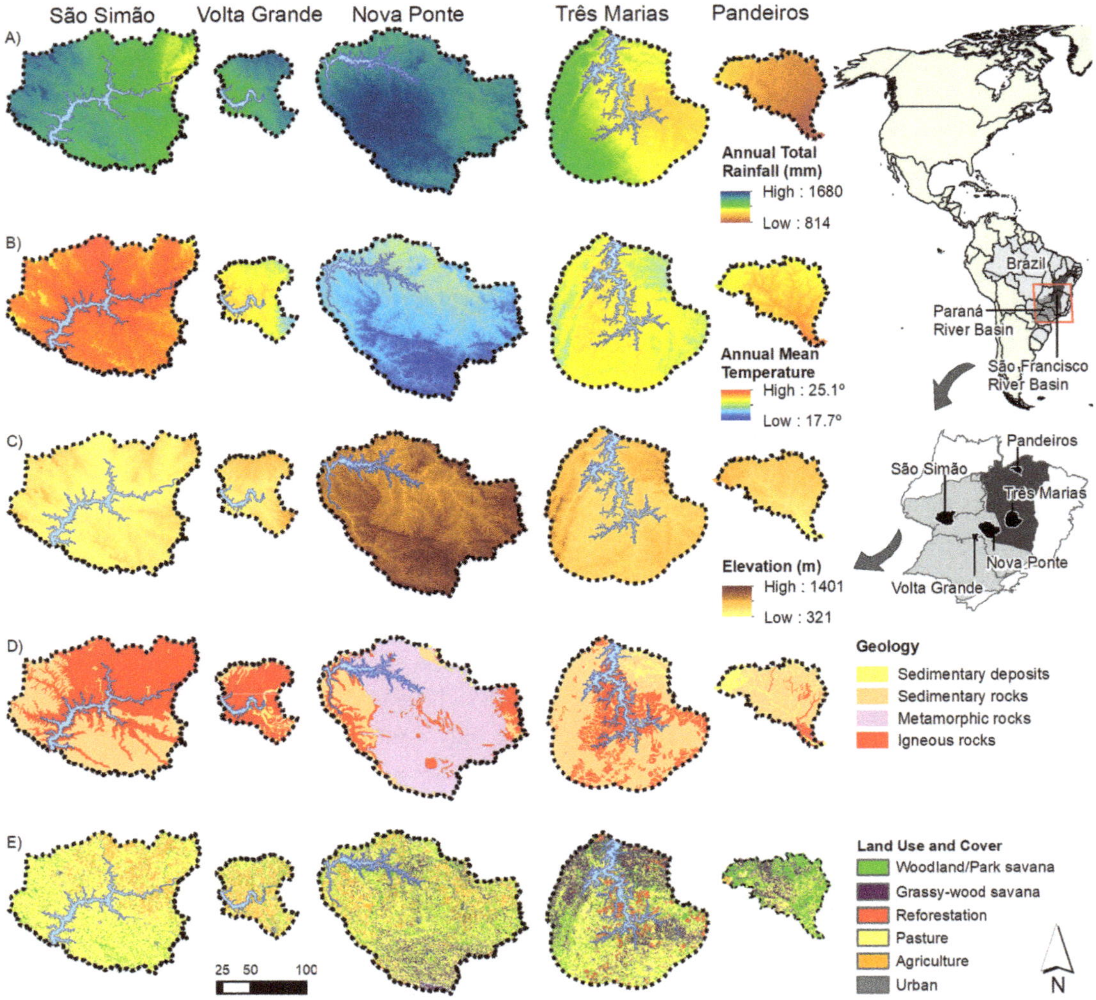

Figure 2. Landscape features of the hydrologic units: **(A)** total annual rainfall, **(B)** mean annual air temperature, **(C)** topography, **(D)** geology, and **(E)** land use and cover.

logical condition (high macroinvertebrate multimetric index scores). However, the sites in São Simão, Três Marias, and especially Volta Grande were dominated by those in poor biological condition (Figure 7).

In general, the correlations between landscape disturbance indicators and biological response indicators were greater for stream sites than for reservoir sites (Table 2). For reservoir sites, percentage of invaders were more strongly positively correlated than taxa richness with the disturbance indices and reservoir taxa richness was significantly negatively correlated with only the hemeroby index. On the other hand, in streams taxa richness was more strongly correlated than macroinvertebrate multimetric index (MMI) scores with the disturbance indices.

Student Assessment of Urban Stream Sites

As expected, local ecosystem condition was strongly correlated with landscape disturbance. The physical habitat index and

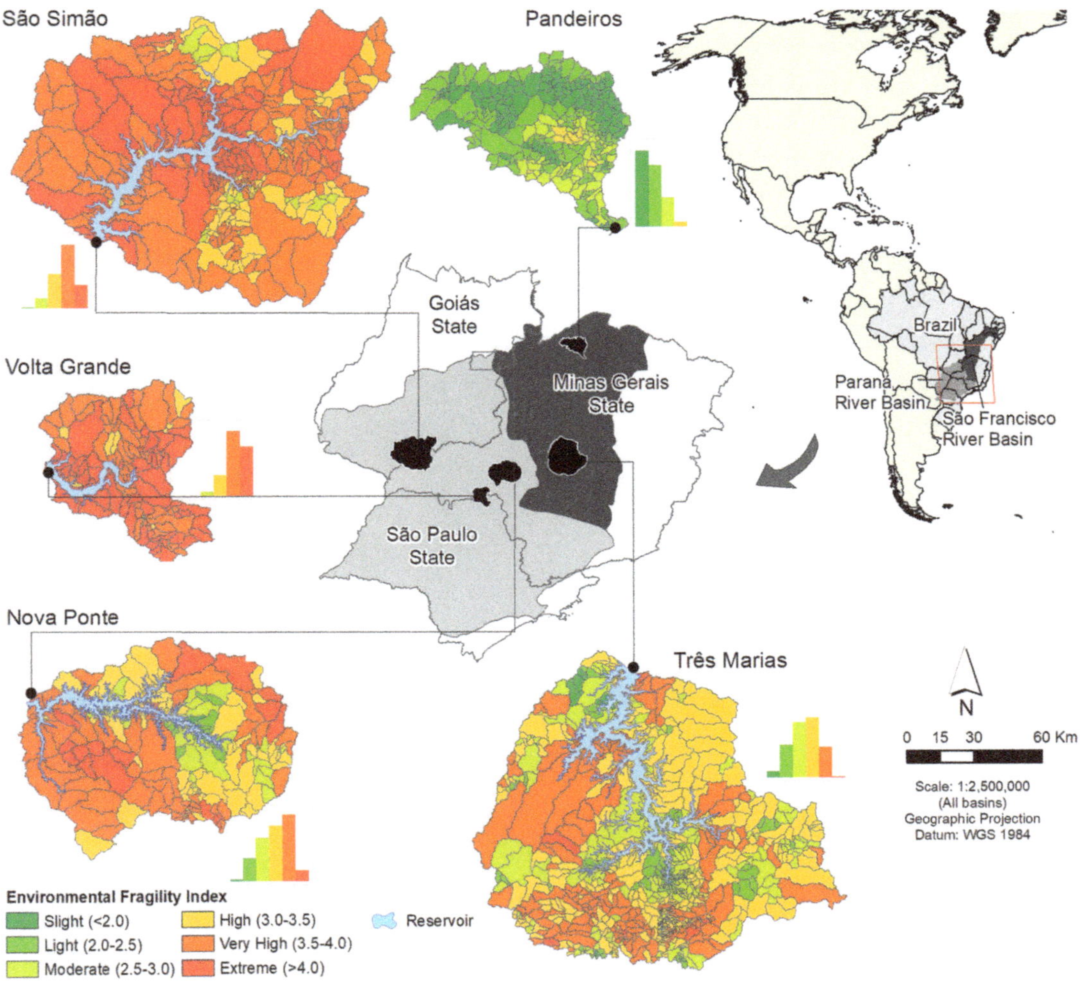

Figure 3. Environmental fragility index for five hydrologic units.

CS-BMWP results were significantly negatively correlated with the three landscape disturbance indices, but the least with the BDI (buffer land use; Table 3). The habitat index scores were more strongly correlated with the LDI, as might be expected because both assess local pollution evidence. The CS-BMWP scores were more strongly correlated with the IDI, suggesting the importance of both local and buffer conditions on macroinvertebrate assemblage composition. The scores of all three disturbance indices, but especially the LDI (incorporating qualitative pollution evidence), were significantly related with sites that violated water quality criteria (Table 4).

Discussion

As a general pattern, our results showed that the landscape metrics we measured are markedly more important for streams than for reservoirs. In streams, landscape metrics were major factors explaining benthic macroinvertebrate taxonomic richness, functional diversity (as indicated by MMI scores; Silva et al. 2017), and spatial distribution (as indicated by comparing maps of landscape disturbance [Figures 2–5] with taxa richness [Figure 6A]).

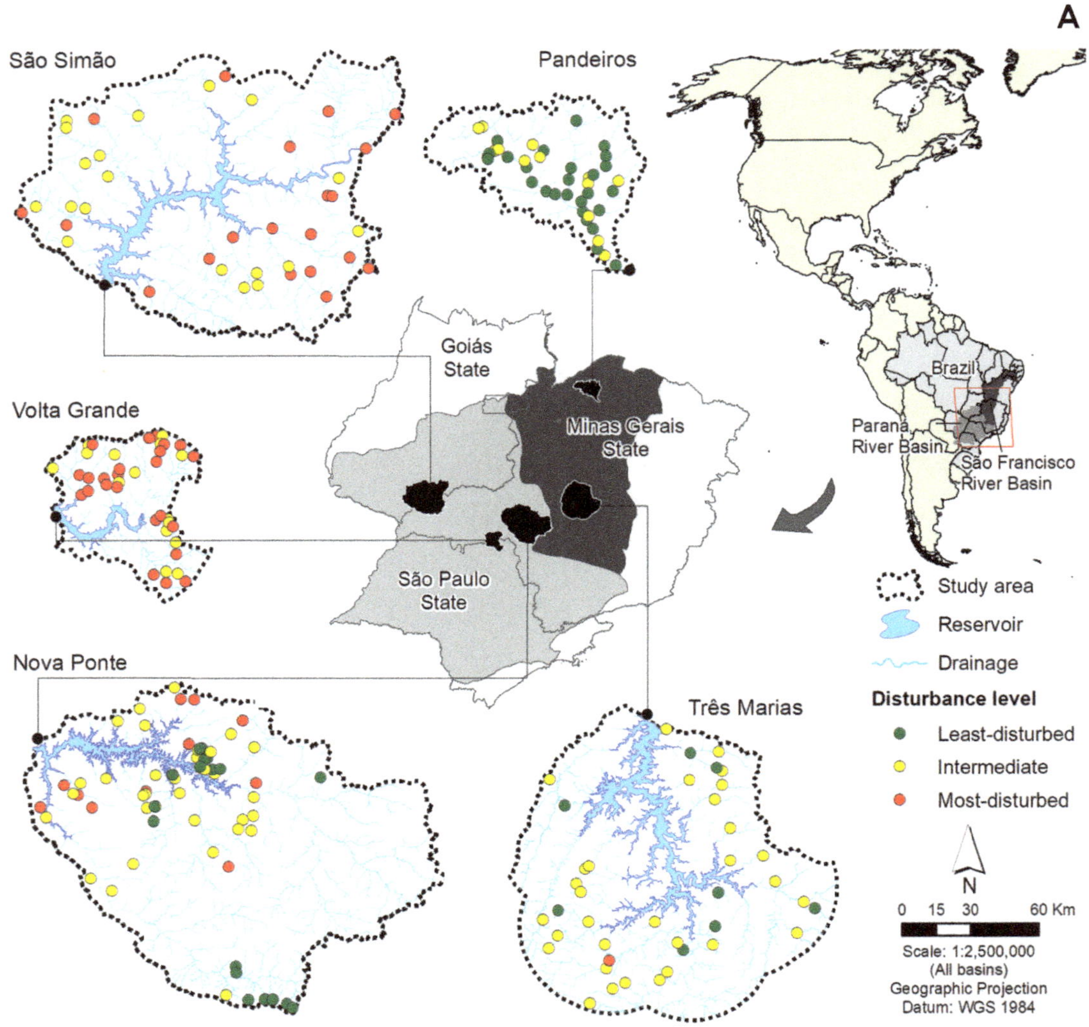

Figure 4. Integrated disturbance index calculated for **(A)** stream sites and **(B)** reservoir sites.

On the other hand, reservoir benthic assemblages, especially the presence and relative abundance of invasive nonnative mollusk species, were shown to be more associated with local metrics.

Our results show that catchment land use can be a useful indicator of lotic ecosystem condition. Clearly, the amount of natural land cover in the catchment is strongly correlated with good ecological condition, whereas the amount of urban and agricultural land cover has the opposite effect (Ligeiro et al. 2013; Macedo et al. 2014a; Silva et al. 2017). This may result from the relationship between land cover and local aquatic habitat quality, as well as the fact that many benthic macroinvertebrates are terrestrial as adults, thereby doubling the effect on their distribution and abundance (Ferreira et al. 2014; Firmiano et al. 2017; Janaina de Brito, Instituto Nacional de Pesquisas da Amazônia, unpublished data). Land-use disturbances can also affect the structure and functioning of lotic ecosystems by altering their energetic pathways

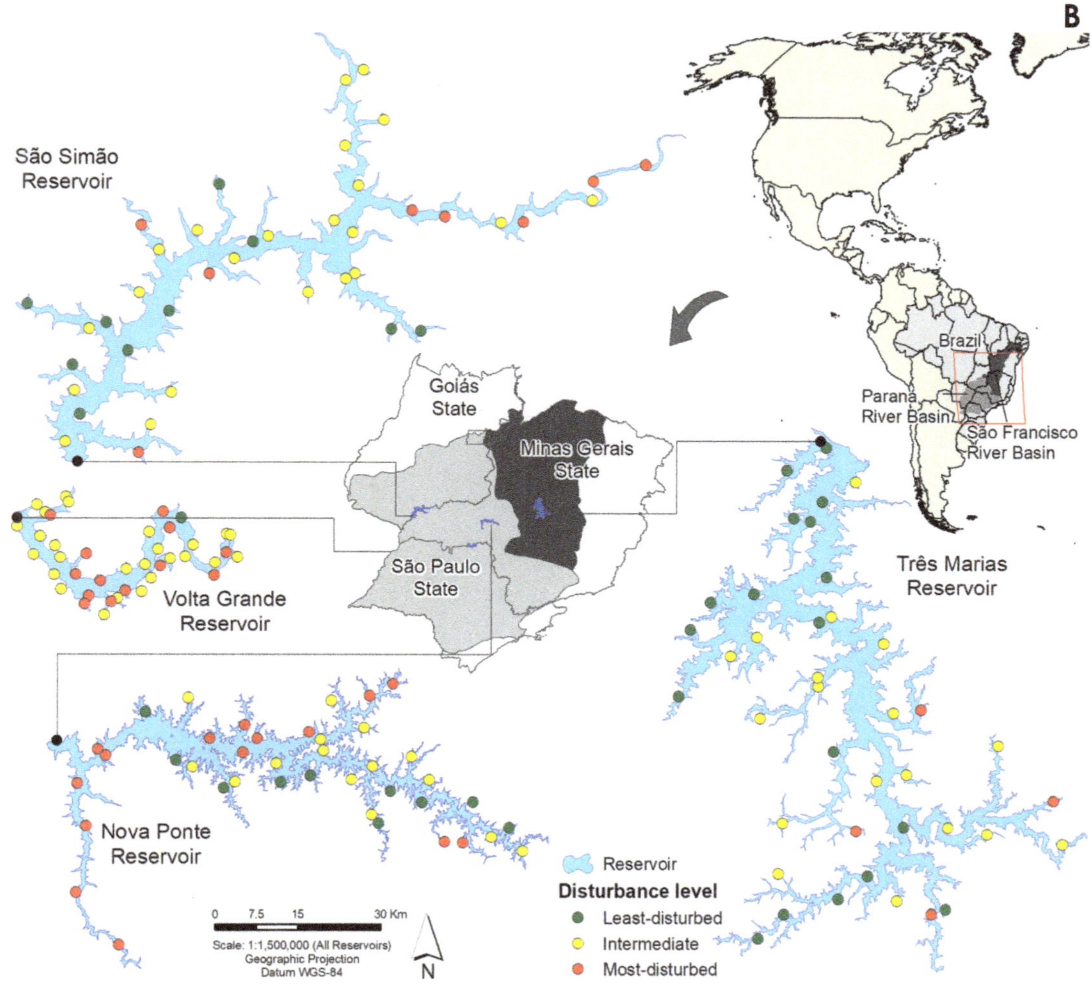

Figure 4. Continued.

(Carvalho et al. 2015; Castro et al. 2016) and reducing functional diversity (Castro et al. 2017; Kai Chen, Nanjing Agricultural University, unpublished data). The stream macroinvertebrate indicators (MMI; taxa richness) in our study showed stronger correlations with land use than did fish species richness in most of the same sites (Pompeu et al 2019, this volume). This is likely because macroinvertebrates are more sensitive to the pressures we measured as well as the more integrated manner in which we assessed those pressures.

Because our studies of reservoirs and their hydrologic units were based on systematic and probabilistic survey designs, we were able to infer our results in a statistically rigorous manner to entire reservoir margins and the entire stream length of each unit and region. Such inferences provide much more accurate ecological assessments than those based on ad hoc site selection (Olsen and Peck 2008; Jiménez-Valencia et al. 2014; Silva et al. 2017, 2018). Using similar designs but a different MMI and reference conditions, Jiménez-Valencia et al. (2014) determined that 62% and 25% of the stream length in the Guapiacu-Macacu River basin (Brazil) were in poor and good condition, respectively. The major stressors were poor physical habitat structure, poor watershed condition, and

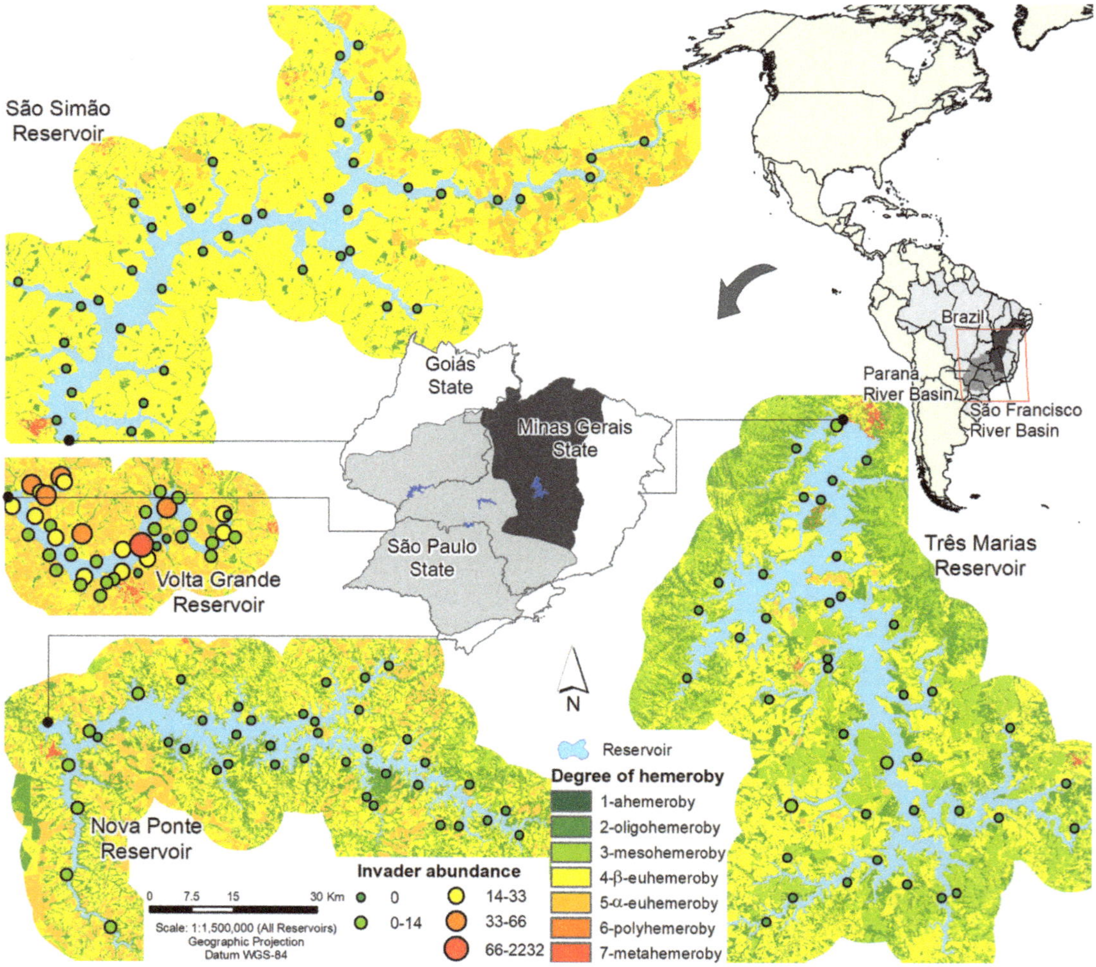

Figure 5. Hemeroby index and invasive mollusk abundance in hydropower reservoir sites.

poor riparian vegetation condition. Using the same MMI we used, Silva et al. (2017) estimated that 27% and 38% of the stream length in the region were in poor and good condition, respectively. Silva et al. (2018) determined that turbidity, percentage of fine sediments, and percentage of agriculture were the key stressors associated with poor MMI scores regionally, but percentage of pasture was also a key stressor in Nova Ponte and Três Marias hydrologic units. Likewise, Paulsen et al. (2008) determined that 42% and 28% of the stream length in the conterminous U.S. were in poor and good condition, respectively. The major stressors were excess nutrients and sediments, both nationally and regionally. In a national assessment, the USEPA (2016) estimated that 31% and 33% of lakes were in most-disturbed and least-disturbed condition, respectively. Again, the major stressors were excess nutrients from agriculture plus lakeshore disturbance. Neither USEPA survey assessed the effect of catchment-scale stressors, but both Jiménez-Valencia et al. (2014) and Silva et al. (2018) found catchment land use to be a significant pressure on their biotic indicators.

Geographic data such as land use and cover, anthropogenic disturbance, precipita-

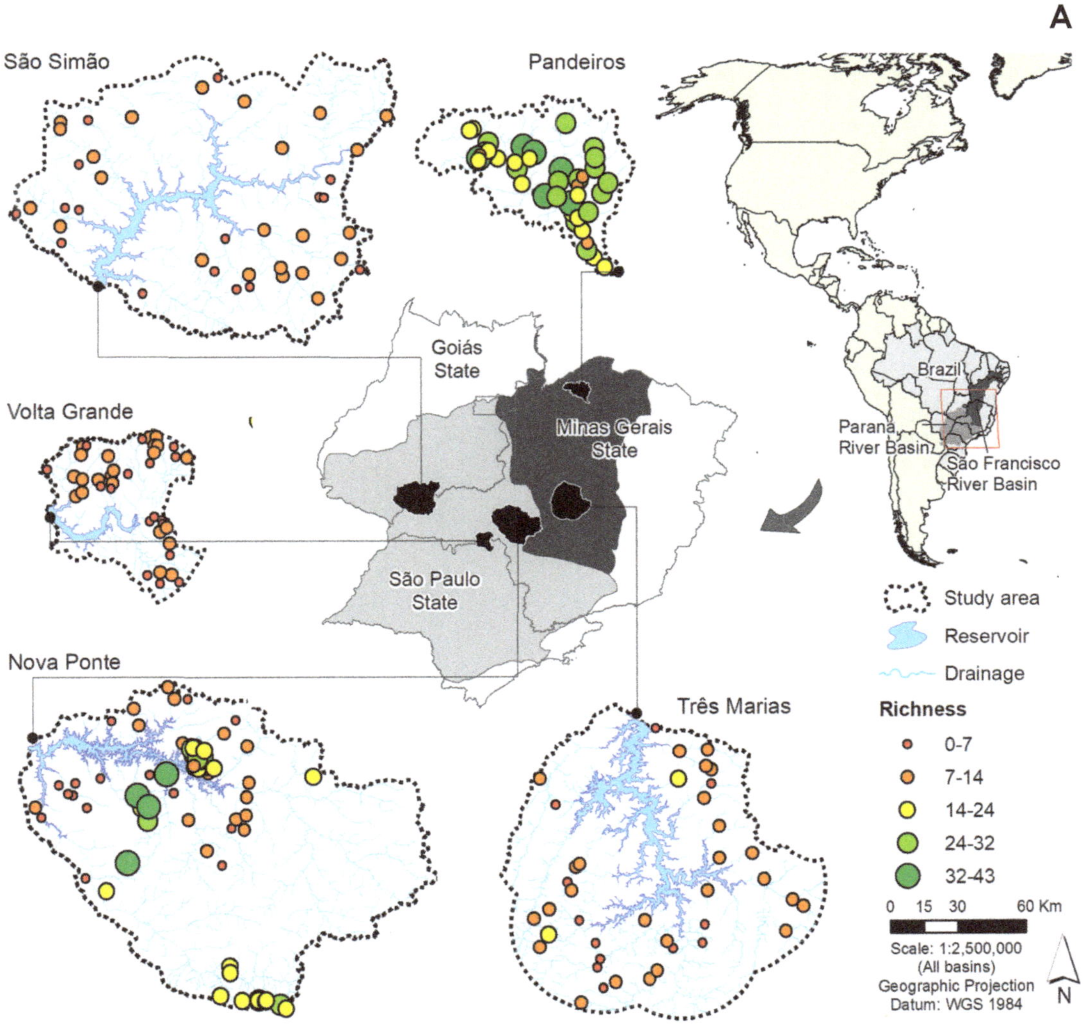

Figure 6. Macroinvertebrate taxa richness in **(A)** wadeable stream sites and **(B)** reservoir sites.

tion, temperature, elevation, and geology are frequently mapped continuously through use of GIS procedures (Figure 2; Wang et al. 2011; Esselman et al. 2013; Macedo et al. 2018). However, most data collected from water body sites have not been accurately extrapolated across large geographic regions in useful ways until relatively recently. Those extrapolations have been facilitated through the use of standardized data collection programs combined with synoptic interpretation of multiple maps (Omernik 1977; Omernik and Powers 1983) or GIS modeling procedures (Hill et al. 2017). Our results offer one more example of linking continuously mapped landscape data with biological responses. Clearer explanations of catchment effects on biota are afforded by relating catchment-scale pressures to site-scale stressors and the latter to biotic indicators through statistical modeling (Marzin et al. 2012; Macedo et al. 2014; Leal et al. 2016; Feld et al. 2018; Leitão et al. 2018).

The environmental fragility results corroborate other research emphasizing the im-

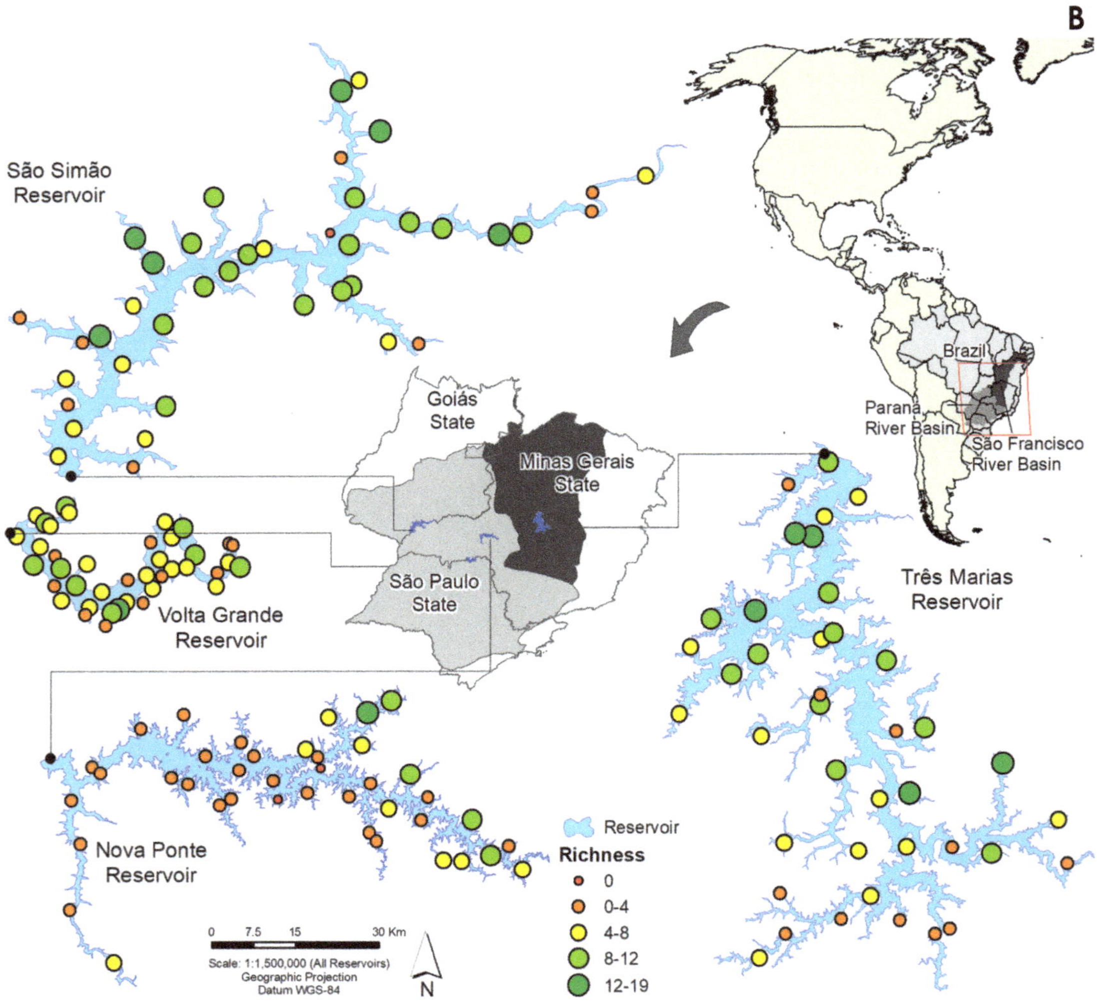

Figure 6. Continued.

portance of maintaining natural vegetation to conserve soil and extend reservoir life in tropical regions and elsewhere (Arias et al. 2011; Liu et al. 2014; Chen et al. 2015). This approach also has value for biodiversity conservation because excess fine sediments are associated with biodiversity losses in streams (Kaufmann et al. 2009; Bryce et al. 2010; Ferreira et al. 2014) and lakes (Lenhardt et al. 2008; Molozzi et al. 2013a). Furthermore, this approach can be used to predict which water bodies are most threatened by future landscape erosion and water body sedimentation (Macedo et al. 2018). Natural cover is the most important component in environmental fragility evaluation, and it reinforces the importance of native vegetation and vegetation cover conservation and restoration. Therefore, the environmental fragility methodology has great potential for application in the management, recovery, and preservation of hydroelectric reservoirs and streams in tropical river basins (Macedo et al. 2018).

Our results showed that invasive nonnative mollusks were a very important factor in the structure and composition of benthic macroinvertebrate assemblages in reservoirs. Most of the impacts on benthic assemblages

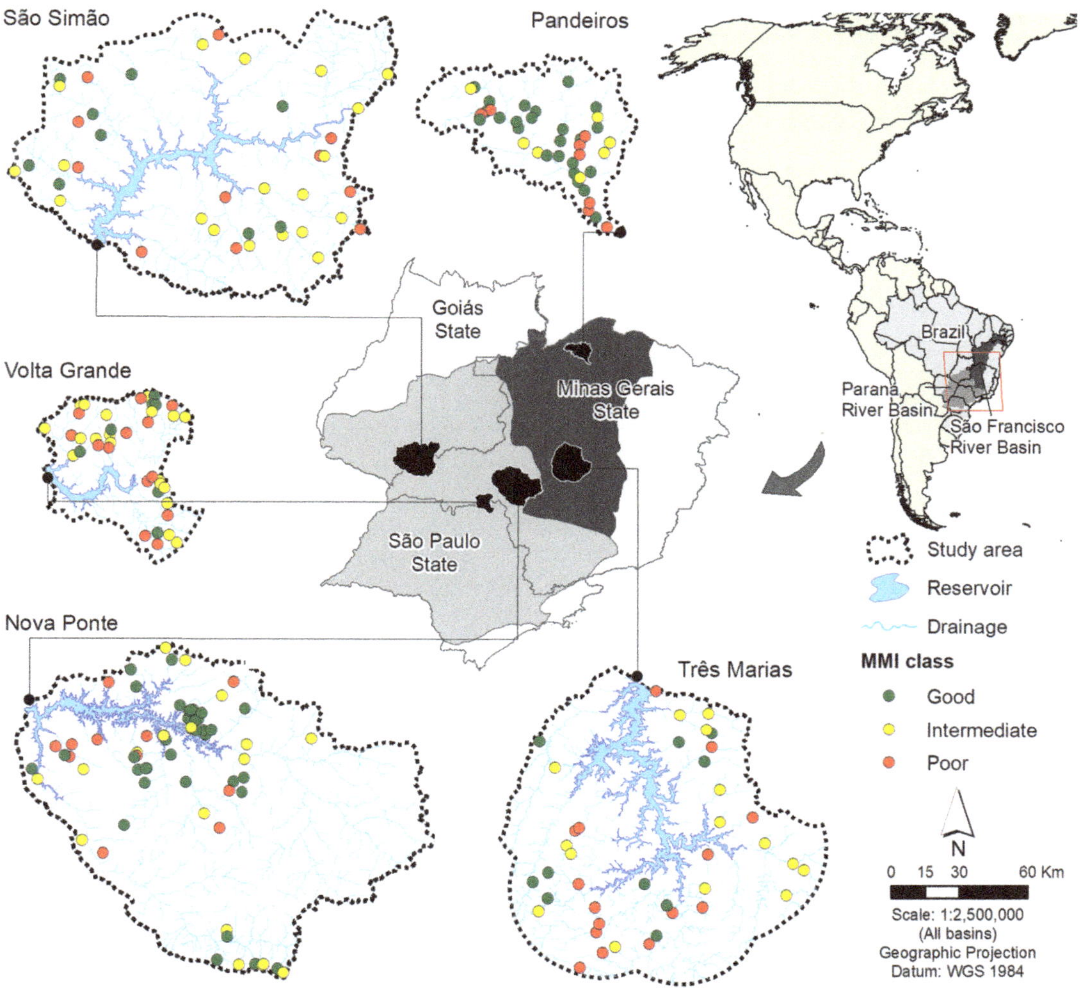

Figure 7. Macroinvertebrate multimetric index (MMI) values for stream sites.

caused by invasive mollusks are a direct result of their function as ecosystem engineers (i.e., species that change the physical habitat structure and trophic balance in the ecosystems where they become established; Jones et al. 1994; Sousa et al. 2009; Zaiko and Daunys 2015; Egan 2018). Hydropower dams are linked to the introduction and establishment of nonnative invasive species in the river systems where they are built (Johnson et al. 2008; White 2014). A recent study (Linares et al. 2018b) showed that even a small run-of-river dam can increase the density of invasive species by many times when compared to free-flowing sites nearby in the same river. There is also a significant correlation between invasive species relative abundance and density with disturbance in reservoirs (Molozzi et al. 2011; de Morais et al. 2017; Anacleto et al. 2018), suggesting that these metrics are very useful indicators of reservoir biological condition.

A common thread in our studies of reservoirs in the Neotropical savanna was that the most abundant invasive species were the gastropod red-rim melania *Melanoides tuberculata* and the bivalves Asian clam *Corbicula fluminea* and golden mussel *Limnoperna fortunei*. All three are com-

Table 1. Reservoir and stream biological condition based on macroinvertebrate taxa richness at 40 sites per reservoir and stream (in parentheses) from Figure 6. Greater than 50% taxa loss indicates poor condition, and the top quintile indicates good condition.

Reservoir (region)	% taxa loss	% top quintile
Nova Ponte	87.5 (73.5)	2.5 (6.5)
São Simão	47.5 (100)	17.5 (0)
Três Marias	60.0 (100)	12.5 (0)
Volta Grande	75.0 (100)	2.5 (0)
Pandeiros	NA (45)	NA (15)
Reservoir mode	Poor	
Streams regionally	86.2 Poor	4.8 Good

Table 2. Correlations between landscape-scale condition indices and biological responses for reservoir and stream sites. Hemeroby = vegetation disturbance index; IDI = integrated disturbance index; EFI = environmental fragility index; MMI = macroinvertebrate multimetric index.

	Reservoir			
	% invaders	*P*-value	Taxa richness	*P*-value
Hemeroby	0.29	<0.001	−0.25	<0.001
IDI	0.29	<0.001	−0.05	0.53
EFI	0.37	<0.001	0.08	0.31
	Stream			
	MMI	*P*-value	Taxa richness	*P*-value
Hemeroby	−0.30	<0.001	−0.60	<0.001
IDI	−0.35	<0.001	−0.61	<0.001
EFI	−0.34	<0.001	−0.52	<0.001

Table 3. Urban habitat index and Citizen Science-Biological Monitoring Working Party (CS-BMWP) tolerance index score correlations with the landscape disturbance indices. LDI = local disturbance index; BDI = buffer disturbance index; IDI = integrated disturbance index.

Variables	Correlation	*P*-value
Habitat × LDI	−0.88	<0.0001
Habitat × BDI	−0.48	0.0007
Habitat × IDI	−0.77	<0.0001
CS-BMWP × LDI	−0.61	<0.0001
CS-BMWP × BDI	−0.57	<0.0001
CS-BMWP × IDI	−0.70	<0.0001

Table 4. Mean (±SE) and generalized linear mode (GLM) results of the disturbance indices for sites violating or not violating national water quality criteria. LDI = local disturbance index; BDI = buffer disturbance index; IDI = integrated disturbance index.

	Water quality		GLM		
Disturbance index	Nonviolated	Violated	*F*	df	*P*-value
LDI	0.78 (±0.16)	1.79 (±0.19)	146.48	47	<0.0001
BDI	202.11 (±25.11)	291.12 (±20.41)	25.35	47	<0.0001
IDI	0.79 (±0.10)	1.32 (±0.07)	19.15	47	<0.0001

mon nonnative invasive species in Brazil. In freshwater ecosystems, mollusks are among the most successful invasive taxa (Ricciardi 2007; Oliveira et al. 2011; Paschoal et al. 2015) because they are often *r*-selected species, exhibiting high fecundity, rapid growth, and broad physiological tolerance to several abiotic factors. Therefore, they are adapted to colonize a wide range of aquatic environments (Nakano et al. 2015). Such characteristics help them to become dominant species, exhibiting high densities and constituting a major fraction of the benthic macroinvertebrate assemblage biomass (Karatayev et al. 2007). These species can affect benthic macroinvertebrate assemblages by shifting reservoir substrates from simple soft bottoms to complex hard surfaces (Molozzi et al. 2013a; Egan 2018).

Opening new energetic pathways is another factor. Previous studies using thermodynamic indicators (Linares et al. 2017, 2018b) showed that invasive nonnative mollusks accessed resources that were not previously available for reservoir benthic assemblages. For instance, South American freshwaters lack powerful native benthic filter feeders that are able to provide a direct link between benthic and water column components of the ecosystem (Karatayev et al. 2007). That role is filled by invasive nonnative bivalves where they become abundant. Those mollusks can shift lake food webs from limnetic to benthic by removing algae and their prey from the water column, thereby resulting in oligotrophication (Evans et al. 2011).

Our urban streams study indicated that ecological monitoring by young citizen scientists was a valuable new educational experience and contributed to their confidence in society and governance. Simplified data collection protocols and analyses offered useful and cost-effective information regarding urban stream ecological condition (França et al. 2019). Presentations to metropolitan authorities encouraged improved governance of environmental quality of urban streams. Therefore, we believe it is wise to invest in such citizen science programs for pedagogical, ecological, and sociological reasons (Stapp 1978). Citizen science information does not substitute for more precise and accurate data regarding the water quality and physical habitat structure of urban streams. Although student monitoring is most useful when supported by higher-level scientists, we found that involving riverine citizens is a critically important investment for engaging entire communities with decision makers to develop and establish more effective water management policies for urban streams.

Future Challenges

The influence of landscape variables on lotic ecosystem function and processes is an important research and management topic. The effects of land use on ecosystem stability and resilience, for example, is information that could be used to inform decision makers and environmental managers regarding the kinds of activities that should be allowed in sensitive areas and across extensive landscapes. Thermodynamic indicators, such as eco-exergy and specific eco-exergy, can be used as proxies for ecosystem stability and resilience (Jørgensen 2007) and have already proven sensitive to local-scale disturbances (Molozzi et al. 2013b; Linares et al. 2017, 2018b). More assessments of ecosystem processes, such as organic matter decomposition, ecosystem metabolism, and primary and secondary production, in response to landscape disturbance, are also needed (de Rezende et al. 2016).

A second important focus is to improve the ability of catchment and landscape-scale indicators to predict site-scale biological responses (Hill et al. 2017) and ecosystem services (Costanza et al. 1997). Studies using landscape-scale indicators have typically predicted markedly less than 50% of biological variation in European rivers (Sály et al. 2011; Marzin et al. 2012) and Brazilian streams (Macedo et al. 2014a; Terra et al. 2015; Leal et al. 2018; Leitão et al. 2018) regardless of the biological indicator, but see Wang et al. (2003) for U.S. streams. This suggests that we need better landscape predictor and biological response indicators, as well as better statistical analyses for linking landscape-scale pressures and the resulting site-scale stressors and biological responses.

A third critical need is robust ecological syntheses (e.g., eco-bioinformatics). Although our studies have methodological consistency, allowing for relatively easy comparisons between different studies, the results are largely independent of one another. It is necessary to unite these disparate data sets into a cohesive whole that can be easily analyzed (Michener 2006). That would enable us to find patterns and a holistic perspective on the effects of land use, dam construction and operation, and other large-scale anthropogenic activities on freshwater ecosystems in the Neotropical savanna. An important first step in this direction is to form an ontological database and a formal model that uses mathematical logic to clarify and define terms/concepts and relationships between them within a domain of interest (Madin et al. 2008). With this, the ecological data collected will be integrated into a single relational database with a semantically and syntactically standardized language. In this way, we will facilitate the comparison of data between different ecological systems, the use of data efficiently for different lines of research, and the availability of the relational database for future studies.

We need national- or continental-scale standard sampling methods also. Our Brazilian studies and those implemented by the Rede Amazônia Sustentável (Gardner et al. 2013) have employed consistent sampling methods and multiple academic disciplines (terrestrial and aquatic ecologists, hydrologists, statisticians, geographers, and economists) at regional scales. However, there are no consistent standard methods for sampling fish and aquatic macroinvertebrates, basic water quality, and physical habitat structure in Brazilian aquatic ecosystems. This greatly hinders data sharing and interpretation—even if multiple Brazilian institutions should desire to do so. Continental-scale programs such as those of USEPA National Aquatic Resource Surveys (Shapiro et al. 2008), the U.S. Geological Survey's National Water Quality Assessment Program (USGS 2018), and the European Union's AQEM (Hering et al. 2004), and FAME (Schmutz et al. 2007) projects offer insights for developing such a Brazilian program. Individuals operating individually and locally cannot hope to fully and rigorously assess the conditions of Brazil's rich aquatic ecosystems (Hughes et al. 2000; Buss et al. 2015).

Finally, it is important for future studies to focus on developing better predictive assessment models that are understandable to the public and decision makers. With global threats like climate change, population and economic growth, and a possible sixth mass extinction looming on the horizon in the Anthropocene (Limburg et al. 2011; Joly 2018; Ripple et al. 2018), ecology, in general, is in dire need of improved predictive tools. Landscape models are especially useful in this regard because of the way humans experience nature and many environmental management actions are made at this scale (Metzger 2001). Structurally dynamic models, ones that account for the changes in the model parameters resulting from the adaptation or the shift in species composition resulting from current changes in the forcing functions (Jørgensen 2006), can be used for predictive assessments. These models can be useful tools for providing reliable scientific bases for environmental management in a rapidly changing global context.

Management Needs and Challenges and Policy Implications

In recent decades, the Cerrado biome has suffered intensive deforestation, 2.5 times higher than that of the Amazon biome (Strassburg et al. 2017), mainly resulting from agricultural expansion. Furthermore, the Cerrado contains the headwaters of major South American rivers (e.g., the São Francisco, Tocantins, and Paraná) and the Guarani Aquifer. Almost 25% of the Brazilian population lives in the Cerrado (IBGE 2010), and because of the land cover changes, today the Cerrado is a major focus of conservation efforts.

Most Brazilian protected areas were created to protect terrestrial biodiversity, neglecting aquatic biodiversity, with dramatic consequences such as the decline of large migratory fishes (Azevedo-Santos et al. 2019) and loss of sensitive macroinvertebrate taxa (Firmiano et al. 2017; Oliveira-Junior et al. 2017; Brito, unpublished data). The current Brazilian Forest Code (Law Number 12.651/2012) mandates that landowners must protect a minimum size of the river riparian zones varying with river width and property size. Streams less than 10 m wide must have at least a 30-m-wide permanently protected buffer on both sides (Metzger 2001). However, those mandated buffers have failed to protect sensitive fish and insect taxa (Firmiano et al. 2017; Oliveira-Junior et al. 2017; Leal et al. 2018; Brito, unpublished data). Clearly, current land-use regulations must be amended if Brazil wishes to conserve its aquatic biodiversity.

Sewage treatment is rare (<44% of the urban population) in Minas Gerais (IGAM 2015) and Brazil (ANA 2017a), meaning that most urban streams and rivers are highly enriched by organic wastes, phosphorus, and nitrogen, thereby eliminating sensitive macroinvertebrate (Silva et al. 2017; França et al. 2019) and fish (Pompeu et al. 2005; Pinto et al. 2006; Terra et al. 2013) species. If Brazil is interested in improving the biological condition and ecosystem services of its waters, it would be wise to follow other nations and install modern waste collection and treatment systems for its cities and industries (Hughes et al. 2019; Yoder et al. 2019).

River fragmentation by dams and reservoirs substantially alters ichthyofaunas and aquatic insects globally (Vörösmarty et al. 2010; Sánchez-Bayo and Wyckhuys 2019) and certainly in Brazil (Agostinho and Júlio 1996; Agostinho et al. 2007; Pompeu et al. 2012). Also, our results indicate that macroinvertebrate richness and composition are fundamentally altered by dams and reservoirs, as others have reported (Molozzi et al. 2013; Linares et al. 2018b). Although hydropower is the major source of electricity in Brazil and reservoirs are the major sources of urban drinking water, the nation would be wise to take advantage of its abundant sunshine and coastal winds for alternative power sources.

Brazil is a beautiful megadiverse country of continental dimensions, where the agribusiness and mining lobbies hold great political and economic power—despite tailings dam failures causing hundreds of human deaths (Reuters 2019). Habitat loss and the fragmentation of freshwater ecosystems are consequences of poor landscape and river management, perhaps even more so than the impacts of global climate change (Tedesco et al. 2013; Jacobson et al. 2019, this volume). Social activism is imperative for countering environmental degradation to restore and maintain the ecological condition of water bodies and their ecosystem services. We believe that citizen science is more than simply translating scientific papers into language understandable by nonscientists and to delivering it to decision makers in management agencies. It must also become a collaborative effort actively involving academia, policymakers, management agencies, and citizens (Hughes et al. 2019). Human and ecological well-being and the future sustainability of our society are dependent on this alliance.

Our results have been used to improve water resource management by decision makers and environmental technicians of the Minas Gerais energy company, state and municipal environmental agencies, watershed committees, city councilors and mayors, and school teachers and student scientists in southeastern Brazil. Our information is also being used in discussions about future environmental investments in the national integrated electrical energy system. In addition, we are working to have our methodologies incorporated in the National Water Agency (ANA) aquatic ecosystem assessment procedures.

Acknowledgments

We thank Bianca de Freitas Terra for inviting us to prepare this chapter. The team of undergraduate and graduate students of the Laboratório de Ecologia de Bentos/ICB-UFMG supported the field activities. MC was awarded National Council for Scientific & Technological Development (CNPq) research productivity grants 303380/2015-2 and by Fundação de Amparo à Pesquisa do Estado de Minas Gerais research grant (PPM 00104-18). DRM received a research grant from Pró-Reitoria de Pesquisa (PRPq) -UFMG (ADRC-05/2016) and a research project grant from CNPq (402907/2016–7). RMH received a Fulbright Brasil Grant. This research has been continually funded by Programa Peixe Vivo/Companhia Energética de Minas Gerais, Programa de Pesquisa e Desenvolvimento Tecnológico do Setor de Energia Elétrica-Companhia Energética de Minas Gerais (P&D Aneel-Cemig GT-487, GT-550, GT-599, and GT-611) and by Fundação de Amparo à Pesquisa de Minas Gerais (APQ-01961-15). This study was financed in part by the Coordenação de Aperfeiçoamento de Pessoal de Ni´vel Superior - Brasil (CAPES) - Finance Code 001. We also thank Bianca de Freitas Terra and three anonymous referees who contributed critical inputs to this chapter.

References

Agostinho, A. A., L. C. Gomes, and F. M. Pelicice. 2007. Ecologia e manejo de recursos pesqueiros em reservatórios do Brasil. [Ecology and management of fishery resources in Brazilian reservoirs.] Eduem, Maringá, Paraná, Brazil.

Agostinho, A. A., and H. F. Júlio, Jr. 1996. Peixes de outras águas: ameaça ecológica. [Fish from other waters: ecological threat.] Ciência Hoje 21(124):36–44.

Agra, J. U. M., R. L. B. Santos, D. R. Macedo, R. M. Hughes and M. Callisto. 2019. Ecoregions and stream types aid understanding ecological variability of Neotropical reference streams. Marine and Freshwater Research 70:594–602.

ANA (Agencia Nacional de Águas). 2017b. Atlas esgotos: despoluição de bacias hidrográficas. [Atlas of sewers: depollution of river basins.] ANA, Brasília, Brazil.

ANA (Agencia Nacional de Águas). 2017a. Hidroweb: sistema de informações hidrológicas. [Hydroweb: hydrological information system. ANA, Brasília, Brazil.

Anacleto, M. J. P., R. Ligeiro, J. E. L. Barbosa, J. Molozzi, and M. Callisto. 2018. Effects of an atypical drought on the benthic macroinvertebrate community in a tropical reservoir. Biota Neotropica [online serial] 18(2):e20170352.

Anderson, D., H. Moggridge, P. Warren, and J. Shucksmith. 2015. The impacts of "run-of-river" hydropower on the physical and ecological condition of rivers. Water and Environment Journal 29:268–276.

APHA (American Public Health Association). 2005. Standard methods for the examination of water and wastewater, 21st edition. APHA, Washington, D.C.

Arias, M. E., T. A. Cochrane, K. S. Lawrence, T. J. Killeen, and T. A. Farrell. 2011. Paying the forest for electricity: a modelling framework to market forest conservation as payment for ecosystem services benefiting hydropower generation. Environmental Conservation 38:473–484.

Azevedo-Santos, V. M., R. G. Frederico, C. K. Fagundes, P. S. Pompeu, F. M. Pelicice, A. A. Padial, M. G. Nogueira, P. M. Fearnside, L. B. Lima, V. S. Daga, F. J. M. Oliveira, J. R. S. Vitule, M. Callisto, A. A. Agostinho, F. A. Esteves, D. P. Lima-Junior, A. L. B. Magalhães, J. Sabino, R. P. Mormul, D. Grasel, J. Zuanon, F. Villela, and R. Henry. 2019. Protected areas: a focus on Brazilian freshwater biodiversity. Diversity and Distributions 25:442–448.

Becker, B., B. O. S. Galhardo, D. R. Macedo, R. M. Hughes, M. Callisto, and G. B. Santos. 2016. Influence of limnological zones on the spatial distribution of fish assemblages in three Brazilian reservoirs. Journal of Limnology 75:156–168.

Braga, F. M., and L. M. Gomieiro. 1997. Análise da pesca experimental realizada no reservatório de Volta Grande, rio Grande (MG-SP). [Analysis of experimental fishing at Volta Grande reservoir, Rio Grande (MG-SP).] Boletim do Instituto de Pesca 24:131–138.

Brazil. 2004. Carta geológica do Brasil ao milionésimo. [Geological map of Brazil to the millionith.] Ministério de Minas e Energia/ CPRM/Serviço Geológico do Brasil, Brasília, Brazil.

Brasil. 2005. Classificação de águas, doces, salobras e salinas do Território Nacional. [Classification of freshwaters and salt waters of the Brazilian National Territory.] Resolução CONAMA N°357, Brasília, Brazil.

Bryce, S. A., G. A. Lomnicky, and P. R. Kaufmann. 2010. Protecting sediment-sensitive aquatic species in mountain streams through the application of biologically based streambed sediment criteria. Journal of the North American Benthological Society 29:657–672.

Buss, D. F., D. Carlisle, T. S. Chon, J. Culp, J. S. Harding, H. E. Keizer-Vlek, W. A. Robinson, S. Strachan, C. Thirion, and R. M. Hughes. 2015. Stream biomonitoring using macroinvertebrates around the globe: a comparison of large-scale programs. Environmental Monitoring and Assessment [online serial] 187:4132.

Callisto, M., W. R. Ferreira, P. Moreno, M. Goulart, and M. Petrucio. 2002. Aplicação de um protocolo de avaliação rápida da diversidade de habitats de atividades de ensino e pesquisa (MG-RJ). [Application of a rapid assessment protocol of habitat diversity in teaching and research activities (MG-RJ).] Acta Limnologica Brasiliensia 14:91–98.

Callisto, M., R. M. Hughes, J. M. Lopes, and M. A. Castro. 2014. Ecological conditions in hydropower basins. Série Peixe Vivo, Belo Horizonte, Minas Gerais, Brazil.

Callisto, M., A. S. Melo, D. F. Baptista, J. F. Gonçalves Junior, M. A. S. Graça, and F. G. Augusto. 2012. Future ecological studies of Brazilian headwater streams under global-changes. Acta Limnologica Brasiliensia 24:293–302.

Carvalho, D. R., D. Castro, M. Callisto, M. Z. Moreira, and P. S. Pompeu. 2015. Isotopic variation in five species of stream fishes under the influence of different land uses. Journal of Fish Biology 87:559–578.

Castro, D. M. P. de, D. R. de Carvalho, P. D. S. Pompeu, M. Z. Moreira, G. B. Nardoto, and M. Callisto. 2016. Land use influences niche size and the assimilation of resources by benthic macroinvertebrates in tropical headwater streams. PLOS (Public Library of Science) ONE [online serial] 11(3):e0150527.

Castro, D. M. P. de, S. Dolédec, and M. Callisto. 2017. Landscape variables influence taxonomic and trait composition of insect assemblages in Neotropical savanna streams. Freshwater Biology 62:1472–1486.

Chen, N., M. Chen, J. Li, N. He, M. Deng, J. Iqbal Tanoli, and M. Cai. 2015. Effects of human activity on erosion, sedimentation and debris flow activity: a case study of the Qionghai Lake watershed, southeastern Tibetan Plateau, China. The Holocene 25:973–988.

Costanza, R., R. D'Arge, R. de Groot, S. Farber, M. Grasso, B. Hannon, K. Limburg, S. Naeem, R. V. O'Neill, J. Paruelo, R. G. Raskin, P. Sutton, and M. van den Belt. 1997. The value of the world's ecosystem services and natural capital. Nature (London) 387:253–260.

de Morais, L., B. de Oliveira Sanches, G. B. Santos, P. R. Kaufmann, R. M. Hughes, J. Molozzi, and M. Callisto. 2017. Assessment of disturbance at three spatial scales in two large tropical reservoirs. Journal of Limnology 76:240–250.

de Rezende, R., S., M. A. S. Graça, A. M. dos Santos, A. O. Medeiros, P. F. Santos, Y. R. Nunes, and J. F. Gonçalves, Jr. 2016. Organic matter dynamics in a tropical gallery forest in a grassland landscape. Biotropica 48:301–310.

Dudgeon, D. 2010. Prospects for sustaining freshwater biodiversity in the 21st century: linking ecosystem structure and function. Current Opinion in Environmental Sustainability 2:422–430.

Egan, D. 2018. The death and life of the great lakes. Norton, New York.

EPE (Empresa de Pesquisa Energética). 2016. Balanço energético nacional. [Brazil's national energy balance.] Ministério de Minas e Energia, Brasília, Brazil.

Esselman, P. C., D. M. Infante, L. Wang, A. R. Cooper, D. Wieferich, Y. P. Tsang, D. J. Thornbrugh, and W. W. Taylor. 2013. Regional fish community indicators of landscape disturbance to catchments of the conterminous United States. Ecological Indicators 26:163–173.

Evans, M. A., G. Fahnenstiel, and D. Scavia. 2011. Incidental oligotrophication of North American Great Lakes. Environmental Science and Technology 45:3297–3303.

Feld, C. K., M. R. Fernandes, M. T. Ferreira, D. Hering, S. J. Ormerod, M. Venohr, and C. Gutierrez-Canovas. 2018. Evaluating riparian solutions to multiple stressor problems in river ecosystems–a conceptual study. Water Research 139:381–394.

Fernandes, G. W., N. P. U. Barbosa, B. Alberton, A. Barbieri, R. Dirzo, F. Goulart, T. J. Guerra, L. P. C. Morellato, and R. R. C. Solar. 2018. The deadly route to collapse and the uncertain fate of Brazilian rupestrian grasslands. Biodiversity and Conservation 27:2587–2603.

Ferreira, W. R., R. Ligeiro, D. R. Macedo, R. M. Hughes, P. R. Kaufmann, L. G. Oliveira, and M. Callisto. 2014. Importance of environmental factors for the richness and distribution of benthic macroinvertebrates in tropical headwater streams. Freshwater Science 33:860–871.

Firmiano, K. R., R. Ligeiro, D. R. Macedo, L. Juen, R. M. Hughes, and M. Callisto. 2017. Mayfly bioindicator thresholds for several anthropogenic disturbances in Neotropical savanna streams. Ecological Indicators 74:276–284.

França, J. S., R. Solar, R. M. Hughes, and M. Callisto. 2019. Student monitoring of the ecological quality of neotropical urban streams. Ambio 48:867–878.

Gardner, T. A., J. Barlow, R. Chazdon, R. M. Ewers, C. A. Harvey, C. A. Peres, and N. S. Sodhi. 2009. Prospects for tropical forest biodiversity in a human-modified world. Ecology Letters 12:561–582.

Gardner, T. & 86 coauthors. 2013. A social and ecological assessment of tropical land uses at multiple scales: the Sustainable Amazon Network. Philosophical Transactions of the Royal Society of Britain B 368(1609):20120166.

Hamada, N., J. L. Nessimian, and R. B. Querino. 2014. Insetos aquáticos na Amazônia brasileira: taxonomia, biologia e ecologia. [Aquatic insects in the Brazilian Amazon: taxonomy, biology, and ecology.] Editora do INPA, Manaus, Amazonas, Brazil.

Hamada, N., J. Thorp, and D. C. Rogers, editors. 2018. Thorp and Covich's freshwater invertebrates, volume 3: keys to Neotropical Hexapoda. Academic Press, London.

Hering, D., O. Moog, L. Sandin, and P. F. M Verdonschot. 2004. Overview and application of the AQEM assessment system. Hydrobiologia 516:1–20.

Hijmans, R. J., S. E. Cameron, J. L. Parra, P. G. Jones, and A. Jarvis. 2005. Very high resolution interpolated climate surfaces for global land areas. International Journal of Climatology 25:1965–1978.

Hill, R. A., E. W. Fox, S. G. Leibowitz, A. R. Olsen, D. J. Thornbrugh, and M. H. Weber. 2017. Predictive mapping of the biotic condition of conterminous U.S. rivers and streams. Ecological Applications 27:2397–2415.

Hughes, R. M., B. L. Bangs, S. V. Gregory, P. D. Scheerer, R. C. Wildman, and J. S. Ziller. 2019. Recovery of Willamette River fish assemblages: successes and remaining threats. Pages 157–184 *in* C. C. Krueger, W. W. Taylor, and S. Youn editors. From catastrophe to recovery: stories of fish management success. American Fisheries Society. Bethesda, Maryland.

Hughes, R. M., S. G. Paulsen, and J. L. Stoddard. 2000. EMAP-Surface Waters: a national, multiassemblage, probability survey of ecological integrity. Hydrobiologia 422–423:429–443.

Hunke, P., E. N. Mueller, B. Schröder, and P. Zeilhofer. 2015. The Brazilian Cerrado: assessment of water and soil degradation in catchments under intensive agricultural use. Ecohydrology 8:1154–1180.

IBGE (Instituto Brasileiro de Geografia e Estatística). 2010. Censo demográfico 2010. [Brazilian demographic census 2010.] Instituto Brasileiro de Geografia e Estatística, Rio de Janeiro, Brazil.

IGAM (Instituto Mineiro de Gestão das Aguas). 2015. Qualidade das águas superficiais de Minas Gerais em 2015. [Minas Gerais' surface water quality by 2015.] Instituto Mineiro das Águas, Belo Horizonte, Minas Gerais, Brazil.

INMET (Instituto Nacional de Metereologia). 2017. Rede hidrometereológica nacional. [Brazilian national hydrometereological network.] INMET, Brasília, Brazil.

Jacobson, P. C., G. J. A. Hansen, L. G. Olmanson, K. E. Wehrly, C. L. Hein, and L. B. Johnson. 2019. Loss of coldwater fish habitat in glaciated lakes of the Midwestern United States after a century of land use and climate change. Pages 141–157 *in* R. M. Hughes, D. M. Infante, L. Wang, K. Chen, and B. F. Terra, editors. Advances in understanding landscape influences on freshwater habitats and biological assemblages. American Fisheries Society, Symposium 90, Bethesda, Maryland

Jiménez-Valencia, J., P. R. Kaufmann, A. Sattamini, R. Mugnai, and D. F. Baptista. 2014. Assessing the ecological condition of streams in a southeastern Brazilian basin using a probabilistic monitoring design. Environmental Monitoring and Assessment 186:4685–4695.

Johnson, P. T., J. D. Olden, and M. J. Vander Zanden. 2008. Dam invaders: impoundments facilitate biological invasions into freshwaters. Frontiers in Ecology and the Environment 6:357–363.

Joly, C. A. 2018. Biodiversity at risk in the Americas. Biota Neotropica [online serial] 18(2):e20180301.

Jones, C. G., J. H. Lawton, and M. Shachak. 1994. Organisms as ecosystem engineers. Oikos 69:373–386.

Jørgensen, S. E. 2006. Application of holistic thermodynamic indicators. Ecological Indicators 6:24–29.

Jørgensen, S. E. 2007. An integrated ecosystem theory. Annals of the European Academy of Sciences 2006–2007:19–33.

Junqueira, V. M., and S. C. M. Campos. 1998. Adaptation of the "BMWP" method for water quality evaluation to Rio das Velhas watershed (Minas Gerais, Brazil). Acta Limnologica Brasiliensia 10:125–135.

Karatayev, A. Y., D. K. Padilla, D. Minchin, D. Boltovskoy, and L. E. Burlakova. 2007. Changes in global economies and trade: the potential spread of exotic freshwater bivalves. Biological Invasions 9:161–180.

Kaufmann, P. R., D. P. Larsen, and J. M. Faustini. 2009. Bed stability and sedimentation associated with human disturbances in Pacific Northwest streams. JAWRA Journal of the American Water Resources Association 45:434–459.

Kaufmann, P. R., P. Levine, E. E. G. Robison, C. Seeliger and D. V. Peck. 1999. Quantifying physical habitat in wadeable streams. U.S. Environmental Protection Agency, EPA/620/R-99/003, Washington, D.C.

Kaufmann, P. R., D. V. Peck, S. G. Paulsen, C. W. Seeliger, R. M. Hughes, T. R. Whittier, and N. C. Kamman. 2014. Lakeshore and littoral physical habitat structure in a national lakes assessment. Lake and Reservoir Management 30:192–215.

Leal, C. G., J. Barlow, T. Gardner, R. M. Hughes, R. P. Leitão, R. MacNally, P. Kaufmann, S. F. B. Ferraz, J. Zuanon, F. R. de Paula, J. Ferreira, J. R. Thomson, G. D. Lennox, E. P. Dary, C. P. Röpke, and P. S. Pompeu. 2018. Is environmental legislation conserving tropical stream faunas? A large-scale assessment of local, riparian and catchment-scale influences on Amazonian fish. Journal of Applied Ecology 55:1312–1326.

Leal, C. G., P. S. Pompeu, T. A. Gardner, R. P. Leitão, R.M. Hughes, P. R. Kaufmann, J. Zuanon, F. R. de Paula, S. F. B. Ferraz, J. R. Thomson, R. MacNally, J. Ferreira, and J. Barlow. 2016. Multi-scale assessment of human-induced changes on Amazonian instream habitat. Landscape Ecology 31:1725–1745.

Leitão, R. P., J. Zuanon, D. Mouillot, C. G. Leal, R. M. Hughes, P. R. Kaufmann, S. Villéger, P. S. Pompeu, D. Kasper, F. R. de Paula, S. F. B. Ferraz, and T. Gardner. 2018. Disentangling the pathways of land use impacts on the functional structure of fish assemblages in Amazon streams. Ecography 41:219–232.

Lenhardt, M., G. Markovic, and Z. Gacic. 2008. Decline in the index of biotic integrity of the fish assemblage as a response to reservoir aging. Water Resources Management 23:1713–1723.

Ligeiro, R., R. M. Hughes, P. R. Kaufmann, D. R. Macedo, K. R. Firmiano, W. R. Ferreira, D. Oliveira, A. S. Melo, and M. Callisto. 2013. Defining quantitative stream disturbance gradients and the additive role of habitat variation to explain macroinvertebrate taxa richness. Ecological Indicators 25:45–57.

Limburg, K. E., R. M. Hughes, D. C. Jackson, and B. Czech. 2011. Population increase, economic growth, and fish conservation: collision course or savvy stewardship? Fisheries 36:27–35.

Linares, M. S., M. Callisto, and J. C. Marques. 2017. Invasive bivalves increase benthic communities complexity in Neotropical reservoirs. Ecological Indicators 75:279–285.

Linares, M. S., M. Callisto, and J. C. Marques. 2018a. Compliance of secondary production and eco-exergy as indicators of benthic macroinvertebrates assemblages' response to canopy cover conditions in Neotropical headwater streams. Science of the Total Environment 613- 614:1543–1550.

Linares, M. S., M. Callisto, and J. C. Marques. 2018b. Thermodynamic based indicators illustrate how a run-of-river impoundment in Neotropical savanna attracts invasive species and alters the benthic macroinvertebrate assemblages' complexity. Ecological Indicators 88:181–189.

Liu, Y. J., T. W. Wang, C. F. Cai, Z. X. Li, and D. B. Cheng. 2014. Effects of vegetation on runoff generation, sediment yield and soil shear strength on road-side slopes under a simulation rainfall test in the Three Gorges Reservoir Area, China. Science of the Total Environment 485- 486:93–102.

Macedo, D. R., R. M. Hughes, W. R. Ferreira, K. R. Firmiano, D. R. O. Silva, R. Ligeiro, P. R. Kaufmann, and M. Callisto. 2016. Development of a benthic macroinvertebrate multimetric index (MMI) for Neotropical savanna headwater streams. Ecological Indicators 64:132–141.

Macedo, D. R., R. M. Hughes, P. R. Kaufmann, and M. Callisto. 2018. Development and validation of an environmental fragility index (EFI) for the Neotropical savannah biome. Science of the Total Environment 635:1267–1279.

Macedo, D. R., R. M. Hughes, R. Ligeiro, W. R. Ferreira, M. A. Castro, N. T. Junqueira, D. R. Oliveira, K. R. Firmiano, P. R. Kaufmann, P. S. Pompeu, and M. Callisto. 2014a. The relative influence of catchment and site variables on fish and macroinvertebrate richness in Cerrado biome streams. Landscape Ecology 29:1001–1016.

Macedo, D. R., P. S. Pompeu, L. Morais, M. A. Castro, C. B. M. Alves, J. S. França, B. O. Sanches, J. U. M. Agra, and M. Callisto. 2014b. Sampling site selection, land use and cover, field reconnaissance, and sampling. Pages 61–83 *in* M. Callisto, R. M. Hughes, J. M. Lopes, and M. A. Castro, editors. Ecological conditions in hydropower basins. Serie Peixe Vivo 3. Companhia Energética de Minas Gerais, Belo Horizonte, Minas Gerais, Brazil.

Mapbiomas. 2018. Collection 2.3 of Brazil's annual series of land cover and land use maps. Available: http://mapbiomas.org/. (July 2018).

Madin, J. S., S. Bowers, M. P. Schildhauer, M. B. Jones. 2008. Advancing ecological research with ontologies. Trends in Ecology and Evolution 23:159–168.

Martins, I., R. Ligeiro, R. M. Hughes, D. R. Macedo, and M. Callisto. 2018. Regionalisation is key to establishing reference conditions for Neotropical savanna streams. Marine and Freshwater Research 69:82–94.

Marzin, A., P. F. M. Verdonschot, and D. Pont. 2012. The relative influence of catchment, riparian corridor, and reach-scale anthropogenic pressures on fish and macroinvertebrate assemblages in French rivers. Hydrobiologia 704:375–388.

Metzger, J. P. 2001. O que é ecología de paisagens? [What is landscape ecology?] Biota Neotropica 1:1–9.

Michener, W. K. 2006. Meta-information concepts for ecological data management. Ecological Informatics 1:3–7.

Molozzi, J., J. S. França, T.L.A. Araujo, T.H. Viana, R.M. Hughes, and M. Callisto. 2011. Diversidade de habitats físicos e sua relação com macroinvertebrados bentônicos em reservatórios urbanos. [Diversity of physical habitats and their relationship with benthic macroinvertebrates in urban reservoirs.] Iheringia Série Zoologia 101(3):191–199.

Molozzi, J., L. U. Hepp, and M. Callisto. 2013a. The additive partitioning of macroinvertebrate diversity in tropical reservoirs. Marine and Freshwater Research 64:609–617.

Molozzi, J., F. Salas, M. Callisto, and J. C. Marques. 2013b. Thermodynamic oriented ecological indicators: application of eco-exergy and specific eco-exergy in capturing environmental changes between disturbed and non-disturbed tropical reservoirs. Ecological Indicators 24:543–551.

Mugnai, R., J. L. Nessimian, and D.F. Baptista. 2010. Manual de identificação de macroinvertebrados aquáticos do estado do Rio de Janeiro. [Manual of identification of aquatic macroinvertebrates of the state of Rio de Janeiro.] Technical Books Editora, Rio de Janeiro, Brazil.

Nakano, D., T. Baba, N. Endo, S. Nagayama, A. Fujinaga, A. Uchida, A. Shiragane, M. Urabe, and T. Kobayashi. 2015. Invasion, dispersion, population persistence and ecological impacts of a freshwater mussel (*Limnoperna fortunei*) in the Honshu Island of Japan. Biological Invasions 17:743–759.

Oliveira-Junior, J. M. B., P. de Marco-Junior, K. Dias-Silva, R. P. Leitão, C. G. Leal, P. S. Pompeu, T. A. Gardner, R. M. Hughes, and L. Juen. 2017. Effects of human disturbance and riparian conditions on Odonata (Insecta) assemblages in eastern Amazon basin streams. Limnologica 66:31–39.

Oliveira, M. D., D. F. Calheiros, C. M. Jacobi, and S. K. Hamilton. 2011. Abiotic factors controlling the establishment and abundance of the invasive golden mussel *Limnoperna fortunei*. Biological Invasions 13:717–729.

Olsen, A. R., and D. V. Peck. 2008. Survey design and extent estimates for the Wadeable Streams Assessment. Journal of the North American Benthological Society 27:822–836.

Omernik, J. M. 1977. Nonpoint source-stream nutrient level relationships: a nationwide study. U.S. Environmental Protection Agency, EPA-600/3-77-105, Washington, D.C.

Omernik, J. M., and C. F. Powers. 1983. Map supplement: total alkalinity of surface waters—a national map. Annals of the Association of American Geographers 73:133–136.

Paschoal, L., D. Andrade, and G. Darrigran. 2015. How the fluctuations of water levels affect populations of invasive bivalve Corbicula fluminea (Müller, 1774) in a Neotropical reservoir? Brazilian Journal of Biology 75:135–143.

Paulsen, S. G., A. Mayio, D. V. Peck, J. L. Stoddard, E. Tarquinio, S. M. Holdsworth, J. Van Sickle, L. L. Yuan, C. P. Hawkins, A. T. Herlihy, P. R. Kaufmann, M. T. Barbour, D. P. Larsen, and A. R. Olsen. 2008. Condition of stream ecosystems in the US: an overview of the first national assessment. Journal of the North American Benthological Society 27:812–821.

Peck, D., A. Herlihy, B. Hill, R. Hughes, P. Kaufmann, D. Klemm, J. Lazorchak, F. McCormick, S. Peterson, P. Ringold, T. Magee, and M. Cappaert. 2006. Environmental Monitoring and Assessment Program—Surface Waters Western Pilot Study: field operations manual for wadeable streams. U.S. Environmental Protection Agency, EPA/620/R-06/003, Washington, D.C.

Pinto, B. C. T., F. G. Araujo, and R. M. Hughes. 2006. Effects of landscape and riparian condition on a fish index of biotic integrity in a large southeastern Brazil river. Hydrobiologia 556:69–83.

Pinto-Coelho, R. M. 2013. Ordenamento da aquicultura no reservatório de São Simão. [Management of aquaculture in São Simão reservoir.] EPAMIG, Belo Horizonte, Minas Gerais, Brazil.

Pompeu, P. S., A. A. Agostinho, and F. M. Pelicice. 2012. Existing and future challenges: the concept of successful fish passage in South America. River Research and Applications 28:504–512.

Pompeu, P. S., C. B. M. Alves, and M. Callisto. 2005. The effects of urbanization on biodiversity and water quality in the Rio das Velhas basin, Brazil. Pages 11–22 *in* L. R. Brown, R. H. Gray, R. M. Hughes, and M. R. Meador, editors. American Fisheries Society, Symposium 47, Bethesda, Maryland.

Pompeu, P. S., C. G. Leal, M. A. Castro, N. Junqueira, D. Carvalho, and R. M. Hughes. 2019. Effects of catchment land use on stream fish richness and assemblage structure in the Brazilian savanna. Pages 303–320 *in* R. M. Hughes, D. M. Infante, L. Wang, K. Chen, and B. F. Terra, editors. Advances in understanding landscape influences on freshwater habitats and biological assemblages. American Fisheries Society, Symposium 90, Bethesda, Maryland.

Ratter, J. A., J. F. Ribeiro, and S. Bridgewater. 1997. The Brazilian Cerrado vegetation and threats to its biodiversity. Annals of Botany 80:223–230.

Reid, A. J., A. K. Carlson, I. F. Creed, E. J. Eliason, P. A. Gell, P. T. J. Johnson, K. A. Kidd, T. J. MacCormack, J. D. Olden, S. J. Omerod, J. P. Smol, W. W. Taylor, K. Tockner, J. C. Vermaire, D. Dudgeon, and S. J. Cooke. 2019. Emerging threats and persistent conservation challenges for freshwater biodiversity. Biological Reviews 94:849–873.

Reuters. 2019. Factbox: Vale tailings dam collapse adds to long list of mining disasters. Available: https://www.reuters.com/article/us-vale-sa-disaster-accidents-factbox/factbox-vale-tailings-dam-collapse-adds-to-long-list-of-mining-disasters-idUSKCN1PO0VI. (February 2019).

Ricciardi, A. 2007. Are modern biological invasions an unprecedented form of global change? Conservation Biology 21:329–336.

Ripple, W. J., C. Wolf, T. M. Newsome, M. Galetti, M. Alamgir, E. Crist, M. I. Mahmoud, and W. F. Laurance. 2018. World scientists' warning to humanity: a second notice. BioScience 67:1026–1028.

Sály, P., P. Takács, I. Kiss, P. Bíró, and T. Erös. 2011. The relative influence of spatial context and catchment- and site-scale environmental factors on stream fish assemblages in a human-modified landscape. Ecology of Freshwater Fish 20:251–262.

Sanches, B. O., R. M. Hughes, D. R. Macedo, M. Callisto, G. B. Santos, B. O. Sanches, R. M. Hughes, D. R. Macedo, M. Callisto, and G. B. Santos. 2016. Spatial variations in fish assemblage structure in a southeastern Brazilian reservoir. Brazilian Journal of Biology 76:185–193.

Sánchez-Bayo, F., and K. A. G. Wyckhuys. 2019. Worldwide decline of the entomofauna: a review of its drivers. Biological Conservation 232:8–27.

Santos, J. P., I. Martins, M. Callisto, and D. R. Macedo. 2017. Relações entre qualidade da água e uso e cobertura do solo em múltiplas escalas espaciais na bacia do Rio Pandeiros, Minas Gerais. [Relationship between water quality and land use and cover at multiple spatial scales in the Pandeiros River basin, Minas Gerais.] Revista Espinhaço 6(2):36–46.

Schmutz, S., I. G. Cowx, G. Haidvogl, and D. Pont. 2007. Fish-based methods for assessing European running waters: a synthesis. Fisheries Management and Ecology 14:369–380.

Shapiro, M. H., S. Holdsworth, and S. G. Paulsen. 2008. The need to assess the condition of aquatic resources in the U.S. Journal of the North American Benthological Society 27:808–811.

Silva, D. R. O., A. T. Herlihy, R. M. Hughes, and M. Callisto. 2017. An improved macroinvertebrate multimetric index for the assessment of wadeable streams in the Neotropical savanna. Ecological Indicators 81:514–525.

Silva, D. R. O., A. T. Herlihy, R. M. Hughes, D. R. Macedo, and M. Callisto. 2018. Assessing the extent and relative risk of aquatic stressors on stream macroinvertebrate assemblages in the Neotropical savanna. Science of the Total Environment 633:179–188.

Sousa, R., J. L. Gutiérrez, and D. C. Aldridge. 2009. Non-indigenous invasive bivalves as ecosystem engineers. Biological Invasions 10:2367–2385.

Stapp, W. 1978. An instructional model for environmental education. Prospects 8:495–507.

Stevens, D. L., Jr., and A. R. Olsen. 2004. Spatially balanced sampling of natural resources. Journal of the American Statistical Association 99:262–278.

Strahler, A. N. 1957. Quantitative analysis of watershed geomorphology. Transactions of the American Geophysical Union 38:913–920.

Strassburg, B. B. N., T. Brooks, R. Feltran-Barbieri, A. Iribarrem, R. Crouzeilles, R. Loyola, A. E. Latawiec, F. J. B. Oliveira Filho, C. A. de M. Scaramuzza, F. R. Scarano, B. Soares-Filho, and A. Balmford. 2017. Moment of truth for the Cerrado hotspot. Nature Ecology and Evolution [online serial] 1(4):0099.

Tedesco, P. A., T. Oberdorff, J. F. Cornu, O. Beauchard, S. Brosse, H. H. Dürr, G. Grenouillet, F. Leprieur, C. Tisseuil, R. Zaiss, and B. Hugueny. 2013. A scenario for impacts of water availability loss due to climate change on riverine fish extinction rates. Journal of Applied Ecology 50:1105–1115.

Terra, B. F., R. M. Hughes, and F. G. Araujo. 2015. Fish assemblages in Atlantic Forest streams: the relative influence of local and catchment environments on taxonomic and functional species. Ecology of Freshwater Fish 25:527–544.

Terra, B. F., R. M. Hughes, M. R. Francelino, and F. G. Araujo. 2013. Assessment of biotic condition of Atlantic Rain Forest streams: a fish-based multimetric approach. Ecological Indicators 34:136–148.

USEPA (U.S. Environmental Protection Agency). 2011. National lakes assessment: field operations manual. USEPA, Washington, D.C.

USEPA (U.S. Environmental Protection Agency). 2016. National lakes assessment 2012: a collaborative survey of lakes in the United States. USEPA, EPA 841-R-16-113, Washington, D.C.

USGS (U.S. Geological Survey). 2018. National Water Quality Assessment (NAWQA) Program. Available: https://water.usgs.gov/nawqa/. (February 2019).

USGS (U.S. Geological Survey). 2005. Shuttle radar topography mission—new products in 2005. USGS, Fact Sheet 2005-3068, Reston, Virginia. Available: https://pubs.usgs.gov/fs/2005/3068/report.pdf. (January 2019).

USGS (U.S. Geological Survey). 2015. SRTMGL1: NASA shuttle radar topography mission global 1 arc second V003 [data set]. USGS, Reston, Virginia. Available: https://lpdaac.usgs.gov/dataset_discovery/measures/measures_products_table/srtmgl1_v003. (January 2019).

Vörösmarty, C. J., P. B. McIntyre, M. O. Gessner, D. Dudgeon, A. Prusevich, P. Green, S. Glidden, S. E. Bunn, C. A. Sullivan, C. R. Liermann, and P. M. Davies. 2010. Global threats to human water security and river biodiversity. Nature (London) 468:334–334.

Walz, U., and C. Stein. 2014. Indicators of hemeroby for the monitoring of landscapes in Germany. Journal for Nature Conservation 22:279–289.

Wang, L., D. Infante, P. Esselman, A. Cooper, D. Wu, W. Taylor, D. Beard, G. Whelan, and A. Ostroff. 2011. A hierarchical spatial framework and database for the national river fish habitat condition assessment. Fisheries 36:436–449.

Wang, L., J. Lyons, P. Rasmussen, P. Seelbach, T. Simon, M. Wiley, P. Kanehl, E. Baker, S. Niemela, and P. M. Stewart. 2003. Watershed, reach, and riparian influences on stream fish assemblagees in the Northern Lakes and Forest Ecoregion, U.S.A. Canadian Journal of Fisheries and Aquatic Sciences 60:491–505.

White, D. S. 2014. The benthic macroinvertebrates of Kentucky Lake, a mainstem reservoir on the Tennessee River, U.S.A. Transactions of the American Entomological Society 140:83–99.

Yoder, C. O., E. T. Rankin, V. L. Gordon, L. E. Hersha, and C. E. Boucher. 2019. Degradation and recovery of Scioto River (Ohio, USA) fish assemblages from presettlement to present-day conditions. Pages 233–265 *in* C. C. Krueger, W. W. Taylor, and S. Youn editors. From catastrophe to recovery: stories of fishery management success. American Fisheries Society, Bethesda, Maryland.

Zaiko, A., and D. Daunys. 2015. Invasive ecosystem engineers and biotic indices: giving a wrong impression of water quality improvement? Ecological Indicators 52:292–299.

American Fisheries Society Symposium 90:303–320, 2019
© 2019 by the American Fisheries Society

Effects of Catchment Land Use on Stream Fish Assemblages in the Brazilian Savanna

Paulo S. Pompeu*
Graduate Program in Applied Ecology, Department of Biology
Federal University of Lavras, Lavras 37200, Brazil

Cecília G. Leal
Forest Hydrology Laboratory, Luiz de Queiroz College of Agriculture
University of São Paulo, Piracicaba 13418, Brazil

Débora R. Carvalho and Nara T. Junqueira
Graduate Program in Applied Ecology, Department of Biology
Federal University of Lavras, Lavras 37200, Brazil

Miriam A. Castro
FAJ Consultoria em Hidrobiologia Sanitária e Ambiental
Belo Horizonte 30730, Brazil

Robert M. Hughes
Amnis Opes Institute, and Department of Fisheries and Wildlife, Oregon State University
2895 SE Glenn, Corvallis, Oregon 97331, USA

Abstract.—The Brazilian savanna (Cerrado) is a biome of global importance with great endemism and environmental heterogeneity, but it is highly threatened and overexploited. Such a set of conditions is a key aspect of freshwater biodiversity and a challenge to our understanding of species-rich regions. Therefore, we investigated the fish diversity patterns and the effects of different land uses on fish assemblage structure in 155 Cerrado stream sites in four hydrological units. We assessed catchment land use and cover upstream of each sample site, where fish were sampled once during the dry season. Stream fish diversity patterns and the effects of different land uses on assemblage structure differed among the four hydrologic units, and in the region as a whole, but high values of beta diversity due to species turnover were consistently observed. We observed low explanation of land use in relation to fish assemblage structure, probably because of the high level of species turnover and large number of rare species. For some units, the most-correlated land uses were anthropogenic, and alien species were positively related to anthropogenic impacts. Our analysis highlights the importance of the heterogeneous composition of the fish fauna in Brazilian savanna streams and the significance of shifting towards protecting or properly managing whole basins and drainage networks.

* Corresponding author: pompeu@ufla.br

Introduction

Freshwater ecosystems are considered more imperiled than terrestrial or marine ecosystems, yet they are poorly protected globally (Dudgeon et al. 2006; Vörösmarty et al. 2010; Reid et al. 2019). The main drivers of these threats are linked to anthropogenic pressures leading to physical habitat degradation, water pollution, flow regulation and water extraction, fisheries overexploitation, and alien species introductions (Strayer and Dudgeon 2010). However, our scientific knowledge of freshwater systems is incomplete, and human-induced changes remain largely unknown and may be underestimated. This situation is likely to be more critical in tropical systems than in temperate ones for two major reasons. First, species composition and distribution are poorly known for most tropical river basins (Dudgeon et al. 2006). Second, original land cover in the hyperdiverse tropics is under rapid and extensive conversion to urban and agricultural uses, with irreversible widespread consequences for natural ecosystems (Barlow et al. 2018).

The Brazilian savanna (Cerrado) is a biome of global importance. It is one of the five South American biodiversity hotspots because of its remarkably rich endemic fauna, yet it is highly threatened and overexploited (Strassburg et al. 2017). The Cerrado is the second largest biome in the neotropics and in Brazil, and it encompasses the headwaters of important Brazilian rivers that host unique faunas and are responsible for the maintenance of critical hydrological processes (Fernandes et al. 2016). Despite its enormous relevance for biodiversity conservation and provision of ecosystem services, the combination of agriculture intensification since the 1960s, infrastructure development, low legal protection, and inadequate conservation incentives has resulted in critical conditions in the Cerrado (Strassburg et al. 2017). So far, it has lost 46% of its native vegetation cover, whereas only 19.8% remains undisturbed (Strassburg et al. 2017), and its major rivers (Paraná, São Francisco, and Tocantins) are extensively dammed (ANEEL 2018). Wherever the effects of land use on small Cerrado streams have been investigated, multiple negative consequences on its fauna were evident (Macedo et al. 2014; Carvalho and Tejerina-Garro 2015; Castro et al. 2018).

The replacement of native vegetation for agriculture, pasture, or urban areas alters the physical structure and energy flow in tropical and temperate stream ecosystems through the modification of the riparian zone, water quality, substrate type, and channel morphology (Allan 2004; Hughes et al. 2006; Leal et al. 2016). Those changes are likely to affect stream condition and consequently biodiversity (Wang et al. 2000; Hughes et al. 2014; Terra et al. 2016; Leitão et al. 2018). Junqueira et al. (2016) found that land-use cover explained about 23% of fish abundance and 19% of occurrence in Cerrado streams, with the most pronounced effects in urban catchments. On the other hand, land use explained 10% of fish taxonomic richness in the same biome (Macedo et al. 2014). Moreover, human-modified Cerrado streams had more homogeneous aquatic macroinvertebrate assemblages, characterized by lower functional richness, lower taxonomic diversity, and lower specialization of sensitive groups (i.e., Ephemeroptera, Plecoptera, and Trichoptera; Castro et al. 2018; Callisto et al. 2019, this volume). Different land uses also resulted in multiple trophic changes in fish assemblages (Carvalho et al. 2015; de Carvalho et al. 2017a). The presence of pasture can positively influence trophic diversity, whereas sugarcane plantations had a negative effect on fish (de Carvalho et al. 2017a; Zeni et al. 2017) and benthic macroinvertebrates (Castro et al. 2016). However, the taxonomic responses of stream fish to different land uses in the Cerrado are still unclear, and most studies have focused on the effects of stream morphology and water quality on fish assemblages (e.g., Araújo and Tejerina-Garro 2007, 2009; Dias and Tejerina-Garro 2010).

Understanding beta diversity patterns, especially the contribution of the spatial turnover and nestedness components, is crucial to our knowledge of biogeographic, ecological, and conservation issues (Baselga and Orme 2012). Whereas common species dominate species richness patterns (Lennon et al. 2004), rare species usually drive beta diversity patterns of entire assemblages (Gering et al. 2003). Tropical streams overall are environmentally heterogeneous and biologically distinct in their fish species composition, evidenced by the extremely high beta diversity between different stream sites and river basins (Leal et al. 2018). Such heterogeneity is a key aspect of freshwater biodiversity and a challenge to our understanding of species-rich regions, often dominated by rare species. However, large-scale studies focusing on land-use effects on fish assemblages in such biodiverse systems are lacking. Therefore, we investigated fish beta diversity patterns and the effects of different land uses on fish assemblage structure in 155 Cerrado stream sites in two major Brazilian river basins and four large hydrologic regions experiencing intensive agricultural activity. We hypothesized that the effects of land uses on fish assemblages would be detectable but weak, mainly because of high species turnover and many rare species.

Methods

Study Area

We studied streams located in the Paranaíba, Grande, São Francisco, and Araguari River basins (Figure 1). In each basin, the sampled sites are located within 40 km of four hydropower reservoirs, namely São Simão, Volta Grande, Três Marias, and Nova Ponte (for convenience, hereafter, sites will be referred to as hydrologic units named after hydropower reservoirs from each basin). Site selection followed the generalized random tessellation stratified sampling design of Stevens and Olsen (2004) and Olsen and Peck (2008). We excluded all tributaries greater than Strahler order three on a digital 1:100,000 scale map because our targets were wadeable perennial streams. The length of each sample site was 40 times its mean wetted width, with a minimum length of 150 m (Kaufmann et al. 1999; Hughes and Peck 2008). We selected 39 sites each in São Simão, Volta Grande, and Três Marias and 38 sites in Nova Ponte (Figure 1). The mean catchment area of the sites was 30.2 km^2, 630 km^2, 27.5 km^2, and 45.2 km^2, respectively.

Land Use

We assessed catchment land use and cover upstream of each sample site through use of screen digitizing of land use and land cover. We extracted catchments from the terrain model of the Shuttle Radar Topographic Mission (USGS 2005). Then we interpreted Landsat TM sensor multispectral imagery (R4G3B2 false-color band combination) in conjunction with fine-resolution images (0.6–5 m spatial resolution, Google Earth data; Google 2010), from the same month of fish sampling. We quantified percentages of five different land covers: natural vegetation, pasture, *Eucalyptus* plantation, agriculture, and urban. Sample sites covered a diverse range of those land uses but with agriculture dominating in Nova Ponte, Volta Grande, and São Simão, whereas pasture dominated in Três Marias (Figure 2).

Fish Sampling

We sampled each site once in September during the dry season as follows: Três Marias in 2010, Volta Grande in 2011, São Simão in 2012, and Nova Ponte in 2013 under Chico Mendes Institute for Biodiversity Conservation (ICMBio in Portuguese) sampling license 10327-1. By sampling during the dry season, we minimized seasonal variation resulting

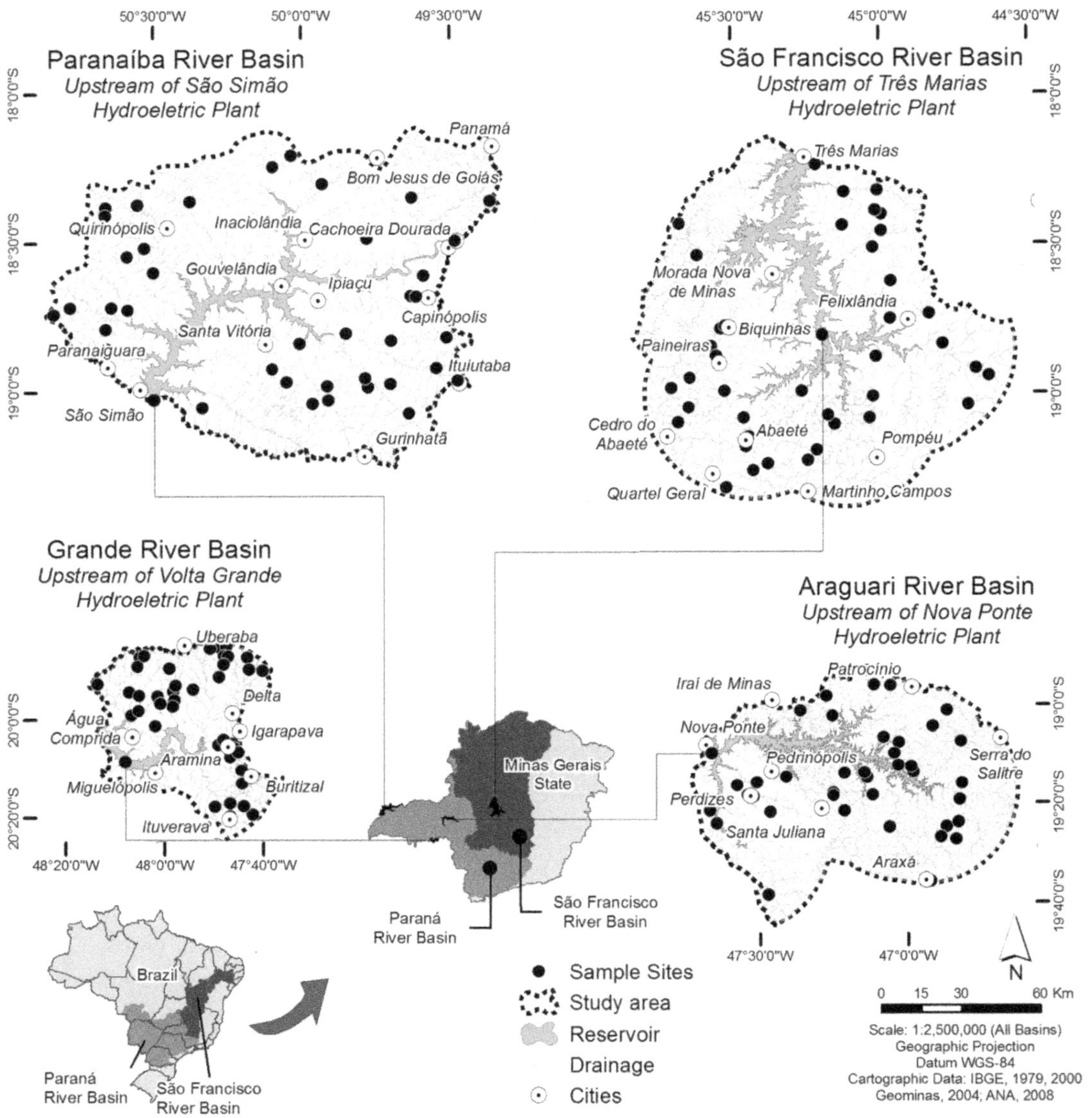

Figure 1. Locations of stream sites randomly arranged upstream from São Simão, Três Marias, Volta Grande, and Nova Ponte reservoirs.

from marked differences in flow intensity, especially those occurring in the wet season. Each site was divided by 11 transects to create 10 sampling sections. The fish sampling effort was standardized in terms of the time and number of devices used. In each section, two or three people sampled fish with two semicircular hand nets (80 cm in diameter, 1 mm mesh) and, whenever possible (large, deep pools), a seine (4.0 m long, 2.0 m high, 5.0 mm between opposite knots) for a period of 12 min, totaling 2 h in each site. Captured specimens were separated by section, labeled, anesthetized, and fixed in 10% formaldehyde. At the laboratory, fish were washed in water, preserved in 70% alcohol, and identified to the lowest possible taxon. Subsequently, they were deposited in the Ichthyology Collection of the Federal University of Lavras (CI-UFLA 302–351; 460–493; 733–767; 804–858).

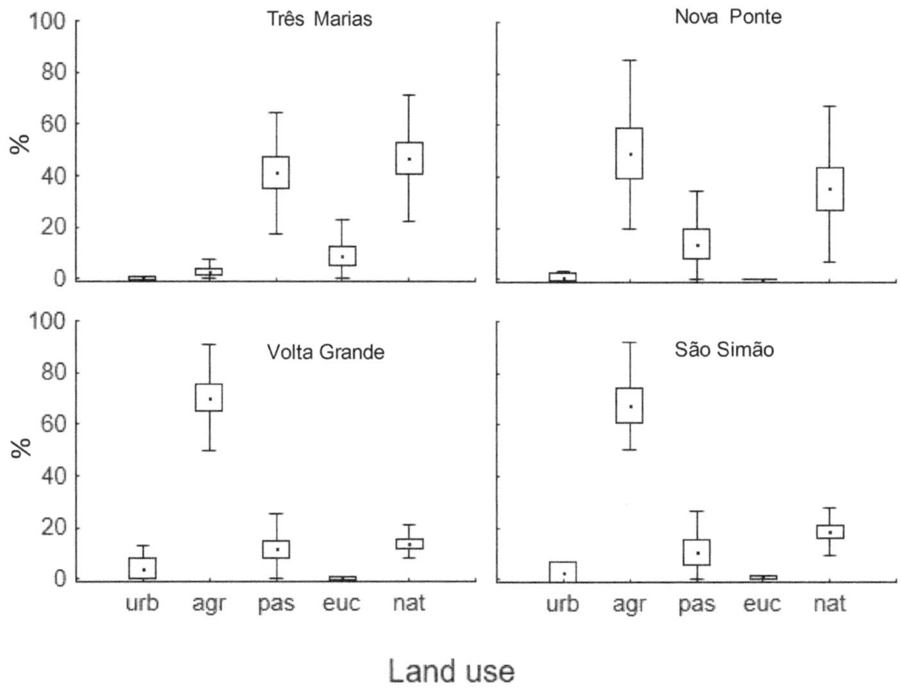

Figure 2. Mean (bar) and 95% confidence interval (whisker) of the different land uses (urb = urban, agr = agriculture, pas = pasture, euc = *Eucalyptus* plantation, and nat = natural vegetation) in the four sampled hydrologic units.

Data Analyses

To analyze the spatial distribution of fish taxonomic richness, we used multiplicative partitioning of Hill numbers (so-called "effective number of species"). Those numbers are a mathematically unified family of richness indices that consider both abundance and species richness and differ by the order q (Jost 2006; Chao et al. 2012, 2014). The analyses were performed considering two spatial scales: each hydrologic unit separately and all four units together. Diversity was calculated considering

$$ {}^{q}R = \left(\sum_{i=1}^{S} p_i^q \right)^{1/(1-q)}, $$

where p_i is the abundance of the ith species in the community, S is the total number of species, and q is the order of the index, which is related to the sensitivity to species relative abundance. When $q = 0$, abundance is not considered, the index simply represents species richness. When $q = 1$, diversity considers each species' relative abundance and is equivalent to the exponential of Shannon's diversity index, a typical measure of species diversity. Last, when $q = 2$, dominant species are weighted more, making the index equivalent to the inverse of Simpson diversity.

Multiplicative diversity partitioning was performed using the 'multipart' function from the R 'vegan' library considering 1,000 simulations (Oksanen et al. 2018; R Core Team 2014). We decomposed the components of β (among-site) diversity to investigate whether variation in species composition across stream sites in each river basin resulted from turnover (species replacement) or nestedness (species loss or gain), using Sørensen (βSOR) and Simpson (βSIM) indices (Baselga and Orme 2012). βSOR accounts for both turnover and nestedness, whereas βSIM considers only turnover.

Therefore, the difference between the two indices gives a representation of the nestedness component (βNES): βNES = βSOR − βSIM. We ran beta decomposition for richness (Hill number 0) using the 'beta.multi' function in R 'betapart' library (Baselga and Orme 2012; R Core Team 2014).

We used distance-based linear models (DistLM) to test the degree to which fish assemblage structure was explained by the proportion of each land use in the respective catchment. To do so, we used Primer+Permanova software (Clarke and Gorley 2006) to perform DistLM for each hydrologic unit and for all units together. We chose Bray–Curtis distance as the similarity index for the matrix to incorporate species abundance information. We used the forward selection of models and adopted a 0.05% significance value. We performed the marginal test for each land use separately, and the best model was the one with greater adjusted R^2 among those that were statistically significant.

We evaluated how the variables selected by the best DistLM models related together, and how species responded to those influences through a distance-based redundancy analysis (DbRDA). DbRDA is a suitable method for semimetric variables, such as Bray–Curtis similarity (Legendre and Anderson 1999). To test those relationships, we used the land uses as vectors and the abundance data of the species to generate the most parsimonious models. Only the species with at least 40% of correlation with one of the DbRDA axes were plotted.

Results

We collected 22,178 individuals representing 138 fish species, 21 families, and six orders (Appendix). In São Simão, richness was greater, with 67 species, followed by Três Marias (58), Volta Grande (40), and Nova Ponte (36). São Simão also had the highest mean site (alpha) richness (10), whereas the other three hydrologic units each had seven species on average. For both single hydrologic units and the entire region, rare species (singletons and doubletons) represented from 33% to 46% of the sampled species (Table 1). Hill's numbers indicated that the higher fish richness in São Simão resulted from a higher number of rare species (Figure 3A). When considering both rare and dominant species, São Simão had greater richness than all the other hydrologic units and Volta Grande had significantly greater richness than Três Marias (Figure 3B). When considering only the dominant species (Figure 3C), São Simão had significantly greater richness than all but Volta Grande, which had significantly greater richness than Nova Ponte and Três Marias. Beta diversity values were high in all regions, mainly because of species turnover (Figure 4).

The predictor variables varied among the four hydrologic units, and when all units were analyzed together, they only weakly explained fish assemblage structure (Table 2). For the entire data set, natural vegetation cover was

Table 1. Fish species richness, singletons, and doubletons for the study regions.

Hydrologic unit	Mean local richness (range)	Regional richness	Singletons (%)	Doubletons (%)
All together	8 (0–20)	138	34 (25)	18 (13)
Três Marias	7 (0–20)	58	17 (29)	10 (17)
Nova Ponte	7 (0–15)	36	7 (19)	5 (14)
Volta Grande	7 (1–14)	40	10 (25)	8 (20)
São Simão	10 (2–20)	67	20 (30)	11 (16)

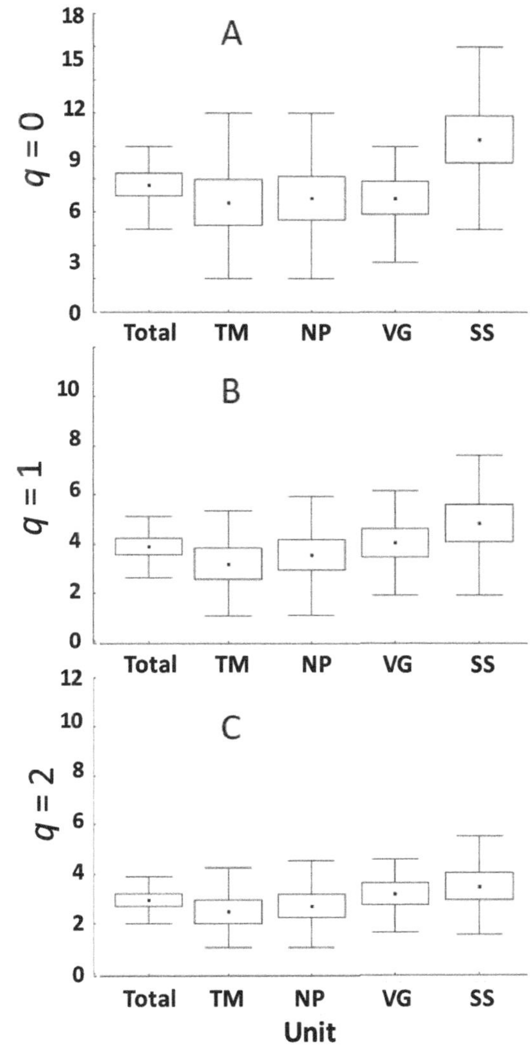

Figure 3. Effective number of species (Hill´s numbers): **(A)** q = 0, species richness; **(B)** q = 1, rare and dominant equally weighted; **(C)** q = 2, dominant species weighted more than rare species. TM = Três Marias, NP = Nova Ponte, VG = Volta Grande, and SS = São Simão.

the main predictor but explained only 4% of fish assemblage structure. *Eucalyptus* plantation and pasture were also included in the best model, raising the explanation to 6.9%. In Três Marias, *Eucalyptus* plantation and pasture comprised the best significant model, explaining 13% of fish assemblage variability among sites. In Nova Ponte, fish assemblage structure was explained by natural vegetation and agriculture in the best composite model (14%). In Volta Grande, the best model explained only 8% of fish assemblage structure and included urban and agriculture. None of the land uses was able to explain fish assemblage structure in São Simão.

There was little consistency among the species that were most correlated with the land uses among the four hydrologic units, and when the regions were analyzed together (Figure 5). In Três Marias, the abundance of the South American darter *Characidium zebra* and the catfish *Microglanis leptostriatus* were positively correlated with pasture. The abundance of the armored catfish *Harttia* sp. and the nonnative characin *Knodus moenkhausii* were positively associated with *Eucalyptus* plantation, whereas the catfish *Pimelodella lateristriga* and the characin *Hyphessobrycon santae* were associated with natural cover. In Nova Ponte, the scrapetooth *Apareiodon ibitiensis* and the characin *Astyanax scabripinnis* were positively correlated with natural vegetation. In Volta Grande, the nonnative Guppy *Poecilia reticulata* was associated with urban land use, whereas the characin *Piabina argenta* was correlated with *Eucalyptus* plantation. In the same region, the characin *Brycon orbignianus* and *Apareiodon ibitiensis* were negatively associated with both land uses. When all four units were analyzed together, only the armored catfish *Hypostomus garmani* was correlated more than 40% with land uses, and it was positively associated with *Eucalyptus* plantation.

Discussion

Stream fish diversity patterns and the effects of different land uses on assemblage structure differed among the four hydrologic units, but high values of beta richness resulting from high species turnover were consistently observed. Similar results were reported by Leal et al. (2018) in a study of fish richness in three

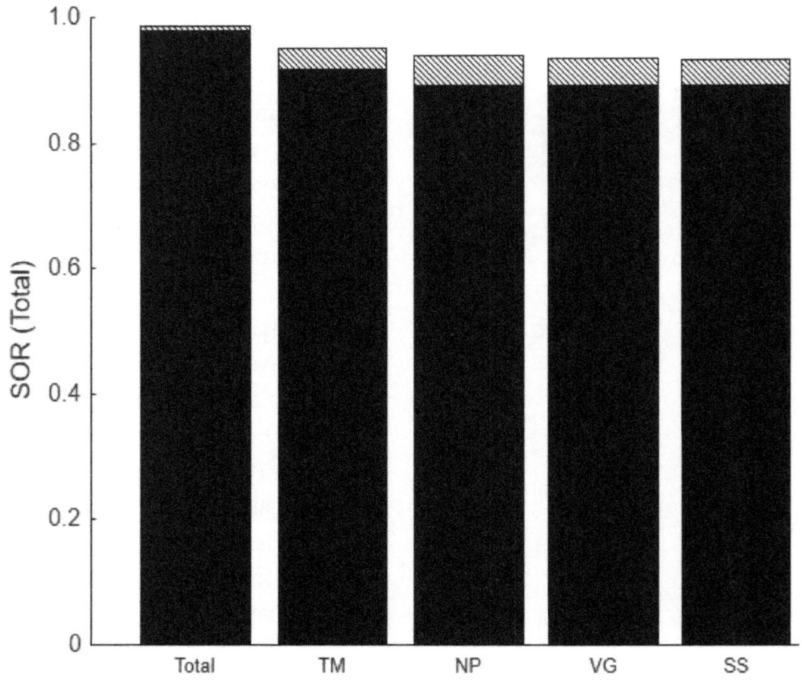

Figure 4. Diversity partitioning for Nova Ponte (NP), Volta Grande (VG), Três Marias (TM), and São Simão (SS) hydrologic units: Sorensen beta diversity index decomposed into nestedness (crosshatched) and turnover (shaded).

Table 2. Relationship among land uses and fish assemblage structure tested by distance-based linear model (DisLM) analysis of Bray–Curtis similarity. For each hydrologic unit, the best significant ($P < 0.05$) model is presented.

Basin/variables	Adjusted R^2	Pseudo-F	P-value
All together			
% natural vegetation	0.04	7.46	<0.01
+ % *Eucalyptus*	0.06	3.66	<0.01
+ % pasture	0.07	3.02	<0.01
Três Marias			
% *Eucalyptus*	0.09	3.38	<0.01
+ % pasture	0.13	1.92	0.04
Nova Ponte			
% natural vegetation	0.09	3.83	<0.01
+ % agriculture	0.14	1.79	0.05
Volta Grande			
% urban	0.05	1.97	0.02
+ % agriculture	0.08	1.16	0.03
São Simão			
No significant model			

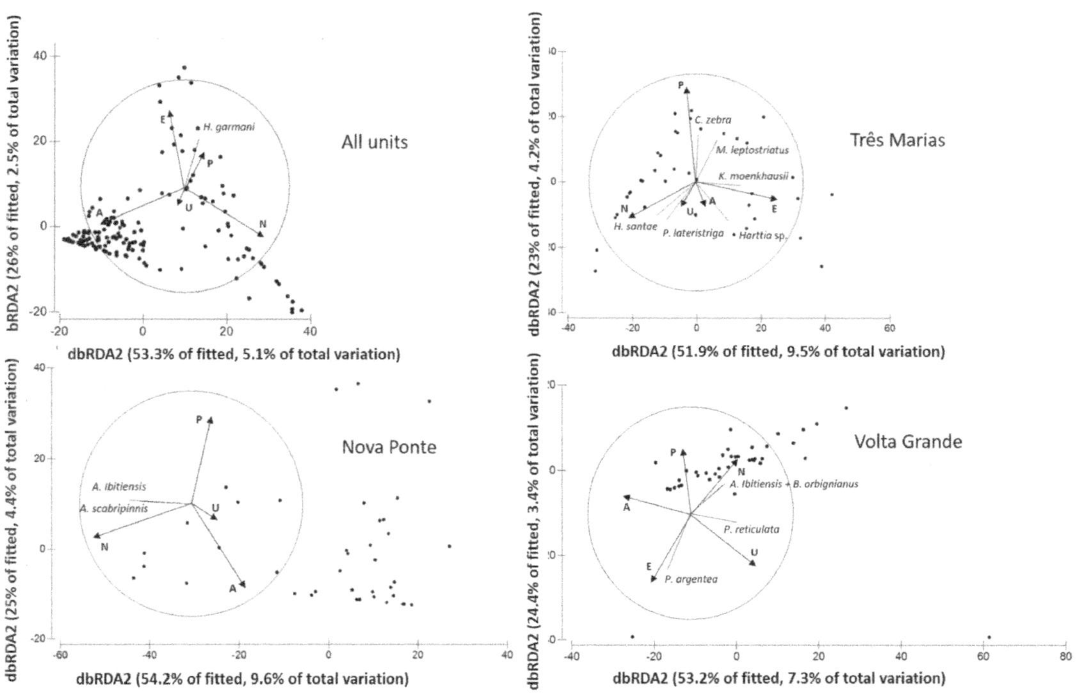

Figure 5. Distance-based redundancy analysis (dbRDA) showing the variables included in the best significant DistLM (distance-based linear model). Land uses are indicated by arrows and the five most correlated species by lines (see Table 2 for correlation values). A = agriculture, E = *Eucalyptus*, N = natural, P = pasture, and U = urban.

Amazon River basins. Especially in the São Simão basin, higher fish richness was associated with a higher number of rare species. We observed a weak effect of land use on fish fauna structure. In addition, alien species were positively related to anthropogenic impacts. In a study in the Brazilian Amazon, Leitão et al. (2018) also reported that the key anthropogenic predictors varied between river basins and predicted only moderate levels of taxonomic and functional richness.

We were able to significantly explain extremely little of the fish assemblage structure patterns (0–14%), with the five land uses selected (Table 1). São Simão was the only unit with no explanation related to land use. Other studies have also reported weak correlations between land uses and fish assemblage structure or fish richness (Allen et al. 1999; Meador and Goldstein 2003; Allan 2004; Sály et al. 2011; Macedo et al. 2014; Leitão et al. 2018). Although in the Neotropics, land-use correlations seem to be as important as the effect of local riparian vegetation or natural covariates (Junqueira et al. 2016; Leal et al. 2018; Leitão et al. 2018). There are several possible reasons for this high level of unexplained variance, including the fact that we did not account for land-use effects at both riparian and catchment spatial scales (see Roth et al. 1996; Wang et al. 2003; Schiff and Benoit 2007) or the influence of past land use (Harding et al. 1998). Species richness may be much more strongly affected by natural differences in catchment location and area, geophysical differences (drainage pattern, geology, slope, elevation), site differences (physical habitat structure, water quality), or all of the above (Allan 2004; Wang et al. 2003; Sály et al. 2011; Macedo et al. 2014). In addition, richness is partly governed by metacommunity dynamics such as differences in species interactions (species sorting, niche separation, and competition), immigration and emigration (Leibold et al. 2004), or both simultaneously, along with habitat heterogeneity (Leibold and Chase 2018). Last, species richness is an assemblage descriptor that does not discriminate among sensitive and tolerant taxa (Kaufmann et al. 2014; Oliveira-Junior et al. 2015, 2017; Firmiano et al. 2017). For all such reasons, as in our study, catchment, channel network, and local environmental predictors, together, typically explain a small proportion of the variability observed in fish species richness or other assemblage condition indicators (Wang et al. 2003; Sály et al. 2011; Macedo et al. 2014; Terra et al. 2016; Leitão et al. 2018).

Our inability to detect any land-use effects on fish assemblage structure in São Simão could be related to low variability in catchment variables (Heino et al. 2008; Roa-Fuentes and Casatti 2017), but such landscape variation was similar to the other units. One other possibility is the large number or rare species collected there, although their percentage was similar to that in the other units. The higher richness in São Simão may be associated with historic biogeographic reasons because the studied units have similar basin areas, precipitation, and elevation. Rarely occurring species add sampling effort noise to fish assemblage studies (Kanno et al. 2009), as well as analytical noise to predictor-response modeling. Therefore, because of the lack of consistent predictor–species relationships, they are often removed before modeling speciose assemblages (Hawkins et al. 2000). However, it seems plausible that many of these rare species could be the most sensitive to anthropogenic impacts, particularly if they are only present in the relatively small number of streams occurring in undisturbed catchments. Although common and rare species are affected by similar environmental factors, rare species also respond to processes operating at smaller spatial and temporal scales, particularly biotic interactions (Siqueira et al. 2012).

Different land uses alter the characteristics of the aquatic environment and may have negative or positive effects on aquatic organisms, depending on the biological characteristics of each species. Their dependence on regional context probably leads to their low abilities in explaining land use versus species abundance when the units were analyzed together. Regionally, even little urban or *Eucalyptus* cover influenced fish assemblages, but in some units, such land uses were absent. The former acts on different hydrology-, geomorphology-, and water-quality-related aspects (Paul and Meyer 2001), whereas the latter represents homogenization and afforestation in a diverse and otherwise grass-dominated biome (Bridgewater et al. 2004).

The replacement of native vegetation by agriculture, pasture, or urban areas can benefit certain species of fish because it promotes changes in resource supplies, type of predominant substrate, flow, and physical and chemical characteristics of the water (Pusey and Arthington 2003; de Carvalho et al. 2017). Unfortunately, there are few studies on the biology of the species correlated with land uses in the different units, but *Characidium zebra* and *Knodus moenkhausii* are the most studied. Both species demonstrate great trophic variation in response to land-use changes in Cerrado streams (Carvalho et al. 2015). This may explain their abilities to successfully occupy stream habitats with low complexity (Ceneviva-Bastos et al. 2010), such as areas experiencing *Eucalyptus* or pasture influence in the Três Marias hydrologic unit.

Nonnative invasive species also have close relationships with changes in aquatic habitats because their potentials to colonize new geographic locations are highly related to environmental quality (e.g., Herborg et al. 2007; Kilroy et al. 2008; Kulhanek et al. 2011). In fact, the presence of a disturbance sometimes is a necessary precursor for their establishment (Hobbs and Huenneke 1992; Minchinton 2002; Marchetti et al. 2004). The Guppy, widely distributed in tropical freshwaters in association with human interventions, is considered an indicator of degraded aquatic environments (de Carvalho et al. 2017b) and was the only one associated with urban streams. The Guppy was introduced in Brazil and other countries with the aim of controlling mosquitoes (Allan and Flecker 1993; Warbanski et al. 2017). Poeciliids exhibit a broad range of food choices (Meffe and Snelson 1989) and will often take the most abundant item available. In addition, because they are ovoviviparous species, their members have an invasive advantage over species of other popular nonnative families. *Knodus moenkhausii*, in turn, is a small characid, originally described from specimens collected in the Paraguay River basin and whose presence in the upper Paraná has an uncertain origin (Langeani et al. 2007). This species has been consistently listed in studies from the upper Paraná (Casatti et al. 2006, 2009; Langeani et al. 2007; Teresa and Casatti 2013; Fagundes et al. 2015), and its high abundance has been a recurrent feature there (Casatti et al. 2006; Fagundes et al. 2015). According to Ceneviva-Bastos and Casatti (2007), *K. moenkhausii* can be considered an opportunistic feeder because of its various feeding tactics and the high variety of consumed items. Such opportunism can be reflected in the abundance of the species, which demonstrates a strong capacity to allocate an important portion of its energy to reproduction, even in environments that are physically degraded by anthropogenic activities (Ceneviva-Bastos and Casatti 2007).

Contrary to the above two species, *Brycon orbignianus* and *Apareiodon ibitiensis* seem to be related to less-disturbed catchments. Both were positively correlated with the presence of natural vegetation and were negatively influenced by urban areas and agriculture in Nova Ponte and Volta Grande. *Brycon orbignyanus*, popularly known as "piracanjuba," is native to the Paraná and Uruguay River basins (Borba et al. 2003). It is a migratory species with high commercial value and is currently classified as critically endangered (Machado 2005). This species is highly dependent on the presence of riparian vegetation for feeding and needs free-flowing rivers with running waters and high flooding for recruitment and reproduction (Panarari-Antunes et al. 2011). *Apareiodon ibitiensis* also prefers running waters and rocky or gravel bottoms and is predominantly herbivorous (Sazima 1980). Therefore, both species tend to be habitat and trophic specialists, and changes in land use can greatly alter their abundances.

When studying small streams in heterogeneous, diverse, and poorly studied regions like the Cerrado, it is challenging to outline research priorities when basic information on species distribution and natural history is still lacking. Nevertheless, further studies that expand our understanding on the effects of land-use changes on stream biodiversity are unquestionably needed. We recommend that similar approaches, as presented here, be performed in other river basins in the Cerrado to further understand the effects of regional and basin contexts. Understanding the factors that underpin this context specificity is a vital step in extrapolating fish responses to land-use change to larger spatial scales (Leal et al. 2018) and can assist with regional conservation planning in the Cerrado. As with the U.S. Environmental Protection Agency's National Aquatic Resource Surveys (Hughes et al. 2000; Shapiro et al. 2008), thoroughly and consistently monitoring a large number of sites over time is needed to determine how time lags and shifting baselines in modified landscapes influence stream fish assemblages, as well as to account for multiple spatial landscape scales.

In this study, we did not include information on functional diversity or species life histories, which can help disentangle the effects of land-use change on stream fish assemblages (Winemiller 1989; Mouillot et al. 2013; Leitão et al. 2018). However, life-history information for Neotropical fish is a huge knowledge gap and a challenge for future studies. Given the amount of unexplained variance we found, we also recognize that more detailed information on land-use change should be included in future work. For instance, history of land use (e.g., land-use intensity, deforestation timing, pattern, and type) and agriculture intensification (mechanization, biocide application) can clarify the impacts of human activities on tropical streams (Leal et al. 2016).

Our study analyzed catchment-scale land-use cover for predicting assemblage structure and included a large biogeographically complex area, which makes our results an important contribution for future studies. Despite the low explanation of land uses on the richness of stream fish assemblages, our analysis highlights the importance of the heterogeneous composition of the fish fauna in Brazilian savanna streams. The high levels of beta diversity and turnover and the great contribution of rare species have important implications for land and water management and fish conservation policies. The existing legislative frameworks focus on protecting riparian vegetation, and the creation of protected areas in Brazil was strongly based on terrestrial biota (Azevedo-Santos et al. 2019). To safeguard the species-rich freshwater biota of streams, conservation actions must shift towards managing or protecting whole basins and drainage networks, as well as improving agricultural practices in already modified landscapes (Leal et al. 2018; Azevedo-Santos et al. 2019).

Acknowledgments

We thank the Universidade Federal de Lavras Laboratório de Ecologia de Peixes team, the Universidade Federal de Minas Gerais Laboratório de Ecologia de Bentos team, the Pontificia Universidade Católica-Minas Zoologia de Vertebrados Postgraduate Program team, and the Centro Federal de Educação Tecnológica de Minas Gerais team for participating in the field sampling and laboratory work. We also thank the landowners where the sites we visited are located. We are grateful to the Companhia Energética de Minas Gerais - Peixe Vivo Program for financing the project and to Coordenação de Aperfeiçoamento de Pessoal de Nível Superior (CAPES), Conselho Nacional de Desenvolvimento Cientifico e Techológico (CNPq), Fundação de Amparo à Pesquisa do Estado de Minas Gerais (FAPEMIG), and Fundação de Amparo àPesquisa do Estado de São Paulo (FAPESP) for scholarship grants. CGL was awarded a Ph.D. (CAPES PDSE 2943/13-1) and a postdoctorate grant (FAPESP 2017/25383-0); NTJ, MAC, DRC, and DFS were awarded Master's program grants; MAS and DFS were awarded scientific initiation grants; and PSP was awarded a research productivity grant (CNPq No. 303548/2017-7) and a Minas Gerais researcher scholarship grant (FAPEMIG PPM-00237/13). RMH was supported by Fulbright Brasil and Amnis Opes Institute. We also thank Bianca de Freitas Terra and three anonymous reviewers for helpful comments on an earlier draft manuscript.

References

Allan, J. D. 2004. Landscapes and riverscapes: the influence of land use on stream ecosystems. Annual Review of Ecology, Evolution, and Systematics 35:257–284.

Allan, J. D., and A. S. Flecker. 1993. Biodiversity conservation in running waters. BioScience 43:32–43.

Allen, A. P., T. R. Whittier, D. P. Larsen, P. R. Kaufmann, R. J. O'Connor, R. M. Hughes, R. S. Stemberger, S. S. Dixit, R. O. Brinkhurst, A. T. Herlihy, and S. G. Paulsen. 1999. Concordance of taxonomic composition patterns across multiple lake assemblages: effects of scale, body size, and land use. Canadian Journal of Fisheries and Aquatic Sciences 56:2029–2040.

ANEEL (Agência Nacional de Energia Elétrica). 2018. Atlas de energia elétrica do Brasil. [Atlas of power production of Brazil.] Available: www2.aneel.gov.br/arquivos/pdf/atlas_par2_cap3.pdf. (October 2018).

Araújo, N. B., and F. L. Tejerina-Garro. 2007. Composição e diversidade da ictiofauna em riachos do Cerrado, bacia do ribeirão Ouvidor, alto rio Paraná, Goiás, Brasil. [Fish diversity and composition in Cerrado streams, Ribeirão Ouvidor basin, upper Paraná River basin, Goiás, Brazil.] Revista Brasileira de Zoologia 24:981–990.

Araújo, N. B., and F. L. Tejerina-Garro. 2009. Influence of environmental variables and anthropogenic perturbations on stream fish assemblages, upper Paraná River, central Brazil. Neotropical Ichthyology 7:31–38.

Azevedo-Santos, V. M., R. G. Frederico, C. K. Fagundes, P. S. Pompeu, F. M. Pelicice, A. A. Padial, M. G. Nogueira, P. M. Fearnside, L. B. Lima, V. S. Daga, F. J. M. Oliveira, J. R. S. Vitule, M. Callisto, A. A. Agostinho, F. A. Esteves, D, Lima, Jr., A. L. B. Magalhães, J. Sabino, R. P. Mormul, D. Grazel, J. Zuanon, F. S. Villela, and R. Henry. 2019. Protected areas: a focus on Brazilian freshwater biodiversity. Diversity and Distributions 25:442–448.

Barlow, J., F. França, T. A. Gardner, C. C. Hicks, G. D. Lennox, E. Berenguer, L. Castello, E. P. Economo, J. Ferreira, B. Guénard, C. G. Leal, V. Isaac, A. C. Lees, C. L. Parr, S. K. Wilson, P. J. Young, and N. A. J. Graham. 2018. The future of hyperdiverse tropical ecosystems. Nature (London) 559:517–526.

Baselga, A., and C. D. L. Orme. 2012. Multiplicative partition of true diversity yields independent alpha and beta components. Methods in Ecology and Evolution 3:808–812.

Borba, M. R., D. M. Fracalossi, L. E. Pezzato, D. Menoyo, and J. M Bautista. 2003. Growth, lipogenesis and body composition of piracanjuba (*Brycon orbignyanus*) fingerlings fed different dietary protein and lipid concentrations. Aquatic Living Resources 16:362–369.

Bridgewater, S., J. A. Ratter, and J. F. Ribeiro. 2004. Biogeographic patterns, β-diversity and dominance in the cerrado biome of Brazil. Biodiversity and Conservation 13:2295–2317.

Carvalho, D. R., D. M. P. Castro, M. Callisto, M. Z. Moreira, and P. S. Pompeu. 2015. Isotopic variation in five species of stream fish under the influence of different land uses. Journal of Fish Biology 87:559–578.

Carvalho, R. A., and F. L. Tejerina-Garro. 2015. The influence of environmental variables on the functional structure of headwater stream fish assemblages: a study of two tropical basins in central Brazil. Neotropical Ichthyology 13:349–360.

Casatti, L., C. P. Ferreira, and F. R. Carvalho. 2009. Grass-dominated stream sites exhibit low fish species diversity and dominance by guppies: an assessment of two tropical pasture river basins. Hydrobiologia 632:273–283.

Casatti, L., F. Langeani, A. M. Silva, and R. M. C. Castro. 2006. Stream fish, water and habitat quality in a pasture dominated basin, southeastern Brazil. Brazilian Journal of Biology 66:681–696.

Castro, D. M. P., D. R. de Carvalho, P. S. Pompeu, M. Z. Moreira, G. B. Nardoto, and M. Callisto. 2016. Land use influences niche size and the assimilation of resources by benthic macroinvertebrates in tropical headwater streams. PLOS (Public Library of Science) One [online serial] 11(3):e0150527.

Castro, D. M. P., S. Dolédec, and M. Callisto. 2018. Land cover disturbance homogenizes aquatic insect functional structure in neotropical savanna streams. Ecological Indicators 84:573–582.

Ceneviva-Bastos, M., and L. Casatti. 2007. Oportunismo alimentar de Knodus moenkhausii (Teleostei, Characidae): uma espécie abundante em riachos do noroeste do Estado de São Paulo, Brasil. [Feeding opportunism of *Knodus moenkhausii* (Teleostei, Characidae): an abundant species in streams of the northwest of the state of Sao Paulo, Brazil.] Iheringia Série Zoologia 97:7–15.

Ceneviva-Bastos, M., L. Casatti, and D. C. Rossa-Feres. 2010. Meso and microhabitat analysis and feeding habits of small nektonic characins (Teleostei: Characiformes) in Neotropical streams. Zoologia (Curitiba) 27:191–200.

Chao, A., C. Chiu, and T. C. Hsieh. 2012. Proposing a resolution to debates on diversity partitioning. Ecology 93:2037–2051.

Chao, A., A. Chao, N. J. Gotelli, T. C. Hsieh, E. L. Sander, K. H. Ma, R. K. Colwell, and A. M. Ellison. 2014. Rarefaction and extrapolation with Hill numbers: a framework for sampling and estimation in species diversity studies. Ecological Monographs 84:45–67.

Clarke, K. R., and R. N. Gorley. 2006. PRIMER v6: user manual/tutorial. PRIMER-e, Auckland, New Zealand.

de Carvalho, D. R., D. M. P. de Castro, M. Callisto, M. Z. Moreira, and P. S. Pompeu. 2017a. The trophic structure of fish communities from streams in the Brazilian Cerrado under different land uses: an approach using stable isotopes. Hydrobiologia 795:199–217.

de Carvalho, D. R., C. G. Leal, N. T. Junqueira, M. A. Castro, D. C. Fagundes, C. B. M. Alves, R. M. Hughes, and P. S. Pompeu. 2017b. A fish-based multimetric index for Brazilian savanna streams. Ecological Indicators 77:386–396.

Dias, A. M., and F. L. Tejerina-Garro. 2010. Changes in the structure of fish assemblages in streams along an undisturbed-impacted gradient, upper Paraná River basin, central Brazil. Neotropical Ichthyology 8:587–598.

Dudgeon, D., A. H. Arthington, M. O. Gessner, Z.-I. Kawabata, D. J. Knowler, C. Lévêque, R. J. Naiman, A.-H. Prieur-Richard, D. Soto, M. L. J. Stiassny, and C. A Sullivan. 2006. Freshwater biodiversity: importance, threats, status and conservation challenges. Biological Reviews of the Cambridge Philosophical Society 81:163–182.

Fagundes, D. C., C. G. Leal, D. R. Carvalho, N. T. Junqueira, F. Langeani, and P. S. Pompeu. 2015. The stream fish fauna from three regions of the upper Paraná River basin. Biota Neotropica 15:1–8.

Fernandes, G. W., M. S. Coelho, R. B. Machado, M. E. Ferreira, L. M de Souza Aguiar, R. Dirzo, A. Scariot, and C. R. Lopes. 2016. Afforestation of savannas: an impending ecological disaster. Natureza & Conservação 14:146–151.

Firmiano, K. R., R. Ligeiro, D. R. Macedo, L. Juen, R. M. Hughes, and M. Callisto. 2017. Mayfly bioindicator thresholds for several anthropogenic disturbances in neotropical savanna streams. Ecological Indicators 74:276–284.

Gering, J. C., T. O. Crist, and J. A. Veech. 2003. Additive partitioning of species diversity across multiple spatial scales: implications for regional conservation of biodiversity. Conservation Biology 17:488–499.

Google. 2010. Google Earth. Google, Mountain View, California. Available: www.google.com/earth/.

Harding, J. S., E. F. Benfield, P. V. Bolstad, G. S. Helfman, and E. B. D. Jones. 1998. Stream biodiversity: the ghost of land use past. Proceedings of the National Academy of Sciences of the United States of America 95:14843–14847.

Hawkins, C. P., R. H. Norris, J. N. Hogue, and J. W. Feminella. 2000. Development and evaluation of predictive models for measuring the biological integrity of streams. Ecological Applications 10:1456–1477.

Heino, J., H. Mykrä, and J. Kotanen. 2008. Weak relationships between landscape characteristics and multiple facets of stream macroinvertebrate biodiversity in a boreal drainage basin. Landscape Ecology 23:417–426.

Herborg, L. M., C. L. Jerde, D. M. Lodge, G. M. Ruiz, and H. J. MacIsaac. 2007. Predicting invasion risk using measures of introduction effort and environmental niche models. Ecological Applications 17:663–674.

Hobbs, R. J., and L. F. Huenneke. 1992. Disturbance, diversity, and invasion: implications for conservation. Conservation Biology 6:324–337.

Hughes, R. M., and D. V. Peck. 2008. Acquiring data for large aquatic resource surveys: the art of compromise among science, logistics, and reality. Journal of the North American Benthological Society 27:837–859.

Hughes, R. M., S. Dunham, K. G. Maas-Hebner, J. A. Yeakley, C. Schreck, M. Harte, N. Molina, C. C. Shock, V. W. Kaczynski, and J. Schaeffer. 2014. A review of urban water body challenges and approaches: (1) rehabilitation and remediation. Fisheries 39:18–29.

Hughes, R. M., S. G. Paulsen, and J. L. Stoddard. 2000. EMAP-Surface Waters: a national, multiassemblage, probability survey of ecological integrity. Hydrobiologia 422–423:429–443.

Hughes, R. M., L. Wang, and P. W. Seelbach, editors. 2006. Landscape influences on stream habitats and biological assemblages. American Fisheries Society, Symposium 48, Bethesda, Maryland.

Jost, L. 2006. Partitioning diversity into independent alpha and beta components. Ecology 88:2427–2439.

Junqueira, N. T., D. R. Macedo, R. C. R. D. Souza, R. M. Hughes, M. Callisto, and P. S. Pompeu. 2016. Influence of environmental variables on stream fish fauna at multiple spatial scales. Neotropical Ichthyology [online serial] 14(3):e150116.

Kanno, Y., J. C. Vokoun, D. C. Dauwalter, R. M. Hughes, A. T. Herlihy, T. R. Maret, and T. M. Patton. 2009. Influence of rare species on electrofishing distance when estimating species richness relationships at stream sites. Transactions of the American Fisheries Society 138:1240–1251.

Kaufmann, P., P. Levine, E. Robison, C. Seeliger, and D. Peck. 1999. Quantifying physical habitat in wadeable streams. U.S. Environmental Protection Agency, EPA/620/R-99/003, Washington, D.C.

Kaufmann, P. R., R. M. Hughes, T. R. Whittier, S. A. Bryce, and S. G. Paulsen. 2014. Relevance of lake physical habitat assessment indices to fish and riparian birds. Lake and Reservoir Management 30:177–191.

Kilroy, C., T. H. Snelder, O. Floerl, C. C. Vieglais, and K. L. Dey. 2008. A rapid technique for assessing the suitability of areas for invasive species applied to New Zealand's rivers. Divers Distributions 14:262–272.

Kulhanek, S. A., L. Brian, and A. Ricciardi. 2011. Using ecological niche models to predict the abundance and impact of invasive species: application to the Common Carp. Ecological Applications 21:203–213.

Langeani, F., R. M. C. Castro, O. T. Oyakawa, O. A. Shibatta, C. S. Pavanelli, and L. Casatti. 2007. Ichthyofauna diversity of the upper Rio Paraná: present composition and future perspectives. Biota Neotropica 7:181–197.

Leal, C. G., J. Barlow, T. A. Gardner, R. M. Hughes, R. P. Leitão, R. Mac Nally, P. R. Kaufmann, S. F. B. Ferraz, J. Zuanon, F. R. de Paula, J. Ferreira, J. R. Thomson, and P. S. Pompeu. 2018. Is environmental legislation conserving tropical stream faunas? A large-scale assessment of local, riparian and catchment-scale influences on Amazonian fishes. Journal of Applied Ecology 55:1312–1326.

Leal, C. G., P. S. Pompeu, T. A. Gardner, R. P. Leitão, R. M. Hughes, P. R. Kaufmann, J. Zuanon, F. R. De Paula, S. F. B. Ferraz, J. R. Thomson, R. Mac Nally, J. Ferreira, and J. Barlow. 2016. Multi-scale assessment of human-induced changes to Amazonian instream habitats. Landscape Ecology 31:1725–1745.

Legendre, P., and M. J. Anderson. 1999. Distance-based redundancy analysis: testing multispecies responses in multifactorial ecological experiments. Ecological Monographs [online serial] 69:1–24.

Leibold, M.A., and J.M. Chase. 2018. Metacommunity ecology. Princeton University Press, Princeton, New Jersey.

Leibold, M. A., M. Holyoak, N. Mouquet, P. Amarasekare, J. M. Chase, M. F. Hoopes, R. D. Holt, J. B. Shurin, R. Law, D. Tilman, M. Loreau, and A. Gonzalez. 2004. The metacommunity concept: a framework for multi-scale community ecology. Ecology Letters 7:601–613.

Leitão, R. P., J. Zuanon, D. Mouillot, C. G. Leal, R. M. Hughes, P. R. Kaufmann, S. Villéger, P. S. Pompeu, D. Kasper, F. R. de Paula, S. F. B. Ferraz, and T. A. Gardner. 2018. Disentangling the pathways of land use impacts on the functional structure of fish assemblages in Amazon streams. Ecography 41:219–232.

Lennon, J. J., P. Koleff, J. J. Greenwood, and K. J. Gaston. 2004. Contribution of rarity and commonness to patterns of species richness. Ecology Letters 7:81–87.

Macedo, D. R., R. M. Hughes, R. Ligeiro, W. R. Ferreira, M. Castro, N. T. Junqueira, D. R. O. Silva, K. R. Firmiano, P. R. Kaufmann, P. S. Pompeu, and M. Callisto. 2014. The relative influence of multiple spatial scale environmental predictors on fish and macroinvertebrate assemblage richness in Cerrado ecoregion streams, Brazil. Landscape Ecology 29:1001–1016.

Machado, A. B. M. 2005. Lista da fauna brasileira ameaçada de extinção: incluindo as espécies quase ameaçadas e deficientes em dados. [List of threatened Brazilian fauna: including near-threatened and data-deficient species.] Fundação Biodiversitas, Belo Horizonte, Brazil.

Marchetti, M. P., T. Light, P. B. Moyle, and J. Viers. 2004. Fish invasions in California watersheds: testing hypotheses using landscape patterns. Ecological Applications 14:1507–1525.

Meador, M. R., and R. M. Goldstein. 2003. Assessing water quality at large geographic scales: relations among land use, water physicochemistry, riparian condition, and fish community structure. Environmental Management 31:504–517.

Meffe, G. K., and F. F. Snelson. 1989. Ecology and evolution of livebearing fishes (Poeciliidae). Prentice Hall, Upper Saddle River, New Jersey.

Minchinton, T. E. 2002. Disturbance by wrack facilitates spread of *Phragmites australis* in a coastal marsh. Journal of Experimental Marine Biology and Ecology 281:89–107.

Mouillot, D., A. J., N. A. J. Graham, S. Villéger, N. W. H. Mason, and D. R. Bellwood. 2013. A functional approach reveals community responses to disturbances. Trends in Ecology and Evolution 28:167–177.

Oksanen, J., F. G. Blanchet, M. Friendly, R. Kindt, P. Legendre, D. McGlinn, P. R. Minchin, R. B. O'Hara, G. L. Simpson, P. Solymos, M. Henry, H. Stevens, E Szoecs, and H. Wagner. 2018. vegan: community ecology package. R package version 2.0-7. Available: https://cran.r-project.org/web/packages/vegan/vegan.pdf. (February 2019).

Oliveira-Junior, J. M. B., P. de Marco-Junior, K. Dias-Silva, R. P. Leitão, C. G. Leal, P. S. Pompeu, T. A. Gardner, R. M. Hughes, and L. Juen. 2017. Effects of human disturbance and riparian conditions on Odonata (Insecta) assemblages in eastern Amazon basin streams. Limnologica 66:31–39.

Oliveira-Junior, J. M. B., Y. Shimano, T. A. Gardner, R. M. Hughes, P. de Marco Junior, and L. Juen. 2015. Neotropical dragonflies (Insecta: Odonata) as indicators of ecological condition of small streams in the eastern Amazon. Austral Ecology 40:733–744.

Olsen, A. R., and D. V. Peck. 2008. Survey design and extent estimates for the wadeable streams assessment. Journal of the North American Benthological Society 27:822–836.

Panarari-Antunes, R. S., A. J. Prioli, S. M. A. P. Prioli, A. S. Galdino, H. F. Julio Junior, and L. M. Prioli. 2011. Genetic variability of *Brycon orbignyanus* (Valenciennes, 1850) (Characiformes: Characidae) in cultivated and natural populations of the upper Paraná River, and implications for the conservation of the species. Brazilian Archives of Biology and Technology 54:839–848.

Paul, M. J., and J. L. Meyer. 2001. Streams in the urban landscape. Annual Review of Ecology and Systematics 32:333–365.

Pusey, B. J., and A. H. Arthington. 2003. Importance of the riparian zone to the conservation and management of freshwater fish: a review. Marine and Freshwater Research 54:1–16.

Reid, A. J., A. K. Carlson, I. F. Creed, E. J. Eliason, P. A. Gell, P. T. J. Johnson, K. A. Kidd, T. J. MacCormack, J. D. Olden, S. J. Ormerod, J. P. Smol, W. W. Taylor, K. Tockner, J. C. Vermaire, D. Dudgeon, and S. J. Cooke. 2019. Emerging threats and persistent conservation challenges for freshwater biodiversity. Biological Reviews of the Cambridge Philosophical Society 94:849–873.

Roa-Fuentes, C. A., and L. Casatti. 2017. Influence of environmental features at multiple scales and spatial structure on stream fish communities in a tropical agricultural region. Journal of Freshwater Ecology 32:281–295.

Roth, N. E., J. D. Allan, and D. L. Erickson. 1996. Landscape influences on stream biotic integrity assessed at multiple spatial scales. Landscape Ecology 11:141–156.

Sály, P., P. Takács, I. Kiss, P. Bíró, and T. Erős. 2011. The relative influence of spatial context and catchment- and site-scale environmental factors on stream fish assemblages in a human-modified landscape. Ecology of Freshwater Fish 20:251–262.

Sazima, I. 1980. Behavior of two Brazilian species of parodontid fishes, *Apareiodon piracicabae* and *A. ibitiensis*. Copeia 1980:166–169.

Schiff, R., and G. Benoit. 2007. Effects of impervious surface at multiple spatial scales on coastal watershed streams. Journal of the American Water Resources Association 43:712–730.

Shapiro, M. H., S. M. Holdsworth, and S. G. Paulsen. 2008. The need to assess the condition of aquatic resources in the US. Journal of the North American Benthological Society 27:808–811.

Siqueira, T., L. M. Bini, F. O. Roque, S. R. Marques Couceiro, S. Trivinho-Strixino, and K. Cottenie. 2012. Common and rare species respond to similar niche processes in macroinvertebrate metacommunities. Ecography 35:183–192.

Stevens, D. L., and A. R. Olsen. 2004. Spatially balanced sampling of natural resources. Journal of the American Statistical Association 99:262–278.

Strassburg, B. B. N., T. Brooks, R. Feltran-Barbieri, A. Iribarrem, R. Crouzeilles, R. Loyola, A. E. Latawiec, F. J. B. Oliveira Filho, C. A. M. De Scaramuzza, F. R. Scarano, B. Soares-Filho, and A. Balmford. 2017. Moment of truth for the Cerrado hotspot. Nature Ecology and Evolution [online serial] 1:0099.

Strayer, D. L., and D. Dudgeon. 2010. Freshwater biodiversity conservation: recent progress and future challenges. Journal of the North American Benthological Society 29:344–358.

Terra, B. F., R. M. Hughes, and F. G. Araújo. 2016. Fish assemblages in Atlantic Forest streams: the relative influence of local and catchment environments on taxonomic and functional species. Ecology of Freshwater Fish 25:527–544.

Teresa, F. B., and L. Casatti. 2013. Development of habitat suitability criteria for Neotropical stream fishes and an assessment of their transferability to streams with different conservation status. Neotropical Ichthyology 11:395–402.

USGS (U.S. Geological Survey). 2005. USGS EROS archive – digital elevation – Shuttle radar topography mission (SRTM). Available: www.usgs.gov/centers/eros/science/usgs-eros-archive-digital-elevation-shuttle-radar-topography-mission-srtm?qt-science_center_objects=0#qt-science_center_objects. (October 2018).

Vörösmarty, C. J., P. B. McIntyre, M. O. Gessner, D. Dudgeon, A. Prusevich, P. Green, S. Glidden, S. E. Bunn, C. A. Sullivan, C. R. Liermann, and P. M. Davies. 2010. Global threats to human water security and river biodiversity. Nature (London) 467:555–561.

Wang, L., J. Lyons, P. Kanehl, R. Bannerman, and E. Emmons. 2000. Watershed urbanization and changes in fish communities in southeastern Wisconsin streams. Journal of the American Water Resources Association 36:1173–1189.

Wang, L., J. Lyons, P. Rasmussen, P. Kanehl, P. Seelbach, T. Simon, M. Wiley, E. Baker, S. Niemela, and P. M. Stewart. 2003. Watershed, reach, and riparian influences on stream fish assemblages in the Northern Lakes and Forest Ecoregion, U.S.A. Canadian Journal of Fisheries and Aquatic Sciences 60:491–505.

Warbanski, M. L., P. Marques, T. C. Frauendorf, D. A. T. Phillip, and R. W. El-Sabaawi. 2017. Implications of Guppy (*Poecilia reticulata*) life-history phenotype for mosquito control. Ecology and Evolution 7:3324–3334.

Winemiller, K. O. 1989. Patterns of variation in life history among South American fishes in seasonal environments. Oecologia 81:225–241.

Zeni, J. O., D. J. Hoeinghaus, and L. Casatti. 2017. Effects of pasture conversion to sugarcane for biofuel production on stream fish assemblages in tropical agroecosystems. Freshwater Biology 62:2026–2038.

Appendix: Number of Collected Species by Order and Family

Table A.1.1. Number of species collected in the orders and families for Três Marias (TM), Nova Ponte (NP), Volta Grande (VG) and São Simão (SS).

Taxon	TM	NP	VG	SS
Characiformes				
Parodontidae	3	2	1	2
Curimatidae	–	–	–	2
Anostomidae	2	1	–	7
Crenuchidae	4	3	3	2
Characidae	19	10	11	17
Acestrorhynchidae	–	1	–	1
Erythrinidae	2	2	2	2
Lebiasinidae	–	–	1	1
Siluriformes				
Cetopsidae	–	–	–	1
Trichomycteridae	2	3	2	–
Callichthyidae	–	1	2	1
Loricaridae	13	4	5	10
Heptapteridae	7	4	5	8
Pseudopimelodidae	1	–	–	–
Auchenipteridae	1	–	1	1
Gymnotiformes				
Gymnotidae	1	1	2	2
Sternopygidae	1	–	–	3
Cyprinodontiformes				
Rivulidae	–	–	–	1
Poeciliidae	1	2	2	1
Synbranchiformes				
Synbranchidae	1	1	1	1
Perciformes				
Cichlidae	–	1	3	4
Total richness	58	36	40	67

American Fisheries Society Symposium 90:321–350, 2019
© 2019 by the American Fisheries Society

A New Regionalization Framework to Quantify How Physiography Mediates the Effect of Land Use on Stream Fishes

JEFFERSON T. DEWEBER[1], LOGAN SLEEZER, AND EMMANUEL A. FRIMPONG*
Department of Fish and Wildlife Conservation, Virginia Tech
310 West Campus Drive, Blacksburg, Virginia 24060, USA

Abstract.—Surrounding land use and cover can have profound effects on the physical, chemical, and biological properties of stream ecosystems. For this reason, changes in land use and cover throughout catchments often have strong effects on stream ecosystems that are particularly interesting to researchers. Additionally, natural physical and climatic, or physiographic, characteristics are important for determining natural land cover and constraining human land use and are also strongly related to stream habitat and biota. Because the physiographic template differs among catchments and is an important mediator of catchment processes, it is important to account for natural physiographic differences among catchments to understand the relationship between land use/cover and stream biota. In this paper, we develop and assess the usefulness of a regional framework, land use/cover distinguished physiographic regions (LDPRs), which is designed for understanding relationships between land use/cover and stream biota while accounting for the physiographic template. We classified hydrologic units into LDPRs based on physiographic predictors of land use and cover for the eastern and western United States through the use of multivariate regression tree analysis. Next, we used case study data to assess the usefulness of LDPRs by determining if the relationships between fish assemblage function and land use/cover varied among classes using hierarchical logistic regression models. Eight physiographic characteristics determined land cover patterns for both the eastern and western United States and were used to classify hydrologic units into LDPR classes. Five commonly used biotic metrics describing trophic, reproductive, and taxonomic groupings of fish species responded in varying ways to agriculture and urban land use across LDPRs in the upper Mississippi River basin. Our findings suggest that physiographic differences among hydrologic units result in different pathways by which land use and cover affects stream fish assemblages and that LDPRs are useful for stratifying hydrologic units to investigate those different processes. Unlike other commonly used regional frameworks, the rationale and methods used to develop LDPRs properly account for the often-confounded relationship

* Corresponding author: frimp@vt.edu
[1] Current address: Fishery Research Station, Argenweg 50/1, Langenargen, Baden-Wuerttemberg 88085, Germany

between physiography and land use/cover when relating land cover to stream biota. Therefore, we recommend the use and refinement of LDPRs or similarly developed regional frameworks so that the varying processes by which human land use results in stream degradation can be better understood.

Introduction

The adverse effects of agricultural and urban land use on streams and stream biota have been well documented by a large and growing volume of literature (Allan 2004; Wagenhoff 2011; Schinegger et al. 2013; Wolter et al. 2016). Land uses refer to the myriad ways in which humans manipulate land to obtain products or services from present natural resources, while land cover can be defined as the natural (vegetation, ice, rock outcrops, etc.) or anthropogenic (buildings, roads, etc.) structures that occur on the Earth's surface. Because human land use plays a major role in modifying and shaping land cover manifestations across the Earth's surface, we use the term "land use/cover" in the remainder of this chapter to refer to both the anthropogenic surfaces created by human land uses and natural surfaces as land cover, unless specifically noted. Whereas researchers now possess a basic understanding of how agricultural and urban land uses can affect stream assemblages, debate still exists as to how the impacts of land use on stream assemblages varies with spatial extent (Downes et al. 2002; Allan 2004; Esselman and Allan 2010). The development of effective catchment management strategies requires an understanding of the processes by which land use/cover disturbances at multiple extents affect habitat and biota. Restoration of urbanized and agricultural streams, for example, hinges upon knowing how the impacts occur, attempting to reverse these processes, and then monitoring the resulting changes (Barton and Farmer 1997; Bernhardt and Palmer 2007).

In addition to uncertainty regarding spatial extents, regional variation in the physiographic template, or the natural physical and climatic characteristics of a region, leads to inconsistent conclusions regarding the relative effects of environmental stressors on streams by different land use/cover types. The physiographic template partially determines stream and river characteristics (Hynes 1975; Montgomery 1999) and, as a result, is related to associated biotic assemblages. Steep slopes can be directly related to flashy flows; soil and bedrock permeability determine groundwater infiltration, which can control base flow; and regional geology places constraints on channel morphology (Hynes 1975; Frissell et al. 1986; Burt 1992; Montgomery 1999). These natural differences in processes are also likely to result in different pathways by which streams interact with their catchments (Montgomery 1999).

A growing number of studies have addressed how basin physiography plays a role in determining the magnitude of the effects of land use/cover on stream biota in streams with differing geologic and temperature regimes (e.g., Utz et al. 2009, 2011; Neff and Jackson 2012). Some of those studies have uncovered important regional differences. For example, Utz et al. (2009) found that aquatic macroinvertebrate diversity was more vulnerable to degradation by urban land use/cover in the Piedmont physiographic region than in the coastal plain of the eastern United States. Regional variation in aquatic ecosystem responses to urban development was also evident in a series of studies aimed at comparing the effects of urban sprawl on streamflow, instream habitat, and physicochemical and biotic properties of catchments containing major metropolitan areas across the United States (Sprague and Nowell 2008; Brown et al. 2009; Cuffney et al. 2010; Hopkins et al. 2015). Neff and Jackson (2012) demonstrated the effects of changing surficial geology on fish species richness and emphasized the role of geology in shaping not only the trophic and reproductive characteristics of fish species present at sites, but also the land use/cover occurring across these geological gradients. In addition, McManamay et al. (2015) showed that natural variations in stream temperature, substrate, and flow characteristics interact with a type of land use disturbance (dam flow regulation) to influence spatial patterns in abundance of fishes with different life history and reproductive strategies. Fish assemblages have also been closely linked to stream gradient, catchment area, temperature, groundwater inputs, and geology (Burton and Odum 1945; Paller 1994; Wang et al. 2003; Esselman and Allan 2010; Neff and Jackson 2012).

Many of these same physiographic characteristics also determine natural land cover and suitable locations for agriculture and urban land uses. Studies have found that slope and elevation are positively correlated with forest and grasslands, presumably because areas of high elevation and steep slopes are less suitable for agricultural land use (Iverson 1988; Hietel et al. 2004; Frimpong et al. 2006; Reger et al. 2007). Agricultural land use is positively associated with organic matter content, percentage clay, and moistness of soil (Iverson 1988; Burgi and Turner 2002; Hietel et al. 2004) and negatively associated with sandy soils (Burgi and Turner 2002). Although there are few studies available regarding constraints on urban land use, Iverson (1988) suggested that urbanization may only be constrained by water availability in a case study of Illinois.

Because physiographic characteristics are strongly related to many stream properties, stream fish assemblage characteristics, and land use/cover signatures, it is important for future studies aimed at quantifying effects of land use/cover changes on aquatic assemblages to account for these underlying characteristics (Allan 2004; Govenor et al. 2019). Several investigators have acknowledged this and included measurements of various landscape characteristics as covariates in their analyses (Fitzpatrick et al. 2001; Snyder et al. 2003; Frimpong et al. 2005; Roy et al. 2007; Feld et al. 2016). A number of different metrics have been used, including catchment area, catchment length, drainage shape, drainage density, stream length, mean channel slope, channel sinuosity, percent slope of the catchment, topographical relief of the catchment, elevation above mean sea level, length-slope factor, permeability, percent sandy surficial deposits, and geology type (Fitzpatrick et al. 2001; Frimpong et al. 2005; Roy et al. 2007; Wilson and Xenopoulos 2008; Marzin et al. 2013; Dala-Corte et al. 2016; Feld et al. 2016). This approach can be informative and shed light on processes that might otherwise go unnoticed. For example, Snyder et al. (2003) concluded that fish assemblages in higher gradient streams showed greater responses to urban land use than those in lower gradient streams. However, the process of selecting and measuring physiographic variables can require extensive time and resources, and comparisons among studies can be difficult because the same variables may be measured differently across studies. Such an approach, then, seems ineffective if the goal is to standardize comparisons and develop management strategies that transcend a single study area.

A second technique that may be useful for accounting for the physiographic template is to stratify a study region based on existing regional frameworks. Many studies account for physiographic characteristics by stratifying the study region through the use of U.S. Environmental Protection Agency (EPA) ecoregions (e.g., Goldstein et al. 2002; Heitke et al. 2004; Dodds and Oaks 2007; Surasinghe and Baldwin 2015; Govenor et al. 2019) or physiographic provinces (Baker et al. 2006; Barker et al. 2006; Utz et al. 2009, 2011). Although existing regional frameworks, such as ecoregions and physiographic regions, are often used in an attempt to control for physiographic differences (Utz et al. 2009; Surasinghe and Baldwin 2015; Govenor et al. 2019), these regional frameworks were not designed for, and thus do not properly deal with, the confounding between physiography and land use/cover. Furthermore, regional land use/cover was used in the delineation of EPA ecoregions, but in a qualitative and subjective manner (Omernik 1987). This makes the use of EPA ecoregions inappropriate for controlling for underlying physical and climatic characteristics alone. A regional framework that is delineated using physiographic predictors of land use/cover rather than by land use/cover itself reduces the likelihood of confounding land use/cover with physiography and may prove more informative for understanding the effects of anthropogenic land uses.

Because of the many ways in which regional frameworks are constructed to calibrate bioassessments, the efficacy and interpretability of such frameworks in describing variation in biotic communities attributable to spatial variation in physical properties differs greatly (Frimpong and Angermeier 2010). The usefulness of those frameworks is also dependent on the goals of individual study designs and the scales at which these studies are conducted. Aside from physiographic provinces and ecoregions, many other regional frameworks have been proposed for stratifying stream bioassessment studies at various scales from global regionalization to classifications of catchments and even individual stream segments (Higgins et al. 2005; Abell et al. 2008; McManamay et al. 2018). However, to our knowledge, no proposed regional framework has specifically been developed to address the confounding effects of physiography and land use on stream biota at a resolution appropriate for describing those patterns across the entire United States.

In this study, we developed a regional framework specifically designed to investigate relationships between land use/cover and stream ecosystems based on two hypotheses: (1) land use/cover relationships with stream biota vary between regions. (2) Those differences are at least partly explained by differences in underlying physiographic attributes. To test the first hypothesis, we first identified physiographic predictors of land use/cover at multiple spatial extents and developed a classification system, land use/cover distinguished physiographic regions (LDPRs), for selecting hydrologic units with relatively homogeneous physiographic templates. Second, we tested whether relationships between stream fish assemblages and land use/cover varied across LDPRs by using generalized linear mixed models with fish assemblages sampled from sites throughout the upper Mississippi River basin. We predicted that if land use/cover relationships with streams varied with underlying physiography and if LDPRs adequately capture these important differences, then metrics describing fish assemblages would have different responses to land cover across LDPRs.

Methods

Hydrologic Unit Framework and Data Description

The hydrologic landscape regions (HLR) data set was developed by the U.S. Geological Survey (USGS) as part of the National Water-Quality Assessment (NAWQA) program for the purpose of grouping streams based on characteristics expected to yield similar hydrologic responses (Wolock et al. 2004).

In the development of HLRs, Wolock et al. (2004) delineated 43,931 small (~200 km^2) subwatershed hydrologic units for the entire United States and then attributed the mean values of 13 physiographic variables to them, which are available in the online version (USGS 2003). Because these variables are potentially important for determining land use patterns, stream characteristics, and stream biota, this data set represented a unique, publicly available source of information suited to our purpose. It is important to note that the hydrologic units delineated by Wolock et al. (2004) include the catchments of tributaries, headwaters, and confluence to confluence stream segments (Omernik et al. 2017). To reduce confusion, we will refer to these as hydrologic units throughout this chapter and reserve the term catchment for all areas upstream of a given point on a stream network.

To ensure spatial representativeness in training data, we selected two stratified random samples of hydrologic units for the eastern and western United States by randomly selecting 10% of the hydrologic units from each of the original 20 HLR classes of Wolock et al. (2004). We developed separate eastern and western U.S. classifications because a large amount of variation explained by differences in potential evapotranspiration (PMPE) resulted in a natural split (see results). In cases where 10% constituted a sample size smaller than 30, either 30 or all available hydrologic units for that HLR class were selected to include as much natural variability in the sample as possible. The HLR data set included a number of small polygons that did not contain mapped streams and are apparently errors resulting from the original hydrologic unit delineation process. To minimize these errors, hydrologic units not containing streams were removed from the selected samples and replaced with hydrologic units containing streams, as determined with the medium resolution (1:100,000) USGS National Hydrography Dataset (USGS 2000). The final number of hydrologic units sampled totaled 2,200 and 1,890 for the eastern and western United States, respectively. We acquired the 30 m resolution National Land Cover Dataset 2001 and collapsed categories into eight broad classes of land use/cover: forest, wetland, shrubland, grassland, agriculture, urban, water, and barren (Homer et al. 2004). We then calculated the percent of each land use/cover class within the sampled hydrologic units and within 30-m buffers around streams. We used ArcGIS 9.2 for all geographic information system analyses.

Delineation of Land Use/Cover Distinguished Physiographic Regions through Multivariate Regression Trees

We developed multivariate regression trees (MRTs) for classifying hydrologic units into LDPRs based on physiographic predictors of land cover using the 'mvpart' package in R (De'ath 2010; R Core Team 2012). Multivariate regression tree analysis assumes that response variables follow a multivariate normal distribution. If all classes of land cover were used in the response matrix, the values for each extent would sum to one, creating a multinomial response and summarily violating the multivariate normal assumption. Therefore, we did not include land cover types that were underrepresented at both the hydrologic unit and riparian extents. Shrubland, barren, and water were excluded for the eastern United States, whereas wetland, barren, and water were excluded for the western United States. The 'mvpart' package creates MRTs by successively splitting the original sample based on predictors that maximize the sums of squared distances between the centroids of the two resulting groups (De'ath 2002). We used the default 10-fold cross-validation procedure to help select an optimal dendrogram (De'ath 2010). Eleven of the 13 physiographic attributes given in the HLR data set (Wolock et al.

2004) were used as predictors of land cover and are listed in Table 1. We did not include hydrologic unit perimeter (PERIMETER, km) in the analysis because there is likely no direct link to land use/cover patterns. We also did not include average annual potential evapotranspiration (PET, mm/year) because of redundancy with other climatic variables. We used MRTs to create alternative LDPR classifications by predicting land use/cover in both the hydrologic unit and a 30-m riparian buffer, and we refer to these as SubwatClass and RipClass, respectively.

The regression trees were restricted to have a minimum final node size of 50 so that the final classification system would contain what we deemed to be a sufficient number of hydrologic units in each class (an approximate minimum of 500) for future investigations. The simplest tree that fell within 1 SE of the minimum cross-validation error was selected each time (De'ath 2010). For each extent, the MRT that appeared most often out of 50 runs (i.e., the modal tree) was selected as the best and final model. We then numbered the SubwatClass and RipClass LDPRs from left to right in order of appearance on the trees and translated these into the geographic information system by following the decisions conveyed regarding break points in the physiographic attributes.

Case Study: Assessing Land Use/Cover Distinguished Physiographic Region Usefulness with Fish Assemblages

We acquired data from fish assemblage samples collected as part of the U.S. Geological Survey (USGS) NAWQA program for the years 1993–2002 (Gilliom et al. 1995; USGS 2010) and the EPA Regional Environmental Monitoring and Assessment Program (REMAP) data for the Mid-Atlantic Highlands sampled between the years 1993–1998 (EPA 1993, 2009). To ensure that fish assemblages in our samples were relatively similar while still spanning a relatively wide geographic range, we selected sites within the freshwater zoogeographic province for the upper Mississippi River basin (region A2A; Maxwell et al. 1995). We chose this province because it encompassed the greatest number of sampling

Table 1. Names and descriptions of physiographic attributes used with multivariate regression tree analysis to delineate land use/cover distinguished physiographic regions. All attributes come from the hydrologic landscape regions data set described by Wolock et al. (2004).

Abbreviation	Description	Units
AREA	Hydrologic unit area	km^2
PMPE	Average annual precipitation minus potential evapotranspiration	mm/year
TAVE	Average annual temperature	°C
AQPERMNEW	Aquifer permeability ranking, ordinal scale of 1–7, lowest to highest	None
SAND	Percent sand in soil	%
PFLATTOT	Percent total flatland	%
PFLATUP	Percent flatland in upland	%
PFLATLOW	Percent flatland in lowland	%
RELIEF	Hydrologic unit relief	m
SLOPE	Average hydrologic unit slope	%

sites in the combined data sets and spans a large area covering multiple LDPRs. Unlike the REMAP data, the NAWQA data set included samples from multiple years at many sites; therefore, we selected the sample from each NAWQA site from the year nearest to 2001 so that fish sampling occurred nearest the year of land cover imagery collection. We delineated catchments for REMAP sites using preprocessed digital elevation models acquired with the National Hydrography Dataset Plus (NHDPlus) after snapping the sites to NHDPlus streams (EPA and USGS 2009). For the NAWQA data, we acquired delineated catchments that were part of the NHDPlus data set. We visually checked to ensure that sampling sites had correctly delineated catchments and that each was located near a stream (i.e., within approximately 100 m); those not meeting these criteria were removed. The initial number of fish sites within the upper Mississippi numbered 201, with 166 sites from the NAWQA data and 35 from the REMAP data. We restricted sites to have catchment areas between 10 and 10,000 km^2 to reduce variation in fish assemblages due to catchment area, and we required a site to have more than 50 individual fish to help ensure that samples were representative of the fish assemblages in the streams.

We initially selected 28 fish metrics that were expected to respond to land use/cover based on previously published studies, hypothesized relationships, and availability of information for preliminary analyses (Table 2). We used the FishTraits database (Frimpong and Angermeier 2009) to classify fish species into guilds, functional groups, and taxonomic groups. All of the selected metrics were calculated as the number of species belonging to the group or displaying the trait divided by the total number of species in the sample. Prior to fitting final models, we used a correlation matrix, visually analyzed scatterplots and fit preliminary models to select several fish metrics that potentially exhibited a relationship with different land use/cover types. We selected the following five metrics for further analyses: proportion of benthic invertivores; proportion of centrarchids; proportion of darters, madtoms, and sculpins; proportion of omnivores; and proportion of substrate generalist species (shown in bold in Table 2). The proportion of darter, madtom, and sculpin species was moderately correlated (r = 0.75) with the proportion of benthic invertivore species because most darter, madtom, and sculpin species are benthic invertivores. There also are a number of cyprinid species that are benthic invertivores, however, so we decided to include both of these metrics in our analysis. Preliminary analysis for the other 23 metrics did not suggest strong relationships with land cover, so we did not use them in further analyses.

We calculated the percentage of urban and agricultural land cover in the catchment of each site and then assigned sites to LDPR SubwatClasses and RipClasses. Because the catchments often span across multiple LDPR hydrologic units, we developed a definition to assign sites while removing those that did not clearly belong to a single LDPR SubwatClass or RipClass. A site was assigned to a SubwatClass or RipClass whenever both the location of the fish sampling site and the majority of the catchment belonged to the same SubwatClass or RipClass, respectively. We removed sites from the analysis belonging to greatly underrepresented SubwatClasses and RipClasses to reduce inequality in sample sizes among classes, and also required sites to belong to both a SubwatClass and RipClass so that the usefulness of those spatial extents could be directly compared.

We fit generalized linear mixed models with a logistic link using the 'lme4' package in R (Bates et al. 2012) to explain each of the selected fish metrics. We used the default Gauss-Hermite approximation to the log like-

Table 2. The list of fish metrics used in preliminary analysis, along with a description and the formula used to define the metric. All metrics were calculated as percent of species (PS) or percent of abundance (PA), and we highlighted metrics chosen for further analysis in bold font. All fish trait information came from the FishTraits data set, and the FishTraits formula gives the exact expression used based on FishTraits field names (Frimpong and Angermeier 2009).

Metric abbreviation	Metric description	Formula
N_Species	Species richness	Sum of all species captured
Abundance	Total number of individuals	Sum of all individuals captured
Simpson	Simpson's diversity index	$1 - \Sigma(n_i/N_t)^2$, where n_i is the abundance of species i and N_t is the total abundance.
BenInv_PS and BenInv_PA	**Benthic invertivore**	Defined as a species that is a specialized benthic feeder and a specialized invertivore, one consuming small invertebrates and larval fish but not macrophytes, detritus, algae, phytoplankton, adult fish, crabs, crayfish, or blood. By this definition, a species may consume eggs or other unidentifiable matter. FishTraits formula: BENTHIC = 1, SURWCOL = 0, "INVLVFSH" = 1, all other food items equal 0 except "EGGS" and "OTHER".
Omni_PS and Omni_PA	**Omnivore**	Defined as a species for which phytoplankton, algae, macrophytes, and detritus were described as consisting of between 25% and 75% of the diet FishTraits formula: 0.25 < [("ALGPHYTO" + "MACVASCU" + "DETRITUS")/ ("ALGPHYTO" + "MACVASCU" + "DETRITUS" + "INVLVFISH" + "FSHCRCRB" + "BLOOD" + "EGGS" + "OTHER")] > 0.75
Gen_PS and Gen_PA	Trophic generalist	Defined as a species whose diet commonly consisted of four or more food items. FishTraits formula: "ALGPHYTO" + "MACVASCU" + "DETRITUS" + "INVLVFISH" + "FSHCRCRB" + "BLOOD" + "EGGS" + "OTHER" ≥ 4
OmniGen_PS and OmniGen_PA	Omnivore and tropic generalist	Defined as a species that is either an omnivore or trophic generalist.
HerbDet_PS and HerbDet_PA	Herbivore and detritivore	Defined as a species whose diet commonly consisted of one or more of phytoplankton, algae, macrophytes, and detritus but did not include any other food items. FishTraits formula: "ALGPHYTO" + "MACVASCU" + "DETRITUS") > 1, "INVLVFISH" + "FSHCRCRB" + "BLOOD" + "EGGS" + "OTHER" < 1
SubIn_PS and SubIn_PA	Substrate indifferent	Defined as a species that is described as either pelogophilic or a bearer. FishTraits formula: A_1_1 or C_1_3_4_C2_4 = 1

Table 2. Continued.

Metric abbreviation	Metric description	Formula
SubGen_PS and SubGen_PA	**Substrate generalist**	Defined as a species described as a phytolithophil, polyphil, ariadnophil, or speleophil not specific to rock cavities. FishTraits formula: A_1_4, A_2_4C, B_2_2, B_2_4, B_2_7B, or B_2_7C = 1
SubInGe_PS and SubInGe_PA	Substrate indifferent and/or generalist	Defined as either a substrate indifferent or a substrate generalist species.
Litho_PS and Litho_PA	Simple lithophil	Defined as a species described as a nonguarding lithophil associated with rock, gravel or sand substrate. FishTraits: A_1_3A or A_1_3B = 1
Speleo_PS and Speleo_PA	Specialized speleophils	Defined as a species described as either a nonguarding or guarding rock cavity speleophil. FishTraits formula: A_2_4A or B_2_7A = 1
CatMox_PS and CatMox_PA	Catostomid or mosostomid	Defined as a species belonging to either the genus *Catostomus* or *Moxostoma*. FishTraits formula: GID = "Catostom" or "Moxostom")
Centrar_PS and Centrar_PA	**Centrarchid**	Defined as a species belonging to the family Centrarchidae. FishTraits formula: FID = "Centrarc"
DrtMdSc_PS and DrtMdSc_PA	**Darter, madtom, and sculpin**	Defined as a species that is a member of one of the following genera: *Etheostoma*, *Percina*, *Noturus*, or *Cottus*. FishTraits formula: GID = "Etheosto," "Percina," "Noturus," or "Cottus"

lihood but used five points per axis to achieve greater accuracy (i.e., nAGQ = 5). Before analysis, catchment area was log-transformed and all predictors were centered by subtracting each value from the mean. Catchment area was included in all models because it is known to be an important predictor of many fish metrics and should be properly accounted for (Burton and Odum 1945). Models with more than one random effect were fit, assuming that random slopes and intercepts were uncorrelated. We developed two sets of models to make two different comparisons of interest to our analysis (see Table 3). First, we compared a simple model where the effect of one land cover (either urban or agriculture) was fixed across all sites (i.e., fixed slope), with a model where the effect could vary across LDPRs (i.e., random slopes). Specifically, we compared models 4 and 2 for urban land cover and models 5 and 3 for agricultural land cover (Table 3). Second, we compared a set of five models (model numbers 2–6 in Table 3) to extend our understanding of the relative and combined effects of agriculture and urban land cover on stream fish assemblages in a mixed modeling framework. We compared all models with sites grouped either by LDPR SubwatClass or RipClass, which made a total of four and eight models for each metric for the first and second comparisons, respectively. We did not make a direct comparison between fixed and random intercept models because the marginal log likelihood estimates of the 'lme4' package cannot be directly compared with log likelihoods from generalized linear models with no random effects. Comparing SubwatClass and RipClass models with only random intercepts and/or fixed slope terms, however, enabled us to indirectly determine whether random intercepts were better supported by one classification scheme and therefore were meaningful.

Before developing all models, we calculated c-hat estimates to check for overdispersion by building the full model in a logistic regression framework with only fixed effects. All c-hat values were between 0.69 and 1.04, suggesting that no corrections were necessary (Anderson 2008). To compare models, we calculated the Akaike information criterion corrected for small sample sizes (AIC_c), its change relative to the lowest AIC_c (ΔAIC_c), and the weight of evidence (WOE) for each model using the 'AICcmodavg' package in R (Mazerolle 2012). The number of parameters reported in the 'lme4' package is always counted as one parameter. In reality, the ef-

Table 3. Formulas used in hierarchical logistic regression analysis of fish responses to land cover. All formulas are part of the larger formula $y_{ij} = \pi_j + \varepsilon_{ij}$ for observation *i* in group *j*. *W* is catchment area, *U* is catchment urban land use, *A* is catchment agriculture land use, α is the intercept, μ_j is the intercept for class *j*, β_1 is effect of *W*, β_2 is the effect *U*, β_3 is the effect of *A*, θ_j is the effect of *U* for group *j*, and Z_j is the effect of *A* for group *j*. Sites were assigned into group *j* based on either land use/cover distinguished physiographic region SubwatClass or RipClass.

Number	Formula
1	$\text{logit}(\pi_{ij}) = \alpha + \mu_j + \varepsilon_{ij}$
2	$\text{logit}(\pi_{ij}) = \alpha + \mu_j + \beta_1 W_i + \beta_2 U_i + \varepsilon_{ij}$
3	$\text{logit}(\pi_{ij}) = \alpha + \mu_j + \beta_1 W_i + \beta_3 A_i + \varepsilon_{ij}$
4	$\text{logit}(\pi_{ij}) = \alpha + \mu_j + \beta_1 W_i + \beta_2 U_i + \theta_j U_i + \varepsilon_{ij}$
5	$\text{logit}(\pi_{ij}) = \alpha + \mu_j + \beta_1 W_i + \beta_3 A_i + Z_j A_i + \varepsilon_{ij}$
6	$\text{logit}(\pi_{ij}) = \alpha + \mu_j + \beta_1 W_i + \beta_2 U_i + \beta_3 A_i + \theta j U_i + Z_j A_i + \varepsilon_{ij}$

fective number of parameters estimated for a random effect in a generalized linear mixed model is unknown but is somewhere between $K = 1$ and $K = n - 1$, where n is the number of groups estimated (Bolker et al. 2009). Not knowing the exact number of parameters can also hinder estimating information criterion, but inferences at the population level (i.e., random effects are significant) can still be made using the estimated marginal log likelihood (Vaida and Blanchard 2005). We show information for all four models for comparison 1, but show only the set of models that explained a combined 95% of the WOE (i.e., the 95% confidence set) for each metric for comparison 2 because of the large number of weakly supported models. We then selected a representative model for each metric by favoring clearly supported models and requiring each extra term to reduce AIC_c by more than 2 and avoid the pretending variable problem (Anderson 2008).

Results

Our preliminary analyses using multivariate regression trees for the entire conterminous United States split the HLR data set roughly into eastern (PMPE) ≥ −7.94 mm/year) and western (PMPE < −7.94 mm/year) halves, with a few exceptions primarily in the Pacific Northwest. The strong split in PMPE was overwhelmingly responsible for the total explained variation, thus obscuring the details in the rest of the tree. As a result, we developed LDPRs separately for the eastern and western United States to have sufficient classification detail in each half, as described below.

Eastern Land Use/Cover Distinguished Physiographic Regions

The selected MRT for hydrologic unit land use/cover in the eastern United States has 10 LDPR SubwatClasses split by five landform and climatic variables (Figure 1). The five physiographic variables explained approximately 55.6% of the total variation in the eastern hydrologic unit land cover data and had a cross-validation error of 47.7%. The most important predictor of land cover was PMPE, explaining 41.4% of the total variance explained by the tree in the first split. The next split in the tree was defined by SLOPE and explained 22.9% of the explained variance. Four of the variables (PMPE, SLOPE, SAND, and TAVE) appear twice in the tree while PFLATTOT only appeared as a predictor in one tree split and explained 4.7% of the explained variance.

The land cover characteristics of the 10 LDPR SubwatClasses are given in Table 4, and the tree splits show how the physiographic template characteristics constrain these different land cover classes. For example, eastern SubwatClass 1 (E-1) is characterized by a mix of forest, wetland, and agricultural land cover and has a humid climate (PMPE ≥ 233.8 mm), flat topography (SLOPE < 0.66%), and sandy soils (SAND ≥ 23.5%). Eastern SubwatClasses E-2 and E-3 have less sandy soils (SAND < 23.5%) and have greater amounts of agriculture land cover. Eastern SubwatClasses E-4 and E-5 are humid with relatively steeper slopes (SLOPE > 0.66%) and are largely dominated by forest land cover. Eastern SubwatClasses E-6–E-10 are typically less humid (PMPE < 9.2) and generally have a larger proportion of grassland than the first five SubwatClasses but differ from each other based on other attributes. Some major differences in average land cover between classes result from percent sand, average temperature, and slope. E-8 is cooler and has much more agricultural activity than E-7, which is typified by a mix of agriculture, forest, and grassland. A sharp contrast in average land cover can also be seen between E-9 and E-10, as higher slopes (SLOPE > 0.68%) in E-10 are characteristics of hydrologic units with more grassland than agriculture.

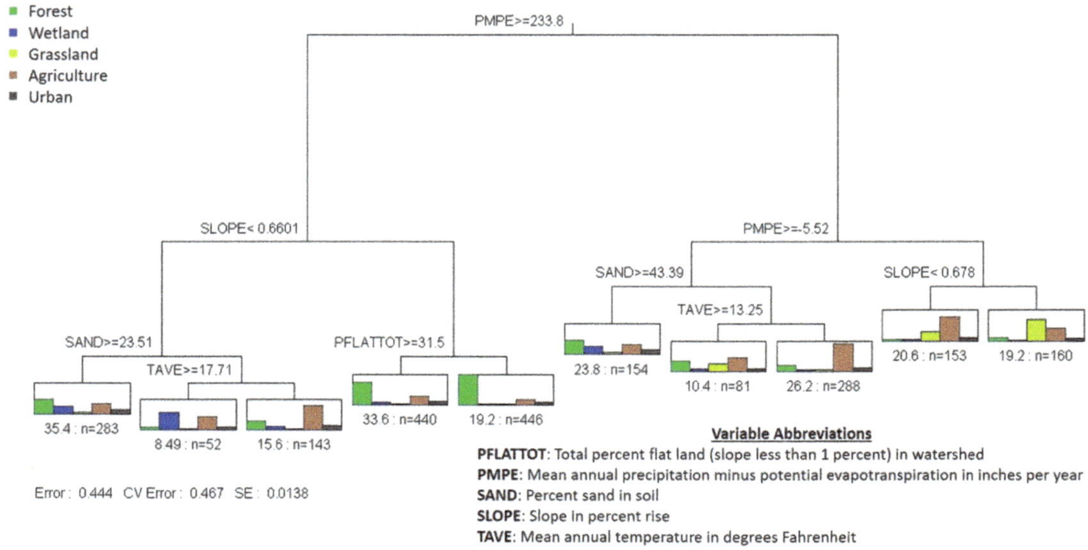

Figure 1. The multivariate regression tree for the eastern hydrologic unit land use/cover distinguished physiographic region SubwatClass shows the relationship between physiographic predictors and percent cover for five land use/cover types at the hydrologic unit extent. The original sample of 2,200 hydrologic units is split successively by physiographic variables, and the branches represent the amount of variance explained by each split. The number (*n*) of hydrologic units in each class is shown below each histogram. Classes were named as E-*X*, where *X* is the number from left to right.

As expected, there are strong spatial patterns for most of the 10 eastern LDPR classes (Figure 2). Not surprisingly, the wetlands dominated SubwatClass E-2 appears to be limited almost entirely to areas near the Mississippi River delta, and the agricultural-dominated Midwest is represented well by SubwatClass E-8. However, some SubwatClasses (such as E-1 and E-6) do not show such strong spatial correlations, but are rather widely distributed. The forested classes (E-4 and E-5) are generally constrained to mountainous or hilly regions.

The 30-m riparian buffer regression tree for the eastern United States (Figure 3) has 11 LDPR classes (RipClasses) determined by five physiographic characteristics: SLOPE, PMPE, MINELE, SAND, and TAVE. This tree explained 47.5% of the variance and had a cross-validation error of 56%. In contrast to the hydrologic unit land cover tree, PMPE was second to third in importance in the model and accounted for only 25.2% of the explained variance. Two landform variables, SLOPE and MINELE, explained 38.4% and 25.6% of the explained variance. TAVE and SAND appeared in the lower branches of the tree and together explained the remaining 10.8%.

The physiographic differences between RipClasses result in land use/cover signatures in a manner similar to the eastern SubwatClasses (Table 4). In general, hydrologic units with SLOPE < 0.69% (eastern RipClasses ERip-1–ERip-7) have more agriculture land cover in riparian buffers than those with greater slopes. This inverse relationship between agriculture and slope can also be seen by comparing ERip-1 with ERip-2. The model also implies that there is an inverse relationship between temperature and riparian agriculture, which distinguishes ERip-3 from

Table 4. Eastern catchment (E) and riparian (ERip) land use/cover distinguished physiographic region (LDPR) names and land cover characteristics. Mean (range) of the percent of total area covered by eight land cover types represent typical land cover signatures associated with physiographic attributes. Shrubland, barren, and water were not used in the classification of LDPRs in the eastern United States.

LDPR	Forest	Wetland	Grassland	Agriculture	Urban	Shrubland	Water	Barren
E-1	34.9	17.5	3.7	32.4	9.9	3.9	4.7	0.9
	(0, 90.8)	(0, 80.2)	(0, 27.2)	(0, 87.2)	(0, 94.4)	(0, 45.9)	(0, 99.8)	(0, 5.2)
E-2	6.5	42.4	1.8	30.7	5.2	6.5	10.1	0.2
	(0, 61.2)	(0, 91.0)	(0, 32.1)	(0, 96.9)	(0, 36.3)	(0, 29.0)	(0, 97.1)	(0, 4.6)
E-3	19.8	7.4	0.9	58.5	8.8	2.2	2.3	0.1
	(0, 95.0)	(0, 47.8)	(0, 9.6)	(0, 94.6)	(0, 68.8)	(0, 21.8)	(0, 21.3)	(0, 3.1)
E-4	55.7	5.7	2.8	20.7	8.3	3.7	2.6	0.2
	(6.2, 97.2)	(0, 36.9)	(0, 20.3)	(0, 86.8)	(0, 84.9)	(0, 22.6)	(0, 32.2)	(0, 6.0)
E-5	74.3	2.2	1.5	12.5	5.9	1.7	1.7	0.3
	(19.4, 99.8)	(0, 26.7)	(0, 19.8)	(0, 58.6)	(0, 44.4)	(0, 20.9)	(0, 32.1)	(0, 6.0)
E-6	33.4	20	5.3	23.6	11.3	1.9	4.1	0.4
	(0, 88.8)	(0, 100)	(0, 33.6)	(0, 78.1)	(0, 93.0)	(0, 19.5)	(0, 30.2)	(0, 11.4)
E-7	25.8	7.3	18.9	33.9	7.2	2.6	4.1	0.1
	(0.5, 69.3)	(0, 89.8)	(0, 75.4)	(0.1, 71.4)	(0.6, 86.8)	(0, 18.3)	(0, 36.2)	(0, 1.7)
E-8	14.3	3.2	4.3	66.7	8.6	0.3	2.5	0
	(0, 85.8)	(0, 77.9)	(0, 85.5)	(1.3, 96.9)	(0.3, 88.0)	(0, 6.7)	(0, 78.4)	(0, 1.8)
E-9	4.1	3.5	21.9	58.5	7.6	1.2	3.0	0.1
	(0, 46.3)	(0, 21.0)	(0, 96.8)	(0, 93.9)	(0, 80.3)	(0, 20.9)	(0, 59.2)	(0, 5.8)
E-10	7.5	1.9	51	31.2	5.3	1.2	1.7	0.2
	(0, 47.6)	(0, 7.1)	(1.8, 95.5)	(0, 91.5)	(0, 91.4)	(0, 45.8)	(0, 21.1)	(0, 6.3)
ERip-1	7.3	41.9	1.8	27.8	6.9	1.1	12.9	0.4
	(0, 59.6)	(0, 100)	(0, 27.9)	(0, 95.7)	(0, 78.9)	(0, 12.6)	(0, 100)	(0, 6.0)
ERip-2	25.8	45.8	2.1	8.3	5.4	2.7	9.6	0.4
	(0, 85.6)	(0, 100)	(0, 14.1)	(0, 51.1)	(0, 100)	(0, 24.2)	(0, 98.6)	(0, 9.8)
ERip-3	13.3	12.2	31.7	28.3	5.4	0.5	8.4	0.3
	(0, 54.5)	(0, 67.4)	(0, 74.7)	(0, 88.7)	(0, 46.8)	(0, 13.4)	(0, 77.3)	(0,16.7)
ERip-4	2.9	11.4	15.9	58.7	4.6	0.2	6.2	0.0
	(0, 33.0)	(0, 52.1)	(0, 82.1)	(9.4, 100)	(0, 18.7)	(0, 4.9)	(0, 46.1)	(0, 1.9)
ERip-5	26.4	37.4	2.9	18.1	4.9	0.7	9.5	0.1
	(0, 80.0)	(1.1, 100)	(0, 12.0)	(0, 79.2)	(0, 47.2)	(0, 7.3)	(0, 59.1)	(0, 1.1)
ERip-6	37.4	17.2	3.1	29.6	4.3	1.9	6.3	0.1
	(0, 100)	(0, 65.7)	(0, 53.3)	(0, 91.7)	(0, 66.3)	(0, 16.8)	(0, 70.6)	(0, 4.2)
ERip-7	19.8	10.6	3.9	52.3	7.7	0.4	5.2	0.0
	(0, 67.7)	(0, 72.3)	(0, 24.0)	(0, 93.6)	(0.5, 63.5)	(0, 15.8)	(0, 62.7)	(0, 2.2)
ERip-8	38.1	12.1	4.5	31.3	5.0	0.7	8.2	0.1
	(0.1, 100)	(0, 76.1)	(0, 64.4)	(0, 87.8)	(0, 57.9)	(0, 8.1)	(0, 48.3)	(0, 3.3)
ERip-9	47.3	26.5	0.9	8.2	4.7	3.1	9.1	0.2
	(0, 86.0)	(0, 85.0)	(0, 12.0)	(0, 61.3)	(0, 51.3)	(0, 14.4)	(0, 100)	(0, 2.7)
ERip-10	62.2	7.7	1.9	13.7	6.5	1.4	6.4	0.2
	(12.4, 99.8)	(0, 79.2)	(0, 23.2)	(0, 81.3)	(0, 76.4)	(0, 18.5)	(0, 58.6)	(0, 16.5)
ERip-11	16.0	9.4	42.4	20.1	3.1	1.0	7.9	0.2
	(0, 71.6)	(0, 50.7)	(0, 96.5)	(0, 98.2)	(0, 65.8)	(0, 28.7)	(0, 99.9)	(0, 5.5)

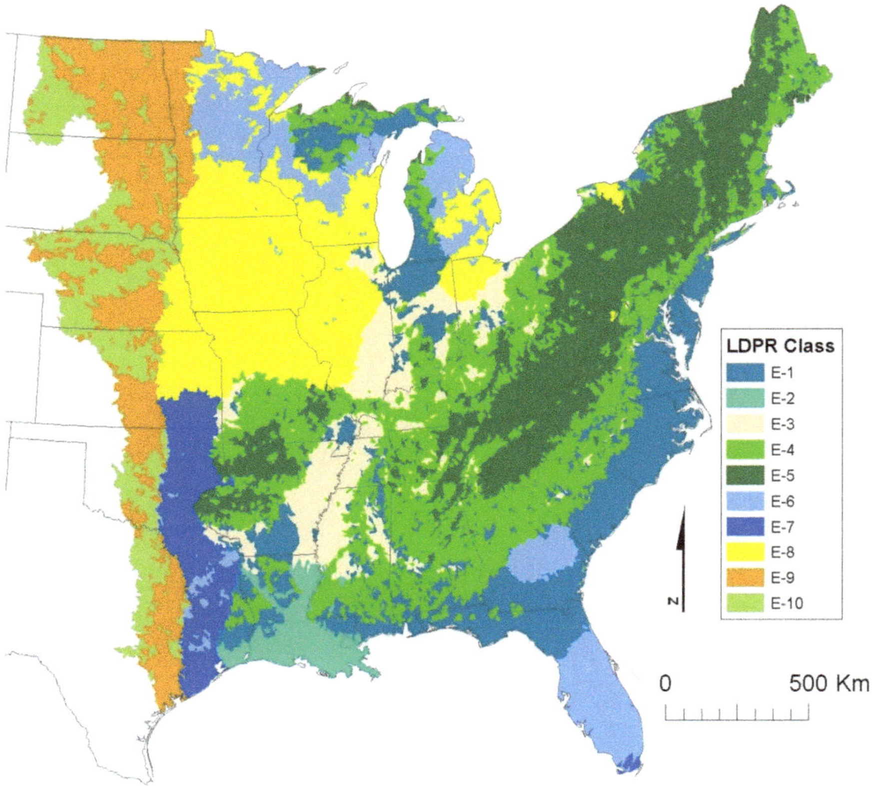

Figure 2. Map showing the distribution of eastern land use/cover distinguished physiographic region (LDPR) SubwatClasses (E), which are hydrologic units classified based on similarities in physiographic attributes that predict hydrologic unit land use/cover for the eastern United States, as described in the text.

ERip-4 and ERip-6 from ERip-7. Hydrologic units with sandy soils (ERip-5) also typically have less agriculture land cover within their riparian buffers.

A map of the different RipClasses (Figure 4) shows less spatial aggregation of hydrologic units than was evident for SubwatClasses. The location of mountainous, forested hydrologic units is still evident, for example, but is much less pronounced than for the SubwatClass regionalization. A close comparison of SubwatClasses and RipClasses reveals that some SubwatClasses have a majority of hydrologic units that are classified into a number of RipClasses, which reveals how different hydrologic unit and riparian land use/cover patterns can prevail under similar physiographic conditions. One clear example is that the largely agriculturally dominated hydrologic units of E-8 may either be typically dominated by riparian agriculture (ERip-7) or have more natural riparian land cover in the form of grasslands and forest (ERip-8).

Western Land Use/Cover Distinguished Physiographic Regions

The regression tree for hydrologic unit land cover in the western United States contains nine LDPR SubwatClasses split by five landform and climatic variables (Figure 5), as distributed in Figure 6. This tree explained 45.7% of the total variation in hydrologic unit land use/cover and had a cross-validation error of 57.9%. Potential evapotranspiration

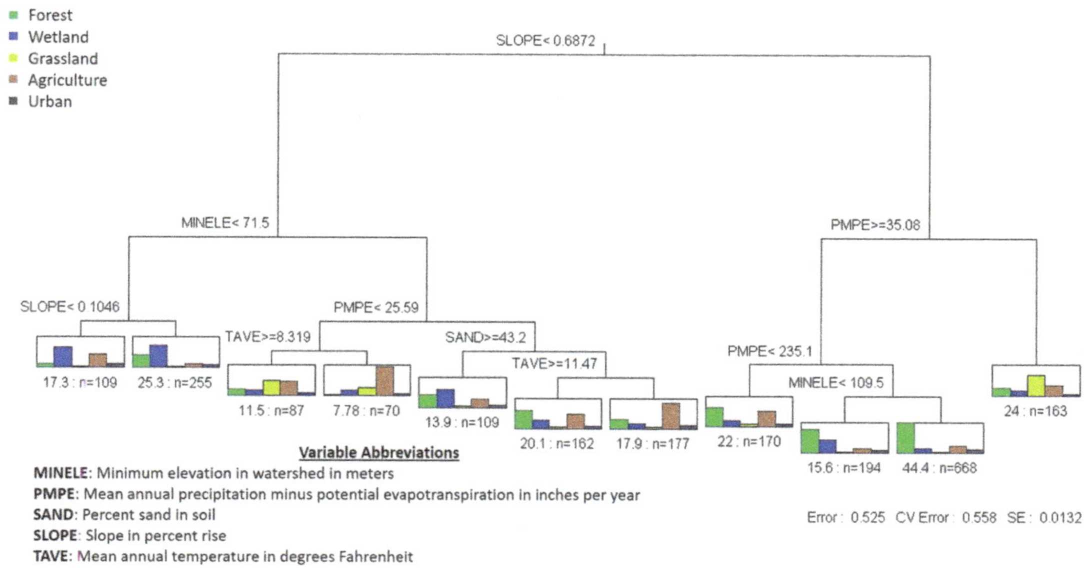

Figure 3. The multivariate regression tree for the eastern land use/cover distinguished physiographic region RipClasses classification shows the relationship between hydrologic unit physiographic predictors and percent cover of five land use/cover types within a 30-m riparian buffer. The original sample of 2,200 hydrologic units is split successively by physiographic variables and the branches represent the amount of variance explained by each split. The number (*n*) of hydrologic units in each class is shown below each histogram. Classes were named as ERip-*X*, where *X* is the number from left to right.

was the most important predictor, explaining 42.2% of the total variance explained by the tree in the first split and appearing two other times. SLOPE was the second most important, explaining 23.7% of the total tree variance, followed by PFLATTOT, and MINELE. In contrast to the results from the eastern United States, PMPE was the only variable to appear more than once and was the only climatic variable present. The 30-m RipClass tree for the western United States has 11 classes of hydrologic units with different land cover signatures separated by four physiographic predictors (Figure 7), shown spatially in Figure 8. In contrast to the eastern United States, the western RipClass model explained more variation (51.8%) than the SubwatClass model and had a lower cross-validation error (52.3%). PMPE was the most important variable and explained 40.1% of the explained variance in the first split and an additional 21.8% in two other branches. SLOPE, the only landform variable appearing, explained 26.3% of the explained variance in three splits, whereas PPT and TAVE explained the remaining 11.7%.

Case Study: Assessing Land Use/ Cover Distinguished Physiographic Regions with Fish Assemblages

The number of fish sampling sites that were eventually included in the analysis was 126 in six LDPR SubwatClasses (E-1, E-3, E-4, E-5, E-6, and E-8) and three RipClasses (ERip-5, ERip-7, and ERip-10). The selected NAWQA and REMAP fish sampling sites are spread across a diverse area and have a relatively even spread across the Upper Mississippi aquatic zoogeographic region A2A (Maxwell et al. 1995; Figure 9). The number of sites within each SubwatClass and RipClass, along with land cover and fish assemblage charac-

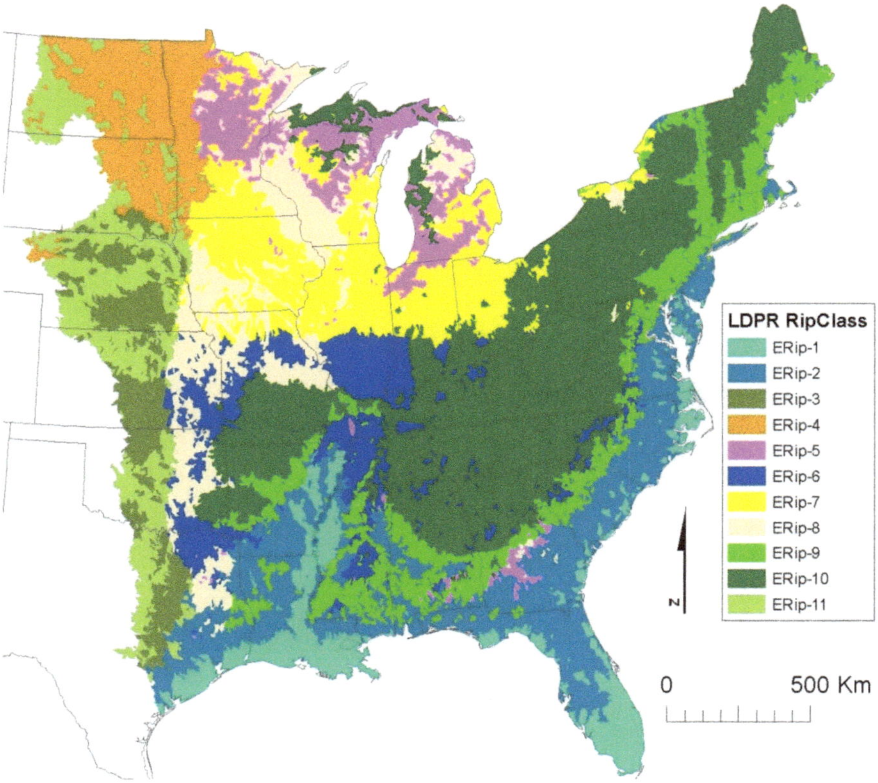

Figure 4. Map showing the distribution of the eastern RipClass (ERip) land use/cover distinguished physiographic regions (LDPRs), which are hydrologic units classified based on similarities in physiographic characteristics that predict land cover within 30-m riparian buffers for the eastern United States.

teristics, are shown in Table 5. Differences in land cover signatures between groups of sites within SubwatClasses and RipClasses were evident, as expected; even so, there was considerable overlap.

SubwatClass models for catchment urban and agriculture land cover were much better supported based on model weight and lower AIC_c scores than RipClass models for all metrics (Table 6). The model weight also clearly supported SubwatClass models with random slopes (models SubwatClass¬_4 and SubwatClass_5) for almost all metrics, and highly supported random slopes in urban effects on benthic invertivores (77%) and agriculture effects on substrate generalists (76%), centrarchids (96%), and omnivores (84%). The AIC_c was also reduced by more than 2 for all SubwatClass models when random slopes were included. In comparison, random slopes were only supported in RipClass models for agriculture effects on centrarchids ($\Delta AIC_c = 7.09$) and substrate generalists ($\Delta AIC_c = 2.92$).

The 95% confidence set for the different biological response metrics included two models for benthic invertivores, two models for substrate generalists, two models for centrarchids, and four models for each of omnivores and darters, madtoms, and sculpins (Table 7). SubwatClass models were again better supported for all metrics, and only four RipClass models appeared for all metrics combined and received no more than 16% of the model weight for any single metric. The most supported of benthic invertivore models was SubwatClass-6, but the reduction in

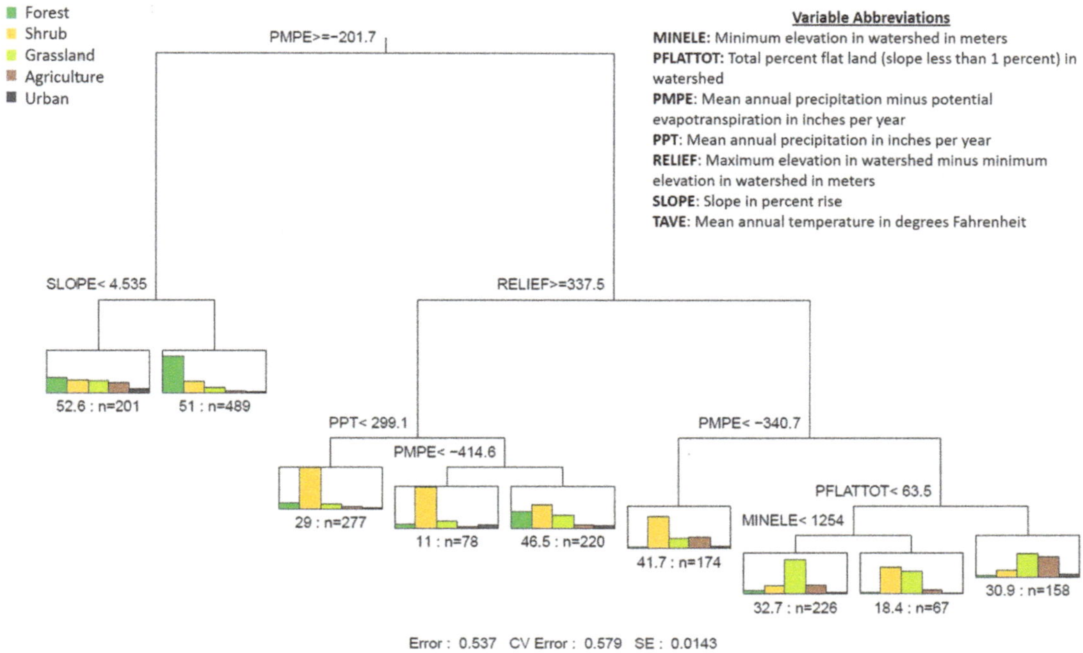

Figure 5. The multivariate regression tree for the western hydrologic units land use/cover distinguished physiographic region SubwatClass shows the relationship between physiographic predictors and percent cover for five land use/cover types at the hydrologic unit extent. The original sample of 1,890 hydrologic units is split successively by physiographic variables, and the branches represent the amount of variance explained by each split. The number (*n*) of hydrologic units in each class is shown below each histogram. Classes were named as W-*X*, where *X* is the number from left to right.

AIC_c was much less than 2. Thus, we selected the simpler SubwatClass_4 as the representative model for this metric. We selected SubwatClass-6 as the best representative model for substrate generalists because it was highly supported (83% model weight) and the less-supported RipClass-6 was equally complex. The selected centrarchid model was SubwatClass-6, which was very highly supported (93% WOE) compared to other models. For darters, madtoms, and sculpins, the more complex, first-ranked SubwatClass-6 had a $\Delta AIC_c = 0.49$ and was not considered. After removing this model, we selected the SubwatClass-4 as the best representative model because the model weight was 2.5 greater than the equally complex RipClass-4. The model results for omnivores suggest that the SubwatClass-6 is too complex based on available data and that either SubwatClass-4 or SubwatClass-5 was best supported, but there was no clear representative model from this group.

Discussion

The multivariate regression tree analysis was a useful method for classifying hydrologic units based on physiographic predictors of land use/cover. In this study, hydrologic units with low slope are typified by high proportions of agricultural land use, which agrees with a number of studies carried out across different spatial locations and extents (Iverson 1988; Hietel et al. 2004; Frimpong et al. 2006; Reger et al. 2007). Sandy soils have been shown in previous studies to constrain agriculture (Burgi and Turner 2002); this is clear-

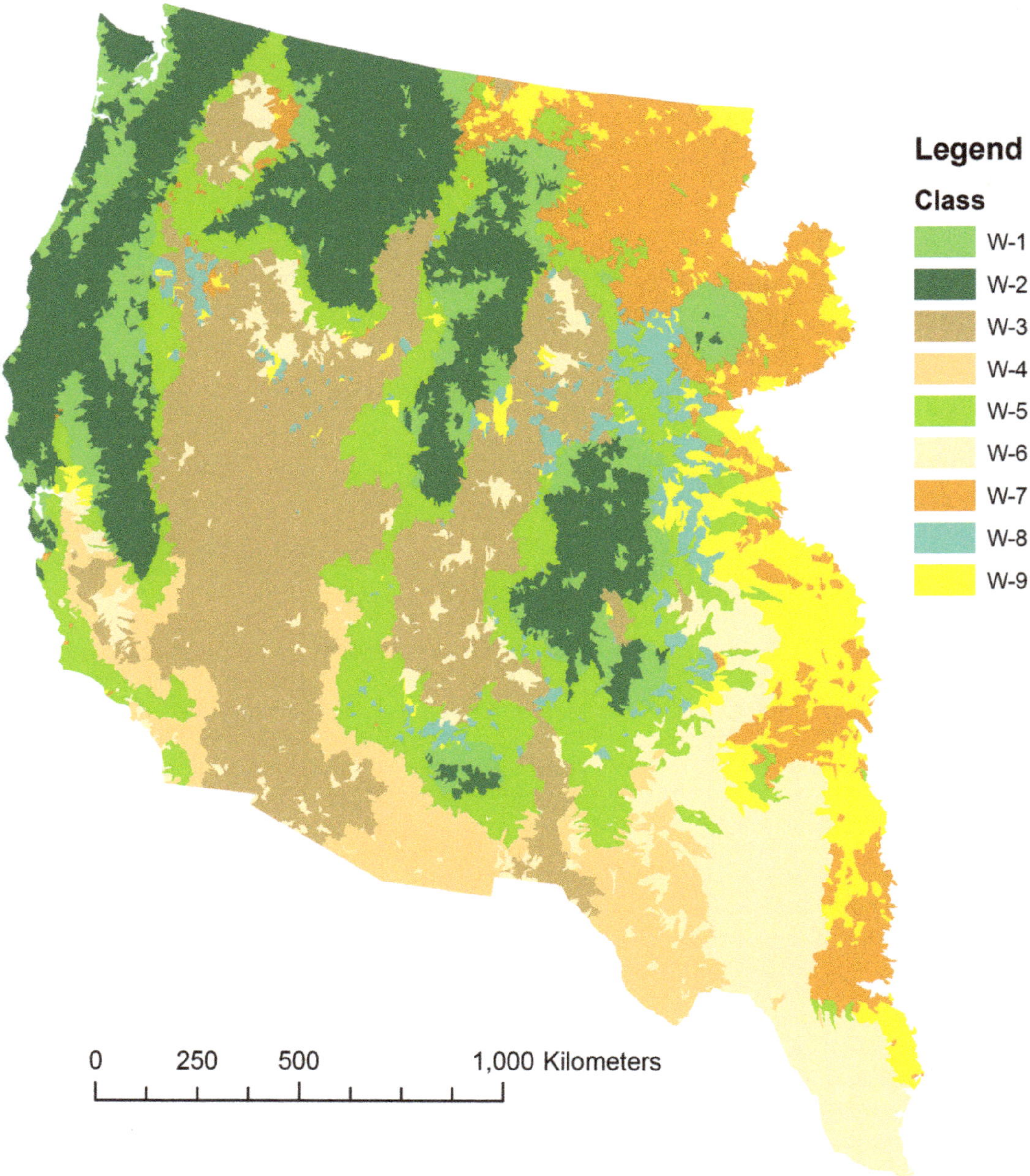

Figure 6. Map showing the distribution of western (W) land use/cover distinguished physiographic region SubwatClasses, which are hydrologic units classified based on similarities in physiographic attributes that predict hydrologic unit land use/cover for the western United States, as described in the text.

ly shown in LDPRs E-1 and E-6, which have relatively more sandy soils and less agricultural land use. Whereas agriculture is highly related to underlying physiography, urban land use differs only moderately between LDPRs, which agrees with the results of Iverson (1988). Despite seemingly low variation, we did find that urban land use at the hydrologic unit extent for the eastern LDPRs ranged from an average of 5.2% in class E-2 to 11.3%

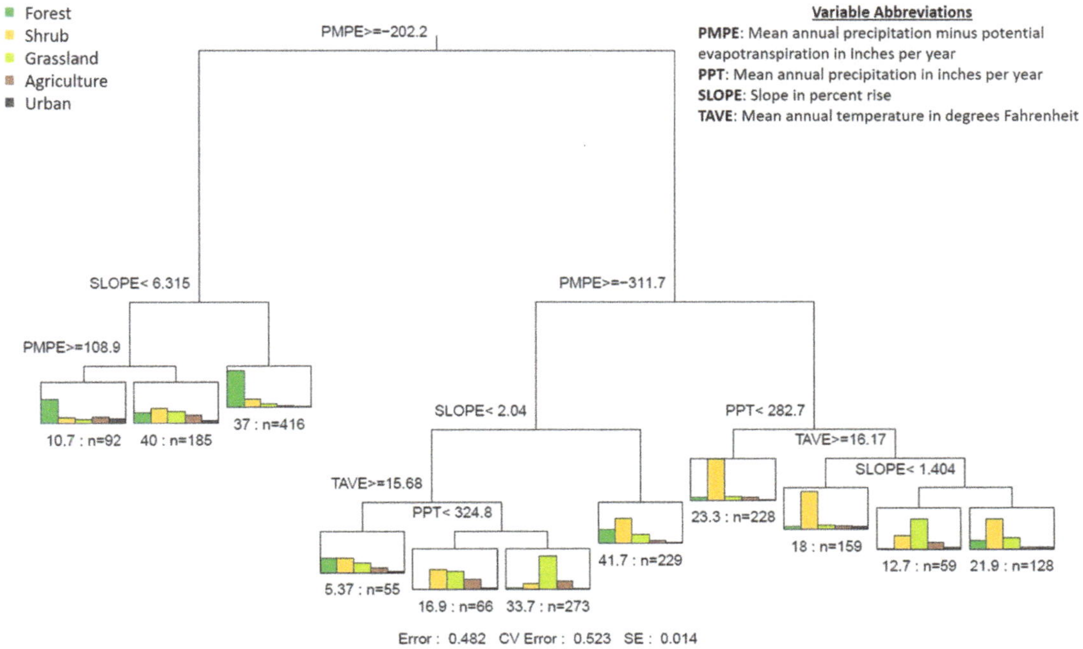

Figure 7. The multivariate regression tree for the western land use/cover distinguished physiographic region RipClasses classification shows the relationship between hydrologic unit physiographic predictors and percent cover of five land use/cover types within a 30-m riparian buffer. The original sample of 1,890 hydrologic units is split successively by physiographic variables, and the branches represent the amount of variance explained by each split. The number (*n*) of hydrologic units in each class is shown below each histogram. Classes were named as WRip-*X*, where *X* is the number from left to right.

in class E-6, which may be a significant level of variation. It could be that variation in urban land use/cover results more from socioeconomic and historical factors that determined where population centers are located than from physiography (Behrens 2007).

Our analysis covers a large extent across which climate varies widely, so it is not surprising that climatic characteristics appear as important predictors of both natural land cover and human land use. Average annual precipitation minus potential evapotranspiration was the most important predictor of land use/cover in this study and appears to primarily determine natural land cover at regional extents. This variable effectively split hydrologic units into those that were characterized by forest, wetland, grassland, or shrubland land cover. Although climatic predictors primarily differentiated natural land cover, we found that hydrologic units in the eastern United States with warmer annual mean temperatures (>13.3°C) had less agricultural land use/cover than hydrologic units with otherwise similar physiography and cooler temperatures.

The delineated LDPRs were found to closely follow regions defined by known land cover patterns, such as the largely agricultural Midwest or mountainous regions that are dominated by forest. The degree to which the classes mirror regions based on land cover is somewhat surprising because the amount of variation in land cover explained by even the best models seems fairly low and further demonstrates the close ties between physi-

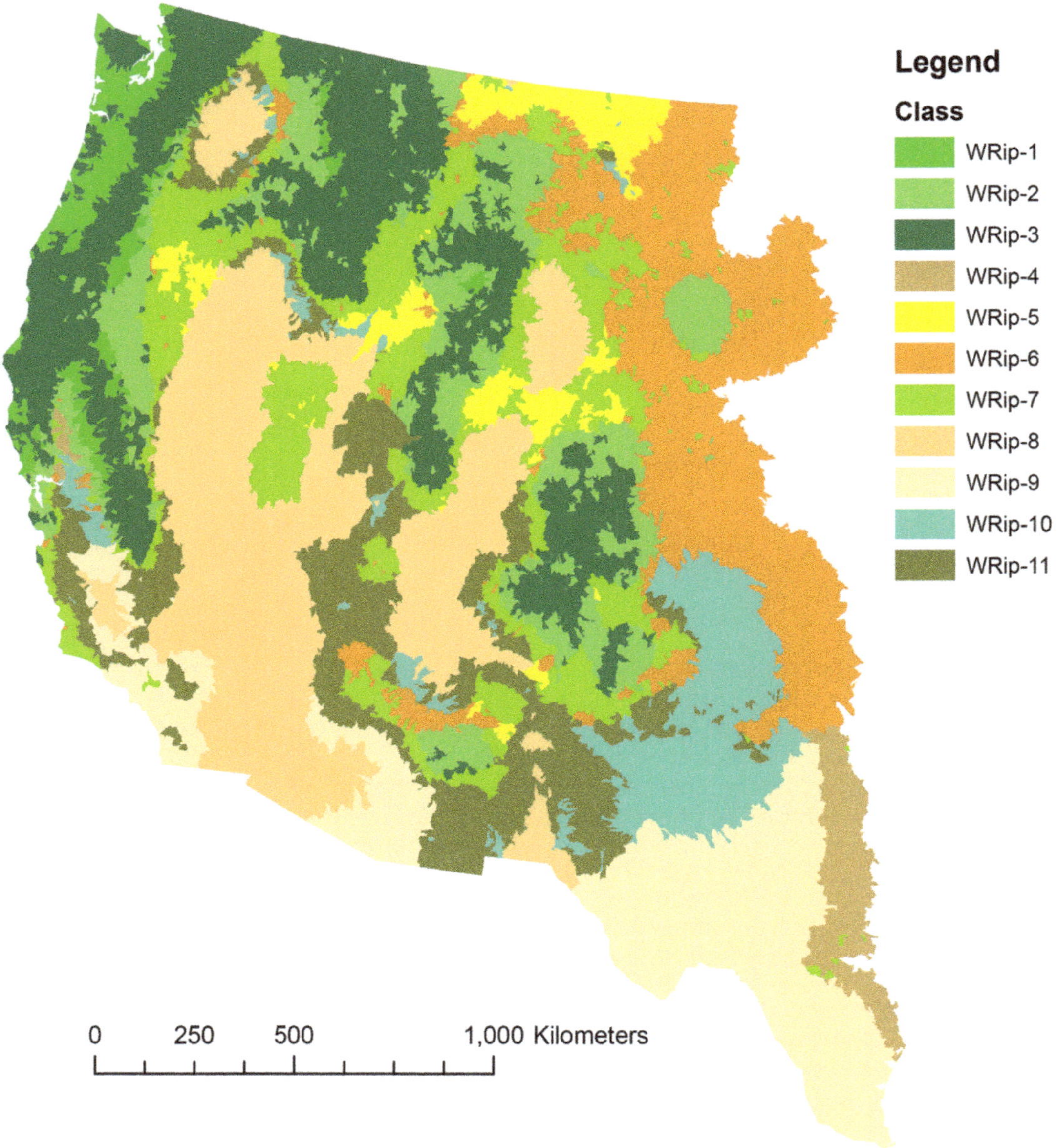

Figure 8. Map showing the distribution of the western RipClass (WRip) land use/cover distinguished physiographic regions, which are hydrologic units classified based on similarities in physiographic characteristics that predict land use/cover within 30-m riparian buffers for the eastern United States.

ography and land cover. Although the low amount of explained variance might at first seem like a drawback and some of this unexplained variance is attributable to unmeasured physiographic variability and sampling error, the unexplained variance in the model is also a useful characteristic of the proposed classifications in that this remaining variation partially represents the actual variation in land cover patterns within regions of similar physiography. If all or most variation in land cover was explained by LDPR boundaries,

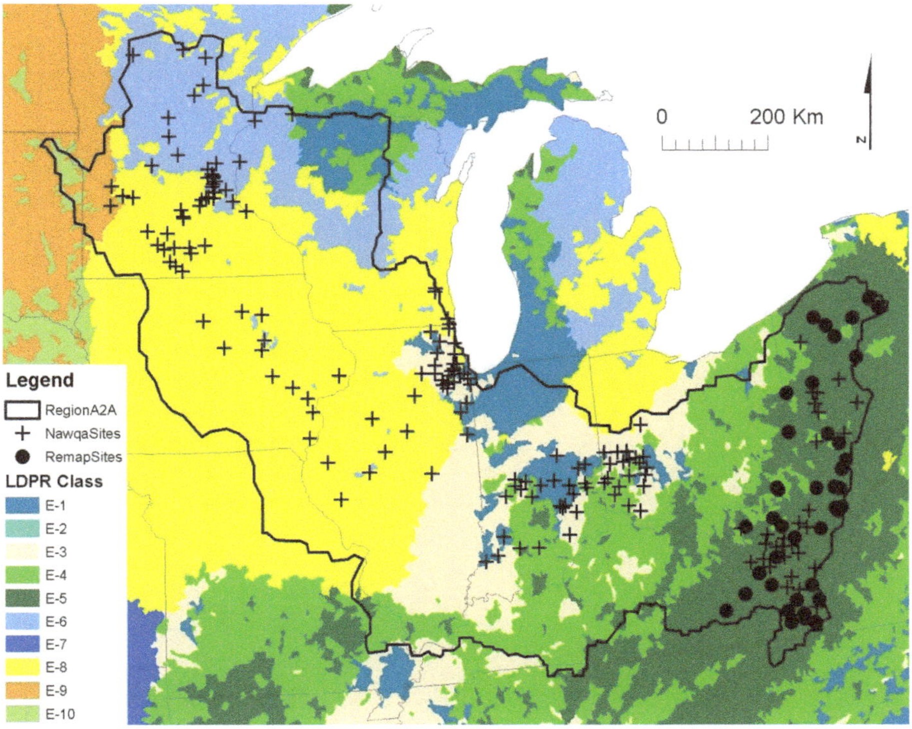

Figure 9. Map showing the location of fish sampling sites from the NAWQA (+) and REMAP (•) data sets within zoogeographic region A2A, the upper Mississippi River basin. State boundaries and land use/cover distinguished physiographic region (LDPR) SubwatClasses (E) are shown in the background.

physiography and land cover would be completely confounded. In that case, there would be no prospect for using this regional framework to better understand the mediating role of physiography in the impacts of land use on stream biota. Because there is variation, investigators can employ LDPRs to study the relationship between a stream response variable of interest and land cover predictors within a similar physiographic template (i.e., a select LDPR) and even compare among sites with different templates (i.e., comparing two or more LDPRs). Two major benefits of this approach are (1) it is far easier than measuring and accounting for physiographic variables on a study-by-study basis, and (2) it can give more interpretable results than stratifying catchments or hydrologic units through the use of existing regional frameworks because the splits in the regression trees convey information about the physiographic template.

Case Study: Assessing Land Use/Cover Distinguished Physiographic Regions with Fish Assemblages

Our results clearly demonstrate that the relationships between urban and agricultural land use/cover and stream fish vary across LDPR SubwatClasses. Such varying responses suggest that the processes linking land use/cover to streams and stream fish differ across LDPRs, supporting the usefulness of this regional framework for future investigations. Whereas we can estimate the variation in random slopes and intercepts, one limitation of the mixed model error structure is that we cannot directly compare the fitted effects among LDPRs because random effects

Table 5. Characteristics of the fish sampling sites grouped by land use/cover distinguished physiographic regions (LDPR) SubwatClass (E) and RipClass (ERip). Mean (range) of the five fish metrics used in models, agriculture and urban land covers, and catchment area (km^2). A description of the fish metric is given in Table 1.

LDPR	*N*	BenInv	Omni	SubGen	Centrar	DrtMdSc	Agriculture	Urban	Catchment area
E-1	13	13.7	56.6	45.5	20.7	17.2	63.8	22.8	957.9
		(0, 25.0)	(41.7, 75.0)	(25.0, 59.3)	(8.3, 29.6)	(0, 25.0)	(0.8, 91.7)	(4.8, 86.7)	(29.1, 5949.1)
E-3	23	13.9	55.7	46.6	20.6	18.2	56.7	32.1	231.2
		(0, 28.6)	(40.0, 75.0)	(22.2, 80.0)	(0, 40.0)	(0, 33.3)	(0.1, 89.9)	(4.4, 96.2)	(33.1, 1664.6)
E-4	11	14.3	58.7	44.5	20.3	18.3	52.0	29.2	133.0
		(0, 30.0)	(40.0, 72.7)	(26.3, 60.0)	(5.3, 31.6)	(0, 31.6)	(0.2, 84.0)	(5.4, 85.6)	(33.5, 824.8)
E-5	42	15.6	51.5	26.3	12.6	22.3	14.4	6.8	526.5
		(0, 40.0)	(20.0, 83.3)	(0, 62.5)	(0, 33.3)	(0, 48.5)	(0, 55.4)	(0.6, 51.8)	(11.6, 9781.7)
E-6	9	6.1	62.7	45.7	16.6	9.7	27.2	28.8	1,679.4
		(0, 12.5)	(48.1, 76.9)	(23.1, 62.5)	(7.7, 25.0)	(5.0, 16.7)	(1.3, 82.7)	(2.6, 81.2)	(39.9, 9986.9)
E-8	28	6.1	59.5	47.6	12.6	14.6	81.0	9.7	713.8
		(0, 21.1)	(41.7, 87.5)	(28.0, 80.0)	(0, 30.0)	(0, 31.6)	(37.2, 91.7)	(4.5, 39.9)	(60.8, 6049.1)
ERip-5	11	6.7	60.1	45.3	16.5	10.6	33.5	25.7	1922.6
		(0, 12.5)	(43.8, 76.9)	(23.1, 62.5)	(7.7, 25.0)	(5.0, 16.7)	(1.3, 82.7)	(2.6, 81.2)	(39.9, 9986.9)
ERip-7	63	10.6	57.6	46.9	17.2	16.7	69.1	20.6	504.3
		(0, 28.6)	(40.0, 87.5)	(22.2, 80.0)	(0, 40.0)	(0, 33.3)	(0.1, 91.7)	(4.4, 96.2)	(29.1, 6049.1)
ERip-10	52	15.2	53.2	29.8	14.1	21.3	21.2	11.4	452.4
		(0, 40.0)	(20.0, 83.3)	(0, 62.5)	(0, 33.3)	(0, 48.5)	(0, 84.0)	(0.6, 85.6)	(11.6, 9781.7)

Table 6. Model results for comparison of models with fixed and random effects. Model name is a combination of how the sites were grouped (by SubwatClass or RipClass) and the number of the model as listed in Table 3. *K* is the estimated number of parameters in the model, LL is the negative log-likelihood, AICcWt is the AIC_c model weight, and CumWt is the cumulative AICcWt.

Metric	Model name	*K*	LL	AIC_c	ΔAIC_c	AICcWt	Cum.Wt
BenInv_PS	SubwatClass_4	5	−60.02	130.08	0	0.77	0.77
	SubwatClass_2	4	−62.24	132.51	2.43	0.23	1
	RipClass_4	5	−69.59	149.21	19.13	0	1
	RipClass_2	4	−70.7	149.42	19.34	0	1
	SubwatClass_5	5	−63.57	137.17	0	0.79	0.79
	SubwatClass_3	4	−65.91	139.85	2.68	0.21	1
	RipClass_5	5	−74.4	158.83	21.67	0	1
	RipClass_3	4	−75.51	159.05	21.88	0	1
SubGen_PS	SubwatClass_4	5	−73.57	157.17	0	0.6	0.6
	SubwatClass_2	4	−75.79	159.6	2.43	0.18	0.78
	RipClass_4	5	−75.23	160.48	3.31	0.12	0.9
	RipClass_2	4	−76.34	160.69	3.52	0.1	1
	SubwatClass_5	5	−81.65	173.34	0	0.76	0.76
	SubwatClass_3	4	−83.94	175.9	2.56	0.21	0.97
	RipClass_5	5	−84.95	179.92	6.58	0.03	1
	RipClass_3	4	−89.5	187.01	13.68	0	1
Centrar_PS	SubwatClass_4	5	−67.42	144.87	0	0.71	0.71
	SubwatClass_2	4	−70.23	148.48	3.61	0.12	0.83
	RipClass_4	5	−69.47	148.98	4.1	0.09	0.92
	RipClass_2	4	−70.58	149.19	4.32	0.08	1
	SubwatClass_5	5	−68.86	147.74	0	0.96	0.96
	SubwatClass_3	4	−73.24	154.5	6.76	0.03	1
	RipClass_5	5	−75.05	160.13	12.39	0	1
	RipClass_3	4	−77.51	163.05	15.31	0	1
DrtMdSc_PS	SubwatClass_4	5	−55.95	121.93	0	0.54	0.54
	RipClass_4	5	−56.9	123.82	1.89	0.21	0.74
	SubwatClass_2	4	−58.17	124.36	2.43	0.16	0.9
	RipClass_2	4	−58.67	125.36	3.43	0.1	1
	SubwatClass_5	5	−61.58	133.19	0	0.72	0.72
	SubwatClass_3	4	−63.81	135.63	2.44	0.21	0.94
	RipClass_5	5	−64.67	139.36	6.16	0.03	0.97
	RipClass_3	4	−65.8	139.62	6.42	0.03	1
Omni_PS	SubwatClass_4	5	−42.69	95.4	0	0.7	0.7
	SubwatClass_2	4	−44.91	97.83	2.43	0.21	0.91
	RipClass_4	5	−45.34	100.7	5.3	0.05	0.96
	RipClass_2	4	−46.45	100.93	5.53	0.04	1
	SubwatClass_5	5	−42.7	95.42	0	0.84	0.84
	SubwatClass_3	4	−45.92	99.85	4.43	0.09	0.93
	RipClass_5	5	−45.87	101.77	6.36	0.04	0.97
	RipClass_3	4	−46.98	101.99	6.57	0.03	1

Table 7. The set of models receiving cumulative 95% of the weight of evidence are shown for each fish metric. The model that we selected as the best representative is shown in bold font (see the text for model selection); no model was selected for Omni_PS. Model name is a combination of how the sites were grouped (by SubwatClass or RipClass) and the number of the model as listed in Table 3. K is the estimated number of parameters in the model, LL is the negative log-likelihood, AICcWt is the AIC_c model weight, and CumWt is the cumulative AICcWt.

Metric	Model name	*K*	LL	AIC_c	ΔAIC_c	AICcWt	Cum.Wt
BenInv_PS	SubwatClass_6	7	−57.8	129.65	0	0.55	0.55
	SubwatClass_4	**5**	**−60.02**	**130.08**	**0.42**	**0.44**	**0.99**
SubGen_PS	**SubwatClass_6**	**7**	**−66.64**	**147.34**	**0**	**0.83**	**0.83**
	RipClass_6	7	−68.26	150.56	3.23	0.16	0.99
Centrar_PS	**SubwatClass_6**	**7**	**−62.37**	**138.8**	**0**	**0.93**	**0.93**
	SubwatClass_4	5	−67.42	144.87	6.07	0.04	0.98
DrtMdSc_PS	SubwatClass_6	7	−53.7	121.45	0	0.45	0.45
	SubwatClass_4	**5**	**−55.95**	**121.93**	**0.49**	**0.35**	**0.8**
	RipClass_4	5	−56.9	123.82	2.38	0.14	0.94
	RipClass_6	7	−55.76	125.57	4.12	0.06	1
Omni_PS	SubwatClass_6	7	−39.48	93.01	0	0.57	0.57
	SubwatClass_4	5	−42.69	95.4	2.39	0.17	0.74
	SubwatClass_5	5	−42.7	95.42	2.41	0.17	0.91
	RipClass_6	7	−41.6	97.25	4.24	0.07	0.98

for each class are not treated as separate parameters in the model (Vaida and Blanchard 2005). Future efforts could fit and compare separate logistic regression models for each LDPR to contrast the response of fish metrics to land cover. Although we did not compare random intercept and fixed intercept models, we did find that random intercept models were always more supported when sites were grouped by SubwatClass than by RipClass. This suggests that there are differences in the baseline proportions of fishes with different traits used in our analysis, and these differences likely have ecological significance representing adaptations to specific habitats. This is not a surprising result because differences in fish assemblages across broad regions with complex physiography are expected and constitute the basis for physiographic regionalization (e.g., Maxwell et al. 1995). For example, higher proportions of fishes reliant upon clean gravel substrates for feeding and reproduction are found in high-gradient mountain streams than in low-gradient streams that are often naturally silty. Such a contrast was evident when comparing the proportion of darter, madtom and sculpin species between mountainous E-5 and relatively flat E-8, and we think that these baseline differences are independent of human land use.

Other researchers have assessed the strength of regional frameworks for explaining biological responses using random groupings (Brenden et al. 2008), arbitrary or neutral regions (Wolock et al. 2004; Frimpong and Angermeier 2010), or site proximity models (Pyne et al. 2007). Such an assessment is unnecessary for this study given the differences between results for the two LDPR hydrologic unit stratification schemes. Models with sites grouped by LDPR SubwatClasses were always better supported than those that used RipClass to group sites, which shows that the criteria used to define regional boundaries are meaningful. This also suggests that LDPR SubwatClasses delineated based on predictors of land cover at the catchment extent may be more useful than RipClasses, for which riparian land cover was predicted, for explaining land cover patterns and investigating the impacts of land cover alterations on streams. If models for the two regionalization schemes were equally supported, one could conclude that the results resulted from the spatial proximity of sites within the groups rather than the boundaries. We did not compare our modeling results with other regional frameworks, but future studies may help to determine how regional frameworks developed to support different objectives compare with LDPRs.

This investigation has provided further support for the need to consider physiographic template differences when investigating the impacts of land use on streams and shows that LDPRs can be used to successfully stratify a study region based on physiography. Because investigations are often conducted in regions believed to be homogenous, there is not a large body of evidence to support these results. Two studies found that responses of stream macroinvertebrate and fish assemblages to urban land use differed between Piedmont and Coastal Plain physiographic regions in Maryland (Utz et al. 2009; Utz et al. 2010). Based on these and our results, we expect that contrasting effects are widespread because of differences in the processes connecting catchments and streams.

Applications

We recommend the use of LDPRs as a catchment stratification tool that can be used in a number of ways. First, control streams that are relatively unimpacted by human land use can be identified because there are catchments within every LDPR that have less-intensive human land use, even in agriculturally dominated areas. Second, environmental monitoring programs aimed at characterizing the streams of a region and identifying major human impacts can stratify catchments through the use of LDPRs. The use of LDPRs would enable streams that are representative of a region with a common physiographic template to be selected, thus increasing land cover gradients within a region and strengthening inferences about stream responses to land use. Last, LDPRs can be used to better understand the role of physiography in the interactions between land use/cover and streams by contrasting the effects of land use/cover between LDPRs.

Frimpong and Angermeier (2010) suggested that regional frameworks that are useful for understanding fish assemblage–land use/cover relationships may also be useful for understanding assemblages of freshwater mussels, macroinvertebrates, and other aquatic taxa because of the ecological and evolutionary relationships among different taxa. We encourage opportunities for further testing the framework with fish from other regions across the United States and against other taxonomic groups. We developed the LDPRs with a relatively coarse-grained data set that represented seamless coverage of the conterminous United States (Wolock et al. 2004). Whereas HLRs have sufficient resolution to represent patterns across the conterminous United States, finer-scale patterns in physiography are likely not represented in this regionalization. The classification could be improved with finer resolution data sets such as the National Hydrography Dataset series (NHD, NHDPlus, NHDPlusV2; www.horizon-systems.com/nhdplus/). With the concept demonstrated, we recommend that more detailed and potentially more accurate classifications within specific regions be pursued with NHDPlusV2 where desired.

Acknowledgments

We thank Dana Infante and three anonymous reviewers for helpful comments on our draft manuscript. J. T. D. would like to sincerely thank Don Orth and Feng Guo for helpful input and guidance towards the completion of this work and his master's thesis. We acknowledge the Edna Bailey Sussman Foundation and Virginia Tech Department of Fish and Wildlife Conservation for providing funding to support this work.

References

Abell, R., M. L. Thieme, C. Revenga, M. Bryer, M. Kottelat, N. Bogutskaya, B. Coad, N. Mandrak, S. Contreras Balderas, W. Bussing, M. L. J. Stiassny, P. Skelton, G. R. Allen, P. Unmack, A. Naseka, R. Ng, N. Sindorf, J. Robertson, E. Armijo, J. V. Higgins, T. J. Heibel, E. Wikramanayake, D. Olson, H. L. López, R. E. Reis, J. G. Lundberg, M. H. Sabaj Pérez, and P. Pauly. 2008. Freshwater ecoregions of the world: a new map of biogeographic units for freshwater biodiversity conservation. BioScience 58:403–414.

Allan, J. D. 2004. Landscapes and riverscapes: the influence of land use on stream ecosystems. Annual Review of Ecology, Evolution, and Systematics 35:257–284.

Anderson, D. R. 2008. Model based inference in the life sciences: a primer on evidence. Springer, New York.

Baker, M. E., D. E. Weller, and T. E. Jordan. 2006. Improved methods for quantifying potential nutrient interception by riparian buffers. Landscape Ecology 21:1327–1345.

Barker, L. S., G. K. Felton, G. K., and E. Russek-Cohen. 2006. Use of Maryland Biological Stream Survey data to determine effects of agricultural riparian buffers on measures of biological stream health. Environmental Monitoring and Assessment 117:1–19.

Barton, D. R., and M. E. D. Farmer. 1997. The effects of conservation tillage practices on benthic invertebrate communities in headwater streams in southwestern Ontario, Canada. Environmental Pollution 96:207–215.

Bates, D., A. Maechler, and B. Bolker. 2012. lme4: linear mixed-effects models using S4 classes. R package version 0.999999-0. Available: http://CRAN.R-project.org/package=lme4. (September 2012).

Behrens, K. 2007. On the location and lock-in of cities: geography vs transportation technology. Regional Science and Urban Economics 37:22–45.

Bernhardt, E. S., and M. A. Palmer. 2007. Restoring streams in an urbanizing world. Freshwater Biology 52:738–751.

Bolker, B. M., M. E. Brooks, C. J. Clark, S. W. Geange, J. R. Poulsen, M. H. H. Stevens, and J. White. 2009. Generalized linear mixed models: a practical guide for ecology and evolution. Trends in Ecology and Evolution 24:127–135.

Brenden, T. O., L. Z. Wang, and P. W. Seelbach. 2008. A river valley segment classification of Michigan streams based on fish and physical attributes. Transactions of the American Fisheries Society 137:1621–1636.

Brown, L. R., T. F. Cuffney, J. F. Coles, F. Fitzpatrick, G. McMahon, J. Steuer, A. H. Bell, and J. T. May. 2009. Urban streams across the USA: lessons learned from studies in 9 metropolitan areas. Journal of the North American Benthological Society 28:1051–1069.

Burgi, M., and M. G. Turner. 2002. Factors and processes shaping land cover and land cover changes along the Wisconsin River. Ecosystems 5:184–201.

Burt, T. P. 1992. The hydrology of headwater catchments. Pages 3–28 *in* P. Calow, and G. E. Petts, editors. The rivers handbook: hydrological and ecological principles. Blackwell Scientific Publications, Oxford, UK.

Burton, G. W., and E. P. Odum. 1945. The distribution of stream fish in the vicinity of

Mountain Lake, Virginia. Ecology 26:182–194.

Cuffney, T. F., R. A. Brightbill, J. T. May, and I. R. Waite. 2010. Responses of benthic macroinvertebrates to environmental changes associated with urbanization in nine metropolitan areas. Ecological Applications 20:1384–1401.

Dala-Corte, R. B., X. Giam, J. D. Olden, F. G. Becker, T. D. F. Gimaráes, and A. S. Melo. 2016. Revealing the pathways by which agricultural land-use affects stream fish communities in South Brazilian grasslands. Freshwater Biology 61:1921–1934.

De'ath, G. 2002. Multivariate regression trees: a new technique for modeling species-environment relationships. Ecology 83:1105–1117.

De'ath, G. 2010. mvpart: multivariate partitioning. R package version 1.6-0. Available: http://CRAN.R-project.org/package=mvpart. (May 2010).

Dodds, W. K., and R. M. Oaks. 2007. Headwater influences on downstream water quality. Environmental Management 41:367–377.

Downes B. J., L. A. Barmuta, P. G. Fairweather, D. P. Faith, M. J. Keough, P. S. Lake. 2002. Monitoring ecological impact: concepts and practice in flowing water. Cambridge University Press, Cambridge, UK.

EPA (U.S. Environmental Protection Agency). 1993. Regional environmental monitoring and assessment program. EPA, EPA/625/R/93/012, Washington D.C.

EPA (U.S. Environmental Protection Agency). 2009. Regional environmental monitoring and assessment program: data. Available: https://archive.epa.gov/emap/archive-emap/web/html/data.html. (May 2019).

EPA (U.S. Environmental Protection Agency) and USGS (U.S. Geological Survey). 2009. NHDPlus user guide. The NHDPlus Team. Available: www.horizon-systems.com/NHDPlus/NHDPlusV1_documentation.php. (May 2019).

Esselman, P. C., and J. D. Allan. 2010. Relative influences of catchment- and reach-scale abiotic factors on freshwater fish communities in rivers of northeastern Mesoamerica. Ecology of Freshwater Fish 19:439–454.

Feld, C. K., S. Birk, D. Eme, M. Gerisch, D. Hering, M. Kernan, K. Maileht, U. Mischke, I. Ott, F. Pletterbauer, S. Poikane, J. Salgado, C. D. Sayer, J. van Wichelen, and F. Malard. 2016. Disentangling the effects of land use and geo-climatic factors on diversity in European freshwater ecosystems. Ecological Indicators 60:71–83.

Fitzpatrick, F. A., B. C. Scudder, B. N. Lenz, and D. J. Sullivan. 2001. Effects of multi-scale environmental characteristics on agricultural stream biota in eastern Wisconsin. Journal of the American Water Resources Association 37:1489–1507.

Frimpong, E. A., and P. L. Angermeier. 2009. FishTraits: a database of ecological and life-history traits of freshwater fishes of the United States. Fisheries 34:487–495.

Frimpong, E. A., and P. L. Angermeier. 2010. Comparative utility of selected frameworks for regionalizing fish-based bioassessments across the United States. Transactions of the American Fisheries Society 139:1872–1895.

Frimpong, E. A., A. L. Ross-Davis, J. G. Lee, and S. R. Broussard. 2006. Biophysical and socioeconomic factors explaining the extent of forest cover on private ownerships in a Midwestern (USA) agrarian landscape. Landscape Ecology 21:763–776.

Frimpong, E. A., T. M. Sutton, B. A. Engel, and T. P. Simon. 2005. Spatial-scale effects on relative importance of physical habitat predictors of stream health. Environmental Management 36:899–917.

Frissell, C. A., W. J. Liss, C. E. Warren, and M. D. Hurley. 1986. A hierarchical framework for stream habitat classification—viewing streams in a watershed context. Environmental Management 10:199–214.

Gilliom, R. J., W. M. Alley, and M. E. Gurtz. 1995. Design of the National Water-Quality Assessment Program: occurrence and distribution of water quality conditions. U..S Geological Survey Circular 1112.

Goldstein, R. M., L. X. Wang, T. P. Simon, and P. M. Stewart. 2002. Development of a stream habitat index for the Northern Lakes and Forests Ecoregion. North American Journal of Fisheries Management 22:452–464.

Govenor, H., L. A. H. Krometis, L. Willis, P. L. Angermeier, and W. C. Hession. 2019. Macroinvertebrate sensitivity thresholds for sediment in Virginia streams. Integrated Environmental Assessment and Management 1:77–92.

Heitke, J. D., C. L. Pierce, G. T. Gelwicks, G. A. Simmons, and G. L. Siegwarth. 2004. Habitat, land use, and fish assemblage relationships in Iowa streams: preliminary assessment in an agricultural landscape. Pages 287–303 *in* R. M. Hughes, L. Wang, and P. W. Seelbach, editors. Landscape Influences on stream habitat and biological assemblages. American Fisheries Society, Symposium 48, Bethesda, Maryland.

Hietel, E., R. Waldhardt, and A. Otte. 2004. Analysing land-cover changes in relation to environmental variables in Hesse, Germany. Landscape Ecology 19:473–489.

Higgins, J. V., M. T. Bryer, M. L. Khoury, and T. W. Fitzhugh. 2005. A freshwater classification approach for biodiversity conservation planning. Conservation Biology 19:432–445.

Homer, C., C. Q. Huang, L. M. Yang, B. Wylie, and M. Coan. 2004. Development of a 2001 national land-cover database for the United States. Photogrammetric Engineering and Remote Sensing 70:829–840.

Hopkins, K. G., N. B. Morse, D. J. Bain, N. D. Bettez, N. B. Grimm, J. L. Morse, M. M. Palta, W. D. Shuster, A. R. Bratt, and A. K. Suchy. 2015. Assessment of regional variation in streamflow responses to urbanization and the persistence of physiography. Environmental Science and Technology 49:2724–2732.

Hynes, H. B. N. 1975. The stream and its valley. Verhandlungen Internationale Vereinigung für Theoretische und Angewandte Limnologie 19:1–15.

Iverson, L. R. 1988. Land-use changes in Illinois, USA: the influence of landscape attributes on current and historic land use Landscape Ecology 2:45–61.

Marzin, A., P. F. M. Verdonschot, and D. Pont. 2013. The relative influence of catchment, riparian corridor, and reach-scale anthropogenic pressures on fish and macroinvertebrate assemblages in French rivers. Hydrobiologia 704:375–388.

Maxwell J. R., C. J. Edwards, M. E. Jensen, S. J. Paustian, H. Parrott, and D. M. Hill. 1995. A hierarchical framework of aquatic ecological units in North America (Nearctic Zone). U.S. Forest Service, General Technical Report NC-176, Washington D.C.

Mazerolle M. J. 2012. AICcmodavg: model selection and multimodel inference based on (Q)AIC(c). R package version 1.26. Available: http://CRAN.R-project.org/package=AICcmodavg. (September 2012).

McManamay, R. A., B. K. Peoples, D. J. Orth, C. A. Dolloff, and D. C. Matthews. 2015. Isolating causal pathways between flow and fish in the regulated river hierarchy. Canadian Journal of Fisheries and Aquatic Sciences 72:1731–1748.

McManamay, R. A., M. J. Troia, C. R. DeRolph, A. O. Sheldon, A. R. Barnett, S. C. Kao, and M. G. Anderson. 2018. A stream classification system to explore the physical habitat diversity and anthropogenic impacts in riverscapes of the eastern United States. PLOS (Public Library of Science) ONE [online serial] 13(6):e0198439.

Montgomery, D. R. 1999. Process domains and the river continuum. Journal of the American Water Resources Association 35:397–410.

Neff, M. R., and D. A. Jackson. 2012. Geology as a structuring mechanism of stream fish communities. Transactions of the American Fisheries Society 141:962–974.

Omernik, J. M. 1987. Ecoregions of the conterminous United States. Annals of the Association of American Geographers 77:118–125.

Omernik, J. M., G. E. Griffith, R. M. Hughes, M. H. Weber, and J. B. Glover. 2017. How misapplication of the hydrologic unit framework diminishes the meaning of watersheds. Environmental Management 60:1–11.

Paller, M. H. 1994. Relationships between fish assemblage structure and stream order in South Carolina coastal plain streams. Transactions of the American Fisheries Society 123:150–161.

Pyne, M. I., R. B. Rader, and W. F. Christensen. 2007. Predicting local biological characteristics in streams: a comparison of landscape classifications. Freshwater Biology 52:1302–1321.

R Core Team. 2012. R: a language and environment for statistical computing. R Foundation for Statistical Computing, Vienna, Austria.

Reger, B., A. Otte, and R. Waldhardt. 2007. Identifying patterns of land-cover change and their physical attributes in a marginal European landscape. Landscape and Urban Planning 81:104–113.

Roy, A. H., B. J. Freeman, and M. C. Freeman. 2007. Riparian influences on stream fish assemblage structure in urbanizing streams. Landscape Ecology 22:385–402.

Schinegger, R., C. Trautwein, and S. Schmutz. 2013. Pressure-specific and multiple pressure response of fish assemblages in European running waters. Limnologica: Ecology and Management of Inland Waters 43:348–361.

Snyder, C. D., J. A. Young, R. Villella, and D. P. Lemarie. 2003. Influences of upland and riparian land use patterns on stream biotic integrity. Landscape Ecology 18:647–664.

Sprague, L. A., and L. H. Nowell. 2008. Comparison of pesticide concentrations in streams at low flow in six metropolitan areas of the United States. Environmental Toxicology and Chemistry 27:288–298.

Surasinghe, T. D., and R. F. Baldwin. 2015. Importance of riparian forest buffers in conservation of stream biodiversity: responses to land uses by stream-associated salamanders across two southeastern temperate ecoregions. Journal of Herpetology 49:83–94.

USGS (U.S. Geological Survey). 2000. National Hydrography Dataset. USGS. Available: http://nhd.usgs.gov/index.html. (December 2006).

USGS (U.S. Geological Survey). 2003. Hydrologic landscape regions of the United States. USGS. Available: https://pubs.er.usgs.gov/publication/ofr03145. (May 2019).

USGS (U.S. Geological Survey). 2010. NAWQA data export: biological community. USGS. Available: http://infotrek.er.usgs.gov/nawqa_queries/biomaster/index.jsp. (April 2006).

Utz, R. M., K. N. Eshleman, and R. H. Hilderbrand. 2011. Variation in physicochemical responses to urbanization in streams between two Mid-Atlantic physiographic regions. Ecological Applications 21:402–415.

Utz, R. M., R. H. Hilderbrand, and D. M. Boward. 2009. Identifying regional differences in threshold responses of aquatic invertebrates to land cover gradients. Ecological Indicators 9:556–567.

Utz, R. M., R. H. Hilderbrand, and R. L. Raesly. 2010. Regional differences in patterns of fish species loss with changing land use. Biological Conservation 143:688–699.

Vaida, F., and S. Blanchard. 2005. Conditional Akaike information for mixed-effects models. Biometrika 92:351–370.

Wagenhoff, A. 2011. Multiple-stressor effects along gradients of deposited fine sediment and dissolved nutrients in streams. Doctoral thesis. University of Otago, Dunedin, New Zealand.

Wang, L. Z., J. Lyons, P. Rasmussen, P. Seelbach, T. Simon, M. Wiley, P. Kanehl, E. Baker, S. Niemela, and P. M. Stewart. 2003. Watershed, reach, and riparian influences on stream fish assemblages in the Northern Lakes and Forest Ecoregion, USA. Canadian Journal of Fisheries and Aquatic Sciences 60:491–505.

Wilson, H. F., and M. A. Xenopoulos. 2008. Landscape influences on stream fish assemblages across spatial scales in a northern Great Plains ecoregion. Canadian Journal of Fisheries and Aquatic Sciences 65:245–257.

Wolock, D. M., T. C. Winter, and G. Mcmahon. 2004. Delineation and evaluation of hydrologic-landscape regions in the United States using geographic information system tools and multivariate statistical analyses. Environmental Management 34:71–88.

Wolter, C., A.D. Buijse, P. Parasiewicz. 2016. Temporal and spatial patterns of fish response to hydromorphological processes. River Research and Applications 32:190–201.

American Fisheries Society Symposium 90:351–383, 2019
© 2019 by the American Fisheries Society

Landscape Effects on Stream Fishes: Broad-Scale Responses to Anthropogenic Land Use across Temperate Mesic Regions of the United States

DARREN J. THORNBRUGH*
U.S. Forest Service, Eastern Regional Office
626 East Wisconsin Avenue, Milwaukee, Wisconsin 53202, USA
and
Department of Fisheries and Wildlife, Michigan State University
1405 South Harrison Road, Suite 318, East Lansing, Michigan 48824, USA

DANA M. INFANTE
Department of Fisheries and Wildlife, Michigan State University
1405 South Harrison Road, Suite 318, East Lansing, Michigan 48824, USA

Abstract.—Anthropogenic activities including urbanization, agriculture, and dams degrade stream habitats and are a dominant reason for global biodiversity declines in fluvial fish assemblages. Declining diversity trends have been well documented in many regions of the world; however, fishes vary regionally in response to anthropogenic land use, resulting from complex relationships between landscape variables and mechanisms controlling stream fish assemblages. To test for differences in regional fish response to anthropogenic land use, we conducted our study across five freshwater ecoregions in the temperate mesic portion of the United States and evaluated data characterizing stream fish assemblages from 10,522 locations across all study freshwater ecoregions. Fishes were summarized by metrics describing assemblage structure, trophic groupings of species, levels of tolerance to anthropogenic stressors, and life history characteristics, with seven metrics used for analyses. Natural and anthropogenic landscape variables were assessed across freshwater ecoregions, and we tested for regionally specific influences of percent catchment urbanization, percent catchment agriculture, and catchment densities of dams and stream-road crossings on stream fishes. We used cascade multivariate regression trees to quantify variance explained in fish metrics by these landscape variables after controlling for influences of natural landscape variables, including catchment area, catchment lithology, and elevation of study sites. Results indicated differences in dominant influences by freshwater ecoregion, as well as differences in the levels of anthropogenic land use influencing fishes within and across freshwater ecoregions. For example, urban land use was the most influential anthropogenic land use in both Appalachian Piedmont and Chesapeake Bay freshwater ecoregions, with fish assemblage metrics showing responses at 10%

* Corresponding author: darren.thornbrugh@usda.gov

and 1% catchment urban land use, respectively. In contrast, dam density in the network catchment was the most influential anthropogenic variable on fish assemblage metrics in both the Laurentian Great Lakes and Middle Missouri freshwater ecoregions. Also, large amounts of agriculture in the catchment was the most influential anthropogenic land use on fish assemblage metrics in the Upper Mississippi freshwater ecoregion. Knowledge of regional differences in the top contributing anthropogenic landscape variables and the levels at which fish assemblages respond to these variables lends insight into mechanisms controlling stream fish assemblages by freshwater ecoregions and can aid in development of region-specific conservation strategies to prevent biodiversity loss from current and future anthropogenic land use.

Introduction

For more than a century, most freshwater fishes have experienced population declines and range reductions globally due to anthropogenic influences in the environment (Dudgeon et al. 2006; Helfman 2007; Vörösmarty et al. 2010). Thirty-nine percent of freshwater fishes are critically imperiled in North America (Jelks et al. 2008). Freshwater fishes are considered one of the most imperiled taxa groups on the planet (Ricciardi and Rasmussen 1999; Cooke et al. 2012). Anthropogenic land use affects fishes by altering streamflows and thermal regimes, as well as through inputs of nutrients, sediments, and toxins, stresses known to degrade habitat for stream fish assemblages (e.g., Dudgeon et al. 2006; Helfman 2007; Jelks et al. 2008). Future modifications to aquatic habitats induced by land cover changes are expected to continue having negative effects on the status of freshwater fishes (Ormerod 2003; Helfman 2007; Jelks et al. 2008), underscoring the need for conservation actions.

In fluvial systems, understanding the extent to which anthropogenic landscapes affect fishes has advanced greatly in the past few decades as a result of many studies that describe relationships between anthropogenic land use in catchments and aquatic habitat quality (e.g., Allan et al. 1997; Wiens 2002; Allan 2004). Two land uses, urbanization and agriculture, and dams have been reported to be particularly detrimental to fish assemblages.

Urban land use is known to be a major source of degradation to stream habitat, in part resulting from increases of impervious surface cover within catchments that alter streamflow, thermal regimes, and stream morphology (Allan and Flecker 1993; Caissie 2006; Olden and Naiman 2010; Carpenter et al. 2011). Additional effects of urban land cover can lead to finer substrates, increased sediment loads, and inputs of excess nutrients and toxins to streams (Paul and Meyer 2001). Collectively, all of these variables can result in declines in biota in streams draining urbanized landscapes (Paul and Meyer 2001; Allan 2004; Walsh et al. 2005), and multiple studies have documented negative effects of urban land use on stream habitats and declines in fish assemblages (e.g., Wang et al. 2000; 2001, Walters et al. 2003; Scott 2006).

Agricultural land use also impairs stream fish habitats. High levels of catchment agriculture have been associated with high levels of suspended sediments and excess nutrients in streams of 20 major river basins throughout the United States (e.g., Meador and Goldstein 2003), with degraded fish assemblage condition in these catchments likely resulting from altered habitats. Catchment agriculture has been associated with reduced stream cover and bank stability and with increased nutrients in Wisconsin streams (Wang et al. 1997). Agriculture has been associated with decreases in fish abundance, species richness, and biotic integrity in both Michigan (e.g., Roth et al. 1996) and Wisconsin (e.g., Wang et al. 1997; Fitzpatrick et al. 2001) streams.

Dams, reservoirs, and other barriers reduce network connectivity for fish (e.g., Hughes et al. 2005; Schinegger et al. 2016; Cooper et al. 2017). They also degrade stream habitat by changing flow, thermal regimes, sediment and wood transport, and water quality (Ligon et al. 1995; Graf 2006; Mbaka and Wanjiru Mwaniki 2015; Wohl et al. 2019). Despite our extensive and advancing understanding of anthropogenic landscape and riverscape effects on stream habitats, declines in stream fishes continue throughout the United States and globally (Allan and Flecker 1993; Dudgeon et al. 2006; Jelks et al. 2008; Strayer and Dudgeon 2010).

Conservation of stream health is further complicated because land use influences on streams vary regionally as a result of the influences of co-occurring natural landscape variables (e.g., differences in climate, lithology, soils, and landforms). Those covariables result in region-specific trends and/or differences in how the types and levels of anthropogenic land use affect stream fishes. Only a few studies have attempted to describe such regional differences in fish responses. For example, Utz et al. (2010) found substantial regional differences in declines of diversity and abundances of stream fishes across physiographic provinces in Maryland, attributing the differences in part to geomorphic characteristics of the physiographic provinces and their ability to buffer effects of urbanization. Catchment-scale anthropogenic disturbance variables better distinguish least- and most-disturbed sites for small streams in the western U.S. mountains than they did in the western U.S. plains (Whittier et al. 2006).

The diversity of fishes, the specific composition of fish assemblages, and land use histories are important variables in accounting for regional effects of catchment anthropogenic stressors. Morgan and Cushman (2005) found differences in fish responses to catchment urban land use in Maryland, in part because of fish assemblage compositional differences between the Eastern Piedmont and Coastal Plains physiographic provinces in Maryland. In a similar study, Meador et al. (2005) showed that fish species richness declined with increasing urbanization at sites in both Boston, Massachusetts and Birmingham, Alabama but attributed observed declines in fish species richness to different variables, including nutrient enrichment, types of habitat alterations, and native and nonnative invasive species compositional differences between the two study areas. In a study of nine metropolitan areas spread across the United States, Brown et al. (2009) found stronger responses of fish assemblages to urbanization in cities replacing natural vegetation than in those replacing agriculture. These findings highlight the observations that fish assemblages can respond differently and be more susceptible to anthropogenic land use depending on location, land use histories, types of anthropogenic stressors, and assemblage composition (Meador et al. 2005; Morgan and Cushman 2005; Brown et al. 2009).

Based on the importance of understanding regional differences in land use effects on conservation and management of fluvial fish, the goal of this study was to characterize region-specific differences in fish assemblage relationships to anthropogenic land use across five freshwater ecoregions in the temperate mesic portion of the United States. We used three approaches to address our goal. First, we used multiple steps, including principle component analysis (PCA), to characterize prominent natural and anthropogenic landscape variables that could potentially affect fish assemblages in the study freshwater ecoregions. Second, we used a redundancy analysis (RDA) to assess the degree to which natural landscape variables versus anthropogenic land use variables explained variance in fish assemblages summarized by select metrics in each study freshwater ecoregion. Finally, we used cascade multivariate regression tree (CMRT) analysis to predicted fish metrics from natural and anthropogenic landscape variables in a manner that specifically allowed us to account for the influence of natural variables and to identify different types and levels of major anthropogenic influences important to fish assemblages in each freshwater ecoregion.

Methods

Study Area

To assess regional differences in the effects of landscape-level variables on fish assemblages, we conducted analyses within five freshwater ecoregions in the United States: Chesapeake Bay, Laurentian Great Lakes, Middle Missouri, Upper Mississippi, and Appalachian Piedmont (Abell et al. 2000, 2008). The study area extends from the western Great Plains to the eastern coastal plains of the United States (Figure 1), and both natural and anthropogenic landscape variables across the study area are highly diverse. The freshwater ecoregions are defined by boundaries of major river basins with similar biogeographic history in their fish faunas (Abell et al. 2000, 2008). Therefore, they represent ecologically relevant units for groupings of fishes based on evolutionary patterns across large landscapes (Maxwell et al. 1995; Matthews 1998; Abell et al. 2000; but see Deweber et al. 2019; Herlihy et al. 2019; both this volume). The Chesapeake Bay freshwater ecoregion includes a partially previously glaciated landscape up to approximately 41 degrees latitude

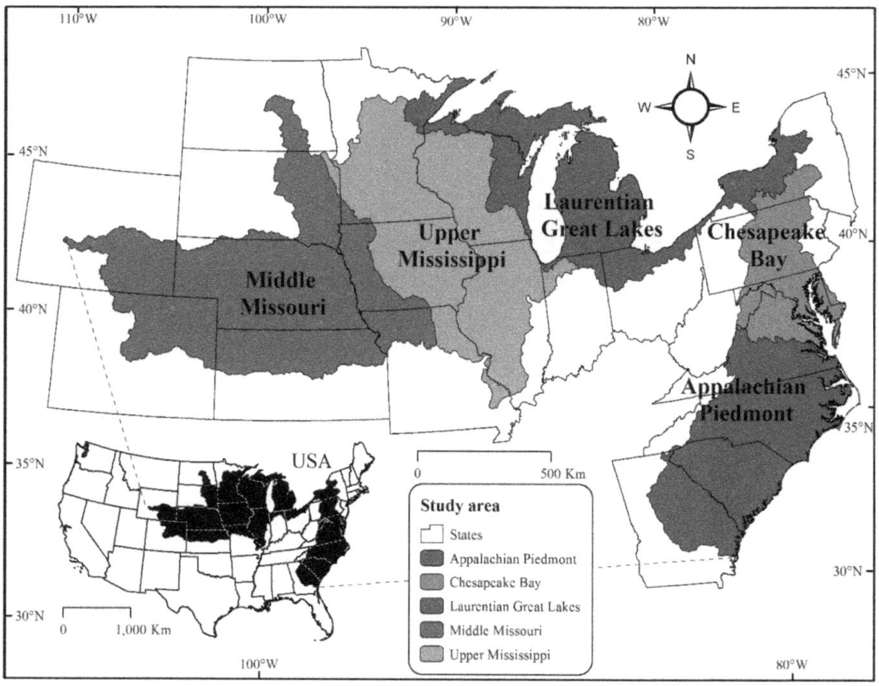

Figure 1. Five freshwater ecoregions of the United States that comprised the study area (Abell et al. 2008).

with areas of high urban land use, and the Laurentian Great Lakes freshwater ecoregion includes a previously glaciated landscape but with a greater amount of forest land cover (Hocutt and Wiley 1986). Both the Middle Missouri and the Upper Mississippi freshwater ecoregions contain areas of intensive agriculture, but the Middle Missouri freshwater ecoregion contains large amounts of grassland. Finally, the Appalachian Piedmont freshwater ecoregion is an unglaciated landscape that includes highlands, piedmont, and coastal plains.

Stream Layer

The 1:100,000-scale National Hydrography Dataset Plus Version 1 (NHDPlusV1) was used as a base layer for geographic representation of stream reaches and their catchments (USEPA and USGS 2005). We define a stream reach as a section of river in the NHDPlusV1 that extends from (1) the stream origin to the first downstream confluence or junction with a lake or reservoir, (2) an upstream confluence or lake/reservoir outflow to the next downstream confluence or lake/reservoir junction, or (3) an upstream confluence or lake/reservoir outflow to the river mouth where it meets with a lake, reservoir, or estuary (Daniel et al. 2015). Two types of catchments were used in analysis based on the NHDPlusV1. Local catchments include all land that drains directly into an individual stream reach without being transported via other fluvial pathways represented by the NHDPlusV1. Network catchments include all land upstream of, and draining into a given reach via fluvial pathways, including the local catchment (Wang et al. 2011a).

Natural Landscape Variables

A variety of natural and anthropogenic landscape variables were summarized at the local catchment and then aggregated to the network catchments of stream reaches in the study freshwater ecoregions, following Tsang et al. (2014; Appendix 1). Natural land cover classes, including water, wetland, barren, forest, shrub, and grassland, were developed from the 2001 National Land Cover Dataset (NLCD; Homer et al. 2004; Appendix 1). Network catchment areas were calculated using a geographic information system. Channel gradient (i.e., slope) and elevation in local catchments were developed from the U.S. Geological Survey (USGS) National Elevation Dataset (Gesch 2007). Surficial lithology metrics were developed from the USGS terrestrial ecosystems surficial lithology of the conterminous U.S. data layer (Soller and Reheis 2004; Cress et al. 2010). Surficial lithology classes were defined by grouping major categories of lithology based on their hydraulic conductivity (e.g., coarse materials that allow rapid water infiltration versus fine materials where infiltration occurs more slowly). Coarse and fine categories were used to compare streams with a more stable hydrology and potentially higher contribution of base flow from groundwater inputs versus streams that may be more prone to variable flow regimes. The base-flow index, a model of hydraulic potential characterizing the percentage of groundwater contribution to streamflow in network catchments, was modeled by the USGS (Wolock 2003; Appendix 1). A range in soil hydrologic groups and soil permeability metrics in local catchments were developed from the U.S. Department of Agriculture Natural Resources Conservation Service (NRCS) State Soil Geographic (STATSGO) database (Schwarz and Alexander 1995). Thirty-year normal mean annual air temperature and mean annual precipitation were summarized for the climatological period from 1981 to 2010 by the PRISM Climate Group (Oregon State University, Corvallis), attributed to catchments of stream reaches, and aggregated to the network catchments (Daly et al. 2008).

Anthropogenic Land Use

Major categories of anthropogenic land cover for urban and agriculture classes from the NLCD were used, including low-, medium-, and high-intensity developed land, open-space developed land, pasture/hay, and cultivated crops (Homer et al. 2004). Individual classes for urban and agricultural land use were summarized into composite metrics within network catchments to aid in meeting assumptions of statistical tests (described below; Table 1). The percentage of impervious surfaces in the network catchment was developed from the 2001 NLCD (Homer et al. 2004). Population density in network catchments was developed from a data layer created by the National Oceanic and Atmospheric Administration (NOAA 2010). To consider cumulative effects of catchment-wide barriers on stream fish, dam density in network catchments was calculated using data from the National Anthropogenic Barrier Dataset (USGS 2013). Density of stream-road crossings, road length, and canals/ditches in catchments were developed from the U.S. Census Bureau TIGER/line files data (USCB 2002).

Fish Data

Data characterizing stream fish assemblages were compiled from state and federal programs, collected from 1990 to 2010, and spatially referenced to corresponding stream reaches of the NHDPlusV1 (Esselman et al. 2013b; Daniel et al. 2015). All fish data were collected using single-pass electrofishing by state and federal programs using methods determined to be comparable for creeks, small rivers, and medium rivers with drainage areas less than 10,000 km^2 (Esselman et al. 2013a; Daniel et al. 2015). Data per reach included abundance measurements for fishes identified to species from assemblage sampling, and all scientific naming was standardized to the Integrated Taxonomic Information System (ITIS 2010). Seven fish metrics were identified for analysis from 22 candidate metrics, after reducing highly correlated metrics (selection process described in Thornbrugh 2014). Selected metrics described trophic groupings of species (Lyons 1992), tolerance levels (Barbour et al. 1999), and life history characteristics (Frimpong and Angermeier 2009). The seven metrics are species richness as well as percentages of intolerant individuals, invertivore individuals, omnivore individuals, lithophilic individuals, rheophilic individuals, and game fish individuals (Table 2; Daniel et al. 2015).

Data Analyses

Landscape variable selection.—Fifty-two variables describing a range of natural and anthropogenic landscape variables were considered for analysis (Appendix 1). All variables were transformed to better meet assumptions of linearity and homoscedasticity. Continuous variables were natural log-transformed; percentage and proportional variables were arcsin square-root transformed. We ran a Pearson correlation on all 52 variables to reduce the number of highly correlated variables (Appendix 2). When a pair of variables was correlated by an absolute value of ≥0.7, one of the two was eliminated from further analysis. To do so, network catchment variables were selected over the same local catchment variables because they were assumed to characterize a broader and more spatially extensive set of influences from landscape variables on streams. Other highly correlated variables were selected based on ease of ecological interpretability. Principal component analysis was then used with 22 remaining natural and anthropogenic landscape variables to explore dimensionality in the data set and to further reduce redundancy (Legendre and Legendre 2012). Landscape metrics were chosen for additional analyses by their weightings on orthogonal axes to retain unique dimensions

Table 1. Descriptive statistics of landscape variables for the five freshwater ecoregions with landscape variable descriptions (i.e., mean, minimum [Min], maximum [Max], and the 10th and 90th percentiles).

	Appalachian Piedmont				
Landscape variable description	Mean	Min	Max	10%	90%
Natural					
Network catchment area (km^2)	155.16	1.42	8,062.60	11.42	180.03
Fine lithology (% network catchment)	21.76	0.00	100.00	0.00	100.00
Mean local catchment elevation (m)	157.30	1.39	937.66	20.42	298.06
Base-flow index (% network catchment groundwater contribution to base flow)	45.95	25.23	78.30	30.54	60.35
Air temperature (°C mean annual local catchment)	15.64	9.53	19.33	14.11	17.52
Area weighted average precipitation (mm mean annual network catchment)	1,215.13	973.82	2,372.90	1,125.53	1,317.59
Anthropogenic land use					
Urban					
Developed, open space (% network catchment)[a]	8.44	0.00	77.33	2.85	18.88
Developed, low, medium, and high intensity (% network catchment)[a]	5.76	0.00	77.97	0.09	16.81
Developed, open space, low, medium, and high intensity (% network catchment)	14.20	0.00	98.62	3.13	37.22
Agriculture					
Pasture/hay (% network catchment)[b]	14.06	0.00	63.49	1.43	32.49
Cultivated crops (% network catchment)[b]	7.27	0.00	68.27	0.00	26.55
Pasture/hay and cultivated crops (% network catchment)	21.34	0.00	72.63	3.83	40.93
Road crossing					
Road crossing (# in network catchment/catchment area [100 km^2])	48.96	0.00	345.30	15.82	89.03
Dams					
Dam count (# in network catchment/catchment area [100 km^2])	3.32	0.00	57.56	0.00	9.06

	Chesapeake Bay				
Landscape variable description	Mean	Min	Max	10%	90%
Natural					
Network catchment area (km^2)	371.17	0.82	8,483.38	3.48	560.87
Fine lithology (% network catchment)	13.22	0.00	100.00	0.00	99.80
Mean local catchment elevation (m)	288.25	2.52	1,050.68	21.71	491.14

Table 1. Continued.

Landscape variable description	Chesapeake Bay				
	Mean	Min	Max	10%	90%
Natural (continued)					
Base-flow index (% network catchment groundwater contribution to base flow)	46.26	26.99	64.02	38.08	54.74
Air temperature (°C mean annual local catchment)	9.57	6.35	14.28	7.00	13.25
Area weighted average precipitation (mm mean annual network catchment)	1,027.70	806.87	1,239.06	914.37	1,108.89
Anthropogenic land use					
Urban					
Developed, open space (% network catchment)[a]	4.55	0.00	39.90	0.95	8.50
Developed, low, medium, and high intensity (% network catchment)[a]	3.61	0.00	83.76	0.01	7.09
Developed, open space, low, medium, and high intensity (% network catchment)	8.15	0.00	98.44	1.27	14.85
Agriculture					
Pasture/hay (% network catchment)[b]	19.55	0.00	69.61	4.84	36.71
Cultivated crops (% network catchment)[b]	12.04	0.00	81.75	0.51	27.43
Pasture/hay and cultivated crops (% network catchment)	31.59	0.00	94.91	7.72	63.49
Road crossing					
Road crossing (# in network catchment/catchment area [100 km^2])	53.99	0.00	588.24	9.40	96.04
Dams					
Dam count (# in network catchment/catchment area [100 km^2])	1.10	0.00	30.53	0.00	3.04

Landscape variable description	Laurentian Great Lakes				
	Mean	Min	Max	10%	90%
Natural					
Network catchment area (km^2)	354.81	0.23	9,533.74	10.27	758.72
Fine lithology (% network catchment)	41.97	0.00	100.00	0.00	100.00
Mean local catchment elevation (m)	277.15	74.56	680.78	180.28	430.92
Base-flow index (% network catchment groundwater contribution to base flow)	45.04	18.69	87.95	21.99	68.45

Table 1. Continued.

Landscape variable description	Laurentian Great Lakes Mean	Min	Max	10%	90%
Natural (continued)					
Air temperature (°C mean annual local catchment)	7.74	2.12	10.56	4.69	9.68
Area weighted average precipitation (mm mean annual network catchment)	929.80	641.04	1,566.84	764.17	1,123.50
Anthropogenic land use					
Urban					
Developed, open space (% network catchment)[a]	5.92	0.00	58.15	0.88	12.08
Developed, low, medium, and high intensity (% network catchment)[a]	5.13	0.00	93.37	0.06	12.57
Developed, open space, low, medium, and high intensity (% network catchment)	11.05	0.00	99.92	1.15	26.31
Agriculture					
Pasture/hay (% network catchment)[b]	10.61	0.00	62.63	0.01	26.51
Cultivated crops (% network catchment)[b]	29.24	0.00	94.26	0.06	79.76
Pasture/hay and cultivated crops (% network catchment)	39.85	0.00	94.26	0.47	83.07
Road crossing					
Road crossing (# in network catchment/catchment area [100 km^2])	48.49	0.00	413.37	8.35	91.15
Dams					
Dam count (# in network catchment/catchment area [100 km^2])	1.13	0.00	73.69	0.00	2.84

Landscape variable description	Middle Missouri Mean	Min	Max	10%	90%
Natural					
Network catchment area (km^2)	764.33	1.57	9,984.84	15.36	2,196.10
Fine lithology (% network catchment)	32.01	0.00	100.00	0.00	100.00
Mean local catchment elevation (m)	591.36	195.32	3,134.90	288.62	1,165.51
Base-flow index (% network catchment groundwater contribution to base flow)	38.90	12.20	84.20	19.19	69.00
Air temperature (°C mean annual local catchment)	10.22	0.19	12.92	8.11	12.29
Area weighted average precipitation (mm mean annual network catchment)	708.96	344.53	1,113.56	441.07	929.31

Table 1. Continued.

Landscape variable description	Middle Missouri				
	Mean	Min	Max	10%	90%
Anthropogenic land use					
Urban					
Developed, open space (% network catchment)[a]	3.88	0.00	34.60	1.75	5.48
Developed, low, medium, and high intensity (% network catchment)[a]	2.06	0.00	82.72	0.04	2.45
Developed, open space, low, medium, and high intensity (% network catchment)	5.93	0.00	92.22	1.99	7.65
Agriculture					
Pasture/hay (% network catchment)[b]	9.92	0.00	80.99	0.00	39.88
Cultivated crops (% network catchment)[b]	41.58	0.00	93.56	2.05	82.94
Pasture/hay and cultivated crops (% network catchment)	51.50	0.00	94.88	4.00	87.86
Road crossing					
Road crossing (# in network catchment/catchment area [100 km^2])	35.90	0.00	254.61	3.63	70.55
Dams					
Dam count (# in network catchment/catchment area [100 km^2])	3.91	0.00	73.31	0.00	11.60

Landscape variable description	Upper Missouri				
	Mean	Min	Max	10%	90%
Natural					
Network catchment area (km^2)	356.01	0.23	9,964.22	9.65	740.40
Fine lithology (% network catchment)	9.49	0.00	100.00	0.00	27.43
Mean local catchment elevation (m)	288.86	114.85	589.09	184.87	382.03
Base-flow index (% network catchment groundwater contribution to base flow)	49.85	7.84	74.87	31.21	66.29
Air temperature (°C mean annual local catchment)	7.57	3.07	13.60	5.48	10.61
Area weighted average precipitation (mm mean annual network catchment)	823.55	556.81	1,190.27	704.84	949.22
Anthropogenic land use					
Urban					
Developed, open space (% network catchment)[a]	4.56	0.00	30.15	2.32	6.81

Table 1. Continued.

Landscape variable description	Upper Missouri Mean	Min	Max	10%	90%
Anthropogenic land use (continued)					
Urban (continued)					
Developed, low, medium, and high intensity (% network catchment)[a]	4.09	0.00	92.83	0.10	7.25
Developed, open space, low, medium, and high intensity (% network catchment)	8.65	0.00	99.46	3.24	13.05
Agriculture					
Pasture/hay (% network catchment)[a]	12.34	0.00	68.33	0.39	27.99
Cultivated crops (% network catchment)[b]	47.87	0.00	98.06	5.14	87.11
Pasture/hay and cultivated crops (% network catchment)	60.21	0.00	98.06	15.54	89.80
Road crossing					
Road crossing (# in network catchment/catchment area [100 km^2])	39.58	0.00	643.09	8.88	69.17
Dams					
Dam count (# in network catchment/catchment area [100 km^2])	0.77	0.00	39.03	0.00	1.88

[a] Combined for cascade multivariate regression tree.

of explained variation in the data and on ease of ecological interpretability. Six natural and eight anthropogenic landscape variables were selected for use in further analysis.

Redundancy analysis.—We partitioned the total variance in fish metrics explained by natural landscape variables, anthropogenic land use variables, and their corresponding interaction terms for each of the five freshwater ecoregions with RDA using CANOCO software (ter Braak and Smilauer 2002). Six natural landscape variables and six anthropogenic land use classes were used in RDA (Appendix 1), and we conducted the analyses regionally to characterize how variance partitions might differ by freshwater ecoregions (Borcard et al. 2011; Legendre and Legendre 2012).

Developing a composite variable to characterize natural landscape influences.—We used PCA to develop a composite variable, or bundled variable, from six natural landscape variables across all five study freshwater ecoregions (Appendix 1; Figure 1). This composite variable was developed for use in multiple wave CMRT analysis (described below) to aid in controlling for natural variability known to influence fish across the five study freshwater ecoregions in a consistent and comprehensive manner (Whittier et al. 2007; Whittier and Van Sickle 2010; Van Sickle 2013).

Cascade multivariate regression tree.—We performed CMRT to quantify regional differences in fish responses to dominant anthropogenic land use after controlling for natural

Table 2. Descriptive statistics (i.e., mean, range, 10th and 90th percentiles) of fish metrics in five freshwater ecoregions with number of sample sites per freshwater ecoregion.

Freshwater ecoregion (# of sites)		Fish richness count	Intolerant % individual	Game fish % individual	Invertivore % individual	Omnivore % individual	Lithophilic % individual	Rheophilic % individual
Appalachian Piedmont (1,083)	Mean	14.68	1.27	28.46	65.38	23.90	49.85	20.25
	Range	1–34	0–95.81	0–100.00	0–100.00	0–93.44	0–100.00	0–97.13
	0.10	8.00	0.00	3.11	40.00	3.71	4.86	0.00
	0.90	22.00	2.59	67.26	88.31	45.74	86.65	48.83
Chesapeake Bay (807)	Mean	9.95	12.67	22.15	35.09	14.12	45.41	19.80
	Range	1–31	0–100.00	0–100.00	0–100.00	0–100.00	0–100.00	0–100.00
	0.10	2.00	0.00	0.00	0.00	0.00	0.00	0.00
	0.90	19.00	37.15	68.91	69.67	38.54	83.00	53.44
Laurentian Great Lakes (3,773)	Mean	11.05	12.53	20.63	38.90	18.10	27.76	29.01
	Range	1–37	0–100.00	0–100.00	0–100.00	0–100.00	0–100.00	0–100.00
	0.10	4.00	0.00	0.00	4.87	0.00	0.00	0.00
	0.90	19.00	43.80	72.48	76.54	45.97	65.08	69.95
Middle Missouri (886)	Mean	10.29	2.60	9.42	35.78	37.79	10.71	42.68
	Range	1–29	0–100.00	0–100.00	0–100.00	0–100.00	0–100.00	0–100.00
	0.10	4.00	0.00	0.00	4.29	2.79	0.00	0.29
	0.90	17.00	3.23	26.51	70.80	81.00	35.31	83.68
Upper Mississippi (3,973)	Mean	12.63	11.75	15.07	42.75	25.97	23.65	24.19
	Range	1–48	0–100.00	0–100.00	0–100.00	0–100.00	0–100.00	0–100.00
	0.10	3.00	0.00	0.00	3.45	0.93	0.00	0.00
	0.90	24.00	37.05	55.25	78.70	60.33	57.96	66.34

landscape influences. A CMRT is based on a tiered analysis of multivariate regression trees (MRTs) and does not require that relationships between explanatory and response variables be linear or that residuals be normally distributed (De'ath 2002; Ouellette et al. 2012). This makes the technique appropriate for large, ecological data sets that do not meet such statistical requirements. The CMRT is conducted in multiple waves where the first wave of data are theorized to have dominant explanatory power over the response data, and a second wave is applied to explain additional variance. The first wave in this CMRT analysis was performed to associate variability in fish metrics with natural landscape influences. Anthropogenic land use metrics (Table 1) were used in the second wave of analysis to describe additional variation explained in the fish metrics after controlling for the natural landscape variable.

Before analysis, fish metrics were centered, and a standardized z-score was calculated for response metrics. To run the CMRT, we used the function 'CascadeMRT' from the package 'mvpart' in R (Therneau et al. 2012; Oksanen et al. 2013; Ouellette and Legendre 2013; R Core Team 2013). This function uses a standard recursive greedy splitting algorithm to fit all MRT models (De'ath 2002). We used the "one standard error" rule for the v-fold cross validation to fully grow the MRTs (rule described by De'ath 2002). The one standard error rule suggests using the simplest model within one standard error of the model that minimizes the cross-validated relative error, with the aim of minimizing risk and complexity of the model. This method picks the simplest best model that minimizes both the cross-validated relative error and the number of splits in the MRT (De'ath 2002; Ouellette and Legendre 2013). The complexity parameter value is the minimum contribution to the R^2 of an MRT for a split to be considered for the package 'mvpart' in R (Ouellette and Legendre 2013; R Core Team 2013). In our analyses, we set complexity parameters high enough in both the first and second waves to identify only the main variables determining variation in species composition, yet values were still considered conservative, ranging between 0.01 and 0.05.

The CMRT analysis was used to characterize influences on stream fishes in a hierarchical manner, with the predominant influence from natural landscape variables accounted for first, followed by influences of anthropogenic land uses. This approach allowed us to separate the explanatory power of two data sets on stream fishes, first controlling for natural landscape variables across the entire study area and second, accounting for anthropogenic landscape variables by freshwater ecoregion (Ouellette et al. 2012). Site scores from the natural composite variable were used in the first wave of the CMRT analysis to compute the main partition of explained variance. Residuals from the first wave of the CMRT analysis were then used in the second wave with the anthropogenic land use variables to compute anthropogenic partitions of additional explained variance based on the theorized controls of riverine ecosystems. This approach provides the ability to characterize the influence of anthropogenic land use variables after accounting for the influence of the natural landscape variables, recognizing that one set might change as a function of the other (Ouellette et al. 2012).

Results

Study Area

The five study freshwater ecoregions covered an area of 2,287,921 km^2 (Figure 1). A total of 10,522 fish sampling locations were used in analyses across the study area, with 1,083 sampling sites in the Appalachian Piedmont, 807 in the Chesapeake Bay, 3,773 in the Laurentian Great Lakes, 886 in the Middle Missouri, and 3,973 in the Upper Mississippi freshwater ecoregions. Natural landscape variables at study sites were highly diverse among and within freshwater ecoregions (Table 1). Mean network catchment area of study sites ranged from 155 km^2 in the Appalachian Piedmont freshwater ecoregion to 764 km^2 in the Middle Missouri freshwater ecoregion. Fine lithology in the network catchment was lowest in the Upper Mississippi freshwater ecoregion (mean of 9%) and highest in the Laurentian Great Lakes freshwater ecoregion (mean of 42%). Mean local catchment elevation was lowest at study sites in the Appalachian Piedmont freshwater ecoregion at 157 m and ranged from <2 m to 3,134 m across catchments in the study area (Table 1). Mean local catchment elevation was highest at study sites in the Middle Missouri freshwater ecoregion at 591 m. Climate varied across sites in the study freshwater ecoregions as well, with lowest mean annual air temperatures in the Upper Mississippi (8°C) and highest mean annual air temperature and precipitation in the Appalachian Piedmont freshwater ecoregions (15°C and 1,215 mm, respectively; Table 1). The lowest mean annual precipitation was in the Middle Missouri freshwater ecoregion at 709 mm.

Study sites across the five freshwater ecoregions also exhibited a wide range in anthropogenic land use (Figure 2). Mean percentage of urban land use in network catchments was highest in the Appalachian Piedmont freshwater ecoregion at 14% and lowest in the Middle Missouri at 6% (Table 1). Mean percentage of agriculture land use in network catchments varied substantially across the study area (Figure 2). It was highest in the Upper Mississippi freshwater ecoregion at 60%, with 10th and 90th percentile values of 16% and 90%, respectively (Table 1). The mean network catchment percent agriculture was lowest in the Appalachian Piedmont freshwater ecoregion with a value of 21% and 10th and 90th percentile values of 4%, and 41%, respectively (Table 1). Density of stream-road crossings in network catchments was highest in the Chesapeake Bay freshwater ecoregion at 54/100 km^2 and lowest in the Middle Missouri freshwater ecoregion at 36/100 km^2. Dam density in network catchments was highest in the Appalachian Piedmont freshwater ecoregion at 4/100 km^2 and lowest in the Upper Mississippi at <1/100 km^2 (Table 1; Figure 2).

Fish metrics used in analyses also varied across the study area sites. Total species richness was highest in the Upper Mississippi freshwater ecoregion (maximum of 48 species collected) and lowest in the Chesapeake Bay freshwater ecoregion (maximum of 31 species collected; Table 2). Mean species richness across the study sites was highest in the Appalachian Piedmont freshwater ecoregion at 15. Mean percent intolerant individuals across study sites was similar in the Chesapeake Bay, Laurentian Great Lakes, and Upper Mississippi freshwater ecoregions at 13%, 13%, and 12%, respectively, but substantially lower in the Middle Missouri and Appalachian Piedmont freshwater ecoregions at 3%, and 1%, respectively (Table 2). Mean percentages of invertivore individuals and lithophilic individuals at study sites were highest in the Appalachian Piedmont freshwater ecoregion at 65% and 50%, respectively. In contrast, invertivores at study sites were lowest in the Chesapeake Bay freshwater ecoregion at 35%, and lithophils at study sites were lowest in the Middle Missouri freshwater ecoregion at 11% (Table 2), which had the highest mean percent of rheophilic individuals at study sites at 43%. Mean percent game fish individuals across study sites was similar in the Appalachian Piedmont, Chesapeake Bay, and Laurentian Great Lakes freshwater ecoregions at 28%, 22%, and 21%, respectively, and lowest in the Middle Missouri freshwater ecoregion at 9% (Table 2).

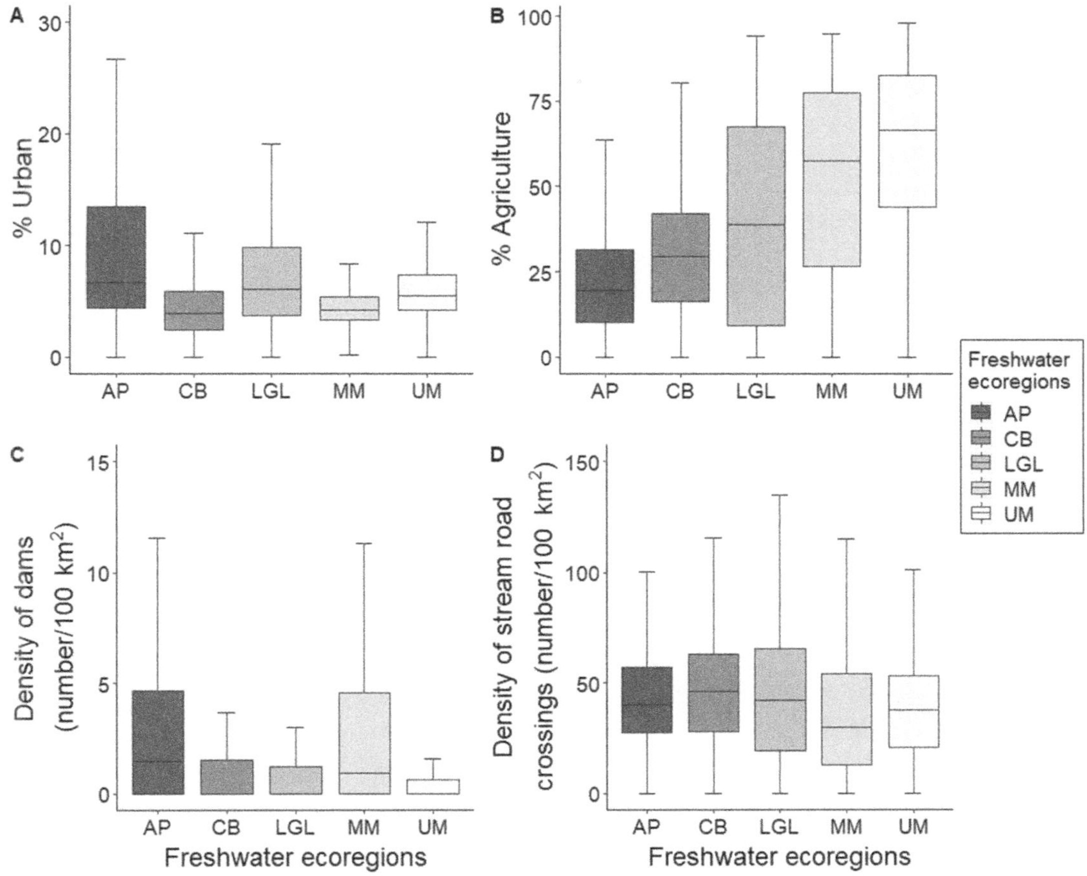

Figure 2. Boxplots of four anthropogenic landscape variables in the network catchment for the five freshwater ecoregions in our study area, where **(A)** is percent urban land use, **(B)** is the percent agriculture land use, **(C)** is the density of dams (number/network catchment area × 100 km^2), and **(D)** is the density of stream road crossings (number/network catchment area × 100 km^2). The black lines in the boxes are median values, the boxes represent 75th and 25th percentiles, whiskers are the highest and lowest values that are within 1.5× the interquartile range of other outliers, and outlier points are not plotted on these graphs. Freshwater ecoregion codes are AP = Appalachian Piedmont, CB = Chesapeake Bay, LGL = Laurentian Great Lakes, MM = Middle Missouri, and UP = Upper Mississippi freshwater ecoregions.

Developing a Composite Variable to Characterize Natural Landscape Influences

Six natural landscape variables were used in PCA to develop a generalized natural composite variable to account for major natural landscape influences in the first wave of the CMRT analysis. The PCA resulted in two axes explaining 60% of the variation across the study sites (Table 3). Axis one, explaining 39% of the variation in the study sites, was weighted positively by air temperature, average precipitation, and fine lithology and was negatively weighted by mean local elevation and base-flow index (Table 3). The first axis of the PCA described the prominent amount of variance representing natural variation in the data, and site scores from axis one were used to create the composite natural variable

Table 3. Principle component analysis results for six natural landscape variables for all study sites across the five study freshwater ecoregions. Variable descriptions and axes weightings are provided. Total amount of variation explained was 59.91%.

	Axis 1	Axis 2
Variance explained (%)	39.10	20.81
Natural landscape variable description		
Air temperature (°C mean annual local catchment)	0.83	–0.01
Area weighted average precipitation (mL mean annual network catchment)	0.68	–0.54
Fine lithology (% network catchment)	0.53	0.55
Network catchment area (km^2)	–0.02	0.64
Base-flow index (% network catchment groundwater contribution to base flow)	–0.58	–0.44
Mean local catchment elevation (m)	–0.77	0.21

for the first wave of the CMRT analysis across all five study freshwater ecoregions.

Variance Partitioning with Redundancy Analysis: Differences across Study Freshwater Ecoregions

Redundancy analysis variance partitioning indicated that natural and anthropogenic landscape variables contributed to between 28% and 42% of total explained variation in fish metrics across the freshwater ecoregions (Figure 3). The maximum variation explained in metrics occurred in the Upper Mississippi freshwater ecoregion (42%), and the minimum variation explained occurred in the Laurentian Great Lakes freshwater ecoregion (28%, Figure 3). Natural landscape variables generally explained the largest proportion of variance in fish metrics as expected, ranging from 12% to 27%, except for the Laurentian Great Lakes freshwater ecoregion, where interactions among the anthropogenic and natural variables explained the highest percentage of variance in fish metrics (13%; Figure 3). Variation explained by anthropogenic land use alone was much less than that explained by natural landscape variables. Approximately 2% of the variation in fish metrics was explained by anthropogenic landscape variables in the Middle Missouri and Upper Mississippi freshwater ecoregions, ranging to a maximum of 5% in the Chesapeake Bay freshwater ecoregion (Figure 3). Interactions between groupings of natural and anthropogenic landscape variables contributed a large proportion of explained variation in fish metrics from 6% in the Chesapeake Bay and Middle Missouri freshwater ecoregions to 13% in the Appalachian Piedmont, Laurentian Great Lakes, and Upper Mississippi freshwater ecoregions (Figure 3).

Predicting Fish Metrics with Cascade Multivariate Regression Trees

Whereas differences in the amount of variance explained in fish metrics occurred among the five freshwater ecoregions, general patterns were similar. Natural and anthropogenic variables combined explained between 7% (Appalachian Piedmont freshwater ecoregion) and 13% (Chesapeake Bay freshwater ecoregion) of the variation in fish metrics (Table 4). The natural landscape composite variable (referred to as main natural) used in the first wave of the CMRT ranged from a low of 6% in the Upper Mississippi freshwater ecoregion to 10% in the Appalachian Piedmont freshwater ecoregion. The

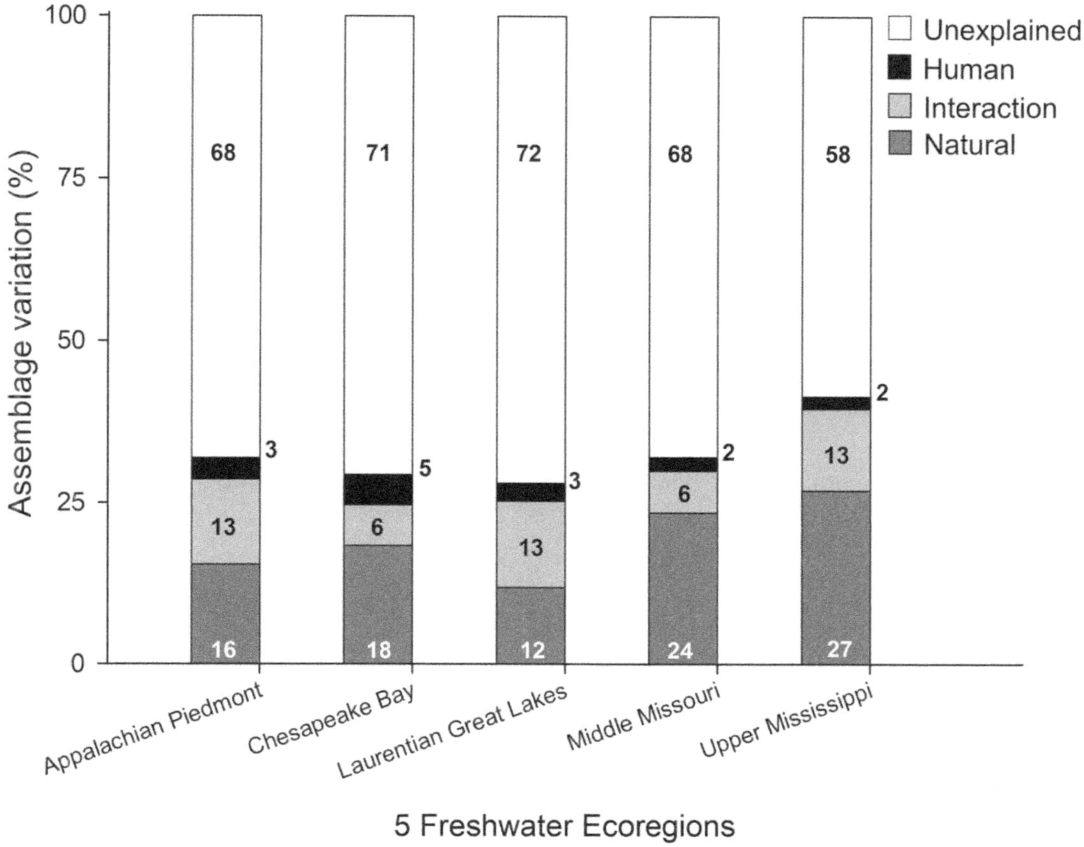

Figure 3. Results from redundancy analysis partitioned variation explained in fish metrics by natural and anthropogenic variables across the five freshwater ecoregions in our study area.

combined variance explained in fish metrics from the two dominant anthropogenic land uses (i.e., anthropogenic 1 + anthropogenic 2) from each drop of the second wave of the CMRT ranged from 3% in the Laurentian Great Lakes freshwater ecoregion to 5% in the Middle Missouri freshwater ecoregion study sites (Table 4).

Regional Influences of Anthropogenic Land Use Variables with Cascade Multivariate Regression Tree Analysis

The second wave of the CMRT, constrained by the composite variable of natural landscape variables, identified the top two anthropogenic land use variables and the amounts of variation in fish metrics that they explained. In the Appalachian Piedmont freshwater ecoregion, the top two contributing land use variables explaining variance in fish metrics were both values of urban land use. The Chesapeake Bay freshwater ecoregion had urban as the top land use and agriculture as a secondary land use. On the other hand, the Laurentian Great Lakes, Middle Missouri, and Upper Mississippi freshwater ecoregions all had network catchment agriculture and network catchment dam density as the top and/or secondary anthropogenic variables describing variance in fish metrics.

In those cases when the top anthropogenic land use metrics were the same across freshwater ecoregions, substantial differences

Table 4. Variable, value, and percent variance explained from the main (natural composite variable), first wave of cascade multivariate regression tree (CMRT), and the top two contributors for (anthropogenic) influences, second wave of CMRT, on fish assemblage metrics across five freshwater ecoregions from CMRT analysis.

Freshwater ecoregion (# of sites)		Main natural	Anthropogenic 1	Anthropogenic 2	Total variance explained
All data (10,522)	Variable	Natural	Agriculture	Agriculture	
	Value	0.12	0.53	0.64	
	% variance explained	0.04	0.02	0.02	0.07
Appalachian Piedmont (1,083)	Variable	Natural	Urban	Urban	
	Value	1.40	0.10	0.38	
	% variance explained	0.10	0.02	0.01	0.13
Chesapeake Bay (807)	Variable	Natural	Urban	Agriculture	
	Value	0.08	0.01	0.00	
	% variance explained	0.06	0.04	0.01	0.11
Laurentian Great Lakes (3,773)	Variable	Natural	Dams	Agriculture	
	Value	0.08	0.05/100 km^2	0.68	
	% variance explained	0.07	0.01	0.01	0.10
Middle Missouri (886)	Variable	Natural	Dams	Agriculture	
	Value	−1.55	0.02/100 km^2	<0.01%	
	% variance explained	0.07	0.03	0.02	0.12
Upper Mississippi (3,973)	Variable	Natural	Agriculture	Dams	
	Value	−0.25	0.54	0.04/100 km^2	
	% variance explained	0.06	0.03	0.00	0.09

in the value of the land use metric causing differences in stream fish assemblage metrics across sampling sites often occurred (Table 4). Percent urban in the network catchment was the most influential anthropogenic land use variable in the Appalachian Piedmont freshwater ecoregion at two values; a high value of 38%, which was above the 90th percentile value of 37%, and a value of 10%, which is just below the mean value of 14% (Tables 1 and 4). Percent urban in the network catchment was also the most influential anthropogenic land use variable in the Chesapeake Bay freshwater ecoregion at a low value of 1%, almost equal to the 10th percentile value of 1% for the freshwater ecoregion. Percent agriculture in the network catchment was the most influential anthropogenic land use variable for the Upper Mississippi freshwater ecoregion and the second most influential anthropogenic land use variable in the Laurentian Great Lakes, Chesapeake Bay, and Middle Missouri freshwater ecoregions at values of 54%, 68%, <1%, and <1%, respectively (Table 4). The values of percent agriculture in the network catchment that fish assemblage metrics were responding to, across all study sites in the CMRT analysis, were substantially different for the Upper Mississippi and Laurentian Great Lakes (i.e., just below the mean value and substantially above the mean value, respectively) versus those in the Chesapeake Bay and Middle Missouri freshwater ecoregions (i.e., below the 10th percentile values; Table 1).

Low dam density in the network catchment was the primary anthropogenic landscape variable describing variation in fish assemblage metrics for the Laurentian Great Lakes and Middle Missouri freshwater ecoregions and a secondary anthropogenic landscape variable in the Upper Mississippi freshwater ecoregion (Table 4). Road crossing density in the network catchment was not found to be one of the two most influential anthropogenic landscape metrics across sites for in any of the five freshwater ecoregions.

Discussion

Our results suggest that effects of anthropogenic land use on stream fishes vary regionally, resulting from complex relationships between landscape variables and mechanisms controlling stream fish assemblages. This study characterized regional differences in types of anthropogenic land use potentially influencing stream fish assemblages across five freshwater ecoregions in the temperate mesic portion of the United States. First, using RDA to test influences of groupings of natural and anthropogenic landscape variables on stream fishes, we found that natural landscape variables generally explained the most variation in each of the five freshwater ecoregions. The amounts of remaining explained variance differed because of differences in contributions of anthropogenic landscape variables and shared variation between natural and anthropogenic landscape variables. The RDA results indicated that in all five freshwater ecoregions, the amount of variance explained by covarying natural and anthropogenic landscape variables was greater than that explained by anthropogenic landscape variables alone. Additional testing for influences of specific anthropogenic variables on stream fish assemblages with CMRT analysis showed regional differences in effects of urban land use, agricultural land use, and dam density on stream fishes. Urban land use was the top anthropogenic variable associated with differences in fish assemblages in the Appalachian Piedmont and Chesapeake Bay freshwater ecoregions, but agricultural land use was the top anthropogenic variable in the Upper Mississippi freshwater ecoregion. In contrast, dam density in network catchments was the top anthropogenic variable in both the Laurentian Great Lakes and Middle Missouri freshwater ecoregions, underscoring the potential influence of multiple dams within network catchments to act as a landscape-scale influence on stream fishes. Besides identifying regional differences in influences of anthropogenic landscape variables on fish assemblage metrics, we also determined that fishes responded to varying severities of landscape disturbances. For example, high percentages of network catchment agriculture (i.e., 54%, and 68%, respectively) were associated with variation in fish assemblage metrics in the Upper Mississippi and Laurentian Great Lakes freshwater ecoregions. However, lower levels (<1%) were associated with variation in fish assemblage metrics in the Chesapeake Bay and Middle Missouri freshwater ecoregions. This approach can be used to assess regional dominance of classes of anthropogenic land use types that are affecting changes in fish assemblages. A better understanding of the top contributing anthropogenic landscape variables and at what level fish assemblages are responding to these variables has implications for management of stream resources and the conservation of stream fish biodiversity.

Variance in Fish Metrics Explained by Anthropogenic and Natural Landscape Variables across the Study Area

By explaining variance in stream fish metrics with groupings of natural landscape variables and anthropogenic land use variables, the RDA indicated regional differences in landscape controls on stream fishes. Total explained variance in fish metrics ranged from 28% in the Laurentian Great Lakes freshwater ecoregion to 42% in the Upper Mississippi freshwater ecoregion, and of that variance, most was attributed to natural landscape variables, which ranged from 12% in the Laurentian Great Lakes freshwater ecoregion to 27% in the Upper Mississippi freshwater ecoregion. Work conducted by Wang et al. (2003) across streams in Minnesota, Wisconsin, and Michigan described similar percentages of variance explained (i.e., 24%) in fish metrics by landscape variables that included stream catchment area, an estimate of groundwater delivery, catchment elevation, longitude, percent catchment urban land use, and percent glacial till. Similarly, studying streams in Louisiana and Mississippi, Alford (2014) described a similar percentage of variance explained (i.e., 24%) in fish species assemblages by catchment variables, including catchment area, elevation, and catchment stream density. An additional insight from the RDA was in the identification of the degree to which natural landscape variables and anthropogenic land use were confounded within our study freshwater ecoregions. In all five freshwater ecoregions, amounts of variance explained by covarying natural and anthropogenic landscape variables were greater than that explained by anthropogenic landscape variables alone. This results from the fact that natural landscape characteristics influence the suitability of landscapes for various anthropogenic land uses (Allan 2004; Scott 2006; Whittier et al. 2006; Deweber et al. 2019). It also underscored the need to explicitly control for natural landscape variables consistently across the five study freshwater ecoregions when developing our natural composite variable for use in CMRT analyses (Whittier and Van Sickle 2010; Van Sickle 2013).

Urban Land Use Effects on Stream Fish

After controlling for the influence of natural landscape variables with our natural composite variable, percent urban land use in network catchments explained the most variance in fish assemblages in the Appalachian Piedmont and Chesapeake Bay freshwater ecoregions at levels of 10%, and 2%, respectively. The lower level detected in the Chesapeake Bay freshwater ecoregion was similar to levels detected in other studies. Yoder et al. (1999) found that declines in sensitive species in Ohio streams occurred at <5% urban land use in stream catchments, resulting in part from degraded substrate, and Miltner et al. (2004) found changes in fish biotic condition in Ohio streams at levels as low as 5% urban land use in stream catchments. Higher levels of urban land use affecting fishes detected in the Appalachian Piedmont freshwater ecoregion (i.e., 10%) were similar to results from Wisconsin. Wang et al. (1997) found decreases in fish index of biotic integrity (IBI) scores in streams with catchment urban land use occurring at 10%; also, Wang et al. (2000, 2001) showed declines in fish species richness and diversity with levels >10% catchment imperviousness. In our study freshwater ecoregions, we used percent urban land use in the network catchments for analysis because it describes more variance in the data set, but percent impervious land cover in network catchments was highly correlated with percent urban land use in network catchments (R^2 = 0.97, Thornbrugh 2014). We observed changes in fish assemblage metrics with specific levels of disturbance, and these levels are consistent with studies in Ohio documenting significant declines in fish IBI scores with impervious land use >15% (Yoder et al. 1999; Miltner et al. 2004). The differences seen in the proportional level of urban land use in the network catchment that fish assemblage metrics are responding to between the Appalachian Piedmont (e.g., 10%) and Chesapeake Bay (e.g., 2%) freshwater ecoregions (and differing proportions between these freshwater ecoregions from the other three study freshwater ecoregions) may be related to different land use mechanisms and differences in fish assemblage composition. There is substantial variability in the proportional amounts of urban land use in the network catchments across our study freshwater ecoregions (Figure 2). Variability in fish response by freshwater ecoregion to the percent of urban land use in the network catchment may result in part from relationships between natural landscape variables (e.g., lithology) and the suitability of the land to be developed for specific types of urban land use (e.g., low-, medium-, and high-intensity urban land use; Utz et al. 2010). Regionally specific types of underlying lithology may exacerbate or buffer effects of urbanization (Utz et al. 2010, 2011). Furthermore, within a region, specific natural landscape characteristics (e.g., topology, lithology, and slope) influence the suitability of land for the development of various anthropogenic land use types (Allan 2004; Scott 2006). Compositional differences in fish assemblages among regions may be more or less susceptible to percent of urban in the network catchment (Meador et al. 2005). For example, regions with a high degree of endemism, or fish habitat specialists may be more susceptible to changes in fish assemblage structure from increases of urban land use than regions where fish assemblage composition is dominated by generalist fish species with a lower degree of endemic or specialist fish fauna. However, the prominence of the urban signal in the Appalachian Piedmont and Chesapeake Bay freshwater ecoregions as opposed to the other three freshwater ecoregions leads us to believe that regional differences may result from landscape characteristics of the freshwater ecoregions. For example, percent of agriculture in network catchments of study sites was low in both the Appalachian Piedmont and Chesapeake Bay freshwater ecoregions in comparison to the other three freshwater ecoregions, allowing the urban signal to prevail over an agricultural signal (Paavola et al. 2006).

In the Appalachian Piedmont freshwater ecoregion, we detected a secondary influence from urban land use on fish assemblages at 38% network catchment urban land use. Fish response at this proportion of urbanization is much higher than typical levels reported in the literature but may be related to low population abundances and/or low fish species diversity resulting from extreme levels of urban land use. Miltner et al. (2004) found an almost complete loss of aquatic life with levels of impervious land cover higher than 27% for stream catchments in Ohio. Again, these landscape variables (i.e., urban land use and impervious land cover) are known to have similar effects on streams and are highly correlated within our study freshwater ecoregions. Influence of the proportion of urban land use in the network catchment on stream fishes at two proportional levels in the Appalachian Piedmont (again, 10% and 38%) may reflect a series of threshold responses in the grouping of fish metrics, indicating a low and high sensitivity of stream fishes to urban land use (Allan 2004; Crawford et al. 2016). At a low proportional level of urban land use in the network catchment, fish assemblages may begin to change from those that would exist in the absence of urban land use in network catchments, with negative effects on fishes resulting from initial habitat degradation. At the higher proportional levels of urban land use in the network catchment, fish assemblage metrics may stop changing because of a loss of specialist species and some functional groups (e.g., invertivores), then level off and become dominated by a few tolerant or generalist fish species (Yoder et al. 1999; Paul and Meyer 2001; Miltner et al. 2004; Crawford et al. 2016). As described previously, Yoder et al. (1999) showed differences in fish assemblages occurring at levels of <5% urban land use in stream catchments and at 10–15% urban land use in stream catchments, and they also described responses at levels >15%, including severe fish faunal degradation from increased nutrients, and toxicity from urbanization in the catchments.

Effects of Agricultural Land Use on Stream Fish

Based on our CMRT results, agriculture was the dominant anthropogenic land use variable influencing fish assemblages in network catchments of the Upper Mississippi freshwater ecoregion at a level of 54%. It was also the secondary anthropogenic land use across study sites influencing fishes in the Laurentian Great Lakes, Chesapeake Bay, and Middle Missouri freshwater ecoregions at levels of 68%, <1%, and <1%, respectively. These results reflect a high (>50%) and low level (<1%) of agriculture land use in network catchments affecting fish assemblages across study freshwater ecoregions. Other studies showed fish assemblages change with high levels of agriculture in the network catchment. Fitzpatrick et al. (2001) described a decline in fish IBI scores with percentage of catchment agriculture >30%, and Wang et al. (1997) found a decline in fish abundance, fish species richness, and fish IBI scores at levels >50%. The low proportional levels of agriculture in network catchments in our study influencing changes in fish assemblage metrics in the Chesapeake Bay and Middle Missouri freshwater ecoregions may result from an initial decline in abundances of fishes, but effects can also be species-specific (Utz et al. 2010). Utz et al. (2010) found that sensitivity of fish species to percent of agriculture in catchments varied by physiographic province, with more species having negative responses to catchment agriculture in the Highlands physiographic province than in the Coastal Plain and Piedmont physiographic provinces. Also, fish sensitive to agriculture may have been reduced in abundances and range in the Coastal Plain and Piedmont physiographic provinces from previous deforestation (i.e., legacy land use effects) to a greater degree than in the Highlands physiographic province. Such legacy effects would lead to compositional differences in fish assemblages and higher sensitivity to agriculture in the Highland physiographic province (Harding et al. 1998; Foster et al. 2003). Similar histories could contribute to differences in the proportional levels of agriculture found to predict fish assemblage metrics at low levels of catchment agriculture. Also, differences may be related to natural landscape variables affecting suitability for types of agriculture (e.g., pasture versus row crop) as in the Sandhills region of the Middle Missouri freshwater ecoregion (Allan 2004).

Effects of Dams in the Network Catchment

A low density of dams in the catchment was the top anthropogenic land use variable influencing fish assemblages across study sites in the Laurentian Great Lakes and Middle Missouri freshwater ecoregions, and secondarily in the Upper Mississippi freshwater ecoregion, indicated by our CMRT results. This supports the idea that dams are acting as a landscape influence and that they may be a dominant anthropogenic landscape variable causing changes in fish assemblages (Cooper et al. 2016, 2017). In the Laurentian Great Lakes freshwater ecoregion, dam density was the top contributing anthropogenic variable that explained differences in groups of study sites based on fish assemblage metrics. Those results in our study are supported by Wang et al. (2011b), who characterized a unique amount of variation in two groupings of fish indicators explained by a set of dam metrics that included dam density in network catchments. Their dam metrics explained 16–19% of unique variation in fish IBI scores and fish habitat preference indicators, respectively. After excluding stream sites that had >60% agricultural or 10% urban land use in their upstream catchments, Wang et al. (2011b) found that most fish assemblage metrics tested had negative associations with dam metrics. Our study showed that even a small number of dams in the network catchment influenced fish assemblages in three out of the five study freshwater ecoregions. In the other two freshwater ecoregions, the proportions of urban and agriculture land use in the network catchments may be driving fish assemblage metric response via multiple functional pathways (e.g., hydrologic regulation, regulation of water chemistry, sediment regulation, habitat provision, etc.) and dominate responses in fish assemblage metrics (Flotemersch et al. 2016; Thornbrugh et al. 2018). Thornbrugh et al. (2018) developed a stress index that conceptualizes the delivery of anthropogenic landscape variables through multiple functional pathways (i.e., hydrologic connectivity, temperature regulation, regulation of water chemistry, etc.) that can degrade total watershed condition. That is not to say that dams in the network catchment are not influencing fish assemblage metrics in the Appalachian Piedmont and Chesapeake Bay freshwater ecoregions, only that they are not one of the top two contributing anthropogenic variables to fish assemblage responses.

Effects of Stream-Road Crossings

Density of stream-road crossings in network catchments was not identified as one of the top two contributing anthropogenic landscape variables affecting fishes in any of our five study freshwater ecoregions. Other studies have shown negative influences on fish movement from stream-road crossings (Warren and Pardew 1998; Schaefer et al. 2003; Mueller et al. 2008; Bouska and Paukert 2010). Januchowski–Hartley et al. (2013) documented the problems that road crossings pose as barriers to fishes in the Great Lakes region and described their potential for having catchment-wide cumulative effects within catchments. The additive effects and the pervasiveness of multiple barriers, including passage ways at stream-road crossings and small and large dams in river catchments, warrant further investigation to quantify their catchment-wide effects on fishes (Perkin and Gido 2012; Cooper et al. 2017). The cumulative effects of stream-road crossing are likely still having an important influence on fish assemblage structure, despite not being one of the top two contributing anthropogenic landscape variables tested.

Research Gaps and Challenges

One challenge with our analysis, as well as with many other landscape-scale analyses occurring over large spatial extents, is the need to control for influences of natural landscape variables on stream fish assemblages to more effectively understand how anthropogenic landscape variables may affect stream fishes (e.g., Daniel et al. 2015; Cooper et al. 2017; Deweber et al. 2019). We attempted to do this in two ways. First, we constrained our analyses within freshwater ecoregions, and second, we developed a composite variable comprised of natural landscape variables. The composite variable described a large percentage of natural variation across the study area (39%), yet this still left a large amount of unaccounted for variation in our data set that likely influenced our ability to fully characterize variation in fish metrics explained by anthropogenic variables in CMRT analysis. A second challenge is the need to consider influences of natural and anthropogenic landscape factors in riparian buffers, as well as in catchments of study streams. Because riparian buffers can mitigate effects of anthropogenic land uses (e.g., Crawford et al. 2016), an accounting of natural and/or anthropogenic landscape features within buffers may have allowed us to more accurately assess the degree to which anthropogenic landscape factors were influencing stream fishes. Addressing both of these challenges are avenues for future work into large-scale studies of landscape influences on stream fishes.

Management Implications and Conclusions

Interregional comparisons into mechanisms controlling regional fish assemblage responses to anthropogenic landscape variables is a fruitful area for additional investigation. A similar CMRT analysis conducted in this study at varying spatial extents could elucidate the specific spatial extent over which anthropogenic landscape variables most affect fish assemblages. Some of the potential regional differences causing different fish assemblage responses may be related to the buffering capacity of lithology and/or topographic landscape variables, as well as natural land covers and the fish beta diversity (Pompeu et al. 2019, this volume) in a region. The types, proportions of natural versus anthropogenic land cover and their spatial arrangement in the catchment have been shown to affect the delivery of runoff and other materials to stream ecosystems (Zimmerman et al. 2003; Blann et al. 2009; Carpenter et al. 2011). Intensity of urban land use, hydrologic connectivity of impervious land cover, specific agricultural practices, and riparian buffers to streams are a few examples of variables that may influence the delivery of materials to riverine ecosystems (Blann et al. 2009; Carpenter et al. 2011). Such variables cause heterogeneous responses in stream biota across regions (Meador and Goldstein 2003; Morgan and Cushman 2005; Utz et al. 2010). Our results and methods can be used to assess regional dominance of classes of anthropogenic land use types that are affecting changes in fish assemblages and can provide insights into levels of anthropogenic land use that are causing change in fish assemblage metrics.

Our study underscored regional differences in dominant anthropogenic land use affecting differences in groups of sites based on fish metrics. If dominant anthropogenic land uses differ across regions, and anthropogenic landscape variables influence fish assemblages at different levels across regions, then these influences should affect relevant management objectives for stream resources. Studies characterizing critical anthropogenic stressors in one region may not be directly applicable to a different region because of region-specific influences. Region-specific influences of anthropogenic landscape variables can be accounted for and incorporated into the planning process to develop realistic management priorities. Future research studies could help further characterize regional differences in anthropogenic land use effects on stream biota, elucidate important mechanisms of influence on stream systems, and have transferable results or components to help meet management priorities across large spatial extents. Assuming that anthropogenic pressures will continue to grow, we need responsible planning, better understanding for future development of landscapes, and land conversion to protect the diversity and complexity of freshwater biota for future generations.

Acknowledgments

This research was conducted with support from the U.S. Geological Survey Aquatic GAP Program and U.S. Fish and Wildlife Service and grants from Michigan State University. This research was also supported by a scholarship from the Austrian Marshall Plan Foundation and a fellowship from the Red Cedar Fly Fishers. We thank graduate committee members, Bianca de Freitas Terra, and three anonymous reviewers for providing feedback that helped to improve earlier versions of the manuscript. We thank the Aquatic Landscape Ecology (ALE) Laboratory at Michigan State University for assistance in developing landscape metrics and Wesley Daniel for assistance in developing the game fish species metric.

References

Abell, R. A., D. M. Olsen, E. Dinerstein, P. T. Hurley, J. T. Diggs, W. Eichbaum, S. Walters, W Wettengel, T. Allnutt, C. J. Loucks, and P. Hedao. 2000. Freshwater ecoregions of North America: a conservation assessment. Island Press, Washington, D.C.

Abell, R., M. L. Thieme, C. Revenga, M. Bryer, M. Kottelat, N. Bogutskaya, B. Coad, N. Mandrak, S. Contreras Balderas, W. Bussing, M. L. J. Stiassny, P. Skelton, G. R. Allen, P. Unmack, A. Naseka, R. Ng, N. Sindorf, J. Robertson, E. Armijo, J. V. Higgins, T. J. Heibel, E. Wikramanayake, D. Olson, H. L. López, R. E. Reis, J. G. Lundberg, M. H. Sabaj Pérze, and P. Petry. 2008. Freshwater ecoregions of the world: a new map of biogeographic units for freshwater biodiversity conservation. BioScience 58:403–414.

Alford, J. B. 2014. Multi-scale assessment of habitats and stressors influencing stream fish assemblages in the Lake Pontchartrain basin, USA. Hydrobiologia 738:129–146.

Allan, J. D. 2004. Landscapes and riverscapes: the influence of land use on stream ecosystems. Annual Review of Ecology, Evolution, and Systematics 35:257–284.

Allan, J. D., D. L. Erickson, and J. Fay. 1997. The influence of catchment land use on stream integrity across multiple spatial scales. Freshwater Biology 37:149–61.

Allan, J. D., and A. S. Flecker. 1993. Biodiversity conservation in running waters. BioScience 43:32–43.

Barbour M. T., J. Gerritsen, B. D. Snyder, and J. B. Stribling. 1999. Rapid bioassessment protocols for use in streams and wadeable rivers: periphyton, benthic macroinvertebrates and fish, 2nd edition. U.S. Environmental Protection Agency, Office of Water, EPA/841-B-99-002, Washington, D.C.

Blann, K. L., J. L. Anderson, G. R. Sands, and B. Vondracek. 2009. Effects of agricultural drainage on aquatic ecosystems: a review. Critical Reviews in Environmental Science and Technology 39:909–1001.

Borcard, D., F. Gillet, and P. Legendre. 2011. Numerical Ecology with R. Springer, New York.

Bouska, W. W., and C. P. Paukert. 2010. Road crossing designs and their impact on fish assemblages of Great Plains streams. Transactions of the American Fisheries Society 139:214–222.

Brown, L. R., M. B. Gregory, and J. T. May. 2009. Relation of urbanization to stream fish assemblages and species traits in nine metropolitan areas of the United States. Urban Ecosystems 12:391–416.

Carpenter, S. R., E. H. Stanley, and M. J. V. Zanden. 2011. State of the world's freshwater ecosystems: physical, chemical, and biological changes. The Annual Review of Environment and Resources 36:75–99.

Caissie, C. 2006. The thermal regime of rivers: a review. Freshwater Biology 51:1389–1406.

Cooke, S. J., C. Paukert, and Z. Hogan. 2012. Endangered river fish: factors hindering conservation and restoration. Endangered Species Research 17:179–191.

Cooper, A. R., D. M. Infante, W. M. Daniel, K. E. Wehrly, L. Wang, and T. O. Brenden. 2017. Assessment of dam effects on streams and fish assemblages of the conterminous USA. Science of the Total Environment 586:879–889.

Cooper, A. R., D. M. Infante, K. E. Wehrly, L. Wang, and T. O. Brenden. 2016. Identifying indicators and quantifying large-scale effects of dams on fishes. Ecological Indicators 61:646–657.

Crawford, S., G. Whelan, D. M. Infante, K. Blackhart, W. M. Daniel, P. L. Fuller, T. Birdsong, D. J. Wieferich, R. McClees-Funinan, S. M. Stedman, K. Herreman, and P. Ruhl. 2016. Through a fish's eye: the status of fish habitats in the United States 2015. National Fish Habitat Partnership. Available: http://assessment.fishhabitat.org/. (April 2019).

Cress, J., D. R. Soller, R. Sayre, P. Comer, and H. Warner. 2010. Terrestrial ecosystems—surficial lithology of the conterminous United States: Scientific Investigations Map 3126, scale 1:5,000,000, 1 sheet. U.S. Geological Survey, Scientific Investigations Map 3126. Available: http://pubs.usgs.gov/sim/3126. (April 2019).

Daly, C., M. Halbleib, J. I. Smith, W. P. Gibson, M. K. Doggett, G. H. Taylor, J. Curtis, and P. P. Pasteris. 2008. Physiographically sensitive mapping of climatological temperature and precipitation across the conterminous United States. International Journal of Climatology 28:2031–2064.

Daniel, W. M., D. M. Infante, R. M. Hughes, Y. Tsang, P. C. Esselman, D. Wieferich, K. Herreman, A. R. Cooper, L. Wang, and W. W. Taylor. 2015. Characterizing coal and mineral mines as a regional source of stress to stream fish assemblages. Ecological Indicators 50:50–61.

De'ath, G. 2002. Multivariate regression trees: a new technique for constrained classification analysis. Ecology 83:1105–1117.

Deweber, J. T., L. Sleezer, and E. A. Frimpong. 2019. A new regionalization framework to quantify how physiography mediates the effect of land use on stream fishes. Pages 321–350 *in* R. M. Hughes, D. M. Infante, L. Wang, K. Chen, and B. F. Terra, editors, Advances in understanding landscape influences on freshwater habitats and biological assemblages. American Fisheries Society, Symposium 90, Bethesda, Maryland.

Dudgeon, D., A. H. Arthington, M. O. Gessner, Z. Kawabata, D. J. Knowler, C. Lévêque, R. J. Naiman, A. Prieur-Richard, D. Soto, M. L. J. Stiassny, and C. A. Sullivan. 2006. Freshwater biodiversity: importance, threats, status and conservation challenges. Biological Reviews 81:163–182.

Esselman, P. C., D. M. Infante, L. Wang, A. R. Cooper, D. Wieferich, Y. Tsang, D. J. Thornbrugh, and W. W. Taylor. 2013a. Regional fish community indicators of landscape disturbance to catchments of the conterminous United States. Ecological Indicators 26:163–173.

Esselman, P., D. M. Infante, D. Wieferich, A. Cooper, L. Wang, W. Taylor, R. Tingley, D. Thornbrugh, J. Ross, and J. Fenner. 2013b. National fish habitat action plan (NFHAP) 2010 community fish data. National Fish Habitat Partnership Data System. Available: www.sciencebase.gov/catalog/item/521519e0e4b06d85e095c50f#. (April 2019).

Fitzpatrick, F. A., B. C. Scudder, B. N. Lenz, and D. J. Sullivan. 2001. Effects of multi-scale environmental characteristics on agricultural stream biota in eastern Wisconsin. Journal of the American Water Resources Association 37:1489–1507.

Flotemersch, J. E., S. G. Leibowitz, R. A. Hill, J. L. Stoddard, M. C. Thoms, and R. E. Tharme. 2016. A watershed integrity definition and assessment approach to support strategic management of watersheds. River Research and Application 32:1654–1671.

Foster, D., F. Swanson, J. Aber, I. Burke, N. Brokaw, D. Tilman, and A. Knapp. 2003. The importance of land-use legacies to ecology and conservation. BioScience 53:77–88.

Frimpong, E. A., and P. L. Angermeier. 2009. Fish traits: a database of ecological and life-history traits of freshwater fishes of the United States. Fisheries 34:487–495.

Graf, W. L. 2006. Downstream hydrologic and geomorphic effects of large dams on American rivers. Geomorphology 79:336–360.

Gesch, D. B., 2007. The national elevation dataset. Pages 99–118 *in* D. Maune, editor.

Digital elevation model technologies and applications: the DEM user's manual, 2nd edition. American Society for Photogrammetry and Remote Sensing, Bethesda, Maryland,

Harding, J. S., E. F. Benfield, P. V. Bolstad, G. S. Helfman, and E. B. D. Jones. 1998. Stream biodiversity: the ghost of land use past. Proceedings of the National Academy of Sciences of the United States of America 95:14843–14847.

Helfman, G. S. 2007. Fish conservation: a guide to understanding and restoring global aquatic biodiversity and fishery resources. Island Press, Washington, D.C.

Herlihy, A. T., J. C. Sifneos, R. M. Hughes, D. V. Peck, and R. M. Mitchell. 2019. Lotic fish assemblage clusters across the conterminous United States and their associations with environmental variables. Pages 385–408 *in* R. M. Hughes, D. M. Infante, L. Wang, K. Chen, and B. F. Terra, editors, Advances in understanding landscape influences on freshwater habitats and biological assemblages. American Fisheries Society, Symposium 90, Bethesda, Maryland.

Hocutt, C. H., and E. O. Wiley, editors. 1986. The zoogeography of North American freshwater fishes. Wiley, New York.

Homer, C., C. Huang, L. Yang, B. Wylie, and M. Coan. 2004. Development of a 2001 national landcover database for the United States. Photogrammetric Engineering and Remote Sensing 70:829–840.

Hughes, R. M., J. N. Rinne, and B. Calamusso. 2005. Historical changes in large river fish assemblages of the Americas: a synthesis. Pages 603–612 *in* J. N. Rinne, R. M. Hughes, and B. Calamusso, editors. Historical changes in large river fish assemblages of the Americas. American Fisheries Society, Symposium 45, Bethesda, Maryland.

ITIS (Integrated Taxonomic Information System). 2010. Integrated Taxonomic Information System [online database]. Available: www.itis.gov. (April 2010)

Januchowski-Hartley, S. R., P. B. McIntyre, M. Diebel, P. J. Doran, D. M. Infante, C. Joseph, and J. D. Allan. 2013. Restoring aquatic ecosystem connectivity requires expanding barrier inventories. Frontiers in Ecology and the Environment 11:211–217.

Jelks, H. L., S. J. Walsh, N. M. Burkhead, S. Contreras-Balderas, E. Díaz-Pardo, D. A. Hendrickson, J. Lyons, N. E. Mandrak, F. McCormick, J. S. Nelson, S. P. Platania, B. A. Porter, C. B. Renaud, J. J. Schmitter-Soto, E. B. Taylor, and M. L. Warren, Jr. 2008. Conservation status of imperiled North American freshwater and diadromous fishes. Fisheries 33:372–407.

Legendre, P., and L. Legendre. 2012. Numerical ecology, 3rd edition. Elsevier, Amsterdam.

Ligon, F. K., W. E. Dietrich, and W. J. Trush. 1995. Downstream ecological effects of dams. BioScience 45:183–192.

Lyons, J. 1992. Using the index of biotic integrity (IBI) to measure environmental quality in warmwater streams of Wisconsin. U.S. Forest Service, General Technical Report NC-149, Saint Paul, Minnesota.

Matthews, W. J. 1998. Patterns in freshwater fish ecology. Chapman and Hall, New York.

Maxwell, J. R., C. J. Edwards, M. E. Jensen, S. J. Paustian, H. Parrott, and D. M. Hill. 1995. A hierarchical framework of aquatic ecological units in North America (Nearctic Zone). U. S. Forest Service, General Technical Report NC-176, Saint Paul, Minnesota.

Mbaka, J. G., and M. Wanjiru Mwaniki. 2015. A global review of the downstream effects of small impoundments on stream habitat conditions and macroinvertebrates. Environmental Reviews 23:257–262.

Meador, M. R., J. F. Coles, and H. Zappia, 2005. Fish assemblage responses to urban intensity gradients in contrasting metropolitan areas: Birmingham, Alabama, and Boston, Massachusetts. Pages 409–422 *in* L. R. Brown, R. H. Gray, R. M. Hughes, M. R. Meador, editors. Effects of urbanization on stream ecosystems. American Fisheries Society, Symposium 47, Bethesda, Maryland.

Meador, M. R., and R. M. Goldstein. 2003. Assessing water quality at large geographic scales: relations among land use, water physicochemistry, riparian condition, and fish community structure. Environmental Management 31:504–517.

Miltner, R. J., D. White, and C. Yoder. 2004. The biotic integrity of streams in urban and suburbanizing landscapes. Landscape and Urban Planning 69:87–100.

Morgan, R. P., and S. E. Cushman. 2005. Urbanization effects on stream fish assemblages in Maryland, USA. Journal of the North American Benthological Society 24:643–655.

Mueller, R. P., S. S. Southard, C. W. May, W. H. Pearson, and V. I. Cullinan. 2008. Juvenile Coho Salmon leaping ability and behavior in an experimental culvert test bed. Transactions of the American Fisheries Society 137:941–950.

NOAA (National Oceanographic and Atmospheric Administration). 2010. Development sprawl impacts on the terrestrial carbon dynamics of the United States. Available: http://www.ngdc.noaa.gov/eog/dmsp/download_sprawl.html. (December 2014).

Oksanen, J., F. G. Blanchet, R. Kindt, P. Legendre, P. R. Minchin, R. B. O'Hara, G. L. Simpson, P. Solymos, M. H. H. Stevens and H. Wagner. 2013. Vegan: community ecology package. R package version 2.0-9. Available: http://CRAN.R-project.org/package=vegan.

Olden, J. D., and R. J. Naiman. 2010. Incorporating thermal regimes into environmental flows assessments: modifying dam operations to restore freshwater ecosystem integrity. Freshwater Biology 55:86–107.

Ormerod, S. J. 2003. Current issues with fish and fisheries: editor's overview and introduction. Journal of Applied Ecology 40:204–213.

Ouellette, M. H., P. Legendre, and D. Borcard. 2012. Cascade multivariate regression tree: a novel approach for modelling nested explanatory sets. Methods in Ecology and Evolution 3:234–244.

Ouellette, M. H., and P. Legendre. 2013. MVPARTwrap: additional features for package mvpart. R package version 0.1-9.2. Available: https://cran.r-project.org/src/contrib/Archive/MVPARTwrap/.

Paavola, R., T. Muotka, R. Virtanen, J. Heino, D. Jackson, and A. Maki-Petays. 2006. Spatial scale affects community concordance among fishes, benthic macroinvertebrates, and bryophytes in streams. Ecological Applications 16:368–379.

Paul, M. J., and J. L. Meyer. 2001. Streams in the urban landscape. Annual Review of Ecology and Systematics 32:333–65.

Perkin, J. S., and K. B. Gido. 2012. Fragmentation alters stream fish community structure in dendritic ecological networks. Ecological Applications 22:2176–2187.

Pompeu, P. S., C. G. Leal, D. R. Carvalho, N. T. Junqueira, M. A. Castro, and R. M. Hughes. 2019. Effects of catchment land use on stream fish assemblages in the Brazilian savanna. Pages 303–320 *in* R. M. Hughes, D. M. Infante, L. Wang, K. Chen, and B. F. Terra, editors. Advances in understanding landscape influences on freshwater habitats and biological assemblages. American Fisheries Society, Symposium 90, Bethesda, Maryland.

Ricciardi, A., and J. B. Rasmussen. 1999. Extinction rates of North American freshwater fauna. Conservation Biology 13:1220–1222.

Roth, N. E., J. D. Allan, and D. L. Erickson. 1996. Landscape influences on stream biotic integrity assessed at multiple spatial scales. Landscape Ecology 11:141–56.

R Core Team. 2013. R: a language and environment for statistical computing. R Foundation for Statistical Computing, Vienna, Austria. Available: http://www.R-project.org.

Schaefer, J. F., E. Marsh-Matthews, D. E. Spooner, K. B. Gido, and W. J. Matthews. 2003. Effects of barriers and thermal refugia on local movements of threatened Leopard Darter, Percina pantherina. Environmental Biology of Fishes 66:391–400.

Schinegger, R., M. Palt, P. Segurado, and S.

Schmutz. 2016. Untangling the effects of multiple human stressors and their impacts on fish assemblages in European running waters. Science of the Total Environment 573:1079–1088.

Schwarz, G. E., and R. B. Alexander. 1995. Soils data for the conterminous United States derived from the NRCS state soil geographic (STATSGO) data base. U.S. Geological Survey, Open-File Report 95-449, Arc 7.0 coverage, Reston, Virginia. Available: https://water.usgs.gov/GIS/metadata/usgswrd/XML/ussoils.xml. (December 2014).

Scott, M. C. 2006. Winners and losers among stream fishes in relation to land use legacies and urban development in the southeastern US. Biological Conservation 127:301–309.

Soller, D. R., and M. C. Reheis, compilers. 2004. Surficial materials in the conterminous United States. U.S. Geological Survey, Open-File Report 03-275, Reston, Virginia.

Strayer, D. L., and D. Dudgeon. 2010. Freshwater biodiversity conservation: recent progress and future challenges. Journal of the North American Benthological Society 29:344–358.

ter Braak, C. J. F. and P. Smilauer. 2002. CANOCO 4.5 reference manual and CanoDraw for Windows. User's guide to Canoco for Windows: software for canonical community ordination. Microcomputer Power, Ithaca, New York.

Therneau, T. M., B. Atkinson, B. Ripley, J. Oksanen, and G. De'ath. 2012. mvpart: multivariate partitioning. R package version 1.6-1. Available: https://CRAN.R-project.org/package=mvpart.

Thornbrugh, D. J. 2014. Regional influence of landscape features and processes on fluvial fish assemblages. Doctoral dissertation. Michigan State University, East Lansing.

Thornbrugh, D. J., S. G. Leibowitz, R. A. Hill, M. H. Weber, Z. C. Johnson, A. R. Olsen, J. E. Flotemersch, J. L. Stoddard, and D. V. Peck. 2018. Mapping watershed integrity for the conterminous United States. Ecological indicators 85:1133–1148.

Tsang, Y., D. Wieferich, K. Fung, D. M. Infante, and A. R. Cooper. 2014. An approach for aggregating upstream catchment information to support research and management of fluvial systems across large landscapes. SpringerPlus 23:589–597.

USCB (U.S. Census Bureau). 2002. Redistricting census 2000 TIGER/ Line Files [machine-readable data files]. U.S. Census Bureau, Washington, D.C. Available: www.census.gov/mp/www/geo/tiger.html. (May 2019).

USEPA (U.S. Environmental Protection Agency) and USGS (U.S. Geological Survey). 2005. National hydrography dataset plus (NHDPlusV1). Available: www.horizon-systems.com/nhdplus/.

USGS (U.S. Geological Survey). 2013. 2012 national anthropogenic barrier dataset (NABD). Available: www.sciencebase.gov/catalog/item/56a7f9dce4b0b28f1184dabd. (May 2019).

Utz, R. M., K. N. Eshleman, and R. H. Hilderbrand. 2011. Variation in physicochemical responses to urbanization in streams between two Mid-Atlantic physiographic regions. Ecological Applications 21:402–415.

Utz, R. M., R. H. Hilderbrand, and R. L. Raesly. 2010. Regional differences in patterns of fish species loss with changing land use. Biological Conservation 143:688–699.

Van Sickle, J. 2013. Estimating the risks of multiple, covarying stressors in the National Lakes Assessment. Freshwater Science 32:204–216.

Vörösmarty, C. J., P. B. McIntyre., M. O. Gessner, D. Dudgeon, A. Prusevich, P. Green, S. Glidden, S. E. Bunn, C. A. Sullivan, C. R. Liermann, P. M. Davies. 2010. Global threats to human water security and river biodiversity. Nature (London) 467:555–561.

Walsh, C. J., A. H. Roy, J. W. Feminella, P. D. Cottingham, P. M. Groffman, and R. P. I. Morgan. 2005. The urban stream syndrome: current knowledge and the search for a cure. Journal of the North American Benthological Society 24:706–723.

Walters, D. M., D. S. Leigh, and A. B. Bearden.

2003. Urbanization, sedimentation, and the homogenization of fish assemblages in the Etowah River basin, USA. Hydrobiologia 494:5–10.

Wang, L., D. Infante, P. Esselman, A. Cooper, D. Wu, W. Taylor, D. Beard, G. Whelan, and A. Ostroff. 2011a. A hierarchical spatial framework and database for the national river fish habitat condition assessment. Fisheries 36:436–449.

Wang, L., D. Infante, J. Lyons, J. Stewart, and A. Cooper. 2011b. Effects of dams in river networks on fish assemblages in non-impoundment sections of rivers in Michigan and Wisconsin, USA. River Research and Applications 27:473–487.

Wang, L., J. Lyons, and P. Kanehl. 2001. Impacts of urbanization on stream habitat and fish across multiple spatial scales. Environmental Management 28:255–266.

Wang, L., J. Lyons, P. Kanehl, R. Bannerman, and E. Emmons. 2000. Watershed urbanization and changes in fish communities in southeastern Wisconsin streams. Journal of the American Water Resources Association 36:1173–1189.

Wang, L., J. Lyons, P. Kanehl, and R. Gatti. 1997. Influences of watershed land use on habitat quality and biotic integrity in Wisconsin streams. Fisheries 22(6):6–12.

Wang, L., J. Lyons, P. Rasmussen, P. Seelbach, T. Simon, M. Wiley, P. Kanehl, E. Baker, S. Niemela, and P. M. Stewart. 2003. Watershed, reach and riparian influences on stream fish assemblages in the Northern Lakes and Forest Ecoregion, USA. Canadian Journal of Fisheries and Aquatic Sciences 60:491–505.

Warren, M. L., Jr., and M. G. Pardew. 1998. Road crossings as barriers to small-stream fish movement. Transactions of the American Fisheries Society 127:637–644.

Whittier, T. R., R. M. Hughes, G. A. Lomnicky, and D. V. Peck. 2007. Fish and amphibian tolerance values and an assemblage tolerance index (ATI) for western USA streams and rivers. Transactions of the American Fisheries Society 136:254–271.

Whittier, T. R., J. L. Stoddard, R. M. Hughes, and G. Lomnicky. 2006. Associations among catchment- and site-scale disturbance indicators and biological assemblages at least- and most-disturbed stream and river sites in the western USA. Pages 641–664 *in* R. M. Hughes, L. Wang, and P. W. Seelbach, editors. Landscape influences on stream habitat and biological assemblages. American Fisheries Society, Symposium 48, Bethesda, Maryland.

Whittier, T. R., and J. Van Sickle. 2010. Macroinvertebrate tolerance values and an assemblage tolerance index (ATI) for western USA streams and rivers. Journal of the North American Benthological Society 29:852–866.

Wiens, J. A. 2002. Riverine landscapes: taking landscape ecology into the water. Freshwater Biology 47:501–515.

Wohl, E., N. Kramer, V. Ruiz-Villanueva, D. Scott, F. Comti, A. Gurnell, H. Piegay, K. Lininger, K. Jaeger, D. Walters, and K. Fausch. 2019. The natural wood regime in rivers. BioScience 69:259–273.

Wolock, D. M. 2003. Base-flow index grid for the conterminous United States, U.S. Geological Survey, Open-File Report 03-263, Reston, Virginia. Available: http://water.usgs.gov/lookup/getspatial?bfi48grd. (March 2012).

Yoder, C. O., R. J. Miltner, and D. White. 1999. Assessing the status of aquatic life designated uses in urban and suburban watersheds. Pages 16–28 in A. Everson, editor. National conference on retrofit opportunities for water resource protection in urban environments. U.S. Environmental Protection Agency, EPA/625/R-99/002, Washington, D.C.

Zimmerman, J. K. H., B. Vondracek, and J. Westra. 2003. Agricultural land use effects on sediment loading and fish assemblages in two Minnesota (USA) watersheds. Environmental Management 32:93–105.

Appendix 1 Evaluated Landscape Variable List

Table A.1.1. List of 52 landscape variables initially evaluated for regional comparison.

Variable code	Landscape variable description	Local catchment	Network catchment
Natural			
Land use (%)			
Water	Water	x	x
Wetland	Wetland	x	x
Barren	Barren land	x	x
Forest	Forest	x	x
Shrub	Shrub	x	x
Grassland	Grassland	x	x
Catchment area			
Areasqkm	Catchment area (km^2)	x	[ab]
Slope			
Slope	Mean slope (degrees)	x	
Elevation			
Elev_mean	Mean elevation (m)	[ab]	x
Lithology			
Fine	Fine lithology (%)		[ab]
Coarse	Coarse lithology (%)		x
Soil hydrology			
Base-flow index	Groundwater contribution to base flow (%)		[ab]
Soilhygrp	Mean soil hydrologic relative rates * 10 (infiltration rates ranged from high = 1 to very slow = 4)	x	
Soilperm	Soil permeability rates (cm/h * 100)	x	
Soilhygrpc	Mean soil hydrologic relative rates * 10 (infiltration rates ranged from high = 1 to very slow = 4)		x
Soilpermc	Mean soil permeability rates (cm/h * 100)		x
Climate			
temp	Mean annual air temperature (°C)	[ab]	
Areawtmap	Area weighted average mean annual precipitation (mm)		[ab]
Anthropogenic			
Land use (%)			
Pasture	Pasture/hay	x	[a]
Crops	Cultivated crops	x	[a]
Pasture_Crops	Pasture/hay and cultivated crops	x	[b]
Developed_open	Open space developed land	x	[a]
Developed_low	Low intensity developed land	x	x
Developed_medium	Medium intensity developed land	x	x

Table A.1.1. Continued.

Variable code	Landscape variable description	Local catchment	Network catchment
Anthropogenic (continued)			
Land use (%) (continued)			
Developed_high	High intensity developed land	x	x
Developed_lmh	Low, medium, and high intensity developed land	x	[a]
Developed_all	All developed land	x	[b]
Imperv	Impervious surfaces		x
Population density			
Popdens	Population density (number/km^2)		x
Roads			
Road_cross	Road crossing (number/100 km^2)	x	[ab]
Road_len	Road density (km/100 km^2)	x	x
Barriers			
Dam	Dam density (number/100 km^2)		[ab]
Canal/Ditch			
Canal_ditch	Canal/ditch (m/km^2)		x

[a] Landscape variables used in the redundancy analysis.
[b] Landscape variables used in the cascade multivariate regression tree analysis.

Appendix 2 Landscape Metrics Pearson Correlation

Table A.2.1. Results of Pearson correlation among landscape metrics; only results with *r* values ≥0.70 or ≤ –0.70 are displayed. Landscape variable names with a "c" at the end are network catchment measures.

Landscape variables	Wetland	Forest	Shrub	Grassland	Forest_c	Developed_open	Developed_low	Developed_med	Developed_high	Developed_lmh	Developed_all	Pasture	Crops	Pasture_Crops	Developed_openc	Developed_lowc	Developed_medc	Developed_highc	Developed_lmhc	Developed_allc	Crop_c	Elev_mean	Soilhygrp	Soilperm	Popdensc	Impervc
Wetland_c	0.75																									
Forest_c		0.82																								
Shrub_c			0.81																							
Grassland_c				0.85																						
Developed_med							0.82																			
Developed_high								0.88																		
Developed_lmh							0.97	0.92	0.82																	
Developed_all						0.79	0.90	0.86	0.77	0.93																
Pasture_Crops													0.90													
Developed_lowc							0.73			0.73	0.71				0.74											
Developed_medc								0.70		0.71						0.90										
Developed_highc																0.83	0.95									
Developed_lmhc							0.71			0.74	0.72				0.72	0.98	0.96	0.90								
Developed_allc											0.74				0.87	0.94	0.91	0.86	0.96							
Pasture_c												0.76														
Crop_c					–0.77								0.85	0.75												
Pasture_Crops_c					–0.74								0.79	0.82							0.92					
Elev_meanc																						0.96				
Soilhygrpc																							0.79			
Soilpermc																								0.83		
Popdensc																0.81	0.78	0.73	0.81	0.79						
Impervc										0.73	0.72				0.75	0.96	0.97	0.93	0.99	0.97					0.81	
Road_lencz															0.74	0.75	0.71		0.74	0.78					0.76	0.76

American Fisheries Society Symposium 90:385–408, 2019
© 2019 by the American Fisheries Society

Lotic Fish Assemblage Clusters across the Conterminous United States and Their Associations with Environmental Variables

Alan T. Herlihy*
Department of Fisheries and Wildlife, Oregon State University
Corvallis, Oregon 97331, USA

Jean C. Sifneos
Department of Statistics, Oregon State University
Corvallis, Oregon 97331, USA

Robert M. Hughes
Amnis Opes Institute, and Department of Fisheries and Wildlife, Oregon State University
2895 SE Glenn, Corvallis, Oregon 97331, USA

David V. Peck
U.S. Environmental Protection Agency, Office of Research and Development
National Health and Environmental Effects Research Laboratory
Western Ecology Division
200 SW 35th Street, Corvallis, Oregon 97333, USA

Richard M. Mitchell
U.S. Environmental Protection Agency, Office of Water
1200 Pennsylvania Avenue, NW, MC 4502T, Washington, D.C. 20460, USA

Abstract.—Between 2008 and 2014, the first two phases of the National Rivers and Streams Assessment sampled fish assemblages in 2,554 stream and river sites across the conterminous United States. Associated physical habitat, water chemistry, and landscape data were also collected. We used cluster analysis to derive fish assemblage clusters. Assemblage clusters were then related to the local and catchment scale environmental data to assess the primary drivers of fish assemblage structure and to predict cluster membership. The results from our study show that whereas variability in fish assemblages is large over the range of stream/river sizes across the conterminous United States, it is possible to divide sites into six clusters that could be defined by specific indicator species and predictable from environmental data using both classification tree analysis and discriminant function analysis. Ordination identified three environmental gradients as the primary drivers of the biological clusters: stream size, temperature, and stream water ionic strength. The biological classification was very reproducible within a sample year based on repeat sample visits to the same sites but was lower

* Corresponding author: alan.herlihy@oregonstate.edu

based on samples taken 4–6 years apart. Biological monitoring is essential for the complete assessment of aquatic ecosystem condition. As the natural variability in aquatic biota is quite high, reporting and analyzing the results of biomonitoring requires a classification framework that minimizes this variability so that expectations of least-disturbed condition and the effects of anthropogenic disturbance can be more clearly defined. Our results showed that biologically derived classes of sites have higher classification strength than either basins or ecoregions. Thus, for the purposes of biomonitoring, it could be advantageous to replace physical regions with a classification based on aquatic biota.

Introduction

Fish assemblage patterns are easily detectable at multiple scales. For example, fish assemblages differ significantly between large and small streams in the same river basin (Hawkes 1975; Hughes and Gammon 1987; Lyons 1996; McGarvey and Hughes 2008; McGarvey and Ward 2008; Ibañez et al. 2009; Pinto et al. 2009). They also differ among ecoregions and basins at state extents (Larsen et al. 1986; Hughes et al. 1987; Rohm et al. 1987; Whittier et al. 1988; Matthews and Robison 1988; Matthews et al. 1992; Changeux and Pont 1995; Lyons 1996; Van Sickle and Hughes 2000). At both landscape extents, fish assemblages differ in association with local environmental conditions as well as geographic proximity (Hughes and Gammon 1987; Matthews et al. 1992; Lyons 1996; McCormick et al. 2000; Van Sickle and Hughes 2000; McGarvey and Hughes 2008; McGarvey and Ward 2008).

Although transitions from one assemblage to another among sites across a region or continent or along a river's course may be obvious, scientists have differed in what causes them. For example, Meek (1891), Gilbert (1980), Hocutt and Wiley (1986), and Matthews (1998) employed river basins to explain patterns in fish distribution at state and national extents. On the other hand, Legendre and Legendre (1984) found climatic, geomorphic, and vegetation patterns more closely associated with fish assemblage patterns than basins in Quebec. Pflieger (1971) used physiographic regions and river size to describe different Missouri fish faunal regions. In the United States and New Zealand, ecoregions are used by many water quality agencies (Biggs et al. 1990; Hughes et al. 1994) for distinguishing fish assemblages. Multiple landscape factors are likely responsible, but chief among them are natural and anthropogenic landscape conditions, river basins, and river size. Also, controlling factors seen locally may not be important over large regions and vice versa (Matthews et al. 1992; Changeux and Pont 1995; Lyons 1996). The answer also varies by location and the spatial extent of the study (Hughes et al. 1994, 2015).

Because of the difficulties and expense of conducting large-extent surveys, there have been many more data-driven studies at local extents than continental extents. However, it is increasingly important to understand, monitor, and assess continental-extent conditions in ecosystems through use of fish assemblages as biological indicators of status and trends (Pont et al. 2006; Meador et al. 2008; Schinegger et al. 2016; USEPA 2016a). Such studies are needed for better allocating funding, improving water and land-use regulations, and focusing conservation efforts. Although previous continental-extent studies of the United States have been attempted by compiling fish assemblage data from various sources, there are inherent problems regarding data availability and method comparability (Herlihy et al. 2006). For example, Herlihy et al. (2006) analyzed fish assemblage data compiled from 23 different state and federal surveys that had been collected through use of different survey designs, sampling methods, and site densities (Herlihy et al. 2006). Furthermore, the available survey data lacked consistently collected physical habitat structure, water quality, and landscape condition data, thereby precluding rigorous analyses of the associations between such data and fish assemblage patterns. Nonetheless, Herlihy et al. (2006) indicated that schemes based on ecoregion, physiography, hydrologic units, and political boundaries had similar mean within-group fish assemblage similarities; therefore, neither supporting nor rejecting the use of any one of those schemes for explaining fish assemblage patterns. In addition, the results were not strongly influenced by removing nonnative fish species and disturbed sites from data analyses, suggesting that the mechanisms responsible for controlling fish assemblage patterns are robust to the effects of nonnative species and anthropogenic disturbances. But the inconsistently collected data used by Herlihy et al. (2006) support the need for a study based on consistently collected fish assemblage and environmental predictor data. Such data are now available from the U.S. Environmental Protection Agency (USEPA).

The USEPA's National Aquatic Resource Surveys began in 2004 and were designed to estimate the condition of surface waters throughout the United States. To address this objective, a large number (~1,000) of randomly selected lakes, streams, rivers, wetlands, or near coastal sites are visited each year during a defined index period (e.g., summer base flow for streams/rivers). Each of the five water body types are visited once every 5 years. For logistic reasons, stream and river sampling were combined into one survey (the National Rivers and Streams Assessment [NRSA]) and done over a 2-year period every 5 years. At each site, a variety of biological-assemblage, water-quality, and physical-habitat data are collected during a 1-d sampling visit. Therefore, the NRSA data provide a unique opportunity to investigate landscape patterns in lotic fish assemblages at a continental extent through use of consistently collected data. In this current analysis, we use NRSA data to develop fish clusters across the United States using cluster analysis. The relationship of the clusters to local- and catchment-extent environmental variables are assessed by a variety of methods including ordination and indicator species analysis. We also predict cluster membership from the environmental data, assess the classification strength of various landscape classifications versus that of the fish clusters, and quantify the repeat visit reproducibility of the fish cluster analysis.

Methods

National Rivers and Streams Assessment

Study overview and design.—During the summers of 2008 and 2009, NRSA field crews made 2,320 sample visits to 2,123 different river and stream sites across the conterminous United States representing approximately 1,231,000 km of flowing water. Using standardized field methods, crews sampled waters as large as the Mississippi River and as small as mountain headwater streams. Sites were selected using a random sampling technique that uses a probability-based design (Herlihy et al. 2000; Stevens and Olsen 2004; see USEPA 2016b for NRSA-specific details). The NRSA target population consists of all streams and rivers within the 48 conterminous states that have flowing water during the study index period (June–September). The target population excludes tidal rivers and streams up to head of salt, and run-of-the-river ponds and reservoirs with greater than a 7-d residence time. Sample sites were selected from a sample frame derived from the medium National Hydrography Dataset (USGS 2013). The sampling design was spatially balanced, explicitly stratified by state, and employed variable site selection probability based on stream order to even out the sample site distribution across stream size-classes. The number of NRSA probability sample sites was set to allow for robust population estimates for each of the nine aggregate ecoregions listed in Table 1. Each probability site included in the survey has a sample weight that represents a known stream or river length in the target population. Thus, the probability site data can be inferred to the cumulative extent and condition of streams and rivers in the conterminous United States by using the sample weights.

A second round of NRSA sampling was conducted during the summers of 2013 and 2014. Crews made 2,293 sampling visits to 2,098 different sites across the conterminous United States. Part of the NRSA design allows for the study of temporal variability and change. Within each sample year, 10% of the sites were randomly selected for a second summer field visit to assess within-year index variability. In addition, about 40% of the probability sites from the 2008–2009 sampling were resampled in 2013–2014 as part of the 2013–2014 probability design. Last, the NRSA data includes information from 497 hand-picked sites that were selected in an attempt to augment the number of least-disturbed reference sites.

Fish data.—Collection methods for fish are described in detail in the NRSA field operations manuals (USEPA 2009, 2013a, 2013b). Collection methods used for NRSA 2013–2014 were essentially unchanged from those used for NRSA 2008–2009 (USEPA 2009). The basic protocol involved setting up a sample site around the randomly chosen sample point and sampling with sufficient effort to characterize the fish assemblage within the site (Hughes et al. 2002; Reynolds et al. 2003). More than 98% of the sites were sampled using backpack or raft/boat electrofishing, the remaining sites were sampled using seines because of very high stream conductivity. Three variants of the basic sampling protocol were used depending on the width of the stream and if it was wadeable. For wadeable streams less than 12.5 m wide, 40 channel widths were sampled for fish. For larger wadeable streams (>12.5 m wide), 500 m or 20 channel widths were sampled (whichever was longer). For boatable (nonwadeable) streams and rivers, at least 20 channel widths were sampled. At large wadeable and boatable sites, sampling continued past the established reach length until 500 individuals were collected (or a reach length equal to 40 channel widths was sampled).

Table 1. Number of unique sample sites sufficiently sampled for fish assemblages in National Rivers and Streams Assessment (NRSA) 2008–2009 and 2013–2014.

Ecoregion	Number of wadeable sites	Number of boatable sites	Total number
Western Mountains (WMT)	150	64	214
Xeric West (XER)	94	75	169
Northern Plains (NPL)	204	30	234
Southern Plains (SPL)	183	21	204
Temperate Plains (TPL)	217	112	329
Upper Midwest (UMW)	126	105	231
Northern Appalachian Mountains (NAP)	188	169	357
Southern Appalachian Mountains (SAP)	262	166	428
Coastal Plains (CPL)	157	231	388
Total NRSA (ALL)	1,581	973	2,554

Fish were tallied and identified in the field, then released alive unless used for fish tissue analyses or vouchers. Voucher specimens were collected if field identification could not be accomplished. Voucher samples of all species collected were also prepared at 10% of sites for each field taxonomist. Voucher samples were sent to an independent taxonomist to evaluate the taxonomic proficiency of each field taxonomist. All names submitted on field data forms were reviewed and revised when necessary to create a listing of nationally consistent common and scientific names. Where possible, taxonomic names (common and scientific) were based on Nelson et al. (2004) and Page et al. (2013). The online database FishBase (www.fishbase.org) served as a secondary source of taxonomic names. In a few cases, a journal article of a newly described species was used. For rare endemic species, we prepared collection maps for each taxon and compared them to published maps in Page and Burr (2011) or alternative Web publications to help confirm the identifications.

Fish data set.—We used the NRSA fish species abundance data from the 2008, 2009, 2013, and 2014 sampling surveys. Not all NRSA sites had fish data; sites were sometimes fishless or were not sampleable because of permit or logistical problems. We only used data from sites where the sampling effort was deemed sufficient to characterize the fish assemblage at the site (USEPA 2016b). Our data set consisted of fish species abundances from 3,560 unique sample visits to 2,663 unique sites. We removed all observations where individuals were not identified to species (e.g., hybrids and unknowns) or were not fish (e.g., amphibians). All told, there were 581 fish species in the data set. We screened the data set to remove rare species by removing species present at fewer than 5% of the total number of sites (507 species removed). Deleting rare species reduces analytical noise and enhances the detection of relationships between assemblage composition and environmental gradients (McCune and Grace 2002). Marchant (2002) also observed that, in general, only common taxa produce sufficiently strong signals for the interpretation of environmental gradients in bioassessments. This also necessitated removing the sites with only rare species present (109 sites). To not overweight the sites that had repeat visits, we only used the data associated with the first sample visit to the site for all statistical analyses (except for the repeat-visit analysis). The final data set was a relative abundance matrix of 2,554 sites by 74 nonrare species that was used for the cluster, ordination, and mean similarity analyses.

Environmental data.—We used 39 environmental variables collected in NRSA to aid in the prediction and interpretation of the fish analyses. The variables assessed were as follows:

Water chemistry: conductivity, dissolved organic carbon (DOC), sulfate, chloride, pH, total phosphorus (TP), total nitrogen (TN), and turbidity.

Physical habitat: geometric mean substrate diameter, % sand + fine substrate, % fine substrate, excess sediment index, total riparian disturbance index, agricultural riparian disturbance index, nonagricul tural riparian disturbance index, % fast water habitat, % pool habitat, mean wetted width, mean thalweg depth, channel slope, riparian canopy cover index, and natural fish cover index.

Natural landscape: mean air temperature, maximum air temperature, latitude, longitude, catchment area, elevation, soil erodibility index, mean annual precipitation, mean annual runoff, and aggregate ecoregion (Table 1).

Land use/land cover: catchment % agriculture, catchment % urban, catchment % wetland, dam influence index, population density, and road density.

For water chemistry, a single grab sample was collected from each sample site at the random selected site in the middle of the reach for wadeable streams, and at the end of the sample site for boatable rivers (USEPA 2009). Samples were shipped by overnight courier to the analytical laboratory. Most of the water samples were analyzed at the USEPA laboratory in Corvallis, Oregon, but each year, data from three to four states were run in state laboratories following the same USEPA laboratory protocols. Sulfate and chloride were measured by ion chromatography, DOC by carbon analyzer, turbidity by nephelometer, TP and TN by acid persulfate digestion and colorimetry, and meters for pH and conductivity.

Physical habitat field methods are described in detail for wadeable streams in USEPA (2013b) and for boatable rivers in USEPA (2013a). Details on the development of the various habitat indices and their calculation are given in USEPA (2016b) and Kaufmann et al. (1999). Briefly, detailed habitat measurements were made at 11 equidistant transects laid out across the same sample site that was sampled for fish assemblages. Riparian cover, fish cover, substrate (pebble counts), canopy density, human influence and bank data were collected at each of the transects using standardized field forms. Between transects, crews collected depth, woody debris, and habitat unit information at regular intervals.

Both natural landscape and land-use data were based on either the randomly selected sample point or the entire catchment. Latitude, longitude, ecoregion location, and elevation data are for the sample point. The remaining landscape and land use/land cover data were catchment averages. Catchment climate, soils, and anthropogenic stressor data were taken from StreamCat (Hill et al. 2016).

Data Analyses

We used Bray–Curtis distance, flexible-beta linkage method (beta = −0.25), and the Agnes function in the cluster library in R version 3.3.2 to cluster the fish relative abundance data. We wanted the final number of clusters to be between about 5 and 25 as a reasonable number for analysis and interpretation. The number of final clusters was determined by a combination of methods. We mostly relied on a method for pruning described in McCune and Grace (2002), which uses indicator species analysis (ISA; Dufrene and Legendre 1997) as an objective criterion for pruning. In addition, other methods for choosing the final number of clusters were examined: plotting average silhouette width versus the number of clusters (Rousseeuw 1987) and plotting mean between-similarity/mean within-similarity versus number of clusters, but our final decision was made primarily by looking at the number of clusters that were considered good choices by the ISA plot and their spatial pattern.

We also used ISA to quantify how indicative individual species were of each of the national fish clusters. Species indicator values were calculated for each cluster as the product of percent faithfulness (all sites in that particular cluster should contain the species) and percent exclusiveness (species should only be in that particular cluster and not others). We reported the percent indication of individual species to each cluster. Values ranged from 0% (no indication) to 100% (perfect indication). A perfect score would indicate that the species is both 100% exclusive and 100% faithful to that particular cluster.

We used the NRSA environmental variables to predict and interpret the fish cluster groups. Descriptive and graphical exploratory statistics were calculated for the environmental variables, and skewed variables were transformed before statistical analysis. As a result, chemistry variables, stream size variables, slope, elevation, precipitation, runoff, population, and road density were all $\log_{10}$ transformed. Fish cover and canopy cover indices were $\log_{10}(x + 1)$ transformed. Percentage variables and other indices were not transformed.

To aid in interpreting the results of the cluster analysis and identify the major environmental correlates with the patterns in fish assemblages, we conducted a nonmetric multidimensional scaling (NMS) ordination of the fish assemblage data. The NMS was conducted using the PC-ORD software package (McCune and Mefford 1999). NMS is a nonparametric ordination technique that is one of the most robust methods for exploring biological assemblage data (McCune 1994; Cao et al. 1996). This method starts off by calculating a matrix of taxonomic similarities between each pair of sites in the database. The similarity matrix is converted into a distance matrix in which distance (dissimilarity) is 1 minus similarity. Nonmetric multidimensional scaling arranges sites in an ordination of few dimensions (axes) by iteratively adjusting between-site distances to represent the ranks of site dissimilarities. We used Bray–Curtis similarity and proportionate abundance data in our ordination. Ordination axis scores were then correlated with the environmental and fish assemblage data.

We used both discriminant function analysis (DFA) and classification tree analysis (CTA) to predict site cluster membership with the environmental data using the MASS and RPART libraries in R. We used all 37 continuous environmental variables and stepwise discriminant function analysis with leave-out-one cross validation and a *P*-value to enter the model and stay in the model of 0.0001. We used the same data in the CTA with the addition of the categorical variables, ecoregion (Table 1), and size-class (wadeable or boatable). To decide upon the number of final nodes in the tree, we looked at the number of nodes suggested by the one standard error rule, the minimum deviance rule, and the default tree in R. The results of the DFA and CTA analyses give an overall percent of how well the clusters were predicted (versus observed) as well as the success rate for predicting individual clusters.

We used mean similarity analysis (Van Sickle 1997) to quantify within-group fish assemblage similarity for a variety of different landscape classifications. Mean similarity analysis calculates the mean similarity between all the pairs of sites within each group and then calculates an overall mean similarity by averaging the individual mean similarities across all the groups in the classification. Mean similarity was calculated using the MRPP function in R in the Vegan library. We assessed five landscape classifications and compared them to the mean similarity of the six fish clusters: the 10 USEPA federal geopolitical regions (which are based on aggregations of states), the 18 U.S. Geological Survey water resource regions/hydrologic units in the conterminous USA (USGS 2013), aggregates of the 18 water resource regions into eight major hydrologic units (Pacific, Closed basins, Colorado, Mississippi, Great Lakes, Hudson Bay, Northern Atlantic, and Southern Atlantic), and two aggregate Omernik level III ecoregion (Omernik 1987) classifications, one with nine groups (see Table 1) and the other that aggregates the nine groups in Table 1 into three groups (WMT+XER, NAP+SAP, CPL+TPL+NPL+SPL+UMW).

The original data set had 898 sites where repeat visits were performed either within-year or between survey years (2008 or 2009 versus 2013 or 2014). To assess the reproducibility of the clustering, cluster analysis was performed on all visits (i.e., treating each sample visit [including repeat visits] as individual sites). We calculated how many of the repeat sites fell in the same cluster for both visits within the same year and for repeat visits 4–6 years after the initial visit.

Results

We used NRSA fish assemblage data from 1,581 wadeable stream sites and 973 boatable river sites (Table 1). There were 587 sites sampled in 2008, 924 sites in 2009, 461 sites in 2013, and 582 sites in 2014. Given that we chose the first visit to a site as the index visit, there were more sites from the earlier 2008–2009 survey. There was a reasonably even distribution of sites across the nine ecoregions with the highest number of sites in the southern Appalachian Mountains and Coastal Plains ecoregions (Table 1). The sites sampled in NRSA covered streams/rivers from 0.1-m-wide headwater streams to the 2.1-km-wide Mississippi River. They ranged in mean thalweg depth from 0.3 to 3,910 cm, in channel slope from 0% to 11.2%, and in elevation from 0 to 3,130 m.

Based on NRSA data and using the 2008–2009 survey design weights to infer results to the entire conterminous U.S. target population, the most commonly occurring fish species in wadeable streams is the Creek Chub *Semotilus atromaculatus*, found in 56.5% of the target length (Table 2). The next five most commonly occurring species, found in more than 25% of the stream length, are Green Sunfish, Bluegill, White Sucker, Central Stoneroller, and Bluntnose Minnow. Because wadeable streams (target length = 1,438,000 km) make up 92% of the total stream/river length (1,561,000 km), these are the same common species by length percentage for all flowing waters. In terms of just the 123,300 km of boatable rivers (Table 2), the most common species are Bluegill (60.0% of river length), followed by Common Carp (49.5%), Largemouth Bass (48.2%), and Channel Catfish (46.8%). All nine of these species are considered tolerant (Whittier et al. 2007).

National Fish Clusters

We ran the ISA for all the different cluster sizes from 2 to 100, and the average *P*-value from the randomization tests averaged across species was calculated and plotted as a function of the number of different clusters. The number of clusters with the smallest *P*-values were considered good choices in terms of species ability to distinguish among clusters. The ISA showed the lowest *P*-values (most significant) for a cluster size of five to seven clusters. Based on looking at the other statistical approaches for setting cluster size and the clusters themselves, we chose six clusters as the size that made the most sense biologically and spatially.

We pruned the national fish cluster analysis to six clusters. Instead of giving the clusters generic cluster numbers or letters, we opted for assigning the five letter descriptive codes based on their geographic and environmental attributes (Table 3). The dendrogram from the cluster analysis (Figure 1) showed the COLD cluster split off first by itself. The next branch split out the RIVER and PLAIN clusters from the SEAST, MWEST, and NORTH clusters. Spatially, the clusters did occur in different locations throughout the United States (Figure 2). The COLD cluster sites were concentrated in the Appalachian Mountains in the east and the various mountain ranges in the west. The RIVER sites were mostly observed along the main-stem rivers of the Mississippi River basin. The SEAST sites were prevalent in the Coastal Plain and Piedmont of the southeastern United States. Similarly, NORTH sites were focused in the northern United States, PLAIN sites were focused in the Great Plains, and MWEST sites in the Midwestern United States (Figure 2).

The Cyprinidae was the most abundant fish family in most of the fish clusters, especially in the PLAIN where they accounted for 77% of the fish in that cluster (Figure 3). The COLD cluster was unique in that it had relatively high relative abundances of salmonids (24%) and cottids (17%); those families were rare in other clusters. Similarly, the RIVER cluster was unique in having a high relative abundance of clupeids (34%), which were rare outside this cluster. The SEAST cluster

Table 2. Presence of the 20 most common stream and river fish species found in National Rivers and Streams Assessment (NRSA) sampling expressed as a percentage of total target wadeable stream length (1,438,000 km) or total target boatable river length (123,300 km). Species are presented in order of decreasing percent length. The population length estimates are calculated using the site weights from the 2008–2009 NRSA panel design.

Species name	% of wadeable stream length	Species name	% of boatable river length
Creek Chub	56.5	Bluegill	60.0
Green Sunfish *Lepomis cyanellus*	44.5	Common Carp *Cyprinus carpio*	49.5
Bluegill *Lepomis macrochirus*	35.4	Largemouth Bass	48.2
White Sucker *Catostomus commersonii*	27.8	Channel Catfish *Ictalurus punctatus*	46.8
Central Stoneroller *Campostoma anomalum*	27.7	Gizzard Shad *Dorosoma cepedianum*	40.5
Bluntnose Minnow *Pimephales notatus*	26.8	Smallmouth Bass *Micropterus dolomieu*	36.5
Yellow Bullhead *Ameiurus natalis*	23.8	Freshwater Drum *Aplodinotus grunniens*	34.9
Blacknose Dace *Rhinichthys atratulus*	22.3	Longear Sunfish	33.0
Largemouth Bass *Micropterus salmoides*	21.2	Spotted Bass *Micropterus punctulatus*	30.3
Johnny Darter *Etheostoma nigrum*	20.0	Longnose Gar *Lepisosteus osseus*	29.9
Longear Sunfish *Lepomis megalotis*	16.7	Green Sunfish	28.6
Fantail Darter *Etheostoma flabellare*	15.5	White Sucker	28.3
Fathead Minnow *Pimephales promelas*	14.5	Golden Redhorse *Moxostoma erythrurum*	24.2
Common Shiner *Luxilus cornutus*	12.7	Rock Bass *Ambloplites rupestris*	24.1
Striped Shiner *Luxilus chrysocephalus*	12.0	Smallmouth Buffalo *Ictiobus bubalus*	24.0
Western Mosquitofish *Gambusia affinis*	11.9	Spotfin Shiner *Cyprinella spiloptera*	23.8
Brook Trout *Salvelinus fontinalis*	11.8	Bluntnose Minnow *Pimephales notatus*	23.5
Pirate Perch *Aphredoderus sayanus*	11.8	Northern Hog Sucker *Hypentelium nigricans*	22.8
Golden Shiner *Notemigonus crysoleucas*	11.2	Flathead Catfish *Pylodictis olivaris*	22.1
Red Shiner *Cyprinella lutrensis*	10.8	Emerald Shiner *Notropis atherinoides*	21.7

Table 3. The six national fish assemblage clusters and a narrative description of the identifying characteristics along with the fish species indicative of each cluster as determined from indicator species analysis. The top three species in each cluster are shown. The values below the cluster ID are the number of sites in the cluster followed by the mean within cluster similarity in parentheses.

Cluster ID	Description	Indicator species	% indication
COLD $n = 490$ (0.17)	Smaller, colder, steeper sites, abundant salmonids and cottids	Rainbow Trout *Oncorhynchus mykiss* Brook Trout Brown Trout *Salmo trutta*	36% 35% 28%
NORTH $n = 494$ (0.14)	Throughout northern USA, colder, wider sites	Smallmouth Bass Rock Bass White Sucker	54% 37% 36%
RIVER $n = 315$ (0.21)	Large river systems, deeper, wider sites, abundant clupeids	Gizzard Shad Freshwater Drum Common Carp	60% 60% 52%
PLAIN $n = 242$ (0.27)	Great Plains, more arid sites, higher ionic strength and erodibility, cyprinids dominant	Fathead Minnow Red Shiner Sand Shiner *Notropis stramineus*	58% 50% 39%
MWEST $n = 397$ (0.24)	Midwestern USA, smaller sites, intermediate slope and substrate size	Creek Chub Central Stoneroller Bluntnose Minnow	60% 44% 28%
SEAST $n = 616$ (0.14)	Southeastern USA, warmer, wettter, lower-elevation sites, abundant centrarchids	Bluegill Largemouth Bass Green Sunfish	48% 30% 25%

had the highest relative abundance of centrarchids (39%), whereas the MWEST had the highest relative abundance of percids (14%) and the NORTH had the highest relative abundance of catostomids (16%).

In terms of indicator species, the COLD cluster was most strongly indicated by the presence of salmonids: Rainbow Trout, Brook Trout, and Brown Trout (Table 3). The RIVER cluster was strongly indicated (60% indication) by both the Gizzard Shad, and Freshwater Drum. Other fish species with high percentage indication values were Creek Chub (60%) for the MWEST cluster, Fathead Minnow (58%) and Red Shiner (50%) for the PLAIN cluster, and Smallmouth Bass (54%) for the NORTH cluster.

The NMS analysis of the national site by fish species matrix (2,554 sites by 74 nonrare fish species) yielded a three-dimensional solution that accounted for 31.4% of the information in the original site dissimilarity

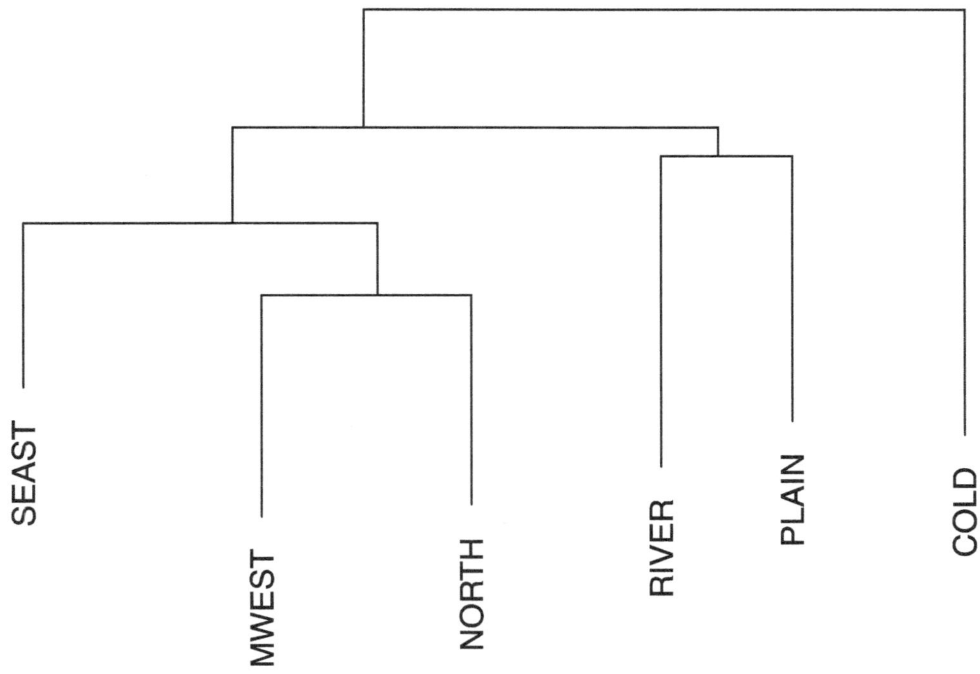

Figure 1. Dendrogram showing the cluster analyses of the National Rivers and Streams Assessment fish database. The dendrogram was pruned to show only the top six clusters.

matrix with a stress of 23.6 (Figure 4). The three NMS axes each explained 10–11% of the information, so no one axis was dominant. Axis 1 was a stream-size axis, being strongly negatively correlated ($r < -0.7$) to catchment area and wetted width (Table 4). Axis 2 was a temperature axis, being positively correlated to air temperature and negatively correlated to latitude and elevation. Axis 3 was a conductivity/ionic strength axis, being positively correlated to conductivity, sulfate, and soil erodibility. In terms of Spearman rank order correlation with fish species relative abundance, axis 1 was negatively correlated with Common Carp ($\rho = -0.47$) and Channel Catfish ($\rho = -0.47$). Axis 2 was positively correlated with both Bluegill ($\rho = 0.45$) and Longear Sunfish ($\rho = 0.40$), whereas axis 3 was positively correlated with Green Sunfish ($\rho = 0.40$) and negatively correlated with Rainbow Trout ($\rho = -0.36$).

Plots showing the location of the six clusters in ordination space (axes 1–2, 1–3, 2–3) are labeled with the above three axis descriptions (Figure 4). The COLD cluster was the most distinctive, falling out on the low end of the temperature, size, and conductivity axes. The RIVER cluster was at the large end of the size axis but intermediate on the other two axes. The PLAIN cluster was at the high end of the conductivity axis, and SEAST was at the high end of the temperature axis.

Box and whisker plots also show the cluster differences in the most discriminatory environmental variables (Figure 5). The MWEST and COLD cluster sites have smaller catchment areas, and the RIVER cluster sites have larger catchment areas than the other clusters. Air temperature is highest in the SEAST and lowest in the COLD and NORTH cluster sites. Conductivity is highest in the PLAIN sites. COLD sites have the steepest slopes and RIVER sites the lowest. The results from the ISA, NMS, and box plots were synthesized to come up with the overall cluster descriptions in Table 3.

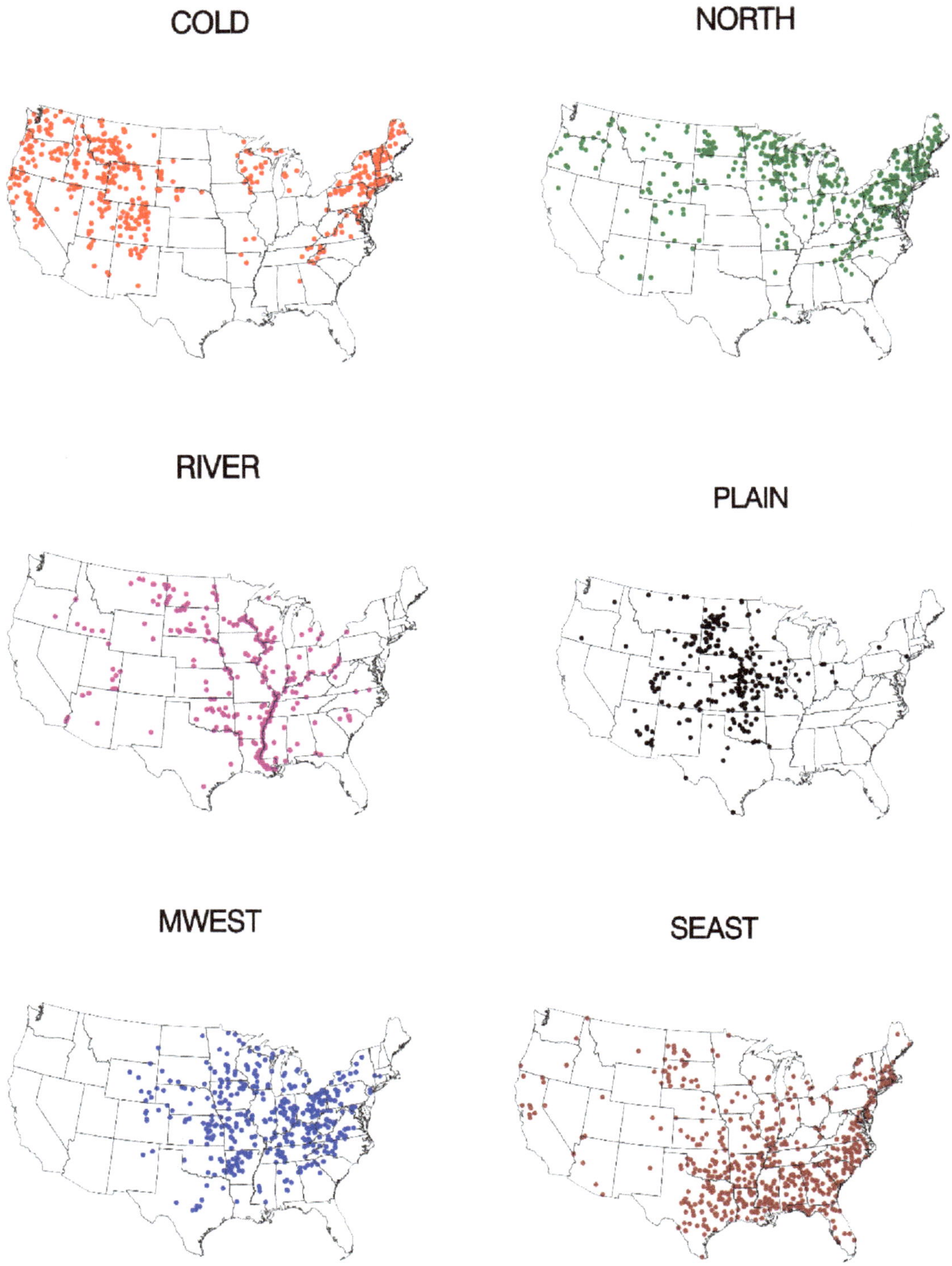

Figure 2. Locations of the sample sites in each of the six fish clusters in the conterminous United States.

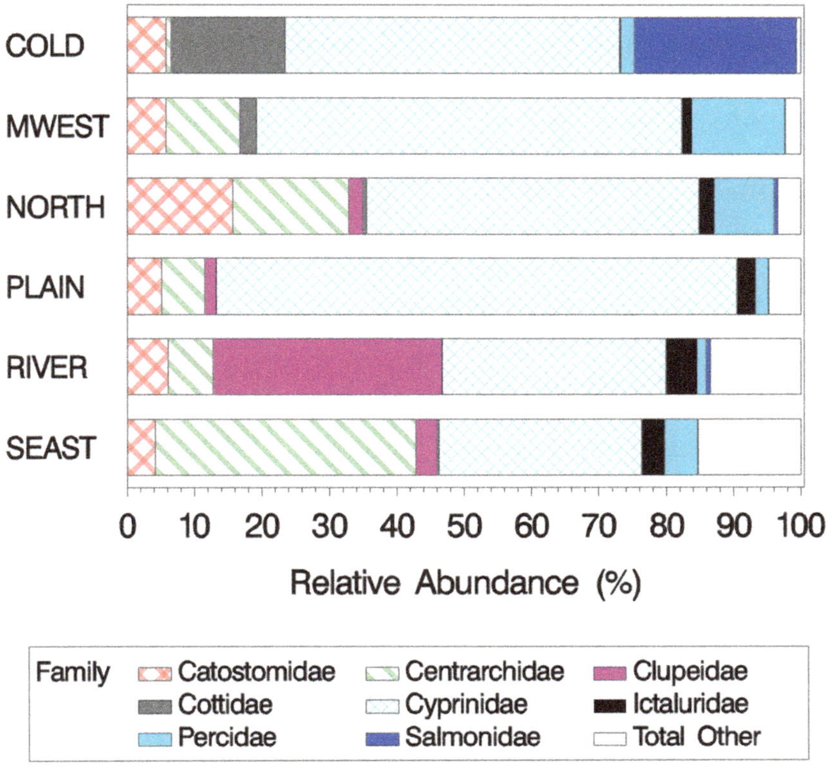

Figure 3. Relative abundance, by family, for all the fish caught in each of the six clusters (as if all the sites in a cluster were combined into one large cluster composite sample). Families are plotted alphabetically from left to right.

Predicting Cluster Membership

The DFA model correctly classified 65.3% of the sites into their correct cluster (Table 5). The COLD cluster was correctly classified most often (77%), and the RIVER (53.6%) and PLAIN (56.1%) clusters had the lowest two correct classification rates. The DFA model included 17 environmental variables: maximum air temperature, catchment area, % fast water habitat, mean air temperature, wetted width, DOC, turbidity, % sand + fine substrate, longitude, sulfate, soil erodibility index, elevation, pH, latitude, % agriculture in catchment, thalweg depth, and % fine substrate.

The CTA produced a 10-node tree to predict the six fish clusters using maximum and mean air temperatures, slope, catchment area, soil erodibility index, runoff, and elevation variables (Figure 6). The first split in the tree was for maximum air temperature. Sites below 29.45°C were classified as COLD if slope was greater than 0.1925%. Sites with lower slopes were classified as either NORTH or RIVER depending on catchment area (>36,250 km^2 were rivers). The CTA correctly classified 60.6% of the sites into their correct cluster (Table 5). The tree did a good job of predicting COLD (73%) and NORTH (77%) sites but a comparatively poorer job in the MWEST (32%) and PLAIN (38%).

Landscape Mean Similarity and Reproducibility

Mean similarity analysis is a way to evaluate the classification strength of different classification schemes (Van Sickle 1997). Classification strength does change as the number of

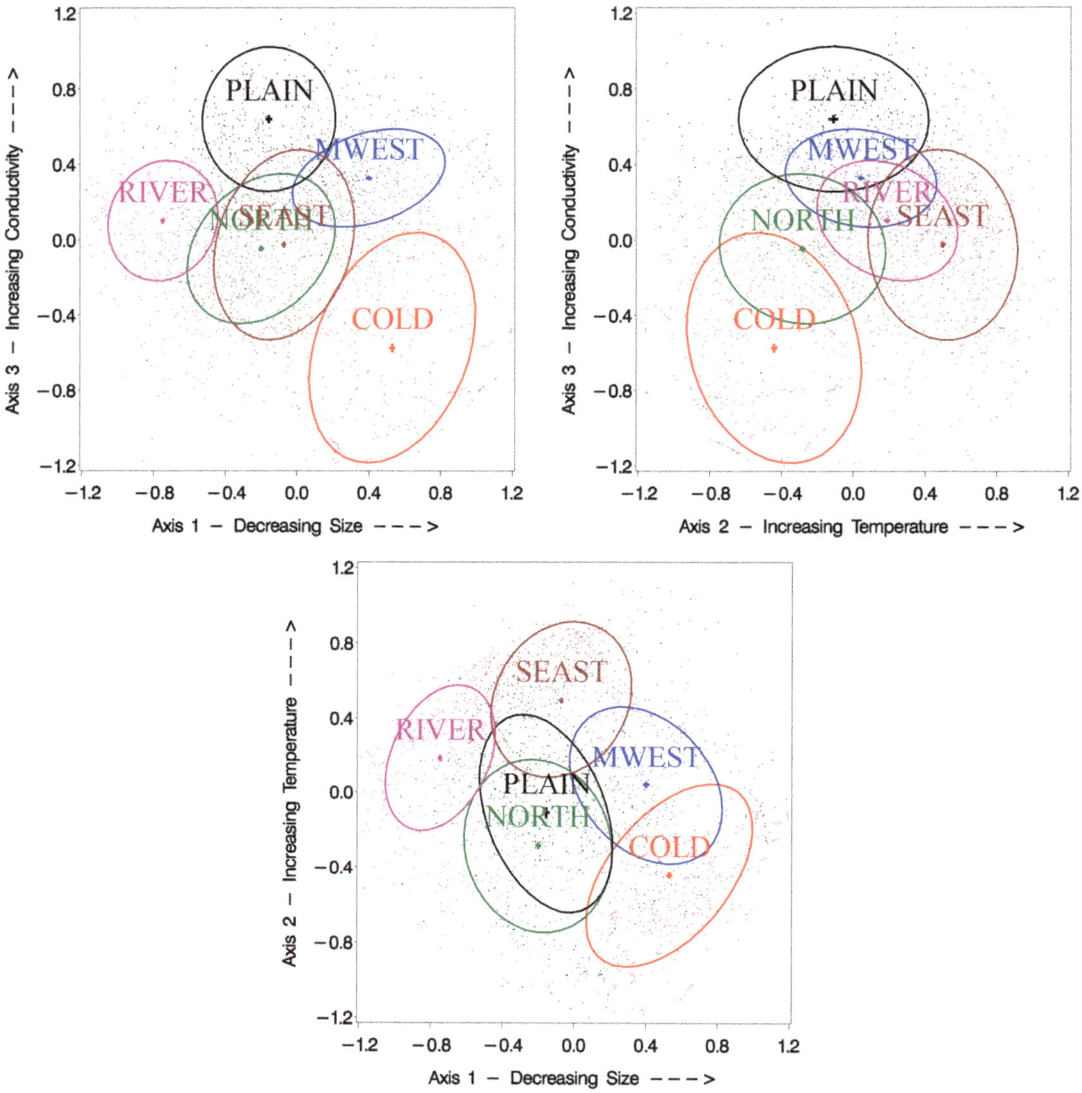

Figure 4. Bivariate plots of the three nonmetric multidimensional scaling fish assemblage ordination axes showing the distribution of the National Rivers and Streams Assessment sample sites (n = 2,554) in each of the six fish clusters in ordination space. To clarify patterns with this site density, ellipses are drawn around each cluster using a 0.5 confidence coefficient, indicating that half the data assuming a bivariate normal distribution is contained within each ellipse. The centroid dot shows the mean axis score for each cluster. The direction of the major environmental correlate with each axis is shown on the axis label.

different groups in the classification changes; the mean within-group similarity of a classification with 1,000 clusters with two sites in each cluster will be much higher than one with two clusters of 1,000 sites each. The within-group similarity for each of the six fish assemblage clusters are given in Table 3. The mean within-group similarity for the fish clusters was 0.182 (Table 6) and represents a theoretical maximum for within-group similarity with six clusters because the cluster analysis used to select the clusters attempts to maximize within-

Table 4. Spearman correlation coefficients for the relationship between nonmetric miltidimensional scaling (NMS) fish assemblage ordination axes scores and environmental variables in the 2008–2014 National Rivers and Streams Assessment data (n = 2,554). Only correlation coefficients greater than ±0.5 are shown.

Variable	NMS Axis 1	NMS Axis 2	NMS Axis 3
Catchment area (km^2)	–0.755	–	–
Mean wetted width (m)	–0.707	–	–
Channel slope (%)	+0.612	–	–
Fast water habitat (%)	+0.506	–	–
Mean air temperature (°C)	–	+0.756	–
Maximum air temperature (°C)	–	+0.725	–
Minimum air temperature (°C)	–	+0.640	–
Site latitude (degrees)	–	–0.682	–
Site elevation (m)	–	–0.520	–
Conductivity (μS)	–	–	+0.560
Catchment soil erodibility factor	–	–	+0.509
Sulfate (μeq/L)	–	–	+0.505

group similarity. The mean similarity between all pairs of sites in the database was 0.064 and represents a theoretical minimum for within-group similarity. Any random grouping of sites would have a similarity near this value. As expected, within-group similarity for the different landscape clusters in Table 7 increases with higher spatial resolution (increasing number of landscape groups). The landscape classification schemes, from political to ecoregional to hydrological, have similar classification strengths after considering the different group sizes. Mean similarities of landscape classification schemes were usually about two-thirds that seen for equivalent-sized biological clusters.

An identical fish assemblage cluster analysis was rerun using all site visits, treating each repeat visit as a separate site to assess the reproducibility of the clustering. In NRSA, about 10% of the sites each year were visited twice within the same summer index period, with the second visit typically occurring 2–6 weeks after the first visit. Also, about 40% of the 2008–2009 NRSA survey sites were revisited in the 2013–2014 NRSA survey (so either 4, 5, or 6 years later). The resulting cluster analysis (not shown) gave very similar-looking clusters as the original six fish clusters shown in Table 3, so we retained the same cluster names for this analysis. Of the 280 sites visited twice within the same year, almost all (91.8%) of them were placed in the same fish cluster (Table 7). The biggest source of misclassification was eight sites that were classified as either NORTH or RIVER and six sites that were classified as either NORTH or SEAST. The correct classification rate was lower for the between-year revisits. Of the 558 sites that were revisited 4–6 years later, 76.3% of them were classified into the same fish cluster (Table 7). The biggest source of misclassification was again 19 sites that were classified as either NORTH or RIVER and 20 sites that were classified as either NORTH or SEAST.

Discussion

We found that six clusters of sites provided a good framework for explaining the major classification of fish assemblage structure across the conterminous United States. Nonmetric multidimensional scaling ordination of the fish assemblage data identified three

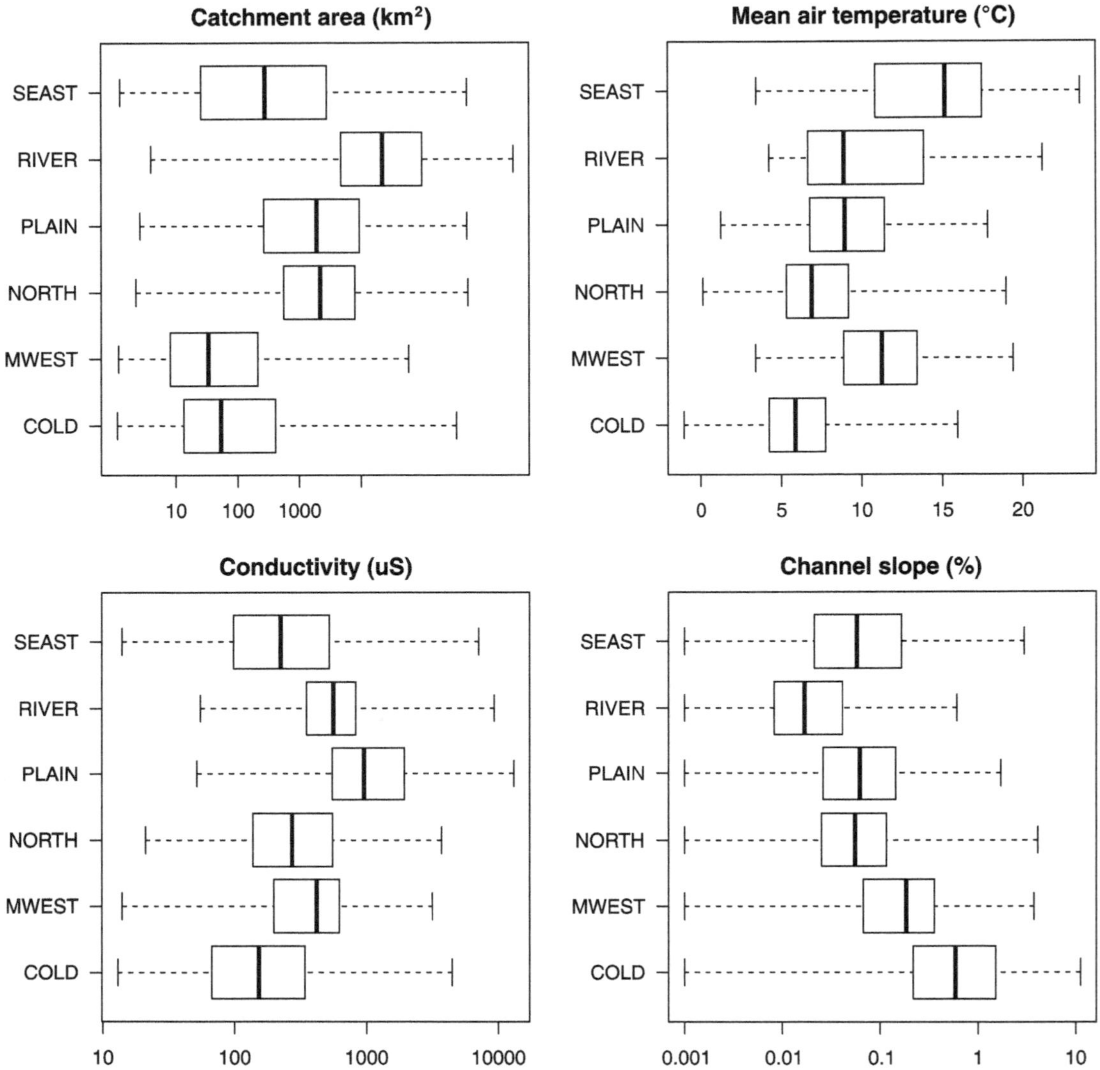

Figure 5. Box and whisker plots showing the distribution of four environmental variables across the sample sites in each of the six fish clusters. Boxes show the interquartile range, the line in the box is the median, and the whiskers show the minimum and maximum.

major axes of roughly equal importance that were extremely useful in interpreting the fish clusters. Each of the NMS axes was strongly related to an environmental gradient. However, the three NMS axes together only explained 31% of the variance in fish assemblage composition. We expect this largely results from trying to explain continental-scale variability with only three axes and not being able to measure all pertinent environmental variables. The first NMS axis was a stream-size axis associated with stream width and catchment area. It clearly separated the RIVER cluster from the other clusters. The RIVER cluster was indicated by fish species found in large river systems. The COLD and MWEST clusters were at the other end of the size axis and represented smaller-sized systems. The second NMS axis was strongly related to variables associated

Table 5. The percent of sites correctly classified into the correct cluster by discriminant function analysis (DFA; *n* = 2,361) and classification tree analysis (CTA; *n* = 2,548). Some sites were missing from each analysis because of missing predictors.

Cluster code	DFA - % correct	CTA - % correct
SEAST	61.9	63.3
MWEST	65.5	32.3
RIVER	53.6	63.5
COLD	77.0	73.2
NORTH	67.0	77.0
PLAIN	56.1	37.8
ALL	65.3	60.6

with temperature and strongly separated the COLD and SEAST clusters. These differences were also reflected in the indicator species of these clusters (warmwater centrarchids versus coldwater salmonids). The third NMS axis was a water conductivity or ionic strength axis. It differentiated the NORTH from the PLAINS clusters that otherwise had similar sizes and temperatures.

There have been few studies that have looked at environmental drivers of fish assemblage structure at continental scales. Melcher et al. (2007) performed a hierarchical cluster analysis of fish assemblages in least-dis-

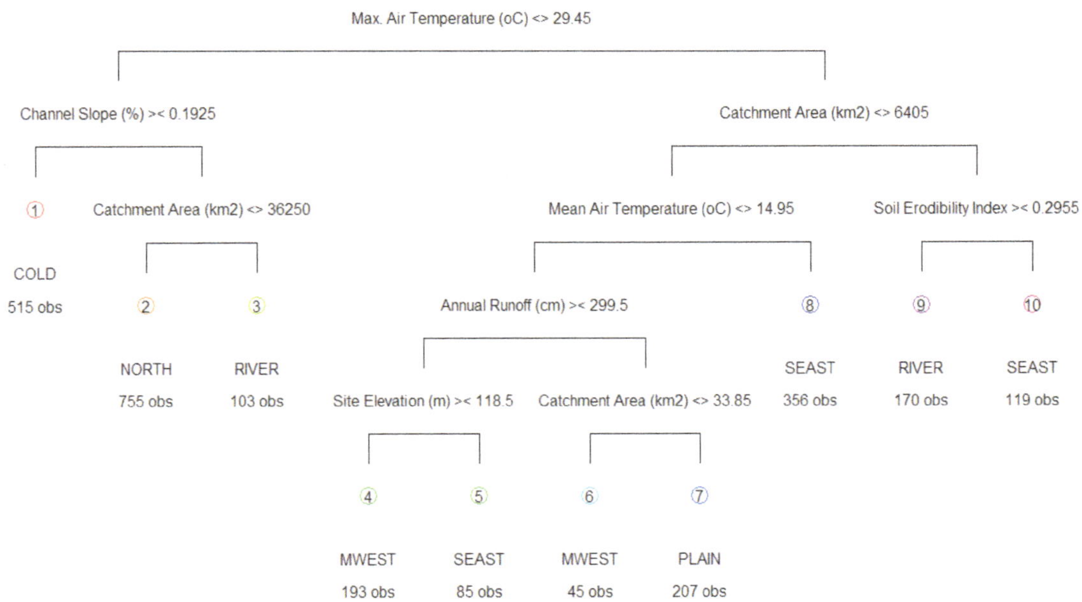

Figure 6. Ten-node classification tree predicting cluster membership from the environmental data. The circled numbers are the terminal tree nodes. Cluster codes predicted by that node are given below the circled number along with the number of sites predicted to occur in that node. The tree can be read like a dichotomous key with the equation at each split indicating that the cases with the lower values go left (<>) or right (><) at the split.

Table 6. Mean within-group Bray–Curtis fish assemblage similarity in different landscape classification schemes. Number of groups is the number of different classes in the classification. For comparison, the mean Bray–Curtis similarity among all possible pairs of sites in the data was 0.064.

Classification scheme	Number of groups	Mean within-group similarity
National Aquatic Resource Surveys fish clusters (this analysis)	6	0.182
Three aggregate Omernik ecoregions	3	0.092
Nine aggregate Omernik ecoregions[a]	9	0.132
U.S. Environmental Protection Agency federal geopolitical regions	10	0.134
Major river hydrologic units	8	0.113
U.S. Geological Survey hydrologic units (two-digit code)	18	0.148

[a] The nine aggregate Omernik ecoregions are listed in Table 1.

Table 7. Reproducibility of the cluster analysis using data from repeat visits to the same sites. A subset of randomly selected National Rivers and Streams Assessment sites were revisited both within years (n = 280) and between years (2009–2009 versus 2013–2014, n = 558). Numbers are the number of sample sites in the cell. Sites along the diagonal indicate that the site was placed in the same cluster for both repeat visits. Overall, 91.8% of the within-year repeat visit sites were placed in the same cluster and 76.3% of the between-year visit sites were placed in the same cluster.

Cluster	SEAST	MWEST	RIVER	COLD	NORTH	PLAIN	Total
Within years							
SEAST	81	–	–	–	–	–	81
MWEST	1	14	–	–	–	–	15
RIVER	3	0	33	–	–	–	36
COLD	0	0	0	43	–	–	43
NORTH	6	0	8	2	69	–	85
PLAIN	1	0	2	0	0	17	20
Total	92	14	43	45	69	17	280
Between years							
SEAST	132	–	–	–	–	–	132
MWEST	14	26	–	–	–	–	40
RIVER	14	0	55	–	–	–	69
COLD	3	4	0	88	–	–	95
NORTH	20	11	19	7	96	–	153
PLAIN	11	9	4	4	12	29	69
Total	194	50	78	99	108	29	558

turbed streams and rivers across Europe and identified 15 European fish assemblage types. They were able to build a DFA model that correctly classified 70% of their sites into the right class, similar to the 65% observed in our DFA model. The European DFA model used seven variables accounting for elevation, stream size, slope, temperature, and spatial position, all variables that were important in the DFA of our fish clusters. Schinegger et al. (2013) also performed a cluster analysis on compiled fish assemblage data from 716 reference sites across Europe and identified four clusters: headwater streams, medium gradient rivers, lowland rivers, and Mediterranean streams. They used CTA to predict cluster membership and found that elevation, air temperature, latitude, and precipitation correctly classified 75% of the sites. Those are similar-looking clusters and drivers to what we observed in the United States, with the exception of an ionic strength/conductivity driver. However, conductivity was not a candidate variable in their CTA.

An important attribute of any clustering or classification scheme is its reproducibility. Is a site placed in the same cluster when it is sampled multiple times? The more reproducible things are, the more useful the classification will be for applied management purposes. The NRSA repeat visit design allowed for an evaluation of the cluster reproducibility. Although no guidelines exist for defining "good" reproducibility, our clustering was highly reproducible when comparing sites revisited twice within the same year (92% of the sites were put in the same cluster both times). Reproducibility among revisits 4–6 years apart was lower (76% of sites were put in the same cluster). Thus, it appears that the clustering is more sensitive to long-term weather changes or interannual changes (wet versus dry years) in fish assemblages than short-term within season changes or crew sampling variability.

Natural landscape drivers of fish assemblage structure can be obscured or confounded by anthropogenic disturbance. This can result from physical/chemical alternations of the environment or the introduction of nonnative species. In our earlier efforts at a national clustering, in addition to running the analysis using all data, we ran the cluster analysis for only least-disturbed reference sites and with and without alien species. Cluster and mean similarity analyses were not strongly influenced by using data subsets that removed nonnative fish species and disturbed sites (Herlihy et al. 2006). In the present analysis, we ran the analysis using all of the data and did not find that landscape- or local-extent human disturbance variables were strongly related to any of the NMS ordination axes. We also ran the cluster analysis using only least-disturbed reference sites (as defined in Herlihy et al. [2008]) and found very similar-looking clusters (data not shown). This strongly suggests that the underlying mechanisms responsible for controlling fish assemblage patterns at the national extent are robust to the effects of nonnative species and anthropogenic disturbances and are mostly driven by natural landscape and local factors.

For a fish classification based on biologically derived clusters to have any transferability, one must define the clusters so that they can be employed by others. It is highly desirable to be able to take site data and determine its cluster membership. We used DFA and CTA as a means of taking observed environmental data to assign cluster membership to a site. Overall, DFA provided slightly higher correct classification rates than CTA (65% versus 61%). There are properties of CTA, however, that make it more practical for general use. The tree output is easy to understand and communicate to others. In addition, classification trees require no distributional assumptions, allow for unspecified interactions between variables, and are invariant to monotonic transformation of the predictor variables; therefore, outliers do not pose a problem. Classification trees can handle mixtures of data types and have provisions for missing values (Clark and Pregibon 1992). For spatial data, mapping the tree provides an easy graphical display and aids interpreting results. Discriminant function analysis results are somewhat more difficult to package for easy use and visualization. With our fish clusters, the NORTH and COLD clusters had a higher correct classification rate than the other four clusters. Presumably, the environmental attributes for these clusters are more distinguishable than the other clusters.

At the national extent, the mean similarity analysis shows that existing landscape classification schemes do not capture the major part of the similarity in fish assemblages that can be obtained by biologically derived fish assemblage clusters (Table 6). It did not matter if the landscape classification was based on river basins, ecoregions, or even geopolitical regions. Smaller extent, local environmental factors appear to play a major role in structuring fish assemblages that appear to be unaccounted for by larger-extent landscape classifications. Our previous national mean similarity analysis with fish assemblage data also found that a wide variety of different landscape classification schemes only accounted for about half of the variation explained by biological fish clusters (Herlihy et al. 2006). Hawkins and Vinson (2000) and Hawkins et al. (2000) suspected that weak landscape classifications occur because aquatic organisms vary independently and continuously along environmental gradients and do not cluster into discrete groups.

Similar lower levels of classification strength in landscape classifications relative to biological classifications as observed in our national study have also been observed in several smaller-extent studies (McCormick et al. 2000; Oswood et al. 2000; Van Sickle and Hughes 2000). Oswood et al. (2000) used published distributions of freshwater fishes of Alaska and found that while some individual classifications worked well, neither hydroregions nor ecoregions alone were sufficient as frameworks for managing aquatic ecosystems across diverse landscapes. Van Sickle and Hughes (2000) examined classification strengths for fish assemblages in western Oregon streams and concluded that while ecoregions and river basins are useful for classifying streams, much of their classification strength may be a result of spatial autocorrelation effects rather than the ecological factors that were used to draw class boundaries. Our results strongly support this observation at the national extent in that ecoregions, basins, and even political units based on state borders all had similar classification strengths.

The results from our study show that whereas variability in fish assemblages is large over the range of stream/river sizes across the conterminous United States, it is possible to divide sites into six biologically based clusters that could be defined by specific indicator species and predictable from environmental data using both CTA and DFA. Nonmetric multidimensional scaling ordination identified three environmental gradients as the primary drivers of the biological clusters: stream size, temperature, and stream water ionic strength. These are the major environmental covariates that need to be taken into account when making any regional extent biomonitoring interpretations of fish assemblage data in the United States. The biological classification was very reproducible within the sampling index period based on repeat sample visits to the same sites, indicating that crew, sampling, and short-term temporal (within-season) differences are not major sources of variability in these kinds of fish bioassessments. Reproducibility was lower, however, based on samples taken 4–6 years apart, suggesting that longer-term (year to year) temporal variability will have more of an impact on fish bioassessments.

Biological monitoring is essential for the complete assessment of aquatic ecosystem condition. As the natural variability in aquatic biota is quite high, reporting and analyzing the results of biomonitoring requires a classification framework that minimizes this variability so that expectations of least-disturbed condition and the effects of anthropogenic disturbance can be more clearly defined. The choice of what classification to use is specific to the study's scope, content, goals, and objectives, so there is no one classification that fits all needs. Our results showed that biologically derived classes of sites have higher classification strength than either basins or ecoregions for lotic fish assemblages. In other words, there are groups of streams defined by fish assemblage composition that have more in common with each other, despite being spread out across the continent, than they have in common with nearby streams in the same landscape class. As noted by Houghton (2003), the failure to consider this disparity may lead to spurious information about human disturbance resulting from natural faunal differences. Thus, for some biomonitoring purposes, it could be advantageous to replace physical regions with a classification based on aquatic biota.

Acknowledgments

The NRSA was planned, funded, and organized by the USEPA's Office of Water and Office of Research and Development and carried out by numerous state, federal, and contractor field crews, information management staff, and laboratory staff whose efforts we gratefully acknowledge. This research was performed while the senior author held a National Research Council Senior Research Associateship award at the USEPA National Health and Environmental Effects Research Laboratory, Western Ecology Division, Corvallis, Oregon. This manuscript has been subjected to agency review and has been approved for publication. The views expressed in this paper are those of the authors and do not necessarily reflect the views or policies of the USEPA. We appreciate the constructive reviews of Stefan Schmutz, Lizhu Wang, and another anonymous reviewer.

References

Biggs, B. J. F., M. J. Duncan, I. G. Jowett, J. M. Quinn, C. W. Hickey, R. J. Davies-Colley, and M. E. Close. 1990. Ecological characterization, classification, and modeling of New Zealand rivers: an introduction and synthesis. New Zealand Journal of Marine Freshwater Research 24:277–304.

Cao, Y., A. W. Bark, and W. P. Williams. 1996. Measuring the responses of macroinvertebrate communities to water pollution: a comparison of multivariate approaches, biotic and diversity indices. Hydrobiologia 341:1–19.

Changeux, T., and D. Pont. 1995. Ichthyogeographic regions and watershed size in the French River Rhône network. Hydrobiologia 300:355–363.

Clark, L. A., and D. Pregibon. 1992. Tree-based models. Pages 377–419 *in* T. J. Hastie and J. M. Chambers, editors. Statistical models in S. Wadsworth and Brooks, Pacific Grove, California.

Dufrene, M., and P. Legendre. 1997. Species assemblages and indicator species: the need for a flexible asymmetrical approach. Ecological Monographs 67:345–366.

Gilbert, C. R. 1980. Zoogeographic factors in relation to biological monitoring of fish. Pages 309–355 *in* C. H. Hocutt and J. R. Stauffer, Jr., editors. Biological monitoring of fish health. Lexington Books, Lexington, Massachusetts.

Hawkes, H. A. 1975. River zonation and classification. Pages 312–374 *in* B. A. Whitton, editor. River ecology. University of California Press, Berkeley.

Hawkins, C. P., R. H. Norris, J. Gerritsen, R. M. Hughes, S. K. Jackson, R. K. Johnson, and R. J. Stevenson. 2000. Evaluation of the use of landscape classifications for the prediction of freshwater biota: synthesis and recommendations. Journal of the North American Benthological Society 19:541–556.

Hawkins, C. P., and M. R. Vinson. 2000. Weak correspondence between landscape classifications and stream invertebrate assemblages: implications for bioassessment. Journal of the North American Benthological Society 19:501–517.

Herlihy, A. T., R. M. Hughes, and J. C. Sifneos. 2006. National clusters of fish species assemblages in the conterminous United States and their relationship to existing landscape classification schemes. pp. 87–112 *in* R. M. Hughes, L. Wang, and P. W. Seelbach editors, Influences of landscapes on stream habitats and biological assemblages. American Fisheries Society, Symposium 48, Bethesda, Maryland.

Herlihy, A. T., D. P. Larsen, S. G. Paulsen, N. S. Urquhart, and B. J. Rosenbaum. 2000. Designing a spatially balanced, randomized site selection process for regional stream surveys: the EMAP mid-Atlantic pilot study. Environmental Monitoring and Assessment 63:95–113.

Herlihy, A. T., S. G. Paulsen, J. V. Sickle, J. L. Stoddard, C. P. Hawkins, and L. L. Yuan. 2008. Striving for consistency in a national assessment: the challenges of applying a reference-condition approach at a continental scale. Journal of the North American Benthological Society 27:860–877.

Hill, R. A., M. H. Weber, S. G. Leibowitz, A. R. Olsen, and D. J. Thornbrugh. 2016. The Stream-Catchment (StreamCat) dataset: a database of watershed metrics for the conterminous United States. Journal of the American Water Resources Association 52:120–128.

Hocutt, C. H., and E. O. Wiley. 1986. The zoogeography of North American freshwater fishes. Wiley & Sons, New York.

Houghton, D. C. 2003. Evaluation of Minnesota geographic classifications based on caddisfly (Trichoptera) data. The Great Lakes Entomologist 36:76–92.

Hughes, R. M., and J. R. Gammon. 1987. Longitudinal changes in fish assemblages and water quality in the Willamette River, Oregon. Transactions of the American Fisheries Society 116:196–209.

Hughes, R. M., S. A. Heiskary, W. J. Matthews, and C. O. Yoder. 1994. Use of ecoregions in biological monitoring. Pages 125–151 *in* S. L. Loeb and A. Spacie, editors. Biological monitoring of aquatic systems. Lewis Press, Boca Raton, Florida.

Hughes, R. M., P. R. Kaufmann, A. T. Herlihy, S. S. Intelmann, S. C. Corbett, M. C. Arbogast, and R. C. Hjort. 2002. Electrofishing distance needed to estimate fish species richness in raftable Oregon rivers. North American Journal of Fisheries Management 22:1229–1240.

Hughes, R. M., A. T. Herlihy, and J. C. Sifneos. 2015. Predicting aquatic vertebrate assemblages from environmental variables at three multistate geographic extents of the western USA. Ecological Indicators 57:546–556.

Hughes, R. M., E. Rexstad, and C. E. Bond. 1987. The relationship of aquatic ecoregions, river basins and physiographic provinces to the ichthyogeographic regions of Oregon. Copeia 1987:423–432.

Ibañez, C., J. Belliard, R. M. Hughes, P. Irz, A. Kamdem-Toham, N. Lamouroux, P. A. Tedesco, and T. Oberdorff. 2009. Convergence of temperate and tropical stream fish assemblages. Ecography 32:658–670.

Kaufmann, P. R., P. Levine, E. G. Robison, C. Seeliger, and D. V. Peck. 1999. Quantifying physical habitat in wadeable streams. U.S. Environmental Protection Agency, Office of Research and Development, EPA 620/R-99/003, Washington, D.C.

Larsen, D. P., J. M. Omernik, R. M. Hughes, C. M. Rohm, T. R. Whittier, A. J. Kinney, A. L. Gallant, and D. R. Dudley. 1986. Correspondence between spatial patterns in fish assemblages in Ohio streams and aquatic ecoregions. Environmental Management 10:815–828.

Legendre, P., and Legendre, V. 1984. Postglacial dispersal of freshwater fishes in the Quebec peninsula. Canadian Journal of Fisheries and Aquatic Sciences 41:1781–1802.

Lyons, J. 1996. Patterns in the species composition of fish assemblages among Wisconsin streams. Environmental Biology of Fishes 45:320–341.

Marchant, R. 2002. Do rare species have any place in multivariate analysis for bioassessment? Journal of the North American Benthological Society 21:311–313.

Matthews, W. J. 1998. Patterns in freshwater fish ecology. Chapman and Hall, New York.

Matthews, W. J., and H. W. Robison. 1988. The distribution of the fishes of Arkansas: a multivariate analysis. Copeia 1988:358–374.

Matthews, W. J., D. J. Hough, and H. W. Robison. 1992. Similarities in fish distribution and water quality patterns in streams of Arkansas: congruence of multivariate analyses. Copeia 1992:296–305.

McCormick, F. H., D. V. Peck, and D. P. Larsen. 2000. Comparison of geographic classification schemes for mid-Atlantic stream fish assemblages. Journal of the North American Benthological Society 19:385–404.

McCune, B. 1994. Improving community analysis with the Beals smoothing function. Ecoscience 1:82–86.

McCune B., and M. J. Mefford. 1999. PC-ORD, multivariate analysis of ecological data. MjM Software Design, Gleneden Beach, Oregon.

McCune, B., and J. B. Grace. 2002. Analysis of ecological communities. MjM Software Design, Gleneden Beach, Oregon.

McGarvey, D. J., and R. M. Hughes. 2008. Longitudinal zonation of Pacific Northwest (U.S.A.) fish assemblages and the species-discharge relationship. Copeia 2008:311–321.

McGarvey, D. J., and G. M. Ward. 2008. Scale dependence in the species-discharge relationship for fishes of the southeastern U.S.A. Freshwater Biology 53:2206–2219.

Meador, M. R., T. R. Whittier, R. M. Goldstein, R. M. Hughes, and D. V. Peck. 2008. Evaluation of an index of biotic integrity approach used to assess biological condition in western U.S. streams and rivers at varying spatial scales. Transactions of the American Fisheries Society 137:13–22.

Meek, S. E. 1891. Report of explorations made in Missouri and Arkansas, conducted during 1889, with an account of the fishes observed in each of the river basins examined. Bulletin of the U.S. Fisheries Commission 13–141.

Melcher, A., S. Schmutz, G. Haidvogl, and K. Moder. 2007. Spatially based methods to assess the ecological status of European fish assemblage types. Fisheries Management and Ecology 14:453–463.

Nelson, J. S., E. J. Crossman, H. Espinosa-Pérez, L. T. Findley, C. R. Gilbert, R. N. Lea, and J. D. Williams. 2004. Common and scientific names of fishes from the United States, Canada, and Mexico, 6th edition. American Fisheries Society, Special Publication 29, Bethesda, Maryland.

Omernik, J. M. 1987. Ecoregions of the conterminous United States. Annals of the Association of American Geographers 77:118–125.

Oswood, M. W., J. B. Reynolds, J. G. Irons, III, and A. M. Milner. 2000. Distributions of freshwater fishes in ecoregions and hydroregions of Alaska. Journal of the North American Benthological Society 19:405–418.

Page, L. M., and B. M. Burr. 2011. A field guide to freshwater fishes of North America north of Mexico, 2nd edition. Houghton Mifflin, Boston.

Page, L. M., H. Espinosa=Pérez, L. T. Findley, C. R. Gilbert, R. N. Lea, N. E. Mandrak, R. L. Mayden, and J. S. Nelson. 2013. Common and scientific names of fishes from the United States, Canada, and Mexico, 7th edition. American Fisheries Society, Special Publication 34, Bethesda, Maryland.

Pflieger, W. L. 1971. A distributional study of Missouri fishes. University of Kansas Publications of the Museum Natural History 20:225–570.

Pinto, B. C. T., F. G. Araujo, V. D. Rodriguez, and R. M. Hughes. 2009. Local and ecoregion effects on fish assemblage structure in tributaries of the Rio Paraíba do Sul, Brazil. Freshwater Biology 54:2600–2615.

Pont, D., B. Hugueny, U. Beier, D. Goffaux, A.

Melcher, R. Noble, C. Rogers, N. Roset, and S. Schmutz. 2006. Assessing river biotic condition at the continental scale: a European approach using functional metrics and fish assemblages. Journal of Applied Ecology 43:70–80.

Reynolds, L., A. T. Herlihy, P. R. Kaufmann, S. V. Gregory, and R. M. Hughes. 2003. Electrofishing effort requirements for assessing species richness and biotic integrity in western Oregon streams. North American Journal of Fisheries Management 23:450–461.

Rohm, C. M., J. W. Giese, and C. C. Bennett. 1987. Evaluation of an aquatic ecoregion classification of streams in Arkansas. Journal of Freshwater Ecology 4:127–140.

Rousseeuw, P. 1987. Silhouettes: a graphical aid to the interpretation and validation of cluster analysis. Journal of Computational and Applied Mathematics 20:53–65.

Schinegger, R., C. Trautwein, and S. Schmutz. 2013. Pressure-specific and multiple pressure response of fish assemblages in European running waters. Limnologica 43:348–361.

Schinegger, R., M. Palt, P. Segurado, and S. Schmutz. 2016. Untangling the effects of multiple human stressors and their impacts on fish assemblages in European running waters. Science of the Total Environment 573:1079–1088.

Stevens, D. L., and A. R. Olsen. 2004. Spatially balanced sampling of natural resources. Journal of the American Statistical Association 99:262–278.

USEPA (U.S. Environmental Protection Agency). 2009. National rivers and streams assessment: field operations manual. USEPA, Office of Water and Office of Environmental Information, EPA 841/B-04/004, Washington, D.C.

USEPA (U.S. Environmental Protection Agency). 2016a. National rivers and streams assessment 2008–2009: a collaborative survey. USEPA, Office of Water and Office of Research and Development, EPA/841/R-16/007, Washington, D.C.

USEPA (U.S. Environmental Protection Agency). 2016b. National rivers and streams assessment 2008–2009 technical report. USEPA, Office of Water and Office of Research and Development, EPA 841/R-16/008, Washington, D.C.

USEPA (U.S. Environmental Protection Agency). 2013a. National rivers and streams assessment 2013/14: field operations manual—non-wadeable. USEPA, Office of Water and Office of Environmental Information, EPA 841/B-12/009a, Washington, D.C.

USEPA (U.S. Environmental Protection Agency). 2013b. National rivers and streams assessment 2013/14: field operations manual—wadeable. USEPA, Office of Water and Office of Environmental Information, EPA 841/B-12/009b, Washington, D.C.

USGS (United States Geological Survey). 2013. National hydrography geodatabase: the national map [online database. Available: https://viewer.nationalmap.gov/advanced-viewer.

Van Sickle, J. 1997. Using mean similarity dendrograms to evaluate classifications. Journal of Agricultural, Biological, and Environmental Statistics 2:370–388.

Van Sickle, J., and R. M. Hughes. 2000. Classification strengths of ecoregions, catchments, and geographic clusters for aquatic vertebrates in Oregon. Journal of the North American Benthological Society 19:370–384.

Whittier, T. R., R. M. Hughes, and D. P. Larsen. 1988. The correspondence between ecoregions and spatial patterns in stream ecosystems in Oregon. Canadian Journal of Fisheries and Aquatic Sciences 45:1264–1278.

Whittier, T. R., R. M. Hughes, G. A. Lomnicky, and D. V. Peck. 2007. Fish and amphibian tolerance values and an assemblage tolerance index for streams and rivers in the western USA. Transactions of the American Fisheries Society 136:254–271.

American Fisheries Society Symposium 90:409–440, 2019
© 2019 by the American Fisheries Society

Multiscale Fish Assemblage Distribution Models to Guide Riverscape Conservation Planning

Joshuah S. Perkin*
Department of Wildlife and Fisheries Sciences, Texas A&M University
534 John Kimbrough Boulevard, College Station, Texas 77843, USA

Juju C. Wellemeyer
Missouri Department of Conservation, Resource Science Division
21999 Highway B, Maitland, Missouri 64466, USA

Jeffrey D. Fore
The Nature Conservancy, Tennessee Chapter
210 25th Avenue North, Suite 810, Nashville, Tennessee 37203, USA

Abstract.—Scale influences the detection of relationships between landscape alterations and stream fishes and therefore has strong implications for conservation planning and implementation. This is especially true in riverscapes because terrestrial landscapes drain into riverscapes in a manner that can be measured at multiple scales. Two commonly employed scales in riverscape ecology and conservation include local catchment (i.e., the area of land draining directly into a segment of stream between two confluences) and network catchment (i.e., the total area of upstream land). We used a multispecies extension of species distribution modeling (i.e., gradient forest) to describe relationships between landscape alterations (measured at local catchment and network catchment scales) and stream fish occurrence patterns in portions of the Mississippi and Tennessee River basins in western Tennessee, USA. Landscape alterations included seven urban or agricultural classes, and densities of roads, road crossings, dams, human population, mines, and confined animal feeding operations. At the network catchment scale, the most influential landscape alterations affecting fish distributions were cultivated crops and pasture/hay land uses, but at the local catchment scale, open-space development, human population density, and road density were most important for describing multispecies fish distributions. Despite these differences, gradient forest model performance measured as explained variation at the species level was consistent between local catchment and network catchment scales. Furthermore, when predictions for unsampled stream segments were mapped across the region, both scales produced consistent patterns in fish assemblages affected by low, medium, and high development or cultivated crops. Our results provide direction to conservation practitioners by identifying regions where limited resources might be allocated to increase efficiencies within two highly altered and taxonomically diverse riverscapes. The framework described here provides a case study for application of new statistical innovations to address

* Corresponding author: jperkin@tamu.edu

conservation challenges and can be used in other landscapes and riverscapes to identify locations where management efforts might be best allocated.

Introduction

Determining the appropriate scale(s) and extents for ecological assessment and application of conservation initiatives is often a challenge (Levin 1992; Boyd et al. 2008). In this chapter, scale encompasses the grain and extent of observation, where grain represents the smallest spatial or temporal resolution and is generally the observational unit and extent represents the spatial dimension or time period over which all observations are made (Wiens 1989). The definitions of these components of scale strongly influence detection of linkages between environmental variables and fish occurrence, abundance, or community composition (Allan et al. 1997; Angermeier and Winston 1998; Kunin 1998; Lammert and Allan 1999). For example, Labbe and Fausch (2000) found that occurrence, population growth rate, and individual growth rate for a single species were each regulated by environmental variables measured at unique spatial scales. Other authors have identified similar scale-dependent regulators of fish populations and assemblages across a range of biomes (Jackson et al. 2001; Chick et al. 2004; Kennard et al. 2007). Consequently, scale has emerged as a focal research area for promoting the conservation of fishes (Fausch et al. 2002; Fagan et al. 2005; Gido et al. 2006). Despite previous advances in the application of scale theory in fish conservation (see overview by Fausch 2010), additional research is required to determine the most appropriate spatial scales for developing preservation or restoration plans for increasingly threated freshwater fauna (Rivers-Moore et al. 2011; Stendera et al. 2012; Erős 2017).

In lotic ecosystems, applied principles of landscape ecology allow for addressing the issue of scale in fish conservation (Frissell et al. 1986; Ward 1998; Wiens 2002). Fausch et al. (2002) reviewed the application of hierarchy theory in riverscapes as a means of connecting basic research on scaling with applied research focused on stream fish conservation. Because riverscapes drain terrestrial landscapes, riverscapes are linked to landscapes through processes mediated by hydrologic connectivity (Pringle 2003; Allan 2004; Burcher et al. 2007). Given these linkages, alterations to land use that result in land-cover transformation degrade ecological integrity and affect the abundance and distribution of fishes (Wang et al. 1997; Wenger et al. 2008). Terrestrial land-cover alterations, such as transformation from natural forest or prairie to urban or agricultural land uses, are among the most widely studied (Wang et al. 2001; Blann et al. 2009); however, these are not the only landscape factors affecting stream fishes. Multiple landscape alterations, including roadways and railways, roadway and railway crossings, confined animal feeding operations (CAFOs), mining runoff, human population density, and large and small dams, are all known to influence stream fishes (Fore et al. 2014; Perkin et al. 2016; Wellemeyer et al. 2018). The capacity of mangers to address the ecological consequences associated with these landscape alterations is scale-dependent. Near-term and local-scale actions such as riparian planting, streambank stabilization, channel redesign, fish passage improvement, and small dam removal are most appropriate when fishes are responsive to fine-scale alterations not affecting entire basins (Kwak and Freeman 2010). Alternatively, long-term and large-scale habitat management actions such as natural flow regime mimicry or coordinated uses of cover crops to reduce soil erosion might be necessary to address landscape alterations associated with fish responses at basin scales (Kwak and Freeman 2010; Strayer and Dudgeon 2010). A growing body of research has identified local and network catchments as relevant scales for assessing landscape–riverscape interactions, where local catchment describes the area draining immediately into a stream segment and network catchment describes the total upstream area draining to a point (Hill et al. 2017; Figure 1). Previous works have demonstrated that landscape alterations can correlate strongest with ecological change in streams at local scales (Wang et al. 2001, 2003), network catchment scales (Roth et al. 1996; Gido et al. 2006), or, in some cases, both scales (Hoeinghaus et al. 2007; Macedo et al. 2014; Leal et al. 2016). Development of management plans aimed at conservation of stream fishes therefore require assessing linkages between landscape alterations and fish ecological integrity at multiple spatial scales, chiefly local catchment and network catchment scales (Closs et al. 2015; Jeffrey et al. 2015; Erős et al. 2018).

A further scale-related challenge in freshwater fish conservation involves the level of organizational complexity used in ecological assessment. Levin (1992) discussed the differential perspectives that come with studying organisms at the levels of individuals, populations, species, assemblages, communities, or ecosystems. Conservation programs have traditionally focused on groups of individuals

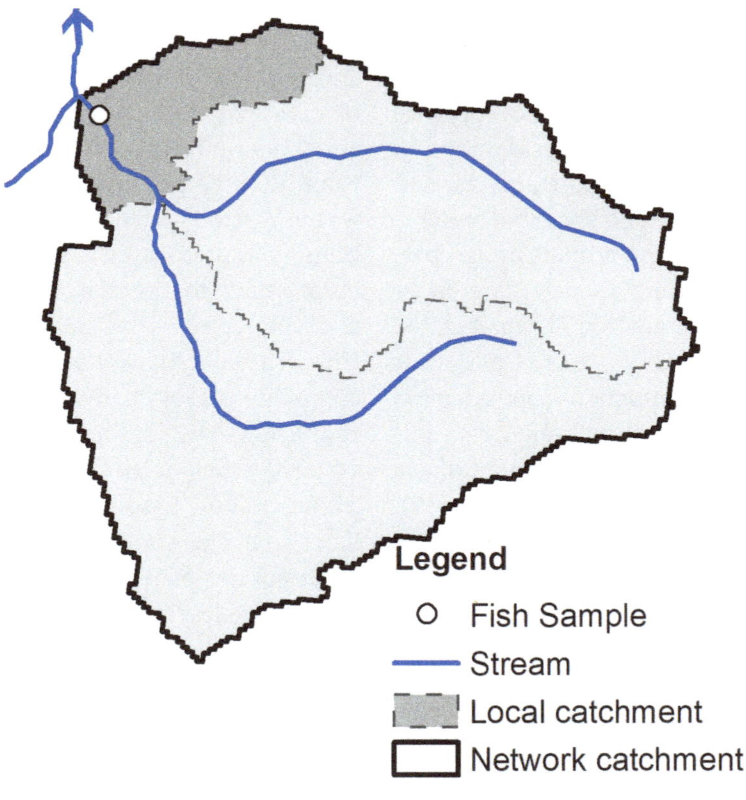

Figure 1. Conceptual diagram illustrating the spatial scales investigated in this study. Fish sampling sites (white circle) were located on stream segments defined as interconfluence sections of stream (blue lines; flow continues in direction of arrow). Each stream segment drains a local catchment (dark gray shading) as well as multiple upstream local catchments (dashed gray lines show outlines) that together are a network catchment (black outline).

within a single population or perhaps multiple populations of a single species (Soulé and Wilcox 1980). This focal level provides convenience in terms of targeted sampling and development of specified and complex models while operating at the level where most change in conservation status is measured (i.e., population decline or extirpation; Olden 2003). Conservation listings typically include species identity and sometimes specific population identities to create species-specific management plans, and this approach has been successful in preserving populations (Simberloff 1998). However, more recent advances in conservation biology advocate multispecies approaches that operate at the level of assemblages or communities rather than populations (Ferrier and Guisan 2006; Maguire et al. 2016). Assemblage-level models (ALMs) of species distributions provide benefits over single-species approaches because they allow for simultaneous consideration of multiple species, including rare species, and assess patterns of co-occurrence that are useful for detecting emergent properties (i.e., patterns not evident in single species approaches) in ecological degradation (Nieto-Lugilde et al. 2018). Furthermore, ALMs allow for identifying the environmental variables most pertinent to management by evaluating multispecies change in occurrence or abundance across ecological gradients (Ellis et al. 2012). In this context, applications of ALMs effectively scale up traditional population- or species-level approaches to assemblage or community levels (Pitcher et al. 2012). Yet ALMs have received relatively little attention in the application of biodiversity conservation (Nieto-Lugilde et al. 2018), and this is especially true of multiscale applications to benefit fish conservation in riverscapes (Leigh and Datry 2017). Given recent calls for scaling up fish conservation to multiple species across entire network catchments through establishment of protected areas such as native fish conservation areas (Williams et al. 2011; Malone et al. 2018), ALMs have potential to contribute to emerging conservation initiatives at broadening spatial scales. Currently, additional research is necessary to determine the appropriate spatial and organizational complexity scales for relating riverscape alterations to stream fish species and assemblages as a means of informing conservation and management planning.

The goal of this study was to develop multiscale (local catchment and network catchment) predictions for stream fish ecological condition measured at two levels (species and assemblage) to inform riverscape conservation planning. Here, we define assemblage as a collection of fishes occurring together in space and time and differentiate this term from community in that only fishes and no other aquatic organisms (e.g., invertebrates, vegetation) are considered (Fauth et al. 1996). To accomplish our goal, we developed three objectives. Our first objective was to assemble data for 13 landscape alterations measured at local and network catchment scales for portions of the Mississippi and Tennessee basins in western Tennessee, USA. These basins are part of a regional fish diversity hotspot characterized by a globally significant level of biodiversity (Warren et al. 2000; Abell et al. 2008). Our second objective was to assess relationships between landscape alterations and the distributions of freshwater fishes compiled at species and assemblage levels. Our third objective was to develop spatially explicit descriptions of the multiscale relationships between landscape alterations and fish biodiversity as a tool for conservation planning across the region. We hypothesized that ecological responses to landscape alterations would not be uniform across the region because of spatial variation in landscape alterations. For example, urban areas should have a unique suite of alterations that affect fishes through pathways that differ from alterations created in agricultural areas (Wang et al. 2001; Feist et al. 2018). We also hypothesized that the magnitude (i.e., measured as model R^2) of ecological responses associated with landscape alterations measured at the local catchment scale would differ from those measured at the network catchment scale. This is expected because some landscape alterations such as urbanization should have stronger localized effects compared to landscape alterations such as agriculture that generally occur over broader spatial extents (Thornbrugh et al. 2018). Tests of these hypotheses have direct implications for informing the locations and scales most appropriate for conservation actions aimed at managing stream fish ecological integrity.

Methods

Study Area

We assessed landscape alterations and fish distributions in the Southeastern Plains and Mississippi Valley Loess Plains ecoregions of Tennessee, USA (Omernik and Griffith 2014). The region is bordered by the Mississippi River to the west, the Tennessee River to the east, the Mississippi State border to the south, and the Kentucky State border to the north (Figure 2). The ecoregions are also separated by the division of the Mississippi and Tennessee basins, a bisection that represents a strong biogeographic barrier to fishes (Etnier and Starnes 1993). Consequently, two subregions were assessed independently: portions of the Mississippi and Tennessee basins. Although the northern and southern borders of the study area are political (rather than biogeographic), all major streams drain either to the west (Mississippi basin tributaries) or the west and east (Tennessee basin tributaries), and therefore most of the local catchments and network catchments assessed, were completely within the study area. Average annual air temperatures are 10–22°C, and average annual rainfall is 136 cm/year across the region (Wellemeyer et al. 2018). The dominant natural land cover was historically deciduous temperate forest, although considerable land-use transformation has occurred to support agriculture and urbanization. Agriculture land use dominates landscapes in the Mississippi Valley Loess Plains ecoregion draining into the Mississippi basin, including both cultivated crops and hay pastures, whereas forest land cover dominates the Southeastern Plains ecoregion draining into the Tennessee River basin (Homer et al. 2015).

Landscape Data

We collected landscape alteration data for 13 variables representing the dominant forms of alteration for the region and for which reliable data were available. Landscape alterations were compiled at local catchment and network catchment scales using a geographic information system, and the protocol is described in detail by Wellemeyer et al. (2018). This protocol included defining stream segments as polylines from the National Hydrologic Dataset (NHD; USEPA and USGS 2012) across the entire region. The NHD is based on a digital elevation model and allows for ascribing the local catchment as the area of land draining immediately into a polyline and the network catchment as all upstream land area draining into the same polyline. We used local catchment and network catchment polygons from the NHD to compile landscape alteration data at each scale independently. Landscape alterations included proportion of the terrestrial landscape covered by cultivated crop (Crop), pasture and hay (Hay), or urban development (five classes: open, Dev1; low intensity, Dev2; medium intensity, Dev3; high intensity, Dev4; and impervious cover, Dev5) and density (#/km^2) of length of paved roads and railroads (Road), roadway and railway crossings (Cross), human population (Pop), dams >3 m (Dam), mining permits (Mine), and CAFO

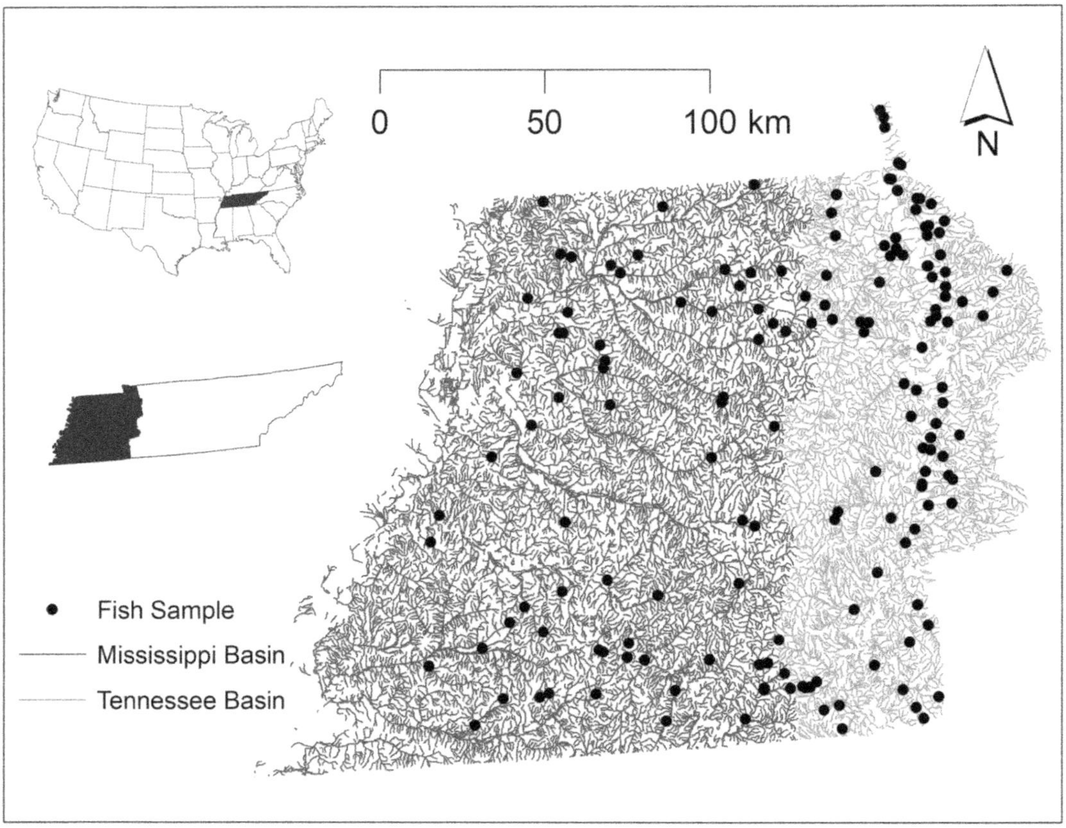

Figure 2. Study area map illustrating Tennessee in the southeastern United States, stream segments, and fish sampling sites in the Mississippi and Tennessee River basins

permits (CAFO) for local catchments and network catchments (Table 1). All land-cover data were obtained from the National Land Cover Dataset (Homer et al. 2015), roadway and railway data were obtained from the U.S. Census Bureau (2014), human population data were obtained from the U.S. Census Bureau (2010), dam locations were obtained from the National Inventory of Dams (USACE 2013), and CAFO and mining permit data were obtained from the Tennessee Department of Environmental Conservation (TDEC 2012).

Fish Data

We used fish occurrence data collected by the Tennessee Wildlife Resources Agency from across the study area during 2000–2015. Those data were collected by using a standardized sampling protocol that consisted of sampling 100–200 m of wadeable stream (i.e., stream order ≤3; Strahler 1957) in an upstream direction using single-pass electrofishing conducted using a single backpack electrofisher (100–200 V, direct current) and a single netter. This level of sampling effort is consistent with recommendations for assessments of ecological condition reported elsewhere (Reynolds et al. 2003; Robinson et al. 2019, this volume). Fishes collected during sampling were held until the entire site was sampled, at which point species were identified and released. Detailed methods used to ensure the quality of these data are given in Wellemeyer et al. (2018) and included retention of only fishes identified to species, assignment of fish collections to NHD polylines us-

Table 1. Landscape alteration magnitude medians (range) for stream segments with fish assemblage data in the Mississippi (MS) and Tennessee (TN) drainages at catchment and watershed spatial scales. Scales at which landscape alterations did not occur are shown as "–". NLCD = National Land Cover Database.

Name	Description	MS local catchment	MS network catchment	TN local catchment	TN network catchment
Crop	Proportion of cultivated crop land cover (NLCD 82)	0.37 (0.00–0.93)	0.31 (0.00–0.92)	0.04 (0.00–0.66)	0.03 (0.00–0.49)
Hay	Proportion of hay and pasture land cover (NLCD 81)	0.05 (0.00–0.49)	0.08 (0.00–0.49)	0.02 (0.00–0.54)	0.05 (0.00–0.53)
Dev1	Proportion of open development land cover (NLCD 21)	0.05 (0.00–0.26)	0.05 (0.00–0.18)	0.03 (0.00–0.16)	0.02 (0.00–0.09)
Dev2	Proportion of low intensity development land cover (NLCD 22)	<0.01 (0.00–0.07)	<0.01 (0.00–0.12)	<0.01 (0.00–0.06)	<0.01 (0.00–0.05)
Dev3	Proportion of medium intensity development land cover (NLCD 23)	<0.01 (0.00–0.06)	<0.01 (0.00–0.04)	<0.01 (0.00–0.04)	<0.01 (0.00–0.02)
Dev4	Proportion of high intensity development land cover (NLCD 24)	<0.01 (0.00–0.02)	<0.01 (0.00–0.02)	<0.01 (0.00–0.01)	<0.01 (0.00–0.01)
Dev5	Proportion of impervious land cover	0.50 (0.00–0.83)	0.52 (0.00–0.90)	0.42 (0.00–0.35)	0.15 (0.00–0.34)
Road	Density of roadways and railways (m/km^2)	1,714 (0–8,465)	1,628 (0–5,399)	1,751 (0–9,200)	1,419 (0–5,831)
Cross	Density of roadway and railway crossings (m/km^2)	0.68 (0.00–33.17)	0.59 (0.00–2.69)	0.56 (0.00–6.21)	0.57 (0.00–2.14)
Dam	Density of dams (#/km^2)	<0.01 (0.00–0.24)	<0.01 (0.00–0.23)	–	<0.01 (0.00–0.06)
Pop	Density of human population (people/km^2)	90 (0–61,615)	12 (0–225)	57 (0–13,809)	10 (0–78)
Mine	Density of mining permits (#/km^2)	–	<0.01 (0.00–0.09)	<0.01 (0.00–0.47)	<0.01 (0.00–0.47)
CAFO	Density of confined animal feeding operation permits (#/km^2)	–	<0.01 (0.00–0.05)	–	<0.01 (0.00–0.07)

ing geographic positioning system coordinates collected during fish sampling, and combining fish occurrence data across multiple collections when the same site was sampled more than once (Gido et al. 2006). We also excluded species that occurred in fewer than five collections. Review of the database and exclusions as noted above resulted in 72 sites (50 species) from the Mississippi basin and 89 sites (62 species) from the Tennessee basin.

Statistical Analyses

We simultaneously modeled species and assemblage levels of organization using four gradient forest models built at two spatial scales for the two ecoregions. Gradient forest (GF) modeling is a multispecies extension of random forest (RF) modeling, a commonly applied tree-based machine learning algorithm. Benefits of RF models include increased classification accuracy when compared with traditional classification trees, the ability to model nonlinearity and threshold responses, and inclusion of complex interactions among predictor variables (Prasad et al. 2006). These models also have advantages over other machine learning models such as artificial neural networks because partial dependence plots give insight into conditioned responses to individual predictor variables (Cutler et al. 2007). An historical limitation to RF modeling was that only a single response variable (here, species) could be modeled, and scaling up to assemblages required stacking multiple independently developed models in a nonintegrated and computationally intensive fashion. This limitation was recently addressed with the advent of GF models (Ellis et al. 2012). In GF modeling, RF models are constructed for each species in a matrix of multiple species, and the relative importance of variables and locations of threshold responses along predictor variable gradients are provided at species and assemblage levels. This approach represents a powerful method for determining how entire assemblages respond to environmental gradients while simultaneously incorporating individual species responses (Pitcher et al. 2011).

We fit GF models in R (R Core Team 2018) using the 'extendedForest' package, an extension of the 'randomForest' package (Liaw and Wiener 2002). We used a matrix of species occurrence data and fit classification-based models separately for each spatial scale and subregion combination (four models total) to test our hypothesis that landscape variables that correlate with fish assemblage structure differ across space and scales. For each model, we used the function 'gradientForest' with fish occurrence data as the response variable, used the 13 landscape alteration variables as predictor variables, constructed 500 trees, and included four randomly selected predictor variables as candidates for each split (Ellis et al. 2012). We used a series of diagnostic plots to assess relationships between landscape alterations and fish occurrences to measure the magnitude of ecological responses to landscape alterations. We plotted overall importance of landscape alteration variables using R^2 weighted importance, cumulative importance to illustrate responses to landscape alteration gradients averaged across all species, and model performance depicting out-of-bag error rate for each species for which the landscape alterations had some explanatory power (Ellis et al. 2012). Values for R^2 weighted importance represent average R^2 values across all species and therefore are generally smaller in magnitude compared with single-species R^2 values, but their relative magnitudes are nonetheless informative and equally interpretable. Finally, we transformed all environmental predictions from the GF model into a common biological scale to illustrate predicted fish assemblage composition based on landscape alteration data in a spatially explicit matter. We used the 'predict' function and landscape alteration data from all sampled and unsampled stream segments to generate biological predictions and then ordinated results using principle components analysis with weighted species scores overlaid (Pitcher et al. 2011). This approach is comparable to direct ordination in that site scores, species scores, and landscape alteration data are simultaneously shown, but here, relationships are shown after accounting for nonlinearities and threshold responses by species. As with most ordination methodologies, a large number of axes are generated, and we retained only the first two for interpretation.

We mapped predictions for fish assemblage composition across the region. We used GF models for each of the two spatial scales in the two subregions to develop predictions for fish assemblage composition at all sampled and unsampled stream segments. We first filtered the data set of unsampled stream segments and removed any segments that had individual landscape alteration magnitudes greater than those in the subset of segments that were sampled for fishes, as well as stream segments with stream orders greater than three. This step was necessary to control for gradient lengths for natural (stream size) and anthropogenic gradients so that predictions were not made for streams outside the domain of our models. This resulted in the removal of highly urbanized stream segments in the Memphis, Tennessee and Jackson, Tennessee areas, as well as all large main-stem rivers. We clustered predictions from each model to improve interpretation using the 'clara' clustering function from the 'cluster' package (Maechler et al. 2018). This clustering technique is capable of handling large numbers of observations (Kaufman and Rousseeuw 1990), including 17,081 segments at the network catchment scale and 15,738 segments at the local catchment scale in the Mississippi basin and 6,961 segments at the network catchment scale and 6,973 segments at the local catchment scale in the Tennessee basin. The number of segments differed among scales within each basin because of exclusions of segments with large-magnitude landscape alterations, which differed across scales. We elected to use four clusters for each model because this number of clusters captured the major gradients in landscape alterations. Finally, we assigned predictions for assemblage composition to polylines from the NHD and mapped predictions as a means of producing spatially structured data useful for conservation planning.

Results

Landscape Alterations

Most landscape alterations occurred in both subregions, with few exceptions. No CAFO permits or mining permits occurred in the sampled catchments in the Mississippi basin, and no dams or CAFO permits occurred within the sampled catchments in the Tennessee basin. The dominant forms of land use in local catchments and network catchments were Crop, Hay, Dev1, and Dev5 (Table 1). The median values for land-use transformation to cultivated crops included more than 30% of the area in the Mississippi basin (0.31–0.37 across scales) but was less intense in the Tennessee basin (0.03–0.04 across scales). Conversely, median values for transformation to hay were largely consistent in the Tennessee (0.02–0.05 across scales) and Mississippi (0.05–0.08 across scales) basins. Development generally occurred to a greater extent in the Mississippi basin compared with the Tennessee basin, including Dev1, Dev5, and Pop.

Environmental Correlates for Fish Distributions

The most influential landscape alterations associated with fish distributions varied by basin and spatial scale. In the Mississippi basin, Crop, Hay, and Dev2 had the highest weighted importance at the network catchment scale, but Dev1, Crop, and Pop became most important when data collection was scaled down to local catchments (Figure 3a). Among all 13 variables, weighted importance was higher at the network catchment scale for eight alterations and higher at the local catchment scale for five alterations. In the Tennessee basin, Crop, Hay, and Cross had the highest weighted importance at the network catchment scale, but Pop, Road, and Dev1 became most important when data collection was scaled down to local catchments (Figure 3b). Among all 13 variables, weighted importance was higher at the network catchment scale for eight alterations and higher at the local catchment scale for five alterations.

Cumulative importance plots highlighted linear (i.e., single species losses or gains) versus threshold (i.e., multiple species losses or gains) changes in assemblage composition and the scales at which these changes were most important for predicting fish assemblage structure. Assemblages changed in approximately linear patterns regardless of scale as the proportion of Crop increased at both network and local catchment scales in the Mississippi basin (Figure 4a), but in the Tennessee basin, Crop was more important at the network catchment scale than the local catchment scale and a threshold response in assemblage change was evident when 20–30% of catchments were cultivated (Figure

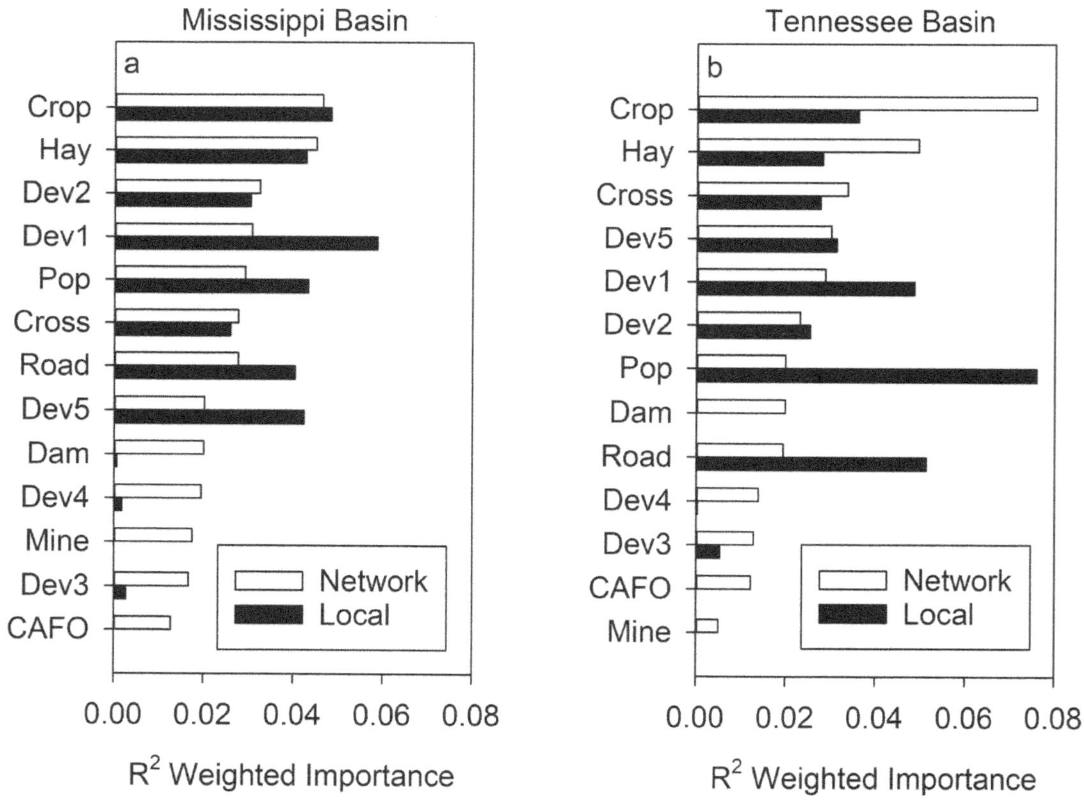

Figure 3. Influence of landscape alterations on stream fish assemblages as measured as R^2 weighted importance for gradient forest models fit to data from the **(a)** Mississippi and **(b)** Tennessee River basins at network catchment (white bars) and local catchment (black bars) scales. Variables are ranked by importance at the network catchment scale; see Table 1 for landscape alteration variable descriptions.

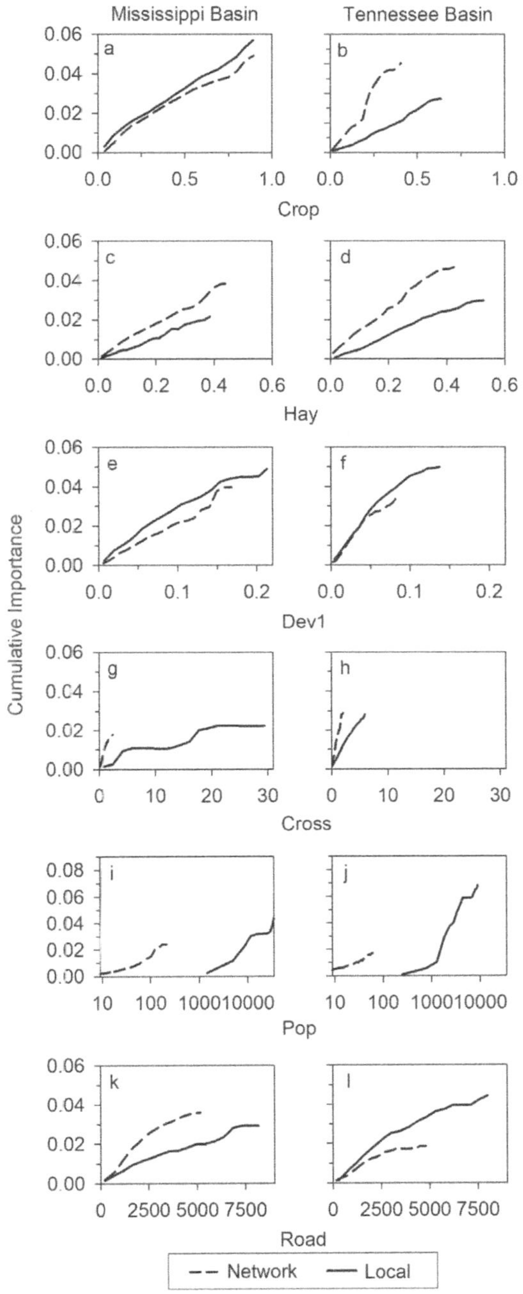

Figure 4. Change in fish assemblage composition across gradients in landscape alterations illustrated as cumulative importance plots from gradient forest models for the Mississippi and Tennessee River basins at network catchment (dashed lines) and local catchment (solid lines) scales. See Table 1 for landscape alteration variable descriptions; note panels **(i)** and **(j)** are shown on log scale.

4b). Assemblages in both the Mississippi and Tennessee basins changed in a linear pattern regardless of scale as proportions of Hay (Figure 4c, 4d) and Dev1 (Figure 4e, 4f) land uses increased, although Hay was more important at the network catchment scale and Dev1 was more important at the local catchment scale. Road crossing density was more important at the network catchment scale across both basins, but the range of magnitudes was smaller at the local catchment scale (Figure 4g, 4h). In the Mississippi basin, threshold changes in assemblage composition occurred when road-crossing densities ranged from 4 to 5 and 18–19 crossings/km^2 at both scales (Figure 4g). Assemblages changed in a curvilinear pattern before plateauing as human population density increased at both network and local catchment scales in the Mississippi and Tennessee basins, but a threshold was evident in the Mississippi basin when 30,000–39,000 people/km^2 were present in local catchments (Figure 4i, 4j). Road density was more important at the network catchment scale in the Mississippi basin (Figure 4k) but more important at the local catchment scale in the Tennessee basin (Figure 4l).

Species and Assemblage Patterns

Models had at least some explanatory power for 47 of 50 species in the Mississippi basin (Table 2) and 61 of 62 species in the Tennessee basin (Table 3). Box plots of random forest model performance were consistent between scales for the Mississippi and Tennessee basins (Figure 5). Among the top 10 species for which models explained the greatest amounts of variation in occurrence, model performance was either greater at the network catchment scale (3 in the Mississippi, 2 in the Tennessee) or was consistent between scales (7 in the Mississippi, 8 in the Tennessee). Among the remaining species in the Mississippi basin, network catchment-scale measurements performed best for 38% (14

Table 2. Common names, scientific names, identification code, random forest model performance measured as 1 – relative error rate (1 – RER), and transformed landscape alteration principal component (PC) scores along PC1 and PC2 for all fishes for which gradient forest models had some explanatory power in the Mississippi River basin at network catchment (NC) and local catchment (LC) spatial scales. Common names marked with an asterisk (*) are listed as imperiled or have limited distributions (<6 drainage units), according to Warren et al. (2000).

			1 – RER		Ordination scores			
Common name	Scientific name	Code	NC	LC	NC PC1	NC PC2	LC PC1	LC PC2
Yellow Bullhead	*Ameiurus natalis*	Amenat	0.20	0.38	0.0034	0.0001	0.0033	0.0038
Pirate Perch	*Aphredoderus sayanus*	Aphsay	0.00	0.02	0.0008	0.0055	0.0012	−0.0004
Bluntface Shiner	*Cyprinella camura*	Cypcam	0.33	0.29	0.0055	0.0032	−0.0108	0.0053
Common Carp	*Cyprinus carpio*	Cypcar	0.46	0.36	0.0162	0.0396	0.0159	0.0172
Red Shiner	*Cyprinella lutrensis*	Cyplut	0.45	0.45	−0.0220	0.0021	−0.0232	−0.0175
Blacktail Shiner	*Cyprinella venusta*	Cypven	0.35	0.23	0.0217	−0.0087	0.0329	0.0002
Gizzard Shad	*Dorosoma cepedianum*	Dorcep	0.37	0.29	0.0076	−0.0114	0.0300	−0.0326
Eastern Creek Chubsucker	*Erimyzon oblongus*	Eriobl	0.36	0.36	−0.0190	−0.0026	−0.0164	0.0032
Grass Pickerel	*Esox americanus*	Esoame	0.25	0.25	0.0042	0.0145	0.0068	0.0192
Bluntnose Darter	*Etheostoma chlorosoma*	Ethchl	0.33	0.41	−0.0093	−0.0117	−0.0161	−0.0163
Slough Darter	*Etheostoma gracile*	Ethgra	0.45	0.45	−0.0118	−0.0013	−0.0090	−0.0056
*Brighteye Darter	*Etheostoma lynceum*	Ethlyn	0.37	0.30	0.0118	−0.0259	0.0042	−0.0199
Johnny Darter	*Etheostoma nigrum*	Ethnig	0.16	0.32	0.0057	0.0023	−0.0008	−0.0036
*Firebelly Darter	*Etheostoma pyrrhogaster*	Ethpyr	0.36	0.45	−0.0193	0.0000	−0.0347	−0.0086
Gulf Darter	*Etheostoma swaini*	Ethswa	0.30	0.30	−0.0212	−0.0085	0.0040	−0.0251
Blackstripe Topminnow	*Fundulus notatus*	Funnot	0.41	0.45	0.0289	−0.0170	0.0164	0.0190
Blackspotted Topminnow	*Fundulus olivaceus*	Funoli	0.23	0.23	0.0038	−0.0027	0.0008	0.0011
Cypress Minnow	*Hybognathus hayi*	Hybhay	0.41	0.41	−0.0149	0.0180	−0.0167	−0.0032
Northern Hog Sucker	*Hypentelium nigricans*	Hypnig	0.44	0.44	−0.0249	−0.0364	−0.0091	−0.0489
Channel Catfish	*Ictalurus punctatus*	Ictpun	0.43	0.37	0.0139	0.0058	0.0204	−0.0038
Green Sunfish	*Lepomis cyanellus*	Lepcya	0.43	0.37	−0.0002	−0.0064	0.0031	−0.0046
Warmouth	*Lepomis gulosus*	Lepgul	0.26	0.22	0.0104	−0.0134	0.0025	−0.0082
Orangespotted Sunfish	*Lepomis humilis*	Lephum	0.22	0.04	0.0021	−0.0038	0.0154	−0.0030
Bluegill	*Lepomis macrochirus*	Lepmac	0.46	0.36	0.0026	−0.0083	0.0046	−0.0043

Table 2. Continued

Common name	Scientific name	Code	1 – RER		Ordination scores			
			NC	LC	NC PC1	NC PC2	LC PC1	LC PC2
Dollar Sunfish	*Lepomis marginatus*	Lepmar	0.36	0.25	0.0156	−0.0192	0.0097	−0.0151
Longear Sunfish	*Lepomis megalotis*	Lepmeg	0.37	0.31	0.0050	−0.0035	0.0117	−0.0017
Redear Sunfish	*Lepomis microlophus*	Lepmic	0.46	0.46	−0.0224	−0.0116	−0.0211	−0.0184
Spotted Gar	*Lepisosteus oculatus*	Lepocu	0.37	0.37	−0.0143	−0.0354	−0.0051	−0.0444
Ribbon Shiner	*Lythrurus fumeus*	Lytfum	0.34	0.30	0.0072	−0.0009	−0.0050	−0.0148
Redfin Shiner	*Lythrurus umbratilis*	Lytumb	0.07	0.00	0.0045	−0.0050	0.0024	0.0048
Spotted Bass	*Micropterus punctulatus*	Micpun	0.30	0.30	0.0021	−0.0020	0.0150	−0.0101
Largemouth Bass	*Micropterus salmoides*	Micsal	0.23	0.27	0.0200	−0.0138	0.0221	−0.0001
Spotted Sucker	*Minytrema melanops*	Minmel	0.45	0.45	−0.0159	0.0009	−0.0336	0.0012
Blacktail Redhorse	*Moxostoma poecilurum*	Moxpoe	0.25	0.33	−0.0144	−0.0012	0.0060	−0.0239
Golden Shiner	*Notemigonus crysoleucas*	Notchr	0.15	0.11	−0.0125	0.0316	−0.0277	0.0100
*Orangefin Shiner	*Notropis ammophilus*	Notamm	0.46	0.36	−0.0394	−0.0336	−0.0273	−0.0269
Emerald Shiner	*Notropis atherinoides*	Notath	0.37	0.43	0.0116	−0.0173	−0.0068	−0.0075
Mimic Shiner	*Notropis volucellus*	Notvol	0.46	0.46	−0.0122	0.0125	−0.0041	0.0312
Tadpole Madtom	*Noturus gyrinus*	Notgyr	0.36	0.52	0.0146	0.0053	−0.0050	0.0266
*Least Madtom	*Noturus hildebrandi*	Nothil	0.25	0.35	−0.0246	0.0092	−0.0226	−0.0015
Brindled Madtom	*Noturus miurus*	Notmiu	0.44	0.44	−0.0153	−0.0189	0.0182	−0.0283
Brown Madtom	*Noturus phaeus*	Notpha	0.17	0.03	−0.0256	−0.0200	−0.0083	−0.0324
Pugnose Minnow	*Opsopoeodus emiliae*	Opsemi	0.46	0.46	−0.0185	−0.0174	−0.0312	−0.0238
Dusky Darter	*Percina sciera*	Persci	0.42	0.42	−0.0159	−0.0085	0.0144	−0.0247
Suckermouth Minnow	*Phenacobius mirabilis*	Phemir	0.36	0.45	0.0547	−0.0310	0.0308	0.0012
Bullhead Minnow	*Pimephales vigilax*	Pimvig	0.30	0.36	−0.0142	0.0135	−0.0173	0.0034
Creek Chub	*Semotilus atromaculatus*	Sematr	0.00	0.13	0.0127	0.0041	0.0071	0.0140

Table 3. Common names, scientific names, identification codes, random forest model performance measured as 1 – relative error rate (1 – RER), and transformed landscape alteration principal component (PC) scores along PC1 and PC2 for all fishes for which gradient forest models had some explanatory power in the Tennessee River basin at network catchment (NC) and local catchment (LC) spatial scales. Common names marked with an asterisk (*) are listed as imperiled or have limited distributions (≤6 drainage units) according to Warren et al. (2000).

			1 – RER		Ordination scores			
Common name	Scientific name	Code	NC	LC	NC PC1	NC PC2	LC PC1	LC PC2
Rock Bass	*Ambloplites rupestris*	Ambrup	0.36	0.18	−0.0168	−0.0176	0.0154	0.0163
Yellow Bullhead	*Ameiurus natalis*	Amenat	0.03	0.31	0.0103	0.0052	−0.0007	−0.0034
Pirate Perch	*Aphredoderus sayanus*	Aphsay	0.27	0.00	0.0005	0.0208	−0.0186	0.0046
Central Stoneroller	*Campostoma anomalum*	Camano	0.08	0.00	−0.0090	−0.0001	−0.0074	0.0065
Largescale Stoneroller	*Campostoma oligolepis*	Campul	0.19	0.21	−0.0035	−0.0093	0.0066	−0.0056
Southern Redbelly Dace	*Chrosomus erythrogaster*	Chrery	0.38	0.45	−0.0146	−0.0201	−0.0100	0.0100
Rosyside Dace	*Clinostomus funduloides*	Clifun	0.16	0.00	−0.0166	−0.0076	0.0030	0.0116
Banded Sculpin	*Cottus carolinae*	Cotcar	0.23	0.04	−0.0177	−0.0131	0.0078	0.0110
Whitetail Shiner	*Cyprinella galactura*	Cypgal	0.46	0.46	0.0017	−0.0044	0.0381	0.0458
Spotfin Shiner	*Cyprinella spiloptera*	Cypspi	0.47	0.36	−0.0226	−0.0009	0.0112	−0.0059
Gizzard Shad	*Dorosoma cepedianum*	Dorcep	0.36	0.36	−0.0074	−0.0190	0.0425	0.0394
Creek Chubsucker	*Erimyzon oblongus*	Eriobl	0.36	0.40	−0.0006	0.0135	−0.0061	−0.0066
Grass Pickerel	*Esox americanus*	Esoame	0.32	0.40	0.0066	0.0166	0.0061	0.0019
Chain Pickerel	*Esox niger*	Esonig	0.47	0.47	0.0054	0.0125	−0.0202	0.0107
Greenside Darter	*Etheostoma blennioides*	Ethble	0.32	0.28	−0.0238	0.0013	0.0064	0.0317
Rainbow Darter	*Etheostoma caeruleum*	Ethcae	0.23	0.00	−0.0179	−0.0079	−0.0011	0.0124
Bluntnose Darter	*Etheostoma chlorosoma*	Ethchl	0.46	0.38	0.0396	0.0091	0.0236	−0.0055
*Fringed Darter	*Etheostoma crossopterum*	Ethcro	0.47	0.47	−0.0251	0.0051	−0.0186	0.0036
Fantail Darter	*Etheostoma flabellare*	Ethfla	0.01	0.12	−0.0065	−0.0027	−0.0043	0.0014
*Saffron Darter	*Etheostoma flavum*	Ethflav	0.26	0.29	−0.0198	−0.0120	0.0107	0.0054
Slough Darter	*Etheostoma gracile*	Ethgra	0.46	0.46	0.0462	0.0099	−0.0082	−0.0190
*Stripetail Darter	*Etheostoma kennicotti*	Ethken	0.47	0.47	−0.0199	−0.0132	0.0258	−0.0321
Johnny Darter	*Etheostoma nigrum*	Ethnig	0.46	0.38	0.0386	−0.0063	−0.0049	−0.0201
*Blackfin Darter	*Etheostoma nigripinne*	Ethnigri	0.46	0.46	−0.0284	0.0089	−0.0159	−0.0052
*Guardian Darter	*Etheostoma oophylax*	Ethoop	0.17	0.25	−0.0020	−0.0089	0.0079	−0.0142
*Redline Darter	*Etheostoma rufilineatum*	Ethruf	0.38	0.38	−0.0254	0.0023	0.0089	0.0017
Orangethroat Darter	*Etheostoma spectabile*	Ethspe	0.26	0.26	−0.0146	−0.0089	−0.0017	−0.0060
*Bandfin Darter	*Etheostoma zonistium*	Ethzoni	0.38	0.27	0.0277	0.0019	−0.0034	−0.0244
Northern Studfish	*Fundulus catenatus*	Funcat	0.28	0.15	−0.0156	−0.0191	0.0215	−0.0129

Table 3. Continued.

Common name	Scientific name	Code	1 – RER		Ordination scores			
			NC	LC	NC PC1	NC PC2	LC PC1	LC PC2
Blackstripe Topminnow	*Fundulus notatus*	Funnot	0.30	0.44	0.0378	0.0118	0.0108	–0.0196
Blackspotted Topminnow	*Fundulus olivaceus*	Funoli	0.15	0.01	0.0137	0.0004	0.0000	0.0003
Western Mosquitofish	*Gambusia affinis*	Gamaff	0.38	0.45	0.0184	0.0119	0.0094	0.0015
Northern Hog Sucker	*Hypentelium nigricans*	Hypnig	0.41	0.11	–0.0072	–0.0216	0.0153	0.0077
Green Sunfish	*Lepomis cyanellus*	Lepcya	0.38	0.31	–0.0017	–0.0019	0.0020	–0.0012
Warmouth	*Lepomis gulosus*	Lepgul	0.24	0.27	0.0082	–0.0030	0.0092	–0.0023
Orangespotted Sunfish	*Lepomis humilis*	Lephum	0.21	0.03	–0.0006	–0.0073	0.0169	–0.0023
Bluegill	*Lepomis macrochirus*	Lepmac	0.44	0.44	–0.0015	–0.0052	0.0028	–0.0020
Dollar Sunfish	*Lepomis marginatus*	Lepmar	0.32	0.28	0.0035	–0.0179	0.0460	0.0081
Longear Sunfish	*Lepomis megalotis*	Lepmeg	0.37	0.27	0.0013	–0.0098	0.0101	–0.0028
Redear Sunfish	*Lepomis microlophus*	Lepmic	0.33	0.33	0.0058	–0.0040	0.0247	0.0142
Striped Shiner	*Luxilus chrysocephalus*	Luxchr	0.30	0.14	–0.0033	–0.0160	0.0135	–0.0003
Scarlet Shiner	*Lythrurus fasciolaris*	Lytfas	0.31	0.35	–0.0128	–0.0048	–0.0096	–0.0009
Ribbon Shiner	*Lythrurus fumeus*	Lytfum	0.47	0.47	0.0735	0.0319	0.0078	–0.0029
Redfin Shiner	*Lythrurus umbratilis*	Lytumb	0.44	0.44	0.0489	0.0163	–0.0120	–0.0120
Smallmouth Bass	*Micropterus dolomieu*	Micdol	0.53	0.24	–0.0244	–0.0017	0.0073	0.0155
Spotted Bass	*Micropterus punctulatus*	Micpun	0.37	0.23	0.0117	–0.0392	0.0385	0.0125
Largemouth Bass	*Micropterus salmoides*	Micsal	0.14	0.17	–0.0024	–0.0037	0.0120	0.0012
Spotted Sucker	*Minytrema melanops*	Minmel	0.37	0.42	0.0116	–0.0017	0.0072	0.0080
Golden Redhorse	*Moxostoma erythrurum*	Moxery	0.28	0.41	–0.0169	–0.0200	0.0444	0.0351
Golden Shiner	*Notemigonus crysoleucas*	Notchr	0.44	0.32	0.0160	–0.0036	–0.0103	–0.0157
Emerald Shiner	*Notropis atherinoides*	Notath	0.38	0.45	0.0123	–0.0079	–0.0035	–0.0011
Bigeye Shiner	*Notropis boops*	Notboo	0.47	0.47	–0.0153	–0.0148	0.0167	–0.0049
Mimic Shiner	*Notropis volucellus*	Notvol	0.32	0.38	0.0211	–0.0025	0.0370	–0.0273
Slender Madtom	*Noturus exilis*	Notexi	0.38	0.44	–0.0034	–0.0299	–0.0102	0.0098
*Least Madtom	*Noturus hildebrandi*	Nothil	0.47	0.47	0.0412	0.0254	–0.0162	–0.0057
Tennessee Shiner	*Notropis leuciodus*	Notleu	0.43	0.43	–0.0114	–0.0351	0.0305	0.0016
Brown Madtom	*Noturus phaeus*	Notpha	0.46	0.46	0.0803	0.0491	–0.0134	–0.0081
Common Logperch	*Percina caprodes*	Percap	0.34	0.12	–0.0212	0.0127	0.0020	0.0049
Bluntnose Minnow	*Pimephales notatus*	Pimnot	0.40	0.40	0.0163	–0.0157	0.0199	0.0045
Blacknose Dace	*Rhinichthys atratulus*	Rhiobt	0.35	0.31	–0.0258	0.0070	–0.0266	0.0158
Creek Chub	*Semotilus atromaculatus*	Sematr	0.00	0.21	–0.0009	0.0047	–0.0021	0.0052

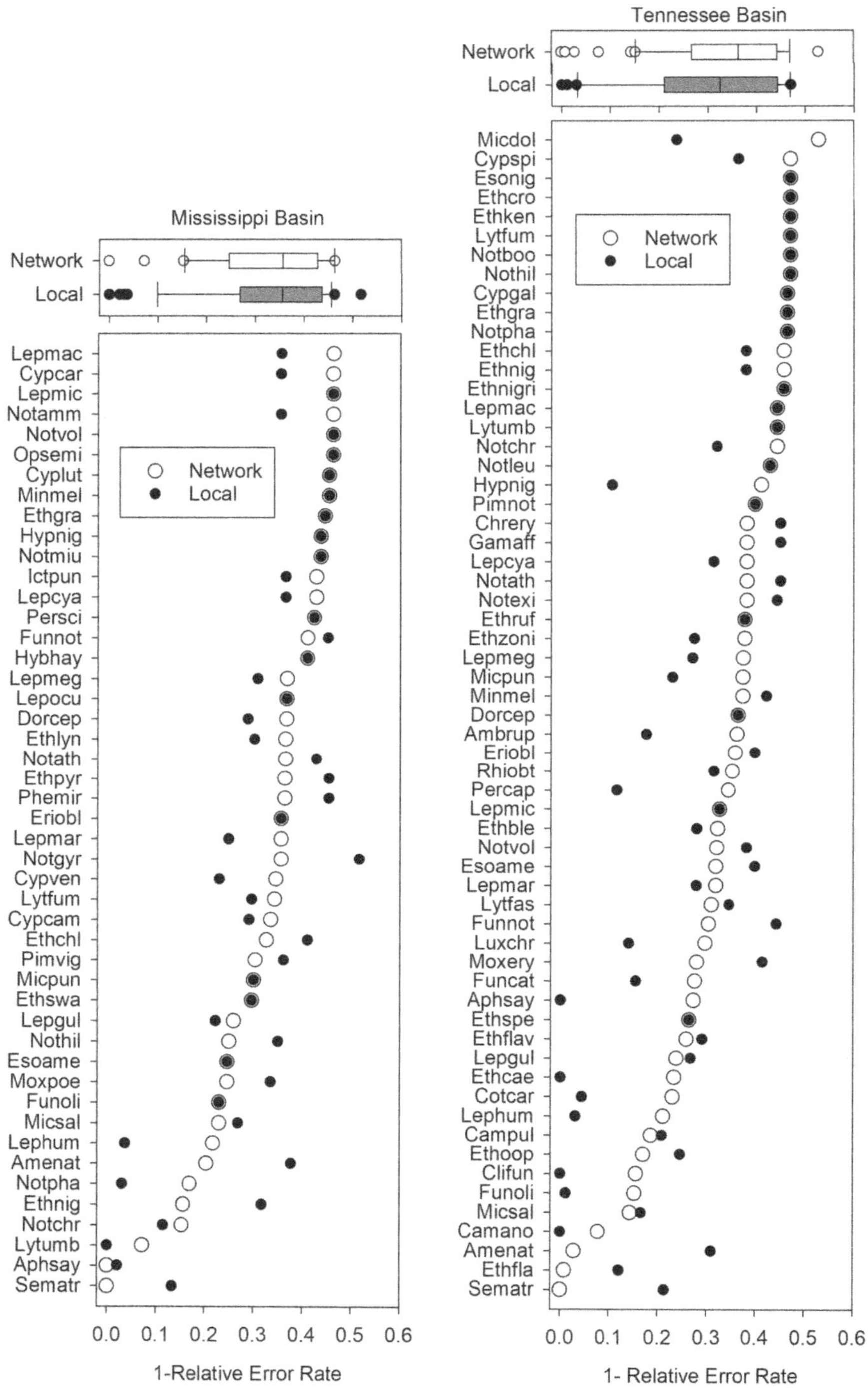

Figure 5. Performance of random forest models developed for species during gradient forest model fitting measured as 1-relative error rate (interpreted as with R^2) for data from the Mississippi and Tennessee River basins at network catchment (white boxes and circles) and local catchment (gray/black boxes and circles) scales. Box plots show the distributions for all species. Only species for which models had some explanatory power are shown; see Tables 2 and 3 for description of species codes.

of 37), local catchment-scale measurements performed best for 38% (14 of 37), and both scales produced equally accurate models for 24% (9 of 37). Among the remaining species in the Tennessee basin, network catchment-scale measurements performed best for 43% (22 of 51), local catchment-scale measurements performed best for 37% (19 of 51), and both scales produced equally accurate models for 20% (10 of 51).

Seventy percent of the variation in assemblage composition predicted by the network catchment GF model in the Mississippi basin was captured in PC1 (46%) and PC2 (24%). The first PC represented an agricultural gradient with catchments characterized by higher proportions of Hay land use arranged negatively and network catchments characterized by higher proportions of Crop land use arranged positively (Figure 6a). The second PC represented a development gradient with network catchments having greater densities or coverages by Road, Dev1, Dev2, Pop, Cross, and Dev 5 arranged positively and network catchments with lower magnitudes for all of those alterations arranged negatively. Along those gradients, predictions for fish assemblage composition clustered as four groups illustrating assemblage variation among catchments with low, medium, and high urban development as well as high Crop coverage (Figure 6b). Weighted species scores illustrated the major components of assemblage change across those gradients, including prevalence of Common Carp, Golden Shiner, and Grass Pickerel in Crop streams and prevalence of Suckermouth Minnow, Blackstripe Topminnow, and Largemouth Bass in heavily developed streams (Figure 6c). Least developed streams were characterized by high occurrences of Firebelly Darter, Gulf Darter, and Brown Madtom.

Sixty-nine percent of the variation in assemblage composition predicted by the local catchment GF model in the Mississippi basin was captured in PC1 (45%) and PC2 (24%). The first PC represented an agricultural gradient, with local catchments characterized by a higher proportion of Hay land use arranged negatively and local catchments characterized by a higher proportion of Crop land use arranged positively (Figure 6d). The second PC represented an urban to rural gradient, with local catchments characterized by development, such as Dev1, Road, Dev5, and Dev2, arranged positively and local catchments with high population density arranged negatively. Along these gradients, predictions for fish assemblage composition clustered into four groups illustrating assemblage variation among local catchments, with low, medium, and high urban development coverage as well as high Crop coverage (Figure 6e). Weighted species scores illustrated the major components of assemblage change across these gradients, including prevalence of Common Carp, Mimic Shiner, and Tadpole Madtom in Crop streams, and Suckermouth Minnow, Blacktail Shiner, and Warmouth in rural streams (Figure 6f). Least developed streams were characterized by high occurrences of Orangefin Shiner, Northern Hog Sucker, and Spotted Gar.

Seventy-three percent of variation in assemblage composition predicted by the network catchment GF model in the Tennessee basin was captured in PC1 (49%) and PC2 (24%). The first PC represented a development gradient, with network catchments characterized by little urban development arranged negatively and network catchments characterized by a higher proportion of Hay, Crop, Dev1, and Dev5 land use arranged positively (Figure 7a). The second PC represented an agricultural gradient, with network catchments characterized by higher proportions of Hay arranged positively and network catchments characterized by higher proportions of Crop land use arranged negatively. Along these gradients, predictions for fish assem-

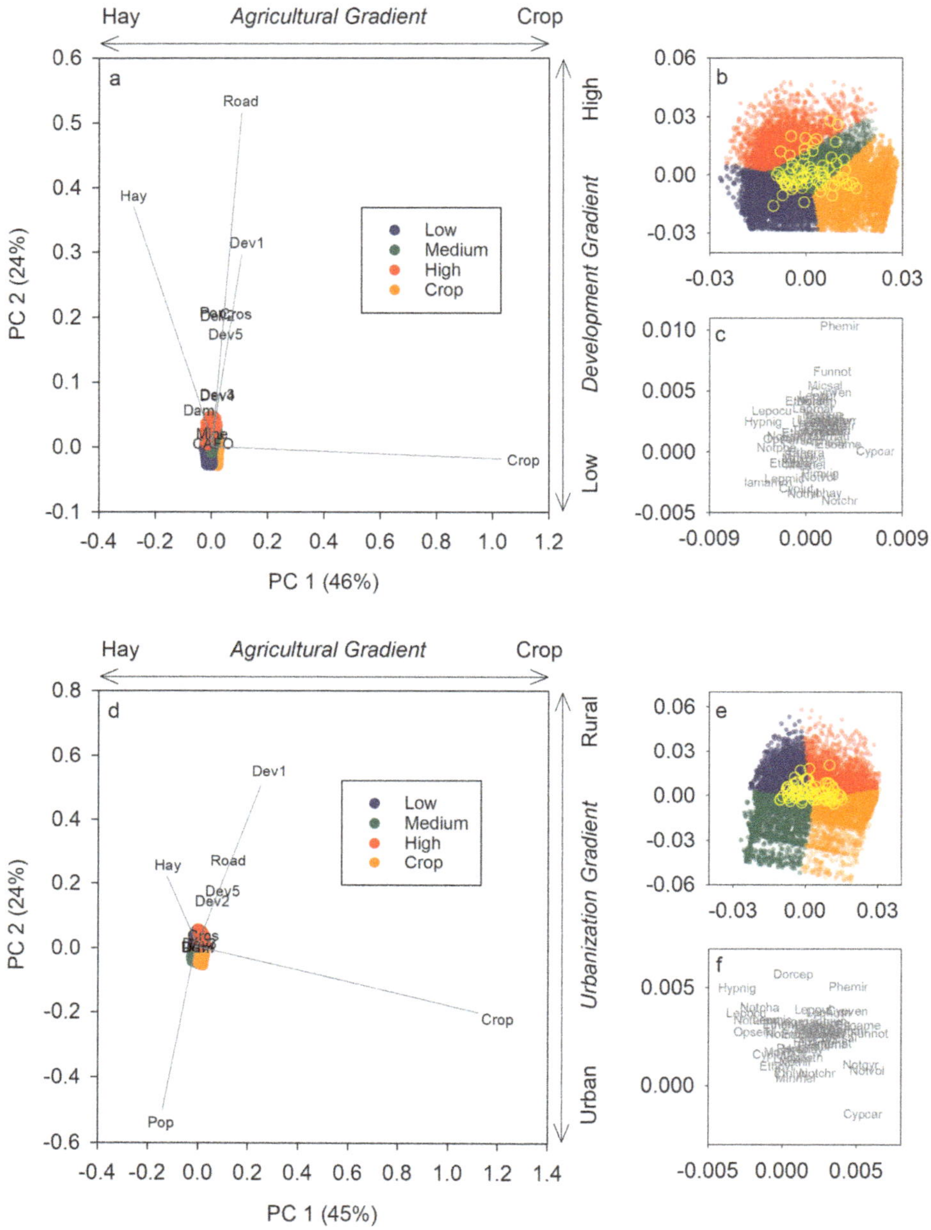

Figure 6. Principle component (PC) plot of biologically transformed landscape alteration variables illustrating predicted fish assemblage composition for stream segments in the Mississippi River basin at **(a–c)** network catchment and **(d–f)** local catchment scales. Panels (a) and (d) illustrate latent PC gradients (% variation explained by each) in predicted assemblage composition developed using gradient forest models, 13 landscape alteration variables (see Table 1 for definitions), and fish assemblage data. Assemblage compositions clustered into four groups illustrate stream segments affected by low (blue), medium (green), or high (red) development as well as cultivated crop land use (orange). Panels (b) and (e) illustrate a close up of the data cloud as well as the distribution of segments with fish assemblage samples (yellow circles) used in model development. Panels (c) and (f) illustrate weighted species scores; see Table 2 for species code descriptions.

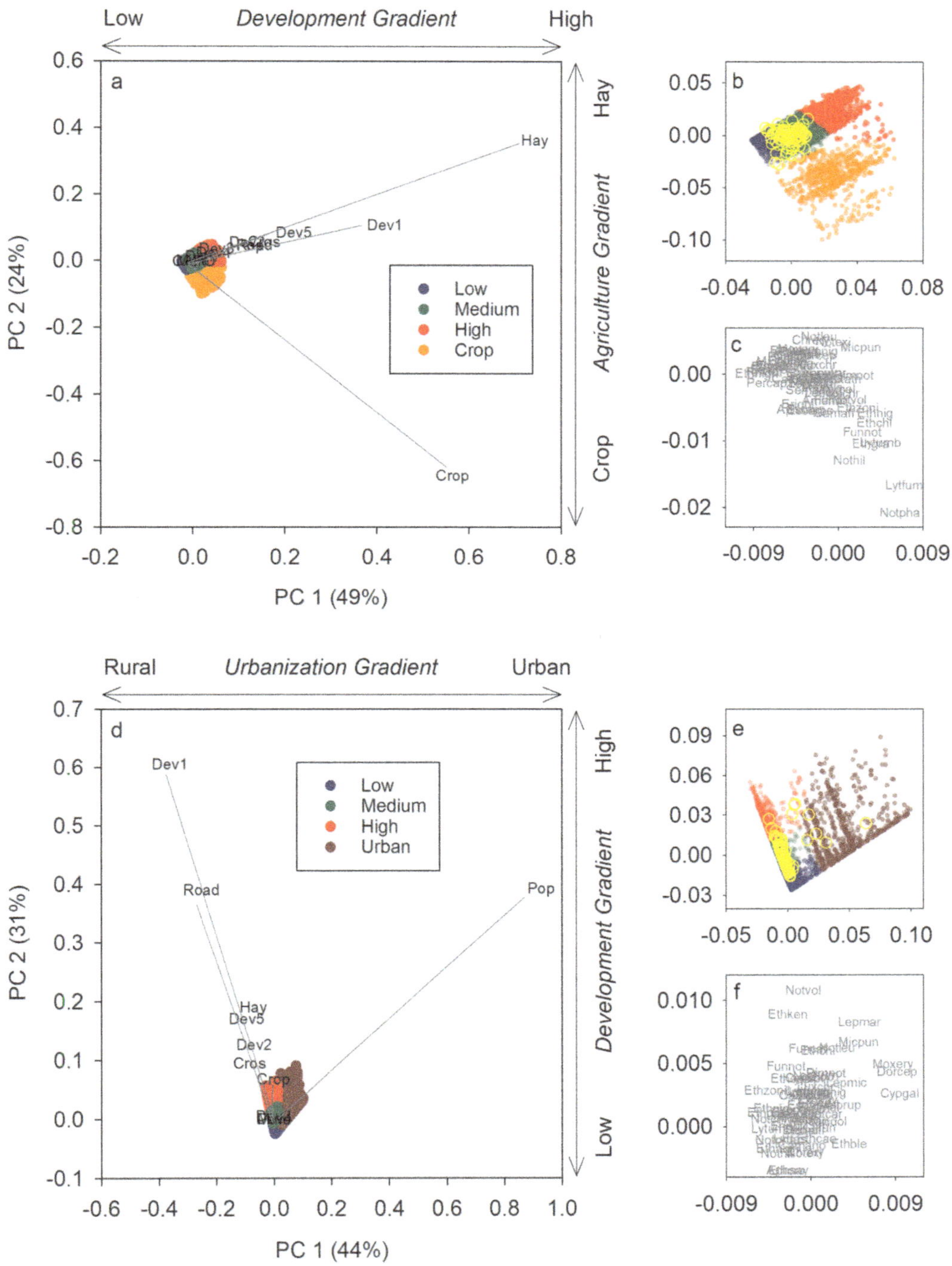

Figure 7. Principle component (PC) plot of biologically transformed landscape alteration variables illustrating predicted fish assemblage composition for stream segments in the Tennessee River basin at **(a–c)** network catchment and **(d–f)** local catchment scales. Panels are as in Figure 6 except that (d) illustrates assemblage compositions clustered into four groups to illustrate stream segments affected by low (blue), medium (green), or high (red) development as well as high human population density (dark red); see Table 3 for species code descriptions.

blage composition clustered into four groups illustrating assemblage variation among network catchments, with low, medium, and high urban development as well as high Crop coverage (Figure 7b). Weighted species scores illustrated the major components of assemblage change across these gradients, including prevalence of Brown Madtom, Ribbon Shiner, and Least Madtom in Crop streams, and Spotted Bass, Slender Madtom, and Bluntnose Minnow in highly developed streams (Figure 7c). Least developed streams were characterized by high occurrences of Blackfin Darter, Blacknose Dace, and Redline Darter.

Seventy-five percent of the variation in assemblage composition predicted by the local catchment GF model in the Tennessee basin was captured in PC1 (44%) and PC2 (31%). The first PC represented an urbanization gradient, with catchments characterized by rural developments such as Dev1, Road, Hay, and Dev5 arranged negatively and urban catchments with higher population density arranged positively (Figure 7 d). The second PC represented a development gradient, with catchments characterized by high proportions of Dev1, Hay, Dev2, and Dev5, as well as high densities of Road and Cross arranged positively and catchments with low values for each of these alterations arranged negatively. Along these gradients, predictions for fish assemblage composition clustered into four groups illustrating assemblage variation among catchments, with low, medium, and high urban development as well as high population density (Figure 7e). Weighted species scores illustrated the major components of assemblage change across these gradients, including prevalence of Mimic Shiner, Stripetail Darter, and Blackstripe Topminnow in rural streams and Gizzard Shad, Whitetail Shiner, and Golden Redhorse in urban streams (Figure 7f). Least developed streams were characterized by high occurrences of Blacknose Dace, Fringed Darter, and Least Madtom.

Mapping Predictions

We mapped the spatial distribution of predicted assemblage compositions across the two basins at network and local catchment scales. At the network catchment scale in the Mississippi basin, a large swath of agriculturally induced alterations to stream fish assemblages was clear, but this swath was interrupted by intact riparian zones along tributaries near larger rivers (Figure 8). To the east and south of the agriculturally dominated area was a region characterized by a mix of low- and medium-development network catchments fragmented by roads. At the network catchment scale in the Tennessee basin, agriculturally affected fish assemblages attenuated in an easterly direction, although assemblages affected by urban development were centered around urban areas.

At the local catchment scale in the Mississippi basin, the same swath of agriculturally influenced fish assemblages was still evident, although a pattern of development-induced alterations to fish assemblages in close proximity to this swath was apparent (Figure 9). In the Tennessee basin, local catchments with minimally altered fish assemblages were distributed to the east, whereas assemblages affected by urban development and human population density increased to the west and along corridors of major roadways.

Discussion

Our study provides strong evidence for spatial variation in relationships between landscape alterations and stream fish assemblages but mixed evidence for an effect of the selected scales on these relationships. Our first hypothesis was that ecological responses to landscape alterations would vary across space because landscape alterations are not spatially uniform. This hypothesis was supported by spatial segregation of fish assemblages affected by low, medium, and high levels of ur-

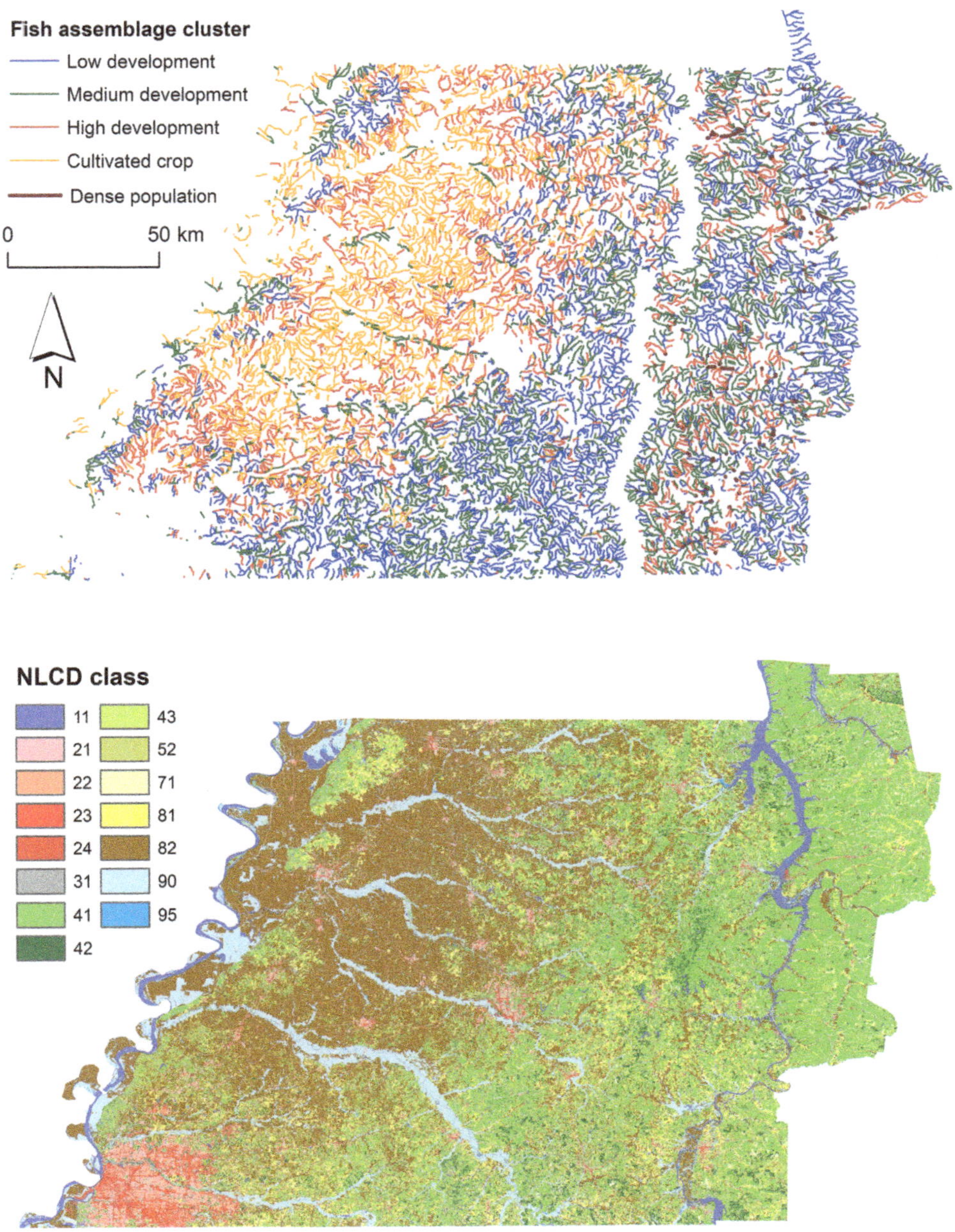

Figure 8. Map of predicted fish assemblage composition based on gradient forest models for stream segments in the Mississippi and Tennessee River basins at the network catchment scale (upper) and land uses from the National Land Cover Database (NLCD; lower). Colors in the upper panel represent predicted fish assemblages clustered into those primarily affected by low development (blue), medium development (green), high development (red), and cultivated crops (orange). Colors in the lower panel are from Homer et al. (2015) and represent land-use classes, including 11 (open water), 21 (developed, open space), 22 (developed, low intensity), 23 (developed, medium intensity), 24 (developed, high intensity), 31 (barren land), 41 (deciduous forest), 42 (evergreen forest), 43 (mixed forest), 52 (shrub/scrub), 71 (grassland), 81 (pasture/hay), 82 (cultivated crops), 90 (woody wetlands), and 95 (emergent herbaceous wetlands).

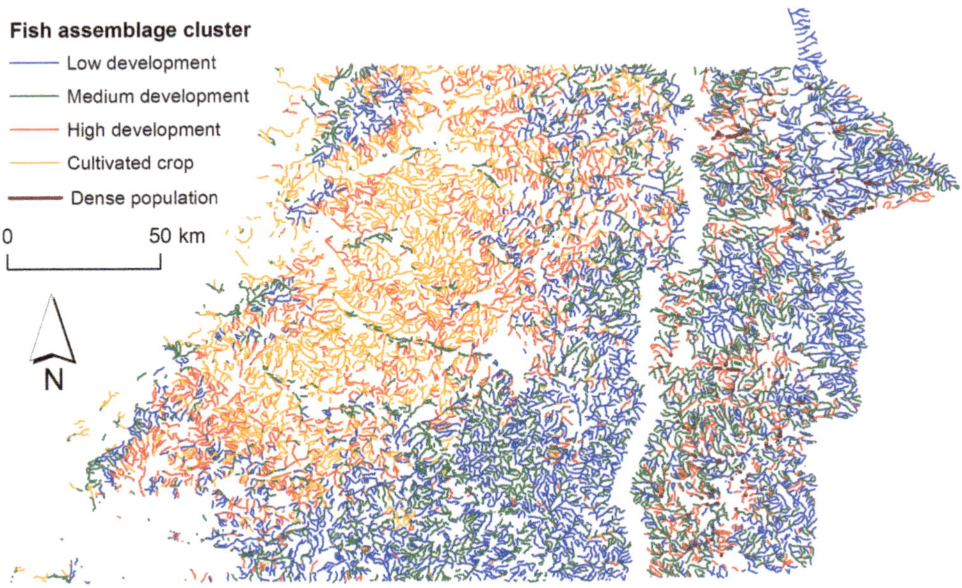

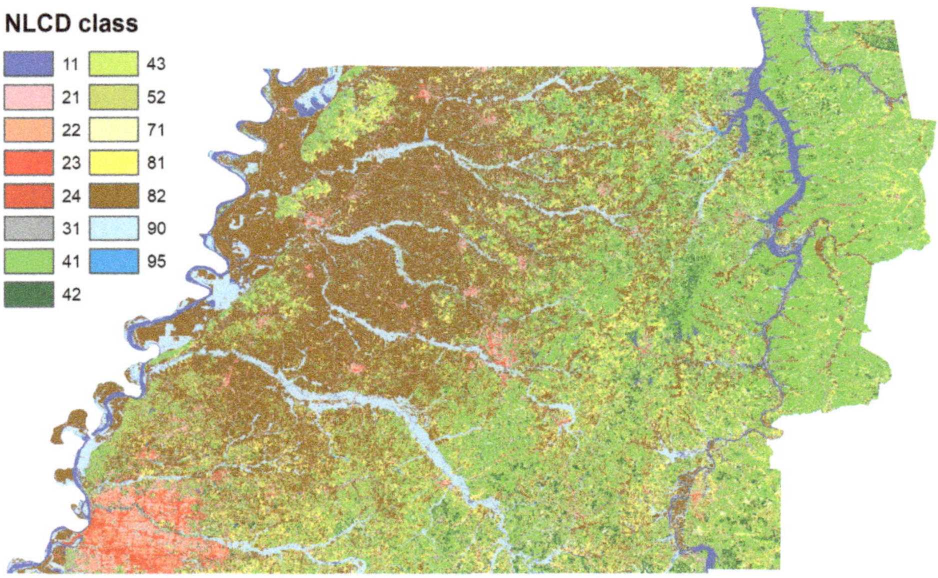

Figure 9. Map of predicted fish assemblage composition based on gradient forest models for stream segments in the Mississippi and Tennessee River basins at the local catchment scale (upper) and land uses from the National Land Cover Database (NLCD; lower). Colors represent predicted fish assemblages clustered into those primarily affected by low development (blue), medium development (green), high development (red), cultivated crops (orange), and high human population density (dark red). Colors in the lower panel are from Homer et al. (2015) and represent land use classes, including 11 (open water), 21 (developed, open space), 22 (developed, low intensity), 23 (developed, medium intensity), 24 (developed, high intensity), 31 (barren land), 41 (deciduous forest), 42 (evergreen forest), 43 (mixed forest), 52 (shrub/scrub), 71 (grassland), 81 (pasture/hay), 82 (cultivated crops), 90 (woody wetlands), and 95 (emergent herbaceous wetlands).

ban development caused by road construction and impervious surfaces as well as agricultural land use. Previous works have demonstrated spatial variation in fish assemblage response to roads (Trombulak and Frissell 2000; Leitão et al. 2018) and agricultural land use (Leitão et al. 2018; Bierschenk et al. 2019; Pompeu et al. 2019, this volume), but the question of how scale might influence these relationships remains a focal research area (Strayer et al. 2003; Kautza and Sullivan 2015; Pease et al. 2015). Our second hypothesis was that the magnitude of ecological responses to landscape alterations would not be consistent when landscape alterations were measured at the network versus local catchment scale. This expectation was based on a growing body of research directed at understanding the relative influences of nearby versus far-upstream landscape alterations on stream fishes (Wang et al. 2001, 2003; Gido et al. 2006; Hoeinghaus et al. 2007; Sály et al. 2011; Marzin et al. 2013; Macedo et al. 2014; Leal et al. 2016; Leitão et al. 2018). Our results only partly supported our second hypothesis because fish assemblage patterns were largely consistent between network and local catchment scales when mapped across the landscape, yet the identities of the most important landscape alterations used in developing spatial predictions and the most responsive fish species varied with scale. Agricultural land uses such as crop and hay were the most important for describing fish assemblage distributions at the network catchment scale, whereas low-intensity development and human population density were most important at the local catchment scale. Combined, our results provide insight into landscape–riverscape linkages that could be included in management planning to advance conservation of freshwater biota (Strayer et al. 2003; Boyd et al. 2008; Kautza and Sullivan 2015).

Including two levels of organizational complexity (species and assemblages) provided inferences not attainable through species- or assemblage-only approaches. The majority of scaling studies in freshwater ecosystems focus on space, time, or a combination of space and time without varying the level of organizational complexity in the response variable. That is, focus is generally on either a single species or assemblage composition. Despite their long history (Caughley 1994), single-species assessments have become less common with the rise of assessments of ecological condition through multimetric indices (Karr 1981; Ruaro and Gubiani 2013; Wellemeyer et al. 2018). However, management of ecological condition cannot ignore the needs of individual species when management plans are developed (Collares-Pereira and Cowx 2004). Consequently, ensemble approaches in which multiple levels of organizational complexity are considered stand the best chance for developing the robust information required for management planning (Olden 2003). The GF models employed here address this need directly by extending single-species RF models into the multispecies realm (Ellis et al. 2012). Our results suggest that multiple species were characteristic of the assemblages inhabiting stream segments affected by a variety of landscape alterations, and these species may represent either indicator or surrogate species. Indicator species allow for targeted monitoring and reduced sampling efforts at the assemblage level to gain the same level of information (Dufrêne and Legendre 1997), whereas surrogate species are defined as those responding in a similar fashion as those of conservation concern but are more easily monitored (Caro and O'Doherty 1999). We found several regionally endemic or listed species that were representative of least-altered stream segments and could be used as ecological indicators (see Tables 2 and 3) as well as nonlisted fishes occupying the same habitats that could be monitored in their place as surrogate species. This promotes efficiencies in resource allocation by reducing the level of effort required to assess or monitor ecological condition and is a level of insight that would not have been possible if we had used a single-species approach. Another benefit of the GF approach is that species for which models have little explanatory power can be identified and inference appropriately constrained. This is a limitation of most direct-ordination assemblage ecology methods that provide species scores, but no insight into how well variation in species occurrence is actually described by the model. Among the species for which GF models explained the most variation in occurrence, local catchment- and network catchment-scale models captured identical levels of variation for most of these species across both basins. These findings suggest consistency in species responses to landscape alterations measured at both local and network catchment scales, a finding reported by some previous researchers (Hoeinghaus et al. 2007; Macedo et al. 2014; Leal et al. 2016). However, other studies suggest that species are responsive to landscape alterations measured at specific scales (Wang et al. 2001, 2003) or across a gradient of scales (Labbe and Fausch 2000). Thus, the question remains: which is the most appropriate scale for measuring landscape alterations? The answer to this question is most likely dependent upon a number of contexts related to scaling.

Scale is "the central problem in biology" (Levin 1992) and remains a challenge for effective management of biodiversity (Boyd et al. 2008). Ecological investigations with a focus on scale are best conducted by maintaining consistent predictor and response variables while varying some component of scale (Allen and Starr 2017). We conducted a study with a focus on scale (sensu Allen and Starr 2017) by maintaining consistent landscape alterations as predictor variables, fish species/assemblage occurrence as response variables, and a fixed spatial extent, but we varied the grain size used to assemble landscape alterations. The two grain sizes we assessed, network catchment and local catchment, are commonly included in multiscale assessments comparing effects of nearby versus distant-upstream landscape alterations (Ward 1998; Wiens 2002; Allan 2004). In theory, nearby landscape alterations should have a greater effect on stream fishes compared with distant alterations if influences are diluted by flow or distance (e.g., Kinsolving and Bain 1993), but this expectation might shift when alterations are classified as point-source versus diffuse-source alterations (Lowrance and Sheridan 1985). Point-source alterations are known to exude strong local-scale influences on stream fishes at the location where the alteration enters the stream (Meyer et al. 2005), but diffuse-source alterations can be more challenging to detect. This is, in part, because diffuse-source alterations operate across broader spatiotemporal scales and do not enter streams at a single location. Agricultural land use is a well-studied form of diffuse-source landscape alteration and was generally of greatest importance at the broader network-catchment grain size in our study. In contrast, urban streams and impervious surface runoff are considered point-source disturbances because water is channeled into streams at specific locations, either through stormwater discharge points or culverts (Walsh et al. 2005). In our study, multiple measures of urban or impervious cover alterations ranked most important at the finer local-catchment grain size, including Pop, Dev1, and Road. In reality, fish distributions are governed by interactions among alterations at multiple spatial and temporal scales (e.g., Harding et al. 1998), and our results suggest that mitigating effects of network-scale agricultural activities and local catchment-scale urban development activities might benefit conservation of stream fishes across the region.

Additional Research and Management Needs

Our study informs conservation efforts by assessing potential fish assemblage composition (and indirectly, ecological condition) and the most influential landscape alterations across a large region of mostly unsampled streams. These results can guide decision making regarding scale-specific prioritization of fish conservation resources and funding allocations. For management goals involving the preservation of currently intact stream segments (i.e., conservation sensu Dauwalter et al. 2010), targeting the stream segments identified here as experiencing low development at either network or local catchment scales might be appropriate. For management goals involving rehabilitating natural conditions (i.e., rehabilitation sensu Dauwalter et al. 2010) through near-term, local-scale actions such as riparian planting, bank stabilization, or channel redesign (see Kwak and Freeman 2010 for other examples), targeting stream segments identified as experiencing medium or high urban development at the local catchment scale might be most appropriate. But not if the network catchment is heavily altered (Hughes et al. 2014). The greater challenge will ultimately involve addressing broadscale improvements in agriculturally dominated network catchments. Agriculturally dominated network catchments were common in portions of our study area and often require substantial conservation practice implementation rates (i.e., >50% of their catchment areas) to reduce stressors impairing fish assemblages (Fore et al. 2017; Scavia et al. 2017). However, in these agriculturally dominated systems, it is likely that combinations of network catchment- and local catchment-scale (e.g., riparian planting) interventions could improve conservation outcomes (Sowa et al. 2016). This is especially true given the large number of stressors that influence stream ecosystems but were not measured in this study (Ormerod et al. 2010). Determining where to target those actions is a management need partly identified in our study. It is only partly identified because although clear patterns existed for relationships between landscape alterations and fish assemblages, we emphasize that the predicted assemblage clusters shown in Figures 8 and 9 are only hypothesized conditions that require validation. This validation is possible through additional research targeting sampling at currently unsampled locations and comparing the observed fish assemblage with predictions generated here. Other future directions for research might include mixing data measured at multiple spatial scales to identify the optimum scale(s) for assessing individual landscape alterations, as well as historical land uses that might have had unmeasured influences on the patterns we observed. For example, Harding et al. (1998) assessed 14 different spatial scales and the temporal influences of land use on several measures of fish and invertebrate diversity and, generally, found that historic catchment-scale land use best predicted present-day assemblages. If we better understand scales at which landscape alterations influence fish assemblages, conservationists could develop more effective conservation strategies. Future studies should seek to derive more explicitly how landscape alterations and management actions are influenced by spatial scale and to identify related economic tradeoffs (Keeler et al. 2012).

Challenges and Policy Implications

This study provides insights about how policies could be improved to more effectively address conservation challenges. In both basins, cultivated crop agriculture had the largest influence on ecological condition at the network catchment scale. In the United States, these lands are nearly all privately owned and implementation of conservation practices on these lands is nearly entirely done on a voluntary basis. Several studies have demonstrated that addressing alterations from cultivated crop agriculture requires high implementation rates of conservation practices, and current cost-share programs are insufficient to address the magnitude of the problem (e.g., Sowa et al. 2016; Fore et al. 2017; Scavia et al. 2017). That is, there is not enough public funding to pay for widespread incentive-based adoption of agricultural conservation practices. Therefore, increasing adoption of conservation practices in these landscapes requires developing and funding new research and technologies that seek to maximize both on-farm economic and environmental outcomes, such that voluntary adoption of these practices is greatly improved. At the local catchment scale, we identified impervious surfaces and road crossings as pressures to ecological condition. Though we did not evaluate specific stresses caused by these alterations, other researchers have found that traditional approaches to managing these types of infrastructure could be improved by developing policies and practices that employ green infrastructure techniques (Shafer et al. 2013; Vogel et al. 2015). For example, road crossings are often barriers to fish passage and can negatively affect fish assemblage structure (Perkin and Gido 2012; Leitão et al. 2018), but new culvert designs promote fish passage while reducing culvert maintenance (Gillespie et al. 2014). In fact, simultaneous ecological and economic benefits associated with culvert replacement are an effective solution to conservation challenges associated with roads (Sleight and Neeson 2018). Similarly, most existing stormwater management infrastructure occurs in highly populated areas and is designed to rapidly move stormwater to lotic systems, which can exacerbate downstream flooding issues in municipalities and often negatively affects fish assemblages (Walsh et al. 2005). Developing policies and funding sources that prioritize stormwater retention and infiltration can reduce many of these ecological effects and benefit urbanized areas that require improved stormwater management (Roy et al. 2008). Finally, the diversity of stream fishes in western Tennessee is still intact and few extinctions have occurred (Etnier and Starnes 1993), meaning that there is still time to rise to the challenge of effective stream fish conservation management in this and other regions. Our work provides examples for how new statistical innovations can be applied to conservation challenges to establish where management efforts might be allocated in the future.

Acknowledgments

This project was funded by the Tennessee Wildlife Resources Agency (TWRA) State Wildlife Grants Program and The Nature Conservancy (TNC) and through support from the Center for the Management, Protection, and Utilization of Water Resources at Tennessee Technological University. Additional support was provided by Texas A&M University and by the National Institute of Food and Agriculture, HATCH Project 1017538. J. Wisby with TNC provided valuable assistance with geographic information systems data and M. Clark and C. Boyd with TWRA provided access to fish occurrence data. We also thank Bob Hughes and three anonymous reviewers for manuscript review.

References

Abell, R., M. L. Thieme, C. Revenga, M. Bryer, M. Kottelat, N. Bogutskaya, B. Coad, N. Mandrak, S. C. Balderas, W. Bussing, and M. L. Stiassny. 2008. Freshwater ecoregions of the world: a new map of biogeographic units for freshwater biodiversity conservation. BioScience 58:403–414.

Allan, D., D. Erickson, and J. Fay. 1997. The influence of catchment land use on stream integrity across multiple spatial scales. Freshwater Biology 37:149–161.

Allan, J. D. 2004. Landscapes and riverscapes: the influence of land use on stream ecosystems. Annual Review of Ecology Evolution and Systematics 35:257–284.

Allen, T. F., and T. B. Starr. 2017. Hierarchy: perspectives for ecological complexity. University of Chicago Press, Chicago.

Angermeier, P. L., and M. R. Winston. 1998. Local vs. regional influences on local diversity in stream fish communities of Virginia. Ecology 79:911–927.

Bierschenk, A. M., M. Mueller, J. Pander, and J. Geist. 2019. Impact of catchment land use on fish community composition in the headwater areas of Elbe, Danube and Main. Science of the Total Environment 652:66–74.

Blann, K. L., J. L. Anderson, G. R. Sands, and B. Vondracek. 2009. Effects of agricultural drainage on aquatic ecosystems: a review. Critical Reviews in Environmental Science and Technology 39:909–1001.

Boyd, C., T. M. Brooks, S. H. Butchart, G. J. Edgar, G. A. Da Fonseca, F. Hawkins, M. Hoffmann, W. Sechrest, S. N. Stuart, and P. P. Van Dijk. 2008. Spatial scale and the conservation of threatened species. Conservation Letters 1:37–43.

Burcher, C. L., H. M. Valett, and E. F. Benfield. 2007. The land-cover cascade: relationships coupling land and water. Ecology 88:228–242.

Caro, T. M., and G. O'Doherty. 1999. On the use of surrogate species in conservation biology. Conservation Biology 13:805–814.

Caughley, G. 1994. Directions in conservation biology. Journal of Animal Ecology 63:215–244.

Chick, J. H., C. R. Ruetz, and J. C. Trexler. 2004. Spatial scale and abundance patterns of large fish communities in freshwater marshes of the Florida Everglades. Wetlands 24:652–664.

Closs, G. P., M. Krkosek, and J. D. Olden. 2015. Conservation of freshwater fishes. Cambridge University Press, Cambridge, UK.

Collares-Pereira, M. J., and I. G. Cowx. 2004. The role of catchment scale environmental management in freshwater fish conservation. Fisheries Management and Ecology 11:303–312.

Cutler, D. R., T. C. Edwards, Jr., K. H. Beard, A. Cutler, K. T. Hess, J. Gibson, and J. J. Lawler. 2007. Random forests for classification in ecology. Ecology 88:2783–2792.

Dauwalter, D. C., W. L. Fisher, and F. J. Rahel. 2010. Warmwater streams. Pages 657–698 *in* W. Hubert and M. Quist, editors. Inland fisheries management in North America, 3rd edition. American Fisheries Society, Bethesda, Maryland.

Dufrêne, M., and P. Legendre. 1997. Species assemblages and indicator species: the need for a flexible asymmetrical approach. Ecological Monographs 67:345–366.

Ellis, N., S. J. Smith, and C. R. Pitcher. 2012. Gradient forests: calculating importance gradients on physical predictors. Ecology 93:156–168.

Erős, T. 2017. Scaling fish metacommunities in stream networks: synthesis and future research avenues. Community Ecology 18:72–86.

Erős, T., J. R. O'Hanley, and I. Czeglédi. 2018. A unified model for optimizing riverscape conservation. Journal of Applied Ecology 55:1871–1883.

Etnier, D. A., and W. C. Starnes. 1993. The fishes of Tennessee. University of Tennessee Press, Knoxville.

Fagan, W. F., C. Aumann, C. M. Kennedy, and P. J. Unmack. 2005. Rarity, fragmentation, and the scale dependence of extinction risk in desert fishes. Ecology 86:34–41.

Fausch, K. D. 2010. Preface: a renaissance in stream fish ecology. Pages 199–206 in K. B. Gido and D. A. Jackson, editors. Community ecology of stream fishes: concepts, approaches, and techniques. American Fisheries Society, Symposium 73, Bethesda, Maryland.

Fausch, K. D., C. E. Torgersen, C. V. Baxter, and H. W. Li. 2002. Landscapes to riverscapes: bridging the gap between research and conservation of stream fishes. BioScience 52:483–498.

Fauth, J. E., J. Bernardo, M. Camara, W. J. Resetarits, Jr., J. Van Buskirk, and S. A. McCollum. 1996. Simplifying the jargon of community ecology: a conceptual approach. The American Naturalist 147:282–286.

Feist, B. E., E. R. Buhle, D. H. Baldwin, J. A. Spromberg, S. E. Damm, J. W. Davis, and N. L. Scholz. 2018. Roads to ruin: conservation threats to a sentinel species across an urban gradient. Ecological Applications 27:2382–2396.

Ferrier, S., and A. Guisan. 2006. Spatial modelling of biodiversity at the community level. Journal of Applied Ecology 43:393–404.

Fore, J. D., S. P. Sowa, D. L. Galat, G. M. Annis, D. D. Diamond, and C. Rewa. 2014. Riverine threat indices to assess watershed condition and identify primary management capacity of agriculture natural resource management agencies. Environmental Management 53:567–582.

Fore, J. D., S. P. Sowa, D. L. Galat, and D. D. Diamond. 2017. Assessing effects of sediment-reducing agriculture conservation practices on stream fishes. Journal of Soil and Water Conservation 72:326–342.

Frissell, C. A., W. J. Liss, C. E. Warren, and M. D. Hurley. 1986. A hierarchical framework for stream habitat classification: viewing streams in a watershed context. Environmental Management 10:199–214.

Gido, K. B., J. A. Falke, R. M. Oakes, and K. J. Hase. 2006. Fish–habitat relations across spatial scales in prairie streams. Pages 265–285 *in* R. M. Hughes, L. Wang, and P. W. Seelbach, editors. Landscape influences on stream habitats and biological assemblages. American Fisheries Society, Symposium 48, Bethesda, Maryland.

Gillespie, N., A. Unthank, L. Campbell, P. Anderson, R. Gubernick, M. Weinhold, D. Cenderelli, B. Austin, D. McKinley, S. Wells, J. Rowan, C. Orvis, M. Hudy, A. Bowden, A. Singler, E. Fretz, J. Levine, and R. Kirn. 2014. Flood effects on road–stream crossing infrastructure: economic and ecological benefits of stream simulation designs. Fisheries 39:62–76.

Harding, J. S., E. F. Benfield, P. V. Bolstad, G. S. Helfman, and E. B. D. Jones. 1998. Stream biodiversity: the ghost of land use past. Proceedings of the National Academy of Sciences of the United States of America 95:14843–14847.

Hill, R. A., E. W. Fox, S.G. Leibowitz, A.R. Olsen, D.J. Thornbrugh, and M. H. Weber. 2017. Predictive mapping of conterminous U.S. rivers and streams. Ecological Applications 27:2397–2415.

Hoeinghaus, D. J., K. O. Winemiller, and J. S. Birnbaum. 2007. Local and regional determinants of stream fish assemblage structure: inferences based on taxonomic vs. functional groups. Journal of Biogeography 34:324–338.

Homer, C. G., J. A. Dewitz, L. Yang, S. Jin, P. Danielson, G. Xian, J. Coulston, N. D. Herold, J. D. Wickham, and K. Megown. 2015. Completion of the 2011 National Land Cover Database for the conterminous United States—representing a decade of land cover change information. Photogrammetric Engineering and Remote Sensing 81:345–354.

Hughes, R. M., S. Dunham, K. G. Maas-Hebner, J. A. Yeakley, C. B. Schreck, M. Harte, N. Molina, C. C. Shock, V. W. Kaczynski, and J. Schaeffer. 2014. A review of urban water body rehabilitation and mitigation challenges and approaches: (1) rehabilitation and remediation. Fisheries 39:18–29.

Jackson, D. A., P. R. Peres-Neto, and J. D. Olden. 2001. What controls who is where in freshwater fish communities—the roles of biotic, abiotic, and spatial factors. Canadian Journal of Fisheries and Aquatic Sciences 58:157–170.

Jeffrey, J. D., C. T. Hasler, J. M. Chapman, S. J. Cooke, and C. D. Suski. 2015. Linking landscape-scale disturbances to stress and condition of fish: implications for restoration and conservation. Integrative and Comparative Biology 55:618–630.

Karr, J. R. 1981. Assessment of biotic integrity using fish communities. Fisheries 6:21–27.

Kaufman, L., and P. J. Rousseeuw. 1990. Finding groups in data. Wiley, Series in Probability

& Mathematical Statistics, Hoboken, New Jersey.

Kautza, A., and S. M. P. Sullivan. 2015. Spatially-dependent human alterations determine fish assemblage composition in a modified river system. River Systems 21:93–108.

Keeler, B. L., S. Polasky, K. A. Brauman, K. A. Johnson, J. C. Finlay, A. O'Neill, K. Kovacs, and B. Dalzell. 2012. Linking water quality and well-being for improved assessment and valuation of ecosystem services. Proceedings of the National Academy of Sciences of the United States of America 109:18619–18624.

Kennard, M. J., J. D. Olden, A. H. Arthington, B. J. Pusey, and N. L. Poff. 2007. Multiscale effects of flow regime and habitat and their interaction on fish assemblage structure in eastern Australia. Canadian Journal of Fisheries and Aquatic Sciences 64:1346–1359.

Kinsolving, A. D., and M. B. Bain. 1993. Fish assemblage recovery along a riverine disturbance gradient. Ecological Applications 3:531–544.

Kunin, W. E. 1998. Extrapolating species abundance across spatial scales. Science 281:1513–1515.

Kwak, T. J., and M. C. Freeman. 2010. Assessment and management of ecological integrity. Pages 353–394 *in* W. A. Hubert and M. C. Quist, editors. Inland fisheries management in North America, 3rd edition. American Fisheries Society, Bethesda, Maryland.

Labbe, T. R., and K. D. Fausch. 2000. Dynamics of intermittent stream habitat regulate persistence of a threatened fish at multiple scales. Ecological Applications 10:1774–1791.

Lammert, M., and J. D. Allan. 1999. Assessing biotic integrity of streams: effects of scale in measuring the influence of land use/cover and habitat structure on fish and macroinvertebrates. Environmental Management 23:257–270.

Leal, C. G., P. S. Pompeu, T. A. Gardner, R. P. Leitão, R. M. Hughes, P. R. Kaufmann, J. Zuanon, F. R. de Paula, S. F. Ferraz, J. R. Thomson, and R. Mac Nally. 2016. Multiscale assessment of human-induced changes to Amazonian instream habitats. Landscape Ecology 31:1725–1745.

Leigh, C., and T. Datry. 2017. Drying as a primary hydrological determinant of biodiversity in river systems: a broad-scale analysis. Ecography 40:487–499.

Leitão, R. P., J. Zuanon, D. Mouillot, C. G. Leal, R. M. Hughes, P. R. Kaufmann, S. Villéger, P. S. Pompeu, D. Kasper, F. R. de Paula, S. F. B. Ferraz, and T. Gardner. 2018. Disentangling the pathways of land use impacts on the functional structure of fish assemblages in Amazon streams. Ecography 41:219–232.

Levin, S. A. 1992. The problem of pattern and scale in ecology: the Robert H. MacArthur award lecture. Ecology 73:1943–1967.

Liaw, A., and M. Wiener. 2002. Classification and regression by random forest. R News 2(3):18–22.

Lowrance, R. L., and J. Sheridan. 1985. Managing riparian ecosystems to control nonpoint pollution. Journal of Soil and Water Conservation 40:87–91.

Macedo, D. R., R. M. Hughes, R. Ligeiro, W. R. Ferreira, M. A. Castro, N. T. Junqueira, D R. Oliveira, K. R. Firmiano, P. R. Kaufmann, P. S. Pompeu, and M. Callisto. 2014. The relative influence of catchment and site variables on fish and macroinvertebrate richness in cerrado biome streams. Landscape Ecology 29:1001–1016.

Maechler, M., P. Rousseeuw, A. Struyf, M. Hubert, and K. Hornik. 2018. Cluster: cluster analysis basics and extensions. R package version 2.0.7-1.

Maguire, K. C., D. Nieto-Lugilde, J. L. Blois, M. C. Fitzpatrick, J. W. Williams, S. Ferrier, and D. J. Lorenz. 2016. Controlled comparison of species-and community-level models across novel climates and communities. Proceedings of the Royal Society B 283:20152817.

Malone, E. W., J. S. Perkin, B. M. Leckie, M. A. Kulp, C. R. Hurt, and D. M. Walker. 2018.

Which species, how many, and from where: Integrating habitat suitability, population genomics, and abundance estimates into species reintroduction planning. Global Change Biology 24:3729–3748.

Marzin, A., P. F. M. Verdonschot, and D. Pont. 2013. The relative influence of catchment, riparian corridor, and reach-scale anthropogenic pressures on fish and macroinvertebrate assemblages in French rivers. Hydrobiologia 704:375–388.

Meyer, J. L., M. J. Paul, and W. K. Taulbee. 2005. Stream ecosystem function in urbanizing landscapes. Journal of the North American Benthological Society 24:602–612.

Nieto-Lugilde, D., K. C. Maguire, J. L. Blois, J. W. Williams, and M. C. Fitzpatrick. 2018. Multiresponse algorithms for community-level modelling: review of theory, applications, and comparison to species distribution models. Methods in Ecology and Evolution 9:834–848.

Olden, J. D. 2003. A species-specific approach to modeling biological communities and its potential for conservation. Conservation Biology 17:854–863.

Omernik, J. M., and G. E. Griffith. 2014. Ecoregions of the conterminous United States: evolution of a hierarchical spatial framework. Environmental Management 54:1249–1266.

Ormerod, S. J., M. Dobson, A. G. Hildrew, and C. R. Townsend. 2010. Multiple stressors in freshwater ecosystems. Freshwater Biology 55(Supplement 1):1–4.

Pease, A. A., J. M. Taylor, K. O. Winemiller, and R. S. King. 2015. Ecoregional, catchment, and reach-scale environmental factors shape functional-trait structure of stream fish assemblages. Hydrobiologia 753:265–283.

Perkin, J. S., and K. B. Gido. 2012. Fragmentation alters stream fish community structure in dendritic ecological networks. Ecological Applications 22:2176–2187.

Perkin, J. S., M. J. Troia, D. C. Shaw, J. E. Gerken, and K. B. Gido. 2016. Multiple watershed alterations influence fish community structure in Great Plains prairie streams. Ecology of Freshwater Fish 25:141–155.

Pitcher, R. C., P. Lawton, N. Ellis, S. J. Smith, L. S. Incze, C. L. Wei, M. E. Greenlaw, N. H. Wolff, J. A. Sameoto, and P. V. Snelgrove. 2012. Exploring the role of environmental variables in shaping patterns of seabed biodiversity composition in regional-scale ecosystems. Journal of Applied Ecology 49:670–679.

Pitcher, C. R., N. Ellis, and S. J. Smith. 2011. Example analysis of biodiversity survey data with R package gradientForest. Available: http://gradientforest.r-forge.r-project.org/biodiversity-survey.pdf. (February 2019).

Pompeu, P. S., C. G. Leal, D. R. Carvalho, N. T. Junqueira, M. A. Castro, and R. M. Hughes. 2019. Effects of catchment land use on stream fish assemblages in the Brazilian savanna. Pages 303–320 *in* R. M. Hughes, D. M. Infante, L. Wang, K. Chen, and B. F. Terra, editors. Advances in understanding landscape influences on freshwater habitats and biological assemblages. American Fisheries Society, Symposium 90, Bethesda, Maryland.

Prasad, A. M., L. R. Iverson, and A. Liaw. 2006. Newer classification and regression tree techniques: bagging and random forests for ecological prediction. Ecosystems 9:181–199.

Pringle, C. 2003. What is hydrologic connectivity and why is it ecologically important? Hydrological Processes 17:2685–2689.

R Core Team. 2018. The R project for statistical computing. R Foundation Vienna, Austria. Available: www.R-project.org. (February 2019).

Reynolds, L., A. T. Herlihy, P. R. Kaufmann, S. V. Gregory, and R. M. Hughes. 2003. Electrofishing effort requirements for assessing species richness and biotic integrity in western Oregon streams. North American Journal of Fisheries Management 23:450–461.

Rivers-Moore, N. A., P. S. Goodman, and J. L. Nel. 2011. Scale-based freshwater conservation planning: towards protecting freshwater biodiversity in KwaZulu-Natal, South Africa. Freshwater Biology 56:125–141.

Robinson, W. A., M. Lintermans, J. H. Harris, and F. Guarino. 2019. A landscape-scale electrofishing monitoring program can evaluate fish responses to climatic conditions in the Murray–Darling River system, Australia. Pages 179–201 *in* R. M. Hughes, D. M. Infante, L. Wang, K. Chen, and B. F. Terra, editors. Advances in understanding landscape influences on freshwater habitats and biological assemblages. American Fisheries Society, Symposium 90, Bethesda, Maryland.

Roth, N. E., J. D. Allan, and D. L. Erickson. 1996. Landscape influences on stream biotic integrity assessed at multiple spatial scales. Landscape Ecology 11:141–156.

Roy, A. H., S. J. Wenger, T. D. Fletcher, C. J. Walsh, A. R. Ladson, W. D. Shuster, H. W. Thurston, and R. R. Brown. 2008. Impediments and solutions to sustainable, watershed-scale urban stormwater management: lessons from Australia and the United States. Environmental Management 42:344–359.

Ruaro, R., and E. A. Gubiani. 2013. A scientometric assessment of 30 years of the index of biotic integrity in aquatic ecosystems: applications and main flaws. Ecological Indicators 29:105–110.

Sály, P., P. Takács, I. Kiss, P. Bíró, and T. Eros. 2011. The relative influence of spatial context and catchment-and site-scale environmental factors on stream fish assemblages in a human-modified landscape. Ecology of Freshwater Fish 20:251–262.

Scavia, D., M. Kalcic, R. L. Muenich, J. Read, N. Aloysius, I. Bertani, C. Boles, R. Confesor, J. DePinto, M. Gildow, J. Martin, T. Redder, D. Robertson, S. Sowa, Y. C. Wang, and H. Yen. 2017. Multiple models guide strategies for agricultural nutrient reductions. Frontiers in Ecology and the Environment 15:126–132.

Shafer, C. S., D. Scott, J. Baker, K.O. Winemiller. 2013. Recreation and amenity values of urban stream corridors: implications for green infrastructure. Journal of Urban Design 18:478–493.

Simberloff, D. 1998. Flagships, umbrellas, and keystones: is single-species management passé in the landscape era? Biological conservation 83:247–257.

Sleight, N., and T. M. Neeson. 2018. Opportunities for collaboration between infrastructure agencies and conservation groups: Road-stream crossings in Oklahoma. Transportation Research Part D: Transport and Environment 63:622–631.

Soulé, M. E., and B. A. Wilcox. 1980. Conservation biology: an evolutionary-ecological perspective. Sinauer Associates, Sunderland, Massachusetts.

Sowa, S. P., M. Herbert, S. Mysorekar, G. M. Annis, K. Hall, A. P. Nejadhashemi, S. Woznicki, L. Wang, and P. J. Doran. 2016. How much conservation is enough? Defining implementation goals for healthy fish communities in agricultural rivers. Journal of Great Lakes Research 42:1302–1321.

Stendera, S., R. Adrian, N. Bonada, M. Cañedo-Argüelles, B. Hugueny, K. Januschke, F. Pletterbauer, and D. Hering. 2012. Drivers and stressors of freshwater biodiversity patterns across different ecosystems and scales: a review. Hydrobiologia 696:1–28.

Strahler, A. N. 1957. Quantitative analysis of watershed geomorphology. Eos, Transactions of the American Geophysical Union 38:913–920.

Strayer, D. L., R. E. Beighley, L. C. Thompson, S. Brooks, C. Nilsson, G. Pinay, and R. J. Naiman. 2003. Effects of land cover on stream ecosystems: roles of empirical models and scaling issues. Ecosystems 6:407–423.

Strayer, D. L., and D. Dudgeon. 2010. Freshwater biodiversity conservation: recent progress and future challenges. Journal of the North American Benthological Society 29:344–358.

TDEC (Tennessee Department of Environment and Conservation). 2012. Water Pollution

Control database. Available: http://tdeconline.tn.gov/dwr/. (March 2019).

Thornbrugh, D. J., S. G. Leibowitz, R. A. Hill, M. H. Weber, Z. C. Johnson, A. R. Olsen, J. E. Flotemersch, J. L. Stoddard, and D. V. Peck. 2018. Mapping watershed integrity for the conterminous United States. Ecological Indicators 85:1133–1148.

Trombulak, S. C., and C. A. Frissell. 2000. Review of ecological effects of roads on terrestrial and aquatic communities. Conservation Biology 14:18–30.

USACE (US Army Corps of Engineers). 2013. National inventory of dams (NID) [online database]. Available: https://nid-test.sec.usace.army.mil/.

U.S. Census Bureau, 2010. United States 2010 census dataset, Washington D.C. Available: www.census.gov/programs-surveys/decennial-census/data/datasets.2010.html. (March 2019).

U.S. Census Bureau. 2014. Topologically integrated geographic encoding and referencing (TIGER). Available: www.census.gov/geo/maps-data/data/tiger.html. (February 2019).

USEPA (U.S. Environmental Protection Agency) and USGS (U.S. Geological Survey). 2012. National hydrography dataset plus (NHDPlus) version 2. Available: www.horizon-systems.com/nhdplus/nhdplusv2_home.php. (February 2019).

Vogel, J. R., T. L. Moore, R. R. Coffman, S. N. Rodie, S. L. Hutchinson, K. R. McDonough, A. J. McLemore, and J. T. McMaine. 2015. Critical review of technical questions facing low impact development and green infrastructure: a perspective from the Great Plains. Water Environment Research 87:849–862.

Walsh, C. J., A. H. Roy, J.W. Feminella, P. D. Cottingham, P. M. Groffman, and R. P. Morgan. 2005. The urban stream syndrome: current knowledge and the search for a cure. Journal of the North American Benthological Society 24:706–723.

Wang, L., J. Lyons, P. Kanehl, and R. Bannerman. 2001. Impacts of urbanization on stream habitat and fish across multiple spatial scales. Environmental Management 28:255–266.

Wang, L., J. Lyons, P. Kanehl, and R. Gatti. 1997. Influences of watershed land use on habitat quality and biotic integrity in Wisconsin streams. Fisheries 22(6):6–12.

Wang, L., J. Lyons, P. Rasmussen, P. Seelbach, T. Simon, M. Wiley, P. Kanehl, E. Baker, S. Niemela, and P. M. Stewart. 2003. Watershed, reach, and riparian influences on stream fish assemblages in the Northern Lakes and Forest Ecoregion, USA. Canadian Journal of Fisheries and Aquatic Sciences 60:491–505.

Ward, J. 1998. Riverine landscapes: biodiversity patterns, disturbance regimes, and aquatic conservation. Biological Conservation 83:269–278.

Warren, M. L., B. M. Burr, S. J. Walsh, H. L. Bart, R. C. Cashner, D. A. Etnier, B. J. Freeman, B. R. Kuhajda, R. L. Mayden, H. W. Robison, S. T. Ross, and W. C. Starnes. 2000. Diversity, distribution, and conservation status of the native freshwater fishes of the southern United States. Fisheries 25(10):7–31.

Wellemeyer, J. C., J. S. Perkin, J. D. Fore, and C. Boyd. 2018. Comparing assembly processes for multimetric indices of biotic integrity. Ecological Indicators 89:590–609.

Wenger, S. J., J. T. Peterson, M. C. Freeman, B. J. Freeman, and D. D. Homans. 2008. Stream fish occurrence in response to impervious cover, historic land use, and hydrogeomorphic factors. Canadian Journal of Fisheries and Aquatic Sciences 65:1250–1264.

Wiens, J. A. 1989. Spatial scaling in ecology. Functional Ecology 3:385–397.

Wiens, J. A. 2002. Riverine landscapes: taking landscape ecology into the water. Freshwater Biology 47:501–515.

Williams, J. E., R. N. Williams, R. F. Thurow, L. Elwell, D. P. Philipp, F. A. Harris, J. L. Kershner, P. J. Martinez, D. Miller, G. H. Reeves, C. A. Frissell, and J. R. Sedell. 2011. Native fish conservation areas: a vision for large-scale conservation of native fish communities. Fisheries 36:267–277.

American Fisheries Society Symposium 90:441–462, 2019
© 2019 by the American Fisheries Society

Ensemble Modelling of Sensitive Stream Fish Species Distributions in Iran: Expanding Knowledge to Aid Species Conservation

Hossein Mostafavi*
Department of Biodiversity and Ecosystem Management
Environmental Sciences Research Institute, Shahid Beheshti University
Shahid Shahriari Square, Daneshjou Boulevard, Shahid Chamran Highway
Tehran 19839, Iran

Ziya Kordjazi
Department of Fisheries, Agricultural and Natural Resources Faculty
Gonbad Kavous University
Post Office Box 163, Fallahi Street, Gonbad Kavous, Iran

Roozbeh Valavi
School of BioSciences, The University of Melbourne
Parkville, Victoria 3010, Australia

Hossein Shafizadeh-Moghadam
Department of GIS and Remote Sensing, Tarbiat Modares University
Jalal AleAhmad, Nasr, Tehran 14115, Iran

Jafar Kambouzia
Department of Agroecology, Environmental Sciences Research Institute
Shahid Beheshti University
Shahid Shahriari Square, Daneshjou Boulevard, Shaid Chamran Highway
Tehran 19839, Iran

Dana M. Infante
Department of Fisheries and Wildlife, Michigan State University
1405 South Harrison Road, Suite 318, East Lansing, Michigan 48824, USA

Abstract.—Species distribution models are important tools for conservation and management of aquatic ecosystems. In this study, nine fish species (Caspian Lamprey *Caspiomyzon wagneri*, *Acanthalburnus urmianus*, *Alburnoides namaki*, *Capoeta buhsei*, Mangar *Luciobarbus esocinus* [also known as *Barbus esocinus*], *Luciobarbus xanthopterus*, *Mesopotamichthys sharpeyi*, *Glyptothorax silviae*, and *Iranocichla hormuzensis*) that are sensitive to habitat changes induced by human activities were predicted by species distribution models throughout rivers in Iran. The fish data used cover several time periods (1970–2000) obtained from data-

* Corresponding author: hmostafaviw@gmail.com

bases originating from field sampling, several museums, and the literature. We considered seven environmental variables, including channel slope, bank-full width, wetted width, elevation, mean air temperature, range of air temperature, and annual precipitation to model distributions of all nine species using an ensemble forecasting approach. Models used included generalized linear models, generalized additive models, classification tree analysis, artificial neural networks, surface range envelopes, boosted regression trees, random forest, multivariate adaptive regression splines, and flexible discriminant analysis. Additionally, we compared known distributions of species with modeled distributions, and we used the models to identify potential habitats for the nine species outside previously sampled areas. True skill statistic for each species was, on average, greater than 0.80 (i.e., excellent). Moreover, whereas surface range envelopes for all species had the lowest performance, random forest and generalized boosting methods had the highest performance. Among species studied, Caspian Lamprey, *Luciobarbus xanthopterus*, and *Mesopotamichthys sharpeyi* were predicted only in basins where they had been previously detected. In contrast, other species (i.e., *Acanthalburnus urmianus*, *Alburnoides namaki*, *Capoeta buhsei*, Mangar, *Glyptothorax silviae*, and *Iranocichla hormuzensis*) were predicted not only in basins with previous records, but also in new basins. These results deepen our understanding of distribution patterns of the studied species in Iran and can be used to guide regional conservation planning, identify critical habitats for threatened species, and inform management and conservation of inland aquatic ecosystems. For this to be effective, a mechanistic framework is needed to untie correlations in potential driving factors. Emerging data sets with fine spatial grain and broad spatial extent will support the transition from correlative models to mechanistic understanding of aquatic invasions.

Introduction

Biodiversity in Iran warrants special consideration (Zehzad et al. 2002; Majnonian et al. 2005; Farashi and Shariati Najafabadi 2017). The wide ranges of geographical and geological conditions, as well as the climatologically varied environments, provide a massive diversity for all taxonomic groups, including plants, mammals, insects, and freshwater fishes. Moreover, the country stretches through three global biomes: Palearctic, Ethiopian, and Oriental (Nalbant and Bianco 1998; Coad 2018). In addition, Iran has a unique evolutionary history, has vast variation in its ecosystems, and is considered a center of origin for many species (Coad 2018). Iran is one of the 34 most important biodiversity hotspots in the world (Marchese 2015) and covers 2.3% of Earth's land surface (www.thecropsite.com/focus/5m/52/biodiversity-protecting-the-natural-world). Overall, a total of 203 freshwater fish species belonging to 104 genera, 28 families, and 17 orders have been recorded in the 19 Iranian drainage basins (Esmaeili et al. 2010; Teimori et al. 2012a, 2012b).

As societies develop, rivers are manipulated to meet human needs. In Iran, rivers have been dammed and straightened, nonnative species (Eastern Mosquitofish *Gambusia holbrooki*, Common Carp *Cyprinus carpio*, and Redbelly Tilapia *Tilapia zillii*) have been introduced that compete with native biota, water is extracted to irrigate crops, and river flows are diverted to develop urban, agricultural, and recreational areas. Because of human demands, freshwater ecosystem health is endangered in many parts of Iran (Mostafavi et al. 2014, 2015). Human activities expose freshwater ecosystems to a wide range of stressors, thereby threatening biodiversity and ecosystem processes. For example, human activities have had profound and usually negative impacts on freshwater ecosystems in Iran (Coad 1980; Esmaeili et al. 2007; Mostafavi et al. 2015). The ultimate effects of human activities in river catchments lead to stressors on the biota and biological processes (Karr and Chu 1998; Schmutz et al. 2000, 2007; Pont et al. 2006, 2009). According to previous studies (Esmaeili et al. 2010; Abdoli 2017) and following the International Union for Conservation of Nature (IUCN) criteria (www.iucnredlist.org), seven Iranian fish species are categorized as endangered, and five as vulnerable. Therefore, modeling freshwater fish distributions is particularly important to implement management and conservation strategies (Dauwalter and Rahel 2008), especially for sensitive species that have experienced declines in their original distributions (Mostafavi et al. 2014; Abdoli 2017; Coad 2018).

Species distribution models (SDMs) are increasingly proposed to support conservation decision making (Guisan and Thuiller 2005). In stream fish ecology, SDMs have been used for different purposes, including assessing the impacts of habitat alteration (Logez and Pont 2011), estimating habitat suitability for species reintroductions (Lek et al. 1996), predicting the likelihood of species invasions (Poulos et al. 2012), examining the influence of scale and geography on relationships between fishes and landscape variables (Pont et al. 2005; Filipe et al. 2013), and identifying areas of persistence for threatened or endangered species (Dauwalter and Rahel 2008).

Various statistical methods are used to model species distributions in the field of freshwater ecology (e.g., Lek et al. 1996; Pont et al. 2005; Buisson et al. 2008). All modeling techniques relate observed distributions of species to environmental variables to identify suitability of habitats where species may not have been previously detected (Austin 2007; Elith and Leathwick 2009). Nevertheless, some scholars (Thuiller et al. 2009b) have demonstrated large discrepancies between different model results; thus, choosing an appropriate approach can be difficult. The results of different models not only depend on the relationship between species occurrence and environmental conditions (e.g., linear or nonlinear), but are also dependent on the data set used, including information on species presence and absence (Elith and Graham 2009). Accordingly, summarizing different model types into an ensemble forecasting approach reduces uncertainty of individual techniques (Araújo and New 2007).

Therefore, this study aims to develop a framework for developing predictive distribution models of nine sensitive fish species (Caspian Lamprey *Caspiomyzon wagneri*, *Acanthalburnus urmianus*, *Alburnoides namaki*, *Capoeta buhsei*, Mangar *Luciobarbus esocinus* (also known as *Barbus esocinus*), *Luciobarbus xanthopterus*, *Mesopotamichthys sharpeyi*, *Glyptothorax silviae*, and *Iranocichla hormuzensis*) as a model for further conservation and management studies in Iran. The objectives of this study include identifying potential distributions of those species across Iran through use of multiple models and comparing performance of those models, determining the importance of environmental variables on the distributions of species, and developing recommendations that can aid in the conservation of these unique fishes.

Material and Methods

Study Area

A flowchart of the modeling process by presence–background data is shown in Figure 1. This study was conducted in Iran (Figure

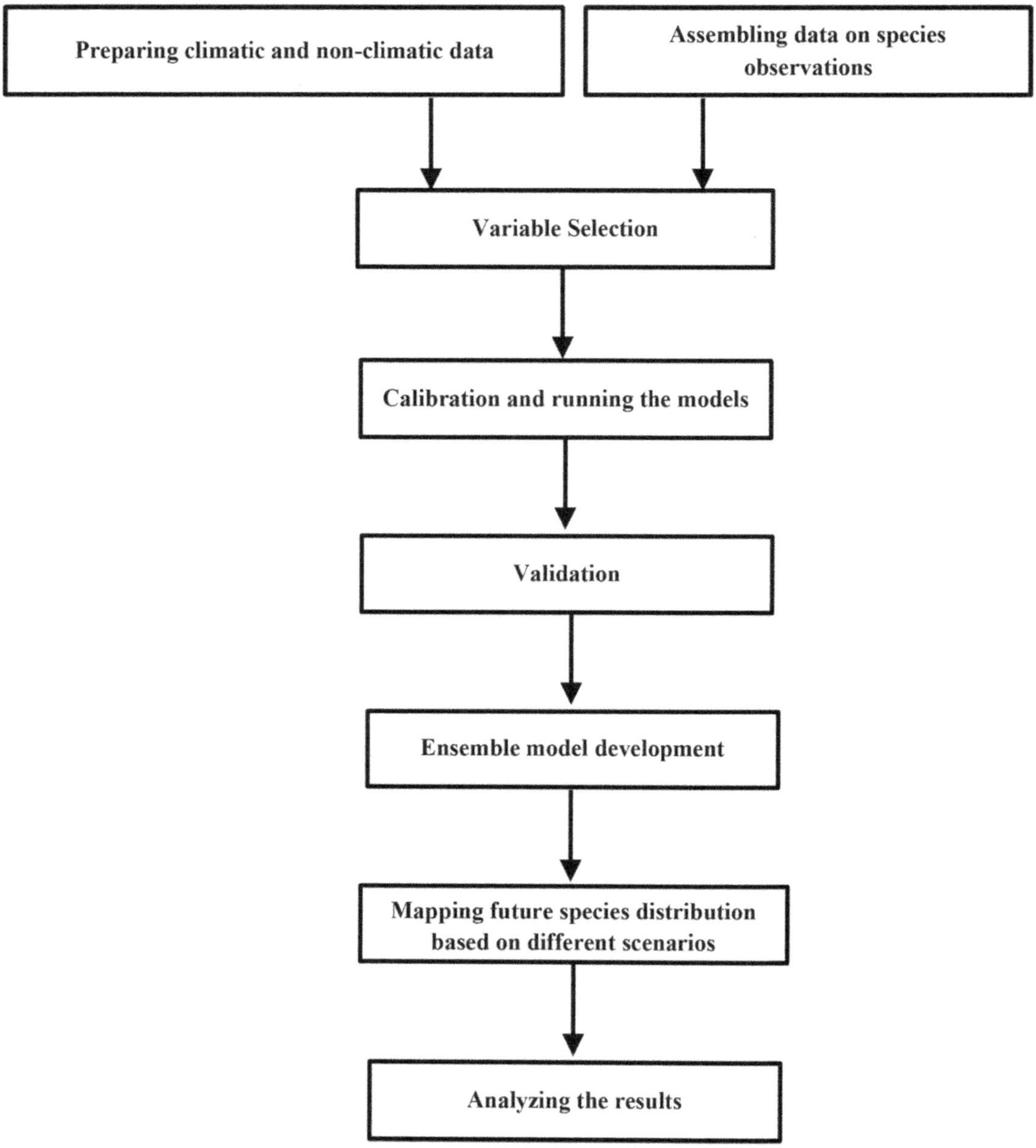

Figure 1. Flowchart of the modeling process by presence–background data.

2), which encompasses 19 hydrologic units (Coad 2018). Iran covers an area of 1,648,195 km^2. It is regarded as the second-largest country in southwest Asia and the 14th-largest country in the world in size (Mostafavi et al. 2015; Coad 2018). More than half of Iran's land consists of mountains, with one-quarter being plains and deserts and less than one-quarter consisting of arable land (Zehzad et al. 2002; Majnonian et al. 2005). The main highlands consist of four distinct mountainous areas: (1) Alborz in the north, which includes the highest elevation in Iran and volcanic Mount Damavand at 5,610 m, located in the center of the Alborz; (2) Kopet-Dagh and the north Khorasan ranges in the north-

Figure 2. Major hydrologic units of Iran (Coad 2018).

east; (3) Zagros in the west and southwest; and (4) Jebal Barez and the Baluchestan mountains, which extend from central areas to the southeast (Zehzad et al. 2002). The majority of Iran occurs at an average altitude of about 1,000 m, and the country is located in the mid-latitude belt of arid and semi-arid regions of the Earth, which cover more than 60% of the country. In this region, precipitation is highly variable in time, space, amount, and duration, and water is the most influential restricting element for biological and agricultural activities. The northern slopes of the Alborz Mountains and the Caspian lowland are the most humid part of this country, which receive 800–2,000 mm annual rainfall, whereas the Dasht-e-Kavir and Dasht-e-Lut deserts are the driest parts of the country with less than 150 mm annual precipitation. The highlands usually receive rainfall from 250 to 800 mm, annually (Zehzad et al. 2002; Majnonian et al. 2005; Akhani et al. 2010; Coad 2018).

Fish Data and Natural Environmental Variables

The nine sensitive fish species were selected according to the conservation status of the species, endemism, and their sensitivity to environmental changes. These species are found in four families, including Petromyzontidae (one species), Cyprinidae (six species), Sisoridae (one species), and Cichlidae (one species) (Table 1). Data used in this study cover several time periods (1970–2000) obtained from databases originating from previous field samplings, several museums, and the literature (Berg 1949; Saadati 1977). Sites where fish occurred are referred to as observed data. Moreover, we excluded sites that had an unclear position in the river network, were sampled outside the temporal period between 1970 and 2000, were stocked, and were located in lakes or wetlands. The primary database contained approximately 1,700 sites that were reduced to 1,090 sites after a detailed quality check concerning the reliability of the biological and spatial information.

After assembling the fish data, we considered four environmental variables at a local extent (elevation [ELE], stream slope [SLO], wetted width [W_WID], and bank-full width [B_WID]), and five regional-extent variables (maximum air temperature [Max_TEM], minimum air temperature [Min_TEM], mean air temperature [A_TEM], the range of air temperature [R_TEM], and annual precipitation [PRE]). These variables have been used in many other studies to understand influences on distributions of fish species (e.g., Oberdorff et al. 2001; Pont et al. 2005; Maloney et al. 2013).

We extracted ELE, W_WID, and B_WID from Google Earth (Version 5; Google Inc.). Rivers in Iran are strongly influenced by climate; therefore, two types of width were recognizable using this approach. W_WID refers to the portion of the stream channel carrying water. B_WID refers to the maximum width that the stream attains, typically marked by changes in vegetation, topography, or texture of sediment. SLO was calculated for a 1-km stretch extending upstream of each site. Climate variables were extracted from WorldClim data (Hijmans et al. 2005, 2007) to characterize annual climate trends based on records for 30 years of monthly means (1970 to 2000) and interpolated at a 30 arc-second grid extent (approximately 1 km at the equator). Climate variables were extracted in a circular buffer (5 km) around each sampling site (Mostafavi et al. 2014), similar to the Catchment Characterization and Modeling database (Vogt et al. 2003, 2007; De Jager and Vogt 2010). Redundancy of environmental variables was tested by Spearman's rank correlation (r). If two vari-

Table 1. List of selected fish species, endemism status, and hydrologic unit collected. International Union for Conservation of Nature (IUCN) status is also included: VU (vulnerable), NT (near threatened), LC (least concern), DD (data deficient), and NE (not evaluated). "+" means endemic and "–" means not endemic

Species	Family	IUCN status	Endemic	Recorded in unit
Caspian Lamprey	Petromyzontidae	NT	–	Caspian Sea
Acanthalburnus urmianus	Cyprinidae	DD	–	Urmia
Alburnoides namaki	Cyprinidae	NE	–	Namak
Capoeta buhsei	Cyprinidae	LC	+	Namak
Mangar	Cyprinidae	VU	–	Gulf
Luciobarbus xanthopterus	Cyprinidae	VU	–	Tigris
Mesopotamichthys sharpeyi	Cyprinidae	VU	–	Tigris
Glyptothorax silviae	Sisoridae	NE	+	Tigris + Gulf
Iranocichla hormuzensis	Cichlidae	NE	+	Hormuz

ables were highly correlated ($r > |0.75|$; Filipe et al. 2013; Mostafavi et al. 2014), one of them was excluded to avoid collinearity, according to our expert judgment and the literature.

Modeling Techniques

This study was based on a heterogeneous data set containing information from several sources. Because of varying sampling methods and investigation targets of compiled original data sets, the absence of species could not be verified in all sites. Accordingly, sites that had no records for species were not considered as actual absences in the modeling procedure. In addition, false absences may decrease the reliability of prediction models (Chefaoui and Lobo 2008); therefore, we used the presence–background modeling approach. This approach involves using a random strategy to create a background data set prior to model calibration. The random selection was repeated 10 times to cover different environmental gradients in the background data set (Thuiller et al. 2009a, 2009b; Barbet-Massin et al. 2012).

In our study, we used the BIOMOD2 package (Thuiller 2003; Thuiller et al. 2009a) in R (R Development Core Team 2018). This tool provided an opportunity to examine methodological uncertainties and maximize predictive performance of the SDMs (Thuiller et al. 2009a, 2009b). We compared the following nine modeling techniques: generalized linear models (GLM), generalized additive models (GAM), classification tree analysis (CTA), artificial neural networks (ANN), surface range envelopes (SRE), generalized boosting method (GBM), random forest (RF), multivariate adaptive regression splines (MARS), and flexible discriminant analysis (FDA). These models come from different families ranging from statistical to machine learning models.

The SRE or bioclimatic envelope model (BIOCLIM) algorithm uses a rectangular envelope that encloses all (or a specified percentile) of the species records in the environmental space (Guisan et al. 2017). It is known as one the simplest modeling approaches that only uses presence data to model species environmental niches. Surface range envelopes assume no interaction between environmental variables and have a simple response shape that is not capable of capturing all the complexity in species–environment relationship (Merow et al. 2014). Based on the number of selected environmental variables, SRE might lead to over- or underfitting (Guisan et al. 2017). A comprehensive comparison study indicated that SRE had low predictive power (Elith et al. 2006).

Classification tree analysis or classification and regression trees use recursive partitioning to capture patterns in the data. In this method, the predictor variable with similar values of the response variable is partitioned recursively, and the response is predicted in each partition (Strobl et al. 2009). Although, CTA has several useful features, including handling missing values, no requirement for data transformation prior to model execution, and incorporating higher order interactions between variables, it has high variance (i.e., a small change in the input data can change the model significantly; Franklin 2010; James et al. 2013). In contrast, RF and GBM algorithms are based on numerous runs of classification and regression trees and they have the capability of CTA but with low variance (i.e., they are more stable with changes in the input data). Both models use ensemble approaches for model construction (RF uses bagging and GBM uses boosting), and they can result in highly accurate predictions (Hastie et al. 2009; Merow et al. 2014; Yates et al. 2018; Valavi et al., in press).

Generalized linear models are parametric linear regressions, meaning that they assume specific distributions (e.g., linear or quadratic) prior to model fitting, and GAM is a nonparametric and nonlinear regression method that is more flexible than GLM (James et al. 2013). Unlike CTA, RF, and GBM, they do not include interaction among the predictors by default. Multivariate adaptive regression splines is a piecewise linear regression that can fit nonlinear models. Unlike GLM and GAM that incorporate global interactions when specified, MARS automatically fits local interaction between the variables (Leathwick et al. 2006). The usefulness of MARS in modeling species distributions has been noted in many studies (e.g., see Leathwick et al. 2005; Elith and Leathwick 2007; Choe et al. 2016).

Artificial neural networks is a machine learning algorithm that mimics the function of biological neurons (Pearson et al. 2002). It can estimate very complex relationship between species and the environment, and it can handle different types of predictor variables with no prior assumptions (Bessa-Gomes and Petrucci-Fonseca 2003). In general, some models such as RF and ANN are quite flexible (i.e., they can have highly nonlinear fitted function), but they might not generalize well (predicting to new areas and\or time), whereas simpler and less flexible models like GLM have lower prediction accuracy but may generalize better (Wenger and Olden 2012). Generalized additive models and MARS results are interpretable, but models such as ANN belong to a black box category in which understanding the structure of the model is not straightforward (Valavi et al., in press).

There can be considerable variation in the spatial prediction of different models, which is one source of uncertainty in the modeling of species distributions; however, combing different models can decrease the uncertainty of modeling prediction (Araújo and New 2007; Grenouillet et al. 2011; Shafizadeh-Moghadam et al. 2018). Ensemble forecasting is a technique of combining the result of different modeling methods (Araújo and New 2007), and it has been successfully applied in predicting species distributions more effectively than single models (Araújo et al. 2005; Breiner et al. 2018; Valavi et al., in press).

Model Calibration, Validation, and Ensemble Forecasting

Model evaluation was based on two criteria: (1) the true skill statistic (TSS), which corresponds to the sum of sensitivity and specificity minus 1 and is independent of prevalence (Lobo et al. 2008). If the TSS index ranges between 0.8 and 1, the model prediction accuracy is classified as excellent or high; if it ranges between 0.6 and 0.8, the accuracy is good; if it ranges between 0.4 and 0.6, the accuracy is considered fair; if it ranges between 0.2 and 0.4, the accuracy is classified as poor, and if it ranges between 0 and 0.2, the accuracy is classified as fail or null (Lobo et al. 2008). Therefore, a TSS index between 0.4 and 1 can be acceptable. Additionally, we applied a (2) cross-validation procedure by randomly splitting the data into calibration (80% of the data) and validation (20%) data sets with 10 repetition runs to assess the stability of model performance (Kuhn and Johnson 2013).

Finally, results of all nine modeling techniques were combined in an ensemble-forecasting framework, as recommended by Araújo and New (2007). We used a mean ensemble approach to combine individual models; this approach was chosen after evaluation of other approaches (e.g., median and confidence intervals). Variable importance was calculated by a permutation procedure used in BIOMOD2, which is independent of the modeling technique (Thuiller et al. 2009a, 2009b). Once the models are trained (i.e., calibrated), a standard prediction is made. Then, one of the variables is randomized and a new prediction is made. The correlation score between that new prediction and the standard prediction is calculated and gives an estimation of the variable importance in the model (Thuiller et al. 2009a, 2009b). After results were generated, we used the software ArcGIS Desktop 10.3 (Esri, Redlands California) to map the predicted distributions of the studied fish species within Iran's 19 major hydrologic units.

Results

Correlation analyses showed that out of nine environmental parameters, seven (B_WID, W_WID, SLO, ELE, A_TEM, R_TEM, and PRE) were independent variables for modeling. Their characteristics are described in Table 2, which emphasizes that our sampling sites cover almost all environmental conditions of Iran. For example, A_TEM varies from 5.5°C to 29.5°C and R_TEM from 6.9°C to 18.5°C. PRE shows extreme variability (53–1,478 mm) because this country receives as much as 1,800 mm/year in the Caspian Sea and as little as 50 mm/year in the eastern deserts (Amiri and Eslamian 2010). ELE ranges between -27 and 2,708 m. SLO shows low to high gradient (0.0–28.0 ‰), and other variables, like B_WID and W_WID, range from small to large rivers (Table 2).

Overall, TSS values ranged from 0.44 to 0.99 for testing validity (Table 3). However, the TSS for each species was excellent, on average (i.e., >0.80). Moreover, whereas SRE for all species had the lowest performance, RF and GBM had the highest performance (Table 3).

The relative importance of the environmental predictors for each species are shown in Table 4. A_TEM, R_TEM, PRE, and ELE were the most important variables for all species, whereas B_ WID, W_WID, and SLO were less important. However, variable importance for all species differed (Table 4). For example, ELE was most important for Caspian Lamprey, R_TEM for Mangar and *Luciobarbus xanthopterus*, A_TEM for *Acanthalburnus urmianus*, *Alburnoides namaki*, and *Capoeta buhsei*, and PRE for *Glyptothorax silviae* and *Iranocichla hormuzensis*.

Among the species studied, Caspian Lamprey, *Luciobarbus xanthopterus*, and *Mesopotamichthys sharpeyi* were predicted only in the recorded hydrologic units (i.e., new sites in the recorded basin) in which they had previously been detected. In contrast, other species (i.e., *Acanthalburnus urmianus*, *Alburnoides namaki*, *Capoeta buhsei*, Mangar, *Glyptothorax silviae*, and *Iranocichla hormuzensis*) were not only predicted in the same recorded hydrologic unit, but also predicted in new units (Table 5; Figures 3–7).

Discussion

According to Guisan et al. (2013), SDMs are one tool for differentiating habitat quality at a rangewide scale, and based on our results, we define critical habitats as those necessary for the persistence, or long-term recovery, of sensitive or threatened fish species. Moreover, SDMs can potentially inform the translocation decision process. Indeed, if translocation is deemed necessary, SDMs can identify potential recipient sites, which may be climate refuges within the current range or sites that are projected to become newly suitable. Additionally, SDMs can be used to identify which local species may be at risk of impact from the

Table 2. Mean and range of environmental variables (bank-full width [B_WID], wetted width [W_WID], stream slope [SLO], elevation [ELE], mean air temperature [A_TEM], range of air temperature [R_TEM], and annual precipitation [PRE]).

Number of sites		B_WID (m)	W_WID (m)	SLO (‰)	ELE (m)	A_TEM (°C)	R_TEM (°C)	PRE (mm)
1,090	Mean:	112.5	35.5	1.6	731	19.1	13.3	384.8
	Range:	1.0–3,539.8	1.0–608.6	0.0–28.0	-27–2708	5.5–29.5	6.9–18.5	53–1,478

Table 3. Evaluation of model performance according to true skill statistic validation index. Modeling techniques: surface range envelopes (SRE), classification tree analysis (CTA), random forest (RF), multivariate adaptive regression splines (MARS), flexible discriminant analysis (FDA), generalized linear models (GLM), generalized additive models (GAM), generalized boosting method (GBM), and artificial neural networks (ANN).

Species	SRE	CTA	RF	MARS	FDA	GLM	GAM	GBM	ANN	Modeling performance (average)
Caspian Lamprey	0.66	0.89	0.98	0.94	0.86	0.94	0.99	0.96	0.96	0.91(Excellent)
Acanthalburnus urmianus	0.71	0.98	0.98	0.98	0.97	0.97	0.98	0.98	0.99	0.95(Excellent)
Alburnoides namaki	0.74	0.75	0.97	0.96	0.84	0.94	0.96	0.97	0.71	0.87(Excellent)
Capoeta buhsei	0.72	0.85	0.93	0.92	0.78	0.92	0.96	0.96	0.85	0.88(Excellent)
Mangar	0.75	0.87	0.99	0.91	0.85	0.88	0.93	0.95	0.9	0.89(Excellent)
Luciobarbus xanthopterus	0.44	0.95	0.99	0.98	0.86	0.94	0.99	0.98	0.97	0.90(Excellent)
Mesopotamichthys sharpeyi	0.64	0.93	0.97	0.95	0.93	0.95	0.95	0.98	0.90	0.91(Excellent)
Glyptothorax silviae	0.65	0.86	0.96	0.75	0.65	0.72	0.98	0.98	0.76	0.81(Excellent)
Iranocichla hormuzensis	0.80	0.95	0.97	0.95	0.94	0.99	0.91	0.97	0.96	0.94(Excellent)
Average	0.68	0.89	0.97	0.93	0.85	0.92	0.96	0.97	0.89	0.90(Excellent)

Table 4. The order of relative importance of environmental variables for each species (bold indicates the three most important variables, respectively). Abbreviations: bank-full width (B_WID), wetted width (W_WID), stream slope (SLO), mean air temperature (A_TEM), annual precipitation (PRE), elevation (ELE), and range of air temperature (R_TEM).

Species	B_WID (m)	W_WID (m)	SLO (‰)	ELE (m)	R_TEM (°C)	A_TEM (°C)	PRE (mm)
Caspian Lamprey	6	5	7	**1**	**3**	4	**2**
Acanthalburnus urmianus	5	7	6	4	**3**	**1**	**2**
Alburnoides namaki	7	5	6	4	**2**	**1**	**3**
Capoeta buhsei	6	5	7	**3**	**2**	**1**	4
Mangar	5	6	7	**3**	**1**	**2**	4
Luciobarbus xanthopterus	6	5	7	**2**	**1**	**3**	4
Mesopotamichthys sharpeyi	7	5	6	**2**	**3**	**1**	4
Glyptothorax silviae	7	5	6	4	**3**	**2**	**1**
Iranocichla hormuzensis	6	5	7	4	**2**	**3**	**1**
Average	6.11	5.33	6.56	3.00	2.22	2.00	2.78

Table 5. Prediction of studied species in the same and new hydrologic unit.

Species	Recorded in unit	Prediction in new unit
Caspian Lamprey	Caspian Sea	_
Acanthalburnus urmianus	Urmia	Upper Tigris
Alburnoides namaki	Namak	Caspian Sea, Urmia, Tigris, Kavir, Hari
Capoeta buhsei	Namak	Caspian Sea, Urmia, Tigris, Kavir, Hari
Mangar	Tigris	Gulf
Luciobarbus xanthopterus	Tigris	_
Mesopotamichthys sharpeyi	Tigris + Gulf	_
Glyptothorax silviae	Tigris + Gulf	Lake Maharlu
Iranocichla hormuzensis	Hormuz	Gulf, Makran + Jaz Murian

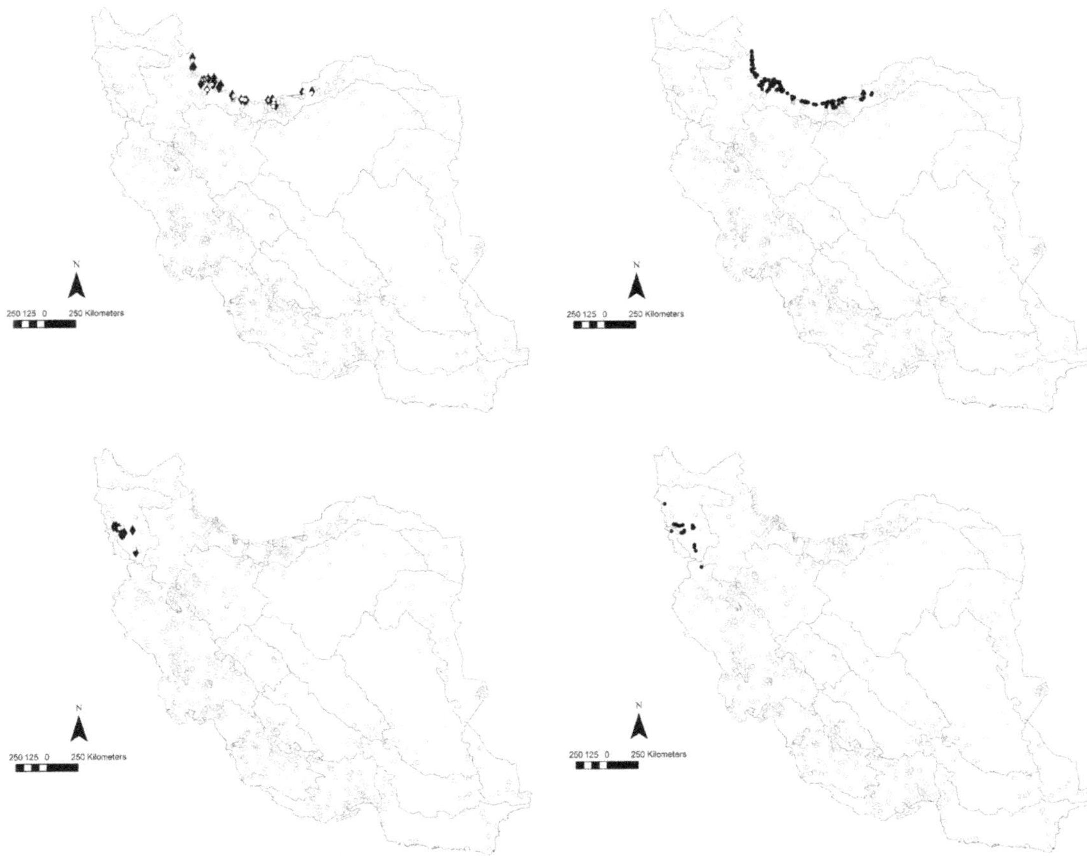

Figure 3. Prediction of Caspian Lamprey (top) and *Acanthalburnus urmianus* (bottom) in the 19 hydrologic units of Iran. Left: observed data, right: prediction (black diamond: presence data/sites, white circle: background data/sites, black circle: predicted sites).

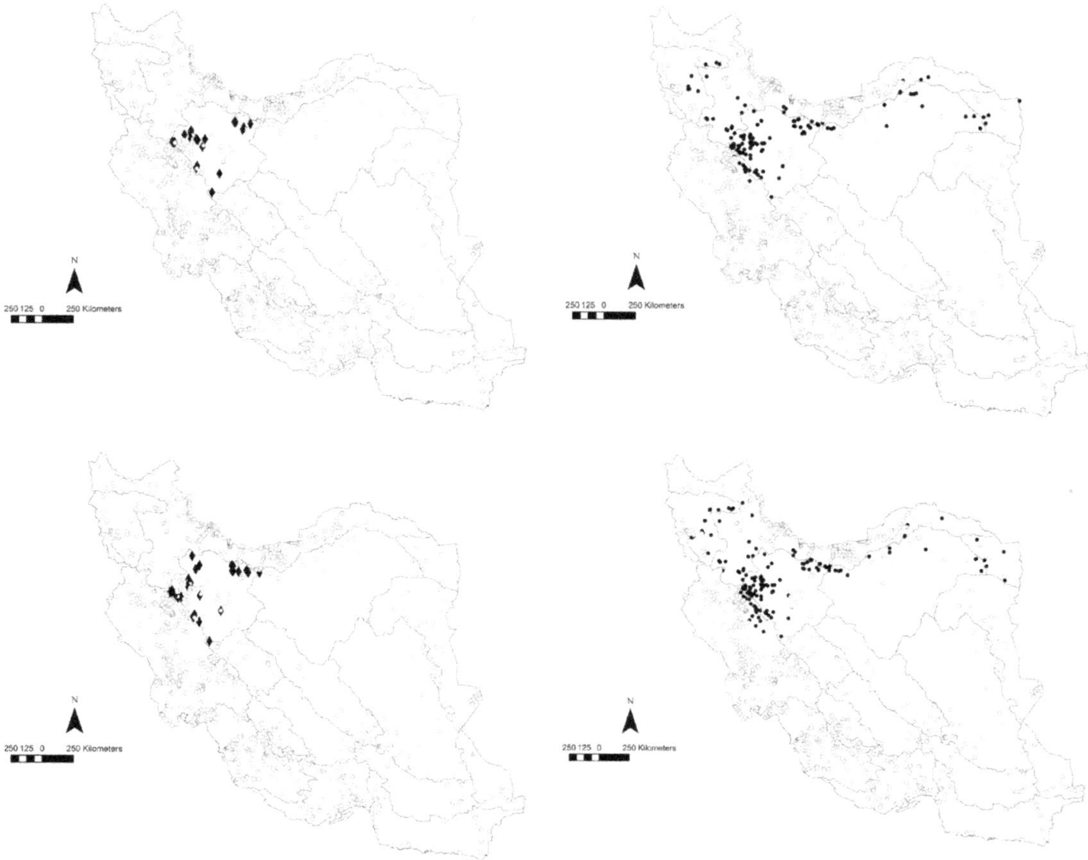

Figure 4. Prediction of *Alburnoides namaki* (top) and *Capoeta buhsei* (bottom) on the 19 hydrologic units of Iran. left: observed data, right: prediction, (black diamond: presence data/sites, white circle: background data/sites, black circle: predicted sites).

introduction of a translocated species through predicted overlapping distributions in the same way as they are used to identify conflict areas between native and invasive species. In addition, SDMs are able to identify sites where species may have been absent because of habitat degradation. Consequently, based on more detailed future studies, effective conservation and restoration measures can be undertaken to maintain and (re)establish species populations. This can be useful for conservation decisions because the habitats of some species, including Caspian Lamprey, Mangar, *Luciobarbus xanthopterus*, and *Mesopotamichthys sharpeyi*, were already degraded by human pressures (Mostafavi et al. 2015).

Model Performance

The results of this study suggested that species distribution modeling can successfully determine the distribution of freshwater fish species in hydrologic units throughout Iran. In addition, the performance of the models used in this study was acceptable according to TSS criteria. Random forest and GBM performed the best while SRE performed the worst for the studied species. However, it is essential to note that for some species, individual models performed better, but using the ensemble modeling approach ensured that the biases and inconsistencies introduced by individual models are directly addressed in the results developed for individual species.

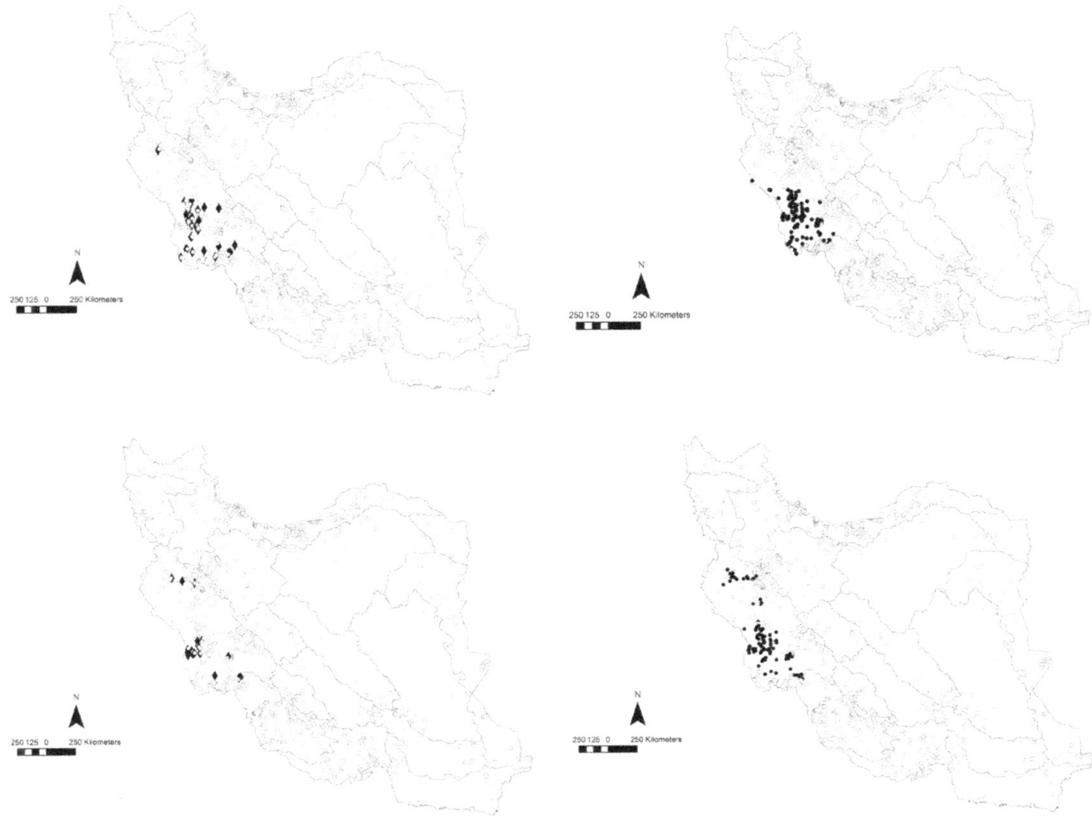

Figure 5. Prediction of Mangar (top) and *Luciobarbus xanthopterus* (bottom) on the 19 hydrologic units of Iran. left: observed data, right: prediction, (black diamond: presence data/sites, white circle: background data/sites, black circle: predicted sites).

In other words, ensemble models make a better prediction and they are more reliable than a single model (Araújo et al. 2005; Marmion et al. 2009).

Evaluation of Selected Variables

A critical component of species distribution modeling is to select and use appropriate environmental variables. These variables can be used to describe geographical and biological conditions of the species and predict correct distribution ranges of the species. It is not possible to identify all variables and factors affecting species niche characteristics in modeling. This might result from deficiencies of ecological knowledge about studied species, a lack of knowledge for recognizing all dimensions of the species niche, and/or scarcity of detailed data. Furthermore, the contrasting sensitivity of the nine selected species to the seven environmental variables leads us to conclude that species of site assemblages have common responses to the hierarchical nature of their environment, which is similar to findings of Pont et al. (2005). Our study confirms that the sensitivity to local- and regional-scale processes is species-specific, and as proposed by Kolasa (1989), all species do not require the same habitat. In this study, we considered both local and regional environmental variables that directly influence the occurrence of fish species because species occurrences are not only constrained by local processes (e.g., slope, river width, and elevation), but

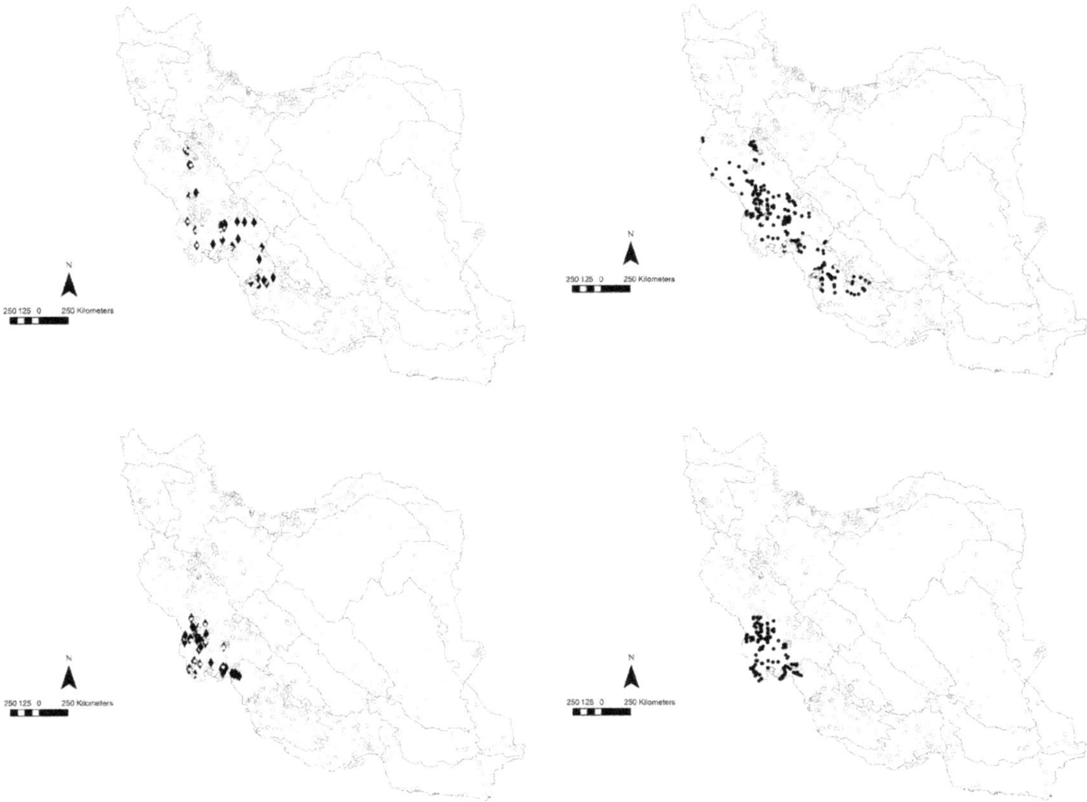

Figure 6. Prediction of *Mesopotamichthys sharpeyi* (top) and *Glyptothorax silviae* (bottom) on the 19 hydrologic units of Iran. left: observed data, right: prediction, (black diamond: presence data/sites, white circle: background data/sites, black circle: predicted sites).

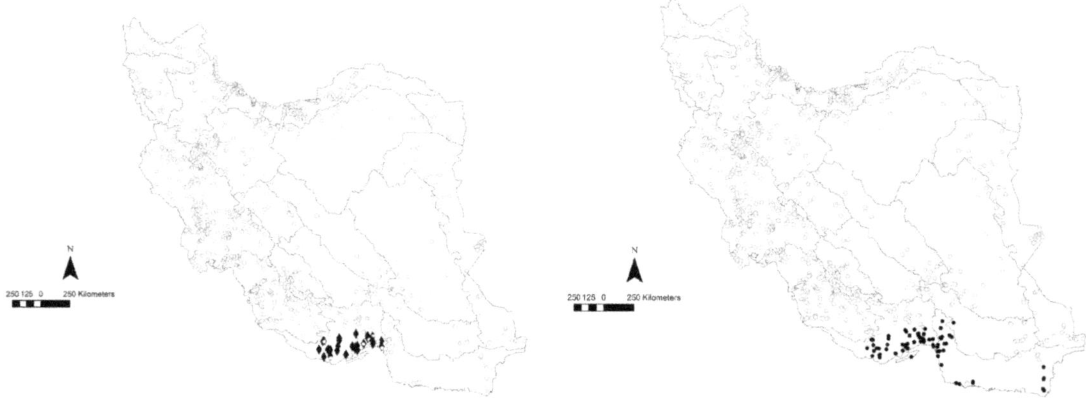

Figure 7. Prediction of *Iranocichla hormuzensis* on the 19 hydrologic units of Iran. left: observed data, right: prediction, (black diamond: presence data/sites, white circle: background data/sites, black circle: predicted sites).

also by factors operating at larger extents, like climate (Tonn et al. 1990; Huston 1999; Pont et al. 2005). The reason is that large-extent factors control the pool of species available locally, whereas small-extent variables limit the number of species that actually occur at a given site.

Temperature appears to be one of the main factors determining the spatial distribution for stream fishes (Buisson et al. 2008; Logez et al. 2012; Herlihy et al. 2019, this volume). Freshwater fishes are ectothermic animals and particularly sensitive to temperature. Temperature fluctuations affect the metabolism, breeding, development, and growth of fishes (Mann 1996). Moreover, because streams and rivers are relatively shallow, turbulent, and well-mixed systems, they easily exchange heat with the atmosphere. Consequently, annual (mean and range) air temperatures were used as a surrogate of water temperature, as was followed by Pont et al. (2005). In our study, mean air temperature was an important variable affecting fish species distributions, and this fact has also been widely reported by other studies (e.g., Pont et al. 2005; Buisson et al. 2008; Abdoli and Naderi 2009; Mostafavi et al. 2014, 2015).

Slope and river width are good descriptors of physical and hydrological conditions at the sites (Oberdorff et al. 2001; Pont et al. 2005) and regionally (Herlihy et al. 2019). Slope is one of the main factors controlling water velocity (Gordon et al. 1992). Importance of water velocity varies with ecology of species, and in this study, it was less important for the species studied, yet it was found to be important for some species in other studies (Filipe et al. 2013; Mostafavi et al. 2014). River width is indicative of both river size and habitat diversity (Matthews 1998) but in our study was not very important. Elevation provides a measure of thermal regime; in our study, there was no correlation between elevation and temperature across the extent of the country because Iran contains different climatic regions. Elevation was important to the prediction of our studied species, which is similar to other studies (Lyons 1989; Álvarez et al. 2017); with increasing elevation, species richness decreases (Lyons and Schneider 1990). Additionally, elevation is negatively correlated with river width (Angermeier and Winston 1998; Álvarez et al. 2017). One could develop independent rationales for expecting relationships between local fish diversity and elevation or between diversity and river width (Angermeier and Winston 1998); however, we found no significant correlation between river width and elevation in our data.

Precipitation was important for all species in our study, which is in line with other studies, such as Logez et al. (2012) and Mostafavi et al. (2014). In fact, precipitation along with slope and elevation influence the dynamics of currents in river systems and create habitat variability along river longitudinal gradients (Bussing 2002; Winemiller et al. 2008).

Comparing Modeling Results with Available Records

Comparing results of the studied species distribution modeling with those of existing reports shows that whereas model predictions and available reports on species distributions agree for most species, some species (i.e., *Acanthalburnus urmianus*, *Alburnoides namaki*, *Capoeta buhsei*, Mangar, *Glyptothorax* silviae, and *Iranocichla hormuzensis*) have a larger predicted range than is currently suggested in the literature (e.g., Abdoli and Naderi 2009; Mostafavi et al. 2014; Abdoli 2017). Two reasons may explain this. First, sampling methods used by researchers are insufficient because almost every year, new species are discovered in remote and mountainous regions of Iran (Teimori et al. 2012a, 2012b; Abdoli 2017) In addition, all species studied in this research were also predicted in the same recorded hydrologic units (i.e., new sites in the recorded unit) in which they were previously detected. The absence of some species in the new predictions for the same hydrologic unit are likely related to different sampling and monitoring methods conducted in diverse habitats and rivers in Iran. Second, these differences can also be attributed to limiting factors that influence distributions of freshwater fish species. Whereas we can consider the distribution of a species based on the relationship between environmental conditions and the species' inherent characteristics in terms of its potential distribution, various natural and artificial factors have constrained the realized distribution of species (Sexton et al. 2009). Because the stream network is typically the only corridor for distribution of fish species, the main limiting natural factor in freshwater ecosystems is the catchment border, by which movement of most species between catchments would be restricted (Schmutz et al. 2000; Pont et al. 2005). Whereas the habitat suitability for a species may extend beyond its main catchment area, distribution of species within the catchment area may be limited by various barriers, especially artificial constructions like dams or natural waterfalls, resulting in limited distribution of a species within a catchment (Davis et al. 2019; Martin et al. 2019; both this volume). Species distribution can also be affected by biological interactions, including competition and predation (Araújo and New 2007). Time is also important for species distributions (Pearson 2007), meaning that despite the availability of suitable environmental conditions and movement corridors, many fish species require time to distribute in their potential distribution areas.

Human Pressure Effects on Fish Species Distribution

Human activities over recent decades have had huge impacts on species occurrences in Iran (Mostafavi et al. 2015). All species studied are very sensitive to environmental changes associated with human activities, some of them have been categorized in the IUCN red list. Therefore, the results of this study can have important implications for conservation activities and management. The modeling framework has the ability to highlight the potential areas of species occurrence and to identify sites where species may have been absent because of habitat degradation. Consequently, based on more detailed future studies, effective conservation and restoration measures can be undertaken to maintain and (re)establish species populations.

Conclusions

Species distribution models are important conservation and management tools because the models can reliably predict areas suitable for species occupation and the species' responses to particular environmental variables can suggest management alternatives. In fact, the definition of feasible actions may be informed by SDMs. When making decisions about where to translocate a sensitive/threatened species or where to target control of an invasive species, SDMs may be used to identify candidate locations as alternative actions that may subsequently be evaluated in greater detail. Information about the costs of management actions, logistical constraints (e.g., distance), or conflicting conservation priorities (e.g., various land ownerships), for example, will ultimately determine the feasibility of different actions, but SDM provides a suite of options (Guisan et al. 2013). Moreover, critical habitats are typically defined as habitats necessary for the persistence, or long-term recovery, of threatened species. In addition, many threatened species have restricted distributions and specific habitat requirements, so decisions to protect critical habitat may need to be made with some urgency to avoid extinction (Martin and Maron 2012). This urgency often leads to protection of minimum amounts of habitat based on occurrence data alone. In such cases, allocating time to collect more data and build accurate SDMs or more complex spatially explicit population models may not necessarily improve predictions—but may delay protective actions. However, deciding to build a simple SDM, or to not build one at all, may overlook some potentially critical habitats for the species (Guisan et al. 2013). Species distribution models are also useful for delineating and establishing protected areas (reserve selection; Margules and Pressey 2000). Finally, the modeling could also provide opportunities for further use (e.g., integration into multimetric indices).

Acknowledgments

Our special thanks are given to Robert Hughes and two anonymous reviewers for their many helpful comments and assistance with the English editing. Roozbeh Valavi is supported by an Australian Government Training Program Scholarship and a Rowden White Scholarship.

References

Abdoli, A. 2017. The field guide of inland water fishes of Iran. Iran-shenasi, Tehran.

Abdoli, A., and M. Naderi. 2009. Biodiversity of fishes of the southern basin of the Caspian Sea. Abzian Scientific Publications, Tehran, Iran.

Akhani, H., M. Djamali, A. Ghorbanalizadeh, and E. Ramezani. 2010. Plant biodiversity of Hyrcanian relict forests, N Iran: an overview of the flora, vegetation, palaeoecology and conservation. Pakistan Journal of Botany 42:231–258.

Álvarez, F. S., W. A. Matamoros, and F. A. Chicas. 2017. The contribution of environmental factors to fish assemblages in the Río Acahuapa, a small drainage in Central America. Neotropical Ichthyology [online serial] 15(3):e170023.

Amiri, M., and S. Eslamian. 2010. Investigation of climate change in Iran. Journal of Environmental Science and Technology 3:208–216.

Angermeier, P., and M. Winston. 1998. Local vs. regional influences on local diversity in stream fish communities of Virginia. Ecology 79:911–927.

Araújo, M. B., and M. New. 2007. Ensemble forecasting of species distributions. Trends in Ecology and Evolution 22:42–47.

Araújo, M. B., R. J. Whittaker, R. J. Ladle, and M. Erhard. 2005. Reducing uncertainty in projections of extinction risk from climate change. Global Ecology and Biogeography 14:529–538.

Austin, M. 2007. Species distribution models and ecological theory: a critical assessment and some possible new approaches. Ecological Modelling 200:1–19.

Barbet-Massin, M., F. Jiguet, C. H. Albert, and W. Thuiller. 2012. Selecting pseudo-absences for species distribution models: how, where and how many? Methods in Ecology and Evolution 3:327–338.

Berg, L. 1949. Presnovodnye ryby Irana i sopredel'nykh stran. [Freshwater fishes of Iran and adjacent countries.] Trudy Zoologicheskogo Instituta Akademii Nauk SSSR 8:783–858.

Bessa-Gomes, C., and F. Petrucci-Fonseca. 2003. Using artificial neural networks to assess wolf distribution patterns in Portugal. Animal Conservation 6:221–229.

Breiner, F. T., M. P. Nobis, A. Bergamini, and A. Guisan. 2018. Optimizing ensembles of small models for predicting the distribution of species with few occurrences. Methods in Ecology and Evolution 9:802–808.

Buisson, L., L. Blanc, and G. Grenouillet. 2008. Modelling stream fish species distribution in a river network: the relative effects of temperature versus physical factors. Ecology of Freshwater Fish 17:244–257.

Bussing, W.A. 2002. Peces de las aguas continentales de Costa Rica. (Freshwater fishes of Costa Rica.) Editorial de la Universidad de Costa Rica, San Jose.

Chefaoui, R. M., and J. M. Lobo. 2008. Assessing the effects of pseudo-absences on predictive distribution model performance. Ecological Modelling 210:478–486.

Choe, H., J. H. Thorne, and C. Seo. 2016. Mapping national plant biodiversity patterns in South Korea with the MARS species distribution model. PLOS (Public Library of Science) ONE [online serial] 11(3):e0149511.

Coad, B. W. 1980. Environmental change and its impact on the freshwater fishes of Iran. Biological Conservation 19:51–80.

Coad, B. W. 2018. Freshwater fishes of Iran. Available: www.briancoad.com/Species%20Accounts/Contents%20new.htm. (April 2019).

Dauwalter, D. C., and F. J. Rahel. 2008. Distribution modelling to guide stream fish conservation: an example using the mountain sucker in the Black Hills National Forest, USA. Aquatic Conservation: Marine and Freshwater Ecosystems 18:1263–1276.

Davis, C. D., C. W. Epps, R. L. Flitcroft, and M. A. Banks. 2019. Beyond isolation by distance: riverscape effects on genetic structure of fall-run Chinook Salmon. Pages 35–56 *in* R. M. Hughes, D. M. Infante, L. Wang, K. Chen, and B. F. Terra, editors. Advances in understanding landscape influences on freshwater habitats and biological assemblages. American Fisheries Society, Symposium 90, Bethesda, Maryland.

De Jager, A., and J. Vogt. 2010. Development and demonstration of a structured hydrological feature coding system for Europe. Hydrological Sciences Journal 55:661–675.

Elith, J., and C. H. Graham. 2009. Do they? How do they? Why do they? On finding reasons for differing performances of species distribution models. Ecography 32:66–77.

Elith, J., C. H. Graham, R. P. Anderson, M. Dudík, S. Ferrier, A. Guisan, R. J. Hijmans, F. Huettmann, J. R. Leathwich, A. Lehmann, J. Li, L. G. Lohmann, B. A. Loiselle, G. Manion, C. Moritz, M. Nakamura, Y. Nakazawa, J. M. M. Overton, A. T. Peterson, S. J. Phillips, K. Richardson, R. Scachetti-Pereira, R. E. Schapire, J. Soberón, S. Williams, M. S. Wisz, and N. E. Zimmermann. 2006. Novel methods improve prediction of species' distributions from occurrence data. Ecography 29:129–151.

Elith, J., and J. Leathwick. 2007. Predicting species distributions from museum and herbarium records using multiresponse models fitted with multivariate adaptive regression splines. Diversity and Distributions 13:265–275.

Elith, J., and J. R. Leathwick. 2009. Species distribution models: ecological explanation and prediction across space and time. Annual Review of Ecology, Evolution, and Systematics 40:677–697.

Esmaeili, H. R., A. Teimori, and A. R. Khosravi. 2007. A note on the biodiversity of Ghadamgah spring–stream system in Fars province, southwest Iran. Iranian Journal of Animal Biosystematics 3:15–23.

Esmaeili, H., B. Coad, A. Gholamifard, N. Nazari, and A. Teimory. 2010. Annotated checklist of the freshwater fishes of Iran. Zoosystematica Rossica 19:361–386.

Farashi, A., and M. Shariati Najafabadi. 2017. Biodiversity hotspots and conservation gaps in Iran. Journal for Nature Conservation 39:37–57.

Filipe, A. F., D. Markovic, F. Pletterbauer, C. Tisseuil, A. De Wever, S. Schmutz, N. Bonada, and J. Freyhof. 2013. Forecasting fish distribution along stream networks: Brown Trout (*Salmo trutta*) in Europe. Diversity and Distributions 19:1059–1071.

Franklin, J. 2010. Mapping species distributions: spatial inference and prediction. Cambridge University Press, Cambridge, UK.

Grenouillet, G., L. Buisson, N. Casajus, and S. Lek. 2011. Ensemble modelling of species distribution: the effects of geographical and environmental ranges. Ecography 34:9–17.

Guisan, A., and W. Thuiller. 2005. Predicting species distribution: offering more than simple habitat models. Ecology Letters 8:993–1009.

Guisan, A., W. Thuiller, and N. E. Zimmermann.

2017. Habitat suitability and distribution models: with applications in R. Cambridge University Press, Cambridge, UK.

Guisan, A., R. Tingley, J. B. Baumgartner, I. Naujokaitis-Lewis, and P. R. Sutcliffe. 2013. Predicting species distributions for conservation decisions. Ecology Letters 16:1424–1435.

Gordon, N. D., T. A. McMahon, and B. L. Finlayson. 1992. Stream hydrology: an introduction for ecologists. Wiley, Chichester, England.

Hastie, T., R. Tibshirani, and J. Friedman. 2009. The elements of statistical learning: data mining, inference, and prediction, 2nd edition. Springer. New York.

Herlihy, A. T., J. C. Sifneos, R. M. Hughes, D. V. Peck, and R. M. Mitchell. 2019. Lotic fish assemblage clusters across the conterminous United States and their associations with environmental variables. Pages 385–408 *in* R. M. Hughes, D. M. Infante, L. Wang, K. Chen, and B. F. Terra, editors. Advances in understanding landscape influences on freshwater habitats and biological assemblages. American Fisheries Society, Symposium 90, Bethesda, Maryland

Hijmans, R., S. Cameron, and J. Parra. 2007. WorldClim version 1.4. Museum of Vertebrate Zoology, University of California, Berkeley. Available: www.worldclim.org. (April 2019).

Hijmans, R. J., S. E. Cameron, J. L. Parra, P. G. Jones, and A. Jarvis. 2005. Very high resolution interpolated climate surfaces for global land areas. International Journal of climatology 25:1965–1978.

Huston, M. A. 1999. Local processes and regional patterns: appropriate scales for understanding variation in the diversity of plants and animals. Oikos 86:393–401.

James, G., D. Witten, T. Hastie, and R. Tibshirani. 2013. An introduction to statistical learning. Springer, New York.

Karr, J. R., and E. W. Chu. 1998. Restoring life in running waters: better biological monitoring. Island Press, Washington D.C.

Kolasa, J. 1989. Ecological systems in hierarchical perspective: breaks in community structure and others consequences. Ecology 70:36–47.

Kuhn, M., and K. Johnson. 2013. Applied predictive modelling. Springer, New York.

Leathwick, J. R., J. Elith, and T. Hastie. 2006. Comparative performance of generalized additive models and multivariate adaptive regression splines for statistical modelling of species distributions. Ecological Modelling 2:188–196.

Leathwick, J. R., D. Rowe, J. Richardson, J. Elith, and T. Hastie. 2005. Using multivariate adaptive regression splines to predict the distributions of New Zealand's freshwater diadromous fish. Freshwater Biology 12:2034–2052.

Lek, S., A. Belaud, P. Baran, I. Dimopoulos, and M. Delacoste. 1996. Role of some environmental variables in trout abundance models using neural networks. Aquatic Living Resources 9:23–29.

Lobo, J., A. Jiménez-valverde, and R. Real. 2008. AUC: a misleading measure of the performance of predictive distribution models. Global Ecology and Biogeography 17:145–151.

Logez, M., P. Bady, and D. Pont. 2012. Modelling the habitat requirement of riverine fish species at the European scale: sensitivity to temperature and precipitation and associated uncertainty. Ecology of Freshwater Fish 21:266–282.

Logez, M., and D. Pont. 2011. Development of metrics based on fish body size and species traits to assess European coldwater streams. Ecological Indicators 11:1204–1215.

Lyons, J. 1989. Correspondence between the distribution of fish assemblages in Wisconsin streams and Omernik's ecoregions. American Midland Naturalist 122:163–182.

Lyons, J., and Schneider, D. W. 1990. Factors influencing fish distribution and community structure in a small coastal river in southwestern Costa Rica. Hydrobiologia 203:1–14.

Majnonian, H., B. H. Kiabi, and M. Danesh. 2005. Reading in zoogeography of Iran (Part 1). Sazman hefazat mohit zist Iran (Department of Environment of Iran), Dayereh Sabz, Tehran.

Maloney, K. O., D. E. Weller, D. E. Michaelson, and P. J. Ciccotto. 2013. Species distribution models of freshwater stream fishes in Maryland and their implications for management. Environmental Modelling and Assessment 18:1–12.

Mann, R. H. K. 1996. Environmental requirements of European non-salmonid fish in rivers. Hydrobiologia 323:223–235.

Marchese, C. 2015. Biodiversity hotspots: a shortcut for a more complicated concept. Global Ecology and Conservation 3:297–309.

Margules, C. R., and R. L. Pressey. 2000. Systematic conservation planning. Nature (London) 405:243–253.

Marmion, M., M. Parviainen, M. Luoto, R. K. Heikkinen, and W. Thuiller.2009. Evaluation of consensus methods in predictive species distribution modelling. Diversity and Distributions 15:59–69.

Martin, B., Y.-P. Tsang, R. W. Tingley, III, H. Clilverd, and D. M. Infante. 2019. Characterizing natural barriers to nonnative stream fauna in Hawai'i. Pages 83–104 *in* R. M. Hughes, D. M. Infante, L. Wang, K. Chen, and B. F. Terra, editors. Advances in understanding landscape influences on freshwater habitats and biological assemblages. American Fisheries Society, Symposium 90, Bethesda, Maryland

Martin, T. E., and J. L. Maron. 2012. Climate impacts on bird and plant communities from altered animal–plant interactions. Nature Climate Change 2:195–200.

Matthews, W. J. 1998. Patterns in freshwater fish ecology. Chapman and Hall, New York.

Merow, C., M. J. Smith, T. C. Edwards, A. Guisan, S. M. McMahon, S. Normand, W. Thuiller, R. O. Wüest, N. E. Zimmermann, and J. Elith. 2014. What do we gain from simplicity versus complexity in species distribution models? Ecography 37:1267–1281.

Mostafavi, H., F. Pletterbauer, B. W. Coad, A. S. Mahini, R. Schinegger, G. Unfer, C. Trautwein, and S. Schmutz. 2014. Predicting presence and absence of trout (*Salmo trutta*) in Iran. Limnologica: Ecology and Management of Inland Waters 46:1–8.

Mostafavi, H., R. Schinegger, A. Melcher, K. Moder, C. Mielach, and S. Schmutz. 2015. A new fish-based multi-metric assessment index for cyprinid streams in the Iranian Caspian Sea basin. Limnologica: Ecology and Management of Inland Waters 51:37–52.

Nalbant, T. T., and P. G. Bianco. 1998. The loaches of Iran and adjacent regions with description of six new species (Cobitoidea). Italian Journal of Zoology 65:109–123.

Oberdorff, T., D. Pont, B. Hugueny, and D. Chessel. 2001. A probabilistic model characterizing fish assemblages of French rivers: a framework for environmental assessment. Freshwater Biology 46:399–415.

Pearson, R. G. 2007. Species' distribution modelling for conservation educators and practitioners. Lessons in Conservation 3:54-89.

Pearson, R. G., T. P. Dawson, P. M. Berry, and P. A. Harrison. 2002. SPECIES: a spatial evaluation of climate impact on the envelope of species. Ecological Modelling 3:289–300.

Pont, D., R. M. Hughes, T. R. Whittier, and S. Schmutz. 2009. A predictive index of biotic integrity model for aquatic-vertebrate assemblages of western U.S. streams. Transactions of the American Fisheries Society 138:292–305.

Pont, D., B. Hugueny, U. Beier, D. Goffaux, A. Melcher, R. Noble, C. Rogers, N. Roset, and S. Schmutz. 2006. Assessing river biotic condition at a continental scale: a European approach using functional metrics and fish assemblages. Journal of Applied Ecology 43:70–80.

Pont, D., B. Hugueny, and T. Oberdorff. 2005. Modelling habitat requirement of European fishes: do species have similar responses to local and regional environmental constraints? Canadian Journal of Fisheries and Aquatic Sciences 62:163–173.

Poulos, H. M., B. Chernoff, P. L. Fuller, and D. Butman. 2012. Ensemble forecasting of potential habitat for three invasive fishes. Aquatic Invasions 7:59–72.

R Core Team. 2018. R: a language and environment for statistical computing. R Foundation for Statistical Computing, Vienna, Austria. Available: www.R-project.org/. (April 2019).

Saadati, M. A. G. 1977. Taxonomy and distribution of the freshwater fishes of Iran. Master's thesis. Colorado State University, Fort Collins.

Shafizadeh-Moghadam, H., R. Valavi, H. Shahabi, K. Chapi, and A. Shirzadi. 2018. Novel forecasting approaches using combination of machine learning and statistical models for flood susceptibility mapping. Journal of Environmental Management 217:1–11.

Schmutz, S., I. Cowx, G. Haidvogl, and D. Pont. 2007. Fish-based methods for assessing European running waters: a synthesis. Fisheries Management and Ecology 14:369–380.

Schmutz, S., M. Kaufmann, B. Vogel, M. Jungwirth, and S. Muhar. 2000. A multi-level concept for fish-based, river-type-specific assessment of ecological integrity. Hydrobiologia 422–423:279–289.

Sexton, J. P., P. J. McIntyre, A. L. Angert, and K. J. Rice. 2009. Evolution and ecology of species range limits. Annual Review of Ecology, Evolution, and Systematics 40:415–436.

Strobl, C., J. Malley, and G. Tutz.2009. An introduction to recursive partitioning: rationale, application, and characteristics of classification and regression trees, bagging, and random forests. Psychological Methods 4:323–348.

Teimori, A., H. R. Esmaeili, Z. Gholami, N. Zarei, and B. Reichenbacher. 2012a. Aphanius arakensis, a new species of tooth-carp (Actinopterygii, Cyprinodontidae) from the endorheic Namak Lake basin in Iran. ZooKeys 215:55–75.

Teimori, A., T. Schulz-Mirbach, H. R. Esmaeili, and B. Reichenbacher. 2012b. Geographical differentiation of *Aphanius dispar* (Teleostei: Cyprinodontidae) from southern Iran. Journal of Zoological Systematics and Evolutionary Research 50:289–304.

Thuiller, W. 2003. BIOMOD—optimizing predictions of species distributions and projecting potential future shifts under global change. Global Change Biology 9:1353–1362.

Thuiller, W., B. Lafourcade, and M. Araujo. 2009a. ModOperating manual for BIOMOD. Available: http://r-forge.r-project.org/scm/viewvc.php/*checkout*/pkg/inst/doc/Biomod%20Manual.pdf?revision=67&root=biomod&pathrev=218. (April 2019).

Thuiller, W., B. Lafourcade, R. Engler, and M. B. Araújo. 2009b. BIOMOD—a platform for ensemble forecasting of species distributions. Ecography 32:369–373.

Tonn, W. M., J. J. Magnuson, M. Rask, and J. Toivonen.1990. Intercontinental comparison of small-lake fish assemblages: the balance between local and regional processes. The American Naturalist 136: 345–375.

Valavi, R., H. Shafizadeh-Moghadam, A. Matkan, A. Shakiba, B. Mirbagheri, and S. H. Kia. In press. Modelling climate change effects on Zagros forests in Iran using individual and ensemble forecasting approaches. Theoretical and Applied Climatology, doi: 10.1007/s00704-018-2625-z.

Vogt, J., R. Colombo, M. L. Paracchini, A. de Jager, and P. Soille. 2003. CMM river and catchment database, version 1.0. European Commission, Joint Research Centre, EUR 20756 EN, Ispra, Italy.

Vogt, J., P. Soille, A. de Jaeger, E. Rimaviciute, W. Mehl, S. Foisneau, K. Bódis, J. Dusart, M. Paracchini, P. Haastrup, and C. Bamps. 2007. A Pan-European river and catchment database. European Commission, Joint Research Centre, EUR 22920 EN, Brussels, Belgium.

Wenger, S. J., and J. D. Olden. 2012. Assessing

transferability of ecological models: an underappreciated aspect of statistical validation. Methods in Ecology and Evolution 2:260–267.

Winemiller, K. O., A. A. Agostinho, and E. P. Caramaschi. 2008. Fish ecology in tropical streams. Pages 107–46 *in* D. Dudgeon, editor. Tropical stream ecology. Academic Press, London.

Yates, K. L, P. J. Bouchet, M. J. Caley, K. Mengersen, C. F. Randin, S. Parnell, A. H. Fielding, A. J. Bamford, S. Ban, A. M. Barbosa, C. F. Dormann, J. Elith, C. B. Embling, G. N. Ervin, R. Fisher, S. Goiuld, R. F. Graf, E. J. Gregr, P. N. Halpin, R. K. Heikkinen, S. Heinanen, A. R. Jones, P. K. Krishnakumar, V. Lauria, H. Lozano-Montes, L. Nannocci, C. Mellin, M. B. Mesgaran, E. Moreno-Amat, S. Mormede, E. Novaczek, S. Oppel, G. Ortuno Crespo, A Townsedn Peterson, G. Rapacciuolo, J. J. Roberts, R. E. Ross, K. L. Scales, D. Schoeman, P. Snelgrove, G. Sundblad, W. Thuiller, L. G. Torres, H. Verbruggen, L. Wang, S. Wenger, M. J. Whittingham, Y. Zharikov, D. Zurell, and A. M. M. Sequeira. 2018. Outstanding challenges in the transferability of ecological models. Trends in Ecology and Evolution 33:790–802.

Zehzad, B., B. H. Kiabi, and H. Madjnoonian. 2002. The natural areas and landscape of Iran: an overview. Zoology in the Middle East 26:7–10.

American Fisheries Society Symposium 90:463–495, 2019
© 2019 by the American Fisheries Society

Advances, Challenges, and Gaps in Understanding Landscape Influences on Freshwater Systems

Dana M. Infante*
Department of Fisheries and Wildlife, Michigan State University
1405 South Harrison Road, Suite 318, East Lansing, Michigan 48824, USA

Lizhu Wang
Great Lakes Regional Office, International Joint Commission
Post Office Box 32869, Detroit, Michigan 48232, USA

Robert M. Hughes
Amnis Opes Institute, and Department of Fisheries and Wildlife, Oregon State University
2895 SE Glenn, Corvallis, Oregon 97331, USA

Kai Chen
Department of Entomology, Nanjing Agricultural University
Weigang #1, Xuanwu District, Nanjing, Jiangsu, 210095, China

Bianca de Freitas Terra
Center for Agrarian and Biological Sciences, State University of Vale do Acaraú
Avenida da Universidade 850, Sobral, Ceará, Brazil

Introduction

Freshwater ecosystems are threatened globally, and conserving them and the organisms they support is challenged in part by the need to account for their connections to the landscape (Sala et al. 2000; Dudgeon et al. 2007; Abell et al. 2017). Natural landscape factors such as geology, topography, and climate determine the ecological potential for a given freshwater ecosystem. Anthropogenic pressures occurring near to and distant from freshwater ecosystems modify habitat and ultimately influence organisms in those systems. Hence, landscape analyses provide frameworks for understanding interactions between landscape characteristics and freshwater conditions, knowledge of how freshwaters may be limited by pressures, and a means for mitigating stressors and conserving freshwaters.

The purpose of this chapter is to describe advances in understanding how landscape influences affect freshwaters and the organisms they support that have occurred since 2006. It is not intended to be a complete review of advances in landscape-aquatic ecology since that time. Instead, we draw from the 19 chapters included in this book as well as other literature to describe progress made in four key areas first mentioned in Wang et al. (2006). These include (1) identifying meaningful spatial units for aquatic systems, (2) determining how humans modify aquatic systems, (3) measuring and understanding how factors occurring at different spatial extents influence aquatic systems, and (4) gathering landscape and aquatic data. This chapter also summarizes remaining challenges and gaps for effectively understand-

* Corresponding author: infanted@msu.edu

ing linkages between landscapes and freshwaters. Finally, we summarize a set of policy recommendations originating from the preceding chapters of this volume.

Advances in Identifying Meaningful Spatial Units for Aquatic Ecosystems

Digital spatial frameworks have now been developed for assessing hierarchical influences of landscape characteristics on streams and lakes. These frameworks classify spatial units (e.g., water body, local catchment, and network catchment). Data are then attributed to spatial units, providing an organized means for analyzing landscape–aquatic linkages at multiple spatial extents. For river systems, a spatial framework has been developed using interconfluence river reaches and their local and network catchments as fundamental study units (Wang et al. 2011). These units are then linked to larger ecological or politically defined spatial units of interest to management. The hierarchical structure used to organize these units facilitates summarizing or analyzing information at varied spatial extents. These ideas were followed in development of StreamCat (Hill et al. 2016), a second national spatial framework that includes stream buffers. This concept has also been expanded to inland lake systems, with the spatial framework including the lake itself, local catchments (land areas draining directly into the lake), and network catchments (areas draining into the lake via inlet river networks) (Wang et al. 2016). A spatial framework has also been described for large lakes (e.g., North American Great Lakes), and its structure accounts for network catchments as well as coastal terrestrial, coastal margin, nearshore/littoral, and offshore zones that encompass an entire lake. Each of the landscape–lake interaction zones is attributed with physicochemical and biological characteristics, data from catchments, and data characterizing terrestrial and coastline conditions (Wang et al. 2015). In Canada, a stream-lake classification is being developed and tested for the province of Ontario (Melles et al. 2014; Allerton et al. 2019, this volume). Collectively, these advances in geospatial frameworks developed for rivers, inland lakes, and large lakes facilitate linking water body units with river or lake catchments, lake zones, lake subbasins, and political boundaries (Wang et al. 2016).

Since 2006, large computer databases describing features of aquatic ecosystems have also been developed. In the United States, national hydrography products (www.usgs.gov/core-science-systems/ngp/national-hydrography) include the National Hydrography Dataset, the National Hydrography Dataset High Resolution, and the Watershed Boundary Dataset. These data sets depict rivers, lakes, coastlines, catchments, and other hydrologic features in a comprehensive, consistent, and comparable manner. Attributing the features represented within these databases with additional information (e.g., climate, geology, and land use for catchments; size, slope, and biological information for water bodies) has yielded national river databases (e.g., Wang et al. 2011; Hill et al. 2016; McManamay and DeRolph 2019). Similar databases have been developed for Austrian rivers (Wimmer et al. 2000), UK rivers (England and Gurnell 2016), European rivers (Rakovec et al. 2018; Schinnegger et al. 2016), and South African aquatic systems (Ollis et al. 2015). A database for the Yangtze River and its tributaries in China is currently under development (Yushun Chen, Institute of Hydrobiology and State Key Laboratory of Freshwater Ecology and Biotechnology, Wuhan, China, personal communication).

Regional, national, and binational lake databases have also been developed. One example is LAGOS-NE, which contains data for 51,101 inland lakes and reservoirs larger than 4 ha from 17 U.S. states (Soranno et al. 2017). That database includes lake location and physical characteristics such as land use, geology, climate, hydrology, and biology organized hierarchically. Another example has been developed for the five North American Great Lakes, which divides the entire 246,000 km^2 of lake area into 30-m grid cells for nearshore zones and into 1.8-km grid cells for offshore zones. These grid cells are attributed with more than 300 physical, chemical, and biological variables where available (Wang et al. 2016; Riseng et al. 2018). A third example is LakeCat (Hill et al. 2018), which includes hundreds of landscape variables for more than 378,000 lakes in the conterminous United States. These large databases have substantially advanced our capability to study landscape–lake system interactions across large areas.

Since 2006, there have also been increased numbers of analyses conducted over large spatial extents (multiple states, countries). In such work, freshwater sites that share similar characteristics are grouped, and then influences of natural factors are quantified before accounting for anthropogenic effects on freshwater systems (e.g., Utz et al. 2010; Esselman et al. 2013; Tang et al. 2017; Daniel et al. 2018; Tingley et al. 2018). Another example occurs in recognizing that responses of freshwater organisms to landscape pressures and stressors often occur nonlinearly (e.g., Allan 2004b; Baker and King 2010; Ellis et al. 2012), an outcome that is sometimes more commonly observed with large versus small sample sizes and in the era of "big data" (Durden et al. 2017).

Since 2006, regional landscape information has been increasingly used by managers. Here, we provide five examples of these extensive efforts in the United States: (1) The National Water Quality Assessment Program (NAWQA) was implemented by the US Geological Survey to assess effects of agriculture and urbanization on waterbodies (Brown et al. 2009; USGS 2014). (2) The National Aquatic Resource Surveys (NARS) are collaborative efforts between the U.S. Environmental Protection Agency (USEPA), states, and tribes to assess the status and trends of all conterminous U.S. rivers, lakes, reservoirs, wetlands, and coastal waters through use of a probability survey design and standard sampling methods (USEPA 2016a, 2016b, 2016c, 2016d). (3) Recently, an assessment of fish habitat in all streams and rivers of the United States was conducted for the National Fish Habitat Partnership (Crawford et al. 2016). (4) Hill et al. (2017) used StreamCat and data from the USEPA's National River and Stream Assessment for predicting the biotic condition of millions of kilometers of streams across the country.

Similar applications have been implemented on other continents. In Europe, the European Union's (EU) Water Framework Directive (Directive 2000/60/EC; http://ec.europa.eu/environment/water/water-framework/index_en.html) has been working to promote good ecological condition for all European water bodies by the year 2027. Additionally, this program encourages sampling of water bodies using comparable approaches in all participating countries, thereby facilitating a landscape perspective in understanding the condition of and limits to aquatic habitats (Pont et al. 2006; Hering et al. 2010). In Australia, the Sustainable Rivers Audit is focused on the waters of the Murray–Darling River basin (Davies et al. 2010; Robinson et al. 2019, this volume). The River Health Program of South Africa monitors and reports on status and emerging problems in South African rivers (Kleynhans 1999; DWAF 2008). In addition, the Ramsar Convention on Wetlands (www.ramsar.org) promotes a landscape perspective as well as international cooperation in the conservation of all wetlands globally, including freshwaters.

Advances in Understanding How Human Activities Modify Aquatic Systems

Quantifying how human activities affect freshwater ecosystems by modifying water body characteristics and biological assemblages is one of the primary applications of a landscape approach (Johnson and Host 2010). Global environmental changes, including land use and climate change, increase stress on aquatic ecosystems through degradation of river habitats, alterations in flow regimes, loss of river connectivity, changes to water quality, and modification of biological assemblages (Januchowski-Hartley et al. 2016; Grill et al. 2019). Landscape pressures (e.g., agriculture, urbanization, and fragmentation) coupled with climate changes from human activities affect aquatic habitats and biological assemblages through processes at multiple spatial extents (Poff 1997; Parrott and Meyer 2012; Cooper et al. 2017). Understanding the landscape and other confounding factors controlling aquatic assemblages remains a critical need for informing management and conservation of freshwater ecosystems.

Both natural and anthropogenic barriers fragment the connectivity of critical habitats (Cooper et al. 2017; Couto and Olden 2018; Grill et al. 2019; Poff 2019), resulting in limits to species distribution and assemblage composition changes (Fullerton et al. 2010). Anthropogenic fragmentation interferes with effective ecological processes that depend on longitudinal river connectivity, including transport of organic and inorganic matter and upstream and downstream movements of aquatic and riparian species (e.g., McIntyre et al. 2016; Grill et al. 2019). Lakes and reservoirs can serve as refuges or barriers, depending on the species (Pompeu et al. 2012; Allerton et al. 2019; Alofs 2019, this volume). The natural fragmentation of waterfalls can also restrict upstream migrations of native and nonnative stream species (Reznick and Ghalambor 2005; Rahel 2013; Martin et al. 2019, this volume). Understanding the relative degrees that human and natural barriers affect species migrations can facilitate research in native species conservation and nonnative species management (Martin et al. 2019).

Climate change and anthropogenic land use are mounting threats to global biological diversity (Sala et al. 2000; Urban 2015), compromising freshwater ecosystem structure and function (Chapin et al. 2000; Comte and Olden 2017). Effects of climate change and land use on river ecosystems include significant changes to stream temperature (Hill et al. 2014; Guse et al. 2015), flow and sediment regimes (Jones 2011; Guse et al. 2015), wood delivery (Wohl et al. 2019), water quality (Allan 2004b; Whitehead et al. 2008), and biota (Dudgeon et al. 2007; Winfield et al. 2016; Isaac et al. 2018). Land surface processes also modulate the severity of temperature, droughts, and other extreme climatic events (Findell et al. 2017). Thus, river ecosystems are threatened by antagonistic or synergistic interactive changes from land use and climatic conditions (Radinger et al. 2016). Such interactions have large impacts on biological and ecological processes and important implications for habitat and landscape management (Oliver and Morecroft 2014). The population dynamics and phenology of aquatic organisms are particularly susceptible to shifting flow regimes and temperatures, both of which are altered by global changes (Falke et al. 2019, this volume; Robinson et al. 2019).

In addition to climate change, there is growing recognition of the linkages between freshwaters and factors originating beyond a water body's catchment. The surface water effects of groundwater connections (Falke et al. 2012; Warner et al. 2012; Perkin et al. 2017) and atmospheric dispersal of contaminants and nutrients (Baker et al. 1991; Peterson et al. 2007; Landers et al. 2010; Meire et al. 2012; Stoddard et al. 2016) have been recognized as important influences on surface waters. Managers and scientists now increasingly acknowledge these broader types of landscape influences on freshwater ecosystems.

Effects of landscape changes on aquatic ecosystem condition involve processes operating at multiple extents and via multiple pathways (Poff 1997; Riseng et al. 2011). The most common approach for conducting environmental assessments is to compare least-disturbed reference conditions for physicochemical habitat and biological assemblages of a site with those of test sites (Stoddard et al. 2006; Wang et al. 2006). When data are limited, risk-based assessments (Cooper et al. 2019, this volume) are a useful alternative to estimate single and cumulative effects with summarized types and intensities of multiple pressures. However, the natural and anthropogenic processes that affect the structure and function of aquatic ecosystems differ regionally (Allan 2004a; Hawkins et al. 2010; Deweber et al. 2019, this volume; Herlihy et al. 2019; Thornbrugh and Infante 2019, this volume). A regionalization framework delineated using physiographic predictors of land use as opposed to physiographic variables not tied to land use may improve our understanding of the effects of anthropogenic land uses (Deweber et al. 2019). Because covarying natural and anthropogenic factors complicate understanding causal systems, it is necessary to identify and evaluate the specific direct and indirect causal paths that link landscape changes to the observed responses of aquatic systems (Baker and Wiley 2009; Riseng et al. 2011). Ecological mechanistic models can be employed to evaluate cause–effect relationships within a set of observed and latent variables. For example, structural equation modeling (Shipley 2000; Leitão et al. 2018), random forest modeling (Cutler et al. 2007), boosted regression tree models (Alofs 2019), gradient forest modeling (Ellis et al. 2012; Tang et al. 2017; Wagenhoff et al. 2017), and Bayesian approaches (Allan et al. 2012) can provide greater insight into influences on conditions in freshwaters.

Advances in Understanding and Measuring How Various Spatial Factors Interactively Influence Aquatic Systems

During recent decades, studies incorporating a landscape approach have increasingly applied landscape ecological principles focusing on extent, pattern, structure, process, connectivity, and interactions to study freshwater systems (Wiens 2002). This is largely a result of the recognition that management and conservation of aquatic habitats and their biota are strongly influenced by various natural and anthropogenic environmental factors at different spatial extents. Previously, Hughes et al. (2006) identified three challenges, including (1) identification of key landscape factors that are most influential and the spatial extents at which they are manifest, (2) the influence of study region size on identifying key landscape factors, and (3) the influence of study region condition (range in variation of influential factors) on the outcomes of the analyses. Since Hughes et al. (2006), considerable effort has been directed toward these challenges in the literature and in this volume. As suggested above, key advances include improved statistical methods, more meaningful spatial extents and levels of resolution, better databases and classification systems, and greater understanding of overall landscape influences. Each of these advances is discussed in the following subsections.

Improved Statistical Methods

Applications of more sophisticated statistical methodologies to landscape–aquatic studies have substantially improved our understanding of how landscape influences freshwater systems. Structural equation modeling, classification and regression decision trees, random forest modeling, gradient forest modeling, neural network modeling, threshold indicator analysis, generalized additive modeling, and Bayesian modeling are examples of such approaches. Although omitted variables are concerns for all statistical analyses, each of these approaches has contributed to our knowledge of landscape influences on freshwater ecosystems regardless of their limitations.

Structural equation modeling focuses on building theoretical models for clarifying the direct and indirect pathways of influences. Structural equation modeling is based on theoretically justified models that are parameterized by minimizing differences between model predictions and observed data (Grace 2008). But with structural equation modeling, there are concerns with reduced importance in lower-order model components and the inaccuracy of some commonly used rules (Tomarken and Waller 2005). Nonetheless, Leitão et al. (2018) was able to explain nearly 30% of fish taxonomic diversity and evenness and functional originality as an indirect function of catchment deforestation in an Amazonian basin compared with random forest analysis, which explained less than 10% (Leal et al. 2018).

Classification and regression decision trees are ideally suited for the analysis of landscape and aquatic data. Classification and regression decision trees explain variation of a single response variable by repeatedly splitting data into more homogeneous groups, using combinations of explanatory variables that may be categorical or numerical. Decision trees tend to have high variance when they use different training and test sets of the same data because they tend to overfit on training data (De'ath and Fabricius 2000). Classification and regression decision trees were used to predict distributions of Michigan stream fish occurrence from landscape variables (Steen et al. 2008) and to predict endemic fish assemblages and species richness in the upper Yangtze River (He et al. 2010). They were also used to map ecological classifications of terrestrial and marine environments by combining spatial, biotic, and abiotic characteristics for environmental planning and monitoring (Pesch et al. 2011). In this book, classification and regression decision trees were used to determine variables most related to dispersal and regional- and lake-extent environment factors to explain Rainbow Smelt *Osmerus mordax* invasion success (Alofs 2019), to classify hydrologic units based on physiographic predictors of land use (Deweber et al. 2019), and to predict fish assemblage clusters across the conterminous United States from environmental variables (Herlihy et al. 2019).

Random forest modeling can be used to build predictive models for both classification and regression problems. It combines many decision trees and trains each one on a slightly different set of observations, splitting nodes in each tree considering a limited number of features. The final predictions of the random forest are made by averaging the predictions of each individual tree. Random forests have very high classification accuracy, the ability to model complex interactions among predictor variables, and the flexibility to perform several types of statistical data analysis (regression, classification, and unsupervised learning; Cutler et al. 2007). Examples of this approach include the study of relationships between landscape and stream fish occurrence patterns (Evans et al. 2011) and land cover classification (Hayes et al. 2014).

Gradient forest analysis is a novel statistical approach that is highly flexible for quantifying the nonlinear shape and magnitude of species compositional turnover across multiple environmental gradients (Ellis et al. 2012; Pitcher et al. 2012). Gradient forest begins with random forest modeling response of each taxon along each gradient, and it then aggregates the collation of split points of random forest models for all species to calculate compositional turnover along each environmental gradient. Gradient forest offers a new capability for exploring relationships between biodiversity and environmental gradients by generating detailed multispecies responses. It is also flexible enough for use of disparate or multiple data sets (e.g., Pitcher et al. 2012; Wagenhoff et al. 2017). Gradient forest modeling can be used to assess the importance of predictor variables (e.g., Tang et al. 2017), define ecosystem thresholds for human activities and environmental pressures (e.g., Roubeix et al. 2016; Samhouri et al. 2017), and support spatial planning, management, and conservation in freshwater and other realms (e.g., Pitcher et al. 2012; Samhouri et al. 2017; Stephenson et al. 2018; Perkin et al. 2019, this volume).

Neural networks are powerful computational data models that are able to capture and represent complex input–output relationships. They do not require users to identify a mathematical model before analysis; instead, they use algorithms that are adjusted sequentially by trial and error during the learning phase (Basheer and Hajmeer 2000). The advantage of neural networks lies in their ability to represent both linear and nonlinear relationships and in their ability to learn these relationships directly from the data being modeled. Examples of this approach include using both artificial neural networks and random forests to predict native fish species richness in Mediterranean rivers (Olaya-Marín et al. 2013) and stream fish abundance in New York streams (McKenna and Johnson 2011). Mostafavi et al. (2019, this volume) used channel slope, bankfull width, wetted width, elevation, mean and range of air temperatures, and annual precipitation to model distributions of nine fish species in Iran by using an ensemble forecasting approach, including neural networks.

Threshold indicator analysis is used to identify transition points or zones of rapid change in biological measures in response to small, continuous increases in a stressor or pressure. Threshold indicator analysis provides a sensitive and precise method for assessing assemblage or taxon thresholds to inform conservation of sensitive, rare, or threatened species; identify reference conditions; and support development of numerical regulatory criteria (Baker and King 2010). Examples of this approach are studies to detect urban land-use levels at which stream assemblage degradation occurred (Brenden et al. 2008), to forecast potential biodiversity loss from human development (Hilderbrand et al. 2010), and to examine impacts of catchment agriculture and urban development on benthic macroinvertebrate, fish, bird, diatom, and plant assemblages in Great Lakes coastal wetlands (Cardoso et al. 2013). A more recent study using threshold indicator analysis in eastern Brazilian Amazonia indicated that thresholds for macroinvertebrate assemblages occurred at 9% and 2% of deforestation at catchment and site-riparian spatial extents, respectively (Janaina de Brito, Instituto Nacional de Pesquisas da Amazônia, unpublished data). She also found that the assemblages were more clearly responsive to forest loss at the catchment/network-riparian extents than at the site extent. This further indicates the value of evaluating landscapes at both spatial extents and suggests that conservation strategies must focus on entire catchments as well as riparian zones.

Generalized additive models are an extension of generalized linear models in which predictor variables can be modeled nonparametrically in addition to linear and polynomial terms for other predictors (Guisan et al. 2002). Tukiainen et al. (2017) used generalized additive models to analyze relationships between land-use intensity and both geodiversity (geological, geomorphological, and hydrological richness) and plant species richness across Finland. Generalized additive models were used in this volume to estimate relative contributions of eutrophication and climate change to loss of coldwater fish habitat for 5,220 stratified Midwestern U.S. lakes (Jacobson et al. 2019, this volume) and to predict fish distributions from physical habitat and climate conditions (Mostafavi et al. 2019; Robinson et al. 2019).

Bayesian modeling is another approach that has been used in improving landscape–aquatic system analysis because Bayesian approaches are able to incorporate prior information. For example, Allan et al. (2012) used Bayesian belief networks to model presumed causal relationships between pressures and response variables. They depicted the chain of causal relationships resulting in environmental impairment and used evidence from expert judgment as well as observational data. Stuber et al. (2017) used a Bayesian method to select spatial scales of predictors by using latent scale indicator variables that are estimated with reversible-jump Markov chain–Monte Carlo sampling. That approach returned accurate descriptions of the explanatory power of multiple spatial scales and unbiased and precise parameter estimates under commonly encountered data limitations, including spatial autocorrelation, effect size, and sample size. Although, the application of Bayesian modeling for understanding landscape influences on freshwaters is still in the exploratory stage, it has great potential to overcome some of the issues that may limit other statistical techniques.

Because each analytical method has its own strength and weakness, combination or sequential uses of these techniques are common practices. An excellent example of such studies is the ensemble forecasting approach for multiple fish species by using generalized linear models, generalized additive models, classification tree analysis, artificial neural networks, surface range envelopes, generalized boosting methods, random forest, multivariate adaptive regression splines, and flexible discriminant analysis (Mostafavi et al. 2019). Herlihy et al (2019) also used cluster analysis, discriminant function analysis, and classification tree analysis sequentially to classify and verify classification correctness of fish assemblages based on landscape factors for streams and rivers across the United States.

More Meaningful Spatial Extents

Advances in understanding the influences of study region sizes and study region conditions can be attributed in part to greater availability of large-extent spatial databases. The influence of study region size on identifying key landscape factors can be addressed by choosing the right size of a study region for answering specific questions for that region. For example, Herlihy et al. (2019) identified stream size, temperature, and stream water ionic strength as the primary drivers of fish assemblage structure in streams for improving biomonitoring sampling design at a national extent. Jacobson et al. (2019) identified Secchi depth, geometry ratio, and mean July air temperature as the drivers of coldwater fish distribution for stratified lakes across three states. Lyons et al. (2010) modeled effects of climate change on 50 fish species distributions through use of 69 environmental variables at a state extent for development of climate-change adaptation strategies. Using appropriate spatial and temporal extents and high-resolution data (Yates et al. 2019, this volume) to answer scientific and management questions is critically important for understanding landscape influences on freshwaters.

Better Classification Systems and Databases

Advances in the development of freshwater system classification, in conjunction with availability of large databases, have also improved our understanding of the influence of study region condition on the outcome of assessments. Although true natural patterns and processes influencing freshwater ecosystems can only be identified and analyzed under undisturbed conditions (Stoddard et al. 2006; Wang et al. 2006; Hughes 2019), aquatic system classification provides a tool to minimize within-class and maximize among-class variation in natural condition measurements. Classification can aid in identifying natural drivers determining habitats and biological assemblages and in grouping their conditions into types. Hence, classification minimizes the influences of natural condition measures, such as elevation, geology, soil, topography, and climate, on the interpretation of landscape anthropogenic impacts on the condition of freshwater systems (Omernik 1987; Omernik et al. 2017). For example, McManamay and DeRolph (2019) developed a multilayered, empirically driven stream classification system for the conterminous United States based on emergent natural variation in six habitat layers of size, gradient, hydrology, temperature, network bifurcation, and valley confinement. Tingley et al. (2018) conducted a classification of all stream reaches in the five main Hawaiian islands using landscape factors, including climate, found to be associated with distributions of native stream species. Results were then integrated with projected climate data to identify climate-sensitive stream reaches (Tingley et al. 2018). Riseng et al. (2018) developed an ecosystem classification for the entire North American Great Lakes using a combination of statistical analyses, published knowledge, and expert opinion on key driving variables and their ecologically relevant thresholds. This classification captured gradients that characterize types of aquatic systems in the Great Lakes and provides a tool for resource inventory and status and trend assessment. Deweber et al. (2019) developed a regional classification framework for understanding relationships between land use and stream biota while accounting for the influence of physiography.

Greater Understanding of Landscape Influences

Recent studies have also improved our knowledge in understanding and measuring how spatial factors interactively influence freshwater habitat and biota. For example, Deweber et al. (2019) recognize that catchment natural settings, such as soil, climate, and topography influence urban and agricultural land developments, which jointly influence aquatic systems. Historical agricultural land uses have been found to be a major influence on the subsequent effects of present-day urbanization on stream biota (Brown et al. 2009). Studies have also enhanced our understanding that natural condition characteristics influence the magnitude of the effects of anthropogenic activities on stream biota in streams with differing hydrologic and temperature regimes (e.g., Utz et al. 2009, 2011). Additionally, we currently acknowledge that influences of factors across catchments are the result of connectivity of resources, energy, and information among social, physical, and biological systems. Such intercatchment influences are carried out by the movements of people, air, water, and groundwater (Grimm et al. 2008; Cox et al. 2019). This is a critically important concept for aquatic landscape ecology, especially under changing climate, atmospheric circulation of contaminants, and natural gas fracking.

Advances in Collecting and Gathering Landscape and Aquatic Data

It has long been clear that large sample sizes are needed for making accurate generalizations about patterns and processes (Tversiky and Kahneman 1971). Both the National Aquatic Resource Surveys of the USEPA (Olsen and Peck 2008) and Bruckerhoff and Gido (2019, this volume) support sampling of at least 50 sites per reporting unit. For trend detection in regional and basin surveys of physical habitat variables, Larsen et al. (2004) recommended sampling 30–50 sites because of the large effect that sample size has on variance calculations. Employing nested survey designs, Callisto et al. (2019, this volume) and Pompeu et al. (2019, this volume) sampled 36–46 sites per hydrologic unit and 155–230 sites in the Cerrado biome, whereas Robinson et al. (2019) sampled 20 sites per catchment and 180 sites in the Murray–Darling River basin. Of course, the ideal number is a function of the regional variability of the study region, the standard errors of the predictor and response variables, the desired confidence intervals of those variables in the reporting, and logistical and funding realities (Stevens and Olsen 2004; Hughes and Peck 2008). Nonetheless, small sample sizes have a high probability of being misleading (Tversky and Kahneman 1971; Larsen et al. 2004; Stevens and Olsen 2004).

Detections of local and regional patterns are also a function of the variability in the study landscape. For example, at the continental spatial extent, Stoddard et al. (2008) and Tang et al (2016) reported that different metrics sufficed for reporting instream condition better than they did for each of nine ecoregions. Across freshwater ecoregions (large hydrologic units that drained multiple ecoregions), Thornbrugh and Infante (2019) reported differences in influences of landscape factors on stream fishes. At a smaller spatial extent, Martins et al. (2018) demonstrated that three different landscape units within a relatively small hydrologic unit contained significantly different channel slopes, leaf-pack cover, and biota at reference sites. Similarly, Agra et al. (2019) determined that reference sites within one ecoregion differed in channel width and slope as well as biota, but this was not evident in a different, less variable ecoregion. In other words, important ecological differences in local variability may not be detected when the analysis is conducted across large spatial extents. This variability supports the use of stream typology (Martins et al. 2018; Agra et al. 2019), assemblage typology (Schmutz et al. 2007; Herlihy et al. 2019), or natural gradient modeling (Oberdorff et al. 2002; Moya et al. 2011; Esselman et al. 2013; Vander Laan and Hawkins 2014; Tang et al. 2016; Chen et al. 2017) when developing biological expectations for fish, macroinvertebrate, and diatom assemblages.

Another aspect of landscape and biotic variability is the capacity to detect the effect of land use change on biotic responses. For example, Heitke et al. (2006) detected little response of row-crop agriculture on stream fish assemblages, presumably because of the pervasiveness of agriculture across Iowa. Pompeu et al. (2019) observed that land use explained little variability in Brazilian fish assemblage structure, likely because of high levels of beta diversity and rare species. However, Brito (unpublished data) and Firmiano et al. (2017) detected the loss of sensitive macroinvertebrate species at low levels of catchment and riparian deforestation through use of threshold indicator analysis. Such species may have been already eliminated from more disturbed landscapes (Leitão et al. 2016).

Collecting good and trustable data is challenging when studying natural systems. It can be much more challenging when studies involve large temporal and spatial extents. Geographic information systems (GIS) have been recognized as a powerful source of large-extent data and can generate innumerable variables. Progress in the use of GIS for understanding landscape influences on aquatic biological assemblages is clear in this book. Allerton et al. (2019) improved hydrologic network metrics that capture essential information about the spatial distribution and size of lakes in aquatic networks, using lake and stream geographic information. Cooper et al. (2019) evaluated two potential future threats to fluvial habitats, mineral mining and climate change, through use of GIS data. Davis (2019, this volume), Callisto et al (2019), Negishi (2019, this volume), Pompeu et al (2019), Jacobson et al. (2019), and Perkin et al (2019) used GIS to determine landscape characteristics (e.g., slope, elevation, and catchment area) and calculate land use pressures. Martin et al. (2019) used GIS to estimate human population density and road length density to characterize human activities that may have facilitated the establishment of nonnative species.

Additionally, this book contains chapters that employ other available data sets. Many of those are from countries that have been building large platforms to store data for applications by researchers and environmental managers. Jacobson et al. (2019) estimated historic climate conditions through use of existing air temperature data from 1895 to 1914. Robinson et al. (2019) used existing flow, temperature, and fish data to predict the effects of climate change on fish assemblages. Cooper et al (2019), Herlihy et al (2019), Thornbrugh and Infante (2019), Deweber et al (2019), and Martin et al (2019) used existing fish databases to evaluate landscape effects on fish assemblages in different temporal scales. Some regions lack computerized data sets; therefore, Mostafavi et al. (2019) obtained fish data from field sampling, museums, and the literature. Although many efforts have been made to compile data from different parts of the world, Asia, South America, and Africa still lack trustable, continental-extent, temporal and spatial databases.

Since 2006, the use of standard methods for biotic, water quality and physical habitat sampling have been adopted and spread worldwide. Strategies for answering questions and assessing freshwater efficiently and rigorously have been produced (Hughes and Peck 2008; Olsen and Peck 2008). In this book, Robinson et al. (2019) suggested a simple electrofishing-only monitoring program for assessing fish assemblages at the landscape scale through use of an efficient randomized survey design. Similar randomized survey designs and standard levels of sampling effort were employed for fish in the United States (Herlihy et al. 2019) and in Brazil for assessing fish (Terra et al. 2013; Pompeu et al. 2019) and macroinvertebrate (Silva et al. 2016; Callisto et al. 2019) assemblages. Bruckerhoff and Gido (2019) quantified how the distribution of sampling sites across landscape gradients affected the responses of stream fish assemblages to these gradients at different sample sizes, indicating the importance of balancing logistic and financial constraints.

Remaining Challenges and Gaps in Linking Landscape Characteristics with Aquatic Physicochemical and Biological Conditions

Data Capturing

In efforts to understand influences of landscape characteristics on freshwaters, many studies have focused on wadeable, perennial streams as well as inland lakes that support game fish species. Other types of systems, including intermittent and ephemeral streams and small inland lakes, may be understudied from a landscape perspective. One reason this occurs is because those systems are not mapped accurately. Challenges with efforts to map both intermittent and ephemeral headwater streams are described in Meyer et al. (2007) and Colvin et al. (2019), and efforts to better account for small inland lakes globally are described in Downing et al. (2006). These challenges in mapping such waters prohibit fully understanding influences of landscape factors on intermittent and ephemeral headwaters and small lakes. This gap is important to bridge because intermittent and ephemeral headwater streams, ponds, and wetlands contribute to the development and maintenance of downstream waters, and they provide critical habitat for many rare and vulnerable organisms (FEMAT 1993; Meyer et al. 2003; Falke et al. 2012; Medley and Shirey 2013). Additionally, small lakes and wetlands dominate the global area covered by freshwaters, and processes occurring in such systems (e.g., carbon, nitrogen, and nutrient processing) may be occurring at much higher rates because of their prominence on the landscape. To better consider relationships between these aquatic systems and the landscape, a better accounting of all types of freshwaters is required.

A second factor limiting our ability to understand influences of landscapes on certain types of freshwaters may result from the absence of easily applied, standardized sampling methods. Nonwadeable rivers, including great rivers of the world, provide a unique set of economic benefits and ecosystems services by supporting recreational and subsistence fisheries, serving as navigation corridors, and processing nutrients (e.g., Tank et al. 2008; Vörösmarty et al. 2010). Because of larger catchment areas, however, habitats in nonwadeable rivers may be subject to more complex stressors than in smaller rivers, and landscape influences may be more difficult to detect because of this. One strategy for landscape stressor detection involves investigating multiple sites or channel habitats throughout riverscapes of nonwadeable rivers (Hughes and Gammon 1987; Oberdorff and Hughes 1992; Angradi et al. 2009; Yoder et al. 2019). Compared to wadeable streams and rivers, however, nonwadeable rivers may be undersampled because natural resource management agencies commonly charged with sampling these habitats may not have protocols or other resources for assessing these systems (e.g., Kaufmann 2000; Wilhelm et al. 2005; Wessell et al. 2008). For all types of freshwaters, standardized sampling leads to many benefits including reducing uncertainty that may result from different gear types, allowing for data collected in different time periods or different regions to be compared, promoting data sharing, and improving communication among data collectors and the public (Bonar and Hubert 2002; Hughes and Peck 2008; Bonar et al. 2009, 2015). Given the growing emphasis of conducting large, landscape-scale studies, the value of data collected with standardized sampling methods becomes even more important to allow for integration of different data sets collected within smaller areas.

Predictive Modeling

Predictive models characterizing physical or biological characteristics of aquatic habitats help us to improve our understanding of influences on freshwater ecosystems, and they allow us to fill in gaps in observations that may occur in space or time (Janssen et al. 2015). Although many advances have been made in this area since 2006, improving accuracy of models to predict physical or biological characteristics of aquatic habitats remains a challenge.

One factor contributing to this challenge is inadequate sampling designs for teasing apart influences of multiple landscape factors operating over multiple spatial extents, as well as spatial autocorrelation that may occur between samples (Bruckerhoff and Gido 2019). This requires use of new analytical techniques (described previously), and it also suggests that larger sample sizes may be required. Ideally, enough samples should be selected from appropriate locations for modeling to represent the entire range of variation in landscape factors over which modelers wish to assess influences on the physical or biological characteristics of interest (Albert et al. 2010). Unfortunately, agencies and researchers frequently lack the capacity to collect sufficient information or to select locations for sampling in a fully statistically strategic manner (Hughes and Peck 2008). This suggests a need for more investment in development of standardized protocols for data collection so that data from different sources may be combined. Additionally, when additional data are not available, modelers must more carefully describe the range of conditions over which their models may be applied.

A second challenge to improving accuracy of predictive models is the need to move from correlative approaches to mechanistic models to better understand how landscape factors affect habitat and biological characteristics of freshwaters. While correlative approaches allow us to predict habitat features and/or patterns of organism distributions or abundances (Poff 1997), they do not fully elucidate how landscapes may be influencing habitat or biota. Following the landscape approach of Allan (2004a), a recommendation for addressing mechanisms is to incorporate factors into modeling that directly influence a habitat feature or an organism (Johnson and Host 2010). These include various measures of physical habitat as well as measurements of interactions among organisms, physiology, life history, phenology, functional traits, and dispersal (Daniel et al. 2018; Leitão et al. 2018; Alofs 2019; Falke et al. 2019; Robinson et al. 2019). Additionally, given the role that flow plays in structuring many features of aquatic environments (Poff 1997), flow data and/or flow models are critical for mechanistic studies to understand landscape influences on habitat and in many cases on organisms. To move forward, a better accounting of streamflow regimes is required (Deweber et al. 2014), as are models that estimate flows from landscape factors (e.g., Miller et al. 2018)

A third challenge to improving predictive modeling is the need to better account for additive, overlapping, and interacting effects of landscape pressures on freshwaters. Landscape pressures, including urbanization, agriculture, and fragmentation, rarely influence freshwaters independently and often interact (Brown et al. 2009; Schinegger et al. 2016). Currently, we have guidance in the literature on approaches for disentangling effects of landscape stressors on freshwaters (e.g., Chariton et al. 2016; Feld et al. 2016; Schinegger et al. 2016; Schinegger et al. 2018). Additionally, Brown et al. (2013) demonstrated the value of accounting for interacting pressures from a management perspective to more effectively mitigate effects of local versus global pressures and stressors. To move forward and given changing climate and the role of historical landscape pressures on current characteristics of freshwaters (e.g., Harding et al. 1998; Van Sickle et al. 2004; Brown et al. 2009; Ross et al. 2019, this volume), accounting for historical, current, and future pressure and stressor interactions should become a common component of studies incorporating a landscape approach.

Related to the challenge of accounting for interacting pressures and stressors, a fourth challenge is the need to better account for covariation between natural landscape factors and human pressures. This need has been described previously in Allan (2004a), in Wang et al. (2006), and in multiple chapters of Hughes et al. (2006). Advances in analytical techniques also provide researchers with greater ability to distinguish natural versus anthropogenic landscape influences on freshwaters. However, in an era of increasingly large data, we have the capacity to account for these patterns across large regions (countries, continents, and globally). It should be possible to develop a standard understanding about large-scale patterns in covariation in key factors, why they occur, and how they may influence aquatic habitats (Utz et al. 2010; Deweber et al. 2019; Thornbrugh and Infante 2019). Such knowledge, documented and tested through multiple studies, would provide benchmark information for future research and assessment efforts.

Future Gaps

Since 2006, the world in which we attempt to understand landscape influences on freshwaters has changed. This stems from actual changes in the threats facing freshwaters and our ability to understand those threats, which has been improved in part because of changes in data availability and technology as well as more evolved scientific understanding. In editing this volume, we recognize some key gaps that should be addressed going forward. While the gaps we describe are not meant to be comprehensive, we hope that by highlighting them, we can draw attention to research needs that others can address to improve understanding of landscape influences on freshwaters into the future.

First, we need a better understanding of landscape influences on aquatic organisms for the simple yet striking reason that species dependent on freshwaters are in decline globally (Dudgeon et al. 2007). Whereas some organisms are fairly well studied (e.g., fish species of socioeconomic importance), we know relatively little about many fishes, not to mention other taxonomic groups like mussels, snails, crayfish, and freshwater insects, including how they may respond to landscape pressures. We need knowledge of organisms in different life stages, including their varied habitat preferences throughout riverscapes, and knowledge of how these organisms interact with one another, both in their native ranges and under novel conditions that may be prompted by changes in climate or other stressors. Use of genetics data is one strategy for addressing these knowledge gaps; investigating genetic differences for populations of organisms occurring across a range of landscape conditions can identify sensitive responses of organisms to stressors (Mulfeld et al. 2014; Scribner et al. 2016; Davis et al. 2019).

A second gap relates to freshwater habitat, which includes physical, chemical and biological characteristics of freshwaters that can influence aquatic organisms (e.g., Jowett 1997; Maddock 1999). Habitat degradation and loss are frequently suggested as leading mechanisms by which landscape stressors affect freshwaters (Helfman 2007; Jelks et al. 2008). Additionally, because of this, managers of freshwaters often consider manipulations of various habitat features as strategies for mitigating negative effects of landscape stressors (Maddock 1999). Despite this recognition, habitat data for freshwaters are frequently not available in many regions or over large landscapes. Habitat data can be difficult or costly to collect, and when they are collected, standardized protocols may be lacking, which hinders integration of multiple data sets (Roper et al. 2010). More investment in collecting or modeling important habitat factors is required (e.g., streamflow, as a "master" habitat variable was described previously in this chapter; Poff 1997). Coupled with that need, more investment in testing effectiveness of habitat manipulations as strategies to achieve conservation objectives is essential to ensure that investments are yielding desired outputs.

A related freshwater gap is the interaction of surface waters and groundwaters, including hyporheic and aquifer waters. Groundwater delivery to freshwaters is widely acknowledged as promoting more stability in streamflows and lake levels and water temperatures than would occur in freshwaters receiving relatively more surface runoff (Dunne and Leopold 1978; Wiley et al. 1997). Groundwater inputs to freshwaters are dictated by landscape features, including land cover, topographic characteristics, soil properties, surficial lithology, and anthropogenic pressures. Because they are out of sight, groundwaters are typically ignored and often polluted by agriculture and industry. Groundwater pumping depletes surface waters (Falke et al. 2012; Perkin et al. 2017; Condon and Maxwell 2019). Agricultural nutrients and biocides and oil, gas, and mineral mines contaminate surface waters via groundwaters (Rupert 2008; Hughes et al. 2016). Nonetheless, regulations pursuant to the Clean Water Act and EU Water Framework Directive leave those waters poorly protected or totally unprotected from depletion or contamination. Increased research is needed that documents the connections between surface waters and both shallow and deep groundwaters in ways that are convincing to water resource management and regulatory agencies.

Continued and improved investigations of landscape influences on freshwaters depends on the availability of landscape-scale data sets that describe natural influences on freshwaters (e.g., groundwater delivery estimates, current climate conditions), anthropogenic stressors (e.g., accurate characterizations of agricultural and urban land uses), and future threats (e.g., projected changes in climate). Despite the fact that, since 2006, development of such data sets has been substantial and supported by agencies and researchers across the globe, important landscape data sets are still not available in a comprehensive and comparable manner, and their omission represents a third gap in our ability to fully account for landscape influences. In a recent effort to assess fish habitat in all streams of the conterminous United States, Crawford et al. (2016) listed some of the key landscape stressors that were not incorporated into their work because of the absence of consistent, comparable data layers for the entire region. These included current stressors like oil and gas extraction activities, fragmentation effects of culverts and small dams on rivers, locations of animal feed lots, effects of logging, measures of grazing intensity, direct modifications to stream channels like dredging and channelization, and other regionally prominent pressures. Historical pressures having legacy effects on the landscape were also missing from this effort and have been cited as missing from other efforts, despite their known impacts on freshwater. These include legacy mining (Woody et al. 2010), beaver trapping (Naiman et al. 1988), channelization (Hughes et al. 2019), nonnative species introductions (Hughes and Herlihy 2012; Egan 2017), elimination of large predators (Ripple and Beschta 2011), and agriculture (Denevan 2011; Mann 2005). The problem with missing disturbances is that in an assessment context, the condition of water bodies may be overestimated, and in a predictive capacity, we fail to accurately understand what may be dictating conditions in freshwater.

Policy Recommendations

As indicated in the Discussion sections of this book's chapters, their authors described policy recommendations that arose from their data analyses and results. In the following paragraphs, these are grouped under the following topics: education and communication, data acquisition and analyses, riverscape management, conservation reserves, land use regulation, and climate change.

Education and Communication

Both Callisto et al. (2019) and Martin et al. (2019) indicated the great need to improve public education regarding effects of land use and species introductions on freshwater ecosystems. Callisto et al. (2019) and França et al. (2019) demonstrated how the increased use of citizen scientist monitoring and reporting could better engage youths and other citizens, as well as local agency managers. Martin et al. (2019) recommended increased public communication regarding the severity of nonnative species impacts on streams. Those proposed communication campaigns included educational curricula for students and public outreach events that involved local citizens with local streams. Not only do such activities increase awareness of the effects of humans on freshwaters and greater interest in protecting those resources, Crouse et al. (2017) also documented the importance of urban greenness and nature recreation in reducing human mortality. Weber and Ringold (2019) documented the importance of engaging the public in determining key indicators for gaining public support for water resource monitoring, conservation, and management. The importance of nature access and nature appreciation study are grounded in biophilia (Wilson 1986) and led to the International Union for Conservation of Nature resolution of a human right to experience a healthy world (IUCN 2012).

Data Acquisition and Analyses

The databases available to landscape ecologists have increased substantially on at least four continents since Hughes et al. (2006). For example, Herlihy et al. (2019) employed data from more than 2,000 sites collected via the USEPA's National Rivers and Streams Assessment and Thornbrugh and Infante (2019) used fish data from thousands of sites compiled from samples collected by state agencies and museums from across the United States (Daniel et al. 2015). Schinegger et al. (2016) assessed landscape-stream patterns in more than 3,000 sites collected from 14 western European nations, which was driven by the EU Water Framework Directive. Robinson et al. (2019) evaluated data from nearly 4,000 river sites obtain in Australia's Murray–Darling basin Sustainable Rivers Audit. Callisto et al. (2019) and Pompeu et al. (2019) analyzed data from 230 and 155 Brazilian tropical savanna sites, respectively. Those studies indicate the importance of funding continental-extent and intensive state-extent ecological surveys employing standard methods and randomized survey designs to facilitate data aggregation and sharing across multiple agencies and programs (Larsen et al. 2007; Mulvey et al. 2009; Callisto et al. 2019; Pompeu et al. 2019; Robinson et al. 2019).

However, increased sampling effort and additional biological surveys are needed (Bruckerhoff and Gido 2019; Callisto et al. 2019; Martin et al. 2019; Pompeu et al. 2019). Yates et al. (2019) argued that agencies in all levels of government should mandate, fund, and ease the acquisition of more-detailed human-activity data. Funding post-hoc power analyses and revisit sampling would also help scientists assess the confidence in their results so that decision makers and stakeholders can make better decisions (Bruckerhoff and Gido 2019; Herlihy et al. 2019). Similarly, assessments of biological, as well as statistical, significance would better inform the public and resource managers regarding the implications of their research (Bruckerhoff and Gido 2019; Callisto et al. 2019). For some biomonitoring, assessment, and reporting purposes, it may be useful for managers to base landscape classifications and funding decisions on patterns in aquatic biota versus terrestrial landscape patterns (Schmutz et al. 2007; Herlihy et al. 2019).

Riverscape Management

It is useful to conceptualize, study, and manage entire riverscapes (Fausch et al. 2002). That thinking includes the need to assess and manage the groundwater, hyporheic, riparian and floodplain effects of land uses, not just the immediate water body effects (Wipfli et al. 2007; Leibowitz et al. 2018; Negishi 2019). This is also true regarding riverscape genetic studies as they pertain to understanding the genetic heterogeneity and persistence of keystone migratory species (Schindler et al. 2010; Davis et al. 2019). Dams and reservoirs (Hughes et al. 2005; Pompeu et al. 2012), as well as road crossings (Perkin and Gido 2012; Hughes and Dunham 2014; Leitão et al. 2018; Perkin et al. 2019), often create barriers to fish passage and can negatively affect fish assemblage structure and diversity. Therefore, a key aspect of riverscape management is the importance of maintaining habitat connectivity to allow adequate rates of dispersal and access to critical habitats required by various species (Nel et al. 2010; Allerton et al. 2019; Davis et al. 2019). Improved culvert designs and placement promote fish passage and reduce culvert maintenance and road damage, thereby offering both conservation and economic benefits (Gillespie et al. 2014; Sleight and Neeson 2018). Unlike road crossings, water, sediment, wood, and fish passage of dams and reservoirs is much more problematic (Poff et al. 2007; Pompeu et al. 2012; Thornbrugh and Infante 2019; Wohl et al. 2019). This means that conservation and rehabilitation are often best served by dam and reservoir removal (Bellmore et al. 2017; Lohan

2018; American Rivers 2019; Brenkman et al., in press) or normalizing as much of the system as possible (Stanford et al. 1996). However, dams and other natural barriers also impede the movements of nonnative species (Martin et al. 2019). Those movements can be further reduced by restricting the movement, release, and harvest of nonnatives (Drake and Mandrak 2014; Alofs 2019), as well as human access to waterways (Martin et al. 2019).

Conservation Reserves

Given the widespread and destructive impacts of anthropogenic activities, several authors called for setting aside more large, protected areas of the landscape for ensuring the persistence of biodiversity (Allerton et al. 2019; Thornbrugh and Infante 2019). Unlike terrestrial-oriented reserves, those concerned with aquatic biodiversity should be based on entire river basins and riverscapes (Callisto et al. 2019; Pompeu et al. 2019), and Mostafavi et al. (2019) recommended using species distribution models as one strategy for delineating and establishing large areas for protecting at-risk fish species. Protecting areas with low current levels of anthropogenic disturbance will increase their resilience to future threats of climate change (Cooper et al. 2019). Similarly, protecting the entire catchments of deep, clear, coldwater lakes is important for conserving those unique and threatened ecosystems (Jacobson et al. 2019). Nations might also focus on locating aquatic conservation reserves in their globally recognized biodiversity hotspots (e.g., Brazil's Amazonia, Atlantic Forest, and Cerrado; China's Southwestern Mountains; and the United States' Southern Appalachian Highlands; Marchese 2015).

Land Use Regulation

Given the difficulty of setting aside large conservation preserves and the knowledge that low levels of catchment devegetation can impair aquatic biota (e.g., Negishi 2019), Callisto et al. (2019) and FEMAT (1993) called for riparian protection zones along water bodies as a means to buffer some of the worst effects of land use. However, Leal et al. (2018) and Brito (unpublished data) reported that current 30-m riparian buffer protections in Brazil are insufficient for protecting sensitive fish and macroinvertebrate taxa in forested Amazon streams. Jacobson et al. (2019) recommended fee-title acquisition and conservation easements for protecting extensive intact forested catchments. Several authors reported on the negative effects of agriculture on streams (Bruckerhoff and Gido 2019; Callisto et al. 2019; Deweber et al. 2019; Perkin et al. 2019; Pompeu et al. 2019; Thornbrugh and Infante 2019), indicating a critical need for regulating agricultural practices to control sediment, nutrient, and pesticide runoff. In agricultural landscapes, vegetated buffer strips, tiling prohibition, and cover crops should be required to reduce those losses (Jacobson et al. 2019; Perkin et al. 2019). Livestock access to riparian areas should be restricted as should livestock numbers on public lands (Beschta et al. 2013). Others have called for reforming agriculture and pasturing by treating land and soil as ecosystems and building soil and biota instead of destroying them (Brown 2018; Crews et al. 2018). Authors found that urban areas markedly degraded streams and recommended that storm water capture should be mandated (Roy et al. 2008; Jacobson et al. 2019; Perkin et al. 2019). Callisto et al. (2019) called for improved wastewater collection and treatment, and Hughes et al. (2014) reported the need for protecting natural land cover and increasing hydrological connectivity and morphological complexity of urban streams.

Climate Change

Cooper et al. (2019) and Falke et al. (2019) discussed the probability that climate change would alter the quality of anadromous fish habitats. Robinson et al. (2019) found that several small-bodied fish species were more frequently collected when conditions were drier whereas common medium- and large-bodied species were less likely collected when flows were lower. To counter some of these effects, Callisto et al. (2019) recommended replacing hydropower with solar and wind power. Although climate change and land use threaten aquatic ecosystem health globally (Oberdorff and Hughes 2019), they are driven by uncontrolled human population and economic growth (Limburg et al. 2011; Ripple et al. 2017); thus, both need to be stabilized and slowly reduced. This also means fundamental restructuring of national and global economies (Green Party US 2019).

Acknowledgment

We thank J. David Allan for his many contributions to the field of aquatic-landscape ecology—particularly the insights that he has provided us via writings and conversations over the past 20 years.

References

Abell, R., B. Lehner, M. Thieme, and S. Linke. 2017. Looking beyond the fenceline: assessing protection gaps for the world's rivers. Conservation Letters 10:384–394.

Agra, J. U. M., R. L. B. Santos, D. R. Macedo, R. M. Hughes, and M. Callisto. 2019. Ecoregions and stream types aid understanding ecological variability of Neotropical reference streams. Marine and Freshwater Research 70:594–602.

Albert, C. H., N. G. Yoccoz, T. C. Edwards, Jr., C. H. Graham, N. E. Zimmermann, and W. Thuiller. 2010. Sampling in ecology and evolution: bridging the gap between theory and practice. Ecography 33:1028–1037.

Allan, J. D. 2004a. Influence of land use and landscape setting on the ecological status of rivers. Limnetica 23:187–198.

Allan, J. D. 2004b. Landscapes and riverscapes: the influence of land use on stream ecosystems. Annual Review of Ecology, Evolution, and Systematics 35:257–284.

Allan, J. D., L. L. Yuan, P. Black, T. O. M. Stockton, P. E. Davies, R. H. Magierowski, and S. M. Read. 2012. Investigating the relationships between environmental stressors and stream condition using Bayesian belief networks. Freshwater Biology 57:58–73.

Allerton, M. L., N. E. Jones, S. J. Melles, and B. J. Schmidt. 2019. Ecologically relevant structural metrics for complex stream–lake networks. Pages 125–139 *in* R. M. Hughes, D. M. Infante, L. Wang, K. Chen, and B. F. Terra, editors. Advances in understanding landscape influences on freshwater habitats and biological assemblages. American Fisheries Society, Symposium 90, Bethesda, Maryland.

Alofs, K. M. 2019. Landscape approaches to understanding invasions in inland lakes: Rainbow Smelt as a case study. Pages 105–123 *in* R. M. Hughes, D. M. Infante, L. Wang, K. Chen, and B. F. Terra, editors. Advances in understanding landscape influences on freshwater habitats and biological assemblages. American Fisheries Society, Symposium 90, Bethesda, Maryland.

American Rivers. 2019. Restoring damaged rivers. Available: www.americanrivers.org/threats-solutions/restoring-damaged-rivers/. (June 2019).

Angradi, T. R., D. W. Bolgrien, T. M. Jicha, M. S. Pearson, B. H. Hill, D. L. Taylor, E. W. Schweiger, L. Shepard, A. R. Batterman, M. F. Moffett, C. M. Elonen, and L. E. Anderson. 2009. A bioassessment approach for mid-continent great rivers: the upper Mississippi, Missouri, and Ohio (USA). Environmental Monitoring and Assessment 152:425–442.

Baker, L. A., A. T. Herlihy, P. R. Kaufmann, and J. M. Eilers. 1991. Acidic lakes and streams in the United States: the role of acidic deposition. Science 252:1151–1154.

Baker, M. E., and R. S. King. 2010. A new method for detecting and interpreting biodiversity and ecological community thresholds. Methods in Ecology and Evolution 1:25–37.

Baker, M. E., and M. J. Wiley. 2009. Multscale control of flooding and riparian-forest composition in lower Michigan, USA. Ecology 90:145–159.

Basheer, I. A., and M. Hajmeer. 2000. Artificial neural networks: fundamentals, computing, design, and application. Journal of Microbiological Methods 43:3–31.

Bellmore, R. J., J. J. Duda, L. S. Craig, S. L. Greene, C. E. Torgersen, M. J. Collins, and K. Vittum. 2017. Status and trends of dam removal research in the United States. Water [online serial] 4(2):e1164.

Beschta, R. L., D. L. Donahue, D. A. DellaSala, J. J. Rhodes, J. R. Karr, M. H. O'Brien, T. L. Fleischner, and C. D. Williams. 2013. Adapting to climate change on western public lands: addressing the ecological effects of domestic, wild, and feral ungulates. Environmental Management 51:474–491.

Bonar, S. A., and W. A. Hubert. 2002. Standard sampling of inland fish: benefits, challenges, and a call for action. Fisheries 27:10–16.

Bonar, S. A., W. A. Hubert, and D. W. Willis. 2009. Standard methods for sampling North American freshwater fishes. American Fisheries Society. Bethesda, Maryland.

Bonar, S. A., N. Mercado-Silva, M. Rahr, and Y. T. Torrey, and A. Cate. 2015. A simple Web-based tool to compare freshwater fish data collected using AFS standard methods. Fisheries 40:580–589.

Brenden, T. O., L. Wang, and Z. Su. 2008. Quantitative identification of disturbance thresholds in support of aquatic resource management. Environmental Management 42:821–832.

Brenkman, S. J. In press. Rapid recolonization and life history responses of Bull Trout following dam removal in Washington's Elwha River. North American Journal of Fisheries Management, doi:10.1002/nafm.10291.

Brown, C. J., M. I. Saunders, H. P. Possingham, and A. J. Richardson. 2013. Managing for interactions between local and global stressors of ecosystems. PLOS (Public Library of Science) ONE [online serial] 8(6):e65765.

Brown, G. 2018. Dirt to soil: one family's journey into regenerative agriculture. Chelsea Green Publishing, White River Junction, Vermont.

Brown, L. R., T. F. Cuffney, J. F. Coles, F. Fitzpatrick, G. McMahon, J. Steuer, A. H. Bell, and J. T. May. 2009. Urban streams across the USA: lessons learned from studies in 9 metropolitan areas. Journal of the North American Benthological Society 28:1051–1069.

Bruckerhoff, L. A., and K. B. Gido. 2019. Assessing site-selection strategies for modeling the influence of landscape factors on stream fish assemblages. Pages 159–177 *in* R. M. Hughes, D. M. Infante, L. Wang, K. Chen, and B. F. Terra, editors. Advances in understanding landscape influences on freshwater habitats and biological assemblages. American Fisheries Society, Symposium 90, Bethesda, Maryland.

Callisto, M., D. R. Macedo, M. S. Linares, and R. M. Hughes. 2019. Multistatus and multispatial scale assessment of landscape effects on benthic macroinvertebrates in the Neotropical savanna. Pages 275–302 *in* R. M. Hughes, D. M. Infante, L. Wang, K. Chen, and B. F. Terra, editors. Advances in understanding landscape influences on freshwater habitats and biological assemblages. American Fisheries Society, Symposium 90, Bethesda, Maryland.

Cardoso, P., F. Rigal, S. Fattorini, S. Terzopoulou, and P. A. Borges. 2013. Integrating landscape disturbance and indicator species in conservation studies. PLOS (Public Library of Science) ONE [online serial] 8(5):e63294.

Chapin, F. S., III, E. S. Zavaleta, V. T. Eviner, R. L. Naylor, P. M. Vitousek, H. L. Reynolds, D. U. Hooper, S. Lavorel, O. E. Sala, S. E. Hobbie, and M. C. Mack. 2000. Consequences of changing biodiversity. Nature 405:234–242.

Chariton, A. A., M. Sun, J. Gibson, J. A. Webb,

K. M. Y. Leung, C. W. Hickey, and G. C. Hose. 2016. Emergent technologies and analytical approaches for understanding the effects of multiple stressors in aquatic environments. Marine and Freshwater Research 67:414–428.

Chen, K., R. M. Hughes, J. G. Brito, C. G. Leal, R. P. Leitão, J. M. B. de Oliveira-Júnior, V. C. de Oliveira, K. Dias-Silva, S. F. B. Ferraz, J. Ferreira, N. Hamada, L. Juen, J. Nessimian, P. S. Pompeu, J. Zuanon. 2017. A multi-assemblage, multi-metric biological condition index for eastern Amazon streams. Ecological Indicators 78:48–61.

Colvin, S. A. R., S. M. P. Sullivan, P. D. Shirey, R. W. Colvin, K. O. Winemiller, R. M. Hughes, K. D. Fausch, D. M. Infante, J. D. Olden, K. R. Bestgen, R. J. Danehy, and L. Eby. 2019. Headwater streams and wetlands are critical for sustaining fish, fisheries, and ecosystem services. Fisheries 2:73–91.

Comte, L., and J. D. Olden. 2017. Climatic vulnerability of the world's freshwater and marine fishes. Nature Climate Change 7:718–722.

Condon, L. E., and R. M. Maxwell. 2019. Simulating the sensitivity of evapotranspiration and streamflow to large-scale groundwater depletion. Science Advances [online serial] 5(6):eaav4574.

Cooper, A. R., D. M. Infante, J. A. Ross, W. M. Daniel, K. Herreman, C. C. Estes, and G. E. Whelan. 2019. Contemporary and future landscape-based anthropogenic disturbances to Alaskan streams: implications for anadromous fish habitat conservation. Pages 229–253 *in* R. M. Hughes, D. M. Infante, L. Wang, K. Chen, and B. F. Terra, editors. Advances in understanding landscape influences on freshwater habitats and biological assemblages. American Fisheries Society, Symposium 90, Bethesda, Maryland.

Cooper, A. R., D. M. Infante, L. Wang, K. Wehrly, and T. Brenden. 2017. Assessment of dam effects on streams and fish assemblages of the conterminous USA. Science of the Total Environment 586:879–889.

Couto, T. B., and J. D. Olden. 2018. Global proliferation of small hydropower plants—science and policy. Frontiers in Ecology and the Environment 16:91–100.

Cox, K. D., G. A. Covernton, H. L. Davies, J. F. Dower, F. Juanes, and S. E. Dudas. 2019. Human consumption of microplastics. Environmental Science and Technology 53:7068–7074.

Crawford, S., G. Whelan, G., D. M. Infante, K. Blackhart, W. M. Daniel, P. L. Fuller, T. Birdsong, D. J. Wieferich, R. McClees-Funinan, S. M. Stedman, K. Herreman, and P. Ruhl. 2016. Through a fish's eye: the status of fish habitats in the United States, 2015. National Fish Habitat Partnership. Available: http://assessment.fishhabitat.org. (June 2019).

Crews, T. E., W. Carton, and L. Olsson. 2018. Is the future of agriculture perennial? Imperatives and opportunities to reinvent agriculture by shifting from annual monocultures to perennial polycultures. Global Sustainability [online serial] 1:e11.

Crouse, D. L., L. Pinault, A. Balram, P. Hystad, P. A. Peters, H. Chen, A. van Dankelaar, R. V. Martin, R. Ménard, A. Robichaud, and P. J. Villeneuve. 2017. Urban greenness and mortality in Canada's largest cities: a national cohort study. Lancet Planet Health [online serial] 1:e289–e297.

Cutler, D. R., T. C. Edwards, Jr, K. H. Beard, A. Cutler, K. T. Hess, J. Gibson, and J. J. Lawler. 2007. Random forests for classification in ecology. Ecology 88:2783–2792.

Daniel, W. M., A. R. Cooper, P. J. Badra, and D. M. Infante. 2018. Predicting habitat suitability for eleven imperiled fluvial freshwater mussels. Hydrobiologia 809:265–283.

Davies, P. E , J. Harris, T. Hillman, and K. Walker. 2010. The Sustainable Rivers Audit: assessing river ecosystem health in the Murray–Darling basin, Australia. Marine and Freshwater Research 61:764–777.

Davis, C. D., C. W. Epps, R. L. Flitcroft, and M. A. Banks. 2019. Beyond isolation by distance: riverscape effects on genetic structure of fall-run Chinook Salmon. Pages 35–56 *in* R. M. Hughes, D. M. Infante, L. Wang, K. Chen, and B. F. Terra, editors. Advances in understanding landscape influences on freshwater habitats and biological assemblages. American Fisheries Society, Symposium 90, Bethesda, Maryland.

De'ath, G., and K. E. Fabricius. 2000. Classification and regression trees: a powerful yet simple technique for ecological data analysis. Ecology 81:3178–3192.

Denevan. W. M. 2011. The "pristine myth" revisited. The Geographical Review 101:576–591.

Deweber, J. T., L. Sleezer, and E. A. Frimpong. 2019. A new regionalization framework to quantify how physiography mediates the effect of land use on stream fishes. Pages 321–350 *in* R. M. Hughes, D. M. Infante, L. Wang, K. Chen, and B. F. Terra, editors. Advances in understanding landscape influences on freshwater habitats and biological assemblages. American Fisheries Society, Symposium 90, Bethesda, Maryland.

Deweber, J. T., Y. P. Tsang, D. Krueger, J. B. Whittier, T. Wagner, D. M. Infante, and G. Whelan. 2014. Importance of understanding landscape biases in USGS gage locations: Implications and solutions for managers. Fisheries 39:155–163.

Downing, J. A., T. T. Prairie, J. J Cole, C. M. Duarte, L. J. Tranvik, R. G. Striegl, W. H. McDowell, P. Kortelainen, N. F. Caraco, J. M. Melack, and J. J. Middelburg. 2006. The global abundance and size distribution of lakes, ponds, and impoundments. Limnology and Oceanography 51:2388–2397.

Drake, D. A. R., and N. E. Mandrak. 2014. Bycatch, bait, anglers, and roads: quantifying vector activity and propagule introduction risk across lake ecosystems. Ecological Applications 24:877–894.

Dudgeon, D., A. H. Arthington, M. O. Gessner, Z.-I. Kawabata, D. J. Knowler, C. Leveque, R. J. Naiman, A.-H. Prieur-Richard, D. Soto, M. L. J. Stiassny, and C. A. Sullivan. 2007. Freshwater biodiversity: importance, threats, status and conservation challenges. Biological Reviews 81:163–182.

Dunne, T., and L. B. Leopold. 1978. Water in environmental planning. W. H. Freeman, San Francisco, California.

Durden, J. M., J. Y. Luo, H. Alexander, A. M. Flanagan, and L. Grossman. 2017. Integrating "big data" into aquatic ecology: challenges and opportunities. Limnology and Oceanography Bulletin 26:101–108

DWAF (Department of Water Affairs and Forestry). 2008. National Aquatic Ecosystem Health Monitoring Programme: River Health Programme implementation manual, version 2. DWAF, Pretoria, South Africa.

Egan, D. 2017. The death and life of the Great Lakes. W.W. Norton, New York.

Ellis, N., S. J. Smith, and C. R. Pitcher. 2012. Gradient forests: calculating importance gradients on physical predictors. Ecology 93:156–168.

England, J., and A. M. Gurnell. 2016. Incorporating catchment to reach scale processes into hydromorphological assessment in the UK. Water and Environment Journal 30:22–30.

Esselman, P. C., D. M. Infante, L. Wang, A. R. Cooper, D. Wieferich, Y. Tang, D. J. Thornbrugh, and W. W. Taylor. 2013. Regional fish community indicators of landscape disturbance to catchments of the conterminous United States. Ecological Indicators 26:163–173.

Evans, J. S., M. A. Murphy, Z. A. Holden, and S. A. Cushman. 2011. Modeling species distribution and change using random forest. Pages 139–159 *in* C. A. Drew, Y. F. Wiersma, and F. Huettmann, editors. Predictive species and habitat modeling in landscape ecology. Springer, New York.

Falke, J. A., L. L. Bailey, K. D. Fausch, and K. R. Bestgen. 2012. Colonization and extinction in dynamic habitats: an occupancy approach for a Great Plains stream fish assemblage. Ecology 93:858–867.

Falke, J. A, B. M. Huntsman, and E. R. Schoen. 2019. Climatic variation drives growth potential of juvenile Chinook Salmon along a subarctic boreal riverscape. Pages 57–82 *in* R. M. Hughes, D. M. Infante, L. Wang, K. Chen, and B. F. Terra, editors. Advances in understanding landscape influences on freshwater habitats and biological assemblages. American Fisheries Society, Symposium 90, Bethesda, Maryland.

Fausch, K. D., C. E. Torgersen, C. V. Baxter, and H. W. Li. 2002. Landscapes to riverscapes: bridging the gap between research and conservation of stream fishes. BioScience 52:483–498.

Feld, C. K., S. Birk, D. Eme, M. Gerisch, D. Hering, M. Kernan, K. Maileht, U. Mischke, I. Ott, F. Pletterbauer, S. Poikane, J. Salgado, C. D. Sayer, J. van Wichelen, and F. Malard. 2016. Disentangling the effects of land use and geo-climatic factors on diversity in European freshwater ecosystems. Ecological Indicators. 60:71–83.

FEMAT (Forest Ecosystem Management Assessment Team). 1993. Forest ecosystem management: an ecological, economic, and social assessment. U.S. Forest Service, National Marine Fisheries Service, U.S. Bureau of Land Management, U.S. Fish and Wildlife Service, National Park Service, and U.S. Environmental Protection Agency, Washington, D.C.

Findell, K. L., A. Berg, P. Gentine, J. P. Krasting, B. R. Lintner, S. Malyshev, J. A. Santanello, and E. Shevliakova. 2017. The impact of anthropogenic land use and land cover change on regional climate extremes. Nature Communications [online serial] 8:989.

Firmiano, K. R., R. Ligeiro, D. R. Macedo, L. Juen, R. M. Hughes, and M. Callisto. 2017. Mayfly bioindicator thresholds for several anthropogenic disturbances in neotropical savanna streams. Ecological Indicators 74:276–284.

França, J. S., R. S. Solar, R. M. Hughes, and M. Callisto. 2019. Student monitoring of the ecological quality of neotropical urban streams. Ambio 48:867–878.

Fullerton, A. H., K. M. Burnett, E. A. Steel, R. L. Flitcroft, G. R. Pess, B. E. Feist, C. E. Torgersen, D. J. Miller, and B. L. Sanderson. 2010. Hydrological connectivity for riverine fish: measurement challenges and research opportunities. Freshwater Biology 55:2215–2237.

Gillespie, N., A. Unthank, L. Campbell, P. Anderson, R. Gubernick, M. Weinhold, D. Cenderelli, B. Austin, D. McKinley, S. Wells, J. Rowan, C. Orvis, M. Hudy, A. Bowden, A. Singler, E. Fretz, J. Levine, and R. Kirn. 2014. Flood effects on road–stream crossing infrastructure: economic and ecological benefits of stream simulation designs. Fisheries 39:62–76.

Grace, J. B. 2008. Structural equation modeling for observational studies. Journal of Wildlife Management 72:14–22.

Green Party US. 2019. Green New Deal—full language. Available: www.gp.org/gnd_full. (June 2019).

Grill, G., B. Lehner, M. Thieme, B. Geenen, D. Tickner, F. Antonelli, S. Babu, P. Borrelli, L. Cheng, H. Crochetiere, H. Ehalt Macedo, R. Filgueiras, M. Goichot, J. Higgins, Z. Hogan, B. Lip, M. E. McClain, J. Meng, M. Mulligan, C. Nilsson, J. D. Olden, J. J. Opperman, P. Petry, C. Reidy Liermann, L. Sáenz, S. Salinas-Rodríguez, P. Schelle, R. J. P. Schmitt, J. Snider, F. Tan, K. Tockner, P. H. Valdujo, A. van Soesbergen, and C. Zarfl. 2019. Mapping the world's free-flowing rivers. Nature 569:215–221.

Grimm, N. B., D. Foster, P. Groffman, J. M. Grove, C. S. Hopkinson, K. J. Nadelhoffer, D. E. Pataki, and D. P. Peters. 2008. The changing landscape: ecosystem responses to urbanization and pollution across climatic and societal gradients. Frontiers in Ecology and the Environment 6:264–272.

Guisan, A., T. C. Edwards, Jr., and T. Hastie. 2002. Generalized linear and generalized additive models in studies of species distributions: setting the scene. Ecological Modelling 157(2–3):89–100.

Guse, B., J. Kail, J. Radinger, M. Schröder, J. Kiesel, D. Hering, C. Wolter, and N. Fohrer.

2015. Eco-hydrologic model cascades: simulating land use and climate change impacts on hydrology, hydraulics and habitats for fish and macroinvertebrates. Science of the Total Environment 533:542–556.

Harding, J. S., E. F. Benfield, P. V. Bolstad, G. S. Helfman, and E. B. D. Jones, III. 1998. Stream biodiversity: the ghost of land use past. Proceedings of the National Academy of Sciences of the United States of America 95:14843–14847.

Hawkins, C. P., J. R. Olson, and R. A. Hill. 2010. The reference condition: predicting benchmarks for ecological and water-quality assessments. Journal of the North American Benthological Society 29:312–343.

Hayes, M. M., S. N. Miller, and M. A. Murphy. 2014. High-resolution landcover classification using Random Forest. Remote Sensing Letters 5:112–121.

He, Y., J. Wang, S. Lek-Ang, and S. Lek. 2010. Predicting assemblages and species richness of endemic fish in the upper Yangtze River. Science of the Total Environment 408:4211–4220.

Heitke, J. D., C. L. Pierce, G. T. Gelwicks, G. A. Simmons, and G. L. Siegworth. 2006. Habitat, land use, and fish assemblage relationships in Iowa streams: preliminary assessment in an agricultural landscape. Pages 207–303 *in* R. M. Hughes, L. Wang, and P. W. Seelbach, editors. Landscape influences on stream habitats and biological assemblages. American Fisheries Society, Symposium 48, Bethesda, Maryland.

Helfman, G. S. 2007. Fish conservation: a guide to understanding and restoring global aquatic biodiversity and fishery resources. Island Press, Washington D.C.

Hering, D., A. Borja, J. Caarstensen, L. Carvalho, M. Elliot, C. K. Feld, A.-S. Heiskanen, R. K. Johnson, J. Moe, D. Pont, A. L. Solheim, and W. van de Bund. 2010. The European Water Framework Directive at the age of 10: a critical review of the achievements with recommendations for the future. Science of the Total Environment 408:4007–4019.

Herlihy, A. T., J. C. Sifneos, R. M. Hughes, D. V. Peck, and R. M. Mitchell. 2019. Lotic fish assemblage clusters across the conterminous United States and their associations with environmental variables. Pages 385–408 *in* R. M. Hughes, D. M. Infante, L. Wang, K. Chen, and B. F. Terra, editors. Advances in understanding landscape influences on freshwater habitats and biological assemblages. American Fisheries Society, Symposium 90, Bethesda, Maryland.

Hilderbrand, R. H., R. M. Utz, S. A. Stranko, and R. L. Raesly. 2010. Applying thresholds to forecast potential biodiversity loss from human development. Journal of the North American Benthological Society 29:1009–1016.

Hill, R. A., C. P. Hawkins, and J. Jin. 2014. Predicting thermal vulnerability of stream and river ecosystems to climate change. Climatic Change 125:399–412.

Hill, R. A., M. H. Weber, R. M. Debbout, S. G. Leibowitz, and A. R. Olsen. 2018. The lake-catchment (LakeCat) dataset: characterizing landscape features for lake basins within the conterminous USA. Freshwater Science 37:208–221.

Hill, R. A., M. H. Weber, S. G. Leibowitz, A. R. Olsen, and D. J. Thornbrugh. 2016. The stream-catchment (StreamCat) dataset: a database of watershed metrics for the conterminous United States. Journal of the American Water Resources Association 52:120–128.

Hill, R. A., E. W. Fox, S. G. Leibowitz, A. R. Olsen, D. J. Thornbrugh, and M. A. Weber. 2017. Predictive mapping of the biotic condition of conterminous U.S. rivers and streams. Ecological Applications 27:2397–2415.

Hughes, R. M. 2019. Ecological integrity: conceptual foundations and applications. In E. Wohl, editor. Oxford bibliographies in environmental science. Oxford University Press, New York. Available: www.oxfordbibliographies.com/view/document/obo-9780199363445/obo-9780199363445-

0113.xml?rskey=Mfte5h&result=21. (June 2019).

Hughes, R. M., and S. Dunham. 2014. Fish passage through urban and rural-residential areas. Pages 93–100 *in* J. A. Yeakley, K. G. Maas-Hebner, and R. M. Hughes, editors. Wild salmonids in the urbanizing Pacific Northwest. Springer, New York.

Hughes, R. M., and J. R. Gammon. 1987. Longitudinal changes in fish assemblages and water quality in the Willamette River, Oregon. Transactions of the American Fisheries Society 116:196–209.

Hughes, R. M., and A. T. Herlihy. 2012. Patterns in catch per unit effort of native prey fish and alien piscivorous fish in 7 Pacific Northwest USA rivers. Fisheries 37:201–211.

Hughes, R. M., and D. V. Peck. 2008. Acquiring data for large aquatic resource surveys: the art of compromise among science, logistics, and reality. Journal of the North American Benthological Society 27:837–859.

Hughes, R. M., F. Amezcua, D. M. Chambers, W. M. Daniel, J. S. Franks, W. Franzin, D. MacDonald, E. Merriam, G. Neall, P. dos Santos Pompeu, and L. Reynolds. 2016. AFS position paper and policy on mining and fossil fuel extraction. Fisheries 41:12–15.

Hughes, R. M., B. L. Bangs, S. V. Gregory, P. D. Scheerer, R. C. Wildman, and J. S. Ziller. 2019. Recovery of Willamette River (Oregon, USA) fish assemblages: successes and remaining threats. Pages 157–184 *in* C. C. Krueger, W. W. Taylor, and S. Youn, editors. From catastrophe to recovery: stories of fishery management success. American Fisheries Society, Bethesda, Maryland.

Hughes, R. M., S. Dunham, K. G. Maas-Hebner, J. A. Yeakley, C. B. Schreck, M. Harte, N. Molina, C. C. Shock, V. W. Kaczynski, and J. Schaeffer. 2014. A review of urban water body rehabilitation and mitigation challenges and approaches: (1) rehabilitation and remediation. Fisheries 39:18–29.

Hughes, R. M., J. N. Rinne, and B. Calamusso. 2005. Historical changes in large river fish assemblages of the Americas: a synthesis. Pages 603–612 *in* J. N. Rinne, R. M. Hughes, and B. Calamusso, editors. Historical changes in large river fish assemblages of the Americas. American Fisheries Society, Symposium 45, Bethesda, Maryland.

Hughes, R. M., L. Wang, and P. W. Seelbach, editors. 2006. Landscape influences on stream habitat and biological assemblages. American Fisheries Society, Symposium 48, Bethesda, Maryland.

Isaac, D. J., C. H. Luce, D. L. Horan, G. L. Chandler, S. P. Wollrab, and D. E. Nagel. 2018. Global warming of salmon and trout rivers in the northwestern U.S.: road to ruin or path through purgatory? Transactions of the American Fisheries Society 147:566–585.

IUCN (International Union for the Conservation of Nature). 2012. Child's right to connect with nature and to a healthy environment. Available: https://portals.iucn.org/library/sites/library/files/resrecfiles/WCC_2012_RES_101_EN.pdf. (June 2019).

Jacobson, P. C., G. J. A. Hansen, L. G. Olmanson, K. E. Wehrly, C. L. Hein, and L. B. Johnson. 2019. Loss of coldwater fish habitat in glaciated lakes of the Midwestern United States after a century of land use and climate change. Pages 141–157 *in* R. M. Hughes, D. M. Infante, L. Wang, K. Chen, and B. F. Terra, editors. Advances in understanding landscape influences on freshwater habitats and biological assemblages. American Fisheries Society, Symposium 90, Bethesda, Maryland.

Janssen, A. B. G., G. B. Arhonditsis, A. Beusen, K. Bolding, L. Bruce, J. Bruggeman, R.-M. Couture, A. S. Downing, J. Alex Elliott, M. A. Frassl, G. Gal, D. J. Gerla, M. R. Hipsey, F. Hu, S. C. Ives, J. H. Janse, E. Jeppesen, K. D. Jöhnk, D. Kneis, X. Kong, J. J. Kuiper, M. K. Lehmann, C. Lemmen, D. Özkundakci, T. Petzoldt, K. Rinke, B. J. Robson, R. Sachse, S. A. Schep, M. Schmid, H. Scholten, S. Teurlincx, D. Trolle, T. A. Troost, A. A. Van Dam, L. P. A. Van Gerven, M.

Weijerman, S. A. Wells, and W. M. Mooij. 2015. Exploring, exploiting and evolving diversity of aquatic ecosystem models: a community perspective. Aquatic Ecology 49:513–548.

Januchowski-Hartley, S. R., L.A. Holtz, S. Martinuzzi, P. B. McIntyre, V. C. Radeloff, and B. M. Pracheil. 2016. Future land use threats to range-restricted fish species in the United States. Diversity and Distributions 22:663–671.

Jelks, H. J., S. J. Walsh, M. M. Burkhead, S. Contreras-Balderas, E. Diaz-Pardo, D. A. Hendrickson, J. Lyons, N. E. Mandrack, F. McCormick, J. S. Nelson, S. P. Platania, B. A. Porter, C. B. Renaud, J. J. Schmitter-Soto, E. B. Taylor, and M. L. Warren, Jr. 2008. Conservation status of imperiled North American freshwater and diadromous fishes, Fisheries 33:372–407.

Johnson, L. B., and G. E. Host. 2010. Recent developments in landscape approaches for the study of aquatic ecosystems. Journal of the North American Benthological Society 29:41–66.

Jones, J. A. 2011. Hydrologic responses to climate change: considering geographic context and alternative hypotheses. Hydrological Processes 25:1996–2000.

Jowett, I. G. 1997. Instream flow methods: a comparison of approaches. Regulated Rivers: Research and Management 13:115–127.

Kaufmann, P. R. 2000. Physical habitat characterization—non-wadeable rivers. Pages 6.1–6.29 in J. M. Lazorchak, B. H. Hill, D. K. Averill, D. V. Peck, and D. J. Klemm, editors. Environmental monitoring and assessment program—surface waters: field operations and methods for measuring the ecological condition of non-wadeable rivers and streams. U.S. Environmental Protection Agency, Cincinnati, Ohio.

Kleynhans, C. J. 1999. The development of a fish index to assess the biological integrity of South African rivers. Water SA 25:265–278.

Landers, D. H., S. M. Simonich, D. Jaffe, L. Geiser, D. H. Campbell, A. Schwindt, C. Schreck, M. Kent, W. Hafner, H. E. Taylor, K. Hageman, S. Usenko, L. Ackerman, J. Schrlau, N. Rose, T. Blett, and M. M. Erway. 2010. The Western Airborne Contaminant Assessment Project (WACAP): an interdisciplinary evaluation of the impacts of airborne contaminants in western national parks. Environmental Science and Technology 44:855–859.

Larsen, D. P., P. R. Kaufmann, T. M. Kincaid, and N. S. Urquhart. 2004. Detecting persistent change in the habitat of salmon-bearing streams in the Pacific Northwest. Canadian Journal of Fisheries Science 61:283–291.

Larsen, D. P., A. R. Olsen, S. H. Lanigan, C. Moyer, K. K. Jones, and T. M. Kincaid. 2007. Sound survey designs can facilitate integrating stream monitoring data across multiple programs. Journal of the American Water Resources Association 43:384–397.

Leal, C. G., J. Barlow, T. Gardner, R. M. Hughes, R. P. Leitão, R. MacNally, P. Kaufmann, S. F. B. Ferraz, J. Zuanon, F. R. de Paula, J. Ferreira, J. R. Thomson, G. D. Lennox, E. P. Dary, C. P. Röpke, and P. S. Pompeu. 2018. Is environmental legislation conserving tropical stream faunas? A large-scale assessment of local, riparian and catchment-scale influences on Amazonian fish. Journal of Applied Ecology 55:1312–1326.

Leibowitz, S. G., P. J. Wigington, Jr., K. A. Scholfield, L. C. Alexander, M. K. Vanderhoof, and H. E. Golden. 2018. Connectivity of streams and wetlands to downstream waters: an integrated systems framework. Journal of the American Water Resources Association 54:298–322.

Leitão, R. P., J. Zuanon, D. Mouillot, C. G. Leal, R. M. Hughes, P. R. Kaufmann, S. Villéger, P. S. Pompeu, D. Kasper, F. R. de Paula, S. F. B. Ferraz, and T. Gardner. 2018. Disentangling the pathways of land use impacts on the functional structure of fish assemblages in Amazon streams. Ecography 41:219–232.

Leitão, R. P., J. Zuanon, S. Villéger, S. E. Williams, C. Baraloto, C. Fortunel, F. P. Mendonça, and D. Mouillot. 2016. Rare species contribute disproportionately to the functional structure of species assemblages. Proceedings of the Royal Society B 283(1828):20160084.

Limburg, K. E., R. M. Hughes, D. C. Jackson, and B. Czech. 2011. Population increase, economic growth, and fish conservation: collision course or savvy stewardship? Fisheries 36:27–35.

Lohan, T. 2018. The Elwha's living laboratory: lessons from the world's largest dam-removal project. The Revelator. Available: https://therevelator.org/elwha-dam-removal/. (June 2019).

Lyons, J., J. S. Stewart, and M. Mitro. 2010. Predicted effects of climate warming on the distribution of 50 stream fishes in Wisconsin, USA. Journal of Fish Biology. 77:1867–1898.

Maddock, I., 1999. The importance of physical habitat assessment for evaluating river health 41:373–391.

Mann, C. C. 2005. 1491: new revelations of the Americas before Columbus. Knopf, New York.

Marchese, C. 2015. Biodiversity hotspots: a shortcut for a more complicated concept. Global Ecology and Conservation 3:297–309.

Martin, B., Y.-P. Tsang, R. W. Tingley, III, H. Cliverd, and D. M. Infante. 2019. Characterizing natural barriers to nonnative stream fauna in Hawai'i. Pages 83–103 *in* R. M. Hughes, D. M. Infante, L. Wang, K. Chen, and B. F. Terra, editors. Advances in understanding landscape influences on freshwater habitats and biological assemblages. American Fisheries Society, Symposium 90, Bethesda, Maryland.

Martins, I., R. Ligeiro, R. M. Hughes, D. R. Macedo, and M. Callisto. 2018. Regionalization is key to establish reference conditions for Neotropical savanna streams. Marine and Freshwater Research 69:82–94.

McIntyre, P. B., C. Reidy Liermann, E. Childress, E. J. Hamann, J. D. Hogan, S. R. Januchowski-Hartley, A. A. Koning, T. M. Neeson, D. L. Oele, and B. M. Pracheil. 2016. Conservation of migratory fishes in freshwater ecosystems. Cambridge University Press, Cambridge, England.

McKenna J. E., Jr., and J. H. Johnson. 2011. Landscape models of Brook Trout abundance and distribution in lotic habitat with field validation. North American Journal of Fisheries Management 31:742–756.

McManamay, R. A., and C. R. DeRolph. 2019. A stream classification system for the conterminous United States. Scientific Data [online serial] 6:190017. Available: www.nature.com/articles/sdata201917. (June 2019).

Medley, C. N., and P. D. Shirey. 2013. Review and reinterpretation of Rio Grande silvery minnow reproductive ecology using egg biology, life history, hydrology and geomorphology information. Ecohydrology 6:491–505.

Meire, R. O., S. C. Lee, Y. Yao, A. C. Targino, J. P. M. Torres, and T. Harner. 2012. Seasonal and altitudinal variations of legacy and current-use pesticides in the Brazilian tropical and subtropical mountains. Atmospheric Environment 59:108–116.

Melles, S. J., N. E. Jones, and B. J. Schmidt. 2014. Aquatic ecosystem classification for Ontario: a technical proposal. Ontario Ministry of Natural Resources, Aquatic Research Series 2013-05, Ottawa.

Meyer, J. L., L. A. Kaplan, J. D. Newbold, D. L. Strayer, C. J. Woltemade, J. B. Zedler, R. Beilfuss, Q. Carpenter, R. Semlitsch, M. C. Watzin, and P. H. Zedler. 2003. Where rivers are born: the scientific imperative for defending small streams and wetlands. Sierra Club and American Rivers, Washington, D.C.

Meyer, J. L., D. L. Strayer, J. B. Wallace, S. L. Eggert, G. S. Helfman, and N. E. Leonard. 2007. The contribution of headwater streams to biodiversity in river networks. Journal of the American Water Resources Association 43:86–103.

Miller, M. P., S. M. Carlisle, D. M. Wolock, and M. Wieczorek. 2018. A database of natural monthly streamflow estimates from 1950 to

2015 for the conterminous United States. Journal of the American Water Resources Association 54:1258–1269.

Mostafavi, H., Z. Kordjazi, R. Valavi, H. Shafizadeh-Moghadam, J. Kambouzia, and D. M. Infante. 2019. Ensemble modelling of sensitive stream fish species distributions in Iran: expanding knowledge to aid species conservation. Pages 441–462 *in* R. M. Hughes, D. M. Infante, L. Wang, K. Chen, and B. F. Terra, editors. Advances in understanding landscape influences on freshwater habitats and biological assemblages. American Fisheries Society, Symposium 90, Bethesda, Maryland.

Moya, N., R. M. Hughes, E. Dominguez, F.-M. Gibon, E. Goita, and T. Oberdorff. 2011. Macroinvertebrate-based multimetric predictive models for measuring the biotic condition of Bolivian streams. Ecological Indicators 11:840–847.

Muhlfeld, C. C., R. P. Kovach, L. A. Jones, R. Al-Chokhachy, M. C. Boyer, R. F. Leary, W. H. Lowe, G. Luikart, and F. W. Allendorf. 2014. Invasive hybridization in a threatened species is accelerated by climate change. Nature Climate Change 4:620–624.

Mulvey, M., R. Leferink, and A. Borisenko. 2009. Willamette Basin rivers and streams assessment. Oregon Department of Environmental Quality, DEQ 09-LAB-016, Portland, Oregon.

Naiman, R. J., C. A. Johnston, and J. C. Kelley. 1988. Alteration of North American streams by beaver. BioScience 38:753–762.

Negishi, J. N. 2019. Does the proportion of crop fields in a small catchment predict coupled benthic–hyporheic invertebrate responses? Pages 255–274 *in* R. M. Hughes, D. M. Infante, L. Wang, K. Chen, and B. F. Terra, editors. Advances in understanding landscape influences on freshwater habitats and biological assemblages. American Fisheries Society, Symposium 90, Bethesda, Maryland.

Nel, J. L., B. Reyers, D. J. Roux, N. D. Impson, and R. M. Cowling. 2010. Designing a conservation area network that supports the representation and persistence of freshwater biodiversity. Freshwater Biology 56:106–124.

Oberdorff, T., and R. M. Hughes. 1992. Modification of an index of biotic integrity based on fish assemblages to characterize rivers of the Seine basin, France. Hydrobiologia 228:117–130.

Oberdorff, T., and R. M. Hughes. 2019. 4.2.3 Freshwater ecosystems. Pages 1083–1091 *in* Y.-J. Shin, A. Arneth, R. R. Chowdhury, and G. F. Midgley, editors. IPBES global assessment on biodiversity and ecosystem services. Chapter 4: plausible futures of nature, its contributions to people and their good quality of life. Intergovernmental Science-Policy Platform on Biodiversity and Ecosystem Services. Available: www.biodiversidad.gob.mx/planeta/internacional/pdf/ipbes_7_inf_1_annex.pdf. (June 2019).

Oberdorff, T., D. Pont, B. Hugueny, and J.-P. Porchers. 2002. Development and validation of a fish-based index for the assessment of 'river health' in France. Freshwater Biology 47:1720–1734.

Olaya-Marín, E. J., F. Martínez-Capel, and P. Vezza. 2013. A comparison of artificial neural networks and random forests to predict native fish species richness in Mediterranean rivers. Knowledge and Management of Aquatic Ecosystems [online serial] 409:07. Available: www.kmae-journal.org/articles/kmae/pdf/2013/02/kmae120133.pdf. (June 2019).

Oliver, T. H., and M. Morecroft. 2014. Interactions between climate change and land use change on biodiversity: attribution problems, risks, and opportunities. Wiley Interdisciplinary Reviews: Climate Change 5:317–335.

Ollis, D. J., J. L. Ewart-Smith, J. A. Day, N. M. Job, D. M. Macfarlane, C. D. Snaddon, E. J. J. Sieben, J. A. Dini, and N. Mbona.2015. The development of a classification system for inland aquatic ecosystems in South Africa. Water SA 41:727–745.

Olsen, A. R., and D. V. Peck. 2008. Survey design and extent estimates for the Wadeable Streams Assessment. Journal of the North American Benthological Society 27:822–836.

Omernik, J. M. 1987. Ecoregions of the conterminous United States. Annals of the Association of American Geographers 77:118–125.

Omernik, J. M., G. E. Griffith, R. M. Hughes, M. H. Weber, and J. B. Glover. 2017. How misapplication of the hydrologic unit framework diminishes the meaning of watersheds. Environmental Management 60:1–11.

Parrott, L., and W. S. Meyer. 2012. Future landscapes: managing within complexity. Frontiers in Ecology and the Environment 10:382–389.

Perkin, J. S., and K. B. Gido. 2012. Fragmentation alters stream fish community structure in dendritic ecological networks. Ecological Applications 22:2176–2187.

Perkin, J. S., K. B. Gido, J. A. Falke, K. D. Fausch, H. Crockett, E. R. Johnson, and J. Sanderson. 2017. Groundwater declines are linked to changes in Great Plains stream fish assemblages. Proceedings of the National Academy of Sciences of the United States of America 114:7373–7378.

Perkin, J. S., J. C. Wellemeyer, and J. D. Fore. 2019. Multiscale fish assemblage distribution models to guide riverscape conservation planning. Pages 409–440 *in* R. M. Hughes, D. M. Infante, L. Wang, K. Chen, and B. F. Terra, editors. Advances in understanding landscape influences on freshwater habitats and biological assemblages. American Fisheries Society, Symposium 90, Bethesda, Maryland.

Pesch, R., G. Schmidt, W. Schroeder, and I. Weustermann. 2011. Application of CART in ecological landscape mapping: two case studies. Ecological Indicators 11:115–122.

Peterson, S. A., J. Van Sickle, A. T. Herlihy, and R. M. Hughes. 2007. Mercury concentration in fish from streams and rivers throughout the western United States. Environmental Science and Technology 41:58–65.

Pitcher, C. R., P. Lawton, N. Ellis, S. J. Smith, L. S. Incze, C. L. Wei, M. E. Greenlaw, N. H. Wolff, J. A. Sameoto, and P. V. Snelgrove. 2012. Exploring the role of environmental variables in shaping patterns of seabed biodiversity composition in regional-scale ecosystems. Journal of Applied Ecology 49:670–679.

Poff, N. L. 1997. Landscape filters and species traits: towards mechanistic understanding and prediction in stream ecology. Journal of the North American Benthological Society 16:391–409.

Poff, N. L. 2019. A river that flows free connects up in 4D. Nature 569:201–202.

Poff, N. L., J. D. Olden, D. M. Merritt, and D. M. Pepin. 2007. Homogenization of regional river dynamics by dams and global biodiversity implications. Procedings of the National Academy of Sciences of the United States of America 14:5732–5737.

Pompeu, P. S., A. A. Agostinho, and F. M. Pelicice. 2012. Existing and future challenges: the concept of successful fish passage in South America. River Research and Applications 28:504–512.

Pompeu, P. S., C. G. Leal, D. R. Carvalho, N. T. Junqueira, M. A. Castro, and R. M. Hughes. 2019. Effects of catchment land use on stream fish assemblage structure in the Brazilian savanna. Pages 303–320 *in* R. M. Hughes, D. M. Infante, L. Wang, K. Chen, and B. F. Terra, editors. Advances in understanding landscape influences on freshwater habitats and biological assemblages. American Fisheries Society, Symposium 90, Bethesda, Maryland.

Pont, D., B. Hugueny, U. Beier, D. Goffaux, A. Melcher, R. Noble, C. Rogers, N. Roset, and S. Schmutz. 2006. Assessing river biotic condition at the continental scale: a European approach using functional metrics and fish assemblages. Journal of Applied Ecology 43:70–80.

Radinger, J., F. Hölker, P. Horký, O. Slavík, N. Dendoncker, and C. Wolter. 2016. Synergistic and antagonistic interactions of future land use and climate change on river fish assemblages. Global Change Biology 22:1505–1522.

Rahel, F. J. 2013. Intentional fragmentation as a management strategy in aquatic systems. BioScience 63:362–372.

Rakovec, O., R. Kumar, J. Mai, M. Cuntz, S. Thober, M. Zink, S. Attinger, D. Schäfer, M. Schrön, and L. Samaniego. 2016. Multiscale and multivariate evaluation of water fluxes and states over European river basins. Journal of Hydrometeorology 17:287–307.

Reznick, D. N., and C. K. Ghalambor. 2005. Selection in nature: experimental manipulations of natural populations. Integrative and Comparative Biology 45:456–462.

Ripple, W. J., and R. L. Beschta. 2011. Trophic cascades in Yellowstone: the first 15 years after wolf reintroduction. Biological Conservation 145:205–213.

Ripple, W. J., C. Wolf, T. M. Newsome, M. Galetti, M. Alamgir, E. Crist, M. I. Mahmoud, and W. F. Laurance. 2017. World scientists' warning to humanity: a second notice. BioScience 67:1026–1028.

Riseng, C., M. Wiley, R. Black, and M. Munn. 2011. Impacts of agricultural land use on biological integrity: a causal analysis. Ecological Applications 21:3128–3146.

Riseng, C. M., K. E. Wehrly, L. Wang, E. S. Rutherford, J. E. McKenna J.R, L. B. Johnson, L. A. Mason, C. Castiglione, T. P. Hollenhorst, B. L. Sparks-Jackson, and S. P. Sowa. 2018. Ecosystem classification and mapping of the Laurentian Great Lakes. Canadian Journal of Fisheries and Aquatic Sciences 75:1693–1712.

Robinson, W. A., M. Lintermans, J. H. Harris, and F. Guarino. 2019. A landscape-scale electrofishing monitoring program can evaluate fish responses to climatic conditions in the Murray–Darling River System, Australia. Pages 179–201 *in* R. M. Hughes, D. M. Infante, L. Wang, K. Chen, and B. F. Terra, editors. Advances in understanding landscape influences on freshwater habitats and biological assemblages. American Fisheries Society, Symposium 90, Bethesda, Maryland.

Roper, B. B., J. M. Buffington, S. Bennett, S. H. Lanigan, E. Archer, S. T. Downie, J. Faustini, T. W. Hillman, S. Hubler, K. Jones, C. Jordan, P. R. Kaufmann, G. Merritt, C. Moyer, and A. Pleus. 2010. A comparison of the performance and compatibility of protocols used by seven monitoring groups to measure stream habitat in the Pacific Northwest. North American Journal of Fisheries Management 30:565–587.

Ross, J. A., D. M. Infante, D. J. Martin, A. Cooper and K. Herreman. 2019. Influences of local and landscape characteristics, including past timber harvest practices, on large wood in streams draining young-growth Southeast Alaska catchments. Pages 203–227 *in* R. M. Hughes, D. M. Infante, L. Wang, K. Chen, and B. F. Terra, editors. Advances in understanding landscape influences on freshwater habitats and biological assemblages. American Fisheries Society, Symposium 90, Bethesda, Maryland.

Roubeix, V., P. A. Danis, T. Feret, and J. M. Baudoin. 2016. Identification of ecological thresholds from variations in phytoplankton communities among lakes: contribution to the definition of environmental standards. Environmental Monitoring and Assessment 188:246.

Roy, A. H., S. J. Wenger, T. D. Fletcher, C. J. Walsh, A. R. Ladson, W. D. Shuster, H. W. Thurston, and R. R. Brown. 2008. Impediments and solutions to sustainable, watershed-scale urban stormwater management: lessons from Australia and the United States. Environmental Management 42:344–359.

Rupert, M. G. 2008. Decadal-scale changes of nitrate in ground water of the United States, 1988-2004. Journal of Environmental Quality 37:S240–S248.

Sala, O. E., F. S. Chapin, J. J. Armesto, E. Berlow, J. Bloomfield, R. Dirzo, E. Huber-Sanwald, L. F. Huenneke, R. B. Jackson, A. Kinzig, and R. Leemans. 2000. Global biodiversity scenarios for the year 2100. Science 287:1770–1774.

Samhouri, J. F., K. S. Andrews, G. Fay, C. J. Harvey, E. L. Hazen, S. M. Hennessey, K. Holsman, M. E. Hunsicker, S. I. Large, K. N. Marshall, and A. C. Stier. 2017. Defining ecosystem thresholds for human activities and environmental pressures in the California Current. Ecosphere [online serial] 8:e01860.

Schindler, D. F., R. Hilborn, B. Chasco, C. P. Boatright, T. P. Quinn, L. A. Rogers, and M. S. Webster. 2010. Population diversity and the portfolio effect in an exploited species. Nature 465:609–612.

Schinegger, R., M. Palt, P. Segurado, and S. Schmutz. 2016. Untangling the effects of multiple human stressors and their impacts on fish assemblages in European running waters. Science of the Total Environment 573:1079–1088.

Schinegger, R., M. Pucher, C. Aschauer, and S. Schmutz. 2018. Configuration of multiple human stressors and their impacts on fish assemblages in alpine river basins of Austria. Science of the Total Environment 616–617:17–28.

Schmutz, S., I. G. Cowx, G. Haidvogl, and D. Pont. 2007. Fish-based methods for assessing European running waters: a synthesis. Fisheries Management and Ecology 14:369–380.

Scribner, K. T., W. H. Lowe, E. Landguth, G. Luikart, D. M. Infante, G. E. Whelan. C. C. Muhlfeld. 2016. Applications of genetic data to improve management and conservation of river fishes and their habitats. Fisheries 41:174–188.

Shipley, B. 2000. Cause and correlation in biology: a user's guide to path analysis, structural equations and causal inference. Cambridge University Press, Cambridge, UK.

Silva, D. R. O., R. Ligeiro, R. M. Hughes, and M. Callisto. 2016. The role of physical habitat and sampling effort on estimates of benthic macroinvertebrate taxonomic richness at basin and site scales. Environmental Monitoring and Assessment 188:340.

Sleight, N., and T. M. Neeson. 2018. Opportunities for collaboration between infrastructure agencies and conservation groups: road-stream crossings in Oklahoma. Transportation Research Part D: Transport and Environment 63:622–631.

Soranno, P. A., L. C. Bacon, M. Beauchene, K. E. Bednar, K. G. Bissell, C. K. Boudreau, M. G. Boyer, M. T. Bremigan, S. R. Carpenter, J. W. Carr, and K. S. Cheruvelil. 2017. LAGOS-NE: a multi-scaled geospatial and temporal database of lake ecological context and water quality for thousands of US lakes. GigaScience [online serial] 6(12):gix101.

Stanford, J. A., J. V. Ward, W. J. Liss, C. A. Frissell, R. N. Williams, J. A. Lichatowich, and C. C. Coutant. 1996. A general protocol for restoration of regulated rivers. Regulated Rivers Research and Management 12:391–413.

Steen, P. J., T. G. Zorn, P. W. Seelbach, and J. S. Schaeffer. 2008. Classification tree models for predicting distributions of Michigan stream fish from landscape variables. Transactions of the American Fisheries Society 137:976–996.

Stephenson, F., J. R. Leathwick, S. W. Geange, R. H. Bulmer, J. E. Hewitt, O. F. Anderson, A. A. Rowden, and C. J. Lundquist. 2018. Using gradient forests to summarize patterns in species turnover across large spatial scales and inform conservation planning. Diversity and Distributions 24:1641–1656.

Stevens, D. L, Jr., and A. R. Olsen. 2004. Spatially balanced sampling of natural resources. Journal of the American Statistical Association 99:262–278.

Stoddard, J. L, A. T. Herlihy, D. V. Peck, R. M. Hughes, T. R. Whittier, and E. Tarquinio. 2008. A process for creating multi-metric indices for large-scale aquatic surveys. Journal of the North American Benthological Society 27:878–891.

Stoddard, J. L., D. P. Larsen, C. P. Hawkins, R. K. Johnson, and R. H. Norris. 2006. Setting expectations for the ecological condition of streams: the concept of reference condition. Ecological Applications 16:1267–1276.

Stoddard, J. L., J. Van Sickle, A. T. Herlihy, J. Brahney, S. Paulsen, D. V. Peck, R. Mitchell, and A. I. Pollard. 2016. Continental-scale increase in lake and stream phosphorous: are oligotrophic systems disappearing in the United States? Environmental Science and Technology 50:3409–3415.

Stuber, E. F., L. F. Gruber, and J. J. Fontaine. 2017. A Bayesian method for assessing multi-scale species-habitat relationships. Landscape Ecology 32:2365–2381.

Tang, T., R. J. Stevenson, and D. M. Infante. 2016. Accounting for regional variation in both natural environment an human disturbance to improve performance of multimetic indices of lotic benthic diatoms. Science of the Total Environment 568:1124–1134.

Tang, T., T. Tang, L. Tan, Y. Gu, W. Jiang, and Q. Cai. 2017. Identifying community thresholds for lotic benthic diatoms in response to human disturbance. Scientific Reports [online serial] 7:4134.

Tank, J. L., E. J. Rosi-Marshall, M. A. Baker, R. O. Hall Jr. 2008. Are rivers just big streams? Using a pulse method to measure nitrogen demand in a large river. Ecology 89:2935–2945.

Terra, B. D. F., R. M. Hughes, and F. G. Araujo. 2013. Sampling sufficiency for fish assemblage surveys of Atlantic Forest streams, southeastern Brazil. Fisheries 38:150–158.

Thornbrugh, D. J., and D. M. Infante. 2019. Landscape effects on stream fishes: broad-scale responses to anthropogenic land use across temperate mesic regions of the United States. Pages 351–383 *in* R. M. Hughes, D. M. Infante, L. Wang, K. Chen, and B. F. Terra, editors. Advances in understanding landscape influences on freshwater habitats and biological assemblages. American Fisheries Society, Symposium 90, Bethesda, Maryland.

Tingley, R. W., D. M. Infante, R. A. MacKenzie, A. R. Cooper, and Y. P. Tsang. 2018. Identifying natural catchment landscape influences on tropical stream organisms: classifying stream reaches of the Hawaiian Islands. Hydrobiologia 826:67–83.

Tingley, R. W., III, D. M. Infante, W. M. Daniel, A. R. Cooper, G. C. Smith, and K. Herreman. 2019. Prioritizing Hawaii's stream habitats to conserve native species with changing climate. Pages 109–134 *in* D. C. Dauwalter, T. W. Birdsong, and G. P. Garrett, editors. Multispecies and watershed approaches to freshwater fish conservation. American Fisheries Society, Symposium 91, Bethesda, Maryland.

Tomarken, A. J., and N. G. Waller. 2005. Structural equation modeling: strengths, limitations, and misconceptions. Annual Review of Clinical Psychology 1:31–65.

Tukiainen, H., J. Alahuhta, R. Field, T. Ala-Hulkko, R. Lampinen, and J. Hjort. 2017. Spatial relationship between biodiversity and geodiversity across a gradient of land-use intensity in high-latitude landscapes. Landscape Ecology 32:1049–1063.

Tversky, A., and D. Kahneman. 1971. Belief in the law of small numbers. Psychological Bulletin 76(2):105–110.

Urban, M. C. 2015. Accelerating extinction risk from climate change. Science 348:571–573.

USEPA (U.S. Environmental Protection Agency). 2016a. National wetlands condition assessment 2011: a collaborative survey of the nation's wetlands. USEPA, EPA-843-R-15-005, Washington, D.C.

USEPA (U.S. Environmental Protection Agency). 2016b. National coastal condition assessment 2010. USEPA, EPA-843-R-15-006, Washington, D.C.

USEPA (U.S. Environmental Protection Agency). 2016c. National rivers and streams assessment 2008-2009: a collaborative survey. USEPA, EPA/841/R-16/007, Washington, D.C.

USEPA (U.S. Environmental Protection Agency). 2016d. National lakes assessment 2012: a collaborative survey of lakes in the United

States. USEPA, EPA841-R-16-113, Washington, D.C. Available: https://nationallakesassessment.epa.gov. (June 2019).

USGS (U.S. Geological Survey). 2014. National Water-Quality Assessment Project. About the program. Available: https://water.usgs.gov/nawqa/about.html. (June 2019).

Utz, R. M., R. H. Hilderbrand, and D. M. Boward. 2009. Identifying regional differences in threshold responses of aquatic invertebrates to land cover gradients. Ecological Indicators 9:556–567.

Utz, R. M., K. N. Eshleman, and R. H. Hilderbrand. 2011. Variation in physicochemical responses to urbanization in streams between two Mid-Atlantic physiographic regions. Ecological Applications 21:402–415.

Utz, R. M., R. H. Hilderbrand, and R. L. Raesly. 2010. Regional differences in patterns of fish species loss with changing land use. Biological Conservation 143:688–699.

Van Sickle, J., J. Baker, A. Herlihy, P. Bayley, S. Gregory, P. Haggerty, L. Ashkenas, and J. Li. 2004. Projecting the biological condition of streams under alternative scenarios of human land use. Ecological Applications 14:368–380.

Vander Laan, J. J., and C. P. Hawkins. 2014. Enhancing the performance and interpretation of freshwater biological indices: an application in arid zone streams. Ecological Indicators 36:470–482.

Vörösmarty, C. J., P. B. McIntyre, M. O. Gessner, D. Dudgeon, A. Prusevich, P. Green, S. Glidden, S. E. Bunn, C. A. Sullivan, C. R. Liermann, and P. M. Davies. 2010. Global threats to human water security and river biodiversity. Nature 467:555–561.

Wagenhoff, A., J. E. Clapcott, K. E. Lau, G. D. Lewis, and R. G. Young. 2017. Identifying congruence in stream assemblage thresholds in response to nutrient and sediment gradients for limit setting. Ecological Applications 27:469–484.

Wang, L., D. M. Infante, P. Esselman, A. Cooper, D. Wu, W. Taylor, D. Beard, G. Whelan, and A. Ostroff. 2011. A hierarchical spatial framework and database for the national river fish habitat condition assessment. Fisheries 36:436–449.

Wang L., D. Infante, C. Riseng, and K. Wehrly. 2016. Advancement of geospatial capability by NRiSD and GLAHF in enhancing aquatic ecosystem research and management. Geoinformatics and Geostatistics: An Overview 4:1–5.

Wang, L., C. M. Riseng, L. A. Mason, E. S. Rutherford, S. Sowa, L. Johnson, J. McKenna, D. Infante, K. Wehrly, C. Castiglione, J. Schaeffer, M. Khoury, M. Roberson, J. Ciborowski, and M. Coscarelli. 2015. A spatial classification and database for management, research, and policy making: the Great Lakes aquatic habitat framework. Journal of Great Lakes Research 41:584–596.

Wang, L., P. W. Seelbach, and R. M. Hughes. 2006. Introduction to landscape influences on stream habitats and biological assemblages. Pages 1–23 *in* R. M. Hughes, L. Wang, and P. W. Seelbach, editors. Influences of landscapes on stream habitats and biological assemblages. American Fisheries Society, Symposium 48, Bethesda, Maryland.

Warner, N. R., R. B. Jackson, T. H. Darrah, S. G. Osborn, A. Down, K. Zhao, A. White, and A. Vengosh. 2012. Geochemical evidence for possible natural migration of Marcellus Formation brine to shallow aquifers in Pennsylvania. Proceedings of the National Academy of Sciences of the United States of America 109:11961–11966.

Weber, M. A., and P. L. Ringold. 2019. River metrics by the public, for the public. PLOS (Public Library of Science) ONE [online serial] 14(5):e0214986.

Wessell, K. J., R. W. Merritt, J. G. O. Wilhelm, J. D. Allan, K. W. Cummins, and D. G. Uzarski. Biological evaluation of Michigan's non-wadeable rivers using macroinvertebrates. Aquatic Ecosystem Health and Management 11:335--351.

Whitehead, P., D. Butterfield, and A. Wade.

2008. Potential impacts of climate change on water quality and ecology in six UK rivers. Hydrology Research 40:113–122.

Wiens, J. A. 2002. Riverine landscapes: taking landcape ecology into the water. Freshwater Biology 47:501–515.

Wiley, M. J., S. L. Kohler, and P. W. Seelbach. 1997. Reconciling landsape and local views of aquatic communities: lessons from Michigan trout streams. Freshwater Biology 37:133–148.

Wilhelm, J. G. O., J. D. Allan, J. J. Wessell, R. W. Merritt, K. W. Cummins. 2005. Habitat assessment of non-wadeable rivers in Michigan. Environmental Management 36:592–609.

Wilson, E. O. 1986. Biophilia: the human bond with other species. Harvard University Press, Cambridge, Massachusetts.

Wimmer, R., A. Chovanec, O. Moog, M. H. Fink, and D. Gruber. 2000. Abiotic stream classification as a basis for a surveillance monitoring network in Austria in accordance with the EU Water Framework Directive. Acta Hydrochimica et Hydrobiologica 28:177–184.

Winfield, I. J., C. Baigun, P. A. Balykin, B. Becker, Y. Chen, A. F. Filipe, Y. V. Gerasimov, A. L. Godinho, R. M. Hughes, J. D. Koehn, D. N. Kutsyn, V. Mendoza-Portillo, T. Oberdorff, A. M. Orlov, A.P. Pedchenko, F. Pletterbauer, I. G. Prado, R. Rösch, and S. J. Vatland. 2016. International perspectives on the effects of climate change on inland fisheries. Fisheries 41:399–405.

Wipfli, M. S., J. S. Richardson, and R. J. Naiman. 2007. Ecological linkages between headwaters and downstream ecosystems: transport of organic matter, invertebrates, and wood down headwater channels. Journal of the American Water Resources Association 43:72–85.

Wohl, E., N. Kramer, V. Ruiz-Villanueva, D. N. Scott, F. Comiti, A. M. Gurnell, H. Piegay, K. B. Lininger, K. L. Jaeger, D. M. Walters, and K. D. Fausch. 2019. The natural wood regime in rivers. BioScience 69:259–273.

Woody, C. A., R. M. Hughes, E. J. Wagner, T. P. Quinn, L. H. Roulsen, L. M. Martin, and K. Griswold. 2010. The U.S. General Mining Law of 1872: change is overdue. Fisheries 35:321–331.

Yates, A. G., J. M. Culp, R. C. Bailey, and P. A. Chambers. 2019. Strengths and weaknesses of data sources for describing exposure of aquatic ecosystems to human activity. Pages 11–33 *in* R. M. Hughes, D. M. Infante, L. Wang, K. Chen, and B. F. Terra, editors. Advances in understanding landscape influences on freshwater habitats and biological assemblages. American Fisheries Society, Symposium 90, Bethesda, Maryland.

Yoder, C. O., E. T. Rankin, V. L. Gordon, L. E. Hersha, and C. E. Boucher. 2019. Degradation and recovery of Scioto River (Ohio, USA) fish assemblages from presettlement to present-day conditions. Pages 233–265 *in* C. C. Krueger, W. W. Taylor, and S. Youn, editors. From catastrophe to recovery: stories of fishery management success. American Fisheries Society, Bethesda, Maryland.

Glossary

Assemblage: a taxonomic subset of a community

Catchment, watershed, network catchment: the entire land area draining to a point on a stream or an entire lake. These are difficult to define in areas with karst geology, deep sand, flat plains, continental glaciation, xeric climate, or interbasin water diversion

Community: all the biota in an ecosystem

Diversity: taxonomic or trait richness and evenness in an assemblage at sites (alpha) or among sites (beta)

Ecological condition: ecosystem status as measured physically, chemically, and biologically

Ecological health: status of intensively managed areas that support ecosystem (including aesthetic) services without degrading future uses in that area or damaging outside areas

Ecological integrity: wild nature untrammeled by humans but self-sustaining by adapting and evolving

Ecoregions, ecological regions: regional patterns in the combination of climate, physiography, geology, soils, potential natural variation, and land use

Hydrologic unit (hydrological unit code [HUC]): land areas comprised of catchments, portions of catchments that drain to stream segments, or multiple catchments draining to a large lake or sea. Approximately half of all hydrologic units at any spatial extent are not catchments or watersheds.

Landscape: any land area that affects, and is affected by, an ecological process or response of interest

Local catchment, subcatchment, local hydrologic unit: land area draining directly to the downstream end of any stream segment, reach, or lake shoreline, but not including the land area draining to the channel above the upstream end of that segment or reach, nor the entire network catchment of a river or stream entering a lake

Pressure: any anthropogenic activity that produces one or more stressors

Reach, segment: a spatial unit of a stream or river that is clearly defined as being between confluences or geomorphic features, such as waterfalls or dams

Regions: regional patterns in a single ecoregional component

Richness: the number of taxa, traits or phylogenies in an assemblage at sites (alpha) or among sites (beta)

Site: the explicit spatial unit to which one or more samples pertain; may be an entire lake or a defined portion of a lake, stream, or river

Spatial extent: the geographic space over which a unit of study extends (ideally depicted by a map with longitudes and latitudes)

Spatial scale: the level of resolution at which the geographic unit is examined (ideally depicted by a scale bar)

Station, Point: the exact location where a sample is taken

Stressor: any immediate chemical, physical, or biological condition that affects biota

Index

B

C

H

I

J

K

L

N

O

P

Q

R

S

T

U

V

W

X

Y